Adsorption by Powders and Porous Solids
Principles, Methodology and Applications

Second edition

Adsorption by Powders and Porous Solids
Principles, Methodology and Applications

Second edition

*F. Rouquerol, J. Rouquerol, K.S.W. Sing, P. Llewellyn and G. Maurin**

Aix Marseille University-CNRS, MADIREL Laboratory, Marseille, France
*University of Montpellier 2, Institute Charles Gerhardt, Montpellier, France

AMSTERDAM • BOSTON • HEIDELBERG • LONDON
NEW YORK • OXFORD • PARIS • SAN DIEGO
SAN FRANCISCO • SYDNEY • TOKYO
Academic Press is an imprint of Elsevier

ELSEVIER

Academic Press is an imprint of Elsevier
The Boulevard, Langford Lane, Kidlington, Oxford OX5 1GB, UK
Radarweg 29, PO Box 211, 1000 AE Amsterdam, The Netherlands
225 Wyman Street, Waltham, MA 02451, USA
525 B Street, Suite 1800, San Diego, CA 92101-4495, USA

Second edition **2014**

Notice
No responsibility is assumed by the publisher for any injury and/or damage to persons or property as a
matter of products liability, negligence or otherwise, or from any use or operation of any methods, products,
instructions or ideas contained in the material herein. Because of rapid advances in the medical sciences,
in particular, independent verification of diagnoses and drug dosages should be made.

British Library Cataloguing in Publication Data
A catalogue record for this book is available from the British Library

Library of Congress Cataloging-in-Publication Data
A catalog record for this book is available from the Library of Congress

For information on all **Academic Press** publications
visit our web site at store.elsevier.com

14   15   16   17   18   10   9   8   7   6   5   4   3   2   1
ISBN: 978-0-08-097035-6

Working together
to grow libraries in
developing countries

www.elsevier.com • www.bookaid.org

Contents

Preface to the First Edition

The growing importance of adsorption (e.g. in separation technology, industrial catalysis and pollution control) has resulted in the appearance of an ever-increasing volume of scientific and technical literature on novel adsorbents and catalysts. Also, various new procedures have been introduced over the past few years for the interpretation of adsorption data – particularly for micropore and mesopore size analysis. It is hardly surprising that it is becoming increasingly difficult to obtain a well-balanced view of the significance of the recent developments in such fields as adsorption energetics, network percolation and density functional theory against the background of the more traditional theories of surface coverage and pore filling.

In writing this book, we have endeavoured to give an introductory survey of the principles, methodology and applications of the adsorption of gases and liquids by powders and porous solids. In particular, we hope that this book will meet the needs of all those students and non-specialists who wish to undertake adsorption measurements. In addition, we believe that certain sections of the book will be of interest to those scientists, engineers and other technologists who are already concerned, either directly or indirectly, with the characterization of finely divided or porous solids.

We are conscious of the fact that few people now have the time or inclination to read a scientific volume from cover to cover. Furthermore, we know that some readers are looking for concise information on the general principles or methodology of adsorption while others are more interested in the adsorptive properties of, say, carbons or oxides. For these reasons, we have not adopted the more orthodox arrangement of material in which the description of each theory is immediately followed by a lengthy discussion of its application. Instead, the format of this book is as follows. First, a general treatment of the theoretical principles, thermodynamics and methodology of adsorption is given in Chapters 1–5. Secondly, the use of adsorption methods for evaluating the surface area and pore size is discussed (also in general terms) in Chapters 6–8. Thirdly, some typical adsorption isotherms and energies obtained with various adsorbents (carbons, oxides, clays, zeolites) are presented and discussed in Chapters 9–12. Finally, our general conclusions and recommendations are summarized in Chapter 13.

Throughout this book, the main emphasis is on the determination and interpretation of adsorption equilibria and energetics. We are not concerned here with the dynamic or chemical engineering aspects of adsorption – both are very important topics which we must leave to other authors! Since we have set out to provide useful guidance to newcomers to adsorption science, our approach is to some extent didactic. In a book of this nature, it would be impossible to achieve a comprehensive review of all aspects of adsorption by powders and porous solids, and the inclusion of material is necessarily selective. By drawing attention to certain publications and overlooking other excellent pieces of research, we risk the displeasure of some members of the international adsorption community. In defence of this approach, we can only plead that the choice of material was dictated by the need to explain and illustrate the general principles which are summarized in Chapter 13.

Many of the ideas expressed in this book have been developed as a result of collaborative research over the past 30–40 years. It would be invidious to name all our co-workers here but they can be identified in the references listed at the end of each chapter. Our cordial thanks are extended to all those authors and publishers who have readily agreed to the reproduction of published material. For the sake of clarity and consistency of units, etc., most figures have been either redrawn or restyled.

We wish to express our grateful thanks to the following people for information supplied privately: D. Avnir, F.S. Baker, F. Bergaya, M. Bienfait, R.H. Bradley, P.J. Branton, P.J.M. Carrott, J.M. Cases, B.R. Davis, M. Donohue, D.H. Everett, G. Findenegg, A. Fuchs, P. Grange, K.E. Gubbins, K. Kaneko, N.K. Kanellopoulos, W.D. Machin, A. Neimark, D. Nicholson, T. Otowa, R. Pellenq, F. Rodriguez-Reinoso, N.A. Seaton, J.D.F. Ramsay, G.W. Scherer, W.A. Steele, F. Stoeckli, J. Suzanne, J. Meurig Thomas, K.K. Unger and H. Van Damme.

Special thanks are due to our colleagues Y. Grillet, R. Denoyel and P.L. Llewellyn for their forbearance and encouragement, to P. Chevrot for the quality of the artwork and to Madame M.F. Fiori for her unfailing dedication in processing the manuscript.

Last but not least we should like to pay tribute to the leadership of Dr. S. John Gregg, whose many contributions to adsorption and surface science were made over a period of more than 60 years.

Preface to the Second Edition

The principal aim of the second edition of this book remains essentially the same as that of the first: to provide an introductory survey of the principles, methodology and applications of the adsorption of gases and liquids by powders and porous solids.

As before, special attention is given to the determination and interpretation of experimental data, particularly in relation to the characterization of adsorbents of technological importance. Over the past 14 years, considerable progress has been made in the development of ordered pore structures and computer modelling of well-defined adsorption systems. However, a number of problems remain unresolved – notably in the assessment of the surface area and pore-size distribution of disordered adsorbents. The discerning reader will detect different shades of opinion in certain chapters and we hope that this will serve to stimulate further discussion and research.

In writing this second edition, we are aware that it is becoming increasingly difficult for the non-specialist to appreciate the scope and limitations of recent advances in adsorption science against the background of 'classical' work. We have therefore endeavoured to summarize and explain the significance of the more important developments in relation to the characterization of the surface properties of porous and non-porous materials.

The new chapter on modelling of physisorption in porous solids should be of particular interest to those experimentalists who wish to understand the purpose and application of computational procedures. Another new chapter is on the adsorptive properties of metal-organic frameworks, which have received much attention in recent years. It will be evident that other parts of the book have been substantially revised; much of the subject matter is now rearranged and expanded so that each chapter is to a large extent self-contained.

We set out to provide useful guidance in the interpretation and application of adsorption methods, and to this end, we have selected particular examples of our own research and other related studies to illustrate the principles expounded in the various chapters. In fact, the amount of new work on adsorption published over the past 14 years is far too extensive to discuss properly in a book of this size. Unfortunately, we have had to omit any reference to a number of excellent research papers by distinguished scientists.

Many of the ideas expressed in the book have originated indirectly from the numerous fruitful – and always enjoyable – discussions with our colleagues and friends. Of course, all directly related pieces of work are indicated by the appropriate references to be found in the list at the end of each chapter. There are other people who deserve special thanks for their help. In addition to the names in the preface to the first edition, we must extend our grateful thanks to the following for their continued help and encouragement: Peter Branton, Donald Carruthers, Renaud Denoyel, Tina Düren, Erich Müller, Alex Neimark, Jehane Ragai, 'Paco' Rodriguez-Reinoso, Randall Snurr, John Meurig Thomas, Petr Nachtigall, Matthias Thommes, Klaus Unger and Ruth Williams.

List of Main Symbols

As far as possible, the notation used here follows the recommendations of the International Union of Pure and Applied Chemistry (Cohen et al., 2007)

a	specific surface area
A	surface area
$a(\text{ext})$	external specific surface area
b	Langmuir adsorption coefficient
B	structural constant D–R (Dubinin–Radushkevich)
B_m	second virial (molar) coefficient
$c(B)$	amount concentration $\left(=\frac{n(B)}{V}\right)$
C	BET constant (Brunauer, Emmett and Teller)
d	molecular diameter or particle diameter
l	single linear dimension
D	D–R constant (Dubinin–Radushkevich)
E	energy
E_o	adsorption molar energy at infinitely low coverage
E_1	adsorption molar energy for the first layer
E_L	liquefaction energy
F	Helmholtz energy defined as $U - TS$
G	Gibbs energy defined as $H - TS$
H	enthalpy defined as $U + pV$
H	distance between the nuclei in the parallel walls of a slit pore
i	intercept
k	Boltzmann constant
k_H	Henry's law constant
K	equilibrium constant
K	Kelvin, SI unit
L	Avogadro constant
m	mass
M	molar mass
n	amount of substance
n	specific surface excess amount
N	number of elementary entities
N	number of layers
p	pressure
$p^{\ominus}$	standard pressure
p°	saturation pressure
P	probability
q	electric charge
Q	heat

r	pore radius
r	curvature radius
R	gas constant
s	slope
S	entropy
t	thickness of multimolecular layer
T	thermodynamic temperature
U	internal energy
V	volume
w	pore width
W	work
x	mole fraction
y	mole fraction
z	distance from surface
α	polarisability
α_S	ratio of two adsorbed amounts which one is used as a reference in the α_S-method
ε	porosity
ε	pairwise interaction energy
φ	potential energy
ϕ	heat flux, defined as dQ/dt
γ	surface tension
$\gamma(S0)$	or γ^s surface tension of the clean solid
$\gamma(SG)$	or γ surface tension of the solid in equilibrium with a gas
$\gamma(SL)$	surface tension of the solid in equilibrium with a liquid
Γ	surface excess concentration defined as n^σ/A
μ	chemical potential
π	spreading pressure, defined as $\gamma(S0) - \gamma(SG)$ or $\gamma^s - \gamma$
ρ	mass density (or volumic mass)
σ	molecular cross-sectional area
τ	distance to the surface in the adsorption space
θ	surface coverage, defined as the ratio of two surface excess amounts, one of which is used as a reference
θ	Celsius temperature

Superscripts

g	gas or vapour
l	liquid
s	solid
aq	aqueous solution
a	adsorbed (in the layer model)
σ	surface excess (in the Gibbs representation)
i	interfacial
$\ominus$	standard
*	pure substance
∞	infinite dilution

Subscripts

m	related to a complete monolayer
p	pore

Any symbol of a state variable (e.g. T, V, p, A, n) can be used as a subscript of a physical quantity to denote that this variable is kept constant (e.g. $G_{T,p}$; $V_{T,p}$; $H_{298.15}$)

Use of operator Δ

The symbol Δ followed by a subscript is used to denote the *change of an extensive quantity* associated with a physical or chemical process. The main subscripts used in this book are:

vap   vaporisation
liq   liquefaction
sub   sublimation
fus   fusion
trs   transition (between phases)
mix   mixing of fluids
sol   solution (of solute in solvent)
dil   dilution (of a solution)
ads   adsorption
dpl   displacement
imm   immersion

Reference

Cohen, E.R., Cvitas, T., Frey, J.G., Holmström, B., Kuchitsu, K., Marquardt, R., Mills, I., Pavese, F., Quack, M., Stohner, J., Strauss, H.L., Takami, M., Thor, A.J., 2007. Quantities, Units and Symbols in Physical Chemistry, third ed. RSC Publishing, Cambridge, UK.

1 Introduction

Françoise Rouquerol*, Jean Rouquerol*, Kenneth S.W. Sing*, Guillaume Maurin**, Philip Llewellyn*

*Aix Marseille University-CNRS, MADIREL Laboratory, Marseille, France
**University of Montpellier 2, Institute Charles Gerhardt, Montpellier, France

---

Chapter Contents

1.1 The Importance of Adsorption

Adsorption occurs whenever a solid surface is exposed to a gas or liquid: it is defined as the enrichment of material or increase in the density of the fluid in the vicinity of an interface. Under certain conditions, there is an appreciable enhancement in the concentration of a particular component and the overall effect is then dependent on the extent of the interfacial area. For this reason, all industrial adsorbents have large specific surface areas (generally, well in excess of $100 \text{ m}^2 \text{ g}^{-1}$) and are therefore highly porous or composed of very fine particles.

Adsorption is of great technological importance, as often stressed (Dabrowski, 2001). Thus, some adsorbents are used on a large scale as desiccants, catalysts or catalyst supports; others are used for the separation or storage of gases, the purification of liquids, controlled drug delivery, pollution control or for respiratory protection. In addition, adsorption phenomena play a vital role in many solid state reactions and biological mechanisms.

Another reason for the widespread use of adsorption techniques is the importance now attached to the characterisation of the surface properties and texture of fine powders such as pigments, fillers and cements. Similarly, adsorption measurements are undertaken in many academic and industrial laboratories on porous materials such

Adsorption by Powders and Porous Solids. http://dx.doi.org/10.1016/B978-0-08-097035-6.00001-2

as clays, ceramics and membranes. In particular, gas adsorption has become one of the most widely used procedures for determining the surface area and pore size distribution of a diverse range of powders and porous materials.

1.2   Historical Aspects

Various phenomena which we now associate with adsorption were known in antiquity. The adsorbent properties of such materials as clay, sand and wood charcoal were utilised by the ancient Egyptians, Greeks and Romans (Robens, 1994). These applications were wide-ranging and included the desalination of water, the clarification of fat and oil and the treatment of many diseases.

It has long been known that certain forms of charcoal can take up large volumes of gas. The earliest quantitative studies appear to have been made by Scheele in 1773 and independently by Priestley in 1775 and the Abbé Fontana in 1777 (Deitz, 1944; Forrester and Giles, 1971). The decolourising properties of charcoal were first investigated by the Russian chemist Lowitz in 1785. The exothermal nature of gas adsorption was noted by de Saussure in 1814. Mitscherlich (1843) suggested that the amount of gas adsorbed in a porous carbon was such that it was probably in the liquid state. This prompted Favre (1854, 1874) to study the 'wetting of solids by gases' and to use adsorption calorimetry to show that the heat of adsorption of various gases on charcoal was larger than the heat of liquefaction, which he explained as due to a higher density in the vicinity of the pore walls. However, it was not until 1879–1881 that the first attempts were made by Chappuis (1879; 1881a,b) and Kayser (1881a,b) to relate the amount of gas adsorbed to the pressure. It was then that Kayser (1881a,b) introduced the term *adsorption* and over the next few years the terms *isotherm* and *isothermal-curve* were applied to the results of adsorption measurements made at constant temperature (see Forrester and Giles, 1971).

It was observed by Leslie in 1802 that heat was produced when liquid was added to a powder. The heat evolved by the immersion of dry sand in water was described by Pouillet in 1822. This exothermic phenomenon became known in France as the 'Pouillet effect'. Gore (1894) recognised that the amount of heat was related to the surface area of the powder, while Gurvich (1915) suggested that it was also dependent on the polarity of the liquid and the nature of the powder.

The first recorded isotherms of adsorption from solution were probably those reported by van Bemmelen in 1881 (Forrester and Giles, 1972). In his investigations of the 'absorptive' power of soils, van Bemmelen noted the importance of the colloidal structure and drew attention to the relevance of the final state (i.e. equilibrium concentration) of the solution in contact with the soil. A number of solute–solid isotherms were determined over the next 20 years including those for the uptake of iodine and various dyes by charcoal and other adsorbents, but many of the investigators still believed that the process involved penetration into the solid structure. Freundlich, in 1907, was one of the first to appreciate the role of the solid surface. He proposed a general mathematical relation for the isotherm, which we now refer to as the Freundlich adsorption equation.

In 1909, McBain reported that the uptake of hydrogen by carbon appeared to occur in two stages: a rapid process of adsorption appeared to be followed by a slow process of absorption into the interior of the solid. McBain coined the term *sorption* to cover both phenomena. In recent years, it has been found convenient to use 'sorption', when it is not possible to make a clear distinction between these two stages of uptake and also to use it to denote the penetration of molecules into very narrow pores (Barrer, 1978).

During the early years of the past century, various quantitative investigations of gas adsorption were undertaken. The most important advances in the theoretical interpretation of gas adsorption data were made by Zsigmondy, Polanyi and Langmuir: their ideas set the scene for much of the research undertaken during the first half of the twentieth century.

In 1911, Zsigmondy pointed out that the condensation of a vapour can occur in very narrow pores at pressures well below the normal vapour pressure of the bulk liquid. This explanation was given for the large uptake of water vapour by silica gel and was based on an extension of a concept originally put forward by Thomson (Lord Kelvin) in 1871 (see Sing and Williams, 2012). It is now generally accepted that *capillary condensation* does play an important role in the physisorption by porous solids, but that the original theory of Zsigmondy cannot be applied to pores of molecular dimensions.

The theory proposed by Polanyi in 1914 was developed from an older idea of long-range attractive forces emanating from the solid surface. The adsorbed layer was pictured as a thick compressed film of decreasing density with increase in distance from the surface. The original 'potential theory' did not give an equation for the adsorption isotherm, but instead provided a means of establishing a 'characteristic curve' – relating adsorption potential to amount adsorbed – for a given system. In spite of its initial appeal, it soon became apparent that the principles underlying the potential theory were not consistent with the emerging treatment of intermolecular forces. However, the concept of a characteristic curve was subsequently modified and adopted by Dubinin and his co-workers in their theory of micropore filling.

The year 1916 brought a radical change in the approach to surface science. In that year, the first of Langmuir's monumental papers appeared (1916, 1917, 1918). Lord Rayleigh's earlier conclusion that certain films of polar oils on water were one molecule thick had not received the attention it deserved and Langmuir's great contribution was to bring together all the available evidence to support the unifying concept of the monomolecular layer (the *monolayer*). He proposed that adsorption on both liquid and solid surfaces normally involved the formation of a monomolecular layer. In retrospect, it is not surprising that the advent of the Langmuir theory produced a renaissance in surface science.

Langmuir's work on gas adsorption and insoluble monolayers prepared the way for more progress to be made in the interpretation of adsorption from solution data. In the light of the Langmuir theory, it seemed logical to suppose that the plateau of a solute isotherm represented monolayer completion and that the monolayer capacity could be derived by application of the Langmuir equation.

Another important stage in the history of gas adsorption was the work of Brunauer and Emmett, which preceded the publication of the Brunauer–Emmett–Teller (BET) theory in 1938. In 1934, Emmett and Brunauer made their first attempt to use low-temperature adsorption of nitrogen to determine the surface area of an iron synthetic ammonia catalyst. They noted that the adsorption isotherms of a number of gases, measured at temperatures at, or near, their respective boiling points, were all S-shaped with certain distinctive features. Others, including Langmuir, had recognised that this type of adsorption was not always restricted to monolayer coverage and an empirical approach was adopted by Emmett and Brunauer (1937) to ascertain the stage at which the multilayer adsorption began. They eventually decided that completion of the monolayer was characterised by the beginning of the nearly linear section of the adsorption isotherm (designated Point B – see Figure 5.2). The surface area was then evaluated from the amount adsorbed at Point B by making the further assumption that the completed monolayer was in a close-packed state. In 1938, the publication of the BET theory appeared to provide a sound basis for the identification of Point B as the stage of monolayer completion and the onset of multilayer adsorption.

It would be difficult to over-estimate the historical importance of the BET theory. For over 70 years, it has continued to attract an enormous amount of attention (see Davis and Sing, 2002). Indeed, the BET method has become a standard procedure for the determination of the surface area of a wide range of fine powders and porous materials (Gregg and Sing, 1982; Rouquerol et al., 1999). On the other hand, it is now generally recognised that the theory is based on an over-simplified model of multilayer adsorption and that the reliability of the BET method is questionable unless certain conditions are prescribed (see Chapters 7 and 14).

There was a growing awareness in the early 1930's that a distinction could be made between physical adsorption (i.e. *physisorption*) in which the van der Waals interactions are involved and chemical adsorption (i.e. *chemisorption*) in which the adsorbed molecules are attached by chemical bonding. Taylor (1932) introduced the concept of 'activated adsorption' which, by analogy with the familiar idea of an energy of activation in chemical kinetics, attempted to explain the marked increase in rate of adsorption with rise in temperature in terms of surface bond formation. The activated adsorption theory aroused a good deal of early criticism and with the subsequent improvement of high vacuum techniques it was established that chemisorption of certain gases like oxygen can take place very rapidly on clean metal surfaces. However, there are other chemisorption systems which do appear to exhibit the features of activated adsorption (Ehrlich, 1988).

In his 1916 paper, Langmuir had stated that with highly porous adsorbents such as charcoal '*it is impossible to know definitely the area on which the adsorption takes place*' and that '*there are some spaces in which a molecule would be closely surrounded by carbon atoms on nearly all sides*'. He concluded that equations derived for plane surfaces were not applicable to adsorption by charcoal. Unfortunately, these observations have been overlooked by many investigators, who have applied the simple Langmuir monolayer equation to adsorption data obtained with zeolites, activated carbons and other adsorbents containing pore of molecular dimensions (i.e. micropores).

The significance of Langmuir's comments was appreciated, however, by Dubinin and his co-workers in Moscow, who put forward additional evidence to show that the mechanism of physisorption in very narrow pores is not the same as that in the wider pores (i.e. mesopores) or on the open surface. Dubinin argued that such micropores are filled at low relative pressure by a volume-filling process. By studying a wide range of activated carbons, he identified three groups of pores of different width: *micropores*, transitional pores (now termed *mesopores*) and *macropores*. This classification has been refined (see Table 1.3), but the principles remain the same as those adopted by Dubinin (1960).

Another Russian scientist who played a leading role in the advancement of the understanding of adsorption mechanisms was A. V. Kiselev. With the help of a large team of co-workers and by making a systematic investigation of various well-defined adsorbents (notably oxides, carbons and zeolites), Kiselev was able to demonstrate that certain specific interactions were involved in the adsorption of polar molecules on polar or ionic surfaces. At the same time, in the UK the specificity of physisorption was under investigation by Barrer – especially in the context of his pioneering work on the properties of the molecular sieve zeolites.

In an early discussion of the adsorption of gases by solids, Rideal (1932) had stressed the fundamental importance of the nature and extent of the solid surface and had pointed out that 'there is a considerable latitude even in the definition of specific surface', so that 'it is more correct to speak of *accessible* surface, denoting the extent of surface which can be reached by the reactant under consideration'. The accessible surface area is indeed a basic concept for the understanding of adsorption. This is sometimes forgotten in favour of a hypothetical 'true' surface area of the adsorbent, which raises a similar problem to that of a cartographer required to evaluate a coastal perimeter. Obviously, the answer must depend on the scale of the map: that is on its resolution and, therefore, on the size of the yardstick used to explore an irregular coastal line.

Over the past few decades, much attention has been given to the application of fractal analysis to surface science. The early work of Mandelbrot (1975) explored the replication of structure on an increasingly finer scale, that is the quality of self-similarity. As applied to physisorption, fractal analysis appears to provide a generalised link between the monolayer capacity and the molecular area without the requirement of an absolute surface area. In principle, this approach is attractive, although in practice it depends on the validity of the derived value of monolayer capacity and the tacit assumption that the physisorption mechanism remains the same over the molecular range studied. In the context of physisorption, the future success of fractal analysis will depend on its application to well-defined non-porous adsorbents and to porous solids with pores of uniform size and shape.

Many new adsorbents have been developed over the past 50 years. Older types of industrial adsorbents (e.g. activated carbons, aluminas and silica gels) are still produced in large quantity, but they are generally non-crystalline and their surface and pore structures therefore tend to be ill-defined and difficult to characterise. There are growing numbers of model adsorbents having intra-crystalline pore structures, such as new zeotypes, aluminophosphates and metal-organic frameworks (MOFs). Great interest is also being shown in the design of other new ordered structures (e.g. mesoporous carbons and oxides) having pores of well-defined size and shape.

Various advanced spectroscopic, microscopic and scattering techniques can now be employed for studying the state of the adsorbate and microstructure of the adsorbent. Also, major advances have been made in the experimental measurement of isotherm and adsorption energy data, in the development of computational approaches to model physisorption and in the application of the density functional theory (DFT).

It has become apparent that the interpretation of adsorption from solution data is often difficult. Although many isotherms have a similar shape to the classical

Langmuir isotherm, they rarely obey the Langmuir equation over an appreciable range of concentration. It is evident that consideration must be given to the competition between solute and solvent, the solvation of solute and, in many cases, the lack of thermodynamic equilibration.

1.3   General Definitions and Terminology

Some of the principal terms and properties associated with adsorption, powders and porous solids are defined in Tables 1.1–1.3. These definitions are consistent with those proposed by the International Union of Pure and Applied Chemistry (IUPAC) (see Haber, 1991; Rouquerol et al., 1994; Sing et al., 1985) and by the British Standards Institution (1958, 1992) and other official organisations (see Robens and Krebs, 1991). New IUPAC recommendations on 'Physisorption of gases, with special reference to the evaluation of surface area and pore size distribution', which are now in preparation, are also taken into account (M. Thommes, private communication).

As noted earlier, the term *adsorption* is universally understood to mean the enrichment of one or more of the components in the region between two bulk phases (i.e. the *interfacial layer* or the *adsorption space*). In the present context, one of these phases is necessarily a solid and the other a fluid (i.e. gas or liquid). With certain systems (e.g. some metals exposed to hydrogen, oxygen or water), the adsorption process is accompanied by *absorption*, that is the penetration of the fluid into the solid phase. As already indicated, one may then use the term *sorption* (and the related terms *sorbent*, *sorptive* and *sorbate*) to embrace both phenomena of adsorption and absorption. This

Table 1.1 Definitions: Adsorption

Term	Definition
Adsorption	Enrichment of one or more components in the vicinity of an interface
Adsorbate	Substance in the adsorbed state
Adsorptive[a]	Adsorbable substance in the fluid phase
Adsorbent	Solid material on which adsorption occurs
Chemisorption	Adsorption involving chemical bonding
Physisorption	Adsorption without chemical bonding
Monolayer capacity	*Either* chemisorbed amount required to occupy all surface sites *or* physisorbed amount required to cover the surface
Surface coverage	Ratio of amount of adsorbed substance to monolayer capacity

[a]Translated into French as 'adsorbable'.

Table 1.2 Definitions: Powders

Term	Definition
Powder	Dry material composed of discrete particles with dimensions less than $\sim$1 mm
Fine powder	Powder with particle size below $\sim$1 μm
Aggregate	Loose, unconsolidated assemblage of particles
Agglomerate	Rigid, consolidated assemblage of particles
Compact	Agglomerate formed by compaction of powder
Acicular	Needle-shaped
Surface area	Extent of the surface assessed by a given method (experimental or theoretical) under stated conditions
Specific surface area	Surface area of unit mass of powder, as assessed under stated conditions
External surface area	(1) Area of external surface of particles, taking account of roughness (i.e. all cavities which are wider than deep), but not any porosity (2) Area outside any micropores
Roughness factor	Ratio of external surface area (1) to area of smoothed envelope around particles
Divided solid	Solid made up of more or less independent particles which may be in the form of a powder, aggregate or agglomerate

is the convention that we shall adopt in the present book. The term sorption is used by some authors to denote the uptake of gas or liquid by a molecular sieve, but we do not favour this practice.

The terms *adsorption* and *desorption* are often used to indicate the direction from which the equilibrium states have been approached. *Adsorption hysteresis* arises when the amount adsorbed is not brought to the same level by the adsorption and desorption approach to a given 'equilibrium' pressure or bulk concentration. The relation, at constant temperature, between the amount adsorbed and the equilibrium pressure, or concentration, is known as the *adsorption isotherm*.

The term *surface area* is frequently used in the current literature in a somewhat ambiguous manner. In principle, we can identify: (a) an *experimentally accessible* surface area, which is available for the adsorption of certain adsorptives, as described in Chapter 7 and (b) a virtual *r-distant surface area*, as defined in Chapter 6. In practice, however, it is often difficult to assess the physical significance of the 'surface area' obtained by the application of a particular experimental procedure. This is a controversial topic, which has attracted different shades of opinion as explained in some detail in various chapters of this book (e.g. in Chapters 6, 7, 10 and 14).

Table 1.3 Definitions: Porous Solids

Term	Definition
Porous solid	Solid with cavities or channels which are deeper than wide
Open pore	Cavity or channel with access to the surface
Interconnected pore	Pore which communicates with other pores
Blind pore[a] (or dead-end pore)	Pore with a single connection to the surface
Closed pore	Cavity not connected to the surface
Void	Space between particles
Micropore	Pore of internal width >2 nm
Mesopore	Pore of internal width between 2 and 50 nm
Macropore	Pore of internal width <50 nm
Nanopore	Pore of internal width less than ~100 nm
Pore size	Pore width (diameter of cylindrical pore or distance between opposite walls of a slit)
Pore volume	Volume of pores determined by stated method
Porosity	Ratio of total *pore* volume to apparent volume of particle or powder
Total porosity	Ratio of volume of *voids and pores* (open and closed) to volume occupied by solid
Open porosity	Ratio of volume of voids and open pores to volume occupied by solid
Surface area	Extent of surface assessed by a given method (experimental or theoretical) under stated conditions
External surface area	(**1**) Area of surface outside all pores (**2**) Area outside micropores
Internal surface area	(**1**) Area of all pore walls (**2**) Area of micropore walls
True density	Density of solid, excluding pores and voids
Apparent density	Density including closed and inaccessible pores, as determined by stated method

[a]French note: in the sense of 'borgne'.

A *powder* is easily recognised as a mass of small dry particles, but the precise definition is inevitably somewhat arbitrary. The term *fine powder* is also used in an imprecise manner, but it seems reasonable to apply it to a material consisting of particles less than about 1 μm (i.e. particles of colloidal dimensions). The unit mass of a fine powder contains a large number of small particles and hence exhibits an appreciable surface area.

For example, in the simplest case of an assemblage of spherical particles, all with the same diameter, d, the *geometrical* specific surface area, a (with the assumption that the surface is completely smooth) is given by the relation

$$a = 6/\rho d \tag{1.1}$$

where ρ is the particle absolute density. Thus, a powder composed of smooth spherical particles of $d = 1$ μm and $\rho = 3$ g cm^{-3} would have a specific surface of 2 m^2 g^{-1}. The same calculation would apply to cubic particles, but in this case d would equal the edge length of the cube.

It is evident that it is more difficult to define particle size if the particle shape is not spherical or cubic. With some other simple geometric forms, a single linear dimension, l_x, may be used to calculate the surface area. In particular, when the particle aspect ratio is sufficiently large, l_x is taken as the *minimum* dimension. Thus, if the particles are thin or long (i.e. plates or rods), it is the thickness which mainly determines the magnitude of the specific surface area (Gregg and Sing, 1982).

Perfect spheres are rare, but spheroidal particles are present in some powders produced at high temperature (e.g. pyrogenic silicas) or by the sol–gel process. The term *sphericity* is useful for some purposes. Sphericity has been defined in various ways, the simplest definition being the ratio of the surface area of a sphere of the same volume as a given particle to the actual surface area of that particle (Allen, 1990).

The individual particles (primary particles) in a fine powder are usually clustered together in the form of *aggregates* or *agglomerates*. Loosely bonded aggregates are unconsolidated and non-rigid, but they may be converted into more rigid, consolidated agglomerates as a result of sintering or ageing. The breakdown, or partial breakdown, of the consolidated material can be achieved by grinding. The process of agglomeration involves the bridging or cementation of particles and should not be confused with *Ostwald ripening*, which involves the growth of larger particles at the expense of smaller ones. It is evident that an agglomerate may be regarded as a 'secondary' particle, which always contains within it some internal surface. In many cases, this internal surface provides the major part of its surface area and the agglomerate then possesses a well-defined pore structure.

The classification of *pores* according to size has been under discussion for many years, but in the past the terms *micropore* and *macropore* have been applied in different ways by physical chemists and other scientists. In an attempt to clarify this situation, the limits of size of the different categories of pores included in Table 1.3 have been proposed by the International Union of Pure and Applied Chemistry (Everett, 1972; Sing et al., 1985). As indicated, the *pore size* is generally specified as the *pore width*, that is the available distance between the two opposite walls. Obviously, pore size has a precise meaning when the geometrical shape is well defined. Nevertheless, for most purposes the limiting size is that of the smallest dimension and this is generally taken to represent the effective pore size.

Micropores and mesopores are especially important in the context of adsorption. Though the use of the term *nanopore* has become fashionable in recent years, we do not recommend its use for pores wider than ~100 nm.

The hypothetical types of pores shown in Figure 1.1 relate to the definitions in Table 1.3. In addition to *closed pores* and *open pores*, we may distinguish between *blind pores* (or dead-end pores) and *interconnected pores*. Pores which are open at both sides of a membrane or porous plug are termed *through pores*.

Porosity is usually defined as the ratio of the volume of pores and voids to the volume occupied by the solid. However, it should be kept in mind that the recorded value of porosity is not always a simple characteristic property of the material, since it is likely to depend also on the methods used to assess both the pore volume and the volume of the solid. The pore volume is usually regarded as the volume of open pores, but it may include the volume of closed pores. Moreover, the recorded value may depend on the nature of the probe molecule or the experimental conditions.

It is not always easy to distinguish between roughness and porosity or between pores and voids. In principle, a convenient and simple convention is to consider that the *roughness* includes all surface irregularise which are wider than they are deep. Depending on the context, the *external surface area* of a solid is either the area including roughness but not pores, or the area outside the micropores. Similarly, the *internal surface area* is either the wall area of all pores or only of micropores. We prefer to regard the porosity as an intrinsic property of the material and to designate void as the space between particles, which is dependent on the conditions of packing (and the particle coordination number).

It is evident that the description of many real porous materials is complicated by a wide distribution of pore size and shape and the complexity of the pore network. To facilitate the application of certain theoretical principles, the shape is often assumed to be cylindrical, but this is rarely an accurate portrayal of the real system. With some materials, it is more realistic to picture the pores as slits or interstices between spheroidal particles. The advances of computational modelling and the application of

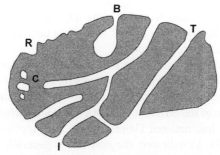

Figure 1.1 Cross section of a hypothetical porous grain showing various types of pores: closed (C), blind (B), through (T), interconnected (I), together with some roughness (R) (Rouquerol, 1990).

percolation theory have made it possible to study the effects of the *connectivity* between pores and their transport properties.

Pore structures can be created in a number of different ways. *Intra-crystalline* pores are an inherent part of certain crystalline structures, for example of zeolites, MOFs and certain clays. These pores are generally of molecular dimensions and are arranged as highly regular networks. A second type of porous material is composed of an assemblage of small particles (as mentioned earlier). The pore structure of the consolidated system (e.g. a xerogel) is mainly dependent on the size and packing density of the primary particles: the process is therefore *constitutive*. A third route is *subtractive* since inherent parts of the original structure are removed to create the pores, for example the thermal decomposition of a hydroxide or carbonate or a leaching process to produce porous glass.

1.4   Physisorption and Chemisorption

Adsorption is brought about by the interactions between the solid and the molecules in the fluid phase. Two kinds of forces are involved, which give rise to either physical adsorption (physisorption) or chemisorption. Physisorption forces are the same as those responsible for the condensation of vapours and the deviations from ideal gas behaviour, whereas chemisorption interactions are essentially those responsible for the formation of chemical compounds.

The most important distinguishing features may be summarised as follows:

a. Physisorption is a general phenomenon with a relatively low degree of specificity.
b. Chemisorbed molecules are linked to reactive parts of the surface and the adsorption is necessarily confined to a monolayer. At high relative pressures, physisorption generally occurs as a multilayer.
c. A physisorbed molecule keeps its identity and on desorption returns to the fluid phase in its original form. If a chemisorbed molecule undergoes reaction or dissociation, it loses its identity and cannot be recovered by desorption.
d. The energy of chemisorption is the same order of magnitude as the energy change in a comparable chemical reaction. Physisorption is always exothermic, but the energy involved is generally not much larger than the energy of condensation of the adsorptive. However, it is appreciably enhanced when physisorption takes place in very narrow pores.
e. An activation energy is often involved in chemisorption and at low temperature the system may not have sufficient thermal energy to attain thermodynamic equilibrium. Physisorption systems generally attain equilibrium fairly rapidly, but equilibration may be slow if the transport process is rate-determining.

1.5   Types of Adsorption Isotherms

1.5.1   Classification of Gas Physisorption Isotherms

The amount of gas adsorbed, n^a, by the mass, m^s, of solid is dependent on the *equilibrium* pressure, p, the temperature, T, and the nature of the gas–solid system. For a given gas adsorbed on a particular solid at a constant temperature we may write:

$$n^a/m^s = f(p)T \tag{1.2}$$

and if the gas is below its critical temperature, it is possible to write

$$n^{a}/m^{s} = f(p/p^{o})T \tag{1.3}$$

where p^{o} is the saturation pressure of the adsorptive at T.

Equations (1.2) and (1.3) represent the *adsorption isotherm* which is the relationship between the amount adsorbed by unit mass of solid and the equilibrium pressure (or relative pressure), at a known temperature. The experimental adsorption isotherm is usually presented in graphical form.

Experimental adsorption isotherms recorded in the literature for many different gas–solid systems have various characteristic shapes. These shapes are important since they provide useful preliminary information about the pore structure of the adsorbent, even before any precise calculations have been carried out. The majority of vapour isotherms (i.e. at sub-critical temperatures) may be divided into nine groups in an extended IUPAC classification (see Figure 1.2). Other shapes are sometimes found and these can usually be explained as a combination of two (or more) of the nine shapes proposed and such an isotherm is then said to be *composite*. The five Types I, II, III, IV and V were similar to those originally proposed by Brunauer, Deming, Deming and Teller (1940), which is usually referred to as the BDDT or Brunauer classification (1945).

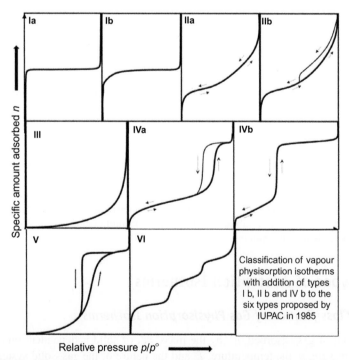

Figure 1.2 Classification of vapour adsorption isotherms combining proposals from IUPAC (Rouquerol et al., 1999; Sing et al., 1985).

The *Type I(a)* and *I (b)* isotherms are reversible and concave to the relative pressure (p/p^o) axis. They rise steeply at low relative pressures and reach a plateau: the amount adsorbed by the unit mass of solid, n^a/m^s approaches a limiting value, as $p/p^o \rightarrow 1$. They are obtained with microporous adsorbents. Enhanced adsorbent–adsorbate interactions occur in micropores of molecular dimensions. A decrease in the micropore width results in both an increase in the adsorption energy and a decrease in the relative pressure at which the micropore-filling occurs. The narrow range of relative pressure necessary to attain the plateau is an indication of a limited range of pore size and the appearance of a nearly horizontal plateau indicates a very small external surface area. Type I(a) corresponds to the filling of narrow micropores, whereas Type I(b) indicates the presence of wider micropores. With all Type I isotherms, the limiting adsorption is dependent on the available micropore volume. We recommend that the term 'Langmuir isotherm' should not be used in the context of physisorption by microporous solids (see Chapter 5).

The *Type II* isotherm is concave to the p/p^o axis, then almost linear and finally convex to the p/p^o axis. This shape is obtained with a non-porous or a macroporous adsorbent, which allows unrestricted multimolecular adsorption to occur at high p/p^o. When the equilibrium pressure equals the saturation vapour pressure, the adsorbed layer becomes a bulk liquid or solid. If the knee of the isotherm is sharp, the uptake at Point B – the beginning of the middle quasi-linear section – is usually considered to represent the completion of the monomolecular layer (the monolayer) and the beginning of the formation of the multimolecular layer (the *multilayer*). The ordinate of Point B gives an estimation of the amount of adsorbate required to cover the unit mass of solid surface with a complete monolayer (i.e. the *monolayer capacity*). If the adsorbent temperature is at, or below, the normal boiling point of the adsorptive, it is not difficult to establish the course of the adsorption–desorption isotherm over the entire range of p/p^o. In the case of a Type II(a) isotherm, complete reversibility of the desorption–adsorption isotherm (i.e. the absence of adsorption hysteresis) is the first condition to be satisfied for 'normal' monolayer–multilayer adsorption on an open and stable surface.

A number of powders or aggregates (e.g. clays, pigments, cements) give Type II isotherms, which exhibit Type H3 hysteresis (see Section 8.6). This isotherm shape is now referred to as Type II(b) – as indicated in Figure 1.2. Here, the narrow hysteresis loop is the result of inter-particle capillary condensation (usually within a non-rigid aggregate).

A *Type III* isotherm is convex to the p/p^o axis over the complete range and therefore has no Point B. This shape is indicative of weak adsorbent–adsorbate interactions on a non-porous or macroporous adsorbent. True Type III isotherms are not common.

The *Type IV(a) and IV(b)* isotherms, whose initial region is closely related to Type II isotherms, level off at high relative pressures with a characteristic saturation plateau, although this may be short and reduced to an inflexion point. They are obtained with mesoporous adsorbents. Type IV(a) isotherms, which are much more common than Type IV(b), exhibit hysteresis loops: the lower (*adsorption*) branch is obtained by the progressive addition of gas and the upper (*desorption*) branch by the progressive withdrawal. The hysteresis loop is associated with the filling and emptying of the mesopores by capillary condensation. Type IV(a) isotherms are common but the exact shape of the hysteresis loop varies from one system to another (see Section 8.6). Type IV(b) isotherms are completely reversible and are given by a few ordered mesoporous structures – notably MCM-41 (see Section 13.2).

The *Type V* isotherm is initially convex to the p/p^o axis and also levels off at high relative pressures. As in the case of the Type III isotherm, this is indicative of weak adsorbent–adsorbate interactions, but here on a microporous or mesoporous solid. A Type V isotherm usually exhibits a hysteresis loop which is associated with pore filling and emptying. Such isotherms are relatively rare.

The *Type VI* isotherm, or stepwise isotherm, is also relatively rare and is associated with layer-by-layer adsorption on a highly uniform surface such as graphitised carbon. The sharpness of the steps is dependent on the system and the temperature.

The above classification is necessarily somewhat idealised since, as already pointed out, many experimental physisorption isotherms have a composite nature and others are more complex than was formerly thought. Some of the more interesting and unusual isotherms will be discussed in subsequent chapters of this book.

1.5.2   Chemisorption of Gases

In contrast to the great variety of adsorption isotherms found in physisorption, *chemisorption* generally gives rise to a simple type of adsorption isotherm, comparable to Type I(a) of physisorption isotherms (see Figure 1.2). The plateau is due to the completion of a chemically bound monolayer. These isotherms may be referred to as *Langmuir isotherms*, even though the mechanism involved may not be strictly in accordance with the Langmuir model. With some systems, and under certain conditions, the slow rate of chemisorption makes it difficult to obtain equilibrium data. Furthermore, the chemisorption reaction may be undetectable at low temperature or pressure and become significant only when the experimental conditions are changed.

1.5.3   Adsorption from Solution

A distinctive feature of adsorption from solution is that it always involves a competition between the solvent and solute which has to be taken into account in any complete analysis of the experimental data. In the case of *adsorption from solution*, the 'apparent adsorption' of a solute at the liquid/solid interface is usually evaluated by measuring the decrease in its concentration when brought into contact with the adsorbent. The adsorption isotherm is then plotted as the apparent adsorption of the solute against the equilibrium concentration.

At low concentrations (i.e. for most practical applications of adsorption), the adsorption from solution isotherms mainly fall into two main types among those listed by Giles and Smith (1974): the *Type L* is concave to the concentration axis (analogous to Type I of the IUPAC classification) and the *Type S* is first convex and then concave to the concentration axis (analogous to Types III or V). A *Type L* isotherm having a long well-defined plateau is generally associated with monolayer adsorption of the solute and minimal competition from the solvent. A *Type S* isotherm is explained by a different balance between the adsorbate–adsorbent and adsorbate–adsorbate interactions: the latter are thought to be responsible for a 'cooperative' adsorption mechanism which produces an upward swing over the first part of the adsorption isotherm.

1.6   Energetics of Physisorption and Molecular Modelling

As a molecule approaches the solid surface, a balance is established between the intermolecular attractive and repulsive interactions. If other molecules are already adsorbed, both adsorbent–adsorbate and adsorbate–adsorbate interactions come into play. It is at once evident that the assessment of the adsorption energy is likely to become relatively complex in the case of a multicomponent system, especially if the adsorption is taking place from solution at a liquid–solid interface. For this reason, in the numerous attempts made to calculate energies of adsorption, most attention has been given to the adsorption of a single component at the gas–solid interface.

It is evident that the adsorption energy is governed by the nature of the adsorption system, that is by the adsorbent/adsorptive pair. There are a few adsorbents which give rise to essentially *non-specific* interactions with a wide range of different adsorbates. We categorise as *non-specific* contributions to the adsorption energy (i) the dispersion attractive term first characterised by London (1930) which arises from the rapid fluctuations in electron density in one atom, inducing an electrical moment in a neighbouring atom, (ii) the short range repulsion term which results from the interpenetration of the electron clouds and (iii) the polarisation term which arises as a result of the close proximity of the adsorbent electric field with the adsorbate. These three contributions labelled as ϕ_D, ϕ_R and ϕ_P respectively are expressed as a function of the distance between the adsorbate and adsorbent atoms (Barrer, 1966) and the corresponding mathematical equations are described in Chapter 6. By the principle of additivity of the pairwise interactions, the adsorption energy thus results from the summation of these energy terms taking place between all atoms of both adsorbate and adsorbent.

The most important non-porous adsorbent of this type is graphitised carbon black, which in its most uniform state has a surface structure composed almost entirely of the graphitic basal planes. The results shown in Figure 1.3 illustrate this behaviour. Here, the adsorption energy at zero coverage, labelled as E_0 is measured from the differential enthalpy of adsorption, $\left[\Delta_{ads} \, h_o \right]$ determined by the chromatographic method at very low surface coverage, as explained in Chapter 3. Note that under these conditions, the resulting adsorption energy can be related to the strength of the adsorbent/adsorbate interactions as we can eliminate or at least neglect the adsorbate/adsorbate interactions. E_0 is plotted against the number of carbon atoms, N_C in the various series of hydrocarbons. It is striking that there is an almost common linear relation between E_0 and N_C and we can also observe that for a given N_C, the values are very similar for the series of alkanes, alkenes and aromatic hydrocarbons (Cao et al., 1991).

By plotting experimental and theoretical low-coverage enthalpies of adsorption as a function of the molecular polarisability, Avgul and Kiselev (1970) also obtained a

$E_0/\text{kJ mol}^{-1}$

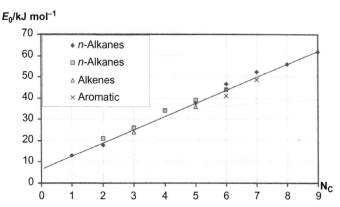

Figure 1.3 'Zero' coverage energy of adsorption versus carbon number, N_C, on graphitised carbon. *n*-Alkanes: diamonds, after Avgul and Kiselev (1970); *n*-alkanes: squares, after Carrott and Sing (1987); alkenes: after Carrott et al. (1989); aromatic: after Avgul and Kiselev (1970).

linear relation for a wide range of polar and non-polar molecules on graphitised carbon black (including noble gases, dimethyl ketone, ethyl ether and a series of alcohols). We may conclude that there is ample evidence to confirm the essentially non-specific nature of the interactions between the surface of graphitised carbon and all types of gas molecules. Such an observation holds true for other porous solids such as the silicalite-1 type zeolite and the MIL-47(V) MOF solid.

The scenario differs when a polar molecule is adsorbed on an ionic or a polar adsorbent surface. In this case, various types of additional *specific* interactions can contribute to the adsorption energy as for instance the field dipole and the field gradient quadrupole energies (Barrer, 1966; see Chapter 6 for more details).

The low-coverage energy data for the adsorption of *n*-hexane and benzene on various solids in Table 1.4 illustrate the importance of the nature of both the adsorbent surface and the adsorbate. Since *n*-hexane is a nonpolar molecule, the value of E_0 depends on the overall *non-specific* interactions with the surface of the adsorbent. Dehydroxylation of a silica surface involves very little change in the surface and therefore no significant difference in the value of E_0 for this adsorbate. However, replacement of the surface hydroxyls by alkylsilyl groups has resulted in a much greater effect. In this case, the weakening of the adsorbent–adsorbate interactions is mainly due to the fact that the surface modification has resulted in a modification of the plausible interacting site. The polarisabilities of benzene and hexane are very similar, but because of its electronic structure benzene exhibits significant specificity in its interaction with ionic or polar surfaces (e.g. hydroxylated silica and barium sulphate). Considerable attention has been given to the specificity associated with hydroxylated silica, but some specific adsorbent–adsorbate interactions are enhanced to an even greater extent by the exposure at the surface of ionic sites as illustrated by the data on $BaSO_4$ and the zeolite NaY (Table 1.4) which both show a higher adsorption

Table 1.4 Differential Enthalpies of Adsorption at Low Coverage $\left(|\Delta_{ads}\dot{h}|/\text{kJ mol}^{-1}\right)$ for n-Hexane and Benzene on Graphitised Carbon, Silica (Hydroxylated, Dehydroxylated and Modified) and Barium Sulphate

Adsorbent	n-Hexane	Benzene	Reference
Graphitised carbon black	42	42	Avgul and Kiselev (1970)
Hydroxylated silica	46	55	Kiselev (1965)
Dehydroxylated silica	48	38	Kiselev (1965)
Trimethylsilylated silica	29	34	Kiselev (1967)
Barium sulphate	47	70	Belyakova et al. (1970)
MIL-47(V)	50.6	43.4	Finsy et al. (2009)
NaY	46.1	62.8	Canet et al. (2005) and Eder and Lercher (1997)

enthalpy for benzene. In contrast, adsorbents with no special adsorption sites for particular adsorbates do not give any significantly enhanced adsorption enthalpies, as is the case for benzene on the MOF type MIL-47(V) solid.

One might expect argon and nitrogen to be similar in their physisorption behaviour since their physical properties are not very different (e.g. molecular sizes, boiling points and polarisabilities). However, the energy data in Table 1.5 show that this is true only if the nitrogen interaction is non-specific, as on graphitised and molecular sieve carbon, silicalite 1, $AlPO_4$-5 and MIL-47(V). The field gradient quadrupole term makes an important contribution when nitrogen is adsorbed on polar or ionic surfaces as hydroxylated silica, NaX, rutile and zinc oxide. In the case of rutile outgassed at $400\,°C$ and the zeolite NaX, the extra-framework cationic sites interact very strongly with nitrogen at low surface coverage.

The results in Table 1.5 also reveal that with some systems the differential enthalpy undergoes a pronounced change with increase in surface coverage whereas in other cases the change is much smaller (at least up to $\theta = 0.5$). Provided that the experimental measurements are made under carefully controlled conditions and that the adsorption systems are well characterised, adsorption energy data can provide valuable information concerning the mechanisms of physisorption. Indeed, the *increase* in the differential enthalpy of adsorption usually observed for an energetically homogeneous adsorbent surface is likely to be due to the attractive interactions between adsorbed molecules ('lateral interactions') becoming more important as the population in the monolayer is increased or as micropore filling approaches completion. The more usual progressive *decrease* in the differential enthalpy is generally to be expected if the adsorbent surface is energetically heterogeneous. An almost flat profile for the differential enthalpy is often related to the interactions between the adsorbate and a mildly energetically heterogeneous adsorbent surface, the significant decrease in the

Table 1.5 Differential Enthalpies of Adsorption $\left(|\Delta_{ads}\dot{h}|/\mathbf{kJ\,mol^{-1}}\right)$ of Argon and Nitrogen at 'Zero' and Half Coverage

Adsorbent	Argon		Nitrogen		Reference
	$\theta \to 0$	$\theta = 0.5$	$\theta \to 0$	$\theta = 0.5$	
Graphitised carbon	10	12	10	11	Grillet et al. (1979)
Hydroxylated silica (mesoporous)	15	9	>20	12	Rouquerol et al. (1979)
Dehydroxylated silica (mesoporous)	15	9	17	11	
Zinc oxide (450 °C)[a]	12	11	21	20	Grillet et al. (1989)
Rutile (150 °C)[a]	13	9	>20	10	Furlong et al. (1980)
Rutile (400 °C)[a]	15	11	30	13	
Molecular sieve carbon	20	15[b]	22	17[b]	Atkinson et al. (1987)
Microporous carbon	21	15[b]	25	15[b]	Rouquerol et al. (1989)
Silicalite I	14	14[b]	15	14[b]	Llewellyn et al. (1993a,b)
H-ZSM5 (Si/Al = 16)	14	14[b]	18	15[b]	
AlPO$_4$-5	11	14[b]	13	14[b]	Grillet et al. (1993)
Sepiolite	14	15	17	15	Grillet et al. (1988)
Attapulgite (130 °C)[a]	15	13	18	17	Cases et al. (1991a,b)
MIL-101(Cr)	16	11	28	13	Llewellyn (2013)
MIL-47(V)	11	11	14.5	14	

$\theta =$ surface coverage.
[a]Outgassing temperature.
[b]$\theta =$ fraction of pore filling.

adsorbent–adsorbate interactions being almost exactly balanced by an increase in the lateral adsorbate–adsorbate interactions.

An important consequence of the additive nature of the molecular interactions is that the entry of adsorbate molecules into pores of molecular dimensions results in an appreciable increase in the adsorption energy over that given by physisorption of the same molecules on the corresponding open surface. In accordance with the theoretical predictions of Everett and Powl (1976) for physisorption in slit-shaped micropores, the energies of adsorption for certain molecular sieve carbons show an approximately twofold enhancement over the corresponding values for graphitised carbon (Table 1.5 and Figure 1.4). It is noticeable that the molecular sieve zeolites NaX and Silicalite also give appreciably higher energies than macroporous silica, but that the degree of enhancement is not as pronounced as for carbon. To explain

E_0/kJ mol^{-1}

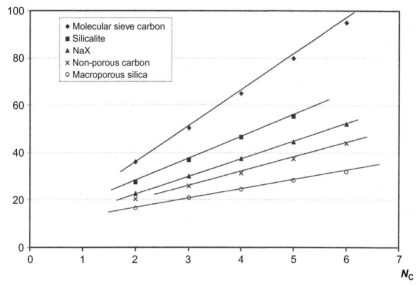

Figure 1.4 'Zero' coverage energy of adsorption of n-alkanes, versus carbon number N_C, on three microporous solids (and two others, for comparison).
Molecular sieve carbon: after Carrott and Sing (1987); silicalite: after Carrott and Sing (1986); NaX and macroporous silica: after Kiselev (1967); non-porous carbon: after Carrott and Sing (1987) and Avgul and Kiselev (1970).

these effects, one must take account the differences in solid structure and pore shape. These aspects are discussed in relation to the major adsorbent systems of carbons, oxides, zeolites and MOFs respectively in Chapters 10–14.

The energetics of physisorption can be complementarily explored using different levels of theory based on either phenomenological consideration or on molecular simulations. The former strategy aims at providing a qualitative interpretation of the experimental data by establishing simple correlations between the thermodynamic properties of interest and the intrinsic properties of both adsorbate and adsorbent without the need to perform accurate and time consuming calculations. As an illustration, in the case of argon adsorption in various X-Faujasites exchanged by mono- and divalent cations, it has been shown that the adsorption differential enthalpy at low coverage follows the same trend as the chemical hardness of the extra-framework cations, which measures their ability to polarise this adsorbate (Maurin et al., 2005). With the aid of fast computing facilities, molecular simulations are nowadays intensively employed as numerical experiments either to elucidate in tandem with the microcalorimetric profile, the adsorption mechanism at the microscopic scale or to predict the adsorption properties of a wide range of solids. This latter point is very helpful for narrowing down the choices of promising adsorbents for some adsorption/separation applications, that will need further experimental investigations. Indeed, modelling tools have now become an integral component of adsorption science. Monte Carlo simulations based on statistical mechanics principle can be employed in various thermodynamics ensemble. To achieve a satisfactory result, the computer runs must be relatively long and the simulated system large enough to overcome statistical uncertainties. Such an approach presents the advantage to be able to treat complex and large systems. The grand canonical (μVT) ensemble, where the chemical potential, volume and temperature are held fixed while the number of particles is allowed to vary, is the natural ensemble that mimics the experimental conditions used in adsorption studies. Prior to performing such calculations, a complete description of the adsorbent is required. This must include details of its solid structure and surface chemistry that can be often obtained by a computational assisted structure determination, which consists of coupling energy minimisation techniques with experimental data extracted from X-ray diffraction. Further, a microscopic model of the adsorbate and the adsorbate/adsorbent interactions via the specification of interatomic potentials (energy terms) must be precisely defined since the validity of the calculations strongly depends on the accuracy of such parameters. For charge-neutral adsorbates, the Lennard-Jones 12-6 potential appropriately parameterised is considered to be a satisfactory starting point for representing the pairwise adsorbate–adsorbent and adsorbate–adsorbate interactions (Lennard-Jones, 1932; Bezus et al., 1978). If the adsorbate molecule has a permanent dipole or quadrupole moment, the system cannot be reasonably described by only Lennard-Jones terms and it is thus necessary to take into account electrostatic interactions. In that case, fixed partial charges usually derived by quantum calculations are assigned to each atom of the adsorbent and the adsorbate, combined with a long-range treatment of the electrostatic interactions.

A critical step usually consists of defining an accurate set of potential parameters for describing the adsorbate/adsorbent interactions. As the parameters taken from generic interatomic potential are not always adequate, a careful parametrisation is required by either *ab initio* cluster calculations combined with a fitting procedure or by empirical derivation. At this stage, low-coverage adsorption enthalpy data are used to refine the evaluation of these parameters. Grand Canonical Monte Carlo (GCMC) simulations can be then performed to calculate the equilibrium thermodynamic properties, including the adsorption isotherm and the enthalpy over the complete range of loading. When one finds a reasonable consistence between the simulated results and the adsorption data, a further step consists of providing a plausible microscopic mechanism for the adsorption process.

Quantum calculations based on a careful choice of basis sets and functionals can be performed either on clusters or when possible periodic models. Such a computational approach while more time consuming than the statistical mechanics based methods, have been widely used for probing the molecular arrangements and the energetics of adsorption at low coverage or pore filling for a wide range of porous solids.

Some of the basic principles of the Monte Carlo/quantum simulations together with the recent adaptations most relevant to the study of adsorption in porous materials are described in Chapter 6. Typical illustrations are also provided to highlight how molecular simulations can not only define the geometrical features (pore size, surface area, pore volume, etc.) but also assist and guide interpretation of the experimental data.

An alternative approach to the molecular simulations is the application of an approximate theory. At present, the most useful theoretical treatment for the estimation of the equilibrium properties is generally considered to be the *density functional theory* (DFT). This involves the derivation of the density profile, $\rho(r)$ of the inhomogeneous fluid at a solid surface or within a given set of pores. Once $\rho(r)$ is known, the adsorption isotherm and other thermodynamic properties, such as the energy of adsorption, can be calculated. The advantage of DFT is its speed and relative ease of calculation, but there is a risk of oversimplification through the introduction of approximate forms of the required functionals (Gubbins, 1997). This method is presented and applied in Chapter 8.

1.7   Diffusion of Adsorbate

Diffusion in porous solids plays a crucial role in most adsorption/separation processes. This issue is particularly important in membrane based applications whose performance strongly depends on both adsorption equilibrium and diffusive transport rates of the adsorbed mixtures (Freeman and Yampolskii, 2010). Further, in applications involving gas storage, besides a large adsorption capacity, a high diffusion rate is also required to avoid a penalising limitation on the charging/discharging time. In the same way, this kinetic parameter is essential for the development of transport based devices.

Several experimental methods have been employed to characterise the *diffusivity* of various guest molecules confined in porous solids. They are usually classified into three categories: (i) 'macroscopic' methods including transient (zero-length-column chromatography) or quasi-state permeation; (ii) 'mesoscopic' techniques such as interference microscopy; (iii) 'microscopic' approaches with quasi-elastic neutron scattering (QENS) or pulse-field gradient nuclear magnetic resonance (PFG-NMR). They differ in the time and length scales the technique is able to probe. For instance, while PFG-NMR allows the exploration of intra-crystalline and inter-crystalline transport phenomena by accessing typical micrometres molecular displacements, QENS probes the nanometer length scale which mainly corresponds to intra-crystalline diffusion. Thus, the diffusivity values reported in the literature for a same solid with different techniques need to be considered with caution: for example, when the crystal contains some defects that can cause some additional transport resistances over long-range distances or when the macroscopic measurements are controlled by external mass and heat transfer resistances. QENS has been particularly useful in providing detailed molecular level information on the diffusion of various adsorbates in porous materials (Jobic and Theodorou, 2007) including the magnitude of the diffusivity and the mechanism for both translational and rotational motions within the crystal. A special application of this technique is to determine the self- (individual motions) or the transport-diffusivity (collective displacements) for the molecules with either incoherent or coherent cross section. Detailed accounts of the relevant aspects of diffusion can be found in the reviews by Kärger and Ruthven (1992), Ruthven (2005), Helmut (2007), Jobic and Theodorou (2007), and Freeman and Yampolskii (2010).

In a complementary way, equilibrium molecular dynamics (MD) simulations are valuable in interpreting the diffusion experiments. This is particularly important when MD simulations are coupled with QENS measurements as they track the same time and length scales. The basic principles of this computational technique consist of solving Newton's equations of motion for the given system. Similarly to the Monte Carlo approach discussed in the previous section, a complete description of the porous solids and the adsorbate/adsorbent interactions via the specification of potentials is required prior to running such calculations. Starting from an initial configuration usually produced by preliminary Monte Carlo simulations, the trajectory of the system consisting of a sequence of positions for the diffusive molecules with respect to time is then generated by integrating the equations of motion numerically over short time steps via appropriate algorithms. These calculations are usually obtained in the canonical (NVT) and micro-canonical (NVE) ensemble. The self- and transport-diffusivities for the guest molecules can then be calculated using Einstein relation (see Demontis and Suffritti, 1997; Jobic and Theodorou, 2007) and further compared to the experimental values extracted from QENS measurements. The basic principles of this modelling tool and some typical illustrations of the QENS–MD synergy applied to porous solids is presented in Chapter 6.

References

Allen, T., 1990. Particle Size Measurement. Chapman and Hall, London.

Atkinson, D., Carrott, P.J.M., Grillet, Y., Rouquerol, J., Sing, K.S.W., 1987. In: Liapis, A.I. (Ed.), Proceedings of the Second International Conference on Fundamentals of Adsorption. Engineering Foundation and American Institute of Chemicals Engineers, New York, p. 89.

Avgul, N.N., Kiselev, A.V., 1970. In: Walker, P.L. (Ed.), Chemistry and Physics of Carbon. Marcel Dekker, New York, p. 1.

Barrer, R.M., 1966. J. Colloid Interface Sci. 21, 415.

Barrer, R.M., 1978. Zeolites and Clay Minerals as Sorbents and Molecular Sieves. Academic Press, London.

Belyakova, L.D., Kiselev, A.V., Soloyan, G.A., 1970. Chromatographia 3, 254.

Bezus, A.G., Kiselev, A.V., Lopatkin, A.A., Du, P.Q., 1978. J. Chem. Soc. Faraday Trans. 74, 367.

British Standards Institution, 1958. British Standard 2955. BSI, London.

British Standards Institution, 1992. British Standard 7591, Part 1. BSI, London.

Brunauer, S., 1945. The Adsorption of Gases and Vapours. Oxford University Press, Oxford.

Brunauer, S., Emmett, P.H., Teller, E., 1938. J. Am. Chem. Soc. 60, 309.

Brunauer, S., Deming, L.S., Deming, W.S., Teller, E., 1940. J. Am. Chem. Soc. 62, 1723.

Canet, X., Nokerman, J., Frere, M., 2005. Adsorption 11, 213.

Cao, X.L., Colenutt, B.A., Sing, K.S.W., 1991. J. Chromatogr. 555, 183.

Carrott, P.J.M., Sing, K.S.W., 1986. Chem. Ind. 360.

Carrott, P.J.M., Sing, K.S.W., 1987. J. Chromatogr. 406, 139.

Carrott, P.J.M., Brotas de Carvalho, M., Sing, K.S.W., 1989. Adsorpt. Sci. Technol. 6, 93.

Cases, J.M., Grillet, Y., François, M., Michot, L., Villieras, F., Yvon, J., 1991a. Clays Clay Miner. 39 (2), 191.

Cases, J.M., Grillet, Y., François, M., Michot, L., Villieras, F., Yvon, J., 1991b. In: Rodriguez, F., Rouquerol, J., Sing, K.S.W., Unger, K.K. (Eds.), Characterization of Porous Solids II. Elsevier, Amsterdam, p. 591.

Chappuis, P., 1879. Wied. Ann. 8, 1.

Chappuis, P., 1881a. Phys. Chim. 178.

Chappuis, P., 1881b. Wied. Ann. 12, 161.

Dabrowski, A., 2001. Adv. Coll. Inerface Sci. 93, 135.

Davis, B.H., Sing, K.S.W., 2002. In: Schuth, F., Sing, K.S.W., Weitkamp, J. (Eds.), Handbook of Porous Solids, Vol. 1. Wiley, Weinheim, p. 3.

Deitz, V.R., 1944. Bibliography of Solid Adsorbents. National Bureau of Standards, Washington, DC.

Demontis, P., Suffritti, G.B., 1997. Chem. Rev. 97, 2845.

de Saussure, N.T., 1814. Gilbert's Annalen der Physik. 47, 113.

Dubinin, M.M., 1960. Chem. Rev. 60, 235.

Eder, F., Lercher, J.A., 1997. Zeolites. 18, 75.

Ehrlich, G., 1988. In: Vansclow, R., Howe, R. (Eds.), Chemistry and Physics of Solid Surfaces VII. Springer Verlag, Berlin-Heidelberg, p. 1 (Chapter1).

Emmett, P.H., Brunauer, S., 1934. J. Am. Chem. Soc. 56, 35.

Emmett, P.H., Brunauer, S., 1937. J. Am. Chem. Soc. 59, 1553.

Everett, D.H., 1972. Pure Appl. Chem. 31, 579.

Everett, D.H., Powl, J.C., 1976. J. Chem. Soc. Faraday Trans. I. 72, 619.

Favre, P.A., 1854. Compt. Rendus Acad. Sci. Fr. 39 (16), 729.

Favre, P.A., 1874. Annales Chim Phys. 5e série. 1, 209.

Finsy, V., Calero, S., Garcia-Perez, E., Merkling, P.J., Vedts, G., De Vos, D.E., Baron, G.V., Denayer, J.F.M., 2009. Phys. Chem. Chem. Phys. 11, 3515.

Forrester, S.D., Giles, C.H., 1971. Chem. Ind., 831.

Forrester, S.D., Giles, C.H., 1972. Chem. Ind., 318.

Freeman, B., Yampolskii, Y., 2010. Membrane Gas Separation. Wiley, New York.

Freundlich, H., 1907. Z. Phys. Chem. 57, 385.

Furlong, N.D., Rouquerol, F., Rouquerol, J., Sing, K.S.W., 1980. J. Chem. Soc. Faraday Trans. I 76, 774.

Giles, C.H., Smith, D., 1974. J. Colloid Interface Sci. 47, 3.

Gore, G., 1894. Phil. Mag. 37, 306.

Gregg, S.J., Sing, K.S.W., 1982. Adsorption, Surface Area and Porosity. Academic Press, London.

Grillet, Y., Rouquerol, F., Rouquerol, J., 1979. J. Colloid Interface Sci. 70, 239.

Grillet, Y., Cases, J.M., François, M., Rouquerol, J., Poirier, J.E., 1988. Clays Clay Miner. 36, 233.

Grillet, Y., Rouquerol, F., Rouquerol, J., 1989. Thermochim. Acta. 148, 191.

Grillet, Y., Llewellyn, P.L., Tosi-Pellenq, N., Rouquerol, J., 1993. In: Suzuki, M. (Ed.), Proceedings of the Fourth International Conference on Fundamentals of Adsorption. Kodansha, Tokyo, p. 235.

Gubbins, K.E., 1997. In: Fraissard, J., Connor, C.W. (Eds.), Physical Adsorption: Experiment, Theory and Applications. Kluwer, Dordrecht, p. 65.

Gurvich, L.G., 1915. J. Russ. Phys. Chim. 47, 805.

Haber, J., 1991. Pure Appl. Chem. 63, 1227.

Helmut, M., 2007. Diffusion in Solids. Springer Series in Solid State Science, Berlin.

Jobic, H., Theodorou, D.N., 2007. Micropor. Mesopor. Mater. 102, 21.

Kärger, J., Ruthven, D.M., 1992. Diffusion in Zeolites and Other Microporous Solids. Wiley, New York.

Kayser, H., 1881a. Wied Ann. Phys. 12, 526.

Kayser, H., 1881b. Wied Ann. Phys. 14, 450.

Kiselev, A.V., 1965. Disc. Faraday Soc. 40, 205.
Kiselev, A.V., 1967. Adv. Chromatogr. 4, 113.
Langmuir, I., 1916. J. Am. Chem. Soc. 38, 2221.
Langmuir, I., 1917. J. Am. Chem. Soc. 39, 1848.
Langmuir, I., 1918. J. Am. Chem. Soc. 40, 1361.
Lennard-Jones, J.E., 1932. Trans. Faraday Soc. 28, 333.
Llewellyn, P.L., Coulomb, J.P., Grillet, Y., Patarin, J., Lauter, H., Reichert, H., Rouquerol, J., 1993a. Langmuir 9, 1846.
Llewellyn, P.L., Coulomb, J.P., Grillet, Y., Patarin, J., André, G., Rouquerol, J., 1993b. Langmuir 9, 1852.
Llewellyn, P.L., 2013. Personal communication.
London, F., 1930. Z. Phys. 63, 245.
Mandelbrot, B.B., 1975. Les Objets Fractals: Forme, Hasard et Dimension. Flammarion, Paris.
Maurin, G., Llewelly, P.L., Poyet, T., Kuchta, B., 2005. Micropor. Mesopor. Mater. 79, 53.
McBain, J.W., 1909. Phil. Mag. 18, 916.
Mitscherlich, E., 1843. Annales Chim. Phys. 3e Série. 7, 15.
Polanyi, M., 1914. Verb. Deutsch Phys. Ges. 16, 1012.
Pouillet, M.C.S., 1822. Ann. Chim. Phys. 20, 141.
Rideal, E.K., 1932. The Adsorption of Gases by Solids. Discussions of the Faraday Society, London, p. 139.
Robens, E., 1994. In: Rouquerol, J., Rodriguez-Reinoso, F., Sing, K.S.W., Unger, K.K. (Eds.), Characterization of Porous Solids III. Elsevier, Amsterdam, p. 109.
Robens, E., Krebs, K.-F., 1991. In: Rodriguez-Reinoso, F., Rouquerol, J., Sing, K.S.W., Unger, K.K. (Eds.), Characterization of Porous Solids II. Elsevier, Amsterdam, p. 133.
Rouquerol, J., 1990. Impact of Science on Society, Vol. 157 UNESCO, Paris p. 5.
Rouquerol, J., Rouquerol, F., Pérès, C., Grillet, Y., Boudellal, M., 1979. In: Gregg, S.J., Sing, K.S.W., Stoeckli, H.F. (Eds.), Characterization of Porous Solids. Society of Chemical Industry, London, p. 107.
Rouquerol, J., Rouquerol, F., Grillet, Y., 1989. Pure Appl. Chem. 61, 1933.
Rouquerol, J., Avnir, D., Fairbridge, C.W., Everett, D.H., Haynes, J.H., Pernicone, N., Ramsay, J.D.F., Sing, K.S.W., Unger, K.K., 1994. Pure Appl. Chem. 66, 1739.
Rouquerol, F., Rouquerol, J., Sing, K.S.W., 1999. Adsorption by Powders and Porous Solids. Academic Press, pp. 51–92.
Ruthven, D.M., 2005. In: Introduction to Zeolite Science and Practice. Studies in Surface Science and Catalysis. Vol. 168, Elsevier, Amsterdam, p. 737.
Sing, K.S.W., Williams, R.T., 2012. Microporous Mesoporous Mater. 154, 16.
Sing, K.S.W., Everett, D.H., Haul, R.A.W., Moscou, L., Pierotti, R.A., Rouquerol, J., Siemieniewska, T., 1985. Pure Appl. Chem. 57, 603.
Taylor, H.S., 1932. In: The Adsorption of Gases by Solids. Discussions of the Faraday society, Royal Society of Chemistry, Cambridge, p. 131.
Thomson (Lord Kelvin), W., 1871. Phil. Mag. 42 (4), 448.
Zsigmondy, A., 1911. Z. Anorg. Chem. 71, 356.

2 Thermodynamics of Adsorption at the Gas/Solid Interface

Françoise Rouquerol, Jean Rouquerol, Kenneth S.W. Sing

Aix Marseille University-CNRS, MADIREL Laboratory, Marseille, France

Chapter Contents

Adsorption by Powders and Porous Solids. http://dx.doi.org/10.1016/B978-0-08-097035-6.00002-4

2.1   Introduction

It was many years ago that the major advances were made in the application of classical thermodynamics to gas adsorption. In particular, the work of Guggenheim (1933, 1940), Hill (1947, 1949, 1950, 1951, 1952), Defay and Prigogine (1951) and Everett (1950, 1972) led to a greatly improved understanding of the fundamental principles involved in the application of thermodynamics to various adsorption systems (see also Young and Crowell, 1962; Ross and Olivier, 1964; Letoquart et al., 1973).

The aim of this chapter is simply to introduce a selection of the most appropriate thermodynamic quantities for the processing and interpretation of adsorption isotherms and calorimetric data, which are obtained by the methods described in Chapter 3. We do not consider here the thermodynamic implications of capillary condensation, since these are dealt with in Chapter 8. Most attention is given to the terminology and the definition of certain key thermodynamic quantities for adsorption of a single gas on a solid adsorbent, whereas the case of gas mixtures is dealt with in Section 2.9.

In order to facilitate this operational approach, we adopt the following conventions and simplifying assumptions:

a. By making use of the *Gibbs dividing surface* (GDS), we define the associated *surface excess* properties, which are directly assessed from experimental measurements, without making any assumptions concerning the state, location or thickness of the adsorbed layer.

b. Since the role of the adsorbent is essentially to provide an adsorption potential, we may picture the adsorption, from the adsorbable gas viewpoint, as a transition from a *3D gas phase* to a *2D adsorbed film*. However, the latter may consist of more than one 2D phase. At equilibrium, the chemical potential of the adsorptive is the same in all phases, and therefore, two independent variables are sufficient to characterize each phase.

c. To process the information provided by the adsorption isotherm data, we adopt the *Helmholtz energy* ($=U-TS$) as the *primary thermodynamic potential* since this is most suitable for experiments carried out at constant temperature, volume and surface area.

d. To derive an equation of state for the adsorbate, it is necessary to select an additional intensive variable. For this purpose, it is convenient to define *the spreading pressure of the adsorbed phase*, which may be regarded as a form of *2D pressure*. The use of the two intensive variables (gas pressure and spreading pressure) also allows us to obtain the integral molar quantities, which are required to compare experimental data with statistical thermodynamic models.

e. We follow the IUPAC recommendations: see Mills et al. (1993) and also Everett (1972) who *inter alia* discourage the use of the imprecise term *heat of adsorption*.

f. We can avoid the usual assumption that the adsorbent always remains inert by embracing the case of when the adsorbent undergoes a deformation and then stores mechanical energy, as observed with carbons (Chapter 10), swelling clays (Chapter 12) or MOF's (Chapter 14). The changes in state functions then hold for the adsorbate/adsorbent system as a whole. For the interpretation of these changes, we must separate the contribution of the adsorbed phase from that of the solid adsorbent. We can make the usual assumption that the latter contribution is zero, or alternatively evaluate a change in the adsorbent structure from experimental data (X-rays, dilatometry, etc.) or by molecular simulation.

2.2 Quantitative Expression of Adsorption of a Single gas

2.2.1 Adsorption up to 1 bar

Without any independent information concerning the structure of the adsorbed layer, we might suppose that the local concentration, $c = dn/dV$, of the adsorbable component decreases progressively with increased distance, z, from the adsorbent surface; at a distance z equal to the thickness t of the adsorbed layer, this concentration reaches the constant value c^g of the gas phase. This form of hypothetical variation of local concentration is shown schematically in Figure 2.1a, where we also identify three zones (I, II and III).

We shall assume that there is no penetration of gas into the solid (i.e. no *absorption*) so that zone I is occupied solely by the adsorbent and therefore the adsorptive concentration in the solid, $c^s = 0$. In zone III, the adsorbable gas is at sufficient distance from the solid surface to have a uniform concentration, c^g, and here $z > t$. In this region, the concentration is dependent only on the equilibrium pressure and temperature. In Figure 2.1a, zone II is the 'adsorbed layer' or *adsorption space*, which is an intermediate region confined within the limits $z = 0$ and $z = t$. Here, the local concentration, c, is higher than the concentration of the gas in zone III and is dependent on z.

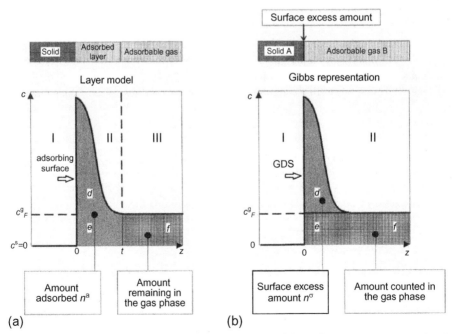

Figure 2.1 The layer model and the Gibbs representation of the surface excess amount (c, local concentration dn/dV of adsorbable gas; z, distance from the surface).

It follows from this simple picture that the volume of the adsorbed layer, V^a, can be expressed as the product of the interfacial area, A, and the thickness, t. Thus

$$V^a = At \tag{2.1}$$

We may define *the amount adsorbed*, n^a, of the substance in the adsorbed layer as

$$n^a = \int_0^{V^a} c \, dV$$
$$= A \int_0^t c \, dz \tag{2.2}$$

In Figure 2.1a, n^a is equivalent to the hatched area $(d + e)$.

The total amount, n, of the adsorbable substance in the whole system can be divided into two parts, the amount adsorbed and the amount remaining in the gas phase:

$$n = A \int_0^t c \, dz + c^g V^g$$

where V^g is the volume occupied by the gas at the concentration c^g, therefore:

$$n^a = n - c^g V^g \tag{2.3}$$

It is evident that the exact evaluation of n^a requires a knowledge of either the exact value of V^g (see Equation 2.3) or of the variation of the local concentration, c, with respect to z (see Equation 2.2). In practise, it is not easy to attain either of these requirements.

To overcome this problem, Gibbs proposed an alternative approach (in his papers published between 1875 and 1878 in the little known '*Transactions of the Connecticut Academy*', fortunately collected later in a single volume; Gibbs, 1928). This approach makes use of the concept of 'surface excess' to quantify the amount adsorbed. Comparison is made with a reference system, which is divided into two zones (I, of volume $V^{s,o}$ and II, of volume $V^{g,o}$) by an imaginary surface – now called the GDS – which is placed parallel and close to the adsorbent surface. The difference between these two surfaces is that the location of the actual adsorbent surface always suffers from some uncertainty, especially if the adsorbent is porous (see the issue of the void volume determination, in Section 3.2.5), whereas the location of the imaginary GDS is, by definition, perfectly defined. Gibbs' objective was indeed the 'precise measurement of the quantities' related to adsorption. The reference system occupies the same total volume V as the real system, so that:

$$V = V^{s,o} + V^{g,o} = V^s + V^a + V^g \tag{2.4}$$

In the reference system, the concentration of the gaseous adsorptive remains constant in the volume $V^{g,o}$, that is, up to the GDS. This is shown in pictorial form in Figure 2.1b. In this model, the *surface excess amount*, n^σ, represented by hatched area d, is defined as the difference between the total amount, n, of the adsorptive (hatched and crossed areas d, e and f) and the amount which would be present in the volume $V^{g,o}$

of the reference system if the final equilibrium concentration c^g would be constant up to the GDS (crossed area $e+f$). Thus:

$$n^\sigma = n - c^g V^{g,o}$$

It is convenient, for the sake of the physical interpretation of the data, to locate the GDS, as much as possible, on the surface which is accessible to the adsorptive used (even if an experimental uncertainty remains about the position of the latter surface), so that $V^{g,o} \approx V^a + V^g$. This is what is assumed in Figure 2.1b where the GDS is located at the same place as the adsorbing surface in Figure 2.1a. The experimental determination of the 'dead-space volume' corresponding to $V^a + V^g$ is examined in Section 3.2.5.

In these conditions:

$$n^\sigma = n - c^g(V^a + V^g) \tag{2.5}$$

Combining with Equation (2.3), one gets:

$$n^a = n^\sigma + c^g V^a \tag{2.6}$$

In Figure 2.1b, the surface excess amount n^σ is represented by hatched area d; whereas the amount adsorbed n^a, which also includes term $c_F^g V^a$ (area e) is represented in Figure 2.1a by area $d+e$.

For experiments carried out up to 1 bar, the quantity $c^g V^a$ is generally negligible in comparison with the quantity n^σ, essentially because the concentration of the gas phase is much lower than that of the adsorbed phase (most often, between 2 and 3 orders of magnitude lower), which would result in the area e being much smaller than shown in Figure 2.1b (i.e. much smaller than area d) and then one can write:

$$n^a \approx n^\sigma \tag{2.7}$$

However, we shall see that at higher pressure, it may become necessary to make some allowance for the difference between these two quantities (see Section 2.2.2).

The amounts n^a and n^σ are extensive quantities, which depend on the extent of the interface. The related 'surface excess concentration', Γ, is an intensive quantity, which is defined as

$$\Gamma = n^\sigma/A \tag{2.8}$$

where the surface area, A, is associated with the mass m^s of the adsorbent. The specific surface area is therefore:

$$a = A/m^s \tag{2.9}$$

What is usually measured and recorded is the specific surface excess amount n^σ/m^s, where

$$n^\sigma/m^s = \Gamma a$$

As we have seen already, n^σ/m^s is dependent on the equilibrium pressure, p, and the adsorbent temperature, T. The usual practise is to maintain constant temperature and then the relation

$$n^\sigma/m^s = f(p)T \tag{2.10}$$

is the adsorption isotherm.

For the sake of simplicity in some later sections of this book and only for pressures <1 bar, where $n^a \approx n^\sigma$, we adopt the *symbol* **n** to denote either the *specific surface excess amount* n^σ/m^s or the *specific amount adsorbed* n^a/m^s. So, we follow a custom which is justified in this pressure range but which should be avoided at higher pressures, as will be explained later in this chapter.

The Gibbs representation provides a simple, clear-cut mode of accounting for the transfer of adsorptive associated with the adsorption phenomenon. The same representation is used to define surface excess quantities assumed to be associated with the GDS for any other thermodynamic quantity related with adsorption. In this way, surface excess energy (U^σ), enthalpy (H^σ), entropy (S^σ) and Helmholtz energy (F^σ) are easily defined (Everett, 1972) as:

$$U^\sigma = U - U^g - U^s \tag{2.11}$$

$$H^\sigma = H - H^g - H^s \tag{2.12}$$

$$S^\sigma = S - S^g - S^s \tag{2.13}$$

$$F^\sigma = F - F^g - F^s \tag{2.14}$$

where U, H, S and F refer to the whole adsorption system at equilibrium (of total volume V and of total solid/gas interface area A), U^s, H^s, S^s and F^s refer to the solid adsorbent (zone I in Figure 2.1b), U^g, H^g, S^g and F^g refer to the gaseous adsorptive.

For the sake of completeness, U^σ and H^σ are separately defined earlier. Actually, when the enthalpy is defined in the standard way ($H = U + pV$), the Gibbs representation results in a useful simplification since the surface excess volume $V^\sigma = 0$, and we can write $H^\sigma = U^\sigma$ (as in Sections 2.4.2 and 2.5.1). Let us stress that any location is admissible for the GDS, provided it is stated. Special locations were proposed by Herrera et al. (2011) and by Gumma and Talu (2010) (see Section 3.5.)

2.2.2   Adsorption Above 1 bar and Much Higher

For gas separation or storage, adsorption at temperatures much lower than ambient is efficient but expensive. This is why, when possible, adsorption at room temperature is preferred in industry. The loss of efficiency due to a higher adsorption temperature must then be balanced by the use of higher pressures, essentially in the 5–50 bar range. In research work, the range is sometimes extended up to 150 bar and even above.

Under these conditions, the concentration of the adsorptive in the gas phase is no longer negligible as compared to its concentration in the adsorbed phase. In Figure 2.1b, area e represents the difference between the amount adsorbed and the

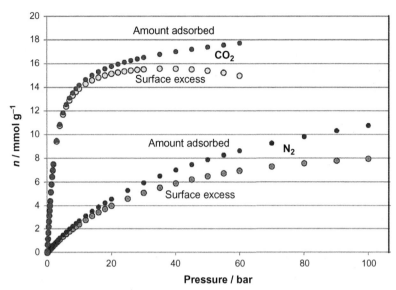

Figure 2.2 Adsorption isotherms of CO_2 and N_2 on a MOF, at 30 °C, plotted in terms of either amount adsorbed n^a or surface excess n^σ.
After Wiersum (2012).

surface excess amount, and it follows that two corresponding adsorption isotherms are appreciably diverging. Thus, area e is not negligible in comparison with area d, and furthermore, it increases with pressure. This is shown in Figure 2.2, for the adsorption of CO_2 and N_2 at 30 °C on a micro-porous MOF (Wiersum, 2012). The procedure used to derive the amounts adsorbed (from the primary data which are the surface excess amounts) is explained further in this section.

Another feature of the CO_2 surface excess isotherm in Figure 2.2 is the maximum (at slightly under 40 bar pressure) which, at first sight, might appear strange. This maximum simply indicates that from that point onwards the concentration in the gas phase increases more rapidly with pressure than the concentration in the adsorption space, which is already high and is approaching a limit.

This is illustrated in Figure 2.3 which shows (i) the concentration in the gas phase (lower curve) and (ii) the mean concentration in the adsorption space (i.e. the mean concentration of the adsorbate, upper curve) versus pressure. The concentration in the gas phase first increases linearly, following the ideal gas law, and then departs from linearity. The mean concentration in the adsorption space follows, over a very limited pressure range, the linear Henry's law and then continues to increase much more sharply than the gas concentration. At higher pressure, the curve of concentration in the adsorption space undergoes pronounced curvature, due to the fact that the adsorbate is approaching a state of maximum concentration (i.e. a condensed state, liquid-like or solid-like). At pressure 2, the slopes of both curves are equal, that is, the condition for a maximum in the surface excess isotherm is fulfilled. At higher pressures, the slope of the adsorbate concentration curve becomes smaller than that of the gas phase concentration curve. The reason is that the compressibility of the

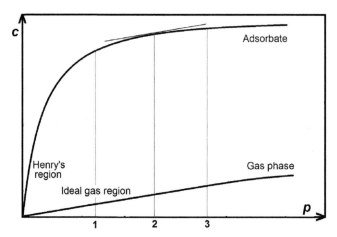

Figure 2.3 Concentrations in the gas phase and the adsorbate (i.e. mean concentration in the adsorption space) versus pressure. At pressure 2, slopes are identical.

gas phase is then higher than that of the adsorbate, whose concentration is approaching the limits for a condensed phase.

This effect of a maximum surface excess – not the total amount adsorbed – is shown in Figure 2.4, the representation being similar to that in Figure 2.1. The three concentration profiles correspond to pressures 1, 2 and 3 of Figure 2.3. One can see that as pressure is increased, the total amount adsorbed, represented by the area $d + e$, increases. At the same time, the surface excess amount, represented by area d, first increases, from 1 to 2, and then decreases, from 2 to 3.

As we saw in Section 2.2.1, the special significance of the surface excess amount is that it avoids any experimental uncertainties and assumptions (e.g. concerning the volume V^g of the gas phase). This is, therefore, a sound way of recording experimental adsorption data and should be systematically used as the first step in reporting and interpreting adsorption isotherms. In the second step, to understand and interpret the results, it is necessary to evaluate the actual amount adsorbed and the space occupied by the adsorbed phase. The distinction between these two steps is essential in

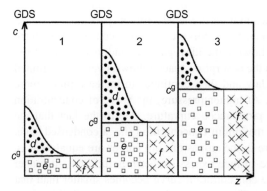

Figure 2.4 Representation of the surface excess amount (area d) and the amount adsorbed (areas $d + e$) around the maximum of a high-pressure surface excess isotherm: before (1), on (2) and after the maximum (3).

high-pressure adsorption work. Although it is generally considered permissible to use the terms 'surface excess amount' and 'amount adsorbed' interchangeably in the pressure range below 1 bar, as we have seen in high-pressure adsorption work, it is necessary to stress the difference between the two amounts.

The terms *absolute amount adsorbed* (Myers et al., 1997) and *total amount adsorbed* (Murata et al., 2001) are found in the literature. The latter term is easier to understand and the adjective 'total' is consistent with the fact that, when the GDS and the actual adsorbing surface are assumed to be close to each other, the amount adsorbed is always larger than the surface excess amount. Of course, determination of the amount adsorbed requires a knowledge of the volume containing the adsorbed layer (i.e. V^a in Figure 2.1a). The term *adsorption space* (Everett, 1972) is of more general significance than 'volume of the adsorbed layer', particularly in relation to micro-pore filling. In the case of micro-porous adsorbents, it is customary to consider that V^a is exactly equal to the micro-pore volume (Pribylov et al., 1991; Quirke and Tennison, 1996; Neimark and Ravikovitch, 1997). This was done to derive the isotherms in Figure 2.2. The micro-pore volume can be evaluated from a standard adsorption isotherm of N_2 at 77 K: if the isotherm is Type I, with a clear horizontal plateau, the height of the plateau directly corresponds to the micro-pore capacity. However, if the isotherm is a combination of Types I and II or I and IV, it is possible to obtain the micro-pore volume by applying an empirical procedure (see Section 9.2). Once this volume V^a is known, it is possible, through Equation (2.4), to derive the volume of the gas phase V^g, provided the volume of the solid V^s was previously determined (by experimental determination or by theoretical evaluation, see Section 3.2.5). Equation (2.3) then allows assessing the amount adsorbed.

Three other quantities are useful in assessing the efficiency of an adsorbent for gas separation or storage, namely:

1. The *working capacity* for a given species and between two stated 'working pressures': this is simply the amount adsorbed between these two pressures. In Figure 2.2, the working capacity between 2 and 20 bar is 12 mmol g^{-1} for CO_2 and 4 mmol g^{-1} for N_2.
2. The *separation factor* for two species 1 and 2, between two given working pressures, is the ratio of the working capacities for these two species. In the above example, the separation factor for CO_2 and N_2 between 2 and 20 bar is equal to $12/4 = 3$.
3. Unlike the previous quantities, the *selectivity* between two species, 1 and 2, is defined for one pressure. It compares, for each species, the amount adsorbed and the amount left in the gas phase. Thus,

$$\text{Selectivity} = \left(n_1^{\text{adsorbed}} / n_1^{\text{in gas phase}} \right) / \left(n_2^{\text{adsorbed}} / n_2^{\text{in gas phase}} \right)$$

2.3 Thermodynamic Potentials of Adsorption

The simplest gas/solid adsorption system is a closed system of volume V and temperature T, containing a mass m^s of solid adsorbent of surface area A and an amount n of a single adsorbable gas.

When the adsorptive gas is brought into contact with the clean adsorbent, part of it leaves the gas phase and becomes adsorbed (i.e. $dn^{\sigma} > 0$). If this adsorption takes place at constant T, V and A, the Helmholtz energy $F_{T,V,A,n}$ is the thermodynamic potential of the adsorption system since this potential attains a minimum value at equilibrium:

$$\left(\frac{\partial F}{\partial n^{\sigma}}\right)_{T,V,A} = 0$$

At equilibrium, the surface excess amount is $n^{\sigma} = n - n^{g}$, where n^{g} is the amount of adsorptive left in the gas phase, up to the GDS.

From Equation (2.14), we arrive at the general condition of equilibrium:

$$\left(\frac{\partial F}{\partial n^{\sigma}}\right)_{T,V,A} = \left(\frac{\partial F^{\sigma}}{\partial n^{\sigma}}\right)_{T,A} + \left(\frac{\partial F^{g}}{\partial n^{\sigma}}\right)_{T,V} + \left(\frac{\partial F^{s}}{\partial n^{\sigma}}\right)_{T,A} = 0 \qquad (2.15)$$

which applies to a system in which the adsorptive is distributed between the adsorbed phase, at the surface excess concentration $\Gamma = n^{\sigma}/A$, and the gas phase, at the concentration $c^{g} = n^{g}/V^{g}$.

In line with Section 2.1f, if we focus our attention on the adsorbate/adsorbent system as a whole and if we call $F^{\sigma,s}$ the Helmholtz energy of the latter, Equation (2.15) becomes:

$$\left(\frac{\partial F}{\partial n^{\sigma}}\right)_{T,V,A,n} = \left(\frac{\partial F^{\sigma}}{\partial n^{\sigma}}\right)_{T,A} + \left(\frac{\partial F^{g}}{\partial n^{\sigma}}\right)_{T,V} + \left(\frac{\partial F^{s}}{\partial n^{\sigma}}\right)_{T,A} = 0$$

Since for a closed adsorption system at equilibrium $dn = dn^{\sigma} + dn^{g} = 0$, we may express the condition of equilibrium in the form

$$\left(\frac{\partial F^{\sigma+s}}{\partial n^{\sigma}}\right)_{T,A} = -\left(\frac{\partial F^{g}}{\partial n^{\sigma}}\right)_{T,V} = +\left(\frac{\partial F^{g}}{\partial n^{g}}\right)_{T,V} \qquad (2.16)$$

The term $\left(\frac{\partial F^{g}}{\partial n^{g}}\right)_{T,V}$ is the chemical potential of the gas, μ^{g}. Similarly, for the equilibrium state of the pair adsorbed phase/adsorbent, which is characterized by variables A and T, we may define a surface excess chemical potential by the relation

$$\mu^{\sigma+s} = \left(\frac{\partial F^{\sigma+s}}{\partial n^{\sigma}}\right)_{T,A} \qquad (2.17)$$

and then using Equations (2.18) and (2.19), we can write:

$$\mu^{\sigma+s} = \mu^{g} \qquad (2.18)$$

Thus, for each equilibrium state, the chemical potential of the adsorbate/adsorbent system is equal to that of the adsorptive in the gas phase. In the case of a single adsorptive, the adsorbed state may be regarded as a one-component phase, which has lost one degree of freedom as compared with the gas. Equation (2.16) indicates that it is sufficient to specify two of the variables which characterize the gaseous adsorptive in order to define the adsorbed state for a particular adsorption system.

In case we assume that the adsorbent does not undergo any deformation, we do not have to consider the chemical potential $\mu^{\sigma+s}$ of the whole adsorbent/adsorbate system but only take account of the chemical potential μ^σ of the adsorbed phase. This simplifies Equations (2.17) and (2.18) into:

$$\mu^\sigma = \left(\frac{\partial F^\sigma}{\partial n^\sigma}\right)_{T,A} \tag{2.19}$$

and

$$\mu^\sigma = \mu^g \tag{2.20}$$

In the measurement of the adsorption isotherm, $\Gamma = f(p)_{T,A}$, the variables T and A are held constant. If we wish to obtain an analytical relation between Γ and p from Equation (2.20), it is necessary to find another adsorption potential. For this purpose, we must characterize the adsorbed phase by another intensive variable.

It is well known that the effect of surface tension is to minimize the area of a liquid surface. From a thermodynamic standpoint, the notion of surface tension, called 'surface stress' for a solid, is a more difficult concept as shown previously by Shuttelworth (1950) and revised by Sanfeld and Steinchen (2000). Its physical significance is more difficult to explain. For our present purpose, we may adopt an analogous definition of the surface tension of a clean solid adsorbent to that for a clean liquid surface.

Thus,

$$\left(\frac{\partial F^s}{\partial A}\right)_{T,V^s} = \gamma_T^s \tag{2.21}$$

which relates the Helmholtz energy, F^s of the solid adsorbent to its surface area A, γ_T^s being the surface tension of the clean solid at the temperature T.

Furthermore, as with a liquid, the surface tension of the adsorbent is reduced as a result of physisorption (Hill, 1968). This lowering of the surface tension is called the *spreading pressure* and is denoted by π so that:

$$\pi = \gamma^s - \gamma \tag{2.22}$$

The spreading pressure π is an intensive variable, which characterizes the adsorbed phase at the given surface excess concentration $\Gamma(=n^\sigma/A)$.

Introduction of the surface tension γ and surface area A in the expressions for H, F and G simply leads to a set of Legendre transforms (Moore, 1972; Alberty and Silbey, 1992) (also known as generalized functions) $\hat{H}, \hat{F}$ and $\hat{G}$. Thus,

$$\hat{H} = H - \gamma A, \quad \hat{F} = F - \gamma A, \quad \hat{G} = G - \gamma A \tag{2.23}$$

We can now characterize the state of equilibrium of the complete adsorption system by its transformed Gibbs energy, $\hat{G}$, defined as

$$\hat{G} = F + pV - \gamma A \tag{2.24}$$

For an adsorption process which would be carried out at constant T, p and γ, the transformed Gibbs energy $\hat{G}_{T,p,\gamma,n}$ would be the thermodynamic potential of adsorption which would attain the minimum value at equilibrium:

$$\left(\frac{\partial \hat{G}}{\partial n^\sigma}\right)_{T,p,\gamma} = 0 \tag{2.25}$$

We may define *the transformed surface excess Gibbs energy* by the relation

$$\hat{G}^\sigma = \hat{G} - G^g - \hat{G}^s \tag{2.26}$$

where $G^g(=n^g\mu^g)$ is the Gibbs energy of the adsorptive remaining in the gaseous state (characterized by T and p), and $\hat{G}^s$ is the transformed Gibbs energy of the solid adsorbent (characterized by T and γ^s), defined as:

$$\hat{G}^s = F^s + pV^s - \gamma^s A \tag{2.27}$$

From Equations (2.23)-(2.27) and taking into account that in the simplified Gibbs approach the excess volume equals zero, we obtain the relation between $\hat{G}^\sigma$ and F^σ:

$$\hat{G}^\sigma = \hat{F} + \pi A = n^\sigma \mu^\sigma \tag{2.28}$$

which becomes $\hat{G}^\sigma = \hat{H}^\sigma - TS^\sigma$ if we define a transformed surface excess enthalpy $\hat{H}^\sigma$ as:

$$\hat{H}^\sigma = U^\sigma + \pi A \tag{2.29}$$

We have seen that, for each equilibrium state, the adsorptive has the same chemical potential in the gas phase and in the adsorbed phase and that, for a given system, only two variables are required to provide a thermodynamic description of each of these phases. Consequently, there is a relation between the two intensive variables π and p.

Let us suppose that under conditions of equilibrium, a reversible change is made in the gas pressure and surface excess concentration of dp and $d\Gamma$, respectively, and that the corresponding change in the spreading pressure is $d\pi$. Since the chemical potential must change by the same amount throughout the system, we may write:

$$\left(\frac{\partial \mu^\sigma}{\partial \pi}\right)_{T,A} d\pi = \left(\frac{\partial \mu^g}{\partial p}\right)_{T,V} dp \tag{2.30}$$

Furthermore, we may express the change in the chemical potential with the spreading pressure in the form of a generalized Gibbs Duhem type equation:

$$d\mu^\sigma = -\frac{S^\sigma}{n^\sigma} dT + \frac{A}{n^\sigma} d\pi \tag{2.31}$$

Equation (2.31) is analogous to that obtained for a pure gas:

$$d\mu^g = -\frac{S^g}{n^g}dT + \frac{V^g}{n^g}dp \tag{2.32}$$

By combining Equations (2.30)–(2.32), we derive, at constant temperature T, the following relation between p and π:

$$\frac{1}{\Gamma}d\pi = v_T^g\,dp \tag{2.33}$$

where $\frac{1}{\Gamma_T}$ is the reciprocal of the surface excess concentration, and v_T^g is the gaseous molar volume.

The value of the spreading pressure π, at any value of Γ, may be calculated by integration of Equation (2.33) between $p=0$ and p, and correspondingly $\pi=0$ and π, after replacing v_T^g by $\frac{RT}{p}$ (assuming the gas to be ideal), so that:

$$\pi = RT\int_0^p \frac{\Gamma}{p}dp \tag{2.34}$$

Equation (2.34) is often referred to as the *Gibbs adsorption equation* where the interdependence of Γ and p is given by the adsorption isotherm. The Gibbs adsorption equation is a surface equation of state which indicates that, for any equilibrium pressure and temperature, the spreading pressure π is dependent on the surface excess concentration Γ. The value of spreading pressure, for any surface excess concentration, may be calculated from the adsorption isotherm drawn with the coordinates $\frac{n}{p}$ and p, by integration between the initial state ($n=0, p=0$) and an equilibrium state represented by one point on the adsorption isotherm.

A mathematical difficulty arises in the evaluation of the integral in the region below the first experimental point ($p=p_1$). First, the exact form of the adsorption isotherm is unknown in this range, and second, the ratio $\frac{n}{p}$ is indeterminate as p tends to zero. A possible solution is to assume a linear variation of n versus p in this lowest part of the adsorption isotherm. By analogy with solutions at infinite dilution, for which Henry proposed his linear law, it is the custom to refer to this linear part of the isotherm as the Henry's law region. Thus,

$$n = k_H \cdot p$$

where k_H is named the *Henry's law constant*.

It then follows that:

$$\int_0^{p_1} \frac{n}{p}dp = kp_1 = (n)_{p_1}$$

The assumption of linearity can be verified most easily by plotting the adsorption isotherm in a semi-logarithmic form, that is, n versus $\ln\{p\}$, where $\{p\}$ is the numerical value of p. An excessive curvature would indicate the need for another form of approximation.

For equilibrium pressures higher than p_1, for which surface excess concentration can be measured, the value of the integral term of Equation (2.34) is obtained by evaluation of the area under the curve $\frac{n}{p} = f(p)$, between pressures p_1 and p.

2.4   Thermodynamic Quantities Related to the Adsorbed States in the Gibbs Representation

By adopting the usual conventions of chemical thermodynamics, we are able to derive from the surface excess chemical potential μ^σ a number of useful surface excess quantities. Our purpose here is to draw attention to the difference between the molar and the differential surface excess quantities.

2.4.1   Definitions of the Molar Surface Excess Quantities

The surface excess chemical potential can be obtained by partial derivation of $\hat{G}^\sigma$ (defined in Equation 2.28) with respect of the surface excess amount n^σ, with the intensive variables T and π remaining constant, so that:

$$\mu^\sigma = \left(\frac{\partial \hat{G}^\sigma}{\partial n^\sigma}\right)_{T,\pi} \tag{2.35}$$

Alternatively, the chemical potential of the adsorbed phase is equal to the transformed Gibbs energy *divided* by the surface excess amount, that is, the transformed *molar* surface excess Gibbs energy, that is:

$$\mu^\sigma = \left(\frac{\partial \hat{G}^\sigma}{\partial n^\sigma}\right)_{T,\pi} = \frac{\hat{G}^\sigma_{T,\Gamma}}{n^\sigma} \tag{2.36}$$

Similarly, for other surface excess thermodynamic quantities, the corresponding molar quantities are as follows:

–the molar surface excess internal energy,

$$u^\sigma_{T,\Gamma} = \frac{U^\sigma_{T,\Gamma}}{n^\sigma} \tag{2.37}$$

–the molar surface excess entropy,

$$s^\sigma_{T,\Gamma} = \frac{S^\sigma_{T,\Gamma}}{n^\sigma} \tag{2.38}$$

–the molar surface excess Helmholtz energy,

$$f^\sigma_{T,\Gamma} = \frac{F^\sigma_{T,\Gamma}}{n^\sigma} \tag{2.39}$$

–the transformed molar surface excess enthalpy,

$$\hat{h}^{\sigma}_{T,\Gamma} = u^{\sigma}_{T,\Gamma} + \frac{\pi}{\Gamma} \tag{2.40}$$

By combining Equations (2.36)-(2.40), we obtain:

$$\mu^{\sigma}_{T,\Gamma} = u^{\sigma}_{T,\Gamma} + \frac{\pi}{\Gamma} - Ts^{\sigma}_{T,\Gamma} \tag{2.41}$$

2.4.2 Definitions of the Differential Surface Excess Quantities

We now return to the definition of the surface excess chemical potential μ^{σ} given by Equation (2.19) where the partial differentiation of the surface excess Helmholtz energy, F^{σ}, with respect to the surface excess amount, n^{σ}, is carried out so that the variables T and A remain constant. This form of partial derivative is generally referred to as *a differential quantity* (Hill, 1949; Everett, 1950). Also, for any surface excess thermodynamic quantity X^{σ}, there is a corresponding differential surface excess quantity $\dot{x}^{\sigma}$. (According to the mathematical convention, the upper point is used to indicate that we are taking the derivative.) So we may write:

–the differential surface excess internal energy,

$$\dot{u}^{\sigma}_{T,\Gamma} = \left(\frac{\partial U^{\sigma}}{\partial n^{\sigma}}\right)_{T,A} \tag{2.42}$$

–the differential surface excess entropy

$$\dot{s}^{\sigma}_{T,\Gamma} = \left(\frac{\partial S^{\sigma}}{\partial n^{\sigma}}\right)_{T,A} \tag{2.43}$$

By combining Equations (2.19), (2.42)-(2.43), we obtain:

$$\mu^{\sigma}_{T,\Gamma} = \dot{u}^{\sigma}_{T,\Gamma} - T\dot{s}^{\sigma}_{T,\Gamma} \tag{2.44}$$

and from Equation (2.29), we get the transformed differential surface excess enthalpy:

$$\hat{\dot{h}}^{\sigma}_{T,\Gamma} = \dot{u}^{\sigma}_{T,\Gamma} + A\left(\frac{\partial \pi}{\partial n^{\sigma}}\right)_{T,A} \tag{2.45}$$

Note that since the surface excess enthalpy, H^{σ}, is defined as $U^{\sigma}+pV^{\sigma}, H^{\sigma}=U^{\sigma}$. Similarly, the differential surface excess enthalpy, $\dot{h}^{\sigma}$ equals the differential surface excess internal energy, $\dot{u}^{\sigma}$.

It is of importance to both experimentalists and theoreticians that the differential and the integral molar excess quantities should not be confused.

2.5 Thermodynamic Quantities Related to the Adsorption Process

If the adsorption isotherm is entirely reversible (i.e. the adsorption and desorption paths coincide), we can assume that thermodynamic equilibrium has been established and maintained over the complete range of relative pressures p/p°. It is then possible to obtain useful thermodynamic quantities from the adsorption isotherm, by applying Equation (2.20), since each point on the isotherm represents a particular adsorbed state defined by one value of Γ (or n) and one value of the equilibrium pressure p (at temperature T).

Since the adsorption is generally carried out when the variables T, V and A are held constant, it is logical to use the differential quantities introduced in Equations (2.19), (2.42) and (2.43). So, taking the differential surface excess Helmholtz energy as the surface excess chemical potential and assuming the gas to be ideal, Equation (2.20) may now be written as:

$$\dot{u}^{\sigma}_{T,\Gamma} - T\dot{s}^{\sigma}_{T,\Gamma} = u^{g}_{T} + RT - Ts^{g}_{T,p} \tag{2.46}$$

where u^{g}_{T} is the molar internal energy of the ideal gaseous adsorptive, which depends only on the temperature T, and $s^{g}_{T,p}$ is the molar entropy of the ideal gaseous adsorptive which depends on T and p:

$$s^{g}_{T,P} = s^{g,\circ}_{T} - R\ln[p] \tag{2.47}$$

Here, $s^{g,\circ}_{T}$ is the molar standard entropy of the ideal gaseous adsorptive at temperature T and at standard pressure $p^{\oplus}$, which is not p° but is equal to either 10^{5} or 101 325 Pa, whereas $[p]$ stands for the ratio $p/p^{\oplus}$.

By combining Equations (2.46) and (2.47), we obtain:

$$\ln[p] = \frac{\dot{u}^{\sigma}_{T,\Gamma} - u^{g}_{T} - RT}{RT} - \frac{\dot{s}^{\sigma}_{T,\Gamma} - s^{g,\circ}}{R} \tag{2.48}$$

2.5.1 Definitions of the Differential Quantities of Adsorption

It is usual to call 'differential quantities of adsorption', the differences between the differential surface excess quantities and the same molar quantity, as in Equation (2.46) or (2.48).

The *differential energy of adsorption*, $\Delta_{ads}\dot{u}_{T,\Gamma}$, is:

$$\Delta_{ads}\dot{u}_{T,\Gamma} = \dot{u}^{\sigma}_{T,\Gamma} - u^{g}_{T} \tag{2.49}$$

Also, the differential energy of adsorption can be regarded as the change of internal energy of the complete adsorption system, produced by the adsorption of an infinitesimal surface excess amount dn^σ, when temperature, volume and surface area are held constant (and assuming the adsorbent to be inert and that its internal energy is not changed). Thus,

$$\Delta_{\text{ads}}\dot{u}_{T,\,\Gamma} = \left(\frac{\partial U}{\partial n^\sigma}\right)_{T,V,A} \tag{2.50}$$

In fact, the differential energy of adsorption may be *directly* obtained by the calorimetric measurement of the heat evolved by adsorption, (see Section 2.7).

We can define the *differential enthalpy of adsorption*, $\Delta_{\text{ads}}\dot{h}_{T,\,\Gamma}$, as:

$$\Delta_{\text{ads}}\dot{h}_{T,\,\Gamma} = \dot{u}^\sigma_{T,\,\Gamma} - u^g_T - RT \tag{2.51}$$

The differential enthalpy of adsorption may be obtained *indirectly* by the isosteric method (see Section 2.6.1). In the past, it was often referred to as the *isosteric heat* and denoted '$-q_{\text{st}}$'. This term is now discouraged and should be replaced by the *isosteric enthalpy of adsorption*. The differential energy and differential enthalpy of adsorption are related by the expression:

$$\Delta_{\text{ads}}\dot{h}_{T,\,\Gamma} = \Delta_{\text{ads}}\dot{u}_{T,\,\Gamma} - RT \tag{2.52}$$

In order to compare energies of adsorption for different adsorptives, it is convenient to evaluate the difference between the differential surface excess internal energy and the molar internal energy of the bulk adsorptive (liquid or solid) at the same temperature, that is, the difference between the differential energy of adsorption and the molar energy of condensation of the liquid ($\Delta_{\text{liq}}u_T$) or of the solid ($\Delta_{\text{sol}}u_T$). Lamb and Coolidge (1920) introduced the term 'net heat of adsorption', which was adopted by Brunauer et al. (1938) (see also Brunauer, 1945), and has been used by many other authors. However, since the original term is somewhat ambiguous, we recommend that it should be replaced by the 'net energy of adsorption'.

We have already seen that the differential surface excess energy is equal to the differential surface excess enthalpy; consequently, we can write:

$$\dot{u}^\sigma_{T,\,\Gamma} - u^l_T = \Delta_{\text{ads}}\dot{h}_T - \Delta_{\text{liq}}h_T \tag{2.53}$$

This difference is the 'net differential enthalpy of adsorption' and could also be called the *net isosteric enthalpy of adsorption*.

The *differential entropy of adsorption*, $\Delta_{\text{ads}}\dot{s}_{T,\,\Gamma}$, is:

$$\Delta_{\text{ads}}\dot{s}_{T,\,\Gamma} = \dot{s}^\sigma_{T,\,\Gamma} - s^g_{T,p} \tag{2.54}$$

The *differential standard entropy of adsorption*, $\Delta_{\text{ads}}\dot{s}^o_{T,\,\Gamma}$, is:

$$\Delta_{\text{ads}}\dot{s}^o_{T,\,\Gamma} = \dot{s}^\sigma_{T,\,\Gamma} - s^{g,\,o}_T \tag{2.55}$$

The differential entropy of adsorption can be readily calculated from the differential enthalpy of adsorption since from Equations (2.46), (2.51) and (2.54), we obtain:

$$\Delta_{\text{ads}} \dot{s}_{T,\Gamma} = \frac{\Delta_{\text{ads}} \dot{h}_{T,\Gamma}}{T} \tag{2.56}$$

It is important not to confuse the differential (or isosteric) enthalpy of adsorption with the *transformed differential enthalpy of adsorption* $\Delta_{\text{ads}} \hat{h}_{T,\Gamma}$, which is derived from Equation (2.45):

$$\Delta_{\text{ads}} \hat{h}_{T,\Gamma} = \ddot{u}^\sigma_{T,\Gamma} + A \left(\frac{\partial \pi}{\partial n^\sigma} \right)_{T,A} - u^g - RT \tag{2.57}$$

so that:

$$\Delta_{\text{ads}} \hat{h}_{T,\Gamma} = \Delta_{\text{ads}} \dot{h}_{T,\Gamma} + A \left(\frac{\partial \pi}{\partial n^\sigma} \right)_{T,A} \tag{2.58}$$

2.5.2   Definitions of the Integral Molar Quantities of Adsorption

The difference between a molar surface excess thermodynamic quantity $x^\sigma_{T,\Gamma}$ and the corresponding molar quantity $x^g_{T,p}$ for the gaseous adsorptive at the same equilibrium T and p is usually called the *integral molar quantity of adsorption* and is denoted as $\Delta_{\text{ads}} x_{T,\Gamma}$:

$$\Delta_{\text{ads}} x_{T,\Gamma} = x^\sigma_{T,\Gamma} - x^g_{T,p}$$

We can thus define the *integral molar energy of adsorption*:

$$\Delta_{\text{ads}} u_{T,\Gamma} = u^\sigma_{T,\Gamma} - u^g_T \tag{2.59}$$

and the *integral molar entropy of adsorption*:

$$\Delta_{\text{ads}} s_{T,\Gamma} = s^\sigma_{T,\Gamma} - s^g_{T,p} \tag{2.60}$$

We may derive the relation between these integral molar quantities of adsorption from Equations (2.20) using the expression of surface excess chemical potential μ^σ given by Equation (2.41) and assuming the gas to be ideal:

$$u^\sigma_{T,\Gamma} + \frac{\pi}{\Gamma} - T s^\sigma_{T,\Gamma} = u^g_T + RT - T s^g_{T,p} \tag{2.61}$$

Then, from Equation (2.40), the *transformed integral molar enthalpy of adsorption* is obtained:

$$\Delta_{\text{ads}} \hat{h}_{T,\Gamma} = u^\sigma_{T,\Gamma} + \frac{\Pi}{\Gamma} - u^g_T - RT \tag{2.62}$$

and therefore:

$$\Delta_{ads}s_{T,\Gamma} = \frac{\Delta_{ads}\hat{h}_{T,\Gamma}}{T} \tag{2.63}$$

The quantity $\Delta_{ads}\hat{h}$ was called the *equilibrium heat of adsorption* by Hill (1949) and Everett (1950), but this is no longer appropriate.

Equation (2.63) relating two integral molar quantities of adsorption must not be confused with the Equation (2.56) relating two differential quantities of adsorption. Note that the differential enthalpy of adsorption is defined without reference to the spreading pressure, whereas this is necessary for the integral molar transformed enthalpy of adsorption in Equation (2.62). It is only in the exceptional case when the differential energy of adsorption does not vary with Γ (e.g. at low coverage on highly homogeneous surfaces) that the differential energy of adsorption equals the integral molar energy of adsorption.

2.5.3 Advantages and Limitations of Differential and Integral Molar Quantities of Adsorption

In any investigation of the energetics of adsorption, a choice has to be made of whether to determine the differential or the corresponding integral molar quantities of adsorption. The decision will affect all aspects of the work including the experimental procedure and the processing and interpretation of the data.

When the main purpose of the gas adsorption measurements is to characterize the adsorbent surface or its pore structure, the preferred approach must be to follow the change in the thermodynamic quantity (e.g. the adsorption energy) with the highest available resolution. This immediately leads to a preference for the differential option, simply because the integral molar quantity is equivalent to the *mean* value of the corresponding differential quantity taken up to a recorded amount adsorbed. Their relationship is indicated by the mathematical form of Equation (2.64), which is explained in the following section.

Numerous examples are given in later chapters of the application of micro-calorimetric measurements. In this chapter, it is sufficient to refer briefly in general terms to the advantages of studying the changes in differential energies or enthalpies of adsorption. Thus, high initial values of $\Delta_{ads}\dot{u}$ or $\Delta_{ads}\dot{h}$ are associated with the interaction of adsorptive molecules with highly active surface sites and/or their entry into narrow micro-pores. A fairly sharp decline in the differential quantity is indicative of pronounced energetic heterogeneity; whereas an increase is generally the result of adsorbate–adsorbate interaction, which in some cases can be related to 2D phase changes. To satisfy the condition for adsorption on a uniform surface, such as the basal plane of pure graphite, one could expect that over an appreciable range of monolayer coverage, the differential energy would either remain constant or gradually increase.

The integral molar quantities are of importance for modelling adsorption systems or in statistical mechanical treatment of physisorption. For example, they are required for comparing the properties of the adsorbed phase with those of the bulk gas or liquid.

As originally pointed out by Hill (1951), the differential quantities are not sufficient to give a complete thermodynamic description of an adsorption system. In order to compare the experimental data with theoretical (e.g. statistical mechanical) values, it is necessary to evaluate integral molar adsorption energies and entropies.

Some experimental techniques are to be preferred for the accurate determination of integral quantities (e.g. from energy of immersion data or a calorimetric experiment in which the adsorptive is introduced in one step to give the required coverage), while others are more suitable for providing high-resolution differential quantities (e.g. a continuous manometric procedure). It is always preferable to experimentally determine the differential quantity directly since its derivation from the integral molar quantity often results in the loss of information.

2.5.4 Evaluation of Integral Molar Quantities of Adsorption

The evaluation of an integral quantity of adsorption, $\Delta_{ads}X$, requires the integration of the corresponding differential quantity between $\Gamma = 0$, and the given surface excess concentration when the variables T, V and A are held constant. The thermodynamic quantity X characterizes the complete adsorption system, and therefore, the integral quantity of adsorption is strictly made up of the three contributions due to the changes in the properties of the adsorbed phase, the gaseous adsorptive and the adsorbent. As before, we shall assume the gas to be ideal and the adsorbent to be rigid. In carrying out the required integration, we must be careful to specify which variables are held constant and also define the limiting equilibrium states. In addition, we must keep in mind that we are dealing with a closed system for which $dn = 0 = dn^\sigma + dn^g$.

2.5.4.1 Integral Molar Energy of Adsorption

Integration of the differential energy of adsorption is quite straightforward from Equation (2.50). Since the gas is ideal, its molar internal energy does not vary with pressure so that:

$$\Delta_{ads}u_{T,\Gamma} = \frac{1}{n^\sigma} \int_0^{n^\sigma} \Delta_{ads}\dot{u}_{T,\Gamma} dn^\sigma \tag{2.64}$$

The integral molar energy of adsorption is, therefore, obtained by integrating the differential energy of adsorption between the limits 0 and n^σ. It equals the mean of the differential energy of adsorption over the range of surface excess concentration from 0 to Γ.

2.5.4.2 Integral Molar Entropy of Adsorption

The integration of the differential entropy of adsorption between 0 and Γ is conveniently carried out with the variables T, V and A held constant:

$$\int_0^{n^\sigma} \left(\Delta_{ads}\dot{s}_{T,\Gamma} \right) dn^\sigma = \int_0^{n^\sigma} \left(\frac{\partial S^\sigma}{\partial n^\sigma} \right)_{T,A} dn^\sigma - \int_0^{n^\sigma} s_{T,p}^g dn^\sigma \tag{2.65}$$

To integrate the second term of the right-hand side of Equation (2.65), it must be taken into account that the molar entropy of the gaseous adsorptive depends on the pressure which varies from 0 to p when n^σ varies from 0 to n^σ (see Equation 2.47). So integration by parts gives:

$$\int_0^{n^\sigma} s^g dn^\sigma = n^\sigma s^g - \int_0^{n^\sigma} n^\sigma ds^g = n^\sigma s^g + R\int_0^{n^\sigma} n^\sigma d\ln[p] \qquad (2.66)$$

From Equations (2.56), (2.65), (2.66), we obtain:

$$\Delta_{ads}s_{T,\Gamma} = \frac{1}{n^\sigma}\int_0^{n^\sigma} \frac{\Delta_{ads}\dot{h}}{T}dn^\sigma + \frac{R}{n^\sigma}\int_0^{n^\sigma} n^\sigma d\ln[p] \qquad (2.67)$$

Comparison of Equations (2.61) and (2.67) shows that the second term of the right-hand side of the Equation (2.67) is due to the spreading pressure, which is related to the interactions between adsorbed molecules, often referred to as the 'lateral interactions'. It is only at very low coverages, where the spreading pressure is negligible, that the integral molar entropy of adsorption is simply dependent on the adsorbate-adsorbent interaction.

In the general case, the integral molar entropy of adsorption is not equal to the mean differential entropy of adsorption over the range of surface excess concentration from 0 to Γ because of the extra term of the right-hand side of Equation (2.67).

2.6 Indirect Derivation of the Quantities of Adsorption from of a Series of Experimental Physisorption Isotherms: The Isosteric Method

2.6.1 Differential Quantities of Adsorption

To obtain differential enthalpies of adsorption from physisorption isothermal data, it is advisable to measure a series of adsorption isotherms at different temperatures.

If we differentiate Equation (2.48) with respect to the adsorption temperature, so that the surface excess concentration Γ(or n) remains constant and assume that $\Delta_{ads}\dot{h}_{T,\Gamma}$ and $\Delta_{ads}\dot{s}_{T,\Gamma}$ do not vary with temperature, we obtain:

$$\left(\frac{\partial}{\partial T}\ln\lfloor p\rfloor\right)_\Gamma = -\frac{\Delta_{ads}\dot{h}_{T,\Gamma}}{RT^2}$$

and therefore:

$$\Delta_{ads}\dot{h}_{T,\Gamma} = R\left(\frac{\partial\ln\lfloor p\rfloor}{\partial(1/T)}\right)_\Gamma \qquad (2.68)$$

We should remember that, here, $[p]$ stands for the ratio $p/p^\ominus$, where $p^\ominus$ is a standard pressure taken as either 10^5 or 101 325 Pa but is not the saturation pressure p^o.

It is evident that Equation (2.68) is analogous to the well-known Clausius–Clapeyron equation for a one-component gas-liquid system. Integration of

Equation (2.68) between the limits of equilibrium pressures and temperatures of p_1, p_2 and T_1, T_2 gives:

$$\Delta_{ads}\dot{h} = -\frac{RT_1T_2}{T_2 - T_1}\ln\frac{p_2}{p_1} \qquad (2.69)$$

The usual method of calculating $\Delta_{ads}\dot{h}$ from a series of adsorption isotherms obtained at different temperatures is to plot $ln\lfloor p\rfloor_\Gamma$ for a given Γ (or n) as a function of $\frac{1}{T}$. This method of calculation of the differential enthalpy of adsorption based on the use of Equation (2.69) is called the *isosteric method*. Although $\Delta_{ads}\dot{h}$ is commonly called *isosteric heat* of adsorption, it is preferable to call it the *isosteric enthalpy* of adsorption. Since *heat* is not a state function, this term should be reserved for the raw calorimetric results. Applying Equation (2.68) at different values of Γ (or n), we can determine the variation of $\Delta_{ads}\dot{h}$ with Γ.

To apply Equation (2.69), one makes use of the equilibrium pressures p_1 and p_2 (measured at temperatures T_1 and T_2), which should not be replaced by *relative* equilibrium pressures (p_1/p_1^0) and (p_2/p_2^0). A simple calculation shows that the use of *relative* pressures would lower the calculated $\Delta_{ads}\dot{h}$ by an amount equal to $\Delta_{vap}h$.

We can verify with Equation (2.68) that $\Delta_{ads}\dot{h} < 0$ since, for physical adsorption, the equilibrium pressure necessary to obtain the surface excess concentration Γ (or n) increases with the adsorption temperature. It follows that $\Delta_{ads}\dot{s}$ is necessarily negative. However, since the differential entropy of adsorption, at constant T, given by Equation (2.56) is directly proportional to the differential enthalpy of adsorption, its calculation is not of great value.

The differential standard entropy of adsorption $\Delta_{ads}\dot{s}^o_{T,\Gamma}$ may also be deduced from Equations (2.55) and (2.56):

$$\Delta_{ads}\dot{s}^o_{T,\Gamma} = \Delta_{ads}\dot{h}_{T,\Gamma} - R\ln[p] \qquad (2.70)$$

As already explained, the application of the isosteric method relies on the principles embodied in Equation (2.68). In practise, the isosteric procedure is applied to at least two adsorption isotherms at different temperatures, which must not be too far apart. A temperature range of, say, 10 K is often considered to be a good compromise. Whenever possible, more than two adsorption isotherms should be measured and the plot of $\log_{10}[p]$ versus $1/T$ checked for linearity. The isosteric method is very sensitive to any error in the measurement of the equilibrium pressure. For this reason, it is not always reliable and should not be used at very low pressure unless the equilibrium pressures are measured with very high accuracy. Systematic comparisons between the values of differential enthalpy determined by the calorimetric and isosteric methods have revealed serious inaccuracies in the isosteric values at low pressures or low surface coverage (Rouquerol et al., 1972; Grillet et al, 1976). It turns out that one must be particularly careful when applying the isosteric method at surface coverage <0.5. A further constraint is that for each constant n^σ, there should be no 2D phase change over the range of temperature studied.

In spite of these limitations, two valuable features of the isosteric method should be recognized:

1. It is a simple and generally applicable procedure for the evaluation of differential enthalpies of adsorption;
2. The accuracy of the method is improved as the pressure is increased; unlike calorimetry, which tends to become less accurate at higher pressures.

2.6.2 Integral Molar Quantities of Adsorption

By making use of Equations (2.47) and (2.63), we can arrive at:

$$\ln[p] = \frac{\Delta_{ads}\hat{h}_{T,\Gamma}}{RT} - \frac{\Delta_{ads}s^o}{R} \tag{2.71}$$

If we differentiate Equation (2.71) with respect to the reciprocal of temperature $1/T$, so that the spreading pressure π remains constant, we obtain:

$$\Delta_{ads}\hat{h}_{T,\Gamma} = R\left(\frac{\partial \ln(p)}{\partial(1/T)}\right)_\pi \tag{2.72}$$

This equation which gives the transformed integral molar enthalpy of adsorption for the adsorption equilibrium characterized by the variables T and Γ (or n), must not be confused with the Equation (2.68) giving the differential enthalpy of adsorption. Indeed, by combining Equations (2.31), (2.44), (2.56) and (2.63) it is easy to obtain the relation between these two enthalpies of adsorption:

$$\Delta_{ads}\hat{h} = \Delta_{ads}\dot{h} + \frac{T}{\Gamma}\left(\frac{\partial \pi}{\partial T}\right)_\Gamma \tag{2.73}$$

Equation (2.72) is not easy to use in the general case in which the spreading pressure is unknown. But in the particular case of stepwise isotherms where there are two adsorbed phases in equilibrium with the gaseous adsorptive (i.e. in the case of an univariant adsorption system), Larher (1968, 1970) showed that the isosteric method may be used with the transition pressure p^n to give integral molar energies u^n and entropies s^n of the 'quasi-layer':

$$\ln \frac{p^n}{p^\infty} = \frac{u^n - u^\infty}{RT} - \frac{1}{R}\left(s^n - s^\infty\right) \tag{2.74}$$

where the standard state is the bulk adsorptive (liquid or solid) characterized by its molar energy, entropy and saturation pressure which he denotes u^∞, s^∞ and p^∞, respectively.

2.7 Derivation of the Adsorption Quantities from Calorimetric Data

Micro-calorimeters are of course well suited for the determination of differential enthalpies of adsorption, as will be commented on in Sections 3.2.2 and 3.3.3 Nevertheless, one should appreciate that there is a big step from the measurement of a heat of adsorption to the determination of a meaningful energy or enthalpy of adsorption. The measured heat depends on the experimental conditions (e.g. on the extent of reversibility of the process, the dead volume of the calorimetric cell and the isothermal or adiabatic operation of the calorimeter). It is therefore essential to devise the calorimetric experiment in such a way that it is the change of state which is assessed and not the mode of operation of the calorimeter.

We shall examine here the two major procedures for gas adsorption calorimetry (see Section 3.3.3). Both procedures make use of a heat-flowmeter micro-calorimeter of the Tian-Calvet type (see Section 3.2.2).

2.7.1 Discontinuous Procedure

The most common calorimetric procedure is the *discontinuous* one where the adsorptive is introduced in successive steps. The calorimetric cell with its contents (adsorbent and adsorptive) must be considered as an *open* system (see Figure 3.15). It is only when the adsorptive is introduced *reversibly* and when the *step is small enough* (so that the amount introduced can be written dn and the pressure increase dp) that the derivation of a differential energy of adsorption (as defined by Equations 2.49 and 2.50) is possible. Under these conditions, and taking into account the internal energy contributed by the gaseous adsorptive, we can write:

$$dU = dQ_{rev} + dW_{rev} + u_T^g dn \qquad (2.75)$$

where dQ_{rev} is the heat reversibly exchanged with the surroundings at temperature T, dW rev is the reversible work of the gas against the external pressure, u_T^g is the molar internal energy of the adsorbable gas at temperature T and dn is the amount of adsorbable gas introduced during a given step.

The calculation of dW_{rev} is easily accomplished (Rouquerol et al., 1980) if one notionally splits the volume of the whole adsorption system into two parts, V_A (external to the calorimetric cell, but in contact with the thermostat) and V_C (located within the calorimetric cell, see Figure 3.15). If we assume a reversible compression of an ideal gas by reduction of volume V_A, the whole systems exchanges work with the surroundings:

$$dW_{rev} = RT \, dn^\sigma + (V_A + V_C)dp \qquad (2.76)$$

where dn^σ is the amount adsorbed during the compression.

The work received by the calorimetric cell is only:

$$dW_{rev}(C) = RT \, dn^{\sigma} + V_C \, dp \tag{2.77}$$

By combining the above equations, we obtain:

$$d(n^g u^g + n^{\sigma} u^{\sigma})_{T,V,A} = dQ_{rev} + RT \, dn^{\sigma} + V_C \, dp + u^g(dn^g + dn^{\sigma}) \tag{2.78}$$

or

$$\left(\frac{dQ_{rev}}{dn^{\sigma}}\right)_{T,A} + V_C \left(\frac{dp}{dn^{\sigma}}\right)_{T,A} = \left[\left(\frac{dU^{\sigma}}{dn^{\sigma}}\right)_{T,A} - u^g - RT\right] = \Delta_{ads}\dot{h}_{T,\Gamma} \tag{2.79}$$

which provides a means for the experimental assessment of the differential enthalpy adsorption, $\Delta_{ads}\dot{h}_{T,\Gamma}$.

One sees that the use of Equation (2.79) requires a knowledge of the following experimental quantities: dQ_{rev} (heat measured by the calorimeter), dn^{σ} (amount adsorbed), dp (increase in equilibrium pressure) and V_C (dead volume of the part of the cell immersed in the heat-flowmeter of the micro-calorimeter; see Figure 3.15). If the conditions of small and reversible introduction of adsorptive are not fulfilled, the quantity assessed by Equation (2.79) can be described as a 'pseudo-differential' enthalpy of adsorption (see Figure 3.16a).

2.7.2 Continuous Procedure

In principle, the *continuous procedure*, where the adsorption takes place continuously and slowly, under quasi-equilibrium conditions, meets the above requirement of reversibility (Rouquerol, 1972). In this experiment, the basic experimental quantities from which one wishes to derive the differential enthalpy of adsorption are the rate of adsorption, f^{σ}, and the corresponding heat flow ϕ.

When the rate of introduction of the adsorptive, $f = \dfrac{dn}{dt}$, is constant (see Section 3.3.2), the rate of adsorption, f^{σ}, is slightly different. Thus:

$$f^{\sigma} = \frac{dn^{\sigma}}{dt} = f - \frac{V_e}{RT_e}\frac{dp}{dt} - \frac{V_C}{RT_C}\frac{dp}{dt} \tag{2.80}$$

where V_e and T_e are the volume and temperature of the external manometric device. The corresponding heat flow is:

$$\phi = \frac{dQ_{rev}}{dt} = \frac{dQ_{rev}}{dn^{\sigma}} \cdot \frac{dn^{\sigma}}{dt} = f^{\sigma} \cdot \left(\frac{dQ_{rev}}{dn^{\sigma}}\right)_{T,A} \tag{2.81}$$

By introducing Equation (2.81) into Equation (2.79), we obtain:

$$\Delta_{ads}\dot{h} = \left(\frac{dQ_{rev}}{dn^{\sigma}}\right)_{T,A} + V_C\left(\frac{dp}{dn^{\sigma}}\right)_{T,A} = \frac{\phi}{f^{\sigma}} + V_C \frac{dp}{dt}\frac{dt}{dn^{\sigma}}$$

$$\Delta_{ads}\dot{h} = \frac{1}{f^{\sigma}}\left(\phi + V_C \frac{dp}{dt}\right) \tag{2.82}$$

It is evident that a knowledge of f^{σ} (rate of adsorption) and ϕ (heat flow) is not enough to derive a continuous curve of differential enthalpy of adsorption. One must also know the dead volume V_C of the calorimetric cell proper and the derivative of the quasi-equilibrium pressure with time. Note that when this derivative is very small (i.e. in the nearly vertical parts of an adsorption isotherm) Equation (2.82) becomes simply:

$$\Delta_{ads}\dot{h} \approx \frac{\phi}{f} \tag{2.83}$$

and under these conditions $f^{\sigma} \approx f$. This means that if f is constant, the recording of heat flow ϕ versus time gives a direct measure of $\Delta_{ads}\dot{h}$ versus amount adsorbed.

The essential requirements for the application of this direct method are a sensitive micro-calorimeter (preferably of the heat-flow type, as described in Section 3.2.2) combined with equipment for the determination of the amount adsorbed. Although the assemblage of the apparatus is somewhat demanding, once the effort has been made the advantages of calorimetry are as follows:

1. In contrast to the isosteric and chromatographic methods, the derivation of differential enthalpies does not involve any assumptions provided that the experiments are carried out sufficiently slowly to ensure equilibrium in the adsorption and gas compression.
2. Accurate thermal data can be obtained, even at low pressures. This method is, therefore, recommended for studying the energetics of micro-pore filling or adsorption on high energy sites.
3. When used along with a continuous measurement of the isotherm, a continuous, high-resolution curve of $\Delta_{ads}\dot{h}$ versus n is obtained. This allows one to detect and characterize sub-steps associated with 2D phase changes (Grillet et al., 1979).

2.8 Other Methods for the Determination of Differential Enthalpies of Adsorption

Our aim in this section is to outline the advantages and limitations of the other experimental procedures available for the determination of the differential enthalpies of adsorption. Both thermodynamic and practical aspects are summarized here, but the latter are discussed in more detail in Chapters 3 and 4.

2.8.1 Immersion Calorimetry

As described in Chapter 4, immersion calorimetry provides an indirect method for determining adsorption energies. The net molar integral energy of adsorption can be obtained from the energy of immersion of the dry solid, but the evaluation of differential energies is more difficult. In this case, a number of independent heat measurements must be made after the progressive pre-coverage of the adsorbent surface. This approach is therefore sample-consuming (since a fresh sample is required for each pre-coverage point) and time-consuming (the weighing, outgassing, etc. may take a day for each point). Moreover, the method is limited to easily condensable adsorptives (mainly water and organic vapours). On the other hand, there are some practical advantages:

1. Provided that sufficient care is taken in the experimentation, the results are more accurate than can be obtained by the isosteric method, especially at low surface coverage (Harkins, 1952; Zettlemoyer and Narayan, 1967).
2. In some respects, the technique is less demanding than gas adsorption calorimetry. For example, the calorimeter and pre-adsorption rig are separate and therefore easy to handle.
3. Condensable vapours pose problems with the other methods. Some of these difficulties (e.g. temperature gradients and spurious condensation) can be avoided by the measurement of energies of immersion.

2.8.2   The Chromatographic Method

The gas chromatographic method is based on the relation between the differential enthalpy of adsorption at 'zero' coverage and the temperature dependence of the Henry's law constant, k_H, as expressed in the form of Equation (5.3). In the low pressure region, where Henry's law applies, the specific retention volume, V_s, is a linear function of k_H (Purnell, 1962; Littlewood, 1970). This relationship makes it possible to use elution chromatography since

$$\Delta_{ads}\dot{h} = RT^2[\partial(\ln\{V_s\})/\partial T]_n \qquad (2.84)$$

The successful application of the method is dependent on a number of requirements (Gravelle, 1978). The chromatographic column must be operating under almost ideal conditions to give sharp, symmetrical peaks, and the carrier gas must not be adsorbed. Ideal gas behaviour is generally a good approximation, but if not, the non-ideality of the adsorptive vapour must be taken into account (Blu et al., 1971).

The advantages of the method are as follows:

1. In principle, the differential enthalpy is determined at very low surface coverage.
2. Measurements are rapid, and with standard commercial equipment, the retention time measurements can be made over a wide range of temperature.
3. Adsorptives with low vapour pressures can be studied.

2.9   State Equations for High Pressure: Single Gases and Mixtures

For adsorption experiments carried out up to 1 bar, it is usually assumed that the gas phase behaves like an ideal gas. For higher pressures, this becomes less and less acceptable and a satisfactory calculation of either the amount left in the gaseous phase (in adsorption manometry, see Section 3.2.1) or its density (to derive the buoyancy in adsorption gravimetry, see Section 3.2.2) requires taking into account the molecular interactions and the limiting volume of the molecules when under compression.

2.9.1 Case of Pure Gases

A number of equations were proposed to relate the pressure p to the thermodynamic temperature T and the gas molar volume V_m. The three listed hereafter are among the easiest to use.

2.9.1.1 The van der Waals Equation (1890)

$$\left(p + a/V_m^2\right)(V_m - b) = RT \tag{2.85}$$

In this equation, the van der Waals constants a (to allow for the molecular interactions) and b (corresponding to the limiting molar volume) are independent of temperature and only dependent on the nature of the gas; they are calculated from the critical temperature T_c and critical pressure p_c (e.g. see Haynes, 2011) by means of the following relationships:

$$a = \frac{R^2 T_c^2}{64 p_c} \quad b = \frac{RT_c}{8 p_c}$$

2.9.1.2 The Redlich–Kwong–Soave Equation

Its initial form (Redlich and Kwong, 1949) was modified by Soave (1972) to become:

$$p = \frac{RT}{V_m - b} - \frac{a\alpha}{V_m(V_m + b)} \tag{2.86}$$

In this equation, constants a, b and α can be calculated from the critical temperature T_c, the critical pressure p_c and the reduced temperature T_r $(=T/T_c)$ by means of the following relationships:

$$a = 0.42747 \frac{R^2 T_c^2}{p_c} \quad b = 0.08664 \frac{RT_c}{p_c}$$

$$\alpha = \left[1 + \left(0.48508 + 1.55171\omega - 0.17613\omega^2\right)\left(1 - \sqrt{T_r}\right)\right]^2$$

where ω, which is called the acentric factor, is a correction factor to take into account the lack of sphericity and the low polarity of the molecule (Poling et al., 2000).

2.9.1.3 The Gasem–Peng–Robinson Equation (2001)

$$p = \frac{RT}{V_m - b} - \frac{a\alpha}{V_m^2 + 2bV_m - b^2}$$

In this equation, constants a, b and α can be calculated from the critical temperature T_c, the critical pressure p_c and the reduced temperature T_r $(=T/T_c)$ by means of the following relationships:

$$a = 0.45724\frac{R^2 T_c^2}{p_c} \qquad b = 0.07780\frac{RT_c}{p_c}$$

$$\alpha = \left[1 + \left(0.37464 + 1.54226\omega - 0.26992\omega^2\right)\left(1 - \sqrt{T_r}\right)\right]^2$$

where ω is the acentric factor already referred above.

Now, when available, the safest means of determining the density of the gas phase at the pressure and temperature of the experiment is of course to measure it experimentally. This is possible in adsorption gravimetry, when using a dedicated sinker of calibrated volume, on which the buoyancy effect is a direct measure of the gas density (see Section 3.2.2). The absolute gas density ρ can then be expressed versus pressure in a polynomial form such as:

$$\rho = Ap + Bp^2 + Cp^3 + Dp^4 + Bp^5 + Ep^5 + Fp^6$$

Results obtained, in that way for CO_2 and CH_4 at 30 °C and up to 50 bar, are reported in Figure 2.5 (Bourrelly, 2006). The corresponding polynomials are:

for CO_2: $A = 1.7276$, $B = 1.191 \times 10^{-2}$, $\qquad C = -2.9567 \times 10^{-4}$, $\qquad D = 1.6513 \times 10^{-5}$,
$\qquad E = -3.2134 \times 10^{-7}$, $F = 2.6628 \times 10^{-9}$

for CH_4: $A = 0.63893$, $B = 6.1066 \times 10^{-4}$, $\qquad C = 3.6945 \times 10^{-5}$, $\qquad D = 1.6439 \times 10^{-6}$,
$\qquad E = 3.4288 \times 10^{-8}$, $F = -2.5702 \times 10^{-10}$

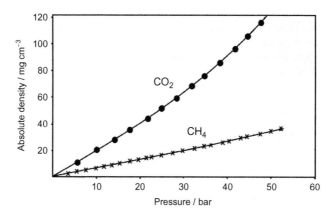

Figure 2.5 Absolute densities of CO_2 and CH_4 at 30 °C: experimental points and corresponding curves fitted with a polynomial equation.
After Bourrelly (2006).

2.9.2   Case of Gas Mixtures

Adsorption of gas mixtures above ambient pressure is of great importance in connection with gas separation or recovery (e.g. H_2, CH_4 in oil refinery exhaust gas). Here, the evaluation of the gas absolute density is more critical than for pure gases, and its direct experimental determination is preferable. However, one can make use of an equation of state, GERG-2004 (Kunz et al., 2007), which was developed in the scope of a project supported by European natural gas companies. This equation, which covers mixtures of up to 18 components, was recently up-dated under the name of the GERG-2008 equation of state to incorporate three more components (Wagner, 2011). The list of the 21 gases and vapours is given in Table 2.1.

The equation is normally applied between 90 and 450 K and 0–35 MPa, but an extended range between 60 and 700 K and 0–70 MPa can also be considered. It applies, *inter alia*, to 210 possible binary mixture combinations. The basic structure of this equation of state is summarized in Figure 2.6. The *departure function* is the function which takes into account the specific behaviour of a mixture of composition X (mole fractions) as compared with its pure components.

Table 2.1 The 21 Gases Covered by the GERG-2008 Equation of State (Wagner, 2011)

Methane	n-Pentane	Hydrogen	Propane	n-Octane	Oxygen	Water
Nitrogen	Isopentane	Hydrogen sulphide	n-Heptane	Ethane	n-Butane	n-Nonane
Carbon dioxide	n-Hexane	Carbon monoxide	Argon	Isobutane	n-Decane	Helium

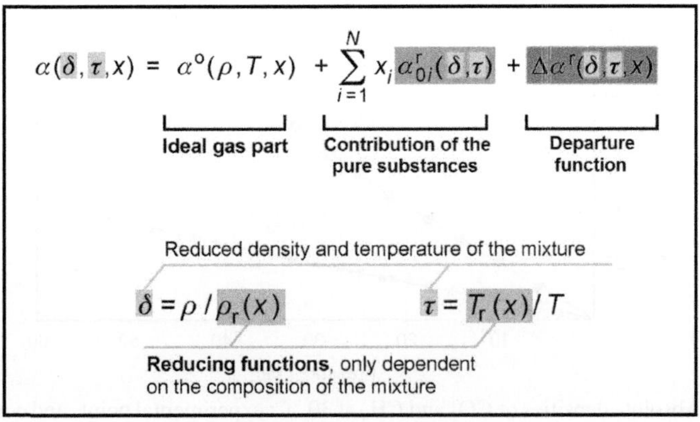

$$\alpha(\delta,\tau,x) = \alpha^{\circ}(\rho,T,x) + \sum_{i=1}^{N} x_i \alpha_{0i}^{r}(\delta,\tau) + \Delta\alpha^{r}(\delta,\tau,x)$$

Ideal gas part   Contribution of the pure substances   Departure function

Reduced density and temperature of the mixture

$$\delta = \rho / \rho_r(x) \qquad \tau = T_r(x)/T$$

Reducing functions, only dependent on the composition of the mixture

Figure 2.6 Basic structure of the GERG-2004 and GERG-2008 equations of state. After Wagner (2011).

References

Alberty, R.A., Silbey, R.J., 1992. Physical Chemistry. John Wiley, New York, p. 108.

Blu, G., Jacob, L., Guiochon, G., 1971. J. Chromatogr. 61, 207.

Bourrelly, S., 2006. Thèse Université de Provence (Marseille). p. 33.

Brunauer, S., 1945. The Adsorption of Gases and Vapors, Physical Adsorption, Vol. I Princeton University Press, Princeton, NJ.

Brunauer, S., Emmett, P.H., Teller, E., 1938. J. Am. Chem. Soc. 60, 309.

Defay, R., Prigogine, I., 1951. Tension Superficielle Et Adsorption. Dunod, Paris.

Everett, D.H., 1950. Trans. Faraday Soc. 46, 453, 942 and 957.

Everett, D.H., 1972. Pure Appl. Chem. 31 (4), 579.

Gibbs, J.W., 1928. Collected Works, Vol. 1 Longmans Green and Co., New York, pp. 221–222.

Gravelle, P.C., 1978. J. Therm. Anal. 14, 53.

Grillet, Y., Rouquerol, F., Rouquerol, J., 1976. Rev. Gen. Therm. 171, 237.

Grillet, Y., Rouquerol, F., Rouquerol, J., 1979. J. Colloid Interface Sci. 70, 239.

Guggenheim, E.A., 1933. Modern Thermodynamics by the Methods of J.W. Gibbs. Methuen, London.

Guggenheim, E.A., 1940. Trans. Faraday Soc. 36, 397.

Gumma, S., Talu, O., 2010. Langmuir. 26 (22), 17013.

Harkins, W.D., 1952. The Physical Chemistry of Surface Films. Reinhold Publishing Corp., New York, p. 268.

Haynes, W.M. (Ed.), 2011. Handbook of Chemistry and Physics. 92nd ed. CRC Press, Boca Raton, Florida.

Herrera, L., Fan, C., Do, D.D., Nicholson, D., 2011. Adsorption. 17, 955.

Hill, T.L., 1947. J. Chem. Phys. 15, 767.

Hill, T.L., 1949. J. Chem. Phys. 17, 507, 520.

Hill, T.L., 1950. J. Chem. Phys. 18, 246.

Hill, T.L., 1951. Trans. Faraday Soc. 47, 376.

Hill, T.L., 1952. Adv. Catal. 4, 212.

Hill, T.L., 1968. Thermodynamics for Chemists and Biologists. Addislon Wesley Publishing Company, Reading, MA.

Hill, T.L., Emmett, P.H., Joyner, L.J., 1951. J. Am. Chem. Soc. 73, 5102.

Kunz, O., Klimeck, R., Wagner, W., Jaeschke, M., 2007. *The GERG-2004 Wide-Range Equation of State for Natural Gases and other Mixtures*, Fortschritt-Berichte VDI Reihe 6, N° 557.

Lamb, A.B., Coolidge, A.S., 1920. J. Am. Chem. Soc. 42, 1146.

Larher, Y., 1968. J. Chim. Phys. Fr. 65, 974.

Larher, Y., 1970. Thèse Université Paris-Sud.

Letoquart, C., Rouquerol, F., Rouquerol, J., 1973. J. Chim. Phys. Fr. 70 (3), 559.

Littlewood, A.B., 1970. Gas Chromatography. Academic Press, New York.

Mills, I., Cvitas, T., Homann, K., Kallay, N., Kuchitsu, K., 1993. Quantities, Units and Symbols in Physical Chemistry, second ed. Blackwell Scientific Publications, London.

Moore, W.J., 1972. Physical Chemistry, fifth ed. Longman, London p. 96.

Murata, K., El-Merraoui, M., Kaneko, J., 2001. J. Chem. Phys. 114, 4196.

Myers, A.L., Calles, J.A., Calleja, G., 1997. Adsorption 3 (2), 107.

Neimark, A.V., Ravikovitch, P.L., 1997. Langmuir 19 (13), 5148.

Poling, B.E., Prausnitz, J.M., O'Connell, J.P., 2000. The Properties of Gases and Liquids, fifth ed. McGraw-Hill, New York.

Pribylov, A.A., Serpinskii, V.V., Kalashnikov, S.M., 1991. Zeolites. 8 (11), 846.

Purnell, H., 1962. Gas Chromatography. Wiley, New York.

Quirke, N., Tennison, S.R.R., 1996. Carbon. 34, 1281.

Redlich, O., Kwong, J.N.S., 1949. Chem. Rev. 44, 233.

Ross, S., Olivier, J.P., 1964. On Physical Adsorption. Interscience Publishers, New York.

Rouquerol, J., 1972. In: C.N.R.S. (Ed.), Thermochimie, vol. 201. Paris, p. 537.

Rouquerol, F., Partyka, S., Rouquerol, J., 1972. In: C.N.R.S. (Ed.), Thermochimie, vol. 201. Paris, p. 547.

Rouquerol, F., Rouquerol, J., Everett, D.H., 1980. Thermochim. Acta 41, 311.

Sanfeld, A., Steinchen, A., 2000. Surf. Sci. 463, 157.

Shuttelworth, R., 1950. Proc. R. Soc. Lond. A 63, 444.

Soave, G., 1972. Chem. Eng. Sci.. 27, 1197.

Wagner, W., 2011. Available from: http://www.thermo.ruhr-uni-bochum.de/en/prof-w-wagner/software.html.

Wiersum, A., 2012. Thèse Aix-Marseille.

Young, D.M., Crowell, A.D., 1962. Physical Adsorption of Gases. Butterworths, London.

Zettlemoyer, A.C., Narayan, K.S., 1967. In: Flood, E.A. (Ed.), The Solid-Gas Interface. Marcel Dekker Inc., New York, p. 152.

3 Methodology of Gas Adsorption

Jean Rouquerol, Françoise Rouquerol

Aix Marseille University-CNRS, MADIREL Laboratory, Marseille, France

Chapter Contents

Adsorption by Powders and Porous Solids. http://dx.doi.org/10.1016/B978-0-08-097035-6.00003-6

3.1   Introduction

The aim of this chapter is to introduce the major experimental procedures for the determination of gas adsorption isotherms and energies of adsorption. These measurements provide the essential physisorption data, which are discussed in other chapters. Indeed, even if a research project is directed towards, say, a spectroscopic study of adsorption, one cannot avoid having recourse to adsorption isotherm data.

It is perhaps worth noting that when a quantity, other than the amount adsorbed or the surface excess amount, is plotted against the equilibrium pressure (or the relative pressure $p/p°$, where $p°$ is the saturation pressure), this quantity should always be clearly defined (e.g. the differential or integral energy of adsorption or a crystalline parameter of the adsorbent). In our opinion, it is important to reserve the expression 'adsorption isotherm' for the relation between the amount adsorbed (or the surface excess amount) and the equilibrium pressure, thereby emphasising its basic character.

Gas adsorption measurements are not difficult to accomplish, especially with the many automated instruments available today on the market, provided that care is taken in their choice and in the selection or design of the full operational procedure, which starts with the outgassing of the adsorbent. Before any measurements are undertaken, it is useful to pose the following questions:

a. What is the purpose of the work?
b. Which technique is most suitable for the particular gas–solid system and the required conditions (i.e. the temperature and range of pressure)?
c. What operational procedure should be followed to obtain data with the desired accuracy and thermodynamic consistency?

Obviously, the choice of technique and experimental conditions (which most often directly influences the duration of the experiment) will depend on whether the measurements are to be used for, say, the routine determination of surface area and pore size, or for fundamental research, or for the acquisition of chemical engineering data. However, if the results are to have any real scientific value, it is essential that they are obtained under carefully controlled and well-defined conditions.

The three different physical quantities that can be utilised for determining a gas adsorption isotherm are pressure, mass and gas flow. They are successively examined in Section 3.2, prior to considering their use – sometimes combined – in the complex case of co-adsorption.

Whatever quantity (i.e. pressure, mass or gas flow) is selected for the determination of the amount adsorbed, one must choose between the two experimental procedures generally referred to as discontinuous (or point-by-point) and continuous (with a continuous recording of the adsorption isotherm, thanks to a slow and continuous introduction of the adsorptive, under quasi-equilibrium conditions (Rouquerol et al., 1988)). We should stress here (as illustrated in Section 3.2.3) that any gas adsorption technique making use of a gas flow, especially with a carrier gas, is not necessarily continuous in the above-mentioned sense, simply because it does not allow a continuous plot of the adsorption isotherm, with an infinity of quasi-equilibrium points.

To evaluate the adsorption energy, it is also necessary to determine either a heatflow or a temperature change, as discussed in Section 3.3.

We shall then examine the crucial issue – although one that is often skipped – of the adsorbent outgassing (Section 3.4) and conclude this chapter with the presentation of the data (Section 3.5), an essential aspect to allow fruitful exchanges within the adsorption community.

3.2   Determination of the Surface Excess Amount (and Amount Adsorbed)

3.2.1   Gas Adsorption Manometry (Measurement of Pressure Only)

This method is based on the measurement of the gas pressure in a calibrated, constant volume, at a known temperature. The beauty of gas adsorption manometry is its economy of resources: determining a gas adsorption isotherm, i.e., plotting the amount adsorbed versus equilibrium pressure, indeed requires, by any other method, measuring separately, with different devices, the amount adsorbed and the equilibrium pressure. Now, gas adsorption manometry carries out both determinations with a single pressure transducer. Probably because of this basic simplicity, this technique is the most widely used.

Practical and historical reasons require us to split this section into two parts, depending on the pressure range of operation.

3.2.1.1   Up to Atmospheric Pressure

The pressure range up to 1 bar is where gas adsorption was first studied and also where most characterisations by adsorption (especially determination of surface area and pore size distribution) are carried out even today.

3.2.1.1.1   Gas Adsorption Volumetry

The expression 'gas adsorption volumetry', which is still employed by some investigators and instrument manufacturers, dates back to the time when adsorption measurements were made with a mercury burette and manometer (as in Figure 3.1).

The technique for determining an adsorption isotherm is often referred to as the 'BET volumetric method' as it was the type of measurement made originally by Emmett and Brunauer (1937) and described by Emmett (1942). However, mercury burettes are

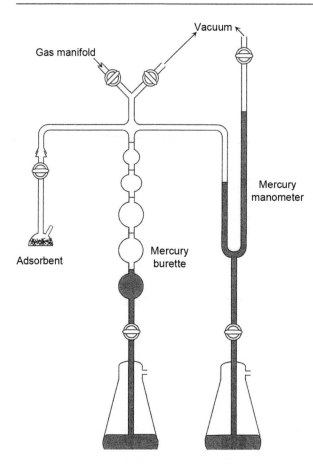

Vacuum

Gas manifold

Adsorbent

Mercury burette

Mercury manometer

Figure 3.1 Original 'BET' apparatus (gas adsorption volumetry).
After Emmett (1942).

generally no longer used, and it is therefore becoming inappropriate to refer to a volumetric procedure when the amount adsorbed is evaluated solely by the change of gas pressure. Along the same lines, many physisorption isotherms are still presented as the volume of gas (STP) adsorbed versus the relative pressure, a presentation that probably appeals to the mind but which has the shortcoming of being closely related to one single type of measurement, adsorption volumetry, and is furthermore obsolete. In a more universal presentation of the isotherms, it is indeed the specific amount adsorbed n (or the specific surface excess amount n^σ) that is plotted versus the relative pressure.

3.2.1.1.2 Simple Gas Adsorption Manometry

A simple modern set-up made of stainless steel (except the adsorption bulb and its stopcock, which is most often made of glass, for reasons of chemical inertia and of optical inspection during and after outgassing) with three valves and a commercial capacitance pressure transducer is shown in Figure 3.2. The 'dosing volume' is within the central cross, i.e., in the connecting tubes between the valves and the pressure transducer, whose volume is also part of it.

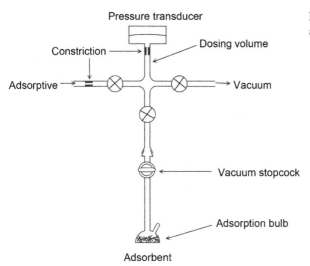

Figure 3.2 Basic gas adsorption manometry.

This technique allows the application of a discontinuous (point-by-point) procedure. For this purpose, within the dosing volume the pressure and temperature of each dose of gas are measured before it is allowed to enter the adsorption bulb. After adsorption equilibrium has been established, the amount adsorbed is calculated from the change in pressure and from the application of either the ideal gas law ($pV = nRT$, where p is the pressure, V the volume accessible to the gas phase, n the amount of gas, R the gas constant and T the thermodynamic temperature) or, alternatively, from a real gas law when required by the gas properties in the pressure and temperature range involved and by the desired accuracy (see Section 3.2.5).

The set-up in Figure 3.2 is relatively easy to build in the laboratory, with bellow-valves and high vacuum metal–metal connections (preferably with an intermediate copper gasket, to ensure a reproducible tightness), especially when one needs to associate it with another piece of equipment such as a calorimeter or a spectrometer of any kind. Because it can be relatively small and light, it also lends itself to being carried to the vicinity of 'large instruments' such as those providing neutrons or bright X-rays, which are useful for the study of the structure of the adsorbed phase or the deformation of the adsorbent under the effect of adsorption.

In reality, most commercial gas adsorption instruments are not far from this simple design, as its simplicity makes their construction, tightness (which is essential) and maintenance easier and cheaper all at once. One easy improvement is the addition, in parallel with the main 1 bar pressure transducer, of a second one, covering a smaller pressure range (i.e. 0–1 mbar or 0–10 mbar) in order to either assess the microporous structure from nitrogen adsorption data (see Chapter 9) or carry out, with reasonable accuracy, krypton adsorption (to determine low surface areas, see Section 7.2.4). Other features to improve the quality of the experimental data are discussed in Sections 3.2.5 and 3.2.6.

3.2.1.1.3   Gas Adsorption Manometry with Intermediate Gas Storage and Measurement

In principle, the preceding set-up has a shortcoming occurring from the addition of the errors made on the independent determination of each dose adsorbed. The greater the number of experimental points, the larger the final error. Such an error is easily detected, in case of a full adsorption–desorption isotherm showing a hysteresis loop, by the fact that, around what should be the lower closure point of the hysteresis loop, the desorption branch either remains above or crosses the adsorption one. The accuracy of most modern commercial equipment usually addresses this issue. Now, in specific set-ups, especially laboratory made, where other constraints (such as the association with another technique, which may impose a large void volume) still limit the accuracy with which each dose is determined, then the following arrangement brings a solution. As represented in Figure 3.3, the set-up features the same elements as in Figure 3.2 and simply requires a reservoir and its corresponding pressure transducer to be added.

One pressure transducer is used to determine the amount of adsorptive remaining in the reservoir, whose volume was previously calibrated, while the second is used to determine the adsorption equilibrium pressure and also the amount of non-adsorbed gas in the central cross and adsorption bulb.

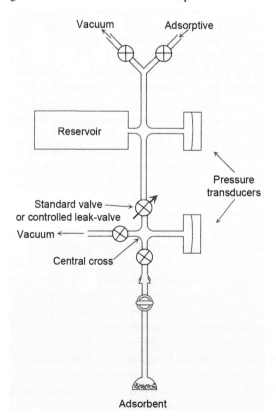

Figure 3.3 Gas adsorption manometry with reservoir and double pressure measurement.

This arrangement gives an integral measurement of the amount adsorbed and avoids the addition of successive errors resulting from a dosing device (Rouquerol and Davy, 1991). It lends itself to either a standard discontinuous procedure or to a continuous one with the help of a controlled-leak valve (Ajot et al., 1991).

3.2.1.1.4 Differential Gas Adsorption Manometry

This technique makes use of two similar bulbs (one with the adsorbent, the other with glass beads to equal the dead volume of the former). In the two forms which were marketed in the past for the sake of industrial control, the amount adsorbed was directly derived from the pressure difference between the two bulbs. This pressure difference was measured:

- Either once, after feeding the two bulbs with the same amount of adsorptive at room temperature, and then immersing them into liquid nitrogen to carry out the adsorption on the sample side, which provides a single point of the adsorption isotherm (Haul and Dümbgen, 1960, 1963), sufficient for a rapid surface area determination (see Section 7.2.2);
- Or continuously, as the adsorptive feeds the two bulbs through two identical capillaries, at a slow and nearly constant flow. As adsorption proceeds, the pressure difference between the two bulbs increases and provides a continuous determination of the amount adsorbed (Schlosser, 1959).

The shortcomings and limitations of these techniques, which are not any more used now, were commented in the previous edition of this book (Rouquerol et al., 1999).

A modern, more sophisticated form of differential gas adsorption manometry (Camp and Stanley, 1991; Webb, 1992) is shown in Figure 3.4. It can be considered as combining the best features of the two ancient set-ups just mentioned. This is a set-up making use of the continuous procedure. The differential assembly allows one to eliminate the dead volume correction, provided that the dead volume of the reference side is properly adjusted and made equal to that of the adsorbent side by means of glass beads, which requires knowing the specific volume of the adsorbent (and also, of course, of the beads). The differential pressure transducer 1, located between the two reservoirs, directly provides at any time the gas consumption difference between the sample side and reference side, with an accuracy which is irrespective of the point reached on the adsorption isotherm. The leak valve above the reference side (on the left) is used to provide a steady, continuous gas flow towards the reference bulb. The differential pressure transducer 2 is used to permanently keep identical the pressures above the beads and the sample, thanks to an appropriate opening of the leak valve above the sample (on the right). The absolute pressure transducer 3 provides the quasi-equilibrium pressure above the sample and is also used to control the opening of the left side leak valve in order to get a linear increase of that pressure versus time.

3.2.1.2 Above Atmospheric Pressure

Initially, a few clever set-ups where devised to allow to make use of glass-made equipment, in spite of the high pressures to be reached. During the last decades, the development of all-stainless steel equipment, easy to build with ready-made high-pressure connectors, tubing and transducers, has made it possible to come back

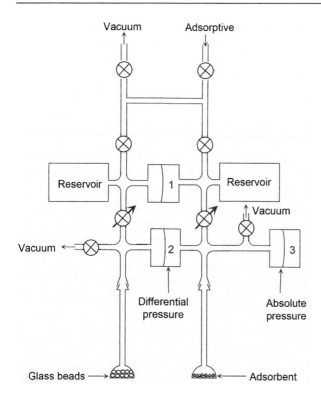

Figure 3.4 Differential adsorption manometry with double reservoir and triple pressure measurement. After Camp and Stanley (1991).

to the same basic design as that used for pressures lower than 1 bar and represented in Figure 3.2. Fittings of the same type as those used for high vacuum may be used up to ca. 150 bar. The main difference is in the internal diameter of the tubing, which, for high pressures, is usually in the 2–4 mm range in order to minimise the volume of the gas phase. Most 'high-pressure' adsorption experiments are not carried out above 50 bar, in part because if one is looking for applications, one should remain within a pressure range acceptable to industry. For pressures up to 1500 bar, special fittings and pressure transducers are also readily available on the market.

Apart from the value of the pressure, the following features are specific of adsorption experiments carried out above atmospheric pressure:

– Need to make use of metallic sample cells, less convenient to clean, fill, observe and submit to high-temperature outgassing than glass-made of silica-made bulbs;
– Difficulties in the sample outgassing, all at once because of the metallic sample cell (as said above) and because of the small-bore tubing, which makes it difficult to reach a good vacuum at the level of the sample. In the absence of any transparent sample cell, its simple evacuation may be problematic when it contains a light adsorbent which tends to spurt out of the cell. An anti-spurting automatic set-up to address this issue is described in Section 3.4.2;
– Need for safety devices, against the risks of explosion and also of leaks, in case harmful or dangerous gases are handled, such as H_2, CO, CH_4 and SH_2, whose adsorption is most studied for practical reasons;

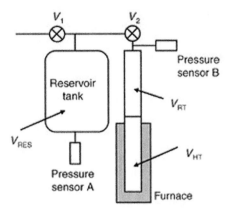

Figure 3.5 Sievert-type apparatus.
After Ichikawa et al. (2005).

-   Major incidence of what is supposed to be the void volume or 'dead space' of the sample bulb upon the final adsorption isotherm: this item is detailed in the general comment about the Gibbs representation (Section 3.5.1) and the amount adsorbed (Section 3.5.2);
-   Finally, it may happen that the overall phenomenon studied is sorption (embracing adsorption and absorption) rather than mere adsorption, especially in the case of H_2 storage by hydrides. In this area, engineers use to call their equipment a 'Sievert-type apparatus'. It can be seen from Figure 3.5 that this equipment has nothing different from an equipment of adsorption manometry.

3.2.1.3   Setting the Parameters for an Automated Experiment of Gas Adsorption Manometry

The automated instruments of gas adsorption manometry available on the market usually aim, first of all, at satisfying the needs of industry by providing the adsorption data in a relatively short time. Nevertheless, the same equipment usually offers a choice of possibilities allowing its adaptation to research work. The choices to be made by the operator are essentially the following:

-   *The sample mass* (see Section 3.2.6).
-   *The extent and quality of outgassing*: is the standard outgassing device and procedure provided with the adsorption equipment satisfactory for the sample under study? Is it not preferable to outgas the sample in a separate rig, with a different heating programme and final temperature (see Section 3.4)? In the latter case, the procedure should include a way to transfer the sample bulb from the outgassing rig to the adsorption equipment with a satisfactory tightness and protection against any further re-adsorption. This may request the use of an intermediate stopcock above the sample bulb or the possibility to introduce a pure gas (i.e. nitrogen, argon or helium) into the bulb before detaching it from the outgassing rig.
-   *The adsorptive*, with a check that the adsorptive properties used in the software are those desired by the operator, especially the parameters of the real gas law used to determine the amount of adsorptive remaining in the dead volume of the sample bulb when at 77 K.

- *The value of the saturation vapour pressure* to be used, either automatically measured by means of a special saturation vapour pressure bulb where a drop of adsorptive is condensed or a value introduced by the experimenter, after assuming, for instance, that the cryogenic liquid is pure enough to have its saturation vapour pressure equal to the surrounding atmospheric pressure.
- *The approximate number of adsorption points* required in each region of the adsorption isotherm. This is strongly dependent on the objective of the experiment and nature of the adsorbent. For instance:
  - if the objective is to determine the BET surface area of a fully unknown adsorbent, then a number of, say, 20 adsorption points is desirable in the $p/p°$ region between 0.01 and 0.3;
  - if the objective is to study the microporosity and apply methods like the t-method, the α_s method or the DFT, the requirement is about the same;
  - if the objective is to get a first, general idea of the adsorption isotherm in order to have an indication about the existence of micropores, mesopores, sheet-like structure and have an order of magnitude of the BET surface area, then a limited number of points (i.e. 20 on the adsorption branch and 10 on the desorption branch) is enough;
  - if the objective is to carry out a careful study of the mesoporosity, it becomes advisable to double the number of desorption points;
  - if the adsorbent is already known and if the objective is, for instance, to check the reproducibility of its synthesis, or follow its ageing, then only 5 adsorption points can do, provided they cover the pressure region previously selected for the application of the BET equation (see Section 7.2.2), thanks to the analysis of a more detailed adsorption isotherm.

- *The conditions of equilibration* of each adsorption point. This depends on the time one is ready to allot to the experiment, on the quality of the adsorption isotherm he wishes and on the presence of narrow micropores which may hinder the diffusion of the adsorptive. Usually, the experimenter is asked to select an 'equilibration interval' which is the time left between two successive checks of the pressure. Equilibration is considered to be achieved when the two successive readings of the pressure differ by less than a minimum pressure step usually fixed by the instrument manufacturer. This pre-selected pressure step is chosen to be larger than the pressure variations expected from the instability of the equipment. Such instability occurs from temperature fluctuations at the level of the cabinet or of the adsorption bulb (i.e. due to either a temperature shift or a level variation of the cryogenic liquid). The best check that the selected conditions of equilibration are satisfactory is to carry out a second experiment with more demanding conditions (i.e. increasing by 50% the equilibration time). If the two curves recorded are satisfactorily superimposed, then the equilibration conditions were satisfactory for both experiments. Otherwise, another experiment should be again carried out with a longer equilibration time, until the superimposition of the curves allows to call them real 'adsorption isotherms', since this term should be reserved to curves only corresponding to a succession of equilibrium points.
- *The upper relative pressure of the adsorption–desorption isotherm.* This is usually conveniently selected between 0.99 and 0.995. With a higher value, there is a risk that any zero shift or sensitivity shift of a pressure transducer (used either for the equilibrium pressure measurement or for the saturation vapour pressure determination) will make it impossible to ever reach that high experimental $p/p°$ value, so that the instrument never switches to the desorption procedure.
- *The conditions of the leak test* at room temperature, a test to be always carried out before the adsorption bulb is brought to the adsorption temperature. The experimenter may be asked to indicate the duration over which he wants the leak test to be performed. Such a test is

especially important if one is to duplicate an adsorption–desorption experiment, which seems to show an anomaly around the expected lower closure point of the hysteresis loop: the absence of closure point or, alternatively, crossing of the adsorption and desorption branches. One should also be aware that the appearance of a residual leak can occur by an unsatisfactory outgassing of a sample with narrow micropores, especially if it was not transferred under vacuum from the outgassing rig to the adsorption apparatus: introducing high-purity nitrogen in the adsorption bulb to protect the sample from humidity during the transfer is not indeed always satisfactory, since it may well produce, under 1 bar, adsorption of nitrogen in the micropores; this nitrogen may then very slowly desorb under vacuum at room temperature and provide a 'leak signal' on the equipment over hours.

3.2.2 Gas Adsorption Gravimetry (Measurement of Mass and Pressure)

3.2.2.1 Up to Atmospheric Pressure

A spring balance for the determination of the amount adsorbed was first used by McBain and Bakr (1926). In its simplest form, the apparatus consists of an adsorbent bucket attached to the lower end of a fused silica spring, which is suspended within a vertical glass tube as in Figure 3.6.

Spring balances are still used in certain research investigations when adsorption equilibration is very slow, e.g., for the study of hysteresis phenomena, or when the nature of the adsorptive (such as SH_2, NH_3, Cl_2 and a number of organic solvents) makes it desirable to use a full-glass equipment in order to avoid interaction with organic insulators or O-rings. For this purpose, it is advisable to replace the mercury manometer with a modern pressure gauge.

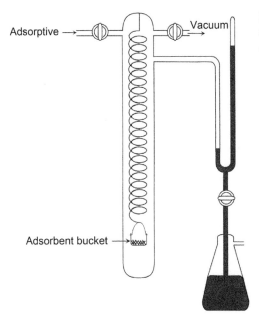

Adsorptive →

Vacuum

Adsorbent bucket →

Figure 3.6 The McBain spring adsorption balance.
After McBain and Bakr (1926).

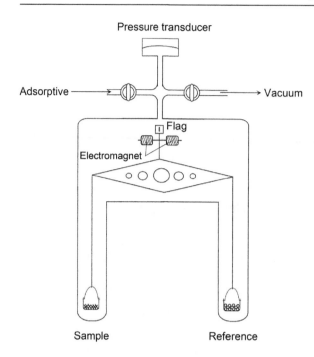

Pressure transducer

Adsorptive → Vacuum

Flag

Electromagnet

Sample   Reference

Figure 3.7 Principle of an electronic, null and symmetrical adsorption balance.

However, in recent years, spring balances have been largely superseded by vacuum electronic microbalances (Eyraud, 1986), first designed and used in adsorption by Eyraud and Eyraud (1955). The essential features of an electronic null microbalance are indicated in Figure 3.7.

The beam sometimes lies on knives but it is more often suspended to a torsion wire, which is especially interesting for adsorption studies as, at the difference between knives, it is insensitive to the presence of the few grains of powder that may leave the sample pan at the time of the evacuation and outgassing. Also, some balances are enclosed in glassware whereas others are all-metal. Some balances are fully symmetrical, which allows compensation for the buoyancy on the moving parts of the balance, provided the reference side is equipped with pan and beads of the same volume as on the sample side.

It is evident that two parallel sets of measurements are required in the application of adsorption gravimetry: that of the amount adsorbed (from the increase in mass) and that of the equilibrium pressure. Adsorption gravimetry happens to be the first technique that allowed an automated recording of a gas adsorption–desorption isotherm. The first approach made use of a discontinuous, stepwise, procedure, with an elegant, fully gravimetric equipment with two identical electronic vacuum balances: one to measure the weight of the adsorbent and the other to measure the gas pressure after the buoyancy on a sinker (Sandstede and Robens, 1970). A second approach made use of the continuous procedure, by introducing the adsorptive through a high-stability leak valve, and allowed determining high-resolution adsorption–desorption isotherms by simply recording the mass uptake versus the quasi-equilibrium pressure (Rouquerol and Davy, 1978). The gravimetric technique is particularly useful for studying the adsorption of a condensable vapour as any uncontrolled condensation on the wall of the balance does not affect the measurement of the amount adsorbed, but obviously condensation on the moving parts of the balance must be avoided by keeping this portion of the balance at a temperature above that of the adsorbent. An interesting combination for adsorption studies is that of gravimetry with heat-flux microcalorimetry (Le Parlouer, 1985). In the context of surface area determination, the sensitivity of a gravimetric procedure can be illustrated in the following manner. A monolayer that covers 10 m^2 of surface would be equivalent to 1 mg of He, 3 mg of N_2, H_2O or C_4H_{10}, 5 mg of Ar and 9 mg of Kr, whereas the reproducibility of commercial electronic balances is usually better than 10^{-2} mg.

The continuous mass signal (a voltage) lends itself to an easy recording, not only during adsorption but also during the outgassing. As the latter takes place *in situ* and as balances lend themselves to a high vacuum, balances are well suited for the study of samples for which a high degree of outgassing is required.

There are a number of potential sources of error in gravimetric measurements and particular attention is drawn to the following:

-   The direct measurement of mass does not eliminate the problem of the evaluation of the dead space. The adsorbent volume correction is now transformed into a *buoyancy correction* (see Section 3.2.5).
-   There is always, in the low-pressure range, *poor thermal transport* between the adsorbent and the surrounding thermostat. Thus, at low temperature (e.g. 77 K) and pressure under, say, 0.1 mbar, the adsorbent temperature is likely to be appreciably above that of the cryostat bath (see Section 3.2.6).
-   After the outgassing stage, the sample bucket is often *electrostatically charged*, together with the glass or silica hang-down tube to which it tends to stick. This problem was sometimes addressed by the use of either a small radioactive source in the bottom of the tube or a grounded, conductive wire netting inside the tube. An efficient way is to ground the sample itself (through metallic bucket, hook, suspension wire, balance beam and torsion wires) and to use a metallic tube.

A different type of extremely sensitive gravimetric technique is based on the effect of change of mass on the resonance frequency of a vibrating quartz crystal (see Figure 3.8).

In this case, the adsorbent must be firmly attached to the crystal. Its area can be as small as a few square centimetres, and mass changes as low as 10^{-2} μg can be detected from the frequency shift (Krim and Watts, 1991). Let us notice that a similar type of equipment is marketed for the study of adsorption on electrodes in solution.

3.2.2.2 Above Atmospheric Pressure

It was possible already several decades ago, starting from stainless steel, null, vacuum microbalances designed for atmospheric pressure, to transform them into balances able to withstand pressures up to 100 bar or even more. These balances allowed to carry out good experiments on the adsorption of pure gases.

More recently, magnetic suspension balances proved to be well suited for this type of work, since they could associate the performance of the best available standard analytical microbalances with the robustness of a relatively small-sized metal cylinder

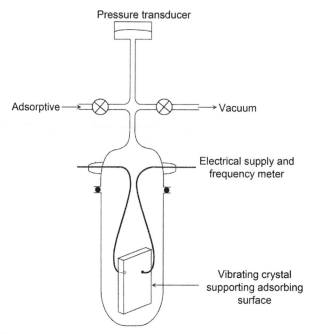

Figure 3.8 Vibrating crystal adsorption balance.

containing the sample pan and the suspension magnet. As we shall see, they are also well suited for the study of co-adsorption. Figure 3.9 shows the details of such a magnetic suspension balance, as devised by Lösch (Dreisbach et al., 1996, 2002).

The tube containing the suspension permanent magnet is surrounded by a set of electromagnets used to detect the position of the latter, whereas another electromagnet, suspended to the analytical microbalance, continuously restores that position, whatever the mass (within certain limits) of the adsorbent in the pan. In high-pressure adsorption gravimetry, the determination of the buoyancy effect is a crucial issue, so the balance can be provided with a 'sinker' of known volume whose apparent weight is permanently reduced by the buoyancy. If one knows the exact volume of the sample, pan and attached suspension system, then one can calculate the buoyancy to be taken into account to correct the apparent mass uptake (see Section 3.2.5).

Such a balance lent itself to the automated determination of adsorption isotherms of pure gases up to 100 bar (De Weireld et al., 1999).

3.2.3 Gas Adsorption with Gas Flow Control or Monitoring

In these techniques, the measurement of the amount adsorbed is derived from the knowledge of a gas flow rate. As already stressed in Section 3.1, we should be careful to distinguish between a gas flow technique and the 'continuous' or 'quasi-equilibrium' procedure, which provides the continuous recording of quasi-equilibrium data. This is because a gas flow does not necessarily involve a slow and continuous

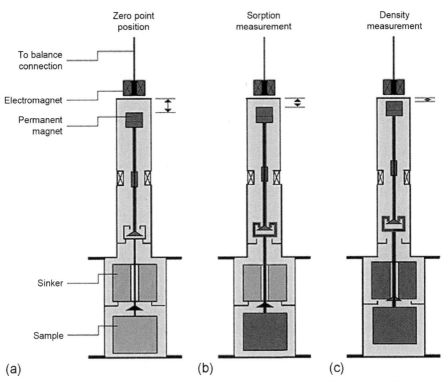

Figure 3.9 Principle of a magnetic suspension high-pressure adsorption balance. Successive weighing of (a) tare alone, (b) tare + sample and (c) tare + sample + sinker.
After Dreisbach et al. (1996).

adsorption. It is true, indeed, that the first two techniques described hereafter (with sonic nozzle or thermal gas flowmeter) lend themselves to a continuous adsorption procedure and, if the conditions of quasi-equilibrium are fulfilled, they provide an adsorption isotherm with an infinity of adsorption points, i.e., with a very high resolution. Conversely, the third technique (desorption under carrier gas, with analysis of outgoing gas stream) still makes use of a constant gas flow but with a discontinuous procedure providing a limited number of adsorption points, whereas the procedure used with inverse gas chromatography (IGC), which was initially discontinuous, is today closer to a continuous one. Let us point out that, for the sake of clarity, the term 'dynamic', sometimes used with these four techniques, should better be avoided, since it can either refer to the simple use of a carrier gas, or to a progressive and continuous adsorption, or to both.

3.2.3.1   Gas Adsorption with Gas Flow Control by a Sonic Nozzle

The distinctive feature of a sonic nozzle is that it can ensure a constant gas flow rate, in spite of an increase in the downstream pressure (which happens in an adsorption

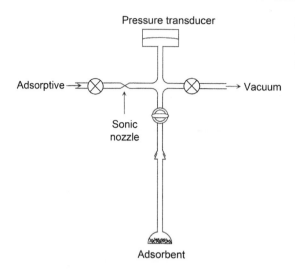

Pressure transducer

Adsorptive →

Sonic
nozzle

Vacuum

Adsorbent

Figure 3.10 Gas adsorption
set-up with a sonic nozzle.
After Rouquerol (1972) and
Grillet et al. (1977a).

experiment as the amount adsorbed increases), provided the upstream pressure remains constant and appreciably higher than the downstream pressure.

This principle was used to devise the simple adsorption apparatus shown in Figure 3.10 (Rouquerol, 1972; Grillet et al., 1977a).

With an upstream pressure of 3 bar, the gas flow rate was shown to remain constant, within 1%, for a downstream pressure of up to 0.4 bar, i.e., high enough to encompass the whole pressure range necessary to apply the BET method with nitrogen at 77 K.

This technique is well adapted for a continuous recording. For any new adsorption system studied, the quasi-equilibrium is checked by two successive experiments whose recordings should be superimposed if the quasi-equilibrium conditions are satisfactory. Experience shows that the flow rates needed are as low as 50–500 μmol h^{-1} (i.e. around 1–10 cm^3(STP) h^{-1}), and most often lower than 200 μmol h^{-1}, which can easily be achieved with a sonic nozzle.

When associated to microcalorimetry, this technique allows to carry out high-resolution gas adsorption calorimetry, with a nearly direct recording of the differential enthalpies of adsorption (see Section 3.3.2).

3.2.3.2   Gas Adsorption with a Thermal Gas Flowmeter (Mass Flowmeter)

An alternative approach is to use a thermal gas flowmeter in association with an automated needle valve, so as to control a constant flow rate, as represented in Figure 3.11 (Pieters and Gates, 1984). Such a thermal gas flowmeter provides a signal which depends on the heat capacity, thermal conductivity and mass flow of the gas; it is usually referred to as a 'mass' flowmeter although there is no direct measurement of mass. This set-up has the advantage of a free choice and setting of the gas flow rate, but its shortcoming is that it does not lend itself to an acceptable stability for flow rates lower than 250 μmol h^{-1} (i.e. 5 cm^3(STP) h^{-1}): as experienced with the nozzle flowmeter, it

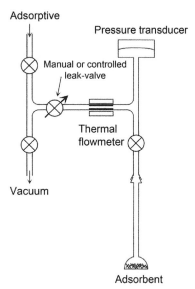

Figure 3.11 Gas adsorption set-up with a thermal gas flowmeter.
After Pieters and Gates (1984).

looks more essential to be able to set a flow rate low enough to ensure good quasi-equilibrium conditions than to have the choice of the setting at the expense of the stability.

3.2.3.3 Desorption Under Carrier Gas, with Analysis of Outgoing Gas Stream

This technique, proposed by Nelsen and Eggertsen (1958) and modified by Atkins (1964) and Karp et al. (1972), is shown in Figure 3.12.

The sample, located in a U-tube, is first outgassed under a stream of carrier gas (e.g. helium), then brought to the adsorption temperature (e.g. 77 K) under a stream of gas mixture (carrier gas + adsorptive at the partial pressure able to provide the desired point on the adsorption isotherm). At the exit from the U-tube, the thermal conductivity of the gas is compared with that of the starting gas mixture, with the help of a catharometer. When equilibrium is reached, i.e., when the signal of the catharometer is stable, then the sample is suddenly warmed (e.g. by immersing the U-tube into water), so that desorption quickly takes place and produces in the catharometer recording a clear peak that can easily be integrated to provide the amount of gas desorbed (provided an appropriate calibration was previously carried out).

If additional points are required, the gas composition is changed and the whole procedure is repeated.

Advantages of this procedure are (a) rapidity (heat exchanges are favoured by the gas mixture flowing at atmospheric pressure, especially when the carrier gas is helium), (b) sensitivity (surface areas as low as 0.5 m^2 can be measured with nitrogen adsorption) and (c) simplicity (it does not require any vacuum facility nor any, or dead,

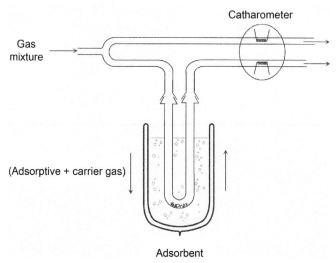

Figure 3.12 Technique of desorption under carrier gas.
After Nelsen and Eggertsen (1958).

volume calibration). The main interest of this procedure is to provide one experimental point for the application of the single point BET determination (see Section 7.2.2). Thus, one adsorptive-carrier gas mixture (for instance, 10% N_2, 90% He) can be stored and used as required. Measurement of equilibrium pressure is not required, but the atmospheric pressure, p_{atm}, should be known. In the above example, the equilibrium pressure is $1/9 p_{atm}$. If possible, the mixture is chosen so that the determination is made within the expected linear range of the BET plot.

The limitations are (a) the carrier gas (usually helium) may be adsorbed in narrow micropores at 77 K, which produces an *apparent increase* in the amount considered to be adsorbed and in the resulting BET surface area; (b) in the presence of somewhat larger micropores, it is the adsorptive, which may not completely and immediately desorb at the temperature of the water bath, that will produce an *apparent decrease* in the amount adsorbed; (c) the determination of successive experimental points involves successive cycles of cooling, flushing with new mixture and heating and (d) a test for adsorption equilibration is not always easy to establish (tailing of the signal).

Because of an apparent analogy with a chromatographic experiment (gas flow, carrier gas, catharometer), there is sometimes a tendency to call the above procedure 'chromatographic', but this should be avoided since no chromatographic effect is studied here and no retention time is measured.

3.2.3.4 Inverse Gas Chromatography

This technique makes use of a chromatographic equipment and procedure, which makes it interesting in a laboratory equipped with chromatography but not with adsorption manometry or gravimetry. It is called 'inverse' simply because instead of studying the composition of a gas mixture with the help of a known chromatographic adsorbent – as normally done in chromatography – the unknown here is the adsorbent, whereas the composition of the gas mixture (usually the carrier gas and a single adsorptive) is known and controlled. Proposed in the early 1940s in view of determining adsorption isotherms (Wilson, 1940; de Vault, 1943; Glueckauf, 1945), inverse frontal chromatography was used especially by the Kiselev School to study catalyst support adsorbents such as silica, alumina or activated carbon (Kiselev and Yashin, 1969) and then progressively improved. There are two main ways to operate: either by *frontal chromatography*, i.e., by continuously injecting the adsorptive in the carrier gas to keep the mixture entering the column at a steady concentration and by recording a breakthrough curve, or by *pulse chromatography*, by injecting a minute amount of adsorptive, which still allows to determine the retention time. Each of these ways can itself be divided into two procedures (Thielmann, 2004):

– A basic procedure where *each experiment provides one point* of the adsorption isotherm; in the case of frontal chromatography, the amount adsorbed is simply derived, for each partial pressure of the adsorptive, from the integration of the breakthrough curve and from the previous determinations of the 'dead time' of the system, the gas flow rate and the sensitivity of the detector. Such a procedure is sound, but time-demanding, and it is not the most popular.
– A much quicker though more sophisticated procedure which relies on a single experiment of the same type as above and on the observation that the shape of the chromatographic signal for a 'characteristic point' of the adsorption isotherm is closely related to the shape of the latter. This procedure is then called either 'frontal analysis of a characteristic point' or 'elution of a characteristic point' (ECP) in case pulse chromatography is used. The theoretical treatment follows the lines proposed early by Roginskii et al. (1960) and Cremer and Huber (1962). The accuracy of ECP was evaluated (Roles and Guiochon, 1992), and although it was mainly used in the Henry region of the isotherm, under the name of infinite dilution IGC, it is considered to provide satisfactory results up to the monolayer coverage. It is then called finite concentration IGC and used to derive the surface area (Balard et al., 2000, 2008).

All procedures above deal with chromatography and involve an elution phenomenon and a retention time. When frontal IGC is modified by a procedure of sudden warming up of the column to produce a quick desorption and to increase the accuracy of the measurement (Paik and Gilbert, 1986), we simply leave the field of chromatography (and of IGC) to enter that of the Nelsen and Eggertsen method presented in the section above.

3.2.4   Gas Co-Adsorption

The main issue in the study of gas co-adsorption is to know, for each adsorption point, the composition of the adsorbed phase and the corresponding partial pressures in the gas phase in order to determine, in parallel, the individual adsorption isotherms for each component of the gas mixture. An additional problem occurs from the fact that co-adsorption is mainly studied in view of gas separation which, for practical reasons, is to be carried out at room temperature, i.e., under elevated pressures.

Gas co-adsorption isotherms are principally determined by adsorption gravimetry and most often with a magnetic suspension balance. Adsorption gravimetry can then be used either alone or associated with adsorption manometry.

The latter arrangement was indeed proposed by Keller et al. (1992) as a simple and safe way to study co-adsorption of two gases, provided their molar masses are sufficiently different. The manometric experiment provides a total amount adsorbed $n_{tot}^{\sigma} = n_1^{\sigma} + n_2^{\sigma}$, whereas the gravimetric experiment provides the total mass adsorbed $m_{tot}^{\sigma} = m_1^{\sigma} + m_2^{\sigma}$. As $n_1^{\sigma} M_1 = m_1^{\sigma}$ and $n_2^{\sigma} M_2 = m_2^{\sigma}$, we have two unknowns, n_1^{σ} and n_2^{σ}, and two equations from which we obtain, for instance:

$$n_1^{\sigma} = \left(m_{tot}^{\sigma} - n_{tot}^{\sigma} \cdot M_2\right)/(M_1 - M_2) \tag{3.1}$$

On practical grounds, some difficulties stem from the large volume of adsorption balances: to accurately measure pressure changes due to adsorption, one must usually place a considerable amount of the adsorbent outside the balance pan (because of the limited loading capacity of the balance) but close enough to be at the same temperature.

Once the mixture of adsorptive is ternary, the above procedure does not hold any more and analytical means must be associated to adsorption gravimetry to determine, for any adsorption point of the isotherm, the composition of the gas phase and, hence, that of the adsorbed phase. The ideal analytical technique would be that which would not withdraw any amount of gas mixture, e.g., a spectroscopic technique simply requesting a circulation of the gas between two windows. Nevertheless, due to problems arising in part from the large pressure range covered, which drastically changes the response of the detector, it does not seem that this approach was much developed. Alternatively, analysis of the gas phase on small samples (say, less than 1% of the total amount present in the balance) was done successfully with either only micro-chromatography (Moret, 2003; Hamon et al., 2008; Ghoufi et al., 2009) or also mass spectrography (Hamon et al., 2008). The set-up shown in Figure 3.13 allows preparing gas mixtures with up to five gases, weighing the gas uptake by the adsorbent placed in the magnetic suspension balance, keeping the gas composition homogeneous (in spite of adsorption) by means of various high-pressure circulators, and determining by micro-chromatography (to use the minimum amount of sample) the composition of the gas phase.

3.2.5 Calibration Procedures and Corrections

3.2.5.1 Calibration of Dosing Volumes

All experiments of gas adsorption manometry rely on the use of a calibrated volume. The basic calibration can be carried out in two ways: either directly or indirectly.

The *direct calibration* implies that this part of the equipment can be isolated, removed and weighed (either evacuated or filled with dry air), filled with an outgassed liquid of known density and then weighed again. A practical limitation comes from the loading capacity of modern analytical balances. It is difficult to find a balance with a sensitivity of 0.1 mg and a loading capacity of more than ca. 300 g. This means that the stainless steel gas storage vessels used in most modern equipment are generally too heavy to be weighed accurately.

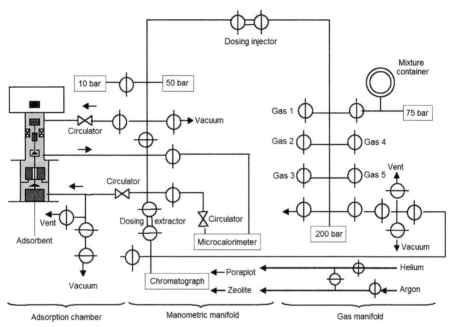

Figure 3.13 Example of a set-up for high-pressure co-adsorption.
After Moret (2003) and Ghoufi et al. (2009).

The indirect calibration implies the use of an external calibrated volume in place of the adsorption bulb. The experiment is similar to that for determining the dead space volume of the adsorption bulb. Maximum accuracy will be obtained if the initial dose of gas admitted to the volume to be calibrated makes use of the full working range of the pressure transducer and if the external calibrated volume is roughly equal to the total volume to be calibrated (i.e. including the pressure transducer and connecting tubing). If there is a difference in temperature, it will be necessary to make a correction for thermal expansion. In the indirect calibration there is inevitably some loss of accuracy in using a separate gas bulb, but there is also the opportunity to use a glass bulb which can be properly cleaned. Any gas bubbles can be seen and withdrawn and also there are no inaccessible regions in the vicinity of metal fittings and valves.

3.2.5.2   Determination of Dead Space Volumes

With most adsorption manometry techniques, it is necessary to know the volumes of two parts of the overall dead space. The first is the *connecting volume* located between the stopcock above the adsorbent bulb and the lowest valve of the dosing volume (see Figure 3.2). The second and more important volume is that of *the dead space within the adsorbent bulb*. Although the connecting volume does not need to be determined for each experiment, its value can be checked in the first stage of the gas expansion calibration procedure.

The determination of the dead space volume of the adsorbent bulb is not quite as straightforward as one might think. It is necessary to consider the following three questions: (i) How do we define the remaining gas volume in relation to the volume occupied by the adsorbent? (ii) What is the most suitable procedure? (iii) If gas expansion is to be used, which gas (e.g. He or N_2) should be adopted and at what temperature?

It follows from the discussion of the quantitative expression of adsorption in Section 2.2 that the most appropriate demarcation between the gas and the adsorbed phase is the Gibbs dividing surface (GDS). This enables us to express the adsorption data in terms of the surface excess and avoid having to determine (or assume) the absolute thickness of the adsorbed layer. As in the Gibbs model the experimenter in principle has the free choice of the GDS position, where should he locate it? The answer may depend on the pressure range. As seen in Section 2.2.1, at pressures below 1 bar, in case the GDS coincides with the probe-accessible surface, one may consider that the surface excess amount n^σ and the amount adsorbed n^a are equal, which, of course, simplifies the interpretation. This is why it is convenient to locate the GDS as closely as possible to the solid surface accessible to the molecules of adsorptive, as suggested by Gibbs himself. By doing so, we facilitate the comparison of adsorption data in the common case when the experimenter does not provide the specific volume (i.e. divided by the sample mass) enclosed by the GDS he has chosen. The case of higher pressures will be dealt with in Section 3.5.1.

Two main routes are available for evaluating the volume up to the probe-accessible surface: the direct one of measuring the volume accessible to a gas, which is not adsorbed at the temperature and pressure of the dead volume determination, and the indirect one of simply subtracting from the volume of the empty bulb the estimated volume of the sample.

The direct route has in the past been preferred and considered to be safer, although this is probably not true when the adsorbent is microporous, as commented on further in this section. It *involves expanding a gas* into the adsorption bulb already containing the adsorbent sample. In the simplest situation, the whole adsorption bulb (up to the top of its stopcock) is maintained at the sample temperature during the subsequent adsorption experiment, whereas all other volumes of the set-up can be considered to be entirely at the controlled ambient temperature. In this case, only one determination must be made, namely that of the volume V_f which is that of the space accessible to the gas in the bulb already filled with the sample. This is acceptable for adsorption experiments at around room temperature, but not for experiments carried out at 77 K. As illustrated in Figure 3.14, there is a considerable temperature change in the CD portion of the bulb neck. In principle, it should be possible to carefully determine the volumes between levels A and B (the bulb proper), C and D (where the gradients lie) and D and F (including the stopcock) and then make a simplifying assumption about the temperature gradients between C and D in order to calculate the amount of gas enclosed.

However, there is another experimental procedure, which can *take into account these temperature gradients* much more accurately – provided they are reproducible – which we can call the procedure of *double imaginary volume*. As we shall see, part of

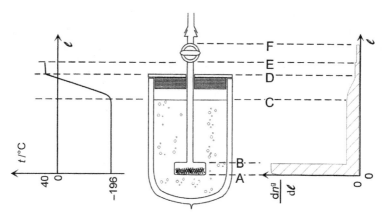

Figure 3.14 Temperature (left) and amount of adsorptive in the gas phase (hatched area, right) as one moves from bottom to top of an adsorption bulb immersed in liquid nitrogen.

this procedure is also suitable if one selects the indirect route explained further. It consists of the following three steps:

1. Expansion of gas from the dosing volume into the empty bulb at ambient temperature (i.e. in the air thermostat) for the determination of the volume V_e of the empty bulb.
2. Again, expansion of gas from the dosing volume but now into the empty bulb immersed in liquid nitrogen, in order to determine the apparent volume V_a which, at ambient temperature, would enclose the same amount of gas as the empty bulb immersed in liquid nitrogen at controlled and reproducible depth, under the same pressure.
3. Finally, expansion of gas from the dosing volume into the bulb filled with adsorbent at ambient temperature for the determination of the void volume V_f left in the bulb by the adsorbent.

From V_a and V_e, we can obtain two imaginary volumes, V_{up} and V_{low}, making up V_e. We can picture V_{up} as a volume totally at room temperature (upper part of the bulb) and V_{low} as a volume totally immersed in liquid nitrogen. In this representation, there is no transition volume with temperature gradients, but V_{up} (at room temperature) and V_{low} (at 77 K) are such as to accommodate exactly the same amount of gas as the empty bulb when immersed in liquid nitrogen during step 2 (with reproducible but unknown temperature gradients). V_{up} does not depend on the presence or absence of the sample. From V_f and V_{up}, we therefore obtain the modified V_{low}, which takes account of the presence of the adsorbent. As V_{up} remains constant, *only step 3 is required for a new sample.*

The gas to be used in the dead space determination must be carefully selected. In the procedure described, step (1) can be carried out with any permanent gas (e.g. helium or nitrogen), whereas for step (2), it is advisable to use a gas with the same virial coefficient B_m as the adsorptive, as B_m and the subsequent correction can vary considerably from one gas to the other (see later in this section). As the measured value of V_a depends on the virial coefficient B_m of the gas used, the simplest procedure is to use the adsorptive itself.

Step (3) is preferably carried out with a gas whose accessibility to the sample is comparable to that of the adsorptive, since we wish to determine a volume up to the probe-accessible surface of the sample, the probe being of course the molecule of adsorptive, not a smaller one like that of helium, for instance, which, in the presence of narrow micropores, could assess a different surface: here again the adsorptive itself, but at a temperature at which it is known not to adsorb, is the best.

As can be seen, helium, contrary to what was once assumed, is not necessarily the best gas to select for the dead space determination. It is sometimes thought that helium allows one to determine the dead space directly at 77 K in the presence of the sample, as it will not adsorb. However, as its virial coefficient is much smaller than that of most adsorptives (see Table 3.2) and because of the possibility of adsorption in micropores (see Chapter 9), its use cannot be recommended. This problem has been discussed by Neimark and Ravikovitch (1997).

The indirect route for determining the dead space volume *makes use of an estimated volume of the adsorbent* sample. This volume can be obtained in two ways:

a. From the theoretical density. This leads to a sample volume which, by definition, excludes all pores of any size (including closed pores and also any micropores inaccessible to the adsorptive).
b. From pycnometric measurements (in a liquid or in a gas) carried out separately. This method takes into account the closed pores, which is an advantage, but may also include in the sample volume those micropores inaccessible to the pycnometric fluid. Like in the direct route above, we expect this fluid to have access to the same pores as the adsorptive. The nature and temperature of the pycnometric fluid must, of course, always be stated.

Both ways have the same advantage of giving a single specific volume of the sample (and therefore a location of the dividing surface) which is kept unchanged from one adsorption bulb to another and from one laboratory to another. It is a sound convention, if the aim is to obtain reproducible measurements and calculations and is consistent with the spirit of the Gibbs representation. It is, for these reasons, certainly well suited for the study of reference materials. Of course, this approach would replace step (3) in the procedure of the double imaginary volume described above and which remains the most accurate to take into account the temperature gradients. Steps (1) and (2) would remain unchanged.

In the case of differential or twin arrangements of adsorption manometry (see Figure 3.4), the dead volume determination is not required, but the volume equalisation and the symmetry of the set-up are essential. The volume equalisation is usually obtained with glass beads on the reference side and sometimes also with adjustable bellows or a piston. The check or adjustment is normally carried out at ambient temperature: the introduction of an identical amount of gas on both sides must result in a zero pressure difference between them.

3.2.5.3 Determination of Buoyancy Correction

The buoyancy correction needed in adsorption gravimetry has the same origin as the dead space correction, in adsorption manometry: it is due to the volume of the sample or, more precisely, to the volume enclosed by the GDS and the resulting change in the apparent amount adsorbed (Rouquerol et al., 1986). This means that any errors – or deliberate changes – in the location of the GDS must have the same effect in gravimetry as in manometry. Here again, we can choose between a direct or an indirect approach and we can cancel part or the whole of this correction by a symmetrical set-up.

The direct determination of the buoyancy is carried out with a gas which does not adsorb at the temperature of the determination: it must be chosen in the same way as the gas for step 3 in the previous section. Here, one should avoid helium: this light gas does not lend itself to the precise determination required as the buoyancy correction is directly proportional to the molar mass of the gas used.

The indirect determination of the buoyancy is obtained by the assessment of the sample volume from its density or by pycnometry, as described in the previous section and with the same implications for the location of the GSD.

A symmetrical balance provides a means of minimising these corrections. To obtain the best compensation, one can adjust the mass and volume on the reference side, for instance, by the use of glass beads and gold wires (Mikhail and Robens, 1983). This compensation is no more accurate than the determination of the buoyancy effect as described just above, but it may allow one to use the balance over a smaller range, where a higher sensitivity is available.

A pertinent question needs to be answered: as the volume of the adsorbed phase increases, do we have to take into account the corresponding increase in buoyancy? (e.g. the buoyancy doubles after saturation of an adsorbent with 50% porosity.) The answer is this: no, provided we want to assess the surface excess mass m^σ. Figure 3.15

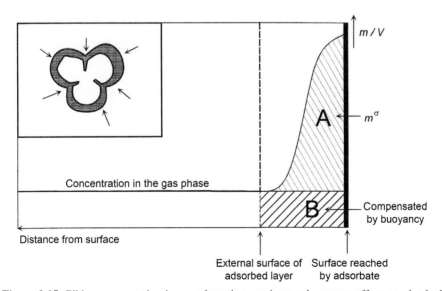

Figure 3.15 Gibbs representation in gas adsorption gravimetry: buoyancy effect on adsorbed layer (top left) makes the balance directly weigh the surface excess mass m^σ, corresponding to area A only.

illustrates indeed that, because of the buoyancy effect, we do not measure the total mass of the adsorbed layer (areas A and B) but simply a surface excess mass (area A only). Thus, adsorption gravimetry and the Gibbs representation are highly compatible (Findenegg, 1997).

Finally, it may be worth considering the effect of buoyancy correction on the magnitude of the error, in the determination of the BET (N_2) surface area. This is indicated in Figure 3.15, for surface areas ranging from 1 to 1000 $m^2 g^{-1}$ and for adsorbent relative densities to 8. The buoyancy is calculated by taking into account the density of nitrogen vapour at 77 K, from 1 to 8. The buoyancy is calculated by taking into account the absolute density of nitrogen vapour at 77 K, at a pressure of 100 mbar, which is assumed to be enough to complete the monolayer. One sees that for a sample of 1 $m^2 g^{-1}$, with a relative density of 2 (for instance, a finely ground quartz), the error amounts to 80%! At the other extreme, for a sample of 1000 $m^2 g^{-1}$, with a relative density of 1.5 (for instance, an active carbon), the error will be as small as 0.1% (Figure 3.16).

3.2.5.4 Pressure Measurement and Corrections

Nowadays, *membrane capacitance pressure transducers* are the most widely used manometers for adsorption studies. A single pressure transducer can be chosen to cover a wide pressure range, but as they lose part of their performances over the lowest few percent of the range, it is usually advisable to have two (e.g. 0–10 and 0–1000 mbar, for N_2 adsorption). It should also be noted that these transducers are fragile and that special care must be taken to avoid sudden pressure changes (this is the reason for the constrictions shown in Figure 3.2) or deposition of powder or condensation of liquid on the membrane.

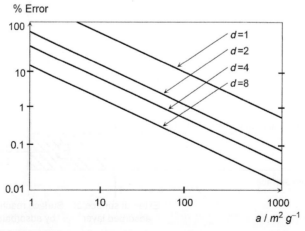

Figure 3.16 Gas adsorption gravimetry: % error on surface area as a result of totally neglecting the buoyancy effect on the adsorbent (assumptions: nitrogen BET, monolayer at 100 mbar, bulk relative density of adsorbent from 1 to 8).

For *the measurement of low pressures* (as required to investigate micropore filling or for krypton or xenon adsorption at 77 K), one may have to take into account the phenomenon of '*thermomolecular flow*' or '*thermal transpiration*'. This happens when two parts of the equipment are maintained at different temperatures and they are connected by a tubing in which the free mean path of the gas molecules is of the order of magnitude of (or above) the tube diameter. The phenomenon results in a steady pressure difference between the ends of the tubing. According to Knudsen (1910), when the mean free path is several times the tube diameter, the pressure ratio is given by the simple relationship:

$$\frac{p_B}{p_G} = \sqrt{\frac{T_B}{T_G}} \tag{3.2}$$

where subscripts B and G stand for 'sample Bulb' and 'pressure Gauge', respectively. Thus, with a sample bulb at 77 K and a pressure gauge at 300 K, the pressure measured by the gauge is nearly twice the true pressure above the sample! In intermediate situations, this effect is smaller and directly depends on the relative size of the mean free path and tube diameter. In these conditions, empirical equations have to be used to estimate the pressure ratio. The most popular equation (Takaishi and Sensui, 1963) employed at present can be written as follows:

$$\frac{p_B}{p_G} = 1 + \frac{\sqrt{T_B/T_G} - 1}{10^5 AX^2 + 10^2 BX + C\sqrt{X} + 1} \tag{3.3}$$

where A, B, C are constants depending on the gas (see Table 3.1) and

$$X = \frac{1.5 p_G D}{T_B + T_G} \tag{3.4}$$

Table 3.1 Thermal Transpiration: Coefficients of the Takaishi and Sensui Equation (3.3), for Various Gases

Gas	A	B	C	Temperature Range/K
Hydrogen	1.24	8.00	10.6	14–673
Neon	2.65	1.88	30.0	20.4–673
Argon	10.8	8.08	15.6	77–673
Krypton	14.5	15.0	13.7	77–673
Methane	14.5	15.0	13	473–673
Xenon	35	41.4	10	77–90
Helium	1.5	1.15	19	4.2–90
Nitrogen	12	10	14	77–195
Oxygen	8	17.5	–	90

In the above equation, the diameter D of the tube (usually the 'neck' of the sample bulb) is expressed in mm, p_G is in mbar and T_B and T_G in kelvins. The results given in the lower part of Table 3.1 (methane and onwards) are considered to be less reliable, especially if they are to be used out of the T_B range over which the measurements were carried out (T_G being at room temperature).

The direct determination of the saturation pressure of the adsorptive has the advantage of providing the real $p°$ and, with nitrogen, of allowing one to calculate to the nearest 0.01 K the adsorption temperature. As the surface layer of a cryogenic liquid tends to become colder (because of evaporation) than the lower part of the liquid (Nicolaon and Teichner, 1968), it is necessary to condense the adsorptive in the bottom of a double-walled ampoule, so that the level at which condensation takes place can be close to that of the adsorbent sample. Measurement of the sample temperature by means of a resistance thermometer is more straightforward, but requires calibration against the saturation vapour pressure thermometer.

3.2.5.5   Correction for Non-ideality

This correction is needed in experiments of adsorption manometry. At the temperatures normally used for physical adsorption, the correction for the non-ideality of real gases generally amounts to several percent. It can be reasonably taken into account by using the first two terms of the virial equation:

$$v_{real}^o = \frac{V}{n} = v_{id}^o + B_m \tag{3.5}$$

where v_{real}^o and v_{id}^o are the real and ideal molar volumes of the gas at a given temperature T and reference pressure (normally 1 bar), respectively, V is the total volume occupied by the given amount of gas and B_m is the second molar virial coefficient, which is usually negative under the conditions of physisorption experiments. Its value for a large number of gases can be found in a critical compilation by Dymond and Smith (1969) and in the thermodynamic tables published by Marsh (1985).

To determine the real amount of gas present in a given volume V, at temperature T and pressure p, it is convenient to use the relation:

$$n = \frac{pV}{T} \times \frac{273.15}{22,711} \left(1 + \frac{\alpha p}{100}\right) \tag{3.6}$$

where V is expressed in cm^3 and p in bar and where $\alpha = -100 B_m / v_{id}^o = -(10 B_m / RT) \times p°$.

Values of α for a number of gases and temperatures used in adsorption studies are given in Table 3.2. They were calculated with the help of the data provided by Dymond and Smith (1969). A detailed study of the incidence of these corrections on the adsorption isotherms and the BET surface areas was made by Jelinek et al. (1990).

Table 3.2 Correction for Non-ideality for a Few Gases Commonly Used in
Adsorption Experiments

Gas		T/K	$a = \%$ Correction
Nitrogen	N_2	77.3	4.0
		90.0	3.1
Argon	Ar	77.3	4.8
		90.0	3.6
Oxygen	O_2	77.3	5.1
		90.0	3.8
Helium	He	77.3	0.2
Carbon monoxide	CO	77.3	1.3
Ammonia	NH_3	273.15	1.5
		298.15	1.2
Neopentane	$(CH_3)_4C$	273.15	4.4
		298.15	3.9
n-Butane	C_4H_{10}	273.15	4.1
		298.15	3.2

An elegant way to avoid this non-ideality correction, especially when operating at high pressures (up to 16.5 MPa) where it can become predominant, was proposed by Bose et al. (1987). In their method, for each equilibrium point, the density of the adsorptive is determined experimentally from its dielectric constant, measured in a gas capacitance cell at same temperature and pressure as the adsorption system studied. The rest of their adsorption procedure is comparable to the discontinuous manometric procedure (see Sections 3.1 and 3.2), the dosing volume being made up of the gas capacitance cell at the temperature of the adsorbent.

In experiments of adsorption gravimetry, the above corrections can be used to calculate the buoyancy at each pressure. Nevertheless, a higher accuracy is obtained by a direct measurement of the buoyancy on a non-adsorbing sinker, either in a blank experiment over the full pressure range or during the adsorption experiment itself in case the balance allows, at any time, a separate weighing of the adsorbent sample and of the sinker (see Figure 3.9).

Finally, it is worth mentioning that a comprehensive study of uncertainties in adsorption manometry and their incidence on the results of the BET and α_s methods was carried out by Badalyan and Pendleton (2003, 2008).

3.2.6   Other Critical Aspects

3.2.6.1   Sample Mass

The first problem to be faced is what amount of adsorbent should be used for the adsorption experiment. This of course depends *inter alia* on the sensitivity of the measuring equipment and the texture and surface properties of the adsorbent.

With modern equipment, reliable measurements are generally obtained with total areas ranging from 20 to 50 m^2 in the adsorption bulb. For materials with specific surface areas below 1 $m^2 \, g^{-1}$, a mass of 10 g or more may be found necessary, but the improvement in sensitivity is counterbalanced by the unavoidable pressure and temperature gradients within the sample.

At the other extreme, with materials of specific surface area above 500 $m^2 \, g^{-1}$, one must be careful not to reduce the mass of sample by too much: *it must remain representative of the batch* of adsorbent and it must be *weighed with an accuracy* consistent with the accuracy provided by the adsorption measurement. For these two reasons, it is usually unwise to use a sample mass under, say, 50 mg. If the full adsorption–desorption isotherm is to be determined, one can be limited by the capacity of adsorptive reservoir or dosing volume or by the automatic control range of the electronic microbalance (typically, between 50 and 100 mg with sensitivity ≥ 1 μg). It therefore often happens that the measurement of one isotherm cannot provide, at the same time, the best determination of the specific surface area and of the full adsorption–desorption isotherm.

The second problem is to obtain a *meaningful mass measurement* in the case of a reactive adsorbent (e.g. one which may absorb or chemisorb H_2O, CO_2). This requires a careful measurement of the mass of adsorbent in its initial state and of the mass change undergone on outgassing – since *the reference mass will be that of the outgassed sample*. This mass change includes the mass of air initially present in the adsorption bulb. This can be calculated from the dead volume of the adsorption bulb and from the absolute density of the air (at a given ambient temperature, pressure and humidity). The adsorbent bulb can also be evacuated at room temperature prior to being weighed, but care must then be taken not to change the sample mass by some desorption. Therefore, this evacuation must be short and not severe: the pressure in the bulb must not be lower than a few mbar to limit water desorption. This should result in an error of less than 1% in the correction term of the mass of air.

3.2.6.2   Temperature (of Various Parts of the Set-Up)

The adsorption temperature is the first, basic datum of an adsorption isotherm. It must be stable over the whole experiment (usually to within 0.1 K): in the case of nitrogen adsorption at 77 K, such a temperature fluctuation would produce a ca. 10 mbar variation in the saturation pressure. In adsorption manometry, the adsorption bulbs are usually immersed in a thermostat or cryostat with which they are in good thermal contact. In adsorption gravimetry, the situation is more complex, because of the poor thermal contact between the sample pan and the surroundings. This is especially critical when the adsorbent is at cryogenic temperatures. A systematic investigation (Partyka and Rouquerol, 1975) has shown that the thermal exchange in a balance

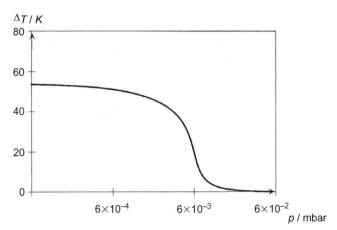

Figure 3.17 Gas adsorption gravimetry: temperature difference between sample and cryostat (at 77 K) as vacuum around sample is progressively reduced.
After Rouquerol and Davy (1978).

can be significantly improved by using a relatively tall (e.g. 80 mm × 10 mm) blackened bucket and by placing black thermal shields above the bucket. Nevertheless, even with these precautions, if the pressure is below 10^{-2} mbar, it is virtually impossible to achieve good thermal equilibrium between the sample and a 77-K cryostat (see Figure 3.17). It follows that it is pointless to wait for equilibrium to be established until the pressure is at least 10^{-2} mbar. This limitation must be kept in mind when adsorption gravimetry is used to study the adsorption of nitrogen by ultramicroporous solids. Above 10^{-2} mbar, the sample is not more than 0.2 K warmer than the surrounding cryostat, but even this is enough to cut off a part of the adsorption isotherm close to saturation.

The cabinet temperature must also be kept constant especially when the adsorptive reservoir is included in the calibrated part of the apparatus (as in Figures 3.3 and 3.4). The pressure transducers are very sensitive to temperature variation, and it is preferable to have the transducer at the same temperature as the rest of the set-up rather than to use individually controlled heaters.

3.2.6.3 Adsorption Equilibrium

By convention, *'adsorption isotherms'* are generally assumed to correspond to a *thermodynamic equilibrium*: if this is not true, the use of the term 'adsorption isotherm' is questionable. The confirmation of adsorption equilibrium is therefore of crucial significance.

Let us first consider *the case of a discontinuous, point-by-point procedure*. An 'absolute' or 'perfect' equilibrium measurement can of course never be attained, since it is limited by the fluctuations (i) of the adsorbent temperature, (ii) of the residual gas temperature and (iii) of the base line of the pressure transducer. As a consequence, after a certain time, the system is as close as it will ever get to true equilibrium.

Nevertheless, this time may be relatively long (e.g. it may take hours to determine each point). The experimenter may then decide to be less demanding in order to save time. A convenient way to operate, which is usually followed by manufacturers of automated instruments (see Section 3.2.1.1.3), is the following:

1. determine the height of the minimum pressure step which can be safely detected (say, 3 or 5 times the fluctuation amplitude observed for the pressure signal);
2. select on 'observation period': equilibrium will be considered to be reached if, over that period, the pressure changes by less than the 'minimum pressure step' determined previously;
3. only then, read and store the data (equilibrium pressure and temperatures).

The longer the pre-selected 'observation period' (usual range: 0.5–5 min), the more demanding the equilibrium requirement.

We should notice that the point-by-point procedure does not always ensure the establishment of a meaningful equilibrium: the adsorptive must be introduced slowly enough to avoid some undesirable scanning within the hysteresis loop. Indeed, when an adsorption system reaches a state corresponding to the adsorption branch of a hysteresis loop, the sudden introduction of a new dose of adsorptive may well favour the initial adsorption on the upper layers of the adsorbent; subsequently, these layers will undergo the desorption of part of their adsorbate, which will be transferred to the lower layers. The resulting experimental point will therefore be located somewhere within the hysteresis loop, instead of on the true adsorption branch. It follows that, in the pressure range of the hysteresis loop, a slow introduction and removal of adsorptive are highly recommended. Maintaining the pressure over the adsorbent at a predetermined level, as done sometimes in automated adsorption gravimetry, can solve this problem provided the size and thickness of the grains and sample bed, respectively, are small enough to make pressure gradients negligible.

Let us now consider *the continuous, quasi-equilibrium procedures*. As the safest and simplest check is to look for the superimposition of two successive adsorption isotherms, this requires one to be able to operate at two different flow rates and, if necessary, to reduce the rate until the test is satisfied. For this reason, it is inadvisable to employ any technique that would not allow the possibility of reducing the flow rate beyond the value finally selected.

3.3 Gas Adsorption Calorimetry

3.3.1 Equipment Available

Probably because of the difficulty to measure heat, over a hundred different types of calorimeters have been described in the literature, of which a number were used to study physisorption systems. Their presentation is much simplified if we distinguish the main categories of calorimeters with the help of Figure 3.18, where a calorimeter is schematically represented as a thermostat **T** (at temperature T_T) surrounding the sample **S** (at temperature T_S) in close thermal contact with its container; the thermal

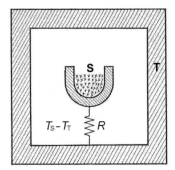

Figure 3.18 Schematic representation of a calorimeter.
After Rouquerol et al. (2007).

connection between the sample and the thermostat is made through the thermal resistance R.

After the heat exchanges between the sample **S** and the thermostat **T**, one easily arrives to the following distinctions (Rouquerol et al., 2007, 2012):

First of all, two broad families of calorimeters can be distinguished:

a. *Adiabatic calorimeters*, whose aim is to *avoid any heat exchange* between the sample and the thermostat.
b. *Diathermal calorimeters*, whose aim is, on the contrary, to *favour these heat exchanges.*

Then, each of these two families can be split into two groups, which can be named 'passive' and 'active', respectively. The passive calorimeters simply rely on the conductivity or insulation of their materials, whereas the active calorimeters make use of electronic temperature control or power compensation to achieve their adiabatic or diathermal character.

We end with the following four groups of calorimeters:

A-1: *Passive adiabatic* calorimeters: here, the adiabaticity is obtained only by increasing the thermal resistance R. There is no control of the temperature difference $T_S - T$. These are the most common calorimeters, also called 'quasi-adiabatic' (after Kubaschewski and Hultgren, 1962) or 'isoperibol' or 'ordinary' and of which the 'water calorimeters' such as those of Thomsen and Berthelot are well known.

A-2: *Active adiabatic* calorimeters: here, the adiabaticity is obtained by a permanent cancellation of the temperature difference $T_S - T_T$, i.e., by permanently adjusting the temperature of the thermostat to that of the sample. This is the most efficient way, so these calorimeters are also called 'true adiabatic'.

B-1: *Passive diathermal* calorimeters, where heat is left to flow to the thermostat and is measured either through a heat-flowmeter or by a phase change taking place in the thermostat.

B-2: *Active diathermal* calorimeters, where the good heat exchange which should normally bring the sample to the temperature of the thermostat is mimicked by an appropriate power compensation (bringing or extracting power, as needed) at the level of the sample.

Let us now comment on the application and suitability of these calorimeters for adsorption or immersion studies.

3.3.1.1   Passive Adiabatic Adsorption or Immersion Calorimetry

The first experiments of gas adsorption calorimetry by Favre (1854) were made at
room temperature with such a calorimeter. More recently, this principle of calorimeter
was adapted for liquid nitrogen temperature by Beebe et al. (1936) and by Kington and
Smith (1964).

In these calorimeters, insulation is necessarily imperfect and the heat leaks must be
evaluated throughout the duration of the experiment by applying Newton's law of
cooling to the ever-changing and continuously monitored temperature difference
$T_S - T_T$. In case of slow, long-lasting phenomena, the heat leak correction can become
predominant. Consequently, this type of calorimeter is not suited for studying thermal
phenomena over a period longer than, say, 30 min. This is, in principle, suitable for
immersion experiments (it was largely used by Zettlemoyer's school) or for discontinuous gas adsorption, but not for experiments of adsorption from solution, sometimes prone to slow displacements and equilibrations. However, the main
drawback is that the experiment is never isothermal: during each adsorption step, a
temperature rise of a few kelvins is common. The corresponding desorption (or lack
of adsorption) must then be taken into account and, after each step, the sample must be
'thermally earthed' so as to start each step at the same temperature. Also, when used in
the form of a water calorimeter, as usual at room temperature for immersion studies,
this calorimeter is not very sensitive because of the comparatively large mass of water
(and in spite of being capable of a high accuracy in the presence of a large and quick
heat evolution such as that occurring in a combustion bomb). In view of these draw-
backs, isoperibol calorimeters are not much used today for studying the energetics of
gas adsorption.

3.3.1.2   Active Adiabatic Adsorption Calorimetry

Here, the heat evolved on adsorption increases the temperature of the sample and its
container (usually a copper cylinder). The heat is prevented from flowing to the
temperature-controlled shield (see Figure 3.19) by an appropriate control of the shield
temperature. Thus, the shield is usually maintained at the same temperature as the
sample container by the use of a differential thermocouple and an external heating
coil. The temperature rise is measured by means of a resistance thermometer attached
to the sample container.

Adiabatic calorimetry is particularly useful for the study of *closed adsorption
systems at low temperatures* (where radiation losses are small) and for temperature
scanning experiments. It is the preferred type of measurement for the determination
of the *heat capacity* of adsorption systems, especially in the temperature range of
4–300 K (Morrison et al., 1952; Dash, 1975). The temperature scan is obtained by
means of the Joule effect applied to the sample container by means of a heating coil.

In some respects, adiabatic calorimetry provides information which is complementary to that provided by heat-flow calorimetry, which is presented just hereafter. The
latter allows a study to be made of the full composition range at constant temperature,
whereas the adiabatic calorimetry study is carried out over a certain range of

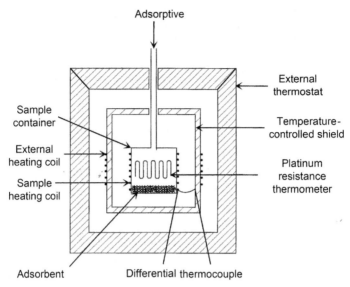

Figure 3.19 Gas adsorption cell (or sample container) in an adiabatic calorimeter.

temperature but with a constant amount of adsorptive in the adsorption cell (of course, this does not mean a constant amount adsorbed, since some desorption occurs on heating). Adiabatic calorimetry allows direct measurements of the heat capacities of adsorbed films, although they are difficult to make accurately because the mass of the film is usually a small fraction of the total mass of the calorimetric cell and sample. Fortunately, phase changes of the adsorbed film give rise to large changes in heat capacity, which are easily detected by adiabatic calorimetry: this technique can be used in the same way as thermal analysis (for 3D systems) to determine phase diagrams of 2D systems (Morrison, 1987). Because of the desorption that takes place on heating, it is necessary to allow for the enthalpy of desorption (e.g. by application of the isosteric method) as it would otherwise lead to under-evaluation of the heat capacity.

3.3.1.3   Passive Diathermal Adsorption or Immersion Calorimetry

This type of calorimetry, which favours the thermal conduction between the sample and the thermostat, has, for adsorption or immersion studies, the double interest of being isothermal (since the thermostat imposes its temperature to the sample) and sensitive. It exists in two main forms.

Phase-change calorimetry. The first instrument of this type, which was also the first to be called a 'calorimeter', was developed in the form of an 'ice calorimeter' by Lavoisier and de Laplace (1783), who weighed the liquid water, and it was then modified by Bunsen (1870), who measured the change of volume. Dewar (1904) devised an elegant adsorption calorimeter at liquid air temperature: the heat was evaluated from the volume of air vapourised. Of course, the temperature of the calorimeter is imposed by the temperature of the phase change. Because of its isothermicity, up to 1960, the Bunsen calorimeter was considered among the best for gas adsorption studies, especially because a version making use of the melting of diphenyl-ether could be used at 27 °C. Nevertheless, these calorimeters are not very easy to operate, they lack adaptability and autonomy (the melting or vapourisation is ensured over a limited time) and cannot be readily automated; therefore, they are mainly of historical interest.

Heat-flow adsorption microcalorimetry. The most important type of isothermal calorimeter in current use is that based on the principle of the heat-flowmeter, which was first applied by Tian (1923) and improved by Calvet and Prat (1958, 1963), who also introduced a differential assembly. The Tian–Calvet thermopile (with up to 1000 thermocouples) is used to channel and measure the heat flux between the sample and the thermostat. Measurement of the heat flux is possible for a mean ΔT smaller than 10^{-6} K. This type of isothermal calorimeter is especially suitable for the study of open systems (i.e. with the introduction and withdrawal of gas or liquid) and is therefore highly recommended for the determination of energies of adsorption either the gas phase, as commented hereafter, or from the liquid phase (see Section 4.3). Because of its sensitivity and stability, it is also well suited for the measurement of energies of immersion (see Section 4.2).

Figure 3.20 shows a gas adsorption bulb inserted in the central part of a Tian–Calvet thermopile. The adsorbent bulb is connected to a device (not represented here) to allow the simultaneous determination of the adsorption isotherm by one of the techniques listed in the previous sections.

Whatever the gas adsorption equipment associated to the microcalorimeter, the assessment of the energy of adsorption requires fulfilling the few conditions hereafter:

a. The adsorptive must be carefully brought to the temperature of the microcalorimeter before entering the adsorption bulb.
b. It must also be introduced very slowly, so that the heat effect corresponding to the gas compression in the calorimeter may be calculated accurately (Rouquerol et al., 1980). This also helps to meet the previous requirement of efficient adsorptive pre-cooling or pre-heating.

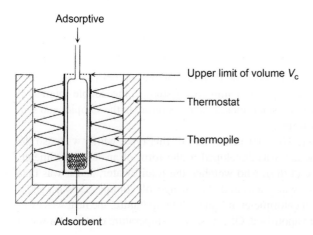

Figure 3.20 Gas adsorption bulb in a Tian–Calvet thermopile.

c. The outgassing must be carried out carefully, bearing in mind that the adsorption bulbs used in a microcalorimeter usually have much longer necks than standard adsorption bulbs. Also, this arrangement may drastically change the actual residual pressure in the immediate vicinity of the sample during outgassing.

A simplified version of the heat-flow gas adsorption microcalorimeter makes use of a single thermal sensor in close-contact with the sample (instead of a multiple integrating heat-flowmeter surrounding the sample). It may be complemented by a second sensor acting as a reference. The thermal sensor can be embedded within the sample, while a continuous flow of carrier gas brings the adsorptive at the desired partial pressure through the bed of adsorbent (Groszek, 1966). This set-up is suitable for preliminary or routine work. In another version, the 'calorimetric bead system' the temperature sensor is embedded in a bead of solid adsorbent, directly cooled by the surrounding gas medium (Jones et al., 1975). These arrangements are simple and sensitive but are not reliable for quantitative measurement.

3.3.1.4   Active Diathermal Adsorption Calorimetry

At a time when the means for recording and integrating the signal from a heat-flowmeter were of limited accuracy, Tian (1923) proposed a way of compensating the major part of the heat liberation by a steady Peltier or Joule effect taking place just against the sample. This idea was refined and adapted to adsorption experiments by Kiselev and his co-workers (Dzhigit et al., 1962). By means of a continuous Joule effect in the vicinity of the adsorbent, they maintained a constant temperature difference ΔT through the heat-flowmeter. As soon as adsorption produced an evolution of heat, the Joule effect was interrupted discontinuously, just enough to keep ΔT unchanged. The heat evolved on adsorption was simply derived from the sum of all non-heating periods. This approach was no longer used when the quality of the recordings allowed an accurate integration.

3.3.2   Calorimetric Procedures

Calorimetry must be associated with a means for determining the amount adsorbed. In principle, any technique could be used to determine the adsorption isotherm, but in practice, the most frequently used for the study of physisorption are (i) the discontinuous, point-by-point, manometric procedure; (ii) the continuous sonic nozzle procedure and (iii) adsorption gravimetry. The former has the advantage of using a conventional assembly (Morrison et al., 1952; Holmes and Beebe, 1961; Isirikyan and Kiselev, 1962; Della Gatta et al., 1972; Gravelle, 1972), whereas the second can provide a continuous measurement of the differential energy of adsorption, with high resolution (Rouquerol, 1972; Grillet et al., 1977b). A fourth procedure can also be used in the case of low-pressure chemisorption, namely the pulse procedure with carrier gas (Gruia et al., 1976).

3.3.2.1 Discontinuous Manometry

As indicated in Section 2.7.1, a discontinuous calorimetric–manometric experiment should be designed to provide the essential experimental data for insertion in the following equation:

$$\Delta_{ads}\dot{h}_{T,\Gamma} = \left(\frac{dQ_{rev}}{dn^{\sigma}}\right)_{T,A} + V_C\left(\frac{dp}{dn^{\sigma}}\right)_{T,A} \tag{3.7}$$

which is, in principle, applicable to very small steps.

Thus, in addition to the data required to determine the surface excess amount (see Section 3.2.1), one needs to know dQ_{rev} (the heat exchanged reversibly during each adsorption step) and V_C (the volume – or dead space – of that part of the adsorption bulb which is located within the calorimetric detector) (see Figure 3.20). V_C is evaluated by liquid weighing or by geometrical considerations and corrected for the sample volume.

Because small steps are time consuming and demanding (in terms of the sensitivity and quality of the base line), the above procedure is normally used to provide a limited number of adsorption steps: the data are then presented in the form of pseudo-differential energies of adsorption, as in Figure 3.21a. A more elaborate way of processing the data involves plotting the integral molar enthalpy of adsorption versus n^{σ} (see Section 2.7.1) followed by the derivation of the differential enthalpies.

3.3.2.2 Continuous Calorimetric Sonic Nozzle Experiment

Figure 3.22 shows the typical assembly of a Tian–Calvet microcalorimeter with a gas adsorption equipment making use of a sonic nozzle flowmeter. For practical reasons, this arrangement is limited to experiments where the final quasi-equilibrium pressure does not exceed 1 bar. It has the advantage of providing a continuous and simultaneous recording of the heat-flow and the quasi-equilibrium pressure versus the amount

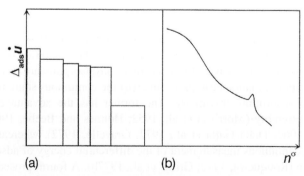

Figure 3.21 Pseudo-differential (a) and differential (b) energies of adsorption, as obtained with a discontinuous or continuous procedure, respectively.

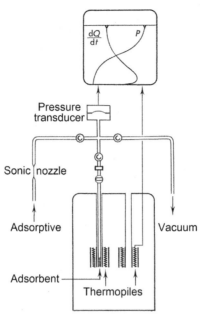

Figure 3.22 Association of Tian–Calvet microcalorimetry with continuous gas flowmetry.

of adsorptive introduced into the system, from which it is possible, as described here-after, to derive a continuous curve of differential enthalpy of adsorption versus amount adsorbed.

As explained in Section 2.7.2, this derivation involves the use of the following equation:

$$\Delta_{ads}\dot{h} = \frac{1}{f^\sigma}\left(\phi + V_C\frac{dp}{dt}\right) \tag{3.8}$$

The 'rate of adsorption' f^σ is determined with the help of Equation (2.80), from the experimental rate of introduction, f, of the adsorptive, the dead volumes V_C and V_E, and their temperatures T_C and T_E, and the slope dp/dt. The only extra information needed is a continuous recording of the heat-flow ϕ from the beginning to the end of the experiment.

The results can be presented in the form of a continuous curve of differential enthalpies of adsorption $\Delta_{ads}\dot{h}$ versus n^σ, as shown in Figure 3.21b, with a resolution that is much higher than obtained by the discontinuous procedure (Figure 3.21a). If the adsorption calorimeter cannot be easily connected to a well-calibrated and well-temperature-controlled adsorption sonic nozzle set-up, or when the adsorption isotherm is difficult to determine (e.g. if very small amounts are adsorbed), there remains the possibility of determining, separately, the adsorption isotherm by any of the discontinuous or continuous procedures described in Sections 3.3.1 or 3.3.2. A rather simple procedure can then be applied to the calorimetric experiment, which does not require any calibration and high stability of the gas flow rate: the simple recordings of quasi-equilibrium pressure and heat-flow versus time are all that are required for the determination of the isotherm and the adsorption energy. If V_C can be calculated (from the dead volume of the empty adsorption bulb and the absolute density of the adsorbent), then one can omit all other calibration steps.

3.4   Adsorbent Outgassing

3.4.1   Aim of the Outgassing

Of particular importance are investigations of the adsorptive properties of powders and porous solids in relation to particular applications. This means that a reproducible initial state of the adsorbent surface should be consistent with the proposed application of the adsorbent. This is why we do not always require a 'perfectly clean' surface, which would usually involve ultra-high vacuum (say, a residual pressure less than 10^{-6} mbar) and high temperatures (say, above 1000 °C). Instead, our aim is (i) to eliminate most of the species physisorbed during storage of the sample (e.g. H_2O, CO_2); (ii) to avoid, during outgassing, any drastic change as a result of ageing, sintering or modification of surface function groups and (iii) to reach a *well-defined, reproducible, intermediate state* which would be suitable for the proposed experiments (e.g. adsorption isotherm measurements or adsorption calorimetry). This state can be attained by an appropriate form of vacuum outgassing or by a gas displacement process.

3.4.2   Conventional Vacuum Outgassing

This type of outgassing is attractive because (i) it is clean by definition; (ii) it allows the operator to use a lower outgassing temperature than would be necessary under atmospheric pressure and static conditions (which may be useful to protect the sample and also the container, especially when in metal) and (iii) it leaves the surface exposed to a vacuum, which is precisely what is required to start most adsorption experiments.

Nevertheless, vacuum outgassing raises a first comment about the reporting of the experimental conditions. While reporting the outgassing conditions, one should resist, indeed, the temptation of stating, for instance, 'outgassed under a vacuum of 10^{-6} mbar', which lets the reader think that the sample was submitted to such a vacuum during its thermal treatment, when no monitoring of the pressure was ever made in the vicinity of the sample. If this is the ultimate vacuum provided by the pump, it is closer to reality to state 'outgassed up to a *final* vacuum of 10^{-6} mbar' since this vacuum will only be reached at the end of the outgassing: as long as outgassing takes place, a pressure gradient from the sample to the pump is needed to evacuate the desorbed species and the actual residual pressure over the sample is unknown.

A second comment deals with the difficulty to easily carry out a satisfactory vacuum outgassing. At first glance, it might seem that this type of outgassing simply needs a sample container, a furnace or oven (with temperature control and, if possible, heating rate or 'ramp' control) and a vacuum (see Figure 3.23).

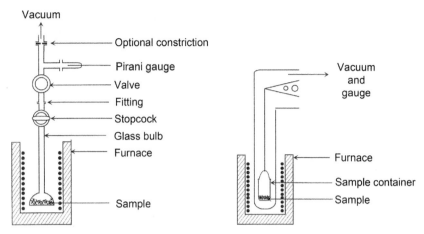

Figure 3.23 Vacuum outgassing from a bulb for adsorption manometry (left) or from the sample container of an adsorption microbalance (right).

In practice, two main problems may have to be faced, namely the spurting of fine powders from the sample container (during the reduction in pressure and increase in temperature) and uncontrolled changes that may occur on heating (a problem that is addressed in Section 3.4.3).

The problem of the *spurting of fine powders* can be solved in several ways. First, spurting can be prevented by using a glass-wool pad (or a glass frit) in the tube above the sample. This practice is simple but adds an uncontrolled barrier between the vacuum line and the region around the sample; it therefore leads to a lack of control in the outgassing. Secondly, spurting can be prevented by manually controlling the opening of the valve connecting the sample to the vacuum line. However, this may be quite time consuming. For example, samples prepared by the sol-gel method may need several hours of slow pumping at room temperature before the stopcock can be completely open. One can then operate in two ways: either by directly observing the sample as the vacuum valve is opened, little by little, or by keeping the gas flow from the sample lower than a pre-selected value that is known, from prior experience, to be safe for the particular sample. Monitoring the gas flow can be achieved by using a Pirani gauge upstream and possibly a constriction (convenient size: 1–3 mm bore and 10 mm length), as shown in the left part of Figure 3.23. When the valve is completely open, some manual control of the heating may still be needed. The same procedure holds for outgassing an adsorbent located in an adsorption balance (see right part of Figure 3.23 where, for simplicity, valve, constriction and gauge are not represented again). Notice that conventional outgassing is often more efficient in a balance (because of the small sample and wide tubings), which can explain differences between adsorption isotherms obtained by either adsorption manometry or adsorption gravimetry, especially on microporous adsorbents.

We found it useful to develop a more complete way of preventing spurting, by introducing two automatic controls, one for opening the vacuum valve and the other

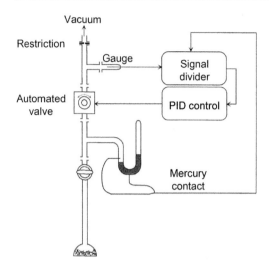

Figure 3.24 Anti-spurting automatic set-up for sample evacuation.

for controlling the sample temperature (see Section 3.4.3). The automatic control of the vacuum valve is similar to that used in the manual procedure and is represented in Figure 3.24. The signal from a Pirani gauge feeds a PID controller acting upon an automated valve (e.g. a standard diaphragm vacuum valve of 10 or 20 mm bore). Experience shows that, on outgassing an adsorbent, two successive stages can be identified: (i) essentially the pumping of air, down to a pressure of, say, 25 mbar and (ii) the desorption proper, which usually involves the removal of water vapour whose flow is responsible for the spurting. Since only a small part of the air pumped comes from the sample itself, this first step can be carried out much more rapidly than the second. A mercury contact control unit can be used to switch the controller from stage (i) to stage (ii). Convenient rates of pumping are, for instance, (i) at $0.5\ dm^3(STP)h^{-1}$ (so that it will require around 15 min) and (ii) $50\ cm^3(STP)h^{-1}$. It takes ca. 5 min to pump 5% physisorbed water from a 100 mg sample, but 1.5 h to pump 10% physisorbed water from a 1 g sample. This set-up is especially useful for outgassing any type of adsorbent prone to spurting (light powder, high water content), especially when located in stainless steel cells (for which there is no possibility of direct visual monitoring), like those needed for high-pressure adsorption experiments.

3.4.3   Controlled Vacuum Outgassing by CRTA

We have just seen how to avoid (either manually or automatically) the spurting of the sample on evacuation, at room temperature. However *when the sample temperature is raised, there is a further risk of spurting and of uncontrolled changes* in the sample itself. The technique for overcoming these problems follows the general principle of controlled rate thermal analysis (CRTA) (Rouquerol, 1970, 1989) where *the heating is directly controlled by the behaviour of the sample itself* (i.e. by a feedback from the sample).

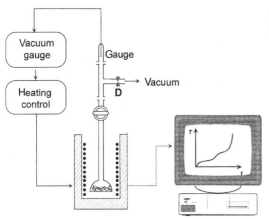

Figure 3.25 Principle of a CRTA set-up for controlled outgassing.

A simple form of CRTA, which is well adapted for outgassing purposes, is presented in Figure 3.25.

The adsorbent bulb is continuously evacuated through a calibrated diaphragm D (convenient size: 1 mm bore, 1 cm long). The pressure drop through the diaphragm is monitored by a simple Pirani gauge. This signal is fed to PID heating control, which heats the sample in such a way as to maintain a constant pressure drop. Thus, both the residual vacuum around the sample (typical values: 5–100 μbar) and the rate of outgassing (typical rates: 1–10 mg lost per hour) are controlled at the same time. Controlling the rate of outgassing means that we also control the pressure and temperature gradients through the sample. They can be lowered at will, by simply reducing the rate of outgassing.

The recorded temperature curve is representative of the thermal path followed by the sample and is equivalent to a thermogravimetry curve. If needed, it can be used for selecting the appropriate outgassing temperature, which should normally correspond to an inflexion point of the curve, indicating that the evolution of the physisorbed species is coming to an end, whereas the thermal decomposition of the sample has not yet started. An advantage of this CRTA curve is also that of its being obtained directly with the sample located in its adsorption bulb. The reproducibility of the heat treatment is easily checked by comparing the temperature curves.

In conventional heating, the sample is linearly brought to a given temperature and, since the sample is far from equilibrium, it is kept at that temperature for a certain time (commonly between 2 and 10 h). This results in a mass–temperature curve that is often far removed from the characteristic curve for the adsorbent (see points 1 and 2 in Figure 3.26).

In some cases, this can invalidate any discussion of the significance of the results. On the contrary, since quasi-equilibrium can be reached at any time during a CRTA experiment, it is possible to stop the heat treatment (and even to quench the sample) at any point of the curve common to all samples (see points 3, 4 and 5 in Figure 3.21). The CRTA approach can be used for the outgassing of a sample located either in a standard adsorption bulb or in the pan of a microbalance (Rouquerol and Davy, 1978).

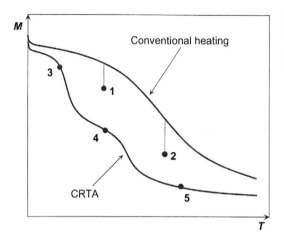

Figure 3.26 Representative points of samples outgassed by conventional heating (1 and 2) or by CRTA (3, 4 and 5).

Apart from outgassing under well-controlled conditions, CRTA also has applications in kinetics, thermal analysis and in the synthesis of highly homogeneous adsorbents. It opened the way to the broader method of sample-controlled thermal analysis, of which it remains, for various practical and theoretical reasons, the most accessible and versatile version (Sorensen and Rouquerol, 2003).

3.4.4 Outgassing with a Carrier Gas

The carrier gas can be any non-reacting gas (e.g. nitrogen, argon or helium) with not more than, say, 100 ppm humidity. To achieve a good tightness against air humidity, metal, polyamide (nylon) or fluoroelastomer (viton) tubes are recommended between the gas bottle and the sample container. The latter can be in the shape of a U-tube, with the sample in the bottom, or in the shape of a standard adsorption bulb into which a long, hollow, needle is introduced and used as the gas inlet.

Heating of the sample is carried out in the same way as in vacuum. In principle, one could again choose between a conventional form of linear heating (followed by a temperature plateau) and a CRTA treatment, the rate of outgassing now being controlled with the help of a highly sensitive, differential gas flowmeter (Rouquerol and Fulconis, 1997). Actually, the main advantage of outgassing with a carrier gas is the simplicity of the technique (no vacuum, no spurting of the sample and good thermal exchange – particularly if the carrier gas is helium). An interesting compromise is to simply monitor the outgassing of the sample by the use of a catharometer (with helium as the carrier gas, it is possible to detect the release of water vapour).

3.5 Presentation of Experimental Data

The adsorption data are to be presented in such a way as to be easily understood, compared and worked on. This implies a few rules to be followed, as indicated hereafter.

3.5.1 Units

It is highly recommended to make use of standardised and broadly accepted quantities and units, i.e., essentially those proposed by IUPAC (Cohen et al., 2007). We should prefer the *surface excess amount* n^σ or the *amount adsorbed* n^a to the 'mass adsorbed' or to the 'volume STP adsorbed', which are related to a specific technique and less easy to compare between themselves. Furthermore, whatever convenient the volume STP can be found for the imagination, it is not part of any system of units and it should be avoided. Also we should prefer *the pascal or the bar* $(=10^5$ Pa) to the mm of mercury (or torr), which, furthermore, does not have the experimental justification it had in the past.

3.5.2 Experimental Conditions

To be meaningful, the adsorption data should always be given together with a description of the essential experimental conditions able to influence the adsorption results: order of magnitude of the *sample mass* and of the *grain size, outgassing* conditions, gas used to determine the dead space volumes, *temperature control* of the sample cell and its tubing, criteria used to check the *adsorption equilibrium.*

3.5.3 Surface Excess Amounts

As detailed in Section 2.2, the safest way to report experimental data on gas adsorption is to do it in surface excess amounts. Nevertheless, any interpretation of the adsorption data, any adsorption isotherm equation, and any modelling and simulation of adsorption always refers to an amount adsorbed, i.e., to the total amount of adsorbate present in the 'adsorption space', where the concentration is different from that in the bulk of the fluid phase (see, for instance, Tolmachev, 2010).

This means that the Gibbs representation must be considered as a *necessary but intermediate step* which should be carried out and reported in such a way as to make feasible the calculation of the adsorbed amount. This makes it *compulsory to provide* the reader with the location of the GDS used for the calculation or, more precisely, with *the volume $V^{S,0}$ enclosed by the GDS* and assumed to be inaccessible to the adsorptive. In the frequent case when the GDS is located with the help of a direct determination of dead volume by gas expansion, the gas, temperature and pressure of this experiment give an indication of the location of the GDS, but the exact information needed is the resulting *specific volume of the sample*. Also, the relationship between the concentration in the gas phase c^g and the pressure used to calculate n^σ (e.g. *ideal or real gas law*) is required.

With these two data, the reader is able to calculate an amount adsorbed n^a after

$$n^a = n^\sigma - c^g\left(V^{S,0} - V^S - V^a\right) \tag{3.9}$$

where he can introduce his assumptions about the actual volume of the solid V^S and of the adsorption space V^a.

It is only under moderate pressures (say, under 1 bar) and when the GDS is assumed to be located exactly on the probe-accessible surface of the adsorbent (which makes $V^{S,0} = V^S$) that the surface excess amount can be considered to equal the amount adsorbed. This is probably the reason why Gibbs suggested placing the GDS close to the solid surface.

Nevertheless, this closeness is not a prerequisite for the application of the Gibbs model and any location of the GDS could, in principle, be envisaged. This is what led Gumma and Talu (2010) to envisage restricting to 0 the volume $V^{S,0}$ enclosed by the GDS. This leads to a well-defined surface excess amount (which they call 'net adsorption'), which does not require any dead volume determination in the presence of the adsorbent.

References

Ajot, H., Joly, J.F., Raatz, F., Russmann, C., 1991. Studies in Surface Science and Catalysis, Vol. 62. Elsevier, Amsterdam, p. 161.

Atkins, J.H., 1964. Anal. Chem. 36, 579.

Badalyan, A., Pendleton, P., 2003. Langmuir. 19 (19), 7919.

Badalyan, A., Pendleton, P., 2008. J. Colloid Interface Sci. 326 (1), 1.

Balard, H., Brendle, E., Papirer, E., 2000. In: Mittal, K. (Ed.), Acid-base Interactions: Relevance to Adhesion Science and Technology. VSP, Utrecht, p. 14.

Balard, H., Maafa, D., Santini, A., Donnet, J.B., 2008. J. Chromatogr. A. 1198–1199, 173.

Beebe, R.A., Low, G.W., Goldwasser, S., 1936. J. Am. Chem. Soc. 58, 2196.

Bose, T.K., Chahine, R., Marchildon, L., St Arnaud, J.M., 1987. Rev. Sci. Instrum. 58 (12), 2279.

Bunsen, R.W., 1870. Ann. Phys. 141, 1.

Calvet, E., Prat, H., 1958. Récents Progrès en Microcalorimétrie. Dunod, Paris.

Calvet, E., Prat, H., 1963. Recent Progress in Microcalorimetry. Pergamon Press, Oxford.

Camp, R.W., Stanley, H.D., 1991. American Laboratory. 9, 34.

Cohen, E.R., Cvitas, T., Frey, J.G., Holmström, B., Kuchitsu, K., Marquardt, R., Mills, I., Pavese, F., Quack, M., Stohner, J., Strauss, H.L., Takami, M., Thor, A.J., 2007. Quantities, Units and Symbols in Physical Chemistry, third ed. RSC Publishing, Cambridge, UK.

Cremer, E., Huber, H., 1962. Gas Chromatogr. Int. Symp. 3, 169.

Dash, J.G., 1975. Films on Solid Surfaces. Academic Press, New York.

Della Gatta, G., Fubini, F., Venturello, G., 1972. In: Thermochimie. Colloques Internationaux du CNRS, Vol. 201. Editions du CNRS, Paris, p. 565.

De Vault, D.J., 1943. J. Am. Chem. Soc. 65, 532.

Dewar, J., 1904. Proc. R. Soc. A. 74, 122.

De Weireld, G., Frère, M., Jadot, R., 1999. Meas. Sci. Technol. 10 (2), 117.

Dreisbach, F., Staudt, R., Tomalla, M., Keller, J.U., 1996. In: LeVan, M.D. (Ed.), Fundamentals of Adsorption. Kluwer Academic Publishers, Dordrecht, p. 259.

Dreisbach, F., Seif, R., Lösch, A.H., Lösch, H.W., 2002. Chem. Ing. Tech. 74 (10), 1353.

Dymond, J.H., Smith, E.B., 1969. The Virial Coefficient of Gases: A Critical Compilation. Clarendon Press, Oxford.

Dzhigit, O.M., Kiselev, A.V., Muttik, G.G., 1962. J. Phys. Chem. 66, 2127.

Emmett, P.H., 1942. Adv. Colloid Sci. 1, 3.

Emmett, P.H., Brunauer, S., 1937. J. Am. Chem. Soc. 56, 35.

Eyraud, C., 1986. Thermochim. Acta. 0040-6031. 100 (1), 223–253. http://dx.doi.org/10.1016/0040-6031(86)87059-9.

Eyraud, C., Eyraud, L., 1955. Laboratoire. 12, 13.

Favre, P.A., 1854. C.R. Acad. Sci. Paris. 39, 729.

Findenegg, G., 1997. In: Fraissard, J. (Ed.), Proceedings of NATO-ASI on Physical Adsorption: Experiments, Theory and Applications. Kluwer Academic Publishers.

Ghoufi, A., Gaberova, L., Rouquerol, J., Vincent, D., Llewellyn, P.L., Maurin, G., 2009. Microporous Mesoporous Mater. 119, 117.

Glueckauf, G., 1945. Nature (Lond.) 156, 748.

Gravelle, P.C., 1972. Advances in Catalysis, Vol. 22. Academic Press, p. 191.

Grillet, Y., Rouquerol, J., Rouquerol, F., 1977a. J. Chim. Phys. 74 (2), 179.

Grillet, Y., Rouquerol, J., Rouquerol, F., 1977b. J. Chim. Phys. 74 (7–8), 778.

Groszek, A.J., 1966. Lubr. Sci. Technol. 9, 67.

Gruia, M., Jarjoui, M., Gravelle, P.C., 1976. J. Chim. Phys. 73 (6), 634.

Gumma, S., Talu, O., 2010. Langmuir. 26 (22), 17013.

Hamon, L., Frère, M., de Weireld, G., 2008. Adsorption. 14, 493.

Haul, R., Dümbgen, G., 1960. Chem. Ing. Tech. 32, 349.

Haul, R., Dümbgen, G., 1963. Chem. Ing. Tech. 35, 586.

Holmes, J.M., Beebe, R.A., 1961. Adv. Chem. Ser. 33, 291.

Ichikawa, T., Tokoyoda, K., Leng, H., Fujii, H., 2005. J. Alloys Compd. 400, 245.

Isirikyan, A.A., Kiselev, A.V., 1962. J. Phys. Chem. 66, 210.

Jelinek, L., Dong, P., Kovats, E., 1990. Adsorpt. Sci. Technol. 7 (3), 140.

Jones, A., Firth, J.G., Jones, T.A., 1975. J. Phys. E. 8, 37.

Karp, S., Lowell, S., Mustacciulo, A., 1972. Anal. Chem. 44, 2395.

Keller, J.U., Staudt, R., Tomalla, M., 1992. Ber. Bunsenges. Phys. Chem. 96 (1), 28.

Kington, G.L., Smith, P.S., 1964. J. Sci. Instr. 41, 145.

Kiselev, A.V., Yashin, Y.I., 1969. In: Gas Adsorption Chromatography. Plenum Press, New York.

Knudsen, M., 1910. Ann. Phys. 31 (210), 633.

Krim, J., Watts, E.T., 1991. In: Mersmann, A.B., Scholl, S.E. (Eds.), Third International Conference on Fundamentals of Adsorption. Engineering Foundation, New York, p. 445.

Kubaschewski, O., Hultgren, R., 1962. In: Skinner, H.A. (Ed.), Experimental Thermochemistry, Vol. II. Interscience Publishers, London, p. 351 (Chapter 16).

Lavoisier, A.L., de Laplace, P.S., 1783. In: Mémoire sur la chaleur. C. R. Académie Royale des Sciences, 28th June.

Le Parlouer, P., 1985. Thermochim. Acta 92, 371.

Marsh, K.N., 1985. TRC – Thermodynamics Tables. Texas Engineering Experimental Station, College Station, TX.

McBain, J.W., Bakr, A.M., 1926. J. Am. Chem. Soc. 48, 690.

Mikhail, R.S., Robens, E., 1983. Microstructure and Thermal Analysis of Solid Surfaces. John Wiley and Sons, Chichester.

Moret, S., 2003. Etude thermodynamique de la co-adsorption de N_2, Ar et CH_4 à 40°C, jusqu'à 15 bar, par des matériaux poreux (thesis). Université de Provence.

Morrison, J.A., 1987. Pure Appl. Chem. 59 (1), 7.

Morrison, J.A., Drain, L.E., Dugdale, J.S., 1952. Can. J. Chem. 30, 890.

Neimark, A.V., Ravikovitch, P.I., 1997. Langmuir. 13, 5148.

Nelsen, F.M., Eggertsen, F.T., 1958. Anal. Chem. 30, 1387.

Nicolaon, G., Teichner, S.J., 1968. J. Chim. Phys. 64, 870.

Paik, S.W., Gilbert, G., 1986. J. Chromatogr. 351, 417.

Partyka, S., Rouquerol, J., 1975. In: Eyraud, C., Escoubès, M. (Eds.), Progress in Vacuum Microbalance Techniques, Vol. 3. Heyden, London, p. 83.

Pieters, W.J.M., Gates, W.E., 1984. US Patent 4 489, 593.

Roginskii, S.Z., Yanovskii, M.L., Lu, P.-C., 1960. Kinet. Catal. 1, 261.

Roles, J., Guiochon, G., 1992. J. Chromatogr. 591, 233.

Rouquerol, J., 1970. J. Therm. Anal. 2, 123.

Rouquerol, J., 1972. In: Thermochimie. Colloques Internationaux du CNRS, Vol. 201. Editions du CNS, Paris, p. 537.

Rouquerol, J., 1989. Thermochim. Acta. 144, 209.

Rouquerol, J., Davy, L., 1978. Thermochim. Acta. 24, 391.

Rouquerol, J., Davy, L., 1991. French Patent on Device for Integral and Continuous Measurement of Gas Adsorption and Desorption, filed 25/10/1991.

Rouquerol, J., Fulconis, J.M., 1997. Private communication.

Rouquerol, F., Rouquerol, J., Everett, D., 1980. Thermochim. Acta. 41, 311.

Rouquerol, J., Rouquerol, F., Grillet, Y., Triaca, M., 1986. Thermochim. Acta. 103, 89.

Rouquerol, J., Rouquerol, F., Grillet, Y., Ward, R.J., 1988. In: Unger, K.K., et al. (Eds.), Characterization of Porous Solids. Elsevier Science Publishers, p. 317.

Rouquerol, F., Rouquerol, J., Sing, K.S.W., 1999. Adsorption by Powders and Porous Solids. Academic Press, pp. 51–92.

Rouquerol, J., Wadso, I., Lever, T.J., Haines, P.J., 2007. In: Brown, M., Gallagher, P. (Eds.), Handbook of Thermal Analysis and Calorimetry. Further Advances, Techniques and Applications, Vol. 5. Elsevier, Amsterdam, pp. 13–54 (Chapter 2).

Rouquerol, J., Rouquerol, F., Llewellyn, P., Denoyel, R., 2012. In: Techniques de l'Ingénieur, Analyse et Caractérisation, Paris, article P 1202, Charles Eyraud 1952.

Sandstede, G., Robens, E., 1970. US patent 3,500,675.

Schlosser, E.G., 1959. Chem. Ing. Tech. 31, 799.

Sorensen, T.O., Rouquerol, J. (Eds.), 2003. Sample Controlled Thermal Analysis (SCTA): Principle, Origins, Goals, Multiple Forms, Applications and Future. Kluwer Academic Publishers, Dordrecht, p. 252, Cohen et al. 2007 green book.

Takaishi, T., Sensui, Y., 1963. Trans. Faraday Soc. 59, 2503.

Thielman, F., 2004. J. Chromatogr. A. 1037, 115.

Tian, A., 1923. Bull. Soc. Chim. Fr. 33 (4), 427.

Tolmachev, A.M., 2010. Prot. Metals Phys. Chem. Surf. 46 (2), 170.

Webb, P.A., 1992. Powder Handling and Processing, Vol. 4(4), p. 439.

Wilson, J.N., 1940. J. Am. Chem. Soc. 62, 1583.

4 Adsorption at the Liquid–Solid Interface: Thermodynamics and Methodology

Jean Rouquerol, Françoise Rouquerol

Aix Marseille University-CNRS, MADIREL Laboratory, Marseille, France

Chapter Contents

Adsorption by Powders and Porous Solids. http://dx.doi.org/10.1016/B978-0-08-097035-6.00004-8

4.1   Introduction

Adsorption at the liquid/solid interface is of great importance in industry and everyday life (e.g. in detergency, adhesion, lubrication, flotation of minerals, water treatment, oil recovery, pigments and particle technology, etc.). Adsorption from solution measurements has been used for many years for the determination of the surface area of certain industrial materials. Immersion microcalorimetry has been applied for the characterization of such materials as clays and activated carbons. The application of the energetics of immersion is based on the observation by Pouillet (1822) that the immersion of an insoluble solid (sand) in a liquid (water) is a measurable exothermic phenomenon. To gain in understanding of the adsorption phenomena at the liquid–solid interface, it is necessary to know not only the characteristic physical properties which also control the adsorption of gases (notably the surface area and porosity) but also the behaviour of the solid in the liquid medium.

The comparison with adsorption at the gas/solid interface is further complicated by the fact that some adsorbents can undergo modifications in their texture either on outgassing or on immersion in a pure liquid or a solution. Under these conditions, the adsorbent is not the same when studied in the presence of a gas or a liquid.

In this chapter, our aim is to give an introductory account of the methodology and underlying thermodynamic principles of adsorption at the liquid/solid interface. We are mainly, but not exclusively, concerned with the characterization of the liquid/solid interface. In this context, there are two relevant topics:

a. the energetics of *immersion* of solids in liquids and
b. isothermal *adsorption from solutions.*

Many attempts have been made to employ immersion microcalorimetry and solution adsorption measurements for the determination of the surface area of porous and nonporous materials, but in our view, insufficient attention has been given to either the basic principles involved or the limitations of the experimental procedures used by many investigators.

Our approach to the thermodynamics of adsorption and immersion remains simple, although rigorous, and close to the experiment. This is why the thermodynamic treatment of the energetics of immersion in Section 4.2 is confined to the simple system of a solid immersed in a pure liquid. Similarly, in Section 4.3, consideration is given only to the adsorption from binary solutions. The thermodynamic nomenclature and definitions proposed here are consistent with the recommendations of IUPAC (Everett, 1972, 1986).

4.2   Energetics of Immersion of Solid in Pure Liquid

4.2.1   Thermodynamic Background

In Chapter 2, we have introduced a number of thermodynamic surface excess quantities (Equations 2.11–2.14) in the case of a simple gas adsorption system involving a single adsorptive. These quantities were expressed as a function of the surface excess amount, n^σ. In the case of the process of immersion of a solid in a pure liquid, the same surface excess quantities can still be defined, and if we are interested in understanding the interfacial phenomenon, it is useful to express them as a function of the surface area. Thus,

$$U^\sigma = A \cdot u^i \tag{4.1}$$

$$H^\sigma = A \cdot h^i \tag{4.2}$$

$$S^\sigma = A \cdot s^i \tag{4.3}$$

$$F^\sigma = A \cdot f^{\,i} \tag{4.4}$$

where A is the area of the solid/fluid interface and u^i, h^i, s^i and f^i (with superscript 'i' which stands for 'interfacial') denote the areal surface excess energy, enthalpy, entropy and Helmoltz energy, respectively (here again, as we are in the Gibbs representation, there is no difference between U^σ and H^σ, because $V^\sigma = 0$). These areal surface excess quantities are characteristic of the nature of the interface, which must be stated.

4.2.1.1   Definition of Immersion Quantities

The energy (or enthalpy) of immersion $\Delta_{imm}U$ (or $\Delta_{imm}H$) is defined (Everett, 1972) as the energy (or enthalpy) change, at constant temperature, when a solid surface is completely immersed in a wetting liquid in which the solid is *insoluble* and *unreactive*. The initial state of the solid surface must of course be specified: the extent of any preliminary outgassing and the nature of surrounding medium (vacuum or vapour of the liquid at a given partial pressure). Each initial state leads indeed to a different surface excess energy U^σ as follows:

$$U^\sigma = U - U^s - U^\ell - U^g \tag{4.5}$$

where U is the internal energy of the whole system in equilibrium, and U^s, U^ℓ and U^g are the internal energies of the bulk solid, liquid and vapour phases, respectively.

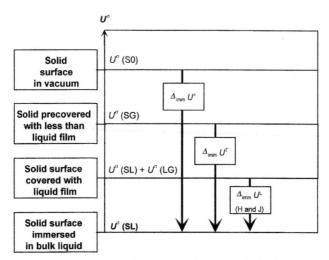

Figure 4.1 Dependence of the energy of immersion $\Delta_{imm}U$ on the initial coverage of the solid surface (H and J stands for Harkins and Jura procedure).

The dependence of U^σ on a few typical states of the system is represented in Figure 4.1.

The upper level corresponds to the clean solid (i.e. in vacuum) whose surface excess energy is denoted as $U^\sigma(S0)$ and, as stated earlier, is proportional to its surface area:

$$U^\sigma(S0) = Au^i(S0) \tag{4.6}$$

The lowest level corresponds to the immersed solid, with a solid/liquid interface whose surface excess energy is now denoted as $U^\sigma(SL)$ with

$$U^\sigma(SL) = Au^i(SL) \tag{4.7}$$

Two intermediate levels deserve consideration, depending on the pre-coverage of the solid surface. If the solid surface has adsorbed less than a liquid film, then the surface excess energy is located at the solid/vapour interface, $U^\sigma(SG)$, and it can be written as follows (always assuming the adsorbent to be inert):

$$U^\sigma(SG) = U^\sigma(S0) + n^\sigma \left(u^\sigma - u^\ell\right) \tag{4.8}$$

where n^σ is the surface excess amount and u^σ the molar surface excess energy. In accordance with Equation (4.1), we can write

$$U^\sigma(SG) = Au^i(SG) \tag{4.9}$$

In Figure 4.1, we see that the surface excess energy of a solid surface covered with a liquid film equals $U^\sigma(SL) + U^\sigma(LG)$. It is useful to separate these two terms by considering that

$$U^\sigma(\text{SL}) = Au^{\text{i}}(\text{SL}) \tag{4.10}$$

at the solid/liquid film interface and that

$$U^\sigma(\text{LG}) = Au^{\text{i}}(\text{LG}) \tag{4.11}$$

at the liquid film/vapour interface.

The maximum energy of immersion, which we designate as $\Delta_{\text{imm}}U^{\text{o}}$, is liberated when the vacuum/solid interface is replaced by the liquid/solid interface. Thus, for the immersion of an outgassed adsorbent, of surface area A, we obtain:

$$\Delta_{\text{imm}}U^{\text{o}} = A\left[u^{\text{i}}(\text{SL}) - u^{\text{i}}(\text{S0})\right] \tag{4.12}$$

where $u^{\text{i}}(\text{SL})$ and $u^{\text{i}}(\text{S0})$ are the areal surface excess energies corresponding to $U^\sigma(\text{SL})$ and $U^\sigma(\text{S0})$ in Figure 4.1. (Note that as the process is exothermic, the $\Delta_{\text{imm}}U$ values are all negative.)

When area A is already covered with a physisorbed layer at surface excess concentration $\Gamma(=n^\sigma/A)$, the energy of immersion becomes

$$\Delta_{\text{imm}}U^\Gamma = A\left[u^{\text{i}}(\text{SL}) - u^{\text{i}}(\text{SG})\right] \tag{4.13}$$

Finally, when the adsorbed layer is thick enough to behave as a liquid film, the energy of immersion, $\Delta_{\text{imm}}U^\ell$, which corresponds to the disappearance of the liquid/gas interface, is simply

$$\Delta_{\text{imm}}U^\ell = -Au^{\text{i}}(\text{LG}) \tag{4.14}$$

The above equations are all based on the internal energy. Similar equations can be written with the enthalpy as *the surface excess enthalpy and energy are identical in the Gibbs representation* when $V^\sigma = 0$ (Harkins and Boyd, 1942). Therefore, the various energies of immersion defined by Equations (4.6)–(4.8) are all virtually equal to the corresponding enthalpies of immersion, that is, ($\Delta_{\text{imm}}H^{\text{o}}$, $\Delta_{\text{imm}}H^\Gamma$ and $\Delta_{\text{imm}}H^\ell$), thus,

$$\Delta_{\text{imm}}H^{\text{o}} = \Delta_{\text{imm}}U^{\text{o}} \tag{4.15}$$

The latter definition of the enthalpy of immersion is that given by Everett (1972, 1986).

Nevertheless, in Section 4.2.1, we shall refer to the energy of immersion which is unambiguous and consistent with our thermodynamic treatment. In the rest of the book, especially when quoting works from the literature, we shall indifferently refer to the energy or the enthalpy of immersion.

In fact, 'energy of immersion' was the term originally used by Harkins in his early papers (Harkins and Dahlstrom, 1930), before resorting to the usual laboratory term of 'heat of immersion'. Although the latter term is still used by a few authors, it is to be discouraged as the heat measured is a raw data which depend on the procedure used and is not directly related to any change in the thermodynamic state of the system: as will be stressed in Section 4.2.2, in practice, the microcalorimetric measurement of the heat exchanged is never equal to the required energy of immersion.

4.2.1.2 Relation Between the Energy of Immersion and the Energy of Gas Adsorption

The process of the immersion of a clean solid surface (which gives rise to $\Delta_{imm}U^o$) can be notionally separated into two successive steps (see Figure 4.2):

1. an adsorption step, during which an amount of liquid n^σ is vaporized from the liquid phase (with a molar energy of vaporization $\Delta_{vap}u$) and adsorbed on the solid surface (with an integral molar adsorption energy $\Delta_{ads}u$, precisely defined for an adsorption process at constant volume [cf. Equation 2.59]), at temperature T and up to the surface excess concentration $\Gamma = n^\sigma/A$ and
2. an immersion step proper, during which the solid with its pre-adsorbed layer is immersed in the liquid.

Let us now consider the case where the first step takes place out of the calorimeter. The corresponding energy change is

$$\Delta_{ads}U + \Delta_{vap}U = n^\sigma \left(\Delta_{ads}u + \Delta_{vap}u\right) = n^\sigma \left(u^\sigma - u^\ell\right) \tag{4.16}$$

The second step, which takes place in the calorimeter, gives rise to an energy of immersion $\Delta_{imm}U^\Gamma$ which is of course smaller than $\Delta_{imm}U^o$ and directly depends on Γ, that is, on the pre-coverage.

If we now add the energy changes for the two steps, we get the following equation:

$$\Delta_{imm}U^o = n^\sigma \left(u^\sigma - u^\ell\right) + \Delta_{imm}U^\Gamma \tag{4.17}$$

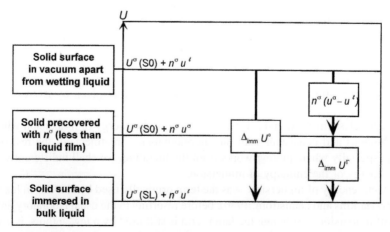

Figure 4.2 Relation between energy of immersion and net molar integral energy of adsorption $n^\sigma(u^\sigma - u^\ell)$.

It thus becomes possible to assess the net molar *integral energy* of adsorption $(u^\sigma - u^\ell)$ from the difference between the energy of immersion of the outgassed adsorbent and that of the adsorbent with a pre-adsorbed surface excess concentration, as was originally pointed out by Hill (1949).

Differentiation of Equation (4.17) with respect to n^σ provides the *differential energy of adsorption*:

$$\Delta_{ads}\dot{u} = \frac{\partial}{\partial n^\sigma}\left(\Delta_{imm}U^\circ - \Delta_{imm}U^\Gamma\right) - \Delta_{vap}u \qquad (4.18)$$

Similarly, we can derive the *differential* (or isosteric) *enthalpy of adsorption* (defined by Equation 2.51):

$$\Delta_{ads}\dot{h} = \frac{\partial}{\partial n^\sigma}\left(\Delta_{imm}U^\circ - \Delta_{imm}U^\Gamma\right) - \Delta_{vap}h \qquad (4.19)$$

It is with this type of equation that, for instance, Micale et al. (1976) was able to check the consistency of the isosteric approach (from gas adsorption isotherms) with immersion calorimetry, for the water–microcrystalline $Ni(OH)_2$ system.

Finally, when integral molar energies of adsorption are directly measured by gas adsorption calorimetry, it is possible to obtain the corresponding integral molar *entropies of adsorption* from Equations (2.65) and (2.66).

If the integral molar entropy of the adsorbed phase is to be compared with the molar entropy of the immersion liquid, it is necessary to express the molar entropy of the gas as a function of the relative pressure and the enthalpy of vaporization. Thus,

$$s^g = s^\ell - R\ln\frac{p}{p^\circ} + \frac{\Delta_{vap}h}{T} \qquad (4.20)$$

The integral molar entropy of adsorption then becomes (as indicated by Jura and Hill, 1952):

$$\left(s^\sigma - s^\ell\right) = \frac{1}{T}\left[\frac{1}{n^\sigma}\left(\Delta_{imm}U^\circ - \Delta_{imm}U^\Gamma\right) + \frac{\Pi}{\Gamma}\right] - R\ln\frac{p}{p^\circ} \qquad (4.21)$$

where Π is the spreading pressure of the adsorbed film characterized by the surface excess concentration Γ, which can be calculated from the adsorption isotherm (cf. Chapter 2).

4.2.1.3 Relation Between the Energies of Immersion and Adhesion

The concept of the energy of adhesion between a liquid and a solid was originally proposed and defined by Harkins (1952) as a means of characterizing the 'dewetting' of a solid to produce a solid surface (in vacuum) and an equal area of liquid surface (in the presence of saturated vapour). The corresponding internal energy change, which we prefer to refer to as the energy of separation, $\Delta_{sep}U$, is given by

$$\Delta_{sep}U = A\left[u^i(S0) + u^i(LG) - u^i(SL)\right] \qquad (4.22)$$

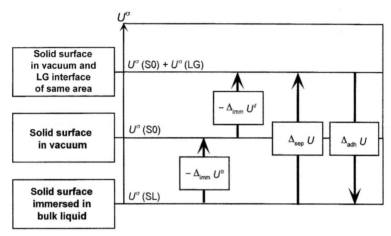

Figure 4.3 Relation between energies of immersion, separation and adhesion.

It follows from Equations (4.12) and (4.14) and Figure 4.3 that

$$\Delta_{\text{sep}} U = -\left[\Delta_{\text{imm}} U^{\circ} + \Delta_{\text{imm}} U^{\ell}\right] \tag{4.23}$$

It is somewhat misleading that Harkins used the term 'energy of adhesion' to denote an 'energy of separation of a liquid from the surface of a solid' (Harkins, 1952). It seems more consistent to define the energy of adhesion $\Delta_{adh}U$, with the same sign as the immersion quantities, so that

$$\Delta_{\text{adh}} U = -\Delta_{\text{sep}} U \tag{4.24}$$

The above relationships are represented in Figure 4.3, where the upper level of surface excess energy corresponds to the equal area of solid/vacuum and liquid/vapour interfaces.

4.2.1.4 Relation Between the Areal Surface Excess Energy and the Surface Tension

As indicated in Chapter 2, the adsorbent surface is characterized by a surface tension γ whose magnitude depends on the nature of the surrounding medium (liquid, gas or vacuum) with which the adsorbent is in equilibrium. The isothermal extension of the surface area A, with no other change in the thermodynamic state of the system of adsorption, results in an increase in dF of the Helmoltz energy of the system. Thus, the surface tension, γ, is defined as follows:

$$\gamma = \left(\frac{\partial F}{\partial A}\right)_{T,V,n^{\sigma}} \tag{4.25}$$

It is convenient to separate the contributions of internal energy U and entropy, S, of the system so that

$$\left(\frac{\partial F}{\partial A}\right)_{T,V,n^\sigma} = \left(\frac{\partial U}{\partial A}\right)_{T,V,n^\sigma} - T\left(\frac{\partial S}{\partial A}\right)_{T,V,n^\sigma} \tag{4.26}$$

and by taking into account Equations (4.1), (4.3) and (4.25) we obtain

$$\gamma = u^i - Ts^i \tag{4.27}$$

As

$$S = -\left(\frac{\partial F}{\partial T}\right)_{A,V,n^\sigma} \tag{4.28}$$

and by making use of Equations (4.25), (4.26) and (4.28), we arrive at

$$\left(\frac{dU}{dA}\right)_{T,V} = \gamma - T\left(\frac{\partial \gamma}{\partial T}\right)_A \tag{4.29}$$

It is noteworthy that Equation (4.29) is analogous to the well-known equation for the isothermal expansion of bulk gas.

$$-\left(\frac{dU}{dV}\right)_T = p - T\left(\frac{\partial p}{\partial T}\right)_V \tag{4.30}$$

where p is the pressure and V the volume. Of course, $(\partial U/\partial V)_p$ is zero in the case of an ideal gas.

The term $(dU/dA)_{T,V,n^\sigma}$ was referred to as the 'total extension energy of the surface per unit increase of area' by Einstein (1901). It is indeed the sum of the work, γ, which must be supplied to extend the surface by unit of area and the heat which must be supplied to carry out this extension reversibly and isothermally.

We can also obtain the areal surface excess entropy from Equations (4.25) and (4.29) as follows:

$$s^i = -\left(\frac{\partial \gamma}{\partial T}\right)_A \tag{4.31}$$

4.2.1.5 Various Types of Wetting

In the preceding sections, we have been mainly concerned with one form of wetting, namely 'immersion'. It is, however, useful to distinguish between various types of wetting, which are illustrated in Figure 4.4.

The four major types of wetting are as follows (the first three following Everett's definitions, 1972):

1. *Immersional* wetting (which we simply call 'immersion' and denote by subscript 'imm') is a process in which the surface of a solid, initially in contact with vacuum or a gas phase, is brought in contact with a liquid without changing the area of the interface. Here, a solid/gas (or solid/vacuum) interface is replaced by a solid/liquid one of same area.

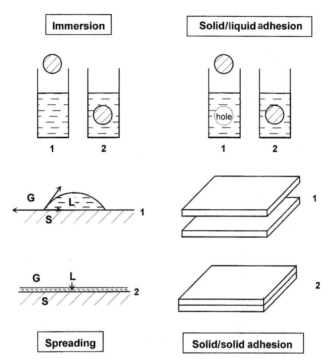

Figure 4.4 Interfaces lost or formed during immersional, adhesional and spreading wetting and during solid/solid adhesion.

2. *Adhesional* wetting (which we shall call 'adhesion' and denote by subscript 'adh') is a process by which an adhesional union is formed between *two pre-existing surfaces* (one of them being solid and the other liquid). Here, two initial interfaces (solid/gas and liquid/gas) are replaced by one (solid/liquid).
3. *Spreading* wetting is a process in which a drop of liquid spreads over a solid substrate (the liquid and solid being previously in equilibrium with the vapour). Here, the solid/vapour interface is replaced by two new interfaces (solid/liquid and liquid/vapour) of same area.
4. *Condensational* wetting is a process in which a clean solid surface (initially in vacuum) adsorbs a vapour up to the formation of a *continuous liquid film*. Here, the solid/vacuum interface is replaced by two new interfaces (solid/liquid and solid/vapour) of same area, as in spreading wetting. The difference between condensational and spreading wetting is in the initial state, the liquid film being formed from a vapour in one case and from a drop in the other case.

At this stage, it may be noted (Jaycock and Parfitt, 1981) that the above types of wetting do not all have the same contact angle requirement ($\theta < 90°$) for spontaneous occurrence. Thus:

a. adhesional wetting requires $\theta < 180°$, which is of course the most general case,
b. immersional wetting requires $\theta < 90°$ (otherwise, external work must be applied), and
c. spreading and condensational wetting require $\theta = 0$.

4.2.1.6 Wettability of a Solid Surface: Definition and Assessment

The concept of wettability of a solid by a liquid is directly related to the wetting processes. This concept is especially useful in the fields of detergency, lubrication or enhanced oil recovery. In the context of the oil industry, proposals were made by Briant and Cuiec (1972) for the experimental assessment of wettability, which was defined in terms of the thermodynamic affinity of a solid surface for a liquid.

According to this approach, wettability is equated to the work exchanged by the immersion system with its surroundings when the process of immersional wetting of a given area of the clean solid surface is carried out reversibly. Thus,

$$\int_{imm} - \left(\frac{\partial F}{\partial A} \right)_{T,V} dA = [\gamma(SO) - \gamma(SL)] A \tag{4.32}$$

where $\gamma(S0)$ and $\gamma(SL)$ are the surface tensions of the solid surface in the presence of vacuum and a liquid film.

The wettability, which we propose to denote $\Delta_{imm}\gamma$, is directly measured by the difference between two surface tensions. This difference was called the 'adhesion tension' by Adamson (1967) and the 'work of immersional wetting per unit area' by Everett (1972). It has also been referred to as the 'wetting tension'. In Equation (4.32), the difference in γ is positive in the case of spontaneous wetting of the solid by the liquid.

It may be useful to distinguish between the initial conditions when the solid is either in vacuum (the wettability will then be denoted as $\Delta_{imm}\gamma^{o}$) or in equilibrium with a gas phase, with a surface excess concentration Γ (the wettability being then denoted as $\Delta_{imm}\gamma^{\Gamma}$)

$$\Delta_{imm}\gamma^{o} = \gamma(S0) - \gamma(SL) \tag{4.33}$$

$$\Delta_{imm}\gamma^{\Gamma} = \gamma(SG) - \gamma(SL) \tag{4.34}$$

The following three main routes are available to assess the wettability. The first method is dependent on the measurement of contact angles, while the second and third make use of energy of immersion and adsorption isotherm data.

(a) Assessment of wettability from the *contact angle*.

A favourable case is when a contact angle θ can be measured between the liquid and solid in equilibrium with the vapour, so that Young–Dupré equation can be applied. As early as 1805, Young considered the possibility of such an equilibrium between surface tensions, but it was only in 1869 that Dupré put it in the well-known form of equation:

$$\gamma(SG) = \gamma(SL) + \gamma(LG) \cos \theta \tag{4.35}$$

and hence:

$$\Delta_{imm}\gamma = \gamma(LG) \cos \theta \tag{4.36}$$

This allows us to derive the wettability simply from θ and the surface tension of the liquid. This method was used, for instance, by Whalen and Lai (1977) in their systematic study of the wetting of modified surfaces of glass.

(b) Assessment of wettability from the *energy of immersion*.

As it is equal in magnitude to the Helmhotz free energy of immersion per unit area (cf. Equation 4.32), the wettability can be written as:

$$\Delta_{\text{imm}}\gamma = -\frac{\Delta_{\text{imm}}F}{A} \tag{4.37}$$

The problem is now to evaluate $\Delta_{\text{imm}}F$. This can be done with help of a useful observation by Briant and Cuiec (1972): these investigators confirmed that, for a number of solid–liquid systems, the following approximate relationship holds:

$$k = \frac{\Delta_{\text{imm}}F}{\Delta_{\text{imm}}U} = \frac{\gamma(\text{LG})}{u^{\text{i}}(\text{LG})} = \frac{\gamma(\text{LG})}{\gamma(\text{LG}) - T(\partial\gamma(\text{LG})/\partial T)} \tag{4.38}$$

This means that the ratio $\Delta_{\text{imm}}F/\Delta_{\text{imm}}U$ can be assessed from the liquid-phase ratio $\gamma(\text{LG})/u^{\text{i}}(\text{LG})$.

Combining Equations (4.37) and (4.38), we obtain

$$\Delta_{\text{imm}}\gamma = -k \cdot \Delta_{\text{imm}}U \tag{4.39}$$

which allows us to make use of the energy of immersion, provided k is known for the immersion liquid. Values of k, calculated by Briant and Cuiec (1972) for liquids commonly used in immersion, are given in Table 4.1.

In fact, the validity of Equation (4.38) is supported by the observation by Robert (1967) that the energies of immersion of carbon blacks in a number of hydrocarbons of the same family (e.g. *n*-alkanes from 7 to 16 carbon atoms) could be ranked exactly in the same order as the 'adsorbability' of these hydrocarbons.

(c) Assessment of wettability from the gas adsorption isotherm

If the solid is covered with a liquid film in equilibrium with saturated vapour p°, the spreading pressure of the film can be derived from Equation (2.22). Thus,

$$\Pi(p^{\circ}) = \gamma(\text{S0}) - (\gamma(\text{SL}) + \gamma(\text{LG})) \tag{4.40}$$

Table 4.1 Calculation of k from Equation (5.38) for a Number of Liquids (Briant and Cuiec, 1972)

Liquid	t °C	$\gamma(\text{LG})$ mJ m^{-2}	$-T(\partial\gamma(\text{LG})/\partial T)$ mJ m^{-2}	$\gamma(\text{LG}) - T(\partial\gamma(\text{LG})/\partial T)$ mJ m^{-2}	k
Water	37	70	50.5	120.5	0.58
	47	68.4	52.1	120.5	0.57
Heptane	55	16	30.5	46.5	0.34
Cyclohexane	43	21.6	36	57.6	0.37
Benzene	35	26.9	40.1	67	0.40
Paraxylene	38	26.5	32.6	59.1	0.45

This expression can be combined with Equation (4.33) to provide the following expression for the wettability of a clean surface:

$$\Delta_{\text{imm}}\gamma^{\circ} = \Pi(p^{\circ}) + \gamma(\text{LG}) \tag{4.41}$$

The spreading pressure $\Pi(p^{\circ})$ can be calculated from Gibbs Equation (2.34) by integration from $p/p^{\circ} = 0$ up to 1. However, this is dependent on the availability of highly accurate data at low p/p°. Briant and Cuiec (1972) evaluated the resulting uncertainty of the wettability to be ca. 10%.

4.2.2 Experimental Techniques of Immersion Microcalorimetry in Pure Liquid

Of the four types of wetting phenomena examined in the previous section, only *immersional wetting* lends itself to direct microcalorimetric measurement: spreading and adhesion experiments would involve too small interfacial areas (say, not more than ca. 100 cm^2), whereas condensational wetting would require measurements up to $p/p^{\circ} = 1$. As we saw in Chapter 3, these are the conditions where accurate measurements of the amounts of gas adsorbed are difficult to achieve. For this reason, we confine the following recommendations to immersion microcalorimetry.

In principle, to carry out immersion microcalorimetry, one simply needs a powder, a liquid and a microcalorimeter. Nevertheless, it was early realized that the heat effects involved are small and the sources of errors and uncertainties are numerous. Many attempts have been made to improve immersion microcalorimetric techniques. Before commenting on this type of experiment, we describe the equipment and procedure which have been found to be of particular value for energy of immersion measurements (Partyka et al., 1979).

4.2.2.1 Recommended Immersion Microcalorimetric Equipment and Experimental Procedure

Figure 4.5 shows the set-up, at a stage which is just prior to the immersion of the solid. Here, the sample (1) is located in a glass bulb (2) with a fragile tip (3). The sample has already been outgassed, out of the microcalorimeter, and has been left either under vacuum or under a given vapour pressure of the immersion liquid. The bulb has been sealed-off and attached, through fitting (4), to a thin glass rod. This assembly has been introduced into the microcalorimeter cell and tightly closed. The glass rod is able to slide down through O-ring 5 (located at least 5 cm above the calorimeter to avoid detecting friction effects). The heat flowmeter (6) (a Tian–Calvet thermopile) surrounds the microcalorimetric cell containing the immersion liquid (7) and the glass bulb.

With the above set-up, the essential steps of a safe and convenient experiment are as follows:

1. Introduction and weighing of adsorbent in immersion bulb.
2. Outgassing of adsorbent with one of the procedures described in Section 3.4.

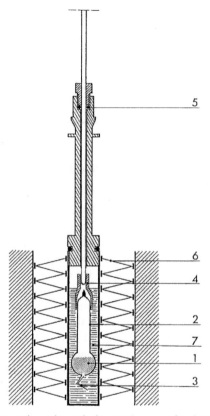

Figure 4.5 Set-up for immersion microcalorimetry (see text for details).

3. If pre-adsorption is required, equilibration of sample with desired relative vapour pressure of immersion liquid.
4. Sealing of ampoule neck at ca. 2 cm above the bulb, if necessary immersed in cooling bath to protect the sample.
5. Determination of weight of sealed bulb.
6. Connection of top of sample bulb to glass rod (3-mm diameter) by means of Teflon plug (4).
7. Introduction of sample bulb (2) into stainless steel calorimetric cell already filled with immersion liquid (7).
8. Tight closure of this set-up, with some lubricant on O-ring 5 and with end of capillary tip of sample bulb ca. 5 mm above the bottom of microcalorimetric cell.
9. Allowance of time for temperature equilibration in the microcalorimeter (this may require 3 h for high sensitivity).
10. Breaking of capillary tip by slowly and gently depressing glass rod until microcalorimetric signal starts to respond.
11. Recording of microcalorimetric signal until it returns to baseline (this usually requires ca. 30 min).
12. Careful removal of sample bulb and broken tip from microcalorimetric cell.

13. Determination of weight of sample bulb filled with immersion liquid (only wiped outside), together with broken tip.
14. Determination of dead volume V of sample bulb from the information gained in steps (5) and (13), knowing mass and density of immersion liquid.
15. Determination of total experimental heat of immersion by integration of the whole micro-calorimetric signal (including the small endothermal peak due to the vaporization of first droplet of immersion liquid into dead volume, just before immersion proper; Partyka et al., 1979).
16. Calculation of correction terms and, finally, of the energy of immersion.

4.2.2.2   Evaluation of the Correction Terms

The two correction terms involved in the above experiment are:

a. The energy of *bulb breaking*, W_b, which is the exothermal sum of the work provided by the operator, plus the energy released by pre-existing stresses in the glass bulb, minus the energy absorbed by the formation of new glass–liquid interfaces. Typical values of W_b for a fine tip of the shape represented in Figure 4.5 are around 5 mJ. The W_b value must be determined by blank experiments with similar bulbs.

b. The energy of *vaporization*, $\Delta_{vap}U$, of the immersion liquid into the dead volume V of the glass bulb, which is endothermal:

$$\Delta_{vap}U = \Delta_{vap}u\frac{(p^o - p)V}{RT} \tag{4.42}$$

where $\Delta_{vap}u$ is the molar energy of vaporization, p is the pressure in the bulb (vacuum or equilibrium pressure) and p^o the saturation vapour pressure of the immersion liquid.

To obtain an impression of the order of magnitude of the correction terms, let us suppose that, in a typical experiment corresponding to the immersion of ca. 2 m^2 of ground sand in water (Partyka et al., 1979) the total heat measured is 1042 mJ. In this case, the above correction terms are ca. -6 and $+76$ mJ, respectively, and therefore,

$$\Delta_{imm}U = 1042\,\text{mJ} - 6\,\text{mJ} + 76\,\text{mJ} = 1112\,\text{mJ} \tag{4.43}$$

If other types of equipment and experimental procedures, which will be commented on hereafter, one may also have to take into account:

- The energy absorbed by the continuing evaporation of the immersion liquid – if the set-up is not airtight.
- The work due to atmospheric pressure when the free level of the liquid is depressed, as it enters the bulb; this term, equal to $V(p_{atm} - p^o)$, is quite small and, in the experiment quoted above, would probably amount to not more than -2 mJ.
- The energy dissipated by stirring, which is often used to ensure a satisfactory dispersion and wetting of the powder.
- And the broad dispersion of the energies of bulb breaking when one does not make use of a brittle end.

4.2.2.3   Critical Aspects of Immersion Calorimetry Techniques

4.2.2.3.1   The Calorimeter

Most immersion calorimetry was carried out with two of the four main categories of calorimeters listed in Section 3.2.1, namely, (i) passive adiabatic calorimeters (also called quasi-adiabatic, isoperibol, or conventional 'temperature rise' type) and (ii) diathermal-conduction microcalorimeters making use of a heat flowmeter. The quasi-adiabatic or isoperibol calorimeters were the only type used until the 1960s: they are easily constructed and are well suited for room temperature operation. Improvements were made in the temperature stability of the surrounding isothermal shield, the sensitivity of the temperature detector: initially a single thermocouple, then a multicouple with up to 104 junctions (Laporte, 1950) and finally a thermistor (Zettlemoyer et al., 1953). Differential or twin mountings were adopted in order to cancel external sources of disturbances (Bartell and Suggitt, 1954; Mackrides and Hackerman, 1959; Whalen, 1961a,b). Nevertheless, the limitations of this type of calorimetry are: (i) the need for rapid experiments (heat effects must last less than a few minutes, in order to minimize the cooling corrections), (ii) the need for complete breaking of the sample bulb and efficient stirring of the powder and consequently, (iii) questionable reproducibility and accuracy due to these two extra heat effects. The availability, in the 1960s, of heat flowmeter microcalorimeters increased the possibilities of the immersion method. The improvements included (i) the much higher sensitivity (at least, 2 orders of magnitude higher), (ii) the long-term stability which allows one to study wetting phenomena over hours (even over days if needed) (iii) to allow more time for the wetting process and hence to let the liquid enter the sample bulb through a thin capillary whose energy of breaking is at least 1 order of magnitude smaller than that of the complete bulb and also (iv) the possibility of avoiding stirring.

4.2.2.3.2   The Sample Container

This is the most critical part of the experiment. The choice must be made between a sealed glass bulb or another type of container. The latter usually look simpler to operate: some are simply closed by a plug or by a cover with a mercury seal and are located above the liquid. The experiment is then started by simply pulling the plug and turning the bucket upside down (Gonzalez-Garcia and Dios Cancela, 1965) or when possible by inverting the whole calorimeter (Vanderdeelen et al., 1972; Nowell and Powell, 1991; Moreno-Pirajan et al., 1996). Another possibility is to insert a pellet of powder, protected on top and bottom by a patch of polymer film, into a tube standing above the liquid. A rod can be used to start the experiment by pushing the pellet out of the tube (Magnan, 1970). Another procedure (Jehlar et al., 1979) is to outgas the powder in a stainless steel cage fitted into the cylindrical cell of a calorimeter. After outgassing, the cage is covered and sealed with mercury. The immersion liquid is poured over the mercury, the cell is closed and made tight by means of a mercury seal and then introduced into the microcalorimeter. When thermal equilibrium is reached, the experiment is started by pulling the cage into the immersion liquid. It is also possible to use a thin metal foil to isolate either the sample (Everett and Findenegg, 1969), or the liquid (Partyka et al., 1975),or to simply use a valve to allow the immersion liquid to flood the sample maintained under vacuum. The limitation of all these systems lies in the *difficulties* encountered either *in the outgassing or in the airtight storage* of the sample during the thermal equilibration of the microcalorimeter (large variations in the recorded energy of immersion can result from a very small pre-adsorption of the vapour). For an energy of immersion to be meaningful, it is essential indeed that the initial state of the sample be well defined, that is, outgassed in a reproducible way. These difficulties are certainly avoided by the use of sealed glass bulbs, which provide the safest operation, but, here again, a choice must be made: either the bulbs are to be completely broken or they are designed with a brittle end which is the only part broken. Complete breakage lends itself to excellent wetting (provided it is assisted by a good stirring) and also to the special case of immersion in a solution. However, the energy of breakage has been reported to range from 57 (Zimmermann et al., 1987) to 3050 mJ (Bartell and Suggitt, 1954), although it is most often in the 200–400 mJ range (Zettlemoyer et al., 1953). The unavoidable uncertainty in these values for the energy of breakage (as the bulbs, most often handmade, lend themselves to large variations of this energy), added to the extra energy introduced by stirring, mean that this approach is mainly justified for immersion in solution, in place of a more sophisticated calorimetric experiment of adsorption from solution as described in Section 4.3.3. For immersion in pure liquids, the brittle end approach provides a considerably higher reproducibility and accuracy. With special care, one can use thin capillary tips 'barely able to sustain the weight of the ampoule' (Everett et al., 1984) giving a heat of breaking as small as ca. 0.2 mJ.

4.2.2.3.3 The Immersion Liquid

To study surface phenomena, the liquid must not dissolve the solid and any chemical interaction must not occur in depth. Features which, depending on the purpose, are worth considering when selecting an immersion liquid are: (i) its wetting properties for the solid under study, (ii) the polarity and molecular size and shape, (iii) its saturation vapour pressure at the immersion temperature and (iv) its corresponding enthalpy of vaporization. One would like the product of these two latter quantities to remain small, so that the vaporization into the dead volume of the bulb does not influence too much the quality of the results. The purity of the immersion liquid is essential: if the latter is non-polar (an alkane, for instance), the least amount of any polar impurity (alcohol, water) can drastically change the immersion microcalorimetry results. This was early realized by Harkins and Dahlstrom (1930), who spent more than two weeks drying their benzene for immersion and the necessary glassware, by making use of 'considerable quantities of sodium wire', of sulphuric acid and of phosphorus pentoxide. The data in Figure 4.6, redrawn from a later paper from the same laboratory (Harkins and Boyd, 1942), show the drastic influence of the residual water concentration in benzene on the enthalpy of immersion of an anatase sample: 20 ppm of water are enough to triple the enthalpy measured! Nowadays, previously degassed 4A zeolite is successfully used as a getter to eliminate any residual water.

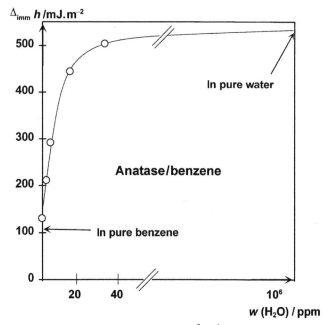

Figure 4.6 Immersion of 1 g TiO$_2$ (anatase) of 9.24 m^2 g^{-1} into 8.74 g of benzene: influence of residual water concentration W on $\Delta_{imm}h$.
After Harkins and Boyd (1942).

4.2.2.3.4 Wetting of the Powder

Evacuating the immersion bulb prior to sealing is probably the most efficient means of achieving the desired wetting after breaking the bulb. Nevertheless, it may happen that a powder is not easily wettable and that it tends to produce aggregates which are not easily wetted. This situation results of course in a lack of good reproductibility. It is often thought that this problem can be overcome by efficient stirring, after complete breakage of the bulb, and that therefore one must be prepared to accept the limitations associated with stirring and complete bulb breakage. However, this is not entirely true: a solution is to use a microcalorimeter able to detect and measure precisely the heat evolution over periods of up to 30 min and in addition to use a bulb with a central constriction: the liquid entering through the broken tip pushes the powder up to the restricted part and percolates through the powder so as to finally fill the upper chamber (Laffitte and Rouquerol, 1970).

4.2.2.3.5 The Dead Volume in the Bulb

This dead volume allows some evaporation of liquid and a corresponding heat absorption. For example, 1 cm^3 of dead volume, when filled with water vapour at 25 °C, absorbs ca. 80 mJ. Depending on the design and filling of the bulb, this dead volume ranges between 0.2 (Everett et al., 1984) and more than 10 cm^3. Two reasons for this dead volume are (i) the fact that, except for highly wettable powders, it is usually preferable to leave the powder loose rather than compact it and (ii) that some distance – and therefore some volume – must be left between the sample and the glass seal to avoid damaging the sample. This distance is typically around 2 cm. For fragile samples, one may use a longer distance or keep the bulb immersed in water or liquid nitrogen during the sealing. In order to restrict the corresponding volume, one can, at will, either introduce a glass rod or limit the diameter of the neck, provided that the conditions for a satisfactory outgassing are still ensured.

4.2.3   Applications of Pure Liquid Immersion Microcalorimetry

If properly used, immersion calorimetry is a versatile, sensitive and accurate technique which has many advantages for the characterization of porous solids and powders. An indication of these possibilities is given in Figure 4.7. The major areas of application are outlined in this section and reference made to specific examples which are discussed in more detail in other chapters.

It should be kept in mind that any change in surface area, surface chemistry or microporosity will result in a change in the energy of immersion. Because immersion calorimetry is quantitative and sensitive and because the technique is not too difficult to apply in its simplest form, it can be used for quality testing. The preliminary outgassing requires the same care, as for a BET measurement, but, from an operational standpoint, energy of immersion measurements are probably less demanding than gas adsorption measurements.

4.2.3.1   Comparison of Surface Areas

For many years, immersion microcalorimetry has been found useful for the routine characterization of fine powders and porous materials such as activated carbons and oxides. In the absence of complicating effects such as micropore filling, it was generally assumed as a first approximation that, *for a given surface chemical*

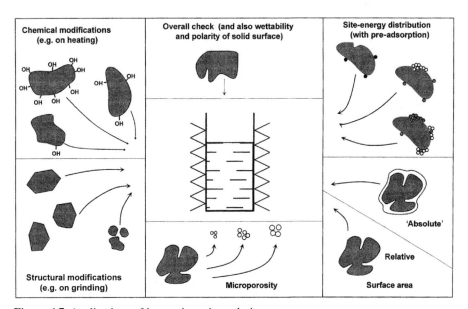

Figure 4.7 Applications of immersion microcalorimetry.

composition, the magnitude of the energy of immersion of the outgassed solid, $\Delta_{imm}U^o$, was proportional to the extent of the available surface A so that

$$\Delta_{imm}U^o = A \cdot \Delta_{imm}u^{i,o} \tag{4.44}$$

where the areal energy of immersion $\Delta_{imm}u^{i,o}$ is characteristic of the nature of a given liquid–solid system. Once it is known (with the help of a sample whose surface area was determined, for instance, by the BET method), one can easily assess a surface area. For that purpose, the technique proposed in Section 4.2.2 simply requires a solid sample in the immersion cell of at least 1 m². This means that, for the routine control of the specific surface areas of a series of similar samples, immersion microcalorimetry is a very useful technique. However, because it is very sensitive to any change in the nature of the surface, the areal energy of immersion in a given liquid is very dependent on the physico-chemical properties of the solid, so that the value of $\Delta_{imm}u^{i,o}$ cannot be used for the assessment of specific surface area of an unknown solid. An elegant way to solve this problem was proposed by Harkins and Jura (1944) under the name of the 'absolute' calorimetric method for the determination of surface areas and is described hereafter.

4.2.3.2 Surface Area of Non-porous Solids: Modified Harkins and Jura 'Absolute' Method

The principle of the method is given in Figure 4.8. In its genuine form (Harkins and Jura, 1944; Harkins, 1952), the outgassed powder (state 1) is brought into equilibrium with the saturation vapour pressure of the immersion liquid in order to form an adsorbed film thick enough (5–7 molecular layers) to hide the specific adsorption sites of the powder and to present an external surface identical in nature to that of the bulk liquid (state 2). The pre-coated powder is then immersed in the liquid located in the calorimeter (state 3). The heat released during the loss of the liquid film/vapour interface is related to its area A. Harkins and Jura made use of a basic relationship which is obtained by integrating Equation (4.29) between the two values of interfacial area $(A,0)$:

$$\Delta_{imm}U^{\ell} = -A\left(\gamma - T\frac{\partial\gamma}{\partial T}\right) \tag{4.45}$$

where γ is the surface tension of the immersion liquid. The immersion energy $\Delta_{imm}U^{\ell}$ has a negative value, and the immersion is an exothermic process. Because A is

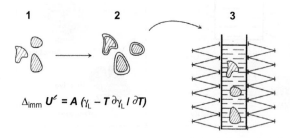

1   **2**   **3**

$\Delta_{imm}\ U^{\ell} = A\ (\gamma_L - T\ \partial\gamma_L\ /\ \partial T)$

Figure 4.8 Principle of Harkins and Jura's 'absolute' calorimetric method for determining the surface area.

obtained *without any assumption concerning the molecular cross-section* of the immersion liquid, Harkins and Jura referred to their method as an 'absolute method'.

The few investigators who have attempted to use the original Harkins–Jura method have encountered a number of inherent difficulties. A major problem is that it is virtually impossible to avoid some inter-particle capillary condensation as $p/p^o \to 1$. This inevitably reduces the extent of the available liquid/vapour interface (Wade and Hackerman, 1960). Moreover, the thickness of a pre-adsorbed film as $p/p^o \to 1$ is highly dependent on the shape, size and roughness of the particles.

By taking advantage of the high sensitivity of Tian–Calvet microcalorimetry (see Chapter 3), it was possible to improve the Harkins–Jura procedure (Partyka et al., 1979). It was found that, with a number of non-porous solids, the pre-adsorption of ca. 1.5 molecular layers of water was sufficient to produce a film/vapour interface with the same energy as a bulk liquid/vapour interface. In other words, this statistical thickness of pre-adsorbed water is enough to effectively 'screen' the adsorbent surface, but small enough to prevent any substantial change in surface area (either by capillary condensation or by increase in the apparent radius of the particles). This is obtained by carrying out the pre-adsorption under a relative pressure of 0.5, instead of 1. With the aid of this modified Harkins–Jura technique and the use of a Tian–Calvet microcalorimeter, it is possible to obtain surface areas which are compared, in Table 4.2, with the corresponding BET (N_2) areas. The large discrepancy found for the processed kaolin is explained by the fact that this sample was modified by the addition of surfactant for the paper industry: the water pre-adsorption used in the Harkins and Jura method is then able to separate sheets of kaolin (leading to 19.2 m^2 g^{-1}), although the latter remain stuck when N_2 adsorption is carried out at 77 K (hence a BET (N_2) area of only 12.1 m^2 g^{-1}).

Table 4.2 Surface Areas Measured on Non-porous Adsorbents by the Modified Harkins and Jura (H$_2$O) Method and by the BET (N$_2$) Method

Sample	HJ (H$_2$O) *area*/m^2 g^{-1}	BET (N$_2$) *area*/m^2 g^{-1}
Aerosil	140	129
Aluminoxid	100	81
Titanoxid	63	57
Gibbsite	27.0	24.0
Gallium hydroxide	21.3	21.0
Kaolin, raw	19.4	19.3
Kaolin, processed	19.2	12.1
Ground quartz	4.2	4.3
Zinc oxide	3.1	2.9
Calcite	0.8	0.6

Source: After Partyka et al. (1979).

Provided that sufficient care is taken, the immersion method is relatively easy to apply and is well suited for the determination of the surface area of insoluble non-porous oxides, ceramics, clays, etc. with specific areas ranging from 1 to 100 m² g⁻¹. This experimental method is one of the very few allowing to assess a surface area *totally independently from the BET method*, without requiring any reference material of known surface area. Hence, its interest to evaluate the validity of the BET method. For the study of hydrophobic surfaces, water becomes unsuitable and penta-nol may be a useful alternative (Partyka et al., 1979).

4.2.3.3   Surface Area of Microporous Carbons

The first question to answer is: how does the presence of micropores change the areal energy of immersion $\Delta_{imm}U/A$? Two answers were brought to this question.

The first answer was based on the calculation made by Everett and Powl (1976) that for the two model cases of slit-shaped and cylindrical micropores, the maximum enhancement of *the adsorption potential* (as compared with that of the flat surface of same nature) should be 2.0 and 3.68, respectively. This maximum occurs when the molecule exactly fits the micropore. Under these conditions, the molecule is covering a much larger area than the 'cross-sectional area' it would cover on a flat surface. Here, the ratios are 2.0 and 3.63, for slit and cylindrical pores, respectively, if one considers on the flat surface a hexagonal compact arrangement. This means that, in the extreme cases of a flat surface on one hand and of a surface completely surrounding the molecular probe on the other hand (i.e. in narrow micropores), the energy of immersion can be considered to be proportional to the area accessible to the probe molecule. This finding was used to propose a simple method to derive *the total* accessible surface area of microporous carbons (Denoyel et al., 1993), after making the two following assumptions:

1. For a given state of the solid surface, the energy of immersion is simply proportional to the surface area available to the immersion liquid for any size and shape of pore. This assumption was later supported by DFT calculations with a slit-shaped pore of varying width (Denoyel, 2004).
2. From the viewpoint of the areal energy of immersion, the behaviour of microporous and external solid surfaces is identical.

The immersion liquid should be chosen so as to limit, as much as possible, any specific interaction with the surface groups and to easily enter the micropores. For activated carbon, benzene is a good choice. One then simply needs to determine the energy of immersion of the sample under study and to compare it with that of a non-porous sample of similar chemical composition and of known surface area (usually determined by the BET (N₂) method), which is used as reference. The areas are assumed to be proportional to the energies of immersion.

So far, this approach has been applied to activated carbons, taking an ungraphitized carbon black Vulcan 3 as the reference, and using the set of immersion liquids listed in the left-hand column of Table 9.1. The results obtained are consistent with the fact that the narrow micropore (or ultramicropore) surface area accessible to a given immersion liquid is dependent on its molecular size. They open a way for the determination of the surface area of narrow microporous solids, a method successfully applied by Gonzalez et al. (1995) and Rodriguez-Reinoso et al. (1997), who were able to show that the presence of oxygen-containing surface groups does not have much effect on the enthalpy of immersion in methanol or water, or, still, by Silvestre-Albero et al. (2001) or Villar-Rodil et al. (2002a,b).

Another answer to the question above was proposed by Stoeckli and Centeno (1997,2005), on the basis of Dubinin's theory. They conclude that the areal enthalpy of immersion is higher in the micropores than on a flat surface and that an additional component of the enthalpy of immersion is due to the liquid filling the micropores between the layers in contact with the pore walls. This raises the question: why, when the pore size exceeds 2 nm, would the areal enthalpy of immersion of the walls suddenly fall down to the value for a flat surface and why would the enthalpy component for the central filling of the micropores suddenly vanish, since, as noticed by Stoeckli and Centeno (2005), a progressive change would be more logical? Also, for each microporous sample, they refer to the total surface area determined by phenol or caffeine adsorption from dilute water solution. Their comprehensive study on a large set of carbons with mean pore sizes ranging from 0.65 to more than 2 nm shows that the phenol or caffeine adsorption method provides areas which are systematically lower than those provided by immersion calorimetry with the help of a single areal enthalpy of immersion. For pores smaller than 0.8 nm, one can explain this discrepancy by the steric hindrance to the formation of a full monolayer of phenol, whose thickness is estimated around 0.41 nm if the molecule lies flat on the surface; for caffeine, a similar hindrance should occur in pores smaller than 1.0–1.2 nm; the amount of phenol adsorbed (and erroneously assumed to form a complete monolayer) logically leads to underestimate the surface area. This explanation holds not only when the mean pore size is lower than 0.8 or 1.2 nm but also as long as the carbon contains both narrow and wide micropores. Alternatively, one can consider that assumption (1) above is implicit (i.e. the energy or enthalpy of immersion is proportional to the area available, irrespective of the pore size). In this case, using immersion calorimetry is not only simpler than determining the amount adsorbed but it also becomes essential for the assessment of the area. Nevertheless, a solvent + a solute is more complex than a pure liquid and may raise diffusion issues in the full range of micropore size and limit the calorimetric signal measured during the time allotted for the measurement, even if a small molecule like phenol gives rise to an important surface diffusion (Ocampo-Perez et al., 2013). Fearing a possible diffusion issue, Bertoncini et al. (2003) waited for 1 week for each equilibrium point of their phenol adsorption isotherms. On the contrary, when a pure liquid of low viscosity and good wetting properties (like, for instance, benzene when used with carbon) is used in immersion calorimetry of an evacuated sample, experience shows that the wetting is completed in a few minutes.

4.2.3.4 Surface Area of Other Porous Adsorbents

Polar adsorbents, such as most oxides, are not amenable to the immersion calorimetry procedures described above (either Denoyel et al., 1993 or Stoeckli and Centeno, 1997) to determine the surface area. This is because the wetting liquids, which are also polar, give rise to specific interactions which modify the areal enthalpy of immersion in an unknown way. Looking for a non-polar liquid able to wet a hydrophilic brought (Rouquerol et al., 2002) to the idea of using *liquid argon*.

A first step in that direction had been done by Chessick et al. (1954) and then by Taylor (1965) in determining the enthalpies of immersion of various solids in *liquid nitrogen*. This was undertaken by simply measuring the amount of gaseous nitrogen vaporized during the experiment, and therefore exactly following the principle of the diathermal, phase change, liquid air calorimeter devised by Dewar (1904). The aim was already to determine the surface area of a solid by means of a liquid which was not expected to interact with any functional groups on the surface. The results obtained with samples of carbon black, calcium silicate, alumina and magnesia seemed to support this view. Nevertheless, this type of calorimeter was not easy to operate because of the difficulty in obtaining a stable rate of boiling and the strong response of the calorimeter to the least change in atmospheric pressure. For example, if say 250 cm^3 of liquid nitrogen were present in the calorimeter, an atmospheric pressure increase of 1 mbar would produce a temperature increase in the liquid bath with a corresponding heat absorption of 8 J. The idea, though, certainly deserved to be reconsidered with the aid of modern technological components.

This is why a technique of immersion calorimetry in liquid nitrogen or liquid argon was devised by Rouquerol et al. (2002), with help of a diathermal, heat flowmeter, low-temperature microcalorimeter normally used for gas adsorption studies at 77 or 87 K. In this arrangement, the microcalorimeter is completely immersed in liquid nitrogen or argon. The bulb containing the evacuated sample is located in the centre of the calorimeter, in a helium atmosphere, and is connected, with help of a glass capillary, to a brittle end immersed in the surrounding liquid. Manual depression of a glass rod breaks the brittle end and lets the liquid nitrogen or argon enter and completely fill the sample bulb. This procedure avoids any disturbance brought by the heat of breakage of the tip (as it is outside the calorimeter proper) or by the introduction of a liquid which would not be exactly at the temperature of the calorimeter. The measurements were carried out on three carbons and three silicas, either non-porous (taken as references for the surface area), mesoporous or microporous. For the three carbons, the enthalpies of immersion in liquid nitrogen and liquid argon were identical within 3%, whereas they differed from 4% to 12% in the case of the three, more polar, silicas. This was easily explained by the permanent quadrupole moment of the nitrogen molecule able to interact with surface hydroxyl groups, whereas the argon molecule practically does not, as shown in earlier work (Rouquerol et al., 1984). Also, the enthalpies of immersion of non-porous carbon and silica in liquid argon differed by 16%: this ruled out the simplifying assumption initially made (Chessick et al., 1954; Taylor, 1965) that the areal enthalpy of immersion was somewhat independent of the chemical nature of the adsorbent, but does not prevent from deriving a meaningful 'immersion surface area', provided that the reference sample is correctly selected.

4.2.3.5 Wettability

We remind that the wettability $\Delta_{imm}\gamma$, adequately called by Everett (1972) 'work of immersional wetting per unit area', is the difference between two surface tensions (see

Equations 4.33 and 4.34). Its determination by immersion calorimetry can involve the application of Equation (4.39), following Briant and Cuiec's method (1972), and is straightforward with immersion liquids for which parameter k is already known: water, heptane, cyclohexane, benzene and paraxylene (cf. Table 4.1). It can also be calculated from $\gamma(LG)$ and its variation with temperature (cf. Equation 4.38), an approach successfully used by Schultz et al. (1977). The last measurements in this area were based on the method proposed by van Oss et al. (1988) and van Oss (2006): works of Zoungrana et al. (1994), Douillard et al. (1995), Medout-Marere et al. (1998) and De Ridder et al. (2013).

A special kind of immersion calorimetry can be used to study the wetting of hydrophobic solids giving with water a contact angle of >90°. Here, water is intruded (into the porous sample located in a microcalorimeter) under a pressure which can amount to 70 MPa (Denoyel et al., 2002, 2004). The simultaneous determination of the work and heat exchanged during an intrusion–extrusion cycle gives access to the apparent contact angle of water with the pore walls (Gomez et al., 2000). Such highly hydrophobic porous solids can find an application in reduced size devices for quick energy storage and delivery, like aircraft dampers (Fadeev and Eroshenko, 1997).

4.2.3.6 Polarity of Solid Surfaces

The polarity of a solid surface can be regarded as the strength of its average electrostatic field F. This field interacts with permanent or induced *dipoles* of adsorbed molecules, whereas its gradient interacts with permanent or induced *quadrupoles*; thus, giving rise to components of the adsorption energy such as $E_{F\mu}$ (with permanent dipoles), E_p (polarization contribution) or $E_{\dot{F}Q}$ (with permanent quadrupoles). Therefore, the selection of the appropriate immersion systems, differing mainly in the molecular dipole moment μ of the immersion liquid, can be expected to provide information on the value of $E_{F\mu}$ as

$$E_{F\mu} = -F\mu \qquad (4.46)$$

This was the basis of the approach by Chessick et al. (1954, 1955) and Zettlemoyer et al. (1958), which involved the use of a series of immersion liquids such as butyl derivatives differing only in their polar groups: 1-butanol, 2-butanol, butanal, 1-aminobutane, 1-chlorobutane, butanoic acid. With the polar surfaces studied (rutile, CaF_2, Aerosil, alumina), an approximately linear relation was found between the energy of immersion and the dipole moment. The slope directly gave the average field strength (for instance, 820 V μm^{-1} on a rutile titanium (IV) oxide) and the intercept gave the average value for the dispersion contribution E_d to the adsorption energy.

The polarization contribution E_p itself could be calculated from F and the polarizability α of the adsorbate, using

$$E_p = -\frac{F^2\alpha}{2} \qquad (4.47)$$

(For a warning about pitfalls to avoid in this type of approach, see Chessick, 1962.)

A similar linear variation in the energy of immersion with the dipole moment was observed by Gonzalez-Garcia and Dios Cancela (1965) when immersing montmoril-lonites in water, ethylene glycol, acetonitrile and dimethyl sulfoxide, although the relationship was more complicated – as is usual with clays – because of the hydration of the exchangeable cations.

Schröder (1979) was able to improve this approach by using a refined expression of the Gibbs free energy of adhesion as a function of the various intermolecular attraction energies between the liquid and the solid. In this way, by using up to 10 different immersion liquids with known parameters, he has calculated the apparent dipole moment and polarizability characterizing the immersion behaviour of various pigment surfaces (e.g. rutile, iron oxide and several phtalocyanines).

As stressed by Jaycock and Parfitt (1981), immersion calorimetry is thus a potential source of fundamental data and is certainly worth further use for this purpose.

4.2.3.7   Surface Modification

One type of chemical modification which has been successfully studied by immersion microcalorimetry in water is the *dehydration and dehydroxylation* of oxides (see Chapter 11). The oxides studied were mainly silica (Brunauer et al., 1956; Young and Bursh, 1960; Whalen, 1961a,b), titania (Wade et al., 1961; Zettlemoyer and Chessick, 1964), alumina (Wade and Hackermann, 1964) and iron oxides (Furuishi et al., 1982; Watanabe and Seto, 1988). The outgassing temperature above which the dehydration was no longer reversible was easily shown from the maximum in the curve of energy of immersion versus outgassing temperature. Also, the influence of dehydroxylation of silicas on their energy of immersion in benzene (which interacts with hydroxyls through its π-electron cloud) but not in cyclohexane (which does not undergo specific interactions) was clearly shown by Whalen (1962), who evaluated the benzene–hydroxyl energy of interaction (from 0.04 to 0.26 μJ site^{-1}, for his three silicas).

Another modification easily assessed by immersion microcalorimetry is the *change in hydrophobicity* of a surface, for example, by oxidizing a graphitized carbon surface: the energy of immersion in water was shown to increase almost linearly with increase in hydrophobicity (Young et al., 1954) and the energy of immersion of hydrophobic and hydrophilic patches was estimated to be 31 and 730 J m^{-2}, respectively (Healey et al., 1955). Lyklema (1995) pointed out that, in the absence of immersion calorim-etry, the notion of surface hydrophilicity–hydrophobicity remains vague. Once the molar enthalpy of immersion in water is assessed it can be readily compared with the value 44 kJ mol^{-1} which corresponds to the enthalpy of condensation of water at room temperature. If higher, the surface is considered to be hydrophilic; if lower, the surface is defined as hydrophobic. The hydrophobic/hydrophilic balance of activated carbon surfaces can also be estimated from the ratio of enthalpies of immer-sion in a non-polar and a polar liquid, like CCl_4 and H_2O (Giraldo and Moreno-Pirajan, 2007).

The *change in nature of the oxidized surface* can be followed with other immersion liquids than water: by increasing the oxygen content of a carbon black (i.e. Spheron 6) up to 12%, Robert and Brusset (1965) obtained an increase in the energy of immersion in methanol from 140 to as much as 390 mJ m^{-2} (practically, the same ratio as that observed with water), whereas the energy of immersion in n-hexadecane remained nearly constant, around 100 mJ m^{-2}.

4.2.3.8   Site-Energy Distribution

This requires a more sophisticated and time-consuming procedure, that is, by following the energy of immersion versus pre-coverage of the sample by the vapour of the immersion liquid: the vapour will first cover the highest available energy sites and hence give a pronounced decrease in the energy of immersion (Zettlemoyer, 1965). Such a curve is shown in Figure 4.9b. Figure 4.9 gives the main types of curve listed by Zettlemoyer and Narayan (1967) for this type of immersion calorimetry experiments with pre-coverage of the sample. Curve (a) is obtained with homogeneous surfaces with respect to the immersion liquid (e.g. chrysotile asbestos in water; Zettlemoyer et al., 1953). Curve (b) is given by heterogeneous surfaces (e.g. most oxides in water) and curve (c) by the immersion in water of hydrophobic surfaces, with only a few hydrophilic sites (e.g. Graphon). Curves (d) and (e) are only in part related to the site-energy distribution: curve (d) is typical of the swelling of a clay (like Wyoming bentonite in water) where the molecules penetrate through platelets of the mineral at certain relative pressures, whereas curve (e) is typical of the progressive filling of micropores and then possibly mesopores. Because of the difficulty of

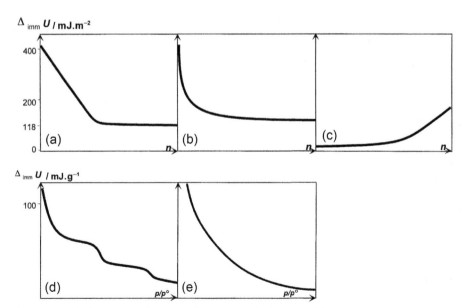

Figure 4.9 Various types of energy of immersion isotherms, (a) homogeneous surface, (b) heterogeneous surface, (c) hydrophobic surface, (d) swelling of stratified adsorbent and (e) filling of porous adsorbent.
After Zettlemoyer and Narayan (1967).

evaluating the true surface area, for these last two systems, the energies of immersion reported are not areal (per m^2) but specific (per g). As the data provided by the above type of immersion calorimetry experiment can be directly related to the results obtained by gas adsorption calorimetry the question arises: which type of experiment is to be preferred? In fact, immersion calorimetry, although time consuming, keeps certain advantages because of the difficulty in handling easily condensable vapours in adsorption manometry.

4.2.3.9   Structural Modifications of the Adsorbent

The surface electrostatic field of a solid strongly depends on the crystallinity of the solid, as can be seen, for instance, by comparing the areal energies of immersion in water of quartz (510 mJ m^{-2}), precipitated silica (300 mJ m^{-2}) and pyrogenic silica (155 mJ m^{-2}) after a similar outgassing at 140 °C (Denoyel et al., 1987). It therefore seems reasonable to consider that there is a direct relationship between the energy of immersion in water and the degree of crystallinity of an oxide such as silica. Furthermore, it is possible to follow, by immersion microcalorimetry, the changes resulting from intensive grinding of powders such as quartz (Wade et al., 1961), titania and alumina (Wade and Hackermann, 1964) or calcite (Cases, 1979).

4.2.3.10   Microporosity

Immersion microcalorimetry is able to characterize microporosity in at least three different ways:

i.   When progressive pre-coverage is used, a curve of the type shown in Figure 4.9e is given, revealing the filling of micropores, even at very low relative pressures of pre-coverage.
ii. When the molecular size of the immersion liquid is close to that of the micropores, a delayed diffusion of the liquid can be immediately detected from the slower response of the microcalorimeter, as was clearly shown by Widyani and Wightman (1982) when immersing a microporous activated carbon in propanol.
iii. When a set of immersion liquids is used, with an appropriate range of molecular sizes and shapes (i.e. flat, like aromatic molecules, or bulky) the energy of immersion is directly dependent on the extent of the molecular penetration into the porous network. This was shown, for example, by Atkinson et al. (1982) and Denoyel et al. (1993) and is developed in Section 9.3.2.

4.2.3.11   Further Comments on the Application of Immersion Microcalorimetry

As we have already seen, it is not difficult to undertake accurate energy of immersion measurements provided that the microcalorimetric technique is carefully selected and that certain precautions are taken. It was pointed out by Chessick (1962) that it is deceptively easy to obtain 'heat of wetting' data, but the results will be of very little scientific value unless steps are taken to ensure that the adsorbent and the liquid have been properly prepared and that the conditions of the experiment meet well-defined thermodynamic requirements: in other words, one must be sure to assess a meaningful enthalpy or energy of immersion between two well-defined states of the adsorbent.

4.3   Adsorption from Liquid Solution

Adsorption from liquid solution is, in some respect, a new world in comparison with adsorption from the gas phase: the fundamental principles and methodology are different in almost all respects.

The simplest system, now involves a binary solution and the composition of the adsorbed layer, is usually unknown. Furthermore, although there is not the large density difference (ca. 3 orders of magnitude) found in the case of gas adsorption, the sizes, shapes and possible configurations of solute molecules cover a much broader range than in gas adsorption. Indeed, from a statistical mechanical standpoint, it may appear that adsorption at the liquid/solid interface is closer to adsorption at the gas/liquid interface than to adsorption at the gas/solid interface (Everett, 1973). A great effort has been made to improve our understanding of these systems which are of major importance in many fields of science and technology.

Major advances in the thermodynamic treatment have been made by Defay and Prigogine (1951), Schay (1970), Schay and Nagy (1961, 1972), Schay et al. (1972), Nagy and Schay (1963), Kipling (1965) and Everett (1972, 1973, 1986). These and other authors have given considerable thought to the presentation and utilization of an experimental data adsorption from solution.

Until relatively recently, the fact that an experimental isotherm necessarily contained composite information concerning the adsorption of the two components of a binary solution was considered to be a major problem. For a rigorous interpretation, it was felt necessary to process the data to obtain the so-called individual adsorption isotherm or separate adsorption isotherm of each component. Actually, this is not at all straightforward and requires the introduction of a number of assumptions relating to the structure of the adsorbed layer. The main problem is of course to know the composition of the adsorbed layer. One assumption often used in the case of volatile components is that introduced by Williams (1913): the solid will adsorb the *same amount of each component from the vapour* in equilibrium with the solution *as from the solution* itself. This of course implies that the adsorbed layer has the same composition at the liquid/solid and gas/solid interfaces and it requires numerous gravimetric measurements from the vapour phase. To limit these measurements to two, a further assumption (Elton, 1951) is that the *adsorption at both interfaces is confined to a monolayer* and that the cross-sectional area of each molecule is independent of coverage, so that it can be derived from the gas adsorption isotherm for each component. This approach was tested by Kipling and Tester (1952) for the adsorption on charcoal from ethanol–benzene mixtures at 20 °C. Their results (see Figure 4.10) illustrate the striking difference between the individual isotherms and the original composite isotherm based on the change of concentration. It is important to recognize that the derivation of the individual isotherms requires an oversimplification in the interpretation of the adsorption process. This is why, by following the IUPAC recommendations (Everett, 1986), we stress the need for a few strict rules on the presentation and interpretation of experimental data. In the next section, it will be shown how the early assumptions may be avoided by the presentation of 'reduced surface excess amounts' versus concentration, which provides a basis for any further theoretical interpretation of the data.

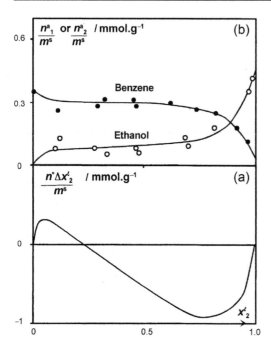

Figure 4.10 Adsorption from ethanol–benzene mixtures onto charcoal at 20 °C, (a) isotherm of 'concentration change' (i.e. of reduced surface excess of ethanol) versus molar fraction of ethanol and (b) corresponding separate isotherms. After Kipling and Tester (1952).

4.3.1 Quantitative Expression of the Amounts Adsorbed from a Binary Solution

4.3.1.1 Scope and Limitations of the Normal Surface Excess Amounts

As for adsorption from the gas phase, one must be able to express unambiguously and quantitatively the observed adsorption phenomenon without reference to any prior knowledge concerning the structure of the adsorbed layer. For this reason, the concept of the Gibbs dividing surface (GDS) and the associated surface excess amounts are just as useful as for gas adsorption.

Figure 4.11 represents the concentration of species 1 and 2 and also the total concentration c (expressed as the amount per unit volume of solution) as one moves away from the adsorbing surface. Figure 4.11a represents a hypothetical adsorption system where the molar volumes of species 1 and 2 are equal, so that any enrichment of the adsorption layer in component 2 (i.e. positive adsorption) is necessarily accompanied by an equal loss of component 1 (i.e. negative adsorption). Figure 4.11b represents another adsorption system, in which the molar volumes of species 1 and 2 are different, so that adsorption also affects the total concentration c in the solution. Although these are representative of real systems, it must be stressed that in most cases these concentration profiles are unknown. Finally, Figure 4.11c gives a Gibbs representation of the simple case where the GDS coincides with the real adsorbing surface. By convention, the concentrations in the solution are taken as constant up to the GDS. The hatched areas represent the amounts apparently lost (for component 2) or gained

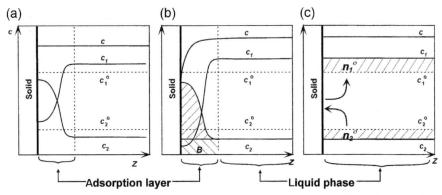

Figure 4.11 Concentrations in a liquid/solid adsorption system (competitive adsorption from binary mixture or solution), (a) real state after preferential adsorption of 2 (molecular volumes and cross-sectional areas of 1 and 2 supposed equal), (b) same as (a), but with different molecular volumes and/or cross-sectional areas for 1 and 2 and (c) Gibbs representation of system (b) for the case where GDS coincides with real adsorbing surface.

(for component 1) by the solution when the adsorption changes the concentrations from c_1^o and c_2^o to c_1 and c_2, respectively. These are the amounts counted as the 'surface excess amounts' on the GDS, one being positive (for component 2) and the other negative (for component 1).

Now, one advantage of the GDS is that it can be located arbitrarily, allowing a change at will of the volume $V^{\ell,o}$ allotted to the liquid phase. Thus, the calculation of the surface excess amount n_i^σ directly follows from its defining equation:

$$n_i^\sigma = n_i - V^{\ell,o} c_i^\ell \tag{4.48}$$

where n_i is the total amount of component i in the system and c_i^ℓ is its concentration in the liquid after adsorption. It is evident that the value of n_i^σ is linearly dependent on the arbitrary value taken for $V^{\ell,o}$. This variation in n_i^σ with the position of the GDS is indicated in Figure 4.12.

Here, the GDS is taken as parallel to the real solid–liquid interface, of area A. Any displacement, Δ, of the GDS normal to the surface (e.g. from position S to position E), produces a change $A\Delta z$ in the volume $V^{\ell,o}$ in Equation (4.48). The corresponding change in n_i^σ is represented in Figure 4.12 by a straight line, its slope being simply $c_i^\ell A$. The strong dependence of n_i^σ on the position of the selected GDS must be recognized in any attempt to provide a standard procedure for reporting the experimental data. For this reason, some investigators have preferred to adopt an alternative approach.

4.3.1.2   The Use of Relative Surface Excess Amounts

The aim now is to replace n_i^σ by another quantity, which would also rely on the existence of a GDS but which would be invariant with respect to the position of the GDS (Guggenheim and Adam, 1933; Defay, 1941). In fact this can be done by simply

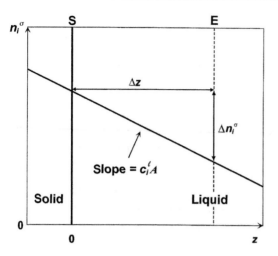

Figure 4.12 Dependence of surface excess amount of component i on location of GDSS: GDS on actual adsorbing surface E: GDS at distance Δz from adsorbing surface.

writing Equation (4.48) for each component, and then eliminating $V^{\ell,\circ}$ from these two equations. In this manner we can obtain an equation of the form

$$n_2^\sigma - n_1^\sigma \frac{c_2^\ell}{c_1^\ell} = n_2 - n_1 \frac{c_2^\ell}{c_1^\ell} \tag{4.49}$$

The variables on the right-hand side are experimental quantities which are independent of the position of the GDS. The left-hand side of the equation must also be independent of the position of the GDS. This function is named '*relative surface excess* of 2 with respect to 1' and is denoted $n_2^{\sigma(1)}$. It is defined as follows:

$$n_2^{\sigma(1)} = n_2^\sigma - n_1^\sigma \frac{c_2^\ell}{c_1^\ell} \tag{4.50}$$

and operationally:

$$n_2^{\sigma(1)} = n_2 - n_1 \frac{c_2^\ell}{c_1^\ell} \tag{4.51}$$

Corresponding equations hold of course for the relative surface excess of 1 with respect to 2. Thus,

$$n_1^{\sigma(2)} = n_1^\sigma - n_2^\sigma \frac{c_1^\ell}{c_2^\ell} = n_1 - n_2 \frac{c_1^\ell}{c_2^\ell} \tag{4.52}$$

In these equations, where the concentrations are expressed as amounts per unit volume, the ratio of concentrations (i.e. c_2^ℓ/c_1^ℓ) can be replaced by the ratio of molar fractions (i.e. x_2^ℓ/x_1^ℓ). Taking into account that $x_1^\ell = 1 - x_2^\ell$ one can then transform Equations (4.50) and (4.49) into

$$n_2^{\sigma(1)} = n_2 - n_1 \frac{x_2^\ell}{x_1^\ell} = n^\circ \frac{\Delta x_2^\ell}{x_1^\ell} \tag{4.53}$$

where $n^\circ = n_1 + n_2$, $n_2 = n^\circ x_2^{\ell,\circ}$ and Δx_2^ℓ is the change $x_2^{\ell,\circ} - x_2^\ell$ due to adsorption.

$n_2^{\sigma(1)}$ is usually divided by m^s (the mass of solid adsorbent) to give the 'specific relative surface excess of 2 with respect to 1'. Alternatively, it can be divided by A to give $\Gamma_2^{(1)}$ the 'areal relative surface excess of 2 with respect to 1'.

4.3.1.3 The Use of Reduced Surface Excess Amounts

Another function invariant with respect to the position of the GDS and named the 'reduced surface excess amount' was derived at by Defay (1941) in a similar way. Starting again from Equation (4.48), we can write the following two equations:

$$n_2^\sigma = n_2 - V^{\ell,\circ} c_2^\ell \tag{4.54}$$

$$n^\sigma = n^\circ - V^{\ell,\circ} c^\ell \tag{4.55}$$

where $n^\sigma = n_1^\sigma + n_2^\sigma$, $n^\circ = n_1 + n_2$ and $c^\ell = c_1^\ell + c_2^\ell$.

Taking into account that c_2^ℓ / c^ℓ can be replaced by x_2^ℓ, we finally obtain the following equation:

$$n_2^\sigma - n^\sigma x_2^\ell = n_2 - n^\circ x_2^\ell = n^\circ \Delta x_2^\ell \tag{4.56}$$

Here again the right-hand side does not depend on anything other than experimentally measurable quantities, and therefore we conclude that the right-hand side does not depend on the position of the GDS. This function is defined as the 'reduced surface excess amount of 2'

$$n_2^{\sigma(n)} = n_2^\sigma - n^\sigma x_2^\ell \tag{4.57}$$

Usually, this quantity is either made 'specific' after division by m^s or 'areal' after division by A. In the latter case, it is denoted by $\Gamma_2^{(n)} = n_2^{\sigma(n)} / A$.

In some instances (especially if the exact molar mass of one component is not known, which often happens with long chain molecules, like those of surfactants and polymers), it is more convenient to think in terms of mass. We then use mass fractions w_1^ℓ or w_2^ℓ (instead of molar concentrations or molar fractions) and arrive at surface excess masses, either relative

$$m_2^{\sigma(1)} = m_2^\sigma - m_1^\sigma \frac{w_2^\ell}{w_1^\ell} = m_2 - m_1 \frac{w_2^\ell}{w_1^\ell} \tag{4.58}$$

or reduced

$$m_2^{\sigma(m)} = m_2^\sigma - m^\sigma w_2^\ell = m_2 - m^\circ w_2^\ell \tag{4.59}$$

4.3.1.4   The Meaning of Relative and Reduced Surface Excess Amounts

Although the relative and the reduced surface excess amounts (or masses) do not
depend on the position of the GDS, there is a special position of the GDS which cancels
the last term of the defining Equations (4.50) and (4.57) and gives a useful idea of the
meaning of these two quantities. For Equation (4.50), this special position of the GDS
is the one for which the surface excess amount n_1^σ of component 1 is zero. We then get

$$n_2^{\sigma(1)} = n_2^\sigma \qquad (4.60)$$

so that the relative surface excess amount of 2 with respect to 1 is equal to the surface
excess of 2 for the GDS which cancels the surface excess of 1. This is represented in
Figure 4.13, which, along the same lines as Figure 4.12, now shows the effect of the
location of the GDS, on the surface excess amounts of components 1 and 2. Here, GDS
'L' provides the special condition for $n_1^\sigma = 0$.

Turning to Equation (4.57), we now require a GDS where n^σ (the whole surface
excess amount $n_1^\sigma + n_2^\sigma$) equals zero. In Figure 4.13, this is the case for GDS 'D', where
$n_2^\sigma = -n_1^\sigma$.

From Equation (4.57), we now obtain

$$n_2^{\sigma(n)} = n_2^\sigma = -n_1^\sigma = -n_1^{\sigma(n)} \qquad (4.61)$$

which shows that the reduced surface excess amounts of 1 and 2 are equal and opposite
in sign. It turns out that this remains true whatever the position of the GDS. As under-
lined by Defay and Prigogine (1951), this representation in terms of the reduced surface excess 'treats the components on an equal footing'. A practical consequence is
that the same curve (either U or S as illustrated in Figure 4.14) provides the adsorption
isotherm for both components. We also note that it is only when the partial molar volumes of 1 and 2, together with their molecular cross-sectional areas, are equal, that for
$n_1^\sigma + n_2^\sigma = 0$ the GDS coincides with the real adsorbing surface.

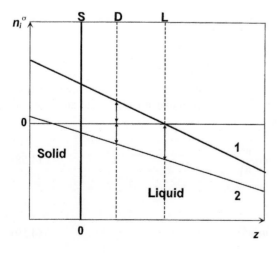

Figure 4.13 Dependence of surface
excess amounts of components 1
and 2 on location of GDSS, GDS on
actual adsorbing surface; D, GDS
for which $n_1^\sigma + n_2^\sigma = 0$; L, GDS for
which $n_1^\sigma = 0$.

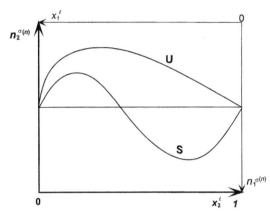

Figure 4.14 The two basic shapes of reduced surface excess (or 'composite') isotherms over the *whole concentration range*: S-shape (S) or inverted U-shape (U). Thick line axes (zero in bottom left): isotherms for component 2. Thin line axes (zero in top right): isotherms for component 1.

By combining Equations (4.53), (4.56) and (4.57), we obtain the following relationship between the relative and reduced surface excess quantities:

$$n_2^{\sigma(1)} = n_2^{\sigma(n)} / x_1^{\ell} \tag{4.62}$$

and

$$\Gamma_2^{(1)} = \Gamma_2^{(n)} / x_1^{\ell} \tag{4.63}$$

and also

$$m_2^{\sigma(1)} = m_2^{\sigma(m)} / w_1^{\ell} \tag{4.64}$$

4.3.1.5 Adsorption Isotherms Expressed in Reduced Surface Excess Amounts

The reduced surface excess amounts, whose use is recommended by the IUPAC (Everett, 1986), offer much more than a form of a precise mathematical accounting of the adsorption experiment: they also offer the most convenient way of reporting the experimental results. In fact, for decades this presentation was intuitively chosen as a way of plotting adsorption data, without any reference to the Gibbs formalism. The quantity plotted to represent adsorption of component 2 was often either in the form of $n^{\circ}\Delta x_2^{\ell}$, which is consistent with Equation (4.56), or $\left[m_2 - m^{\circ}w_2^{\ell}\right]$, which is consistent with Equation (4.59).

The isotherm obtained is generally termed a 'composite isotherm' or an 'an isotherm of apparent adsorption' or less often an 'isotherm of concentration change'. The term 'composite' refers to the fact that this single isotherm contains information about the adsorption of both components 1 and 2.

This is shown in Figure 4.14, which gives the two most important shapes of reduced surface excess (or 'composite') isotherms (S-shape and inverted U-shape) *for completely miscible liquids*. Depending on the axis chosen, one obtains the isotherm for component 2 (*thick* line axis) or for component 1 (*thin* line axis). These shapes were the basis for a more detailed classification given by Schay and Nagy (1961).

When the adsorption is studied *from dilute solutions* (where by convention, we shall consider that component 1 is the solvent), the experimental isotherms are still plotted, strictly speaking, in terms of reduced surface excess. Nevertheless, as $x_1^{\ell} = 1 - x_2^{\ell} \approx 1$, one sees, from Equations (4.53), (4.56) and (4.62) that

$$n_2^{\sigma(1)} \approx n_2^{\sigma(n)} \approx n_2^{\sigma} \qquad\qquad (4.65)$$

so that any one of these three quantities can be plotted, to represent the adsorption data.

Although up to 18 different shapes have been distinguished in a detailed classification (Giles et al., 1960), two main shapes are of special interest, as shown in Figure 4.15: a Langmuir or L-shape and an S-shape with an inflexion in the low-concentration region. If higher concentrations can be attained, the reduced surface excess eventually reaches a maximum, before decreasing as indicated in Figure 4.14.

As the reduced and relative surface excess isotherms convey composite information on the adsorption of the two components, there is a strong incentive to determine the individual (or 'separate') isotherms, that is, the adsorbed amount n_2^{a} (or n_1^{a}) versus concentration, mole fraction or mass fraction. It will be recalled that this implies some assumptions about the thickness, composition and structure of the adsorbed layer, and therefore is not to be recommended for reporting adsorption from solution data in a standard form. Indeed, this second step is already part of the theoretical interpretation of the adsorption mechanisms.

4.3.2 Quantitative Expression of the Energies Involved in Adsorption from Solution

4.3.2.1 Definition of Energies and Enthalpies of Adsorption from Solution

Let us consider the case of a solution with solvent (1) and solute (2). Its composition will be preferably given as a molality, that is, the amount n_2 of solute associated with 1 kg of solvent (this quantity is of course invariant with temperature). Following the

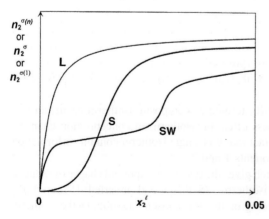

Figure 4.15 Typical shapes of surface excess isotherms from *dilute solutions* (Langmuir or L-shaped, S-shaped, StepWise).

IUPAC recommendation (Mills et al., 1993) the molality of the solute will be denoted b_2 to avoid any confusion with m_2, the mass of solute. The amount of solute contained in a mass m of solution of molality b_2 is therefore,

$$n_2 = \frac{b_2 \cdot m}{1000 + M_2 b_2} \tag{4.66}$$

where M_2 is the molar mass of the solute (in g mol^{-1}). In the solution of molality b_2, each component has a partial molar enthalpy:

$$h_i^\ell = \left(\frac{\partial H^\ell}{\partial n_i} \right)_{T,p,A,n_j \neq n_i} \tag{4.67}$$

where H^ℓ is the total enthalpy of the solution.

We recall that for condensed phases (either liquid or adsorbed) we are generally able to equate the molar enthalpy and the molar internal energy. Here we shall use enthalpies, simply because they are more common in the literature, but in the following definitions h could be replaced by u.

Just as for gas adsorption, we can define a partial differential enthalpy of adsorption of a component, $\Delta_{ads}\dot{h}_i$, which would correspond to the adsorption, from a solution of molality b_i, of an infinitesimal amount of component i, dn_i, on a solid surface already covered with solute at a reduced surface excess concentration $\Gamma_i^{(n)}$

$$\Delta_{ads}\dot{h}_i = \dot{h}_i^{\sigma(n)} - h_i^\ell(b_i) \tag{4.68}$$

where $\dot{h}_i^{\sigma(n)}$ is the reduced surface excess enthalpy of the component i

$$\dot{h}_i^{\sigma(n)} = \left(\frac{\partial H^\sigma}{\partial n_i^{\sigma(n)}} \right)_{T,p,A,n_j \neq n_i} \tag{4.69}$$

H^σ being the total surface excess enthalpy.

Now, in case of adsorption of a finite amount of component i, there is a resulting change in the molality of the solution: it follows that the partial enthalpy of component i not only changes in the solution but also at the adsorbing surface. Rather than integrating Equation (4.68) with an ever-changing initial state, it turns out to be more convenient to refer to a constant reference state comprising a clean solid surface and a pure liquid for the solvent (molar enthalpy $h_1^*(\ell)$) or an infinitely diluted state for the solute (molar enthalpy $h_2^\infty(\ell)$). We can define a standard integral molar enthalpy of adsorption for each component, so that we may write for the solvent

$$\Delta_{ads}h_1^o = h_1^{\sigma(n)} - h_1^*(\ell) \tag{4.70}$$

and for the solute

$$\Delta_{ads}h_2^o = h_2^{\sigma(n)} - h_2^\infty(\ell) \tag{4.71}$$

4.3.2.2   Definition of Displacement Enthalpies (and Energies)

The enthalpies of adsorption as already defined are adequate for an experiment when a clean adsorbent is immersed in the solution, but they are not suited for other experiments, which are usually more accurate, where the adsorbent is initially immersed in the pure solvent. When the pure solvent is replaced by a solution of molality b_2, adsorption of the solute can only take place by displacement of the solvent. The word 'displacement' is used here to indicate that the adsorption of the amount $n_2^{\sigma(n)}$ of solute produces the desorption of a corresponding (but usually not equal) amount of solvent. Following Kiraly and Dekany (1989), let us call r the amount of solvent displaced (i.e. desorbed) by one mole of solute, therefore producing a change in composition of the solution and giving an enthalpy of mixing, $\Delta_{mix}H$, which is part of the overall heat effect, Q_{exp}, measured during the experiment. It is preferable not to include $\Delta_{mix}H$ (which is not a property of the interface) in the definition of the enthalpy of displacement which is therefore:

$$\Delta_{dpl}H_{1,2} = Q_{exp} - \Delta_{mix}H \tag{4.72}$$

4.3.2.3   Definition of the Enthalpies (and Energies) of Mixing

In principle, the overall enthalpy of dilution can be considered to include one term due to the solute and another one due to the solvent. Thus,

$$\Delta_{mix}H = n_2^\ell \left[h_{2f}^\ell - h_{2i}^\ell \right] + n_1^\ell \left[h_{1f}^\ell - h_{1i}^\ell \right] \tag{4.73}$$

where the subscripts i and f refer to the initial and final states. For a dilute solution, say with a mass fraction $\leq 1\%$, the term due to the solvent can be omitted. $\Delta_{mix}H_2$ must be measured in separate experiments, by the addition of a standard solution to the pure solvent.

4.3.3   Basic Experimental Methods for the Study of Adsorption from Solution

We may divide the experimental techniques available for the study of adsorption from solution into three main categories: (i) for the determination of adsorption isotherms, (ii) for the measurement of the energies involved and (iii) for the provision of extra information on the properties of the adsorbed layer.

4.3.3.1   Methods for Determining the Amounts Adsorbed

A first distinction must be made between the methods which use one fresh sample for each point on the adsorption isotherm (i.e. immersion methods) and those using a single sample through which the solution of increasing concentration is allowed to flow (i.e. flow-through methods). A critical outline of most of these methods is given by Everett (1986).

4.3.3.1.1   Immersion Methods

These are the oldest and the easiest to apply with conventional bench-type equipment but they may suffer, as we shall see, from a lack of accuracy or from a large sample consumption.

In the standard immersion method, the dry sample is immersed in the solution (see Figure 4.16a)

In a slightly different version, (Rouquerol and Partyka, 1981), the sample is initially covered with the pure solvent (protected from any contact with ambient atmosphere) before receiving an appropriate dose of mother solution, (see Figure 4.16b). This can be done in a calorimeter (Taraba, 2012).

Equilibration may take between 1 min and more than a day, in a thermostatted bath, with continuous and slow tumbling. The suspension, still kept at the controlled temperature, is allowed to settle, which may take one full day, or, more frequently, the suspension is centrifuged (taking care, in the case of polymers, not to produce a measurable concentration gradient) and the supernatant is then pipetted and analysed. The analysis

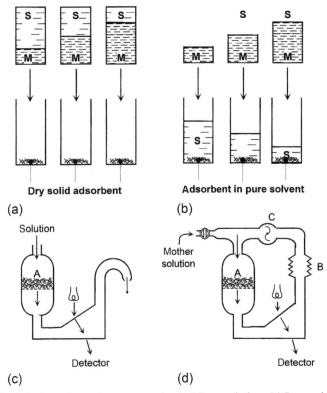

Figure 4.16 Methods to determine amount adsorbed from solution. (a) Immersion method 1: dry samples covered with solution made of various proportions of pure solvent S and mother solution M. (b) Immersion method 2: samples protected by pure solvent receive various amounts of mother solution. (c) Open-flow method with refractometric detector. (d) Circulation method (A: adsorbent packing; B: expansion bellows or vessel; C: circulating pump).

may involve differential refractometry, UV or IR spectroscopy (the former mainly for aqueous solutions, the latter for organic solutions) COT analysis, colorimetry (for dye adsorption), surface tension measurements or, still multichannel electrospray ionization mass spectrometry to determine individual isotherms of each component when the solution contains more than one solute (Benko et al., 2013). As mentioned in Section 5.3.1, this immersion method has the advantage that it directly gives the reduced surface excess amounts. As one experiment only provides one point of the adsorption isotherm, it is usual to undertake a number of simultaneous measurements (each requiring a fresh sample) in order to cover the desired portion of the adsorption isotherm.

A possible way to increase the accuracy of this immersion approach is to use the *slurry method* and to analyse a weighed sample of the slurry in the bottom of the test tube, instead of analysing the supernatant (Nunn et al., 1981). One then simply makes use of Equation (4.51), the operational expression of the relative surface excess of the solute with respect to the solvent. Here, n_1 and n_2 are the total amounts of solute and solvent in the sample of slurry (either adsorbed or in solution) and c_2^{ℓ} and c_1^{ℓ} their concentrations in the solution. If one uses a liquid/solid ratio large enough to avoid any measurable change in concentration on adsorption, then c_2^{ℓ} and c_1^{ℓ} are simply the concentrations in the starting solution. The measurement is accurate provided the quantitative analysis of the slurry, which involves measuring the total amounts of 2 and 1 and also the weight of the dry solid, can itself be carried out accurately. However, this does not solve the problem of sampling: the immersion method relies on the assumption that the various samples used (one for each point of the adsorption isotherm) are all representative of the solid studied.

4.3.3.1.2   Flow-Through Methods

In these methods, only one sample is used and is successively brought into equilibrium with solutions of increasing concentration. Most frequently, the sample is initially equilibrated with the pure solvent. Each increase in concentration produces an adsorption of solute by displacement of solvent from the solid surface. The changes in concentration produced by adsorption are usually monitored by the same techniques as those used in the immersion methods, except that they are now located on-line. A general requirement of these methods is that the sample can form a permeable bed and does not block the filter: this usually requires a grain size over, say, 2 μm and prevents any tendency to gel formation. In the *open-flow method* (see Figure 4.16c), at each step, a fresh solution of constant concentration continuously flows through the sample and is analysed at the outlet (Schay et al., 1972; Sharma and Fort, 1973). The outlet concentration (c_{out}) is lower than the inlet concentration (c_{in}) as long as adsorption equilibrium is not reached. Integration of the difference in concentration between the inlet and outlet over the mass $m (=m_1 + m_2)$ or volume V of the solution flowing through the sample gives the increase in reduced surface excess amount of 2 during this adsorption step. Thus,

$$\Delta n_2^{\sigma(n)} = \int_0^m \Delta w_2 \mathrm{d}m \tag{4.74}$$

where $w_2 = m_2/(m_1 + m_2)$
or

$$\Delta n_2^{\sigma(n)} = \int_0^V \Delta c_2 dV \tag{4.75}$$

where $c_2 = n_2/V$

Because this method makes use of standard HPLC chromatographic equipment (pump, sample cell with filters, on-line detectors), there is a tendency to call it 'chromatographic' or 'frontal chromatographic. We suggest that this tendency should be resisted: it is misleading as there is no chromatographic effect or phenomenon. In the *circulation method*, the same solution is continuously passed through the sample, also with continuous monitoring of the concentration, until equilibrium is reached (see Figure 4.16d) (Kurbanbekov et al., 1969; Ash et al., 1973). Although this method requires a less conventional set-up than the immersion or open-flow methods, it presents the following advantages:

i. Conservation of solution, which is important to save not only the solute (e.g. specially synthesized surfactants or proteins) but also the solvent (e.g. high-purity alkanes).
ii. Straightforward study of the temperature dependence of adsorption (without any addition of solution).
iii. The possible incorporation of a null procedure.

The principle of the *null procedure* (Nunn and Everett, 1983) is to restore, by injection of an appropriate dose of initial solution, the concentration prior to adsorption. This is done at each adsorption step. A refinement is to arrange that the sample cell is by-passed during the injection of a new dose of solution until the determination of a new amount concentration c_2^j. The flow through is re-started and, as the concentration decreases, it is restored to its initial value c_2^j by addition of a volume ΔV^a of a solution of amount concentration c_2^a. The determination of the increase in reduced surface excess amount of 2 simply requires a knowledge of the void volume V_m of the by-passed section and the concentration c_2^{j-1} of the former adsorption equilibrium. Thus,

$$\Delta n_2^{\sigma(n)} = \Delta V^a \left(c_2^a - c_2^j \right) - V_m \left(c_2^j - c_2^{j-1} \right) \tag{4.76}$$

One point of interest in this procedure is that it does not require the knowledge of any other volume than V_m. Also, the accuracy of the determination of $n_2^{\sigma(n)}$ does not depend on the calibration of the detector. A calibration is required of course for the presentation of the adsorption isotherm, that is, $n_2^{\sigma(n)}$ versus c_2^j.

4.3.3.2 Methods for Determining Adsorption Energies

In principle, the following methods are available for the determination of energies, associated with adsorption from solution:

i. The 'isosteric method', based on the temperature dependence of adsorption.
ii. The immersion calorimetry method.
iii. The batch calorimetry method.
iv. The flow-through calorimetry method.

4.3.3.2.1 The Isosteric Method

This method can be considered, for adsorption from solution, in a similar manner as for gas adsorption (see Section 2.6.1). For example, by equating the chemical potentials of component 2 (the solute) in its adsorbed state and the liquid phase, by keeping the specific amounts adsorbed constant and by considering a dilute solution, so that the activity can be replaced by a molality ($b_2 = 1000n_2/m_1$), we obtain

$$\Delta_{ads}\dot{h}_2 = -RT^2 \left(\frac{\partial (\ln b_2)}{\partial T} \right)_{n_2^{\sigma(n)}, n_1^{\sigma(n)}} \tag{4.77}$$

In principle, this method of calculating the differential enthalpy of adsorption could be applied to two adsorption isotherms determined at different temperatures, but the following restrictions must be kept in mind:

  i. The above equation only holds for dilute solutions.
  ii. The structures of the adsorbed phase and the solution are assumed to be unchanged over the temperature range considered.
  iii. Both surface excess amounts must also remain unchanged; this necessarily holds when using reduced surface excess amounts by the application of Equation (4.77), as $n_2^{\sigma(n)} = -n_1^{\sigma(n)}$ (see Equation 4.61) so that it is enough to maintain a constant $n_2^{\sigma(n)}$. This does not hold, however, when relative surface excess amounts or simple surface excess amounts are used. For all these reasons, one should be cautious in the application of the isosteric method (see Lyklema, 1995).

4.3.3.2.2 The Immersion Calorimetry Method

Although the method of immersion calorimetry in solution is directly related to immersion calorimetry in pure liquids, its application to solutions is much less rewarding. The reason is that if one uses the standard sealed glass bulb technique one must now break the bulb completely and stir to obtain a homogeneous solution after adsorption. The resulting heat of breaking and stirring then limits the accuracy of the microcalorimetric measurements, whereas the required sensitivity and reproducibility are usually higher than for a standard immersion microcalorimetry experiment in a pure liquid. It follows that, like the isosteric method, this approach to the energetics of adsorption from liquid solutions is not widely used.

4.3.3.2.3 The Batch Calorimetry Method

In this method, the adsorbent is initially kept in suspension in the pure solvent by means of continuous stirring. The solution is then introduced by successive doses which eventually fill the sample cell (see Figure 4.17).

A particular advantage of this method is that it is suitable for any type of powder, including those of grain size much smaller than 10 μm, which cannot be studied in calorimetry with the flow-through techniques because of the too large heat of friction. It is also suitable for kinetic studies, by using the microcalorimeter as a simple detector of the rate of adsorption after the rapid introduction of each new dose.

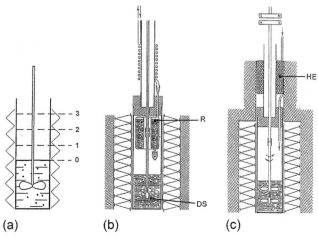

Figure 4.17 Batch calorimetry method. (a) Principle: aliquots of mother solution successively added to suspension of solid (initially in pure solvent). (b) Set-up with disc stirrer (DS) and internal reservoir (R) of mother solution (Rouquerol and Partyka, 1981). (c) Set-up with magnetic stirring and heat exchanger (HE) for mother solution (Nègre et al., 1985).

On the other hand, the requirements of the method are:

- (i) A sample cell large enough (e.g. 20–100 cm^3) to accommodate the stirring device and to receive several successive doses of solution.
- (ii) An efficient upstream heat exchanger so that the doses of added solution are, say, within 10^{-3} K of the temperature of the microcalorimeter.
- (iii) A stirring device able to keep the powder in suspension with a minimum heat evolution.
- (iv) A high calorimetric sensitivity, as the energies of displacement to be measured are usually very small.
- (v) A previous determination of the adsorption isotherm by one of the methods listed in Section 4.3.3.

Two operational arrangements fulfilling the above requirements are represented in Figures 4.17b and c. For convenience, both are incorporated in a Tian–Calvet microcalorimeter with large cells (i.e. ca. 100 cm^3).

The first device uses a disc stirrer (up and down movement) and cancels any temperature difference between the added solution and the adsorbent by placing both the adsorbent and the solution reservoir in top part of the microcalorimetric cell (Rouquerol and Partyka, 1981).

The second device uses a propeller which is given, say, every 10 s, a fast half-turn by means of a hindered magnetic transmission which, also, damps the vibrations from the motor. Here, a long heat exchanger (2-m long coil) allows the solution to reach the microcalorimetric cell at the right temperature. In this set-up, the full volume of the cell is available for progressive filling (Nègre et al., 1985; Rouquerol, 1985).

Both arrangements can be used from 25 to 200 °C (the upper temperatures being useful for studying adsorption from organic solutions) and their thermal stability permits adsorption measurements to be made over several hours. Typically, 0.5 g of adsorbent is kept in suspension in 15 g of pure solvent, whereas the added solution is introduced by steps of 5 g by means of a peristaltic pump operating at a flow rate of 60 mg min^{-1}. The amount of heat Q_{exp} measured during each step must be corrected for the enthalpy of mixing, $\Delta_{mix}H$, as indicated in Equation (4.72), and referred to the change in surface excess reduced amount, $\Delta n_2^{\sigma(n)}$ in order to obtain the corresponding change in integral enthalpy of displacement of solvent 1 by solute 2, $\Delta_{dpl}H_{1-2}$. At the end of each adsorption step, the molality of the solution is required in order to derive $\Delta_{mix}H$ and the reduced surface excess amount $n_2^{\sigma(n)}$ to derive $\Delta n_2^{\sigma(n)}$. This is done by reference to a previously determined adsorption isotherm $n_2^{\sigma(n)}=f(b_2)$ and from the knowledge of the initial amount of solvent and of the successive increments of solution introduced. The calculation can be conveniently presented in a graphical from (Nègre, 1984) and the principle, in the form proposed by Trompette (1995), is explained in Figure 4.18. In Figure 4.18, for a given total amount of solute n_2 introduced into the microcalorimetric cell, the line I gives the dependence of the reduced surface excess amount $n_2^{\sigma(n)}$ on the equilibrium molality b_2. In order to plot line I, we first consider the extreme case of $b_2=0$, that is all the solute is considered to be adsorbed so that

$$\frac{n_2^{\sigma(n)}}{m^s} = \frac{b_{2_0} \cdot \Delta m_i}{1000 + b_{2,0}M_2} \cdot \frac{1}{m^s} \tag{4.78}$$

where $b_{2,0}$ is the molality of the added solution, Δm_i is the mass of the first increment of solution, M_2 is the molar mass of the solute and m^s is the mass of the adsorbent.

The other extreme of line is determined by considering that $n_2^{\sigma(n)}=0$: then all the solute is considered to remain in solution. Then

$$b_2 = \frac{1000 b_{2_0} \Delta m_i}{(m_0 + \Delta m_i)\lfloor 1000 + b_{2,0}M_2 \rfloor - b_{2,0}m_0} \tag{4.79}$$

where m_0 is the initial mass of the pure solvent.

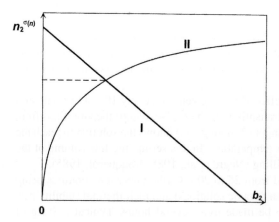

Figure 4.18 Graphical calculation of amount adsorbed in batch calorimetry experiment after introducing amount n_2 of solute. Curve II is the adsorption isotherm. For straight line I, see the text. Crossing represents final adsorption equilibrium. After Trompette (1995).

Curve II is simply the adsorption isotherm: the crossing provides the actual values of $n_2^{\sigma(n)}$ and b_2 after the introduction of Δm_i of solution. A blank experiment with the same set-up provides the curve of $\Delta_{\mathrm{mix}}H$ versus b_2.

A batch microcalorimetric experiment, very similar to the one just described, is possible with a simplified diathermal heat flowmeter type of calorimeter, which is less versatile than the Tian–Calvet microcalorimeter (especially in its temperature range and ultimate sensitivity), but of a simpler design. In the 'Montcal' microcalorimeter (Partyka et al., 1989), the Tian–Calvet thermopile with up to 1000 thermocouples is replaced by a few thermistors.

4.3.3.2.4 The Flow-Through Calorimetry Method

Flow-through adsorption microcalorimetry is not as versatile as the batch procedure. The grain size must be over, say, 20 μm to avoid additional heat effects. Also, the system may require a long time to reach equilibrium (especially in the low-concentration range). Furthermore, the technique may need larger amounts of solution than the batch procedure. On the other hand, it has several advantages.

  i. The final equilibrium molality, pH or ionic strength for each adsorption point can be selected as these are imposed by the incoming solution.
 ii. The amount adsorbed can be determined on-line by simply installing the appropriate detector at the outlet of the microcalorimeter (see the flow-through method for determining the adsorption isotherm);
iii. Both desorption as well as adsorption can be studied, which is practically impossible with the batch procedure.

Groszek (1966) early developed a simple flow-through adsorption calorimeter, which is somewhat similar to a DTA system (because of its single-point temperature detector) and was therefore well suited for the detection of thermal effects and for screening experiments.

To obtain meaningful results demands more sophisticated equipment, however. A heat flowmeter microcalorimeter is normally used for this purpose. Such a microcalorimeter, specially designed for liquid-flow adsorption and for the complementary determination of $\Delta_{\mathrm{mix}}H$ is represented in Figure 4.19. The enthalpy of mixing $\Delta_{\mathrm{mix}}H$ corresponds to the transient decrease in molality which takes place around the adsorbent before the molality of the added solution is eventually restored. If this change in molality is continuously recorded at the outlet of the microcalorimetric cell, there is no difficulty in carrying out the integration, provided the enthalpies of mixing alone have been measured separately (Liphard et al., 1980; Denoyel et al., 1982).

4.3.4 Applications of Adsorption from Solution

The numerous technological applications of adsorption from solution include liquid purification, the stabilization of suspensions, ore flotation, soil science, adhesion, liquid chromatography, detergency, enhanced oil recovery, lubrication and last but not least, a basic help to the life sciences (e.g. adsorption by cell membranes, blood

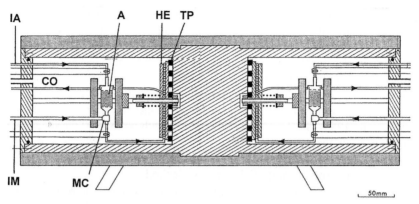

Figure 4.19 Twin liquid-flow adsorption and mixing microcalorimeter. IA, inlet for adsorption; IM, inlet for mixing; MC, mixing cell; CO, common outlet; A, adsorbent; HE, heat exchanger; TP, thermopile.
After Denoyel et al. (1982)

vessels, bones, teeth, skin, eyes and hair). A prerequisite for a successful application of adsorption from solution is of course to have some understanding of the mechanisms involved.

4.3.4.1   Adsorption (and Displacement) Mechanisms

We should start with a warning, adequately developed by Konya and Nagy (2013) about the temptation of deriving an adsorption mechanism (and even thermodynamic data) from the simple fit of the experimental data with a standard isotherm equation. For instance, the Langmuir equation should, in principle, be restricted to adsorption on homogeneous surfaces, with no interaction between adsorbed molecules, whereas the Freundlich equation assumes heterogeneous surface and possible interaction between the adsorbed species. Moreover, one should keep in mind that what is experimentally determined is a 'surface accumulation' which can result from either competitive adsorption (between solvent and solute), or ion exchange or, still, mere precipitation. We shall deal hereafter essentially with competitive adsorption.

The following few examples will serve to illustrate the various parameters able to influence the adsorption mechanisms. Other aspects of this complex area can be found in a comprehensive review by Lyklema (1995).

4.3.4.1.1   pH and Surface Charge
The pH of the solution and the resulting surface charge of the adsorbent have a major influence on the adsorption mechanism, even for the adsorption of non-ionic species. This is strikingly seen in the adsorption of the non-ionic polyoxyethylenic surfactants on various surfaces of silica: quartz, macroporous precipitated silica and pyrogenic silica (Denoyel et al., 1987). For instance, at pH 7, the surface excess density corresponding to the final plateau of the adsorption isotherms increases more than threefold from quartz to precipitated silica: this was shown to result from the much larger

proportion of dissociated silanols in the case of quartz. At pH 2, which is the zero point of charge, the surface excess density is the same for all silicas and has its maximum value, since all silanols are undissociated. This is an example of the indirect but important effect of surface charge on adsorption. This work has also revealed that the crystallinity of the silica surface affects its adsorbing behaviour through the enhancement of surface charge density, pH (Iler, 1979). Conversely, Golub et al. (2004) showed and studied how the adsorption of a cationic surfactant could change the surface charge on rutile and aerosol. As expected, influence of pH on the adsorption by porous silica in the 3.6–10 range was shown to be much larger for a cationic surfactant than for a non-ionic (Kharitonova et al., 2005). Similar phenomena can be used to follow the increase in disorder of solid surfaces on grinding crystalline samples (Cases, 1979).

4.3.4.1.2 Surface Oxidation and Chemistry

The role of surface oxidation of carbons is known to be essential in their adsorbing behaviour in solution, essentially either because of the polarity and acidity of the resulting surface chemical groups or because of their steric role in the accessibility of the micropores (Mattson et al., 1969; Franz et al., 2000; Haydar et al., 2003; Santiago et al., 2005; Guedidi et al., 2013). After a time when the porous texture of the carbons was given, the essential role in physical adsorption processes in solution, the part of the surface chemistry is being more and more taken into account (Rodriguez-Reinoso, 1998; Radovic et al., 2001; Nevskaia et al., 2004; Su et al., 2010; Figueiredo et al., 2011; Yavari et al., 2011). The competing role of surface chemistry and small size micropores was also evidenced for adsorption of valeric acid by carbons (El-Sayed and Bandosz, 2004).

4.3.4.1.3 Surface OH Groups or H_2O Molecules

The presence of surface OH groups or H_2O molecules can play a primary role in adsorption. For example, a microcalorimetric study of the adsorption of stearic acid, from heptane solution, on ferric oxide, (Husbands et al., 1971), revealed that pre-adsorbed water-enhanced adsorption of stearic acid. When adsorption takes place from a dry organic liquid, residual surface water may act as a special agent. This was shown for the adsorption of a silane-coupling agent (γ-amino-propyl-triethoxy-silane) on silica covered with water molecules for $\theta \leq 1$ (Trens and Denoyel, 1996). By the simultaneous determination of adsorption isotherms and the enthalpies of displacement (of heptane by various silanes) it was demonstrated that the amine function was able to displace some of the surface water and make it available for the hydrolysis of the silane into trisilanol, whereas the residual water was able to promote the formation of siloxane bonds between the trisilanol molecules and the surface.

4.3.4.1.4 Chain Length

With the aid of adsorption flow-microcalorimetry, Groszek (1965, 1970) studied the effect of chain length on the behaviour of normal alcohols and the corresponding acids when adsorbed from n-heptane on iron oxides. It was found that the behaviour of *n*-butanol and *n*-octadecanol was quite different: the former was strongly adsorbed in a single stage up to an apparent saturation level, whereas the latter was adsorbed in two stages. In the first stage, the molecules of *n*-octadecanol appeared to lie flat on the surface and, in the second stage, to stand vertically away from the surface. The scale of the microcalorimetric signal was enhanced as the length of the solvent molecule approached that of the solute. Groszek concluded that the stability of the adsorbed layer was increased by the formation of a mixed film (solvent + solute) made up of molecules of similar size.

It is in the field of polymer adsorption that the chain length plays a major role. A microcalorimetric procedure has been used to determine the fraction of 'train' segments in contact with the surface (see Figure 4.20). This approach is based on the assumption that the 'loops' and 'tails' are too far away from the surface to contribute to the measured $\Delta_{dpl}\dot{h}$, and also that the displacement enthalpy of the monomer is an adequate reference (Cohen-Stuart et al., 1982; Killman et al., 1983; Denoyel et al., 1990). This method was found to work well with polyethylene glycol (in the relative molar mass range 400–4,000,000) on macroporous silica gel (Trens and Denoyel, 1993).

In spite of the very small $\Delta_{dpl}\dot{h}$ per mole of segment (ca. 3 kJ mol^{-1}; M of segment: 44 g mol^{-1}), it was shown that (i) at low coverage, irrespective of its length, the molecules adsorb in a flat conformation; (ii) as coverage increases, the molecules tend to stand up; (iii) the molecules must contain more than, say, 5–10 segments to give rise to a complete loop and a second train and (iv) at coverages above 0.5, polymers with more than 500 segments can adsorb only by displacing previously adsorbed molecules.

4.3.4.1.5 Sample Modifications During Adsorption

Finally, one must be aware of possible modifications of the sample during the adsorption process itself. This is what was shown to happen when adsorbing a non-ionic surfactant (nonyphenoloxyethylene, with 9–10 ethoxy groups) on kaolin in the presence of 1% NaCl at 40 °C: in this case, one step only was visible in the normal L-shaped adsorption isotherm whereas two steps were seen in the microcalorimetric recording. The first was attributed to the displacement of water by the surfactant and the second to a partial opening and hydration of the sheet-like structure of kaolin under the action of surfactant and salt (Rouquerol and Partyka, 1981).

A more common situation is the effect of adsorption on the stability of a suspension. This may be favourable (by modifying, for example, the surface charge of the particles, or by increasing the distance between particles) or unfavourable. Such an

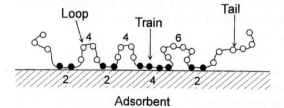

Figure 4.20 Example of a train–loop–tail configuration of an adsorbed chain.
After Fleer et al. (1993).

example of the latter effect was found with non-ionic surfactant TX-100 and silica suspensions. It was found that the adsorption produced micelle-like aggregates which underwent flocculation by a bridging mechanism (Giordano-Palmino et al., 1994). This was strongly supported by the fact that the fastest flocculation occurred at half coverage, for which the theory predicted an optimum amount of bridging flocculation (La Mer, 1966; Kitchener, 1972).

4.3.4.2 Assessment of Surface Area

Adsorption from solution measurements have been used for many years for the routine determination of the surface area of certain porous materials such as activated carbons. This is still a major application, especially when the adsorbent cannot be outgassed or when the analytical techniques available in a laboratory are designed for the study of solutions.

For example, adsorptives as diverse as iodine (Kipling, 1965; Puri and Bansal, 1965; Molina-Sabio et al., 1985; Fernandez-Colinas et al., 1989), p-nitrophenol (Giles and Nakhwa, 1962; Lopez-Gonzalez et al., 1988), salicylic acid (Fernandez-Colinas et al., 1991a,b), various surfactants (Somasundaran and Fuerstenau, 1966), methylene blue (since a long time, see for instance Tewari (2001)) and ethylene oxide chains (for study of expandable clay (Shen, 2002)) have been used for that purpose in dilute aqueous solution. Organic molecules in hydrocarbon solutions have included lauric acid (de Boer et al., 1962). An interesting procedure is that used by Stoeckli et al. (2001a, 2001b) and Centeno and Stoeckli (2010) when using phenol adsorption from 0.4 M water solution: they carry out adsorption by immersion calorimetry and convert the enthalpy of immersion into an area by considering that the areal enthalpy of immersion of a porous carbon into the phenol solution is 109 mJ m^{-2}. Caffeine can also be used in the same way, with carbons, with an areal enthalpy of immersion of 113 mJ m^{-2} (Stoeckli, 1995). Further comments on this approach are given in Section 4.2.3.

Solid/solution adsorption systems have been found to give a variety of isotherm types, as classified by Giles et al. (1960). Many exhibit the characteristic Type I shape (see Figure 1.4), with the equilibrium concentration, c, replacing $p/p°$. In these cases, over a limited range of c, it is usually possible to apply an empirical *equation of Langmuir form*

$$n/n_L = bc/(1 + b\,c) \qquad (4.80)$$

where n is the amount of solute adsorbed (per gram of adsorbent) at c, n_L is the amount adsorbed (per gram of adsorbent) at the plateau and b is an empirical constant. Nevertheless, one should notice that the only real justification for applying such an equation is the case when one (i) does not have enough experimental points to reach the plateau and (ii) knows from prior information on the adsorption system that a plateau does exist. Otherwise, when a Type I isotherm is available, the height of the plateau directly provides n_L. Moreover, this similarity in Type I shape does not mean that the mechanisms involved in solution/solid and gas/solid adsorption are the same. When the adsorbent is microporous, a Type I isotherm may be associated, as for gaseous physisorption, with micropore filling. With some other adsorbents, the plateau of a Type I solute isotherm appears to correspond to monolayer completion in accordance with the classical Langmuir interpretation. But there are a number of reasons why this approach does not always provide a reliable assessment of the overall surface area of an adsorbent. Indeed, solute molecules are generally subject to more specific interactions with the surface than the gas molecules conventionally used (N_2, Ar). An added complication is that adsorbent surfaces are generally heterogeneous. Consequently, the apparent molecular area of the solute is likely to depend on the chemical nature of the exposed sites. Furthermore, the structure of the adsorbed phase may depend not only on the interaction between the adsorbed molecules (e.g. hydrogen bonding) but also on the incorporation of solvent.

As adsorption from solution is a competitive phenomenon between solvent and solute, one easily expects it to be more difficult to interpret than the adsorption of a single gas.

The *role of the solvent* was demonstrated by de Boer and his co-workers (1962). In their study of the adsorption of lauric acid on activated alumina, it was found that solvent competition was most pronounced with diethyl ether, rather less so with benzene and to a negligible extent with pentane. In the latter case, the plateau of the lauric acid isotherm extended over a very wide range of concentration. This type of isotherm is associated with very high adsorption affinity and is generally due to relatively strong adsorbent–adsorbate interactions. However, the solute monolayer is unlikely to be in a close-packed state, the molecular area being dependent on the surface chemistry. Similarly, it has been found (see Gregg and Sing, 1967) that the apparent molecular area of adsorbed dye molecules is not constant from one type of surface to another. It follows that the problem for the measurement of surface areas by adsorption from solution is to choose a solute which is preferentially adsorbed on a particular type of surface and is able to cover all of that surface at low concentration.

Adsorption from solution is often used to characterize the adsorptive capacity of microporous carbons. As we will see in Chapter 10, measurements made with such solutes as iodine (Fernandez-Colinas et al. 1989) or salicylic acid (Fernandez-Colinas et al., 1991a,b) have revealed that it is possible to analyse adsorption from solution data by an extension of the α_s method and to evaluate both the micropore volume and the external surface area.

4.3.4.3   Assessment of Pore Size

Assessment of microporosity from adsorption of organic molecules of different shapes and sizes is examined in Section 9.3.2.

References

Adamson, A.W., 1967. The Physical Chemistry of Surfaces, second ed. Interscience, New York.
Ash, S.G., Brown, R., Everett, D.H., 1973. J. Chem. Thermodyn. 5, 239.
Atkinson, D., Mc Leod, A.I., Sing, K.S.W., Capon, A., 1982. Carbon. 20 (4), 339.
Bartell, F.E., Suggitt, R.M., 1954. J. Phys. Chem. 58, 36.
Benko, M., Puskas, S., Kiraly, Z., 2013. Adsorption. 19, 71.
Bertoncini, C., Raffaelli, J., Fassino, L., Odetti, H.S., Bottani, E.J., 2003. Carbon. 41, 1101.
Briant, J., Cuiec, L., 1972. Comptes-Rendus du 4ème Colloque ARTEP, Rueil-Malmaison, 7–9 Juin 1971. Technip, Paris.

Brunauer, S., Kantro, D.L., Weise, C.H., 1956. Can. J. Chem. 34, 1483.

Cases, J.M., 1979. Bull. Minéral. 102, 694.

Centeno, T.A., Stoeckli, F., 2010. Carbon. 48, 2478.

Chessick, J.J., 1962. J. Phys. Chem. 66, 762.

Chessick, J.J., Young, G.J., Zettlemoyer, A.C., 1954. Trans. Faraday Soc. 50, 587.

Chessick, J.J., Zettlemoyer, A.C., Healey, F.H., Young, G.J., 1955. Can. J. Chem. 33, 251.

Cohen-Stuart, M.A., Fleer, G.J., Bijsterbosch, B.H., 1982. J. Colloid Interface Sci. 90, 321.

de Boer, J.H., Houben, G.M.M., Lippens, B.C., Meij, W.D., Walgrave, W.K.A., 1962. J. Catal. 1, 1.

Defay, R., 1941. Des diverses façons de définir l'adsorption, Mém. Ac. R. Belg. Cl. Sci., Brussels.

Defay, R., Prigogine, I., 1951. Tension Superficielle et Adsorption. Desoer-Dunod, Liège-Paris.

Denoyel, R., 2004. Nanoporous Materials, Series on Chemical Engineering, Vol. 4. p. 727.

Denoyel, R., Rouquerol, F., Rouquerol, J., 1982. In: Rochester, C. (Ed.), Adsorption from Solution. Academic Press, London.

Denoyel, R., Rouquerol, F., Rouquerol, J., 1987. In: Liapis, A.I. (Ed.), Fundamentals of Adsorption. Engineering Foundation, American Institute of Chemical Engineers, New York, p. 199.

Denoyel, R., Durand, G., Lafuma, F., Audebert, R., 1990. J. Colloid Interface Sci. 139 (1), 281.

Denoyel, R., Fernandez-Colinas, J., Grillet, Y., Rouquerol, J., 1993. Langmuir. 9, 515.

Denoyel, R., Beurroies, I., Vincent, D., 2002. J. Therm. Anal. Calorim. 70, 483.

Denoyel, R., Beurroies, I., Lefevre, B., 2004. J. Petrol. Sci. Eng. 45, 203.

de Ridder, D.J., Verliefde, A.R.D., Schoutteten, K., van der Linden, B., Heijman, S.G.J., Beurroies, I., Denoyel, R., Amy, G.L., van Dijk, J.C., 2013. Carbon. 53, 153.

Dewar, J., 1904. Proc. R. Soc. A (Lond.). 74, 122.

Douillard, J.M., Zoungrana, T., Partyka, S., 1995. J. Petrol. Sci. Eng. 14, 51.

Einstein, A., 1901. Ann. der Physik. 4, 513.

El-Sayed, Y., Bandosz, T.J., 2004. J. Colloid Interface Sci. 273, 64.

Elton, G.A.H., 1951. J. Chem. Soc. 2958, .

Everett, D.H., 1972. Pure Appl. Chem. 31 (4), 579.

Everett, D.H., 1973. Specialist Periodical Reports. Colloidal Science, Vol. 1. The Chemical Society, London, p. 51.

Everett, D.H., 1986. Pure Appl. Chem. 58 (7), 967.

Everett, D.H., Findenegg, G.H., 1969. J. Chem. Thermodyn. 1, 573.

Everett, D.H., Powl, J.C., 1976. J. Chem. Soc. Faraday Trans. I. 72, 619.

Everett, D.H., Langdon, A.G., Maher, P., 1984. J. Chem. Thermodyn. 16, 98.

Fadeev, A.Y., Eroshenko, V., 1997. J. Colloid Interface Sci. 187, 275.

Fernandez-Colinas, J., Denoyel, R., Rouquerol, J., 1989. Adsorpt. Sci. Techol. 6, 18.

Fernandez-Colinas, J., Denoyel, R., Grillet, Y., Vandermeersch, J., Reymonet, J.L., Rouquerol, F., Rouquerol, J., 1991a. In: Mersmann, A.B., Scholl, S.E. (Eds.), Fundamentals of Adsorption III. Engineering Foundation, New York, p. 261.

Fernandez-Colinas, J., Denoyel, R., Rouquerol, J., 1991b. In: Rodriguez-Reinoso, F., Rouquerol, J., Sing, K.S.W., Unger, K.K. (Eds.), Characterization of Porous Solids II. Elsevier, Amsterdam, p. 399.

Figueiredo, J.L., Sousa, J.P.S., Orge, C.A., Pereira, M.F.R., Orfao, J.J.M., 2011. Adsorption. 17, 431.

Fleer, G.J., Cohen-Stuart, M.A., Scheutjens, J.M.H.M., Cosgrove, T., Vincent, B., 1993. Polymers at Interfaces. Chapman and Hall, London, p. 31.

Franz, M., Arafat, H.A., Pinto, N.G., 2000. Carbon. 38, 1807.

Furuishi, R., Ishii, T., Oshima, Y., 1982. Thermochim. Acta. 56, 31.

Giles, C.H., Nakhwa, S.N., 1962. J. Appl. Chem. 12, 266.

Giles, C.H., Mac Ewan, T.H., Nakhwa, S.N., Smith, D., 1960. J. Chem. Soc. 3973.

Giordano-Palmino, F., Denoyel, R., Rouquerol, J., 1994. J. Colloid Interface Sci. 165, 82.

Giraldo, L., Moreno-Pirajan, J.C., 2007. J. Therm. Anal. Calorim. 89, 589.

Golub, T.P., Koopal, L.K., Sidorova, M.P., 2004. Colloid J. 66, 38.

Gomez, F., Denoyel, R., Rouquerol, J., 2000. Langmuir. 16, 3474.

Gonzalez, M.T., Sepulveda-Esoribano, A., Molina-Sabio, M., Rodriguez-Reinoso, F., 1995. Langmuir. 11, 2151.

Gonzalez-Garcia, S., Dios Cancela, G., 1965. Studia Chemica I, Salamancap. 37.

Gregg, S.J., Sing, K.S.W., 1967. Adsorption, Surface Area and Porosity, first ed. Academic Press, London.

Groszek, A.J., 1965. Chem. Ind. 482.

Groszek, A.J., 1966. Lubrication Sci. Technol. 9, 67.

Groszek, A.J., 1970. ASLE Trans. 13, 278.

Guedidi, H., Reinert, L., Levêque, J.M., Soneda, Y., Bellakhal, N., Duclaux, L., 2013. Carbon. 54, 132.

Guggenheim, E.A., Adam, N.K., 1933. Proc. R. Soc. 139, 218.

Harkins, W.D., 1952. The Physical Chemistry of Surface Films. Reinhold Publishing, Division, New York, p. 262.

Harkins, W.D., Boyd, G.E., 1942. J. Am. Chem. Soc. 64, 1195.

Harkins, W.D., Dahlstrom, R., 1930. Ind. Eng. Chem. 22, 897.

Harkins, W.D., Jura, G., 1944. J. Am. Chem. Soc. 66, 1362.

Haydar, S., Ferro-Garcia, M.A., Rivera-Utrilla, J., Joly, J.P., 2003. Carbon. 41 (3), 387.

Healey, F.H., Yu, Y.F., Chessick, J.J., 1955. J. Phys. Chem. 59, 399.

Hill, T.L., 1949. J. Chem. Phys. 17, 520.

Husbands, D.I., Tallis, W., Waldsax, J.C.R., Woodings, C.R., Jaycock, M.J., 1971. Powder Technol. 5, 31.

Iler, R.K., 1979. The Chemistry of Silica. Wiley, New York.

Jaycock, M.J., Parfitt, G.D., 1981. Chemistry of Interfaces. John Wiley, Chichester.

Jehlar, P., Romanov, A., Biros, P., 1979. Thermochim. Acta. 28, 188.

Jura, G.J., Hill, T.L., 1952. J. Am. Chem. Soc. 74, 1598.

Kharitonova, T.V., Ivanova, N.I., Summ, B.D., 2005. Colloid J. 67 (2), 242.

Killman, E., Korn, M., Bergmann, M., 1983. In: Ottewill, R.H., Rochester, C.H., Smith, A.L.S. (Eds.), Adsorption from Solution. Academic Press, London, p. 259.

Kipling, J.J., 1965. Adsorption from Solutions of Non-Electrolytes. Academic Press, London.

Kipling, J.J., Tester, D.A., 1952. J. Chem. Soc. 4123.

Kiraly, Z., Dekany, I., 1989. J. Chem. Soc. Faraday Trans. I 85, 3373.

Kitchener, J.A., 1972. Brit. Polym. J. 4, 27.

Konya, J., Nagy, N.M., 2013. Adsorption. 19, 701.

Kurbanbekov, E., Larionov, O.G., Chmutov, K.V., Yudelevich, M.D., 1969. Russ. Phys. Chem. 43, 916.

La Mer, V.K., 1966. Discuss. Farday Soc. 42, 248.

Laffitte, M., Rouquerol, J., 1970. Bull. Soc. Chim. Fr. 3335.

Laporte, F., 1950. Ann. Phys. 5, 5.

Liphard, M., Glanz, P., Pilarski, G., Findenegg, G.H., 1980. Progr. Colloid Polym. Sci. 67, 131.

Lopez-Gonzalez, J., de, D., Valenzuela-Calahorro, C., Navarrete-Guijosa, A., Gomez-Serrano, V., 1988. An. Quim. 84B, 47.

Lyklema, J., 1995. Fundamentals of Interface and Colloid Science. I Fundamentals. II. Solid-liquid Interfaces. Academic Press, London.

Mackrides, A.C., Hackerman, N., 1959. J. Phys. Chem. 63, 594.

Magnan, R., 1970. Am. Ceram. Soc. Bull. 49 (3), 314.

Mattson, J.S., Mark Jr., H.B., Malbin, M.D., Weger Jr., W.J., 1969. J. Colloid Interface Sci. 31 (1), 116.

Medout-Marere, V., Malandrini, H., Zoungrana, T., Douillard, J.M., Partyka, S., 1998. J. Petrol. Sci. Eng. 20, 223.

Micale, F.J., Topic, M., Cronan, C.L., Leidheiser Jr., H., Zettlemoyer, A.C., 1976. J. Colloid Interface Sci. 55 (3), 540.

Mills, I., Cvitas, T., Homann, K., Kallay, N., Kuchitzu, K., 1993. Quantities, Units and Symbols in Physical Chemistry. IUPAC, Blackwell Scientific Publication, London, p. 42.

Molina-Sabio, M., Salinas-Martinez de Lecea, C., Rodriguez-Reinoso, F., Peunte-Ruiz, C., Linares-Solano, A., 1985. Carbon. 23, 91.

Moreno-Pirajan, J.C., Giraldo, G.L., Gomez, O.A., 1996. Thermochimica Acta. 290, 1.

Nagy, L.G., Schay, G., 1963. Acta Chim. Acad. Sci. Hung. 39, 365.

Nègre, J.Cl., 1984. Thèse Université de Provence, Marseille.

Nègre, J.Cl., Denoyel, R., Rouquerol, F., Rouquerol, J., 1985. Actes des Journées de Calorimétrie et d'Analyse Thermique (J.C.A.T.), Marseille.

Nevskaia, D.M., Castillejos, E.-L.E., Guerrero, A., Munoz, V., 2004. Carbon. 42, 653.

Nowell, D.V., Powell, M.W., 1991. J. Therm. Anal. 37, 2109.

Nunn, C., Everett, D.H., 1983. J. Chem. Soc. Faraday Trans. I. 79, 2953.

Nunn, C., Schlechter, R.S., Wade, W.H., 1981. J. Coll. Interface Sci. 80, 598.

Ocampo-Perez, R., Leyva-Ramos, R., Sanchez-Polo, M., Rivera-Utrilla, J., 2013. Adsorption. http://dx.doi.org/10.1007/s10450-013-9502-y.

Partyka, S., Rouquerol, F., Rouquerol, J., 1975. In: Proceedings of the 4th International Conference on Chemical Thermodynamics of IUPAC, Montpellier, CRMT Marseille, vol. 7, p. 46.

Partyka, S., Rouquerol, F., Rouquerol, J., 1979. J. Colloid Interface Sci. 68 (1), 21.

Partyka, S., Keh, E., Lindheimer, M., Groszek, A., 1989. Colloid. Surf. 37, 309.

Pouillet, M.C.S., 1822. Ann. Chim. Phys. 20, 141.

Puri, B.R., Bansal, R.C., 1965. Carbon. 3, 227.

Radovic, L.R., Moreno-Castilla, C., Rivera-Utrilla, J., 2001. In: In: Radovic, L.R. (Ed.), Chemistry and Physics of Carbon, Vol. 27. Dekker, New York, p. 227.

Robert, L., 1967. Bull. Soc. Chim. Fr. 7, 2309.

Robert, L., Brusset, H., 1965. Fuel. 44, 309.

Rodriguez-Reinoso, F., 1998. Carbon. 36, 159.

Rodriguez-Reinoso, F., Molina-Sabio, M., Gonzalez, M.J., 1997. Langmuir. 13, 2354.

Rouquerol, J., 1985. Thermochim. Acta. 95, 337.

Rouquerol, J., Partyka, S., 1981. J. Chem. Tech. Biotechnol. 31, 584.

Rouquerol, J., Rouquerol, F., Grillet, Y., Torralvo, M.J., 1984. In: Myers, A., Belfort, G. (Eds.), Fundamentals of Adsorption. Engineering Foundation, New York, p. 501.

Rouquerol, J., Llewellyn, P., Navarrete, R., Rouquerol, F., Denoyel, R., 2002. Stud. Surf. Sci. Catal. 144, 171.

Santiago, M., Stüber, F., Fortuny, A., Fabregat, A., Font, J., 2005. Carbon. 43, 2134.

Schay, G., 1970. In: Everett, D.H., Otterwill, R.H. (Eds.), Surface Area Determination. Butterworth, London, p. 273.

Schay, G., Nagy, L.G., 1961. J. Chim. Phys. 140.

Schay, G., Nagy, L.G., 1972. J. Colloid Interface Sci. 38 (2), 302.

Schay, G., Nagy, L.G., Racz, G., 1972. Acta Chim. Acad. Sci. Hung. 71, 23.

Schröder, J., 1979. J. Coll. Interface Sci. 72 (2), 279.

Schultz, J., Tsutsumi, K., Donnet, J.B., 1977. J. Colloid Interface Sci. 59 (2), 272.

Sharma, S.G., Fort, T., 1973. J. Coll. Interface Sci. 43, 36.

Shen, Y.H., 2002. Chemosphere. 48, 1075.

Silvestre-Albero, J., Gomez de Salazar, C., Sepulveda-Escribano, A., Rodriguez-Reinoso, F., 2001. Colloids Surf. A. 187–188, 151.

Somasundaran, P., Fuerstenau, D.W., 1966. J. Phys. Chem. 70, 90.

Stoeckli, F., 1995. In: Patrick, J. (Ed.), Porosity in Carbons – Characterization and Applications. Arnold, London, p. 67.

Stoeckli, F., Centeno, T.A., 1997. Carbon. 35 (8), 1097.

Stoeckli, F., Centeno, T.A., 2005. Carbon. 43, 1184.

Stoeckli, F., Lopez-Ramon, M.V., Moreno-Castilla, C., 2001a. Langmuir. 17, 3301.

Stoeckli, F., Lopez-Ramon, M.V., Hugi-Clearly, D., Guillot, A., 2001b. Carbon. 39, 1115.

Su, F., Lu, C., Hu, S., 2010. Colloids Surf. A: Physicochem. Eng. Asp. 353, 83.

Taylor, A.G., 1965. Chemistry and Industry, 2003.

Taraba, B., 2012. J. Therm. Anal. Calorim. 107, 923.

Tewari, B.B., 2001. Russ. J. Gen. Chem. 71 (1), 33.

Trens, P., Denoyel, R., 1993. Langmuir. 9, 519.

Trens, P., Denoyel, R., 1996. Langmuir. 12, 2781.

Trompette, J.L., 1995. Thèse Université des Sciences et Techniques du Languedoc, Montpellier.

Vanderdeelen, J., Rouquerol, J., Baert, L., 1972. In: Thermochimie, Coll. Intern. CNRS, Paris.

van Oss, C.J., 2006. Interfacial Forces in Aqueous Media, second ed. CRC Press, Boca Raton.

van Oss, C.J., Good, R.J., Chaudhury, M.K., 1988. Langmuir. 4, 884.

Villar-Rodil, S., Denoyel, R., Rouquerol, J., Martinez-Alonso, A., Tascon, J.M.D., 2002a. Carbon. 40, 1376.

Villar-Rodil, S., Denoyel, R., Rouquerol, J., Martinez-Alonso, A., Tascon, J.M.D., 2002b. J. Colloid Interface Sci. 252, 169.

Wade, W.H., Hackerman, N., 1960. J. Phys. Chem. 64, 1196.

Wade, W.H., Hackermann, N., 1964. Adv. Chem. Ser. 43, 222.

Wade, W.H., Hackermann, N., Cole, H.D., Meyer, D.E., 1961. Adv. Chem. Ser. 33, 35.

Watanabe, H., Seto, J., 1988. Bull. Chem. Soc. Japan. 61, 3067.

Whalen, J.W., 1961a. J. Phys. Chem. 65, 1676.

Whalen, J.W., 1961b. Adv. Chem. Ser. 33, 281.

Whalen, J.W., 1962. J. Phys. Chem. 66, 511.

Whalen, J.W., Lai, K.Y., 1977. J. Colloid Interface Sci. 59 (3), 483.

Widyani, E., Wightman, J.P., 1982. Colloids Surf. 4, 209.

Williams, A.M., 1913. Medd. k. Veteskapsakad. Nobelinst. 2, 27.

Yavari, R., Huang, Y.D., Ahmadi, S.J., 2011. J. Radioanal. Nucl. Chem. 287, 393.

Young, G.J., Bursh, T.P., 1960. J. Colloid Interface Sci. 15, 361.

Young, G.J., Chessick, J.J., Healey, F.H., Zettlemoyer, A.C., 1954. J. Phys. Chem. 58, 313.

Zettlemoyer, A.C., 1965. Ind. Eng. Chem. 57, 27.

Zettlemoyer, A.C., Chessick, J.J., 1964. Adv. Chem. Ser. 43, 88.

Zettlemoyer, A.C., Narayan, K.S., 1967. In: In: Flood, E.A. (Ed.), The Solid-Gas Interface, Vol. 1. Marcel Dekker, Inc., New York, p. 158.

Zettlemoyer, A.C., Young, G.J., Chessick, J.J., Healey, F.H., 1953. J. Phys. Chem. 57, 649.

Zettlemoyer, A.C., Chessick, J.J., Hollabaugh, C.M., 1958. J. Phys. Chem. 62, 489.

Zimmermann, R., Wolf, G., Schneider, H.A., 1987. Colloids Surf. 22, 1.

Zoungrana, T., Douillard, J.M., Partyka, S., 1994. J. Therm. Anal. 41, 1287.

5 Classical Interpretation of Physisorption Isotherms at the Gas–Solid Interface

Kenneth S.W. Sing, Françoise Rouquerol, Jean Rouquerol

Aix Marseille University-CNRS, MADIREL Laboratory, Marseille, France

Chapter Contents

Adsorption by Powders and Porous Solids. http://dx.doi.org/10.1016/B978-0-08-097035-6.00005-X

5.1   Introduction

The objectives of this chapter are to review the classical theories of physisorption at the gas–solid interface and the application of empirical isotherm equations. It is not our intention to give a comprehensive survey of physisorption theories; instead, our aim is to provide sufficient information to enable the newcomer to surface science to appreciate the advantages and limitations of procedures, which are still widely used for the analysis of experimental data. Our selection of the theoretical material is necessarily somewhat arbitrary in view of the vast literature on physisorption. The decision to include a particular concept or equation is based on either its historical importance or its current usage. Thus, a few equations were considered worthy of inclusion in this chapter, although for our purpose they do not merit further discussion or application in subsequent chapters. A more detailed account of the historical development is to be found in the books by McBain (1932), Brunauer (1945) and Young and Crowell (1962).

Capillary condensation is not discussed in this chapter since it involves the formation of a liquid-like meniscus. This aspect of physisorption by the filling of mesopores is dealt with in Chapter 8 in relation to the characterization of mesoporous materials and evaluation of the mesopore size distribution. Also, the mechanisms involved in the filling of micropores will be mainly dealt with in Chapter 9.

5.2   Adsorption of a Pure Gas

5.2.1   Equations Related to the Gibbs Adsorption Equation: Description of the Adsorbed Phase on Available Surface or in Micropores

As explained in Chapter 2, the adsorbed phase can be characterized by the spreading pressure π, which corresponds to the lowering of the surface tension of the adsorbent due to adsorption (see Equation 2.22). This spreading pressure is an intensive quantity whose value depends on the surface excess concentration $\Gamma(=n^{\sigma}/A)$. Its dependency on the equilibrium pressure p is given by the *Gibbs adsorption equation* (Equation 2.34). This equation allows one to derive the relationship between quantities π and Γ from an experimental adsorption isotherm.

5.2.1.1   Ideal Adsorbed Phase

5.2.1.1.1   Henry's Law
We saw in Chapter 2 that, at pressures small enough, the specific surface excess amount n varies linearly with pressure p, and then one can apply Henry's law:

$$n = k_{H}\, p \qquad\qquad (5.1)$$

where k_{H} is the Henry's law constant, whose value depends on the units chosen for n and p. Under these conditions, the Gibbs adsorption Equation (2.34) becomes:

$$\pi a = nRT \qquad\qquad (5.2)$$

where a is the specific area of the adsorbent.

This relationship, which should hold for any adsorption system at very low surface excess concentration, can be considered as the state equation of an ideal two-dimensional (2D) adsorbed gas. The simplest interpretation of the behaviour of the adsorbed phase in this lowest pressure range is to suppose that the adsorbate molecules are independent of each other.

It is possible to obtain the differential enthalpy of adsorption at 'zero' coverage, $\Delta_{\text{ads}}\dot{h}_0$, from the variation of the Henry's law constant with temperature, since from Equation (2.68):

$$\Delta_{\text{ads}}\dot{h}_0 = RT^2 (\partial(\ln\{k_H\})/\partial T)_n \tag{5.3}$$

The evaluation of $\Delta_{\text{ads}}\dot{h}_0$ is achieved by plotting $\ln\{k_H\}$ against $1/T$, provided that $\Delta_{\text{ads}}\dot{h}_0$ remains constant over the temperature range studied. This is the enthalpy of adsorption on the most active sites of the adsorbent (k_H is related to the gas–solid interaction).

In the simplest case of the adsorption of a spherical nonpolar molecule on a smooth energetically uniform surface, k_H is given by the configurational integral:

$$k_H = \frac{a}{kT} \int_V [\exp(-\phi(z)/kT) - 1] \, dz \tag{5.4}$$

where $\phi(z)$ is the potential energy of adsorption expressed as a function of the distance z of the adsorbed molecule from the surface plane. In this case, it is assumed that ϕ is independent of the location in the xy plane (i.e. the plane parallel to the surface), but a position vector must be introduced in the more general treatment.

In the special case of the physisorption of a simple nonpolar gas by a 'homogeneous' microporous solid, as Everett (1970) pointed out, we may relate the Henry's law constant, k_H, to the potential energy of adsorption, ϕ, by an expression analogous to Equation (5.4). Assuming that ϕ is essentially constant within each pore, we have:

$$k_H = \frac{v_p}{RT} \left[\exp\left(-\frac{\phi}{kT}\right) - 1 \right] \tag{5.5}$$

where v_p is the accessible specific pore volume.

Since primary micropore filling is associated with high adsorption affinity, adsorption isotherm measurements at very low fractional pore filling are most easily made at elevated temperature.

Unfortunately, the energetic heterogeneity exhibited by many microporous adsorbents results in isotherm curvature, even at the lowest recorded pressures. However, there are a few crystalline molecular sieves which have remarkably regular tubular pores and which have been found to give linear nitrogen isotherms over wide ranges of pressure at temperatures of around 300 K (Reichert et al., 1991). In these cases, the Henry's law behaviour is consistent with a nearly constant adsorption energy over an appreciable range of fractional pore filling.

5.2.1.2   Non-ideal Adsorbed Phase

5.2.1.2.1   Virial Equations

Although the initial Henry's law region is well documented with some physisorption isotherms, there are others which deviate from linearity at the lowest pressures so far recorded (Rudzinski and Everett, 1992). Convex curvature with respect to the adsorption axis in this region is likely to be due to the effect of either surface heterogeneity or microporosity. In particular, specific adsorbent–adsorbate interactions are often associated with energetic heterogeneity, and it is not surprising to find that the corresponding isotherms tend to be non-linear at low surface coverage. In some cases, a small knee is evident at very low p/p^o and the isotherm becomes more linear over a range of higher p/p^o. If adsorption energy data are also available, it can be established whether the knee is associated with adsorption on a small high-energy fraction of the surface or whether adsorbate–adsorbate interactions are detectable over the linear range.

A number of different empirical equations have been proposed to allow for the deviations of physisorption isotherms from Henry's law (see Brunauer, 1945; Rudzinski and Everett, 1992). An approach which is analogous to that used in the treatment of imperfect gases and non-ideal solutions is to adopt a virial treatment. Kiselev and his co-workers (Avgul et al., 1973) favoured the form:

$$p = n \exp\left(C_1 + C_2 n + C_3 n^2 + \cdots\right) \tag{5.6}$$

where the coefficients (C_1, C_2, C_3) are characteristic constants for a given gas–solid system and temperature. An alternative, and in some respects more useful, linear form (Cole et al., 1974) is:

$$\ln\{n/p\} = K_1 + K_2 n + K_3 n^2 + \cdots \tag{5.7}$$

Thus, by extrapolation of the virial plot of $\ln\{n/p\}$ versus n, it is possible to obtain k_H, since:

$$k_H = \lim_{p \to 0} (n/p) \tag{5.8}$$

The advantage of using Equation (5.7) is that the linearity of the semi-logarithmic plot generally extends well above the Henry's law limit, and therefore the evaluation of k_H by extrapolation is more reliable.

The virial treatment provides a general method of analysing the low-coverage region of an adsorption isotherm and its application is not restricted to particular mechanisms or systems. If the structure of the adsorbent surface is well defined, virial treatment also provides a sound basis for the statistical mechanical interpretation of the adsorption data (Pierotti and Thomas, 1971; Steele, 1974). As indicated above, K_1 in Equation (5.7) is directly related to k_H and therefore, under favourable conditions, to the gas–solid interaction.

In the case of microporous adsorbents, if the isotherm curvature is not too great, the simple virial plot of $\ln\{n/p\}$ versus n provides a useful means of obtaining k_H. Although the linear range of these plots is generally confined to very low values of n, the evaluation of k_H is achieved by extrapolation and the application of Equation (5.6) (Cole et al., 1974; Carrott and Sing, 1989).

Avgul and Kiselev (1970), Barrer (1978) and others (Ruthven, 1984) used equations of the virial form to analyse physisorption data obtained with molecular sieve zeolites. In their investigations of the adsorption of the noble gases and lower hydrocarbons, Avgul et al. (1973) were able to demonstrate that, at low pore filling/surface coverage, Equation (5.6) can be applied to the isotherms on both

X-type zeolites and graphitized carbons. From an empirical standpoint, this confirms the utility of the virial analysis of adsorption data, but it does not resolve the problems involved in the interpretation of the higher coefficients (Rudzinski and Everett, 1992).

The higher coefficients in Equations (5.6) and (5.7) are more complex and depend on 'mixed' interactions between the adsorbed molecules and with the adsorbent. Although a good deal of attention has been given to the difficult problems associated with surface heterogeneity (Rudzinski and Everett, 1992), the theoretical interpretation of the derived virial coefficients remains speculative.

5.2.1.2.2 The Hill-de Boer Equation

By picturing the adsorbed monolayer as a 2D imperfect gas, one may apply a 2D form of the van der Waals equation in which the gas pressure is replaced by the spreading pressure and the volume by the surface area (see de Boer, 1953; Young and Crowell, 1962; Ross and Olivier, 1964; Gregg and Sing, 1967). By combining this with the Gibbs adsorption equation (Equation 2.34), de Boer (1968) obtained the equation:

$$p = \frac{\theta}{k_H(1 - \theta)} \exp\left(\frac{\theta}{1 - \theta} - k_2\theta\right) \tag{5.9}$$

where θ is the surface coverage, k_H is the Henry's law constant and k_2 is a second empirical constant. A similar form of equation was proposed independently by Hill, and therefore Equation (5.9) is generally known as the Hill-de Boer equation.

Equation (5.9) may be rearranged to give the linear form:

$$k_2\theta + \ln k_H = \frac{\theta}{1 - \theta} + \ln\left(\frac{\theta}{1 - \theta}\right) - \ln\{p\} \equiv W(p, \theta)$$

Accordingly, the plot of W against θ should provide a means of obtaining k_H and k_2. However, since $\theta = n/n_m$, this obviously requires prior knowledge of the monolayer capacity, n_m.

We have seen already that the surface of graphitized carbon black is energetically remarkably homogeneous, and therefore, we would expect it to be a suitable substrate for testing the validity of the Hill-de Boer equation. A number of investigators reported (Ross and Olivier, 1964; Broekhoff and van Dongen, 1970) an appreciable range of fit of Equation (5.9), although generally this is not found above $\theta = 0.5$ (Sing, 1973).

Broekhoff and van Dongen (1970) suggested that it is not appropriate to use the Brunauer–Emmett–Teller (BET) monolayer capacity to calculate θ. Instead, they proposed that appropriate values of the three adjustable parameters (θ, k_H and k_2) should be selected to obtain the best fit. When applied in this manner, Equation (5.9) was claimed to be a useful empirical relation and applicable to a wide range of adsorption isotherms on graphitized carbon.

Several attempts have been made to extend the application of the Hill-de Boer equation to heterogeneous surfaces, following the approach originally adopted by Ross and Olivier (1964). The surface is pictured as an assembly of small uniform patches.

The Hill-de Boer equation is used to describe the form of the isotherm on each 'homo-tattic' patch and the weighted sum is then employed to give the composite isotherm for the heterogeneous surface. This approach has led to extensive interest in the effects of surface heterogeneity on the physisorption of gases (Rudzinski and Everett, 1992).

5.2.2 The Langmuir Theory

5.2.2.1 The Langmuir Model: Ideal Localized Monolayer Adsorption

Langmuir proposed various mechanisms of adsorption, all based on the idea of a limited number of adsorption sites, giving rise to a surface 'chemical combination' (Langmuir, 1918). These mechanisms include cases where (i) there is only one kind of adsorbing site, (ii) there is more than one kind of adsorbing site, (iii) the adsorbing surface is amorphous and presents a continuum of adsorbing sites, (iv) each site can accommodate more than one molecule, (v) the adsorption is dissociative and (vi) multilayer adsorption takes place.

Nevertheless, what is usually referred to as the Langmuir model is case 1: that is, 'adsorption on a plane surface having only one kind of elementary space and in which each space can hold only one adsorbed molecule'. It is evident that Langmuir's original model did not allow for either porosity or physisorption. However, the Langmuir model provided a starting point for the development of the BET treatment and of other more refined physisorption isotherm equations. It is therefore appropriate to consider briefly the mechanism of gas adsorption originally proposed by Langmuir (1916, 1918).

The original derivation of Langmuir equation (Langmuir, 1916) is a kinetic one. The adsorbent surface is pictured as an array of N^s equivalent and independent sites for localized adsorption (one molecule per site).

The fraction of sites occupied by N^a molecules is $\theta = \frac{N^a}{N^s}$.

From the kinetic theory of gases, the rate of adsorption is dependent on the pressure and the fraction of the bare sites $(1 - \theta)$. The rate of desorption is dependent on θ and on the energy of activation E (i.e. equivalent to an energy of adsorption expressed as a positive quantity). The equilibrium is obtained for the values of θ and p for which the rates of adsorption and desorption are equal. Thus, the net rate of adsorption is zero:

$$\frac{dN^a}{dt} = \alpha p(1 - \theta) - \beta\theta \exp\left(-\frac{E}{RT}\right) = 0 \tag{5.10}$$

where α and β are characteristic constants for the given gas–solid system.

If, in an ideal case, the probability of desorption of an adsorbed molecule from the surface is independent of the surface coverage (i.e. there are no lateral interactions between the adsorbed molecules), the value of E is constant for a particular adsorption system. Then, Equation (5.10) is applicable over the complete range of monolayer coverage. By rearrangement and simplification of Equation (5.10), we arrive at the familiar Langmuir isotherm equation:

$$\theta = bp/(1 + bp) \tag{5.11}$$

where b, the 'adsorption coefficient', is exponentially related to the positive value of the energy of adsorption, E:

$$b = K \exp(E/RT) \tag{5.12}$$

the pre-exponential factor, K, being equal to the ratio of the adsorption and desorption coefficients, α/β. Alternatively, b may be regarded as a function of the enthalpy and entropy of adsorption (Everett, 1950; Barrer, 1978).

It is evident that Equation (5.11) is of a very general mathematical form (i.e. a hyperbolic function). At low θ, it reduces to Henry's law; at high surface coverage, a plateau is reached as $\theta \to 1$, which corresponds to the completion of the monolayer. Other equations of the same mathematical form as Equation (5.11) have been derived from a classical thermodynamic standpoint (see Brunauer, 1945) and by application of the principles of statistical mechanics (Fowler, 1935).

Equation (5.11) is usually applied in the linear form:

$$p/n = 1/(n_m b) + p/n_m \tag{5.13}$$

where n is the specific amount of gas adsorbed at the equilibrium pressure p and n_m is the monolayer capacity (as before, $\theta = n/n_m$).

Although the Langmuir model is more strictly applicable to an ideal form of chemisorption, there are a few exceptional physisorption systems, giving well-defined plateaux, which can be attributed to monolayer completion. Examples are shown in Figure 5.1a. Here, the surface of an oxidized aluminium foil is sufficiently well screened by the adsorption of the two straight-chain alcohols, $n\text{-}C_3H_7OH$ and $n\text{-}C_5H_{11}OH$, to exhibit 'gas phase autophobicity' (Blake and Wade, 1971) so that multilayer development is energetically inhibited.

This is confirmed in Figure 5.1b by the extensive range of linearity of the Langmuir plots given by the two n-alcohol isotherms. Langmuir plots with short ranges of linearity are obtained with many other systems. However, a second and more important requirement of the theory is that the energy of adsorption should be independent of surface coverage. Thirdly, the differential entropy of adsorption should vary in

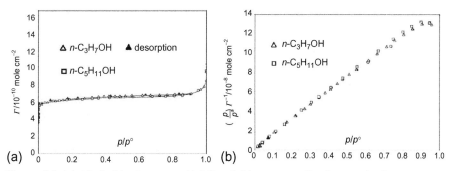

Figure 5.1 (a) Alcohol isotherms on Al foil and (b) corresponding Langmuir plots. After Blake and Wade (1971).

accordance with the ideal localized model (Everett, 1950). That no real physisorption system has been found to satisfy all these requirements is not surprising in view of the complexities noted here and in subsequent chapters. Furthermore, nearly all of the large number of Type I isotherms in the literature are associated with micropore filling rather than monolayer coverage.

5.2.2.2   The Langmuir Equation

It is apparent that the mathematical form of Equation (5.11) is consistent with the general shape of a Type I physisorption isotherm (as shown in Figure 1.10) whose most distinctive feature is a long, horizontal plateau extending up to $p/p^o \rightarrow 1$.

In the context of gas separation (see Yang, 1987) or for some other purposes, it may be expedient to apply an equation of the Langmuir form, but we consider it advisable to treat this as an empirical relation. Thus, we may write:

$$\frac{n}{n_{\mathrm{L}}} = \frac{bp}{1 + bp} \tag{5.14}$$

where the parameters n_{L} and b are to be regarded as empirical constants within stated ranges of p and T.

We should stress again that a Type I isotherm and a linear Langmuir plot do not by themselves imply conformity with the Langmuir model. This is why, in this book, we refer to the application of the Langmuir *equation* rather than the Langmuir *theory*, to make it clear that the use of this mathematical equation does not imply that the assumptions of the theory are justified.

5.2.3   Multilayer Adsorption

5.2.3.1   The Brunauer–Emmett–Teller Theory

In the 1930s, it became evident that the physical adsorption of a vapour is not restricted to monomolecular surface coverage if the relative pressure, p/p^o, is increased above a certain level. In 1937, Emmett and Brunauer came to the empirical conclusion that the beginning of the middle almost linear section of a Type II isotherm (Point B in Figure 5.2) was the point most likely to correspond to monolayer completion.

In attempting to determine the surface area of an iron synthetic ammonia catalyst, Emmett and Brunauer (1937) measured the adsorption isotherms of a number of different gases at or near their respective boiling points (N_2 at $-196\,°C$; O_2, Ar and CO at $-183\,°C$; CO_2 at $-78.5\,°C$; $n\text{-}C_4H_{10}$ at $0\,°C$). Hypothetical surface areas were calculated from each of four characteristic points (A, B, C and D in Figure 5.2) on the assumption that one of these would correspond to a close-packed monolayer. It was found that the closest agreement between the various notional values of surface area was given by the uptakes at Point B and this appeared to be consistent with other experimental evidence – notably the appreciable decrease in differential energy of adsorption in the vicinity of Point B.

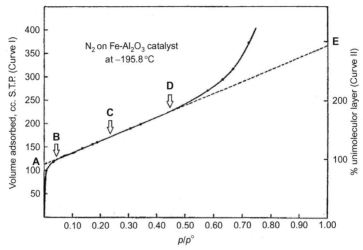

Figure 5.2 Characteristic points on a Type II adsorption isotherm. After Emmett and Brunauer (1937).

By introducing a number of simplifying assumptions, Brunauer et al. (1938) were able to extend the Langmuir mechanism to multilayer adsorption and obtain an isotherm equation (the BET equation), which has the Type II character. The original BET treatment involved an extension of the Langmuir kinetic theory of monomolecular adsorption to the formation of an infinite number of adsorbed layers at the saturation pressure, p^o.

According to the BET model, the adsorbed molecules in one layer can act as adsorption sites for molecules in the next layer and, at any pressure below the saturation vapour pressure p^o, fractions of the surface ($\theta_0, \theta_1, \theta_2, \ldots, \theta_i$) are covered by $0, 1, 2, \ldots, i$ layers of adsorbed molecules (θ_0 of course, represents the fraction of bare surface). It follows that the adsorbed layer is envisaged not to be of uniform thickness, but instead to be made up of random stacks of molecules.

If it is assumed that at equilibrium, characterized by the pressure p, the fractions of bare and covered surface, θ_0 and θ_1, remain constant, we can equate the rate of condensation on the bare surface to the rate of evaporation from the first layer:

$$a_1 p \theta_0 = b_1 \theta_1 \exp\left(-\frac{E_1}{RT}\right) \qquad (5.15)$$

where a_1 and b_1 are adsorption and desorption constants for the first layer and E_1 is the positive value of the so-called energy of adsorption in the first layer. It is assumed that a_1, b_1 and E_1 are independent of the quantity of adsorbed molecules already present in the first layer; that is, as in the Langmuir mechanism, no allowance is made for lateral adsorbate–adsorbate interactions.

In the same way, at the equilibrium pressure p, the fractions of the surface $\theta_2, \theta_3, \ldots, \theta_i$ must also remain constant and we may therefore write:

$$a_2 p \theta_1 = b_2 \theta_2 \exp\left(-\frac{E_2}{RT}\right) \tag{5.16}$$

$$a_3 p \theta_2 = b_3 \theta_3 exp\left(-\frac{E_3}{RT}\right)$$

$$\vdots \qquad \qquad \vdots \tag{5.17}$$

$$\vdots \qquad \qquad \vdots$$

$$\vdots \qquad \qquad \vdots$$

$$a_i p \theta_{i-1} = b_i \theta_i \exp\left(-\frac{E_i}{RT}\right) \tag{5.18}$$

where θ_{i-1} and θ_i represent, respectively, the fractions of surface covered by $i-1$ and i layers; a_i and b_i are adsorption and desorption constants and E_i is the energy of adsorption in the i-th layer.

The sum of the fractions of surface equals unity:

$$\theta_0 + \theta_1 + \cdots + \theta_i + \cdots = 1 \tag{5.19}$$

Moreover, the total adsorbed amount can be expressed as:

$$n = n_m [1\theta_1 + 2\theta_2 + \cdots i\theta_i + \cdots] \tag{5.20}$$

In principle, each adsorbed layer has a different set of values of a_i, b_i and E_i, but the derivation of the BET isotherm equation is dependent on two main assumptions:

a. In the second and all higher layers, the energy of adsorption E_i has the same value as the liquefaction energy, E_L, of the adsorptive (i.e. $E_2 = E_i = E_L$);
b. The multilayer has infinite thickness at $p/p° = 1$ $(i = \infty)$.

Let:

$$\frac{b_2}{a_2} = \frac{b_3}{a_3} = \cdots \frac{b_i}{a_i} = g \tag{5.21}$$

where g is a constant, since all the layers (except the first) have the same properties. We can now express $\theta_1, \theta_2, \ldots, \theta_i$ in terms of θ_0:

$$\theta_1 = y\theta_0, \quad \text{where } y = \frac{a_1}{b_1} p \exp\left(\frac{E_1}{RT}\right) \tag{5.22}$$

$$\theta_2 = x\theta_1, \quad \text{where } x = \frac{p}{g} \exp\left(\frac{E_L}{RT}\right) \tag{5.23}$$

$$\theta_3 = x\theta_2 = x^2\theta_1 \tag{5.24}$$
$$\vdots \qquad \vdots \qquad \vdots$$

$$\theta_i = x^{i-1}\theta_1 = yx^{i-1}\theta_0 \tag{5.25}$$

We may define a constant C:

$$C = \frac{y}{x} = \frac{a_1}{b_1} g \exp\left(\frac{E_1 - E_L}{RT}\right) \tag{5.26}$$

then:

$$\theta_i = Cx^i\theta_0 \tag{5.27}$$

So, we can write:

$$\frac{n}{n_m} = \sum_{i=1}^{\infty} i\,\theta^i = C\sum_{i=1}^{\infty} i\,x^i \cdot \theta_0 \tag{5.28}$$

Taking account of the value of the sum of an infinite geometric progression:

$$\sum_{i=1}^{\infty} x^i = \frac{x}{1-x} \tag{5.29}$$

and of the value of the term $\displaystyle\sum_{i=1}^{\infty} i\,x^i$:

$$\sum_{i=1}^{\infty} i\,x^i = \frac{x}{(1-x)^2} \tag{5.30}$$

and since $\theta = 1 - \displaystyle\sum_{1}^{\infty}\theta_i$:

we obtain from Equations (5.27) and (5.28):

$$\frac{n}{n_m} = \frac{Cx}{(1-x)(1-x+Cx)} \tag{5.31}$$

If at the saturation vapour pressure p^o, the adsorbed layer is assumed to be of infinite thickness, it follows that $x=1$ and (from Equation 5.23) that $x=p/p^o$ so that:

$$\frac{n}{n_m} = \frac{C(p/p^o)}{(1-p/p^o)(1-p/p^o + C(p/p^o))} \tag{5.32}$$

When n/n_m is plotted against p/p^o at constant C, Equation (5.32) gives a curve of Type II shape (see Figure 5.3), provided that $C>2$. It is evident that the shape of the knee (i.e. in the vicinity of $n/n_m=1$) is dependent on the value of C, becoming sharper as C is increased. When $C<2$, but still positive, Equation (5.32) gives a curve with no inflexion point and having the general shape of a Type III isotherm.

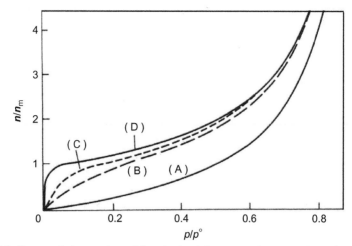

Figure 5.3 Curves of n/n_m against p/p^o, calculated from Equation (5.32) for different values of C: (A) $C=1$, (B) $C=10$, (C) $C=100$, and (D) $C=10,000$. From Gregg and Sing (1982).

Equation (5.32) can be written in the usual linear form:

$$\frac{p}{n(p^o - p)} = \frac{1}{n_m C} + \frac{C-1}{n_m C} \cdot \frac{p}{p^o} \tag{5.33}$$

This 'linear transformed BET equation' provides the basis for the BET plot of experimental isotherm data in the form of $p/[n(p^o - p)]$ versus p/p^o.

The constant C, which is strictly given by Equation (5.26), was assumed to be exponentially related to E_1 by the simplified equation:

$$C \approx \exp\left(\frac{E_1 - E_L}{RT}\right) \tag{5.34}$$

Here, E_1 is defined as a positive quantity and interpreted by Brunauer (1945) as the 'average heat of adsorption on the *less active* part of the adsorbing surface'. Originally $E_1 - E_L$ was known as the 'net heat of adsorption' (Lamb and Coolidge, 1920). It is now recommended that the more general term *net molar energy of adsorption* should be adopted (cf. Chapter 2).

An alternative and elegant derivation of the BET equation is by a statistical mechanical treatment (Hill, 1946; Steele, 1974). The adsorbed phase is pictured as a lattice gas: that is, molecules are located at specific sites in all layers. The first layer is localized and these molecules act as sites for molecules in the second layer, which in turn acts as sites for molecules in the third layer and so on for the higher layers. As the surface is assumed to be planar and uniform, it follows that all surface sites are identical. It is also assumed that the occupation probability of a site is independent of the occupancy of neighbouring sites. This is equivalent to the assumption that there are no lateral interactions between adsorbed molecules. In accordance with the BET model, the probability for site occupation is zero unless all its underlying sites are occupied. Furthermore, it is assumed that it is only the molecular partition function for the first layer which differs from that for molecules in the liquid state.

The statistical thermodynamic treatment of the BET theory has the advantage that it provides a satisfactory basis for further refinement of the theory by, say, allowing for adsorbate–adsorbate interactions or the effects of surface heterogeneity. By making the assumptions outlined above, Steele (1974) has shown that the problems of evaluating the grand partition function for the adsorbed phase could be readily solved. In this manner, he arrived at an isotherm equation, which has the same mathematical form as Equation (5.33). The parameter C is now defined as the ratio of the molecular partition functions for molecules in the first layer and the liquid state.

As mentioned earlier, the mathematical form of Equation (5.32) gives a curve having the general shape of a Type II isotherm, which has also been called an S-shaped, or sigmoid, isotherm (Brunauer, 1945). This is illustrated in Figure 5.4 in which the results for the adsorption of nitrogen are also plotted in the reduced form of n/n_m for a number of non-porous samples of silica and alumina (Gregg and Sing, 1982).

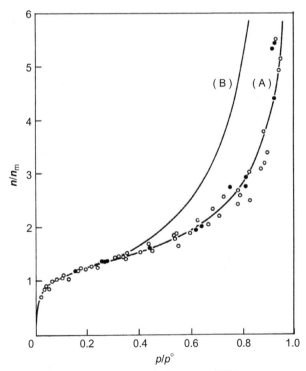

Figure 5.4 Curve (A): adsorption isotherms of nitrogen at 77 K on non-porous silicas (a (BET) from 2.6 to 11.5 m^2 g^{-1}) and aluminas (a (BET) from 58 to 153 m^2 g^{-1}). Curve (B): adsorption isotherm calculated from Equation (5.32), with C values of 100–200. From Gregg and Sing (1982).

It is evident that at $p/p^o > 0.3$, the theoretical BET curve (B) deviates widely from the common experimental curve (A).

If the adsorption at saturation is restricted to a finite number of layers, N, the BET treatment leads to a modified equation which includes this additional parameter (see Chapter 7). Naturally, in the special case when $N = 1$, the extended BET equation corresponds to the Langmuir equation, with p/p^o in place of p in Equation (5.14) and C replacing b.

As we have seen, the BET model is based on the following assumptions: (i) that adsorption takes place on an array of fixed sites, (ii) that the sites are energetically identical and (iii) that no allowance is made for lateral interactions between adsorbed molecules. Extensive adsorption calorimetric measurements (Beebe and Young, 1954; Isirikyan and Kiselev, 1961; Grillet et al., 1979) and simulation studies (Nicholson and Parsonage, 1982; Seri-Levi and Avnir, 1993) on a wide variety of gas–solid systems have shown that real physisorption systems do not behave in this simple way. Most porous and non-porous adsorbents are known to be energetically heterogeneous. Also the evidence for adsorbate–adsorbate interactions on more uniform surfaces is firmly established. Indeed, the inclusion of the lateral interactions appears to flatten the BET stacks into more realistically shaped islands (Seri-Levi and Avnir, 1993). Furthermore, it is generally recognized that the parameter C cannot provide a reliable evaluation of E_1.

In spite of the weakness of the underlying theory, the BET *equation* remains the most widely used of all adsorption isotherm equations. It is extensively used for determining the surface area of porous and non-porous adsorbents. The reasons for the universal application of the *BET method* are discussed in Chapter 7.

5.2.3.2 Multilayer Equations

An extension to the BET model was put forward by Brunauer, Deming, Deming and Teller (BDDT) in 1940. The BDDT equation contains four adjustable parame ters and was designed to fit the isotherm Types I–V. From a theoretical standpoint, the BDDT treatment appears to offer very little more than the original BET theory and nowadays the cumbersome equation is rarely applied to experimental data.

Several other attempts have been made to modify the BET equation in order to improve the agreement with isotherm data in the multilayer region. Brunauer et al. (1969) pointed out that the BET assumption of an infinite number of molecular layers at saturation pressure is not always justified. By replacing p by kp, where k is an additional parameter with a value less than unity, they arrived at the following equation, which has the same form as that originally proposed by Anderson (1946):

$$\frac{kp}{n(p^o - kp)} = \frac{1}{n_m C} + \frac{(C-1)}{n_m C} \frac{kp}{p^o} \tag{5.35}$$

On an empirical basis, this Anderson–Brunauer equation can be applied to some isotherms (e.g. nitrogen and argon at 77 K on various non-porous oxides) over a much wider range of p/p^o than the original BET equation.

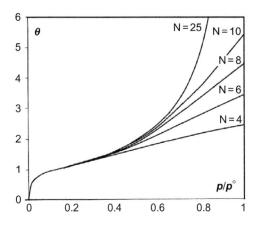

Figure 5.5 Adsorption isotherms calculated for a limiting number of molecular layers ranging from 4 to 25, with the assumption that $C = 100$ (Rouquerol et al., 2003).

These and other modifications of the original BET equation (see Young and Crowell, 1962) have not turned out to be very useful. The main purpose of the equation is for the determination of *surface area*, which does not require any adsorption data above ca. $p/p^{o} = 0.4$ (see Chapter 7). Figure 5.5 shows that changing the limiting number of molecular layers in the BDDT equation only alters the portion of the adsorption isotherm above $p/p^{o} \approx 0.4$.

Conversely, the determination of the *mesopore size distribution* requires a knowledge of the multilayer thickness up to the saturation pressure (see Chapter 8), and for this purpose, a more empirical approach is generally adopted.

When the adsorbate reaches a thickness of several molecular layers, the effects of surface heterogeneity are considerably reduced. If the temperature is not too low, some – but not all – multilayers appear to undergo a continuous increase in thickness as the pressure approaches saturation and bulk behaviour is gradually developed (Venables et al., 1984). With such systems, it seems reasonable to assume that the molar entropy of the thick multilayer is the same as that of the bulk liquid. In that case, the course of the isotherm would be determined solely by the energy of adsorption.

By assuming that the dispersion interactions are dependent on z^{-3} (i.e. by integration of the additive r^{-6} terms), Hill (1952) arrived at the multilayer isotherm equation:

$$\ln(p^{o}/p) = k/\theta^{3} \qquad (5.36)$$

where k is a constant for a given gas–solid system.

Because of the experimental difficulty of making accurate measurements at very high p/p^{o}, it is not easy to verify this expression. However, a more general equation:

$$\ln(p^{o}/p) = k/\theta^{s} \qquad (5.37)$$

in which the parameter s is a non-integer, has been found to be more widely applicable. Although it was first proposed by Halsey (1948), Equation (5.37) is now known as the Frenkel–Halsey–Hill (FHH) equation.

For nitrogen adsorbed at 77 K, this equation becomes:

$$t/\text{nm} = 0.354(5/\ln(p^\circ/p))^{1/3} \tag{5.38}$$

The thickness of the nitrogen multilayer adsorbed at 77 K has also been assessed by application of the Harkins and Jura (1944) equation in the form:

$$\frac{t}{\text{nm}} = \left(\frac{13.99}{0.034 - \log_{10}(p/p^\circ)}\right)^{1/2} \tag{5.39}$$

In a molecular simulation and theoretical study of argon on graphite, Steele (1980) put forward modified forms of FHH equation, which appeared to be applicable over an extensive range of multilayer coverage.

Nitrogen isotherms (at 77 K) on non-porous oxides and carbons have been found to give linear FHH plots over a fairly wide range of p/p°, corresponding to ca. 1.5–3 molecular layers (Carrott and Sing, 1989). With these systems, the individual values of s are remarkably constant. Other adsorptives, such as hydrocarbons and water vapour, give values of s which appear to depend on the surface structure of the adsorbent (see Chapters 10 and 11).

Rudzinski and Everett (1992) drew attention to a possible relation between the value of s and the degree of surface heterogeneity. The interesting suggestion is made that an increase in heterogeneity is likely to extend the range of influence of the surface on the multilayer structure, consequently causing a reduction in the value of s. However, more evidence is required to test this hypothesis.

In spite of its empirical origin, the FHH equation does provide a means of identifying the different effects of pore filling (Carrott et al., 1982). Capillary condensation in mesopores necessarily restricts the range of linearity of the FHH plot and tends to reduce the value of s. Although micropore filling may not significantly affect the linearity of the FHH plot, it does lead to an increase in s.

5.2.4 The Dubinin–Stoeckli Theory: Filling of Micropores

Dubinin pioneered the concept of micropore filling. His approach was based on the early potential theory of Polanyi, in which physisorption isotherm data were expressed in the form of a temperature-invariant 'characteristic curve'. The essential parameter is the quantity E defined by the expression:

$$E = RT \ln(p^\circ/p) \tag{5.40}$$

where E was originally termed the adsorption potential (see Brunauer, 1945), but as explained in Chapter 2, we should more strictly regard E as a difference in chemical potential.

Dubinin followed Polanyi in presenting the characteristic curve for a given adsorbent in the form of the function $E = f(v)$ as illustrated in Figure 5.6, where v is the liquid volume adsorbed (assuming the adsorbate to have the bulk liquid density at the operational temperature). In place of the Polanyi picture of

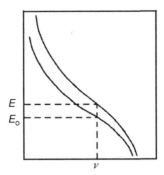

Figure 5.6 Hypothetical characteristic curves for two vapours.
After Lowell et al. (2004).

layer-by-layer adsorption, Dubinin developed the idea of the volume filling of micropores.

If two vapours fill the same available micropore volume, it is assumed that ratio of their adsorption potentials, $\beta = E/E_o$, is constant for the given microporous adsorbent. The scaling factor β was termed the 'affinity coefficient', with E_o taken as the reference value, as indicated in the two hypothetical characteristic curves in Figure 5.6.

In 1947, Dubinin and Radushkevich put forward an equation for the characteristic curve in terms of the fractional filling, v/v_p, of the micropore volume, v_p. This is based on the assumption that the micropore size distribution is Gaussian so that:

$$\frac{v}{v_p} = \exp\left[-kE^2\right] \tag{5.41}$$

where k is another characteristic parameter.

The isotherm equation is obtained by combining Equations (5.40) and (5.41) and the introduction of the scaling factor, β, it becomes:

$$\frac{v}{v_0} = \exp\left[-(RT\ln(p^o/p))^2/(\beta E_o)^2\right] \tag{5.42}$$

where E_o is now termed the 'characteristic energy'.

Another parameter, the 'structural constant', B, is defined a:

$$B = 2.303(R/E_o)^2 \tag{5.43}$$

Rearrangement of Equation (5.42) gives the Dubinin–Radushkevich, DR, equation in its usual form:

$$\log_{10}\frac{v}{v_0} = -D\log_{10}^2\left(\frac{p^o}{p}\right) \tag{5.44}$$

where D is given by the equation:

$$D = 0.434B(T/\beta)^2 \tag{5.45}$$

According to Equation (5.44), a linear relationship should be obtained between $\log_{10}$ (n) and $\log_{10}^2(p^\circ/p)$. In fact, a number of microporous carbons have been found to give linear DR plots over a wide range of p/p° (Dubinin, 1966), but in many other cases, the linear region is restricted to a very limited range of low p/p° (Gregg and Sing, 1982; Carrott et al., 1987). DR plots obtained with zeolites are generally non-linear over virtually the complete isotherm (Dubinin, 1975).

The mathematical relationship between the DR and the Langmuir-BET equations was examined by Gregg and Sing (1976). Thus, DR plots were constructed from a set of hypothetical Langmuir isotherms for different values of the BET parameter, C. As expected, most of the hypothetical DR plots were curved, but an almost linear relation was obtained with $C = 18$ and $D = 0.18$. This analysis revealed that it is possible to obtain linearity with both Langmuir and DR plots for certain micropore-adsorptive systems, but of course this does not validate the applicability of either model. Furthermore, a limited range of linearity of a DR plot is not always associated with micropore filling (Kaganer, 1959).

To allow for the deficiencies of the DR equation, Dubinin and Astakhov (1970) put forward a more general form of characteristic curve:

$$v/v_0 = \exp\left[-(A/E)^N\right] \tag{5.46}$$

where N is another empirical constant.

Dubinin (1975) reported values of N between 2 and 6. Some molecular sieve carbons and zeolites gave $N = 3$. However, in view of the empirical nature of N, it is not surprising to find that usually the 'best' values are not integers. The particular value of N may also depend on the range of the isotherm and the operational temperature.

To overcome these difficulties, Stoeckli (1977) and Stoeckli et al. (1979) suggested that the original DR equation only holds for a carbon with a narrow micropore size distribution. According to this view, the overall isotherm on a heterogeneous microporous solid is made up of the sum of the contributions from the different groups of pores. Thus:

$$v = \sum_j v_{0,j} \exp\left[-B_j(T/\beta)^2 \log_{10}^2(p^\circ/p)\right] \tag{5.47}$$

where $v_{0,j}$ represents the pore volume of the j-th group.

For a continuous distribution, the summation is replaced by integration and then:

$$v(y) = \int_0^\infty f(B) \exp[-By] dB \tag{5.48}$$

where $f(B)$ is a micropore size distribution function and:

$$y = (T/\beta)^2 \log_{10}^2(p^\circ/p) \tag{5.49}$$

By assuming a Gaussian pore size distribution, Stoeckli was able to simplify Equation (5.47) and obtain an isotherm equation in which the distribution function, $f(B)$, was expressed in an analytical form (Huber et al., 1978; Bansal et al., 1988). In principle, $f(B)$ provides an elegant basis for relating the micropore size distribution to the adsorption data. However, it must be kept in mind that the validity of the approach rests on the assumption that the DR equation is applicable to each pore group and that there are no other complicating factors such as differences in *surface* heterogeneity.

5.2.5 Type VI Isotherms: Phase Changes in Physisorbed Layers

Since physisorbed molecules interact with each other, we would expect to find 2D states and phase changes which are analogous to those found in 3D condensed matter (Gregg, 1961). However, we must keep in mind that the structure of a physisorbed layer is dependent, not only on the adsorbate–adsorbate interactions, but also on the magnitude and disposition of the adsorbent–adsorbate interactions.

It can be seen that 2D phase changes are unlikely to be easily detectable unless the adsorbent surface is chemically and physically uniform. The early literature contains a number of claims for the existence of isotherm discontinuities, which appeared to be associated with 2D phase changes. Many of these discontinuities were subsequently found to be spurious and were probably due to faulty technique (see Young and Crowell, 1962). However, the adsorption of CH_4 on MoS_2 was first shown by Bonnetain et al. (1952) to give a genuine stepwise isotherm. The existence of first-order 2D phase changes was also firmly established by the work of Ross and Clark (1954) in which isotherms of C_2H_6 on cubic crystals of NaCl were measured at different temperatures.

Most of the well-documented investigations of 2D phase changes have featured the adsorption of Ar, Kr and Xe on the basal (0001) face of graphite (see Dash, 1975; Suzanne and Gay, 1996), but detailed work has been undertaken also on a number of other systems such as Ar, Kr and CH_4 on layered halides (Larher, 1992) and cubic crystals of MgO (Coulomb et al., 1984). In addition, phase diagrams have been constructed for the adsorption of certain polar molecules on graphite (Terlain and Larher, 1983).

In the systematic investigations undertaken by Thomy and Duval (1970) and Thomy et al. (1972) of krypton adsorption on exfoliated graphite, a series of isotherms was determined over the temperature range of 77–100 K. Stepwise multilayer character was displayed up to four molecular layers, as can be seen in Figure 5.7. Each vertical 'riser' (at constant p/p^o) can be regarded as a phase transition between one adsorbed layer and the next higher layer.

In the sub-monolayer range, three distinctive regions were identified and attributed by Thomy and Duval to 2D 'gas', 'liquid' and 'solid' phases. These measurements provided the first unambiguous evidence for the existence of sub-steps in the monolayer region of a stepwise, Type VI, isotherm.

To illustrate the interpretation of such sub-steps, the monolayer isotherms for the adsorption of Xe on $FeCl_2$ (Larher, 1992) are shown in Figure 5.8. At temperatures below 99.57 K, there is a single vertical step, which corresponds to the transformation of 2D gas to the solid. Very little further compression of the monolayer is possible before its completion at Point B. A smaller sub-step becomes apparent at temperatures

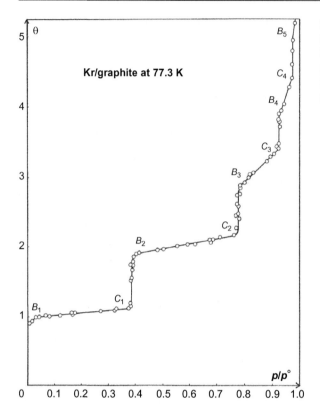

Figure 5.7 Complete adsorption isotherm of krypton on exfoliated graphite at 77.3 K. Courtesy Thomy et al. (1972).

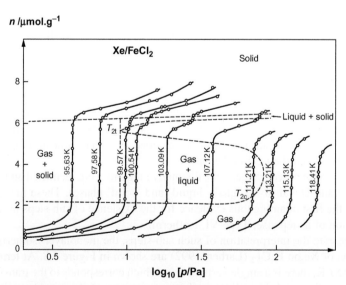

Figure 5.8 Adsorption isotherms of xenon on $FeCl_2$.
Courtesy Larher (1992).

above 99.57 K. As a result of the careful studies of Thomy and Duval, and Larher, the consensus interpretation is that this small sub-step represents a first-order transition between the 2D liquid and solid phases. It is evident that in the case of the Xe/FeCl$_2$ system, 99.57 K is the 2D triple point.

2D phase diagrams are often displayed in the form of $\ln\{p\}$ against $1/T$ (at a constant specific amount adsorbed), which provides a convenient way of indicating the conditions for the coexistence of two phases (see Figure 5.9). Indeed, the application of the Phase Rule indicates that when two adsorbed phases coexist in equilibrium, the system has one degree of freedom: therefore, at constant temperature, the pressure must also remain constant. The coexistence of three adsorbed phases requires the loss of the one degree of freedom and the system then becomes invariant at the 2D triple point.

The limiting temperature for 2D condensation is the 2D critical temperature, T_{2c}, which marks the upper limit for the first-order transition associated with the formation of a 'liquid' monolayer. According to Larher (1992), for the Xe/FeCl$_2$ system in Figure 5.3, $T_{2c} = 112$ K.

Larher (1992) discussed certain aspects of the 2D phase behaviour of different systems. If the adsorbate–adsorbate interaction is not perturbed to any significant extent by the adsorbent, the law of corresponding states should apply to 2D systems. In that case, we would expect the ratio of the 2D and 3D critical temperatures, T_{2c}/T_{3c}, to be constant. It turns out that values close to $T_{2c}/T_{3c} = 0.39$ are obtained for Ar, Kr and Xe on certain layered chlorides (e.g. NiCl$_2$) and also for Ar, and Xe on the basal face of graphite. Larher concludes that the corresponding value of T_{2c} (i.e. $0.39\ T_{3c}$) can be regarded as an 'ideal' critical temperature for 2D condensation on a smooth planar surface.

Higher values of T_{2c}/T_{3c} in the range 0.50–0.55 were reported for the adsorption of Ar on some other dihalides such as CdBr$_2$ and FeI$_2$ (Larher, 1992). In these cases, there appeared to be a somewhat larger degree of incompatibility between the lattice parameters of the crystalline surface and the densest plane – the (111) plane – of the rare gas

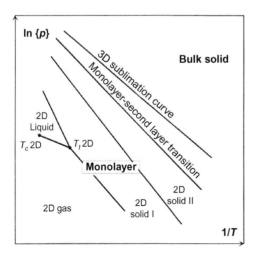

Figure 5.9 Phase diagram for physisorbed rare gas atoms or molecules on uniform solid surfaces such as the basal plane of graphite.
After Suzanne and Gay (1996).

crystal. Larher suggests that, within a certain range of size incompatibility, the most useful quantitative description of monolayer condensation is by a lattice gas model.

Various particle scattering, electron and neutron diffraction and electron spectroscopic techniques have been used to study the structure of physisorbed monolayers (Chiarello et al., 1988; Block et al., 1990; Layet et al., 1993; Bienfait et al., 1997). Heat capacity measurements have also provided strong evidence for the development of different 2D solid structures. Epitaxial monolayers were reported for some systems (see Dash, 1975) in which the adsorbed atoms are arranged in regular patterns in registry with adsorbent structure, as in Figure 5.10.

A number of different 'commensurate' and 'incommensurate' monolayer structures have been identified (Suzanne and Gay, 1996). In the important case of the graphite (0001) face, the commensurate hexagonal structure can be compared with 3D (111) plane of the fcc structure of the rare gases. The lattice mismatch is small for Xe and Kr and much larger for Ar and Ne. The compression of the monolayer from a localized epitaxial state to the close-packed state was suggested as the origin of sub-steps given by Xe and Kr on graphite (Price and Venables, 1976). The argon sub-step at 77 K is more likely to be due to a transformation from the 2D liquid to the solid (as in Figure 5.6). The latter system is discussed in more detail in Chapter 10.

Knowledge of the effective molecular area is of particular importance when gas adsorption is used for the determination of surface area (see Chapter 7). It is often assumed that the completed monolayer is in a 'liquid' close-packed state, but it is now apparent that we must question the soundness of this assumption. The study of phase transitions and monolayer structures leads us to the conclusion that the degree of molecular packing is dependent on the adsorption system (both adsorbent and adsorptive) and the operational conditions of pressure and temperature (Steele, 1996).

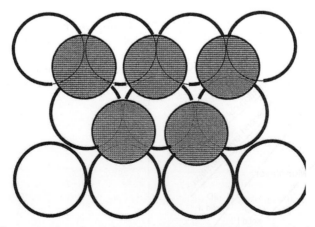

Figure 5.10 Representation of an epitaxial monolayer (black circles), that is, in registry with the adsorbent structure.

5.2.6 Empirical Isotherm Equations

Many applications of physisorption (e.g. industrial gas separation or storage, pollution control) require the interpolation or extrapolation of experimental equilibrium data. Even with the aid of computer-aided techniques for curve fitting, for the analysis of chemical engineering data, it is often useful to set up the isotherm for each component in the form of a relatively simple equation. For this purpose, the Langmuir or DR equations have been adopted on an empirical basis within prescribed limits of pressure and temperature. A few other frequently used empirical isotherm equations are briefly described in this section.

5.2.6.1 Freundlich Equation

The first and best known empirical equation was proposed by Freundlich (1926) in the form

$$\boldsymbol{n} = kp^{1/m} \tag{5.50}$$

where k and m are constants $(m > 1)$.

According to Equation (5.50), the plot of $\ln\{\boldsymbol{n}\}$ against $\ln\{\boldsymbol{p}\}$ should be linear. In general, activated carbons give isotherms which obey the Freundlich equation in the middle range of pressure (Brunauer, 1945), but the agreement is usually poor at high pressures and low temperatures. These limitations are partly due to the fact that the Freundlich isotherm does not give a limiting value of $\boldsymbol{n}$ as $p \to \infty$.

5.2.6.2 Sips (or Langmuir–Freundlich) Equation

It is possible to achieve an improved fit at higher pressures by a combination of the Freundlich and Langmuir equations (Sips, 1948),

$$\boldsymbol{n}/\boldsymbol{n}_{\mathrm{L}} = (kp)^{1/m} / \left[1 + (kp)^{1/m} \right] \tag{5.51}$$

where $\boldsymbol{n}_{\mathrm{L}}$ is the limiting adsorption capacity. Equation (5.51) has been applied as a 'generalized Freundlich' isotherm to multisite occupancy by long-chain hydrocarbons (Rudzinski and Everett, 1992). However, as in the case of the Freundlich isotherm itself, the Sips equation does not reduce to Henry's law as $p \to 0$.

5.2.6.3 Toth Equation

Another empirical variant is the Toth equation (1971):

$$\boldsymbol{n}/\boldsymbol{n}_L = p/(b + p^m)^{1/m} \tag{5.52}$$

which also contains three adjustable parameters ($\boldsymbol{n}_{\mathrm{L}}$, b and m) but has the advantage that it appears to give the required limits for both $p \to 0$ and $p \to \infty$. Thus, although it was originally proposed for monolayer adsorption (Toth, 1962), the Toth equation actually gives a more extensive range of fit when applied to Type I isotherms (Rudzinski and Everett, 1992).

5.2.6.4 Langmuir Multisite Equation

Although, as we have seen, Langmuir's name is connected with a model based on a single type of adsorption sites, he envisaged five other adsorption mechanisms, including that of adsorption on a surface containing several kinds of adsorption sites (Langmuir, 1918). A similar mechanism was proposed by others (see, for instance, Rudzinski and Everett, 1992) and several authors have pointed out that the following expression, written for each kind of site i, can give quite good agreement with experimental data (Chowdhury et al., 2012; Hamon et al., 2012):

$$\theta = b_i p / (1 + \Sigma b_i p) \tag{5.53}$$

Another approach, which is commonly used, involves a form of the Dual-Site Sips equation (Bloch et al., 2012; McDonald et al., 2012; Plaza et al., 2012).

5.2.6.5 Jensen–Seaton Equation

Jensen and Seaton (1996) drew attention to the effect of adsorbate compressibility *at high pressures* (say, above 100 bar). Thus, when the adsorption isotherm is plotted in the form of amount adsorbed, it should *not* reach a plateau but should instead continue to rise (but not if the surface excess amount is used, see Section 2.2.2) They also allowed for the Henry's law requirement at low pressure. By taking account of these two conditions, Jensen and Seaton proposed a semi-empirical isotherm equation limited by two asymptotes: a lower one reflecting the Henry's law at low pressure and an upper one reflecting the compressibility κ of the adsorbate at high pressure. Their equation is:

$$n^a = Kp \left[1 + \left(\frac{Kp}{a(1 + \kappa p)} \right)^c \right]^{-1/c} \tag{5.54}$$

where K is the Henry's law constant and c is a positive empirical constant. If the compressibility of the adsorbate $\kappa = 0$, the equation takes the form of a Toth equation and if, in addition, $c = 1$, it becomes an empirical Langmuir equation.

Equation (5.54) provides a good fit for the systems for which it was developed, that is, particular microporous adsorbents up to high pressure.

5.2.6.6 Combination of Local Isotherms

There is a growing interest in the presentation of physisorption isotherms in a generalized integral form. This approach was first applied to physisorption in the submonolayer region (Adamson et al., 1961), but much of the current interest is centred on the analysis of micropore filling isotherms. An apparent advantage is that it provides a means of constructing a series of model isotherms by systematically combining various hypothetical 'local' isotherms with either notional or computed energy distribution functions.

The starting point is an expression for the overall adsorption, $n(p, T)$, in the form:

$$n(p,T) = n_0 \int \theta(p,T,E) f(E) \mathrm{d}E \tag{5.55}$$

where n_0 represents either the monolayer capacity or the micropore capacity, $\theta(p, T, E)$ represents the local isotherm corresponding to the adsorption energy E and $f(E)$ is the distribution function for the adsorption energy over a specified range of energy.

Stoeckli (1993) pointed out that the Dubinin–Astakhov equation (5.43) can be derived from Equation (5.55), but Mc Enaney (1988) and others (e.g. Jaroniec et al., 1997) have drawn attention to the difficulty in arriving at an unambiguous interpretation of the energy distribution function. A comprehensive review of the significance and application of Equation (5.55) was given by Rudzinski and Everett (1992).

5.3   Adsorption of a Gas Mixture

Although academic work has been largely focused on the mechanisms of pure gas adsorption, many related industrial applications involve gas mixtures and therefore complex co-adsorption phenomena. The variety of components and concentrations makes experimental screening extremely time demanding. This is why predictive models of co-adsorption are most welcome, especially if they are based on the extensive literature data for pure gas adsorption. The most commonly used models are briefly described in this section.

5.3.1   Extended Langmuir Model

In this model, the basic assumptions of the Langmuir theory are applied to each component: localized adsorption on a set of identical adsorption sites, with same energy of adsorption for all molecules of a given gas and with no interaction between the adsorbed molecules.

As in the derivation by Langmuir, the rate of *adsorption* $R_{a,i}$, for a particular component, i, is assumed to be proportional to its partial pressure in the gas phase and to the fraction of vacant sites:

$$R_{a,i} = k_{a,i}\, p_i \left(1 - \sum_{j=1}^{N} \theta_j \right) \tag{5.56}$$

where θ_j is the fractional coverage of each species j, p_i is the partial pressure of species i in the gas phase and $k_{a,i}$ is its adsorption rate constant.

An interesting simplification is that the rate of *desorption* of component i is assumed to be proportional to only its fractional loading, that is, to be independent from the fractional loadings of the other species (after the Langmuir model, there are no interactions between adsorbed molecules):

$$R_{d,i} = k_{d,i}\,\theta_i \tag{5.57}$$

Equating the rates of adsorption and desorption and summing for all the components yields the following equation for the total fractional coverage θ_T in terms of the partial pressures of all the components:

$$\theta_T = \frac{\sum_{j=1}^{N} b_j p_j}{1 + \sum_{j=1}^{N} b_j\,p_j} \tag{5.58}$$

where b_j is the ratio of the adsorption and desorption rate constants, just as in the single gas model ($b_j = k_{a,j}/k_{d,j}$) and is a measure of the component's affinity with the adsorbent. It is unaffected by the presence of other species in the mixture and can therefore be taken from the pure gas model.

Once the total fractional loading has been found, the fractional loading for each individual component can be calculated from:

$$\theta_i = \frac{b_i p_i}{1 + \sum_{j=1}^{N} b_j\,p_j} \tag{5.59}$$

or

$$n_i^a = n_{m,i}^a\,\frac{b_i p_i}{1 + \sum_{j=1}^{N} b_j\,p_j} \tag{5.60}$$

in terms of the amount adsorbed per unit mass of adsorbent.

This equation is known in the literature as the Extended Langmuir (EL) equation and is the simplest model that predicts multicomponent adsorption as it requires only the affinity constants b_i and the monolayer capacities $n_{m,i}^a$ for the pure components. Thermodynamic consistency would require the same monolayer capacity $n_{m,i}^a$ for all species (i.e. $n_{m,i}^a = n_{m,j}^a$). However, when this is not the case and when one is ready to accept a somewhat imperfect fit, it is still apparently possible to estimate a mean monolayer capacity from:

$$\frac{1}{n_m^a} = \sum_{i=1}^{N} \frac{x_i}{n_{m,i}^a} \tag{5.61}$$

where x_i is the mole fraction of the adsorbed component, i:

$$x_i = \frac{n_i^a}{\sum_{j=1}^{N} n_j^a} \tag{5.62}$$

Conversely, if the monolayer capacities are identical for all components, one can assess each affinity constant in order to improve the fit of the corresponding individual isotherm (Qiao et al., 2000).

The EL equation has the advantages of (i) requiring only the appropriate data for adsorption of the pure components and therefore easy to obtain and (ii) being easy to use. Its limitations come (i) from the limited pressure range in which the fit with the experimental isotherm is at its best and (ii) from the fact that a good fit requires that the saturation capacity is the same for all the adsorbed species. Among various improvements, the 'Multi-region EL' proposed by Bai and Yang (2001) takes into account differences in saturation capacities, which are found with many systems.

5.3.2   Ideal Adsorbed Solution Theory

The other most popular model for gas co-adsorption was developed by Myers and Prausnitz (1965) under the name of 'Ideal Adsorbed Solution Theory' (IAST). Like the EL model, it only needs previous knowledge of the adsorption isotherms for single gases, but these two approaches have not much else in common. The novelty in Myers and Prausnitz approach is that the equilibrium between an adsorbed phase and a gas phase is considered to be similar to the equilibrium between a liquid solution and a vapour phase.

In the simplest IAST model, the first assumption is to consider the adsorbate as behaving like *an ideal solution* in equilibrium with the gas phase. The resulting equilibrium equation is analogous to Raoult's law for vapour–liquid equilibria, that is, the concentration of each component in the gas phase is proportional to its concentration in the adsorbed phase. A second assumption is that the spreading pressure of each component π_i, as defined by the Gibbs adsorption equation, is the same as that of the mixture (i.e. π). These two assumptions lead to the following two equations:

$$p.y_i = x_i \, p_i^o(\pi) \tag{5.63}$$

$$\frac{A \, \pi}{RT} = \frac{A \, \pi_i}{RT} = \int_0^{p_i^o} \frac{n_i^a}{p_i} \, dp_i \tag{5.64}$$

where x_i and y_i are the mole fractions in the adsorbed and fluid phase, respectively, and $p_i^o(\pi)$ is the hypothetical pressure, in the gas phase, of the pure component that gives the spreading pressure π. Since two such equations are needed for each component, it follows that for N components $2N$ equations are needed. The number of unknowns is $2N+1$ (Nx_i, Np_i^o and π). The additional equation required to solve these equations is the mole balance:

$$\sum_{i=1}^{N} x_i = 1 \tag{5.65}$$

This set of equations can be solved, either analytically or more frequently numerically, to give the mole fraction x_i and the hypothetical pure component pressure in the gas phase p_i^o. The corresponding amount $n_i^{a,o}$ of pure component in the adsorbed phase is derived from the pure component isotherm equation. A great advantage of IAST, as compared with the EL method, is that it accommodates any type of equation for the pure component isotherm (e.g. Langmuir, Freundlich or Toth). Finally, the total amount adsorbed n_T^a can be calculated by:

$$\frac{1}{n_T^a} = \sum_{i=1}^{N} \frac{x_i}{n_i^a}$$
(5.66)

whereas the amount adsorbed of component i is given by:

$$n_i^a = x_i n_T^a$$
(5.67)

When the behaviour of the adsorbed phase cannot be considered to be ideal, it is possible to take into account the surface heterogeneity by a local application of IAST followed by an integration over all types of adsorption sites: this is the *Heterogeneous IAST* (Valenzuela et al., 1988). One can also calculate the activity coefficients of the equivalent 'real adsorbed solutions', with help of the Real Adsorbed Solution Theory (Dunne and Myers, 1994). These models and others are presented and discussed in detail by Do (1998), Heymans (2011) and Wiersum (2012).

5.4 Conclusions

In view of the complexity of physisorption mechanisms and the heterogeneity of most solid surfaces and pore structures, it is not surprising to find that all the theoretical models summarized in this chapter have practical limitations of one kind or another. Furthermore it must be re-emphasized that the range of fit of a particular equation is not enough by itself to establish the validity of the underlying theory.

In the range of very *low pressure*, adsorption takes place on the most active sites on the surface or within very narrow pores. At somewhat higher pressures, the less active sites are occupied and/or the wider micropores are filled. The molecular interactions are dependent on the adsorption system, surface composition and the micropore structure. Therefore, no relatively simple theoretical treatment can be expected to fit the entire adsorption isotherm or to be applicable equally well to both Type I and Type II isotherms.

The determination of the energy of adsorption is the most direct way of studying surface heterogeneity, but as adsorption calorimetry is experimentally more demanding than the measurement of the isotherm, this approach has inevitably attracted less attention. However, as will become evident in subsequent chapters, there is much to be gained by employing the two experimental techniques in combination.

For the characterization of adsorbents, the BET equation must still be regarded as the most important of all the physisorption equations considered in this chapter. At first sight, it is a curious paradox that the apparent success of the BET *method* for surface area determination does not depend on the validity of the BET *theory*. As will be illustrated in Chapter 7, its simplicity makes the BET method convenient and easily applicable to any adsorption system. However, although the method has come to be accepted as an internationally approved procedure, certain conditions must be satisfied before the significance and reproducibility of the derived BET areas can be established. For this purpose, empirical methods of isotherm analysis are extremely useful. These procedures are introduced in Chapters 7–9.

The pragmatic approach adopted in subsequent chapters involves the identification and characterization of the various mechanisms of adsorption and pore filling. By making use of standard data on well-defined adsorbents and proceeding step by step, we can extract the maximum amount of useful information without having to rely completely on any of the oversimplified models discussed in this chapter.

References

Adamson, A.W., Ling, I., Datta, S.K., 1961. Adv. Chem. Ser. 33, 62.

Anderson, R.B., 1946. J. Am. Chem. Soc. 68, 686.

Avgul, N.N., Kiselev, A.V., 1970. In: Walker, P.L. (Ed.), Chemistry and Physics of Carbon, vol. 6. Marcel Dekker, New York, p. 1.

Avgul, N.N., Bezus, A.G., Dobrova, E.S., Kiselev, A.V., 1973. J. Colloid Interface Sci. 42, 486.

Bai, R., Yang, R.T., 2001. J. Colloid Interf. Sci. 239 (2), 296.

Bansal, R.C., Donnet, J.B., Stoeckli, H.F., 1988. Active Carbon. Marcel Dekker, New York.

Barrer, R.M., 1978. Zeolites and Clay Minerals as Sorbents and Molecular Sieves. Academic Press, London p. 117.

Beebe, R.A., Young, D.M., 1954. J. Phys. Chem. 58, 93.

Bienfait, M., Zeppenfeld, P., Vilches, O.E., Palmari, J.P., Lauter, H.J., 1997. Surf. Sci. 377–379, 504.

Blake, T.D., Wade, W.H., 1971. J. Phys. Chem. 75, 1887.

Bloch, E.D., Queen, W.L., Krishna, R., Zadrozny, J.M., Brown, C.M., Long, J.R., 2012. Science (Washington, DC) 335, 1606.

Block, J.H., Bradshaw, A.M., Gravelle, P.C., Haber, J., Hansen, R.S., Roberts, M.W., Sheppard, N., Tamaru, K., 1990. Pure Appl. Chem. 62, 2297.

Bonnetain, L., Duval, X., Letort, M., 1952. C. R. Acad. Sci. Fr. 234, 1363.

Broekhoff, J.C.P., Van Dongen, R.H., 1970. In: Linsen, B.G. (Ed.), Physical and Chemical Aspects of Adsorbents and Catalysts. Academic Press, London, p. 63.

Brunauer, S., 1945. The Adsorption of Gases and Vapours. Princeton University Press, Princeton.

Brunauer, S., Emmett, P.H., Teller, E., 1938. J. Am. Chem. Soc. 60, 309.

Brunauer, S., Deming, L.S., Deming, W.L., Teller, E., 1940. J. Am. Chem. Soc. 62, 1723.

Brunauer, S., Skalny, J., Bodor, E.E., 1969. J. Colloid Interface Sci. 30, 546.

Carrott, P.J.M., Sing, K.S.W., 1989. Pure Appl. Chem. 61 (11), 1835.

Carrott, P.J.M., Mc Leod, A.J., Sing, K.S.W., 1982. In: Rouquerol, J., Sing, K.S.W. (Eds.), Adsorption at the Gas-Solid and Liquid-Solid Interface. Elsevier, Amsterdam, p. 403.

Carrott, P.J.M., Roberts, R.A., Sing, K.S.W., 1987. Carbon. 25 (6), 769.

Chiarello, R., Coulomb, J.P., Krim, J., Wang, C.L., 1988. Phys. Rev. B. 38, 8967.

Chowdhury, P., Mekala, S., Dreisbach, F., Gumma, S., 2012. Micropor. Mesopor. Mater. 152, 2866.

Cole, J.H., Everett, D.H., Marshall, C.T., Paniego, A.R., Powl, J.C., Rodriguez-Reinoso, F., 1974. J. Chem. Soc. Faraday Trans. I. 70, 2154.

Coulomb, J.P., Sullivan, T.J., Vilches, O.E., 1984. Phys. Rev. B. 30 (8), 4753.

Dash, J.G., 1975. Films on Solid Surfaces. Academic Press, New York.

de Boer, J.H., 1953. The Dynamical Character of Adsorption. Clarendon Press, Oxford.

de Boer, J.H., 1968. The Dynamical Character of Adsorption. Oxford University Press, London p. 179.

Do, D.D., 1998. Adsorption Analysis: Equilibria and Kinetics. Imperial College Press, London.

Dubinin, M.M., 1966. In: Walker, P.L. (Ed.), Chemistry and Physics of Carbon. Marcel Dekker, New York, p. 51.

Dubinin, M.M., 1975. In: Cadenhead, D.A. (Ed.), Progress in Surface and Membrane Science, vol. 9. Academic Press, New York, p. 1.

Dubinin, M.M., Astakhov, V.A., 1970. Adv. Chem. Ser. 102, 69.

Dubinin, M.M., Radushkevich, L.V., 1947. Proc. Acad. Sci. USSR. 55, 331.

Dunne, J., Myers, A.L., 1994. Chem. Eng. Sci. 49, 2941.

Emmett, P.H., Brunauer, S., 1937. J. Am. Chem. Soc. 59, 1553.

Everett, D.H., 1950. Trans. Faraday Soc. 46, 453p. 942 and 957.

Everett, D.H., 1970. In: Everett, D.H., Ottewill, R.H. (Eds.), Surface Area Determination. Butterworth, London, p. 181.

Fowler, R.H., 1935. Proc. Cambridge Phil. Soc. 31, 260.

Freundlich, H., 1926. Colloid and Capillary Chemistry. Methuen, London p. 120.

Gregg, S.J., 1961. The Surface Chemistry of Solids. Chapman and Hall, London.

Gregg, S.J., Sing, K.S.W., 1967. Adsorption, Surface Area and Porosity, 1st ed. Academic Press, London.

Gregg, S.J., Sing, K.S.W., 1976. In: Matijevic, E. (Ed.), Surface and Colloid Science, vol. 9. Wiley, New York, p. 231.

Gregg, S.J., Sing, K.S.W., 1982. Adsorption, Surface Area and Porosity. Academic Press, London.

Grillet, Y., Rouquerol, F., Rouquerol, J., 1979. J. Colloid Interface Sci. 70, 23.

Halsey, G.D., 1948. J. Chem. Phys. 16, 93.

Hamon, L., Heymans, N., Llewellyn, P., Guillerm, V., Ghoufi, A., Vaesen, S., Maurin, G., Serre, C., De, W.G., Pirngruber, G.D., 2012. Dalton Trans. 41, 4052.

Harkins, W.D., Jura, G., 1944. J. Am. Chem. Soc. 66, 1366.

Heymans, N., 2011. Thesis. Université de Mons, Mons, Belgium.

Hill, J.L., 1946. J. Am. Chem. Soc. 68, 535.

Hill, J.L., 1952. Adv. Catal. 4, 211.

Huber, U., Stoeckli, H.F., Houriet, J.P., 1978. J. Coll. Int. Sci. 67 (2), 195.

Isirikyan, A.A., Kiselev, A.V., 1961. J. Phys. Chem. 65, 601.

Jaroniec, M., Kruk, M., Choma, J., 1997. In: Mc Enaney, B., Mays, T.J., Rouquerol, J., Rodriguez-Reinoso, F., Sing, K.S.W., Unger, K.K. (Eds.), Characterization of Porous Solids IV. Royal Society of Chemistry, London, p. 163.

Jensen, C.R.C., Seaton, N.A., 1996. Langmuir. 12, 2866.

Kaganer, M.G., 1959. Zhur. Fiz. Khim. 33, 2202.

Lamb, A.B., Coolidge, A.S., 1920. J. Am. Chem. Soc. 42, 1146.

Langmuir, I., 1916. J. Am. Chem. Soc. 38, 2221.

Langmuir, I., 1918. J. Am. Chem. Soc. 40, 1361.

Larher, Y., 1992. In: Benedek, G. (Ed.), Surface Properties of Layered Structures. Kluwer, Dordrecht, p. 261.

Layet, J.M., Bienfait, M., Ramseyer, C., Hoang, P.N.M., Girardet, C., Coddens, G., 1993. Phys. Rev. B Condens. Matter. 48 (12), 9045.

Lowell, S., Shields, J.E., Thomas, M.A., Thommes, M., 2004. Characterization of Porous Solids and Powders: Surface Area, Pore Size and Density. Kluwer, Dordrecht p. 31.

Mc Bain, J.W., 1932. The sorption of gases and vapours by solids. Routledge, London.

McDonald, T.M., Lee, W.R., Mason, J.A., Wiers, B.M., Hong, C.S., Long, J.R., 2012. J. Am. Chem. Soc. 134 (16), 7056.

Mc Enaney, B., 1988. Carbon. 26, 267.

Myers, A.L., Prausnitz, J.M., 1965. Chem. Eng. Sci. 20, 549.

Nicholson, D., Parsonage, N.G., 1982. Computer Simulation and the Statistical Mechanics of Adsorption. Academic Press, London and New York, 398pp.

Plaza, M.G., Ribeiro, A.M., Ferreira, A., Santos, J.C., Hwang, Y.K., Seo, Y.K., Lee, U.H., Chang, J.S., Loureiro, J.M., Rodrigues, A.E., 2012. Micropor. Mesopor. Mater. 153, 178.

Pierotti, R.A., Thomas, H.E., 1971. In: Matijevic, E. (Ed.), Surface and Colloid Science. Wiley-Interscience, New York, p. 93.

Price, G.L., Venables, J.A., 1976. Surf. Sci. 59 (2), 509.

Qiao, S., Wang, K., Hu, X., 2000. Langmuir. 16, 1292.

Reichert, H., Muller, U., Unger, K.K., Grillet, Y., Rouquerol, F., Rouquerol, J., Coulomb, J.P., 1991. In: Rodriguez-Reinoso, F., Rouquerol, J., Sing, K.S.W., Unger, K.K. (Eds.), Characterization of Porous Solids II. Elsevier, Amsterdam, p. 535.

Ross, S., Clark, H., 1954. J. Am. Chem. Soc. 76, 4291.

Ross, S., Olivier, J.P., 1964. On Physical Adsorption. Wiley-Interscience, New York.

Rouquerol, F., Luciani, L., Llewellyn, P., Denoyel, R., Rouquerol, J., 2003. Techn. l'Ingén. Trai. Anal. Caractér. P1050, 1.

Rudzinski, W., Everett, D.H., 1992. Adsorption of Gases on Heterogeneous Surfaces. Academic Press, London.

Ruthven, D.M., 1984. Principles of Adsorption and Adsorption Processes. Wiley-Interscience, New York.

Seri-Levi, A., Avnir, D., 1993. Langmuir. 9, 2523.

Sing, K.S.W., 1973. Colloid Science I. The Chemical Society, London p. 30.

Sips, R., 1948. J. Chem. Phys. 16, 490.

Steele, W.A., 1974. The Interaction of Gases with Solid Surfaces. Pergamon, New York, p. 131.

Steele, W.A., 1980. J. Colloid Interface Sci. 75, 13.

Steele, W.A., 1996. Langmuir. 12, 145.

Stoeckli, H.F., 1977. J. Colloid Interface Sci. 59, 184.

Stoeckli, H.F., 1993. Adsorpt. Sci. Tech. 10, 3.

Stoeckli, H.F., Houriet, J.P., Perret, A., Huber, U., 1979. In: Gregg, S.J., Sing, K.S.W., Stoeckli, H.F. (Eds.), Characterization of Porous Solids. Society of Chemical Industry, London, p. 31.

Suzanne, J., Gay, J.M., 1996. In: Unertl, W.N. (Ed.), Hanbook of Surface Science. Richardson, N.V., Holloway, S. (Series Eds.), Physical Structure, Vol. 1. North-Holland Elsevier, Amsterdam, 503pp.

Terlain, A., Larher, Y., 1983. Surf. Sci. 125, 304.

Thomy, A., Duval, X., 1970. J. Chem. Phys. Fr. 67, 1101.

Thomy, A., Regnier, J., Duval, X., 1972. Thermochimie, Colloques Internationaux, vol. 201. CNRS, Paris p. 511.

Toth, J., 1962. Acta Chim. Acad. Sci. Hung. 35, 416.

Valenzuela, D.P., Myers, A.L., Talu, O., Zwiebel, I., 1988. AIChe J. 34, 397.

Venables, J.A., Seguin, J.L., Suzanne, I., Bienfait, M., 1984. Surf. Sci. 145, 345.

Wiersum, A., 2012. Thesis, Université d'Aix-Marseille, Marseille, France.

Yang, R.T., 1987. Gas Separation by Adsorption Processes. Butterworths, London.

Young, D.M., Crowell, A.D., 1962. Physical Adsorption of Gases. Butterworths, London, p. 124.

6 Modelling of Physisorption in Porous Solids

Guillaume Maurin

University of Montpellier 2, Institute Charles Gerhardt, Montpellier, France

Chapter Contents

Adsorption by Powders and Porous Solids. http://dx.doi.org/10.1016/B978-0-08-097035-6.00006-1

6.1   Introduction

Computational chemistry has now become an integral component of the adsorption science applied to porous solids. The rapid developments of mathematical methods and computational software and hardware offer a unique platform to assist the experimentalist from the birth of new porous materials to the characterization of their adsorption performances. Molecular modelling that embodies all the methods which model a system at the molecular level using interatomic potential, also currently called 'force field', is thus able to first construct a microscopic representation of the more complex porous solids followed by a careful geometrical characterization of their porosities. Such computational tools that are further used to simulate the adsorption capacities, enthalpies and selectivities for a wide variety of gas/porous solid systems to guide interpretation of the experimental data become increasingly predictive. Rather than blindly screening a large collection of existing or hypothetical porous materials for a given adsorption-related application, large-scale computational strategy based on advanced modelling techniques can substitute experimental high-throughput approaches to save costly and time-demanding measurements. The molecular modelling is also frequently intertwined with the quantum chemical approach which treats the system at the electronic level to provide a deep understanding of the microscopic adsorption mechanism experimentally intractable. Rational approach using advanced statistical methods is nowadays also pursued for identifying structure–adsorption property relationships for a range of adsorbents.

This chapter is written for non-experts in computational chemistry, with the aim to highlight how modelling tools can be valuable not only to assist but also to guide the experimentalists throughout the characterization of the porous solids and the determination of their adsorption and diffusion properties. This strong experimental-modelling interplay is illustrated by typical examples that have been arbitrarily selected from thousands of relevant studies existing on this topic, to cover the main questions that can be addressed. We have deliberately made the choice to introduce briefly the fundamental theory behind each major computational method aided by a minimal number of mathematical equations with the aim to provide just enough information to explain the terminology employed along the text. The readers are referred to specialized books and review articles provided in each section to get more technical details. The computational approaches are also critically evaluated by emphasizing their validities and limitations which hopefully will help the readers to be more confident to assess the pertinence of the simulations reported in this literature in order to select the most reliable ones for further comparison with their own experimental data. A list of software is also provided, some of them being easy to operate in particular for the estimation of the accessible surface area and free volume of the porous solids. It is expected that this chapter will assist the readers in gaining a basic orientation in this field to use such computational tools not like a hazardous 'black box'.

6.2   Microscopic Description of the Porous Solids

The first pre-requirement to predict the adsorption properties of a porous solid consists of building a realistic atomistic description of the adsorbent. The complexity of such a task strongly depends on whether (i) the considered adsorbent is crystalline or not, (ii) its framework is characterized by a chemical disorder or the positions of each atom type are well defined and (iii) its morphology/topology is experimentally controlled or not. It is convenient to categorize the adsorbent as follows.

6.2.1   Crystalline Materials

The inorganic *zeolites* belong to this class of solids (IZA, 2013). While their crystal structures are well established from X-ray diffraction data, these solids show a first complexity with a chemical disorder in their frameworks induced, for instance, by the substitution of aluminium atoms by silicon atoms for the aluminosilicate forms, which can be hardly treated by usual experimental techniques such as nuclear magnetic resonance (NMR) spectroscopy. A statistical approach based on Monte Carlo (MC) techniques is commonly used to properly describe this feature (Maurin et al., 2001). It consists of generating several representative configurations obeying the Lowenstein's rule (Lowenstein, 1954), that is, exclusion of the possibility of alumina as nearest neighbours, which are then selected by combining experimental NMR data and energetic criteria (Maurin et al., 2001). Another aspect consists of modelling via MC simulations the location of the extra-framework cations and their distribution within the porosity of these materials when the crystallographic sites and their occupancies are not experimentally available (Vitale et al., 1997). It has also been evidenced that the adsorption of certain molecules induces a significant redistribution of the extra-framework cations (Beauvais et al., 2004; Maurin et al., 2006). Such a phenomenon is nowadays taken into account in the majority of the computations by allowing these cations to be mobile along the whole adsorption process. This holds also true for the *Clays* for which an additional difficulty sometimes arises from their swelling behaviours upon adsorption that renders the consideration of a fixed value for the interlayer distance highly questionable (Salles et al., 2011). In this latter case, a special attention is paid to implement specific force fields that accurately account for the swelling of the framework. These interatomic potentials typically include bond stretching, bond bending and torsional motions with parameters either available in the literature or derived via different computational approaches (Van Duin et al., 2001). The hybrid inorganic/organic *Metal Organic Framework* materials are also classified as crystalline solids, some of them combining the complexity of the two solids mentioned above, with a chemical disorder sometimes enhanced by a non-periodic distribution of the organic ligands that link together the inorganic node and with a highly flexible character upon adsorption called breathing which is associated to spectacular unit cell volume changes (Férey et al., 2011). This uncommon structural feature needs also to be modelled via sophisticated force fields (Salles et al., 2008a).

Modelling tools involving energy minimization techniques are currently deployed to identify the crystal structures of such porous solids benefiting from a restricted set of experimental information (NMR, IR, EXAFS, etc.) related to local or long-range parts of the materials (Mellot-Draznieks, 2007; Devic et al., 2010). Such a computational-assisted structure determination strategy opens new horizons in the structure resolution able to complement the *ab initio* prediction approach that remains inefficient for some complex crystalline porous solids.

6.2.2 Non-Crystalline Materials

The *silica mesoporous* solids (MCM-41, SBA-15, etc.) belong to this class of adsorbents. Here, the description of a realistic atomistic model requires much computational effort to be deployed as there is a wide range of existing topologies/morphologies such as hexagonal, cylindrical and ellipsoidal pores (Coasne and Pellenq, 2004). The complexity of such a system relies on the surface roughness and defects (constriction and tortuosity), parameters that both vary depending on the sample considered, leading to a drastic impact on its adsorption properties. Microscopic models for such porous solids are usually built using either reconstruction (tailor-made to reproduce the structural features of the real sample) or mimetic approaches (Schumacher et al., 2006). Indeed, on-lattice MC simulations of surfactant–solvent–silica systems that mimic the synthesis conditions are usually employed to generate realistic pore models for these solids (Siperstein and Gubbins, 2003). Other approaches consider a block of the crystalline cristobalite that is carved out in such a way to define a model with the desired topology/morphology features and pore diameter (Coasne et al., 2006). The chemical composition/disorder of the pore wall is then adjusted to reproduce the experimental density of the hydroxyl groups present and to mimic the amorphous nature of the silica surface. A similar strategy is also employed for generating models for *pore glasses* such as the Vycor (Levitz, 1998; Gelb, 2002). Finally, the *porous active carbon* can be modelled in a first approximation as a slit-shaped carbon graphite pore of a constant width, the pore surface being represented by several graphene sheets separated by a distance corresponding to those observed in the graphite (Billemont et al., 2010). A reverse MC approach can also be employed to reconstruct microscopic models for real carbons including porous saccharose cokes that quantitatively match Transmission Electron Transmission images or diffractograms at low angles (Jain et al., 2005; Figure 6.1).

6.3   Intermolecular Potential Function

6.3.1 General Expression of the Pairwise Adsorbate/Adsorbent Interactions

Once the step of defining a microscopic structure model for the adsorbent is achieved, appropriate forms and parameters of a mathematical function need to be selected with the aim to reproduce as fairly as possible the potential energy of the adsorbate/

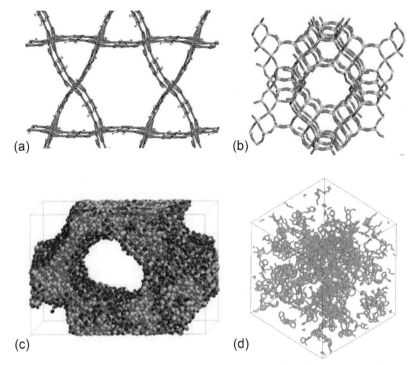

(a)   (b)

(c)   (d)

Figure 6.1 Illustration of the microscopic structural description of porous crystalline: (a) MIL-68(Al) MOF, (b) faujasite zeolite and non-crystalline solids, (c) mesoporous silica SBA-15 and (d) active carbon. (For colour version of this figure, the reader is referred to the online version of this chapter.)
Courtesy: (c) Bhattacharya et al. (2009) and (d) Jain et al. (2005).

adsorbent interactions. This so-called force field function and parameter sets are issued from the literature or derived using *ab initio* calculations or semi-empirical approaches. *To avoid any ambiguity, the readers should bear in mind that the terms 'force field' and 'interatomic potentials' currently manipulated by the modellers in the literature and employed along the chapter refer to an energy term.*

It is customary to assume that the pairwise interactions are additive and can be modelled by an analytical potential energy function expressed by Equation (6.1):

$$\varphi = \varphi_D + \varphi_R + \varphi_P + \varphi_E \tag{6.1}$$

where the former three terms of this expression, φ_D, the dispersion; φ_R, the repulsion; and φ_P, the polarization interactions, occur for every adsorbate/adsorbent system and can consequently be considered as 'non-specific' interaction energy terms, while the latter electrostatic term, φ_E, is classified as a 'specific' interaction energy term since it is involved in only certain adsorbate/adsorbent system.

Of these 'non-specific' interactions, the dispersion interaction energy φ_D is usually represented by Equation (6.2):

$$\varphi_D(r) = -\frac{A}{r^6} \tag{6.2}$$

where r is the distance between the adsorbate and the adsorbent atoms and A is a dispersion force constant characteristic of the atoms pair. This relation only takes into account the dipole–dipole interactions; however, a sum over further excited states can be considered in some systems and then Equation (6.2) can be rewritten as Equation (6.3):

$$\varphi_D(r) = -\frac{A_6}{r^6} - \frac{A_8}{r^8} - \frac{A_{10}}{r^{10}} \tag{6.3}$$

where the term r^{-6} describes the dipole–dipole interactions, r^{-8} the dipole–quadrupole interactions and r^{-10} the quadrupole–quadrupole and dipole–octupole interactions.

This dispersion energy term acts at longer distances than the repulsion contribution which originates from the overlap of the two electron clouds when the atoms approach one another. This repulsion contribution is expressed by the exponential Born–Mayer function or more generally by the term r^{-12} (see Equation 6.4).

$$\varphi_R(r) = B \cdot \exp(-br) \quad \varphi_R(r) = \frac{B}{r^{12}} \tag{6.4}$$

where B and b are constant characteristics of the adsorbate/adsorbent pair.

In most of the adsorbate/adsorbent interatomic potentials reported in the literature, the dispersion and repulsion terms are associated and converted to a Lennard–Jones (LJ) 12–6 potential energy type function (Equation 6.5):

$$\varphi(r) = \frac{B}{r^{12}} - \frac{A}{r^6} \tag{6.5}$$

which is more usually expressed in the form of Equation (6.6),

$$\varphi(r) = 4\varepsilon \left[\left(\frac{\sigma}{r} \right)^{12} - \left(\frac{\sigma}{r} \right)^6 \right] \tag{6.6}$$

where σ and ε are the finite distance at which the potential vanishes and the depth of the potential well, respectively (see Figure 6.2).

This dispersion–repulsion interaction can also be represented by a Buckingham potential function the equation of which is

$$\varphi(r) = B \cdot \exp(-br) - \frac{A}{r^6} \tag{6.7}$$

The polarization interactions arise as a result of the close proximity of the adsorbent electric field with the adsorbate. The resulting polarization energy is given by Equation (6.8):

$$\varphi_P = -\frac{\alpha \cdot E^2}{2} \tag{6.8}$$

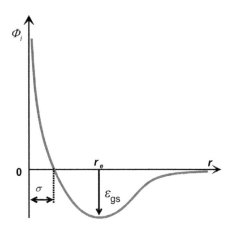

Figure 6.2 Plot of the Lennard–Jones potential function.

where E is the electric field created by the adsorbent and α is the dipole polarizability of the adsorbate. Such a contribution is growing considered in the recent interatomic potential reported in the literature.

Finally, in the general approximation of a monopole–monopole (charge–charge) interaction, the electrostatic energy term is expressed by Equation (6.9):

$$\varphi_{\mathrm{E}} = \frac{q \cdot q'}{4\pi\varepsilon_o r} \tag{6.9}$$

where q and q' are the charges carried by atoms of the adsorbate and the adsorbent, respectively, and ε_o is the vacuum permeability. Additional monopole/dipole or dipole–dipole interactions need in some cases to be considered and lead to higher-order contributions in r^{-2} and r^{-3}, respectively.

6.3.2 Common Strategy for 'Simple' Adsorbate/Adsorbent System

Following the approach initiated by Kiselev (Bezus et al., 1978), an effective LJ 12–6 potential is most often considered to describe the pairwise adsorbent/adsorbate interactions in a wide variety of porous materials including carbons, silicas, zeolites and MOFs. In this Kiselev-type potential that has been widely employed for its simplicity and computational efficiency (Fuchs and Cheetham, 2001), the repulsion/dispersion term acts between LJ sites centred on either all atoms or the most presumably interacting ones of the adsorbent and the adsorbate. Typically, for a purely siliceous zeolite, the LJ sites are only placed on the oxygen atoms of the framework as they concentrate the major part of the interaction between the lattice and the adsorbates. The LJ parameters for all atoms of the adsorbent are usually taken from generic force fields available in the literature, more particularly Universal Force Field (UFF; Rappé et al., 1992), DREIDING (Mayo et al., 1990) and OPLS (Jorgensen et al., 1996). Such popular force fields that contain LJ parameters for the majority of the atoms of the periodic table are revealed surprisingly well transferable for treating the adsorption in a wide class of porous solids. Regarding the atoms of the adsorbate, the corresponding LJ parameters are usually obtained from a fit of the experimental vapour–liquid equilibrium data as, for instance, in the well-known TraPPE force field (Martin and Siepmann, 1998). In such force fields, the adsorbates can be further modelled as either rigid or flexible molecules. The Lorentz–Berthelot combining rules (Allen and Tidesley, 1987) are primarily applied to calculate the cross-terms corresponding to the adsorbent/adsorbate interactions. Another strategy that we can call the 'chemical engineering approach' consists of readjusting the LJ parameters of the simplest existing potential models to match the experimental adsorption data (Dubbeldam et al., 2004). Such a fitting procedure should be really taken with caution as it strongly depends on the quality of the samples and the accuracy of the measurements while it usually suffers from a poor transferability of the potential parameters from one system to another.

When the adsorbate has a permanent dipole moment, an additional electrostatic term is considered to capture the coulombic interactions between all the atoms of the adsorbent and the adsorbate. In this case, either integer or partial (non-integer, also called net atomic) charges are attributed to all the atoms of the adsorbent. *One should bear in mind that estimating the partial charges can be computationally expensive via quantum calculations (Mulliken population analysis (Mulliken, 1955) or electrostatic potential fitting (Heinz and Suter, 2004)); however very recently, several authors have shown that this can be achieved much faster with a reasonable accuracy by means of suitable semi-empirical charge equilibration methods including the charge equilibration (QEq) strategy (Rappé and Goddard, 1991).* This has been revealed of particular interest for screening the adsorption properties of a large library of porous materials.

Such parameterized adsorbate/adsorbent interactions usually associated with a rigid model for the adsorbent are successful for systems implying relatively usual adsorption behaviours. Indeed, this approach is recommended to obtain a rapid prediction of the thermodynamic properties of 'simple' adsorbate/adsorbent system for which the transferability of the interatomic parameters is reasonable and the charges are relatively well described by the low-computational cost, semi-empirical approaches. Simple adsorbents usually include porous materials that do not contain specific adsorption sites such as silicas and zeolites in their siliceous forms, porous carbons and a wide range of MOFs.

However, such a strategy usually fails to accurately capture the adsorption in more complex adsorbents which contain, for instance, chemical defects at the mesoporous silica surface, extra-framework cations in zeolites/clays or coordinative unsaturated sites (cus) in MOFs, or which show some flexibility degree of their frameworks such as in narrow pores/windows zeolites or breathing MOFs. In that case, (i) a more dedicated force field either maintaining the same analytical expression but with a specific parameterization using quantum calculations or empirical fitting, or including additional contributions as, for instance, the PN-TrAZ (transferable for adsorption in zeolites) (Pellenq and Nicholson, 1994) potential applied for zeolites and mesoporous silica which incorporates the polarization energy and higher order in the repulsion/dispersion term and/or (ii) a flexible treatment of the framework via additional intramolecular interaction terms (Salles et al., 2008a) are required.

The next sections provide some typical examples where the use of a Kiselev-type potential combined with generic force fields is not appropriate.

6.3.3 Cases of More 'Complex' Adsorbate/Adsorbent System

6.3.3.1 Quantum-Derived Force Field for the CO_2/Na^+ Faujasite Zeolite

The successful simulation of the CO_2 adsorption in the faujasite-type zeolite containing Na^+ as extra-framework cations requires an accurate description of the interatomic potential between CO_2 and the Na^+ known as the preferential adsorption sites in this system that cannot be achieved by generic force fields. Indeed, the parameterization of such a pairwise interaction can be realized using *quantum* calculations which usually adopt the following procedure. A cluster centred around a Na^+ fixed at its crystallographic site is first cut from the periodic crystal structure of the faujasite. A preliminary quantum calculation using appropriate functional and basis set is then carried out in order to define the optimized geometry of the CO_2–zeolite cluster (Maurin et al., 2005). This optimized geometry provides a suitable starting configuration for generating a potential energy curve using a series of single point energy calculations. The carbon dioxide molecule is then displaced from the cations to a distance between 0.1 and 1.0 nm, at increments of 0.01 nm, the geometry being constrained in order to maintain the previously defined equilibrium ($O=C=O\ldots Na^+$) angle (see Figure 6.3).

At each increment, a single point energy calculation is performed in order to produce the energy profile. This potential energy curve is then fitted by the combination of a coulombic contribution and a Buckingham/LJ potential term to extract the interatomic potential parameters for the considered pair. A similar procedure can then be followed to derive the energy profile for Na^+—$C(CO_2)$ starting from a configuration obtained by a 90° rotation of the previously optimized geometry (see Figure 6.3). Further to this, the resulting parameters are usually tested and validated by comparing the simulated energetic data for the CO_2/faujasite system and those collected by microcalorimetry at low coverage.

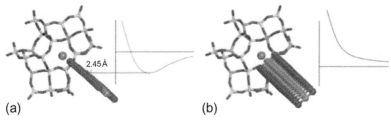

(a)   (b)

Figure 6.3 Schematic representation of the CO_2–Na^+/zeolite geometry considered for the calculation of the potential energy curve corresponding to the Na^+–$O(CO_2)$ (a) and Na^+–$C(CO_2)$ (b) interactions (Maurin et al., 2005). (For colour version of this figure, the reader is referred to the online version of this chapter.)

6.3.3.2 Ab Initio *Potential Energy Surface for the CH₄/cus in the HKUST-1 MOF*

It has been evidenced in different MOFs containing cus (MIL-100(Cr,Fe), MIL-127 (Fe), HKUST-1, etc.) that the generic force fields fail to correctly describe the interactions between the adsorbate molecules and these cus that usually govern the first stage of the adsorption. To address such a limitation, Chen et al. (2011) have recently proposed an alternative strategy based on the determination of the adsorbate/adsorbent potential energy surface (PES) using sophisticated quantum calculations. This was applied for the CH_4/HKUST-1 type MOF system and consisted of placing the carbon atom of a single methane molecule in a random configuration on specific grid points in order to further calculate the interactions energies with the HKUST-1 framework at the quantum level. Such an approach clearly shows the advantages to avoid the fitting procedure described above, sometimes source of ambiguity and inaccuracy, allowing a direct implementation of the PES in MC software to further estimate the thermodynamic properties of the system of interest. In the selected example, this PES is validated by reproducing well the occupancy of the different adsorption sites by CH_4 in the HKUST-1 structure experimentally identified by *in situ* X-ray diffraction measurements. Such a strategy has also been developed to accurately capture the specific interactions at play between the highly important propane/propylene gas mixture and the same HKUST-1 material (Fischer et al., 2012).

6.3.3.3 *Flexibility of the Adsorbent Upon Adsorption: Case of the MOF-Type MIL-53(Cr)*

In contrast to what is predominantly observed for numbers of porous solids (zeolites, mesoporous silica, etc.), characterized by only a tiny rearrangement of the skeleton upon adsorption (contraction/expansion of a few percent in unit cell volume), the scenario is much more complex when one aims to investigate the adsorption properties of a highly flexible adsorbent as, for instance, some clays or MOFs. In that case, such a magnitude of flexibility is expected to drastically impact the adsorption/diffusion properties of these solids. A specific force field is thus required for treating the intra-framework interactions of the host structure that, combined with an accurate description of the adsorbate/ adsorbent interactions, is able to capture substantial degrees of framework flexibility of the porous solids and further reproduce their resulting unusual adsorption behaviour. Some of the existing MOFs such as DMOF-1 (Grosch and Paesani, 2012) and MIL-53 (Férey et al., 2011) belong to this class of materials showing a drastic unit cell volume change upon adsorption of various gases. While high computational cost *ab initio* calculations have been deployed for deriving the necessary force field parameters of the intra-framework energy terms (bond stretching, bond bending, torsion, etc.) for certain MOFs (Tafipolsky et al., 2007), some other authors privileged a less time consuming semi-empirical approach such as for the MIL-53(Cr) solid (Salles et al., 2008a). Indeed starting from an initial set of potential parameters issued from generic force fields, a systematic refinement protocol was conducted with the criteria to be able to accurately reproduce the experimental vibrational properties of the

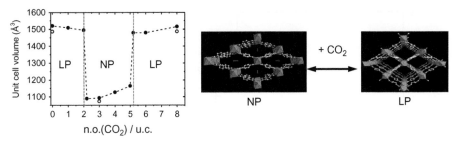

Figure 6.4 Left: evolution of the unit cell volume of the MIL-53(Cr) as a function of CO_2 loading at 300 K: molecular dynamics (full symbols) and *in situ* XRD (empty symbols). Right: illustration of the structural transition from a narrow (NP) and a large (LP) pore forms of MIL-53(Cr) upon CO_2 adsorption (Salles et al., 2008a). (For colour version of this figure, the reader is referred to the online version of this chapter.)

framework via an energy minimization approach at 0 K. This derived force field was further combined with the one selected for representing CO_2 and implemented in Molecular Dynamics (MD) simulations to follow the structural transformations of the porous solid along the whole adsorption process. It allows a successful capture of the two-step structural switching from a large pore (LP) to a NP form, as induced by CO_2 adsorption at ambient temperature, conclusions that were validated by a good agreement with *in situ* XRD observations (Figure 6.4).

6.4 Characterization Computational Tools

6.4.1 Introduction

Once a reliable microscopic description of their structures is defined, molecular simulations based on MC integration algorithm can be applied to characterize porous materials by data such as the *accessible specific surface area* (a_{acc} defined hereafter), pore volume (v_{pore}) and pore size distribution (PSD). These computational tools implemented in Fortran program available in commercial software or free of charge on different webpages can be easily employed by the experimentalists in routine to quickly assess the quality of a sample in terms of activation, thermal stability, etc. Obtaining much smaller experimental BET area and/or pore volume than the theoretical values means that either some remaining solvent and unreacted species are still present within the porosity of a given sample or some structural/textural degradation of the material occurs. It thus indicates that a further optimization of the activation/synthesis procedure is required prior to measure its adsorption properties. Such an approach can also provide valuable information on the expected adsorption performance of the sample for a given adsorbate by establishing a correlation between the most important characteristics of the porous materials and their adsorption capacities. Indeed, it has been established that the adsorption uptakes for some gases in a

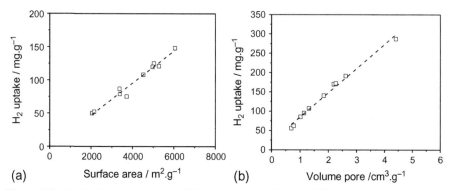

Figure 6.5 H_2 adsorbed amount at (a) 30 bar versus accessible specific surface area (a_{acc}) and at (b) 120 bar versus specific pore volume v_{pore} in the isoreticular series of IRMOFs at 77 K. *Courtesy:* Frost et al. (2006).

large variety of microporous solids including zeolites, porous carbons and MOFs correlate very well with a_{acc}, calculated using a probe diameter of same size than the adsorbate of interest, and v_{pore} at intermediate and high pressure, respectively. A typical illustration is provided in Figure 6.5 for the adsorption of hydrogen in the isoreticular series of IRMOF solids (Frost et al., 2006). Such a relation was further used to evaluate the potential of these materials to meet the targets for hydrogen and methane storage only based on the estimation of a_{acc} and v_{pore}.

Such simple and fast computational methods, therefore, appear as an additional characterization tool in the portfolio of the experimentalist for easily judging the quality of a sample. The basic principles of this computational strategy are summarized in the following section.

6.4.2   Accessible Specific Surface Area

In this section and in the rest of this chapter, we shall use the term 'accessible specific surface area' with its usual meaning among modellers (Leach, 2001), which is different from its initial meaning in the adsorption community (Rideal, 1930), as commented in Chapter 1. In simulation, the calculation of the accessible specific surface area is purely based on the geometric topology of the adsorbent and calculated from a simple MC integration technique where the centre of mass of the probe molecule with hard sphere (of radius r) is 'rolled' over the framework surface. As shown in Figure 6.6, this 'accessible specific surface area' (that we can also call 'r-distant surface area') is the surface covered by the centre of mass, that is, a virtual surface located at distance r from the framework surface. In such a method, a nitrogen-sized probe molecule is randomly inserted around each framework atom of the adsorbent and the fraction of the probe molecule without overlapping with the other framework atoms is then used to calculate the accessible specific surface area. A 0.36-nm value is most commonly considered for the nitrogen probe diameter (Walton and Snurr, 2007), while the LJ size

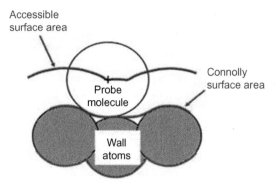

Accessible surface area

Probe molecule

Wall atoms

Connolly surface area

Figure 6.6 Schematic representations of the accessible specific surface area (a_{acc}) and the Connolly specific surface area (a_{Conn}). (For colour version of this figure, the reader is referred to the online version of this chapter.)
Courtesy: Düren et al. (2007).

parameters of the framework atoms are usually taken from generic force fields available in the literature. This latter choice can be arbitrarily done as the value of a_{acc} is only slightly affected from one force field to another.

Such a geometric approach shows a considerable advantage to have a low computational cost (only a few seconds once the microscopic structure model is known) while maintaining a good accuracy.

Another surface area used by the modellers is the '*Connolly specific surface area*' (a_{Conn}), which is calculated from the bottom instead of the centre of mass of the probe molecule (see Figure 6.6.) and which corresponds to the accessible (or 'probe-accessible') specific surface area of the experimentalists. The latter surface area has been assessed to be less appropriate than the accessible specific surface area, as defined above, to characterize the majority of porous solids from the viewpoint of adsorption capacity (Bae et al., 2010). Based on a simple square model (Figure 6.7) on which a probe molecule can adsorb either on the outside or on the inside, Düren et al. (2007)

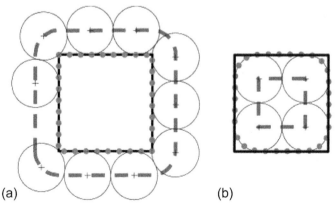

(a)   (b)

Figure 6.7 Accessible specific surface area (line) and Connolly specific surface area (circle) calculated for a probe molecule rolling over the outside (a) and the inside (b) of a simple square model. (For colour version of this figure, the reader is referred to the online version of this chapter.)
Courtesy: Düren et al. (2007).

have clearly demonstrated that while the Connolly specific surface area would lead to the same adsorption capacity in both cases, the accessible specific surface area leads to more realistic adsorption capacities as it makes a clear distinction between the two situations, a_{acc} being much smaller for the adsorption in the inside. This conclusion remains also valid for 3D materials where the molecules can adsorb in the edges (outside curvature) and in the corners (inside curvature) of cavities.

The recent literature seems to show that, in spite of their different origins and calculations, the accessible specific surface area (a_{acc}) and the *experimental BET specific surface area* (a_{exp} (*BET*)) usually compare well for good quality sample. Reservations have to be made in the most frequent case when the adsorbent is microporous since, as explained in Chapter 7, the 'BET monolayer content' then includes the full content of the micropores, where the molecules of adsorbate are far from covering the same area as in the monolayer of an open surfaces. Depending on the relative size of the molecules and micropores, the BET area can be either smaller (in narrow micropores) or larger (in wide micropores) than the probe-accessible or Connolly surface area. In contrast, as visible in Figure 6.7b, the accessible surface area in micropores is always smaller than the Connolly surface area. It was also shown that the above geometric approach to calculate a_{acc} was inappropriate for characterizing the surface areas of solids with extremely small micropores in the range of 0.3–0.45 nm that cannot be or only hardly accessible by the nitrogen-sized probe molecule. A typical illustration is provided in Chapter 14 for the zirconium dicarboxylate MIL-140 MOF-type solid (Guillerm et al., 2012). In that case, the only possible way to get an accurate description of the surface area consists of calculating the adsorption isotherms via grand canonical Monte Carlo (GCMC) simulations (described below) that allow a full exploration of the available surface since the atoms are treated as soft LJ spheres (vs. hard sphere in the geometrical approach) prior to apply the BET method to this simulated isotherm.

It can be generalized that for microporous solids, it makes more sense to compare the *experimental BET specific surface area* a_{exp} (*BET*) with the *BET specific surface area calculated from the simulated isotherm*, a_{sim} (*BET*), than with a_{acc}.

Further, as the accessible specific surface area strongly depends on the probe size, when this geometric method is used as a predictive tool for assessing the potential of a porous solid for the storage of a given adsorbate using correlations between the accessible specific surface area and the adsorption uptake, a probe molecule with a diameter corresponding to the one of the gas of interest is considered. Figure 6.5 provides an illustration of such a relationship between the H_2 uptake of a series of porous solids and their accessible specific surface areas calculated with H_2 (a probe size of 0.28 nm).

6.4.3 Pore Volume/PSD

The *accessible volume* of a porous solid can be estimated by modelling using two different routes. The easiest one is based on a similar geometric method as mentioned above which consists of using a probe molecule with a diameter of 0 nm to determine the volume of the porous solid that is not occupied by the atoms of the framework. One obtains what is usually called the *'free volume'*. Another way consists of using the *thermodynamic approach* first proposed by Myers and Monson (Myers and

Monson, 2002), which mimics the experimental conditions to rigorously evaluate the *'pore volume'*.

The calculated specific *pore volume* is thus expressed by Equation (6.10):

$$v_{\text{pore}} = \frac{1}{m} \int e^{-E/k_B T} \mathrm{d}r \tag{6.10}$$

where m is the mass of a representative sample of porous solid and E is the gas–solid potential energy for a single helium atom at the reference temperature of 298 K.

The integration of this equation over the entire void of the solid requires numerical MC technique. Such a methodology also implemented in some existing code available on the web requires the description of the energy term E, which is most often modelled with Helium as a LJ fluid with the corresponding parameters ($\sigma = 0.258$ nm, $\varepsilon/k_B = 10.22$ K) (Talu and Myers, 2001) interacting with all the framework atoms of the porous solids that are described as soft LJ spheres with parameters issued from various generic force field. Both approaches lead to very similar results for the majority of porous solids, the geometric method showing the advantage to be much faster. However, once again, the thermodynamic estimation of the pore volume is an alternative way to circumvent the limitations of the geometric method discussed above for characterizing small microporous solids. As the pore volume correlates well the uptake of a given adsorbate at high pressure, a rapid estimation of this characteristic can be conducted for a large library of solids to further predict the best candidates for a specific application.

Further, in complement to the approaches described in Chapter 9 for the pore size analysis, the readers are referred to the computational method developed by Bhattacharya and Gubbins (2006). This is based on a constrained non-linear optimization which allows the determination of the maximum radii of test particles at random points inside the pore cavity. The PSD, which is defined as the statistical distribution of the radius of the largest sphere that can be fitted inside the pore at a given point, is then obtained by sampling the test particle radii using an MC integration scheme. Once the structure model for the porous solid is known, such a geometric approach has been revealed to be fast and accurate to estimate the PSD for a wide range of solids including active carbons, mesoporous silica, zeolites and MOFs. Here again this method implemented in some code available on the web can be routinely employed by the experimentalists to get a first idea on the PSD profile they can expect from their newly synthesized samples.

6.5 Modelling of Adsorption in Porous Solids

6.5.1 GCMC Simulations

6.5.1.1 Basic Principles

MC techniques have been widely employed to efficiently calculate the equilibrium properties of porous solids using a stochastic (non-time dependent) method. In a

MC simulation, we generate a large sequence of configurations using random numbers (in reference to games of chance, a popular attraction in the capital of Monaco), each being accepted or rejected according to a certain probability. Such a categorized sampling method further allows the computation of average properties that can be directly compared to those experimentally observed. Indeed, MC simulations are also considered as 'numerical' or 'computer' experiments.

The aim of this section is to discuss in general terms the use of MC techniques for the simulation of the most common thermodynamic properties related to the adsorption of single gas component and mixtures, including the adsorption isotherm, the adsorption enthalpy and the selectivity towards the understanding of the microscopic mechanism at play. For a more thorough technical exposition of the methodology, the readers are referred to specialized books (Nicholson and Parsonage, 1982; Allen and Tildesley, 1987; Frenkel and Smit, 2002).

In this context, force field-based MC simulations (this means that we know exactly all the interactions between the atoms of the system as described in Section 6.3) are predominantly conducted in the grand canonical ensemble where the chemical potential μ, the volume V and the temperature T are fixed. The experimental analogy of such a thermodynamic ensemble consists of placing in a reservoir the adsorbent in equilibrium with the adsorptive species at a given chemical potential and temperature. This ensemble presents the advantage to allow the number of particles to fluctuate, rendering possible an estimation of the amount of adsorbed molecules for fixed values of the temperature and the chemical potential by averaging over the course of the simulation. These calculations can then be directly compared to the experimental data obtained by gravimetry/volumetry/manometry measurements where the equilibrium conditions, T and μ of the gas inside and outside the adsorbent, are exactly mimicked in the grand canonical ensemble. The imposed chemical potential is usually calculated from an equation of state that represents the ideal or non-ideal behaviour of the gas of interest at the gas-phase temperature and pressure and alternatively using the Gibbs ensemble formulation.

Starting with an initial configuration randomly generated, the MC simulations consist of several millions of random moves that allow an efficient sampling of the selected ensemble. During the GCMC simulation, the considered moves include translational and rotational displacements of the molecules, while attempts to insert and to remove molecules ensure a variation in their number.

The corresponding random moves are accepted or rejected with appropriate criteria.

A translation/rotational displacement is accepted with the following probability (see Equation 6.11):

$$P = \min\{1, \exp(-\beta \Delta U)\} \tag{6.11}$$

where ΔU is the change in the total potential energy and $\beta = 1/kT$. The attempt is thus accepted if ΔU is negative or if the magnitude of the potential energy change is lower than a random number ranging between 0 and 1. This criterion is based on the Metropolis algorithm (Metropolis et al., 1953), at the core of all the MC simulations.

In a creation trial where a selected adsorbate molecule is placed in a random position and orientation, the probability of an acceptance of the new configuration is given by Equation (6.12):

$$P = \min\left\{1, \frac{\beta f V}{N+1} \exp(-\beta \Delta U)\right\} \tag{6.12}$$

where f is the fugacity of the gas-phase adsorptive.

Similarly, for a deletion step where a molecule is randomly removed, the new configuration is accepted with the probability given by Equation (6.13):

$$P = \min\left\{1, \frac{N}{\beta f V} \exp(-\beta \Delta U)\right\} \tag{6.13}$$

In the case of gas mixtures, an identity change, commonly called swap, trial is also employed to obtain a faster convergence. This move that consists of converting one randomly selected molecule of type A to type B, with A and B being two different components of the mixture, is accepted with the criteria fixed by Equation (6.14):

$$P = \min\left\{1, \frac{f_B N_A}{f_A (N_B + 1)} \exp(-\beta \Delta U)\right\} \tag{6.14}$$

where f_A and f_B are the fugacities of the components A and B in the gas-phase adsorptive, respectively, and N_A and N_B are the number of molecules.

Using this methodology, a set of configurations that converge towards the specified chemical potential and temperature is generated. The simulation usually takes several millions of step to equilibrate from its original random starting point. The evolution of the total energy over the number of MC steps is usually plotted in order to control the equilibrium conditions. Acceptance rates for each possible trial have to be judiciously adjusted in order to approach the equilibrium in a most efficient way. Usually, the acceptance rate is fixed to be about 0.4–0.5. For accurate statistical results, the steps performed prior to equilibration have to be excluded in the analysis and the averages of interest are calculated over several millions of configurations.

One should bear in mind that such a procedure is valid for treating the adsorption of simple molecules including CO_2, CH_4, N_2, H_2, He, etc. in a porous solid which does not undergo any significant structural change upon adsorption as the adsorbent is maintained fixed during the simulation. For more complex adsorbates (long chain alkanes, xylenes, etc.), which show a high degree of flexibility or/and large size, various statistical bias techniques have been developed to increase the acceptance rate of some trial moves that speeds up the MC simulation by allowing a more efficient sampling of the configurational space. As a typical illustration, the configurational bias technique based on a scheme derived by Rosenbluth (Frenkel and Smit, 2002) consists of regrowing the molecule, atom by atom within the porosity instead of inserting the entire molecule. Such a bias was widely employed for accurately capturing the adsorption behaviour of long chain alkanes in zeolites (Fuchs and Cheetham, 2001). Another popular approach is the cavity bias proposed by Mezei (1980), in which the insertion trials are only attempted in the accessible regions of the pores that can accommodate a molecule. Acceptance rules are associated with each biased move ensuring the generation of new configurations within the appropriate Boltzmann distribution. Further, if the adsorbent shows drastic structural modification upon adsorption as, for instance, the breathing MOFs or some flexible narrow pores/windows zeolites, the standard MC approach fails and it is recommended to include MD steps in a hybrid osmotic Monte Carlo (HOMC) scheme to allow an efficient sampling of the volume fluctuation of the adsorbent (Ghoufi and Maurin, 2010). The details of all these bias and more sophisticated techniques are out of the scope of this chapter and the readers are again referred to a more specialized literature (Nicholson and Parsonage, 1982; Allen and Tildesley, 1987; Frenkel and Smit, 2002).

The adsorption isotherm for either single gas component or mixtures is the thermodynamic data most commonly extracted from a GCMC simulation by running a series of calculations at different fixed chemical potentials (or pressure/fugacity). The readers should bear in mind that *the amounts of adsorbed molecules obtained from the simulation correspond to total (also called 'absolute') amounts, that is, to the total amounts within the pores (n_{total}), while more generally in experiment it is the surface excess amount n^σ (n_{excess}) that is measured* (see Chapter 2). Indeed the simulated (total) and experimental (surface excess) data need to be converted into each other via Equation (6.15)

$$n^\sigma = n_{total} - v_{pore}\rho_{gas} \tag{6.15}$$

where v_{pore} and ρ_{gas} are the pore volume and the gas-phase density, respectively, to ensure a consistent comparison.

Another property of interest is the differential molar enthalpy of adsorption $\Delta_{ads}h$ which is commonly evaluated through the fluctuations over the number of molecules in the system N and the internal energy U by means of Equation (6.16):

$$\Delta_{ads}h = RT - \frac{\langle U \cdot N \rangle - \langle U \rangle \langle N \rangle}{\langle N^2 \rangle - \langle \langle N \rangle \rangle^2} \tag{6.16}$$

where the brackets $\langle \cdots \rangle$ denote an average in the grand canonical ensemble.

This method assumes an ideal behavior of the gas phase and in practice long GCMC simulations are required to get a reasonable statistics for the averages in this equation, more particularly when applied in the limit of low coverage. Alternatively, more elaborated methods have been recently introduced for allowing a more efficient determination of this energetics based on a revised Widom's test particle approach employed in the canonical ensemble where the number of molecules, the volume and the temperature held fixed (Vlugt et al., 2008). Note that the Henry coefficient is also accurately determined using this strategy. The so-obtained data can be directly compared to the differential adsorption enthalpy issued from microcalorimetry. The interest of simulating the enthalpy is also to decompose the resulting value into two components, the adsorbate–adsorbate and adsorbate–adsorbent, in order to further explain the trend of the experimental enthalpy as a function of the loading. The integral molar enthalpy of adsorption can also be calculated from MC simulations. The readers can refer to the recent paper reported by Do et al. (2011) for full details on the equations employed.

Usually, once a good agreement is obtained between the simulation and the experiment for both macroscopic thermodynamic data, that is, isotherm and enthalpy, the following step consists of analysing in detail the molecular configurations at equilibrium, namely, snapshots, and extracting radial distribution functions calculated from this large sampling to gain a further understanding of the microscopic adsorption mechanism. This structural analysis is sometimes also realized in the canonical ensemble for a fixed volume, temperature and number of molecules.

Finally, from the GCMC simulations realised for a mixture of gases, it is possible to determine the separation ability of the porous material for component A relative to component B via the estimation of the corresponding selectivity, labelled as $S_{A/B}$ which is calculated using Equation (6.17):

$$S_{A/B} = \frac{x_A}{x_B} \cdot \frac{y_B}{y_A} \tag{6.17}$$

where x_i and y_i correspond to the mole fractions of component i in the adsorbed and bulk gas phases, respectively.

While GCMC calculations for the simplest adsorbate/adsorbent systems can be routinely realized using commercial software, the readers should pay attention that all the features required for accurately treating the adsorption behaviour of complex molecules such as the bias described above are implemented to avoid the generation of questionable results. In this latter case, we recommend that the experimentalists referred to more appropriate academic codes including Music, CADSS, Raspa, etc.

The following sections will provide through appropriate illustrations how GCMC simulations can be a valuable tool for the experimentalist not only to predict the optimal adsorption/separation performances of the porous solids but also to elucidate the microscopic mechanism sometimes at the origin of unusual adsorption behaviours.

6.5.1.2   Simulation/Prediction of the Adsorption Performances

6.5.1.2.1   Single Component Adsorption

GCMC simulations are most often employed to preliminary-simulate the adsorption isotherms of a series of gases in a given porous solid once a judicious choice of the microscopic models for both the adsorbent and the adsorbate, and the adsorbate/adsorbent force field parameters (see Sections 6.2 and 6.3) has been made. A further comparison between the calculated total amount of adsorbed molecules and the corresponding experimental data needs to be analysed with caution. Indeed, the simulations assume an ideally activated porous material which is not always true from an experimental standpoint. At first sight, the large discrepancy observed, for instance, for the adsorption of CH_4 in the terephtalate MIL-68(Al) MOF material (Yang et al., 2012b) characterized by the coexistence of hexagonal and triangular channels (Figure 6.8) would be attributed to a failure of the simulations. However, a careful characterization of the experimental samples using infra-red spectroscopy evidenced that the narrow triangular channels still contain residual organic compounds not

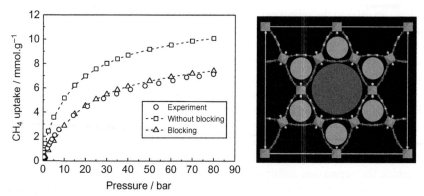

Figure 6.8 Left: comparison between the simulated adsorption isotherms for CH$_4$ with blocking/unblocking the triangular pores and the experimental manometric data at 303 K. Right: view of the crystalline structure of the MIL-68(Al), the circles denoting the triangular and hexagonal channels (Yang et al., 2012b). (For colour version of this figure, the reader is referred to the online version of this chapter.)

removed during the activation protocol that render this region inaccessible for the adsorbate. Based on this information, the simulation of the isotherm was repeated by blocking the triangular channels and led to a very good agreement with the experimental adsorption manometry measurements. It is noticeable that such an accordance is obtained using generic force fields available in the literature, CH$_4$ being treated with the single LJ site model (TraPPE force field; Martin and Siepmann, 1998) while the MIL-68(Al) LJ parameters were taken from the DREIDING force field (Mayo et al. 1990). In such a porous solid that does not contain any strong specific adsorption sites (chemical defects, grafted function, cus, etc.), it was evidenced that other standard force field such as UFF (Rappé et al., 1992) also leads to very similar results. Indeed, as mentioned in Section 6.3, a selection of force field parameters issued from the literature reproduces reasonably well the interactions for simple 'adsorbate/adsorbent' systems. This conclusion also holds true even for other adsorbates including CO$_2$ and H$_2$S that are expected to interact strongly with the MIL-68(Al) framework via electrostatic interactions. It was also observed that the method used for extracting the charges (Mulliken or Electrostatic Potential, ESP) for the MIL-68(Al) framework does not significantly impact the isotherms. This conclusion can be transferable to all porous solids whose surfaces do not show any specific adsorption sites. In contrast, it is clearly stated that once a polar function such as —NH$_2$ is grafted on the organic node of this MOF, the ESP charges much better accurately describe the interactions at play. This observation is expected to be generalized to any types of functionalized porous solids.

In spite of their different meaning in the presence of micropores (cf. Section 6.4.2), the comparison of the experimental BET specific surface area with the accessible surface area calculated from the crystal structure is often considered as a useful starting point. Based on standard force field for both adsorbate and adsorbent, the amount of adsorbed ethane in the terephtalate MIL-53(Cr) was, for instance, over-predicted

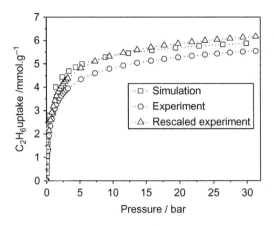

Figure 6.9 Left: Comparison between the simulated adsorption isotherm and the experimental raw and rescaled gravimetric data for ethane in MIL-53(Cr) at 303 K (Rosenbach et al., 2010).

(Figure 6.9) as the a_{BET} of the investigated sample was significantly lower than the theoretical accessible surface value (1350 vs. 1540 m^2/g) (Rosenbach et al., 2010). Nevertheless, once the simulated adsorption isotherm is available, it is more consistent to compare the BET areas calculated on the experimental and simulated isotherm, respectively. In the instance just reported, the authors made use of a scaling factor corresponding to the $a_{sim}(BET)/a_{exp}(BET)$ ratio to get a fair agreement between simulation and experiment, as shown in Figure 6.9.

These two latter examples emphasize that characterizing experimentally the investigated sample is a crucial pre-requisite prior to analysing the comparison between experimental and simulated data. Indeed, it is rather tricky for the modellers to select the most representative experimental data available in the literature to be compared with the simulations as the adsorption uptake of a porous solid varies considerably with the quality of the sample and its mode and extent of outgassing (an aspect often underestimated). This choice is even more critical when the experimental isotherm is fitted to refine the force field parameters. Such a procedure conducted using isotherms collected on a poorly defined sample would lead to force field parameters with unphysical meaning and obviously without any possible transferability to another porous solid. Behind this careful definition and characterization, if the discrepancy between experiments and simulations still persists, this can also come from a gradual structural change of the solid upon adsorption and *in situ* X-ray diffraction measurements can be helpful. If not, this can be attributed to an inefficiently of the simulations and different adsorbate/adsorbent force field parameters and set of charges need to be tested before envisaging a time-consuming derivation of specific force field via quantum calculations.

Beyond a direct comparison with experimental data, GCMC simulations can be used as a predictive tool to assess the optimal adsorption performances of new crystalline porous solids for a wide variety of gases including those with high environmental and economic interests such as H_2, CO_2 and CH_4. The idea behind is to computationally define for each application the best materials for which a special effort will be deployed for optimizing the synthesis/activation protocol for further

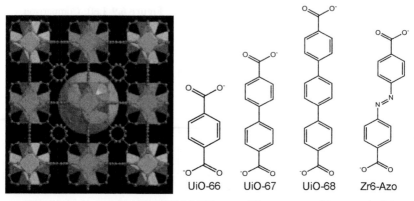

Figure 6.10 Illustration of the UiO-66(Zr) MOFs crystalline structure. The organic linkers for the extended series of Zr-MOFs. (For colour version of this figure, the reader is referred to the online version of this chapter.)

experimental validation. As an illustration, the CH_4 and CO_2 storage capacities have been predicted for an extended isoreticular series of Zr-MOF (Yang et al., 2012a), which derives from the emerging 3D UiO-66(Zr) solid where the initial terephtalate linker is substituted by longer organic spacer (see Figure 6.10).

This modelling approach first requires a computational-assisted structure determination strategy conducted in tandem with X-ray powder diffraction (XRPD) experiments to propose plausible structure candidates for each investigated Zr-MOF. Such a procedure can be generally employed for crystallized porous solids. Indeed, starting with the parent UiO-66(Zr), the crystallographic structure for each Zr-MOF analogue is preliminary built using a ligand replacement strategy based on the unit cell parameters deduced from XRPD experiments and the resulting structures are further refined using a quantum-based geometry optimization procedure. Primarily, one of the listed materials, here the parent UiO-66(Zr), is selected as a model material since its experimental and simulated S_{BET} and v_{pore} features concur very well. A very good agreement between the isotherms simulated for CH_4 and CO_2 in this material and the corresponding experimental data allows a validation of the computational approach (microscopic structure model, force field parameters, partial charges, etc.) to further confidently carry out the explorations for the whole set of Zr-MOFs (see Figure 6.11). UiO-67(Zr) was predicted to show a storage capacity for CH_4 (146 cm^3 (STP)/cm^3) at 35 bar that is only slightly below the target set by the U.S. department of Energy (DOE) for such an application (180 cm^3 (STP)/cm^3). This MOF-type material was found to be particularly promising as it outperforms other conventional adsorbents such as the activated carbon Maxsorb and the zeolite 13X.

In the meantime, UiO-68(Zr) was predicted to show the highest CO_2 adsorption capacity (333 cm^3 (STP)/cm^3) at 40 bar which considerably exceeds the performances of the usual activated carbons and zeolites. The working capacity, defined as the difference in the capacities between the adsorption and desorption pressures (1 bar was chosen here for the desorption), was simulated to be very high and remains similar to

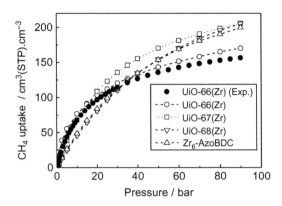

Figure 6.11 Simulated adsorption isotherms for CH_4 in the Zr-MOFs at 303 K. The experimental data are also provided for UiO-66(Zr) as a comparison (Yang et al., 2012a).

its storage capacity (313 vs. 333 cm^3 (STP)/cm^3). Such a favourable behaviour is usually related to an adequate non-rectangular shape of the corresponding simulated CO_2 adsorption isotherm which is not obtained for other hybrid materials, that is, the cus-containing MOFs as, for instance, the MIL-100(Cr) (Hamon et al., 2012), which show working capacities that are drastically decreased by ~50% compared to their storage capacities due to their very high affinity of CO_2 at low loading. Further, a relatively low predicted CO_2 adsorption enthalpy (-20 kJ/mol[1]) compared to the most common adsorbent zeolite 13X (-45 kJ/mol[1]) suggests that such a Zr-MOF can be potentially regenerated without requiring costly operating conditions. With this in hand, a careful optimization of the activation procedure is deployed with a preliminary target to be as close as possible to the theoretical accessible surface area. Once this step is achieved, lab-scale adsorption measurements are realized to validate the predictions prior to pave the way for the uses of such porous solids in physisorption-based processes.

6.5.1.2.2   Gas Mixture Adsorption

As mentioned above, apart from the adsorption isotherms and enthalpies that can be calculated for different gas mixture compositions and compared to the corresponding experimental data, the selectivity of a porous solid which measures its separation ability for a given gas mixture is the major thermodynamic property of interest. The adsorbate/adsorbent force field parameters and the set of charges for the adsorbent (when required) are usually preliminary validated by a good agreement between simulated and experimental single component adsorption isotherms and enthalpies. Once this step is successfully achieved, as mentioned above, most commonly the Lorentz Bertholot mixing rule is employed for the standard gas mixture to describe the cross adsorbate/adsorbate LJ interactions. As a typical illustration, such a protocol has been followed to explore the CO_2/CH_4 selectivity of a series of functionalized UiO-66(Zr) with different polar groups (—Br, —NH_2, —NO_2, —$(CF_3)_2$, —$(OH)_2$, —SO_3H and —CO_2H) grafted on the terephtalate linker (Yang et al., 2011c). The pre-combustion removal of CO_2 from its gas mixture with CH_4 is indeed of great economic and technology importance in the treatment of low-quality natural gas such as

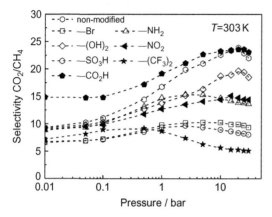

Figure 6.12 Simulated selectivities for CO_2 over CH_4 from their equimolar gas mixture at 303 K in the non-modified UiO-66(Zr) and its various functionalized forms, as a function of the bulk pressure (Yang et al., 2011c).

biogas and landfill gases. In that case, CH_4 and CO_2 were treated by the standard TraPPE (Martin and Siepmann, 1998) and EPM2 (Harris and Yung, 1995) force fields, respectively, while the LJ parameters for all atoms of the UiO-66s were taken from the generic UFF (Rappé et al., 1992) force field and their charges extracted using the ESP scheme.

Figure 6.12 reports the CO_2/CH_4 selectivity for an equimolar mixture simulated in the modified UiO-66(Zr) as a function of the bulk pressure. It has been predicted that the functionalization with $-SO_3H$ and $-CO_2H$ groups leads to even higher selectivities compared to the grafting of $-NH_2$ functions that is generally considered to improve the performance of various porous solids for such mixture separation. Further, the resulting selectivities for both solids that range from 17.0 to 23.0 are at least comparable with those reported for the most commonly used zeolite faujasite 13X (Cavenati et al., 2004) in Pressure Swing Adsorption applications. These materials were also predicted to involve medium-ranged CO_2 adsorption enthalpy values (~ -30 kJ/mol) suggesting a potential regenerability under mild conditions.

Using the same strategy, the CO_2/N_2 separation performances of a series of MOF-type materials have been simulated at 303 K, considering the standard TraPPE force field for representing N_2. Figure 6.13 shows the selectivities at 303 K and 1 bar for the binary CO_2-N_2 mixture with a bulk composition of 15–85 corresponding to the typical operational condition for CO_2 separation from flue gas emitted from the power plants. It has been predicted that UiO-66(Zr)-2(CO_2H) shows the best selectivity (~ 95) for such a gas mixture. A synthesis effort has been further orientated towards the preparation of this functionalized form. This predicted selectivity was further supported by volumetric adsorption measurements realized under the same conditions.

One should bear in mind that the validation of such predictions is much more complicated than for the single gas adsorption as the multi-component adsorption isotherms are relatively scarce in the literature due to the complexity of the experimental protocol. Indeed, the simulated data issued from molecular simulations are most often compared to the results obtained from simple thermodynamic models such as the ideal adsorbed solution theory (IAST) introduced by Myers and Prausnitz

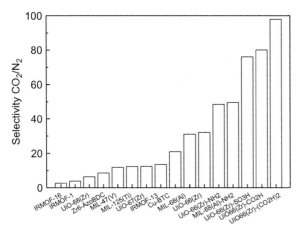

Figure 6.13 Simulated
selectivities for CO_2 over N_2
from a gas molar mixture
(15/85) at 303 K and 1 bar in a
series of MOFs.

(1965) that predict the mixture adsorption behaviour based solely on the consideration of the single component isotherms or more elaborated models such as the real adsorbed solution theory (RAST) one (Yun et al., 1996).

One should notice that such a comparison is also helpful to validate the choice of a thermodynamic model to be applied for predicting the multi-component behaviour of a class of porous solid. A typical example is provided in Figure 6.14 for the co-adsorption of the binary mixture CO_2/CH_4 in the MOF-type MIL-47(V) (Llewellyn et al., 2013). The very good agreement obtained between the amount adsorbed of CO_2 and CH_4 predicted by GCMC simulations and both IAST (Myers and Prausnitz, 1965) and Wilson-VST (Suwanayuen and Danner, 1980) macroscopic models allows the validation of such macroscopic models to preliminary examine the separation properties of this family of MOFs.

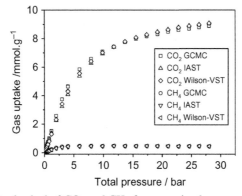

Figure 6.14 Amount adsorbed of CO_2 and CH_4 for an equimolar gas mixture CO_2/CH_4 at 303 K in MIL-47(V): GCMC simulations, IAST and Wilson-VST macroscopic models (Llewellyn et al., 2013).

6.5.1.3   Understanding of the Microscopic Adsorption Mechanism

Beyond the prediction/simulation of the adsorption isotherms and enthalpies, GCMC simulations are valuable to gain further understanding of the microscopic mechanism at play and complement usually very well different types of experimental *in situ* measurements including infra-red spectroscopy and X-ray/neutron diffraction. Such a microscopic level analysis is usually assumed to be reliable only when one finds a reasonable agreement between the simulated and experimental macroscopic properties (isotherms, enthalpies, etc.). As a first typical example, we can consider the adsorption of CO_2 in the DaY, NaY and NaX faujasite zeolites that differ by their Si/Al ratios (Maurin et al., 2005). While a standard force field was used to describe the dealuminated DaY form modelled as a purely siliceous Y faujasite, the two other cation-containing zeolites were treated using a specific force field as detailed in Section 6.3. Starting with a fair experimental-simulation agreement for both the adsorption isotherms and enthalpies at low coverage (Figure 6.15), a careful analysis of the snapshots generated by the GCMC simulations for a wide range of pressure was achieved by constructing 2D density probability and/or plotting radial distribution functions between atoms pair of interest. It was shown that while CO_2 molecules are more or less homogeneously distributed within the supercage of the purely DaY in the whole range of pressure due to the absence of specific adsorption sites, the scenario significantly differs for the cation-containing faujasites where CO_2 strongly interact with Na^+ present in the supercages via a linear ($Na^+...O=C=O$) geometry. Such preferential arrangements have been validated with the conclusions drawn from *in situ* infra-red measurements realized under the same conditions.

Elucidating the adsorption mechanism further allows the interpretation of the adsorption enthalpy profile issued from microcalorimetry measurements performed in a wide range of pressure. Indeed, in DaY, the progressive increase in the adsorption enthalpy with the gas loading (or pressure), also reproduced by the GCMC simulations, was assigned to the interaction between CO_2 and the homogeneous energetic DaY surface giving rise to an almost constant CO_2/adsorbent interaction whatever

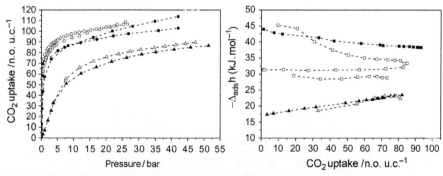

Figure 6.15 Comparison between the simulated (solid symbols) and experimental (empty symbols) adsorption isotherms (left) and differential molar enthalpies (right) for carbon dioxide at 303 K in DaY (triangle symbols), NaY (square symbols) and NaX (circle symbols) (Maurin et al., 2005).

the gas pressure. Combined with the general increase in the adsorbate/adsorbate energetic contribution when the CO_2 loading increases due to a shortening of the distances between adsorbates, it results in an increase in the differential molar enthalpy with the CO_2 pressure. In contrast, the flat enthalpy profile obtained for NaY is attributed to a balance between an increase in the adsorbate/adsorbate interactions and a slight decrease in the CO_2/adsorbent interaction consecutive of a progressive expansion of the coordination sphere around the single possible Na^+ adsorption sites. Finally, the decreasing enthalpy evolution for NaX is due to the energetic heterogeneity of its surface with the existence of two distinct Na^+ sites which are sequentially occupied along the adsorption process. Their differences in terms of affinity for CO_2 are large enough to lead to a drastic decrease in the adsorbate/adsorbent interaction when the CO_2 loading increases, which is not compensated by the increase in the CO_2/CO_2 interaction energy term.

The readers should bear in mind that while a good accordance between the simulated and the experimental adsorption enthalpies in the whole range of pressure usually ensures an accurate description of the adsorption mechanism that is most often validated by in situ *characterization techniques, this is not necessarily true with only a reasonable overall agreement for the isotherms.* The adsorption of CH_4 in the HKUST-1 MOF containing copper as coordinative unsaturated metal sites is a typical illustration (Chen et al., 2011). Indeed, while the generic force field UFF (Rappé et al., 1992) for representing the MOF framework combined with various microscopic models for CH_4 leads to a GCMC-simulated adsorption isotherm at 77 K in fair agreement with the experimental data in terms of shape and maximum uptake (see Figure 6.16), the step being only shifted towards a lower CH_4 concentration (40 molecules/u.c. vs. 85 molecules/u.c.), it dramatically fails to reproduce the microscopic adsorption mechanism evidenced by *in situ* X-ray diffraction experiments. It was further shown that the implementation of a quantum chemical-derived PES in the GCMC-scheme allows a better reproduction of the position of the step in the experimental adsorption isotherm (see Figure 6.16). It further leads to a much more precise description of the preferential adsorption sites for CH_4 within the MOF

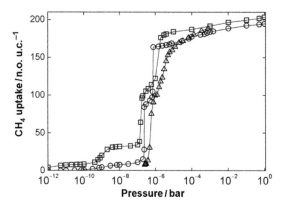

Figure 6.16 Comparison between the GCMC-simulated (UFF force field and quantum chemical potential energy surface with square and circle symbols, respectively) and experimental (triangular symbols) adsorption isotherms for methane at 77 K in HKUST-1.
Courtesy: Chen et al. (2011).

porosity, more particularly around the cus, regions that were not experienced by CH_4 using the generic UFF force field (Rappé et al., 1992). Indeed it is clearly emphasized that one can roughly reproduce the adsorption isotherm using models/force fields which are further not able to capture the mechanism behind.

Computational methods can also play a crucial role in order to deepen the understanding of the adsorption mechanism at the atomic level in complex adsorbent systems that are suspected to undergo concomitant structural changes. A flexible behaviour of a microporous solid upon adsorption leads to an unusual presence of steps or plateau in the isotherm as for MFI and silicalite-1 zeolites (Snurr et al., 1994; Jeffroy et al., 2011) with various halocarbons adsorbates or more spectacularly for the series of breathing MOFs (see Chapter 14). In such a situation, the approximation of a rigid framework for the adsorbent with atoms fixed at positions derived by X-ray diffraction for the 'empty' solid is not anymore valid and, as mentioned in Section 6.3, a specific flexible force field is required to accurately capture the magnitude of the structural modifications upon adsorption. This complex adsorption behaviour needs to be considered with a special attention as the standard GCMC simulations are not appropriate. A more elaborated and time-consuming approach consists of combining MD and GCMC techniques in an HOMC scheme (see specialized references for more details; Allen and Tildesley, 1987; Frenkel and Smit, 2002). The implementation of MD moves is crucial to allow an efficient sampling of the volume fluctuations for the framework. This is particularly true when the volume variations reach spectacular values as it is the case for the MIL-53(Cr) solid upon CO_2 adsorption, the transition between a NP and a LP form implying a unit cell volume change of $\sim$40% (Férey et al., 2011). Such a computational effort combined with a reliable force field for describing the spectacular breathing behaviour of the MIL-53(Cr) solid (see Section 6.3) is the unique way to capture quantitatively (transition pressure, amount adsorbed) the stepwise shape of the experimental CO_2 adsorption isotherm in the whole range of pressure as shown in Figure 6.17. Apart from the reproduction of the adsorption isotherm and the elucidation of the preferential location of the adsorbed molecules (see Chapter 14) also usually obtained for non-flexible

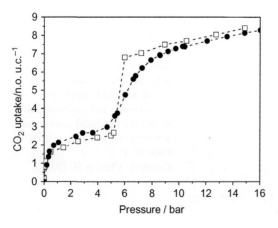

Figure 6.17 Comparison between the hybrid osmotic Monte Carlo simulated (empty squares) and experimental (full circles) adsorption isotherms for carbon dioxide at 303 K in MIL-53(Cr) (Ghoufi et al., 2012).

framework from standard GCMC, this more sophisticated approach allows to get insight into the physics behind such as the drastic structural change of the adsorbent. Indeed, it has been demonstrated that there is a critical interplay between CO_2 and MIL-53(Cr), their interactions inducing first a soft mode in the host framework which is a crucial pre-requirement for further initiating the structural transition of this porous solid (Ghoufi et al., 2012).

6.5.1.4   Screening of a Large Series of Porous Solids

In complement to high-throughput combinatorial chemistry methods and adsorption measurements, there is actually a growing interest to develop large-scale computational tools not only to design novel porous solids but also to assess the adsorption/separation performances of an extensive library of existing and hypothetical porous solids. Such an approach aims not only to avoid the deployment of time-consuming experiments, both for the synthesis of the samples and the characterization/adsorption measurements, but also to probe the adsorption performances of porous solids under severe conditions difficult to be attained in lab-scale experiments as, for instance, their abilities to capture highly risky gases including CO or H_2S. The development of such a large-scale predictive computational method is still at its early stage. As a typical illustration, recent attempts have been able to successfully generate more than hundred thousand hypothetical MOFs and further scan their adsorption abilities for methane storage applications (Wilmer et al., 2011). With the hypothetical MOFs in hand, an efficient screening protocol was employed to determine their methane storage capacities under operating conditions (35 bar, 298 K). This was realized in three steps: the first one consisted of roughly estimating the adsorption uptake via very short GCMC cycles to narrow down the list of promising materials, while the second and the third stages were conducted with an increasing number of GCMC cycles to refine the predicted property for the 5% best MOFs of each previous stage (see Figure 6.18). It came

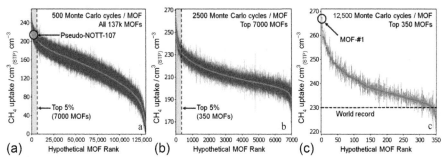

Figure 6.18 Illustration of the three stages of the screening protocol to identify the best MOFs for methane storage at 35 bar, 298 K. (a) GCMC predictions for ~137,000 MOFs using short simulations and (b and c) refinement of the predictions for the top 5% MOFs of the subsequent stages using longer simulations to decrease the statistical errors. (For colour version of this figure, the reader is referred to the online version of this chapter.)
Courtesy: Wilmer et al. (2011)

from this study that a Cu-containing MOF is predicted to store up to $267 \, cm^3(STP)/cm^3$ of methane, well above the U.S. DOE target ($180 \, cm^3(STP)/cm^3$) and the performance of other porous solids.

A large-scale computational strategy was also employed for screening a wide variety of porous MOFs ($\sim$500 materials) for the selective adsorption of a binary CO_2/N_2 mixture (Haldoupis et al., 2012). In contrast to CH_4, simulating the CO_2/N_2 selectivity is more complex as it requires an accurate description of the adsorbate-MOF framework electrostatic interactions. Indeed, semi-empirical approaches such as the charge equilibration PQeq method (Wilmer and Snurr, 2011) is usually favoured as a preliminary step for examining thousands of porous solids in order to avoid the time-consuming quantum chemical charge calculations while maintaining a reasonable accuracy. The Henry's constants for both gases were first simulated at 303 K for all MOFs to define a short list of promising materials with an infinite dilution CO_2/N_2 selectivity higher than 100. Using this criterion, the predicted properties of the 11 best MOFs were refined carefully at finite dilution using both GCMC and MD simulations based on a more accurate description of the partial charges. Two MOFs emerged from this study with high CO_2/N_2 selectivities at 1 bar/303 K (binary mixture CO_2/N_2: 15/85) ranging from 200 to 270 making these porous materials very promising for the purification of flue gas.

While considerable progress has been achieved since the last few years, there are still efforts under deployment in order to set up new algorithms able to systematic screen several thousands of porous solids in reasonable time scales.

6.5.2   Quantum Chemical Calculations

6.5.2.1   Basic Principles

In contrast to the force field-based modelling approaches described in the previous sections that do not explicitly consider any electron in the system, the quantum chemical calculations aim to provide a mathematical description of the behaviour of all the electrons of the explored system. This requires the resolution of the Schrödinger equation that cannot be achieved exactly for any system other than the hydrogen atom. Indeed, the field of quantum chemistry relies on approximate solutions that need to be accurate enough for allowing a favourable comparison between the simulations and the corresponding experimental data. Such a computational approach first requires a microscopic model to represent the porous solids that can be either the periodic structure of the crystalline solids or a cluster judiciously selected to mimic the local environment of the structure where the adsorption phenomenon predominantly occurs. The quantum mechanical method further needs to be specified for describing the electrons present in the model. Several *ab initio* and density functional theory methods can be employed, their reliability strongly depending on the size of the selected microscopic model. In practice, the highly accurate *ab initio* methods such as the Coupled Cluster approach are only applicable for clusters with a very limited number of atoms. At the opposite, density functional theory calculations based on an empirical correction of the long-range dispersion term are computationally less demanding than most of the *ab initio* methods and are thus widely employed for treating the porous solids in a periodic approach. The readers can refer to some highly detailed books and articles on the theoretical foundations of quantum chemistry and the description of the different possible methods (Sauer, 1989; Sauer et al., 1994; Szabo and Ostlund, 1996; Koch and Holthausen, 2000; Sholl and Steckel, 2009).

Basically, while the force field modelling approach is based on an appropriate choice of the interatomic force field parameters, the quantum mechanical method relies on a critical selection of the most reliable basis set into which the orbitals or plane waves are expanded and/or the exchange correlation energy functional, to ensure a qualitative agreement between the simulated and experimental property of interest. *Indeed, one should bear in mind that the quantum chemical strategy is not more accurate than the force field approach, but both methods are fully complementary.* A typical illustration is provided in Section 6.3.3 for the HKUST-1/CH_4 system where a reliable description of the interaction between the adsorbate and the cus of the MOF was first derived at the quantum level and further transferred to a much faster computational GCMC approach to explore the adsorption property of the solid. Another example of such interplay is the use of hybrid or embedded model strategy to sometimes treat the adsorption in porous solids where the system is partitioned into two parts, the inner core usually delimiting the adsorption sites of the solid and the adsorbed species treated at the quantum level, while the rest of the atoms outside this region are described by force field-based approach.

The quantum chemical calculations performed on either cluster or periodic models are nowadays widely employed to provide a complete microscopic description of the adsorption process both geometrically and energetically. This is encouraged by the development of high computing power and efficient quantum chemical software all available with commercial licenses (Crystal09, Gaussian 09, Molpro 2009, VASP 5.2). Indeed, a full geometry optimization of the considered adsorbate/adsorbent allows the determination of not only the positions of the confined molecules but also the local or long-range structural changes of the host porous solids upon adsorption (redistribution of the extra-framework cations in zeolites, rearrangement of small to large unit cell volume changes in zeolites and MOFs, etc.). Such predictions are usually validated by a direct comparison with X-ray/neutron diffractions, EXAFS or infra-red measurements. Based on these geometry optimized structures, it is further possible to simulate two quantities of upmost importance in the field of adsorption:

1. the vibrational frequencies of the adsorbed species. In tandem with infra-red spectroscopy, quantum chemical calculations can gain insight not only into the strength of the adsorbate/ adsorbent interactions and the geometric arrangements of the adsorbates (nature of the interacting atom types, characteristic orientations/distances of the adduct, coordination numbers, etc.) but also into the characterization of the interacting adsorption sites as, for instance, its acid/basic Brönsted or electron donor/acceptor natures or its oxidation state.
2. the adsorbate/adsorbent interaction energy ΔE^{el} (usually called the binding energy) which is obtained by the difference between the energy of the adduct E(porous solids/adsorbate) and those of its single constituents, that is, E(porous solids) $+ E$(adsorbate). The adsorption enthalpy can be further calculated by first correcting ΔE^{el} with the zero point vibration

energy (ΔZPE) usually obtained by using the harmonic approximation, which refers to the interaction energy at the absolute zero $\Delta U(0) = \Delta E^{el} + \Delta ZPE$. This term $\Delta U(0)$ can be further corrected by the thermal effect to relate to the adsorption enthalpy $\Delta H(T)$ measured at a given temperature. Indeed, a thermal energy contribution (ΔE^{Te}) needs to be included which is often derived for ideal gas where $1/2\,RT$ is added for each translational or rotational degree of freedom lost upon gas adsorption, the contribution of the solid with and without adsorbate being neglected. One further needs to account for the pV term ($\Delta H = \Delta U + pV$) which is usually taken from ideal gas equation $pV = RT$.

Indeed the integral molar adsorption enthalpy can be defined by Equation (6.18).

$$\Delta H(T) = \Delta E^{el} + \Delta ZPE + \Delta E^{Te} - RT \qquad (6.18)$$

The resulting values can be directly compared to those determined from microcalorimetry measurements, the simulation-experiment agreement strongly depending on the accuracy of the quantum chemical method/model employed.

The following section provides some typical examples of comparison between adsorption properties issued from quantum-based calculations and experiments.

6.5.2.2 Illustrations

As a first typical example, the interaction between CO_2 and the series of alkali cation-exchanged faujasite Y was investigated by means of density functional theory cluster calculations (Plant et al., 2006). A fair microscopic representation of the zeolite was ensured by the selection of a large 200-atom cluster cut out from the NaY crystal structure, which is centred on the interacting cation site and includes the whole supercage of the faujasite. Such an excised cluster generates dangling bonds that were saturated with hydroxyl groups pointing away from each other to avoid unrealistic interactions. A preliminary validation step of the microscopic model (cluster) and quantum mechanical method (here density functional theory based on the use of the functional PW91 and the double numerical basis) consists of comparing the structure parameters issued from the geometry optimized clusters in the absence of adsorbates, where the positions of both cations and zeolite framework atoms are allowed to fully relax, with those collected by X-ray/neutron diffraction measurements for the corresponding periodic zeolite solids. Here, this step was achieved by a good agreement between the experimental and simulated geometric features for each cation-exchanged form including the M^{n+} (M = Li, Na, K, Cs) – framework oxygen distances and the Al—M^{n+}—Si angles, both used to evaluate the deviation of the cation positions from the plane of the six-membered ring windows of the supercage (see Figure 6.19). The geometries for each M^{n+}-cluster in the presence of CO_2 were further optimized. It was shown that the adsorbate molecules form a quasi-linear ($M^{n+} \cdots O{=}C{=}O$) adduct for all cations, the characteristic M^{n+}—$O(CO_2)$ distances increasing from Li^+ to Cs^+ consistent with infra-red measurements performed on similar CO_2/cation-containing zeolites. Further, while the adsorption induces a significant displacement of Li^+ from the six-membered ring windows, the other cations mainly remain in their initial crystallographic positions.

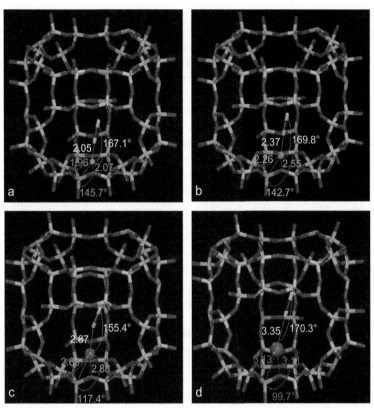

Figure 6.19 Optimized loaded clusters with one CO_2 molecule issued from cluster density functional theory calculations: Li^+ (a), Na^+ (b), K^+ (c) and Cs^+ (d). Both the geometric parameters for M^{n+}-zeolite (M^{n+}—O(zeolite)) distances and the characteristic angles Si—M^{n+}—Al) and for M^{n+}—CO_2 adduct (M^{n+}—O (CO_2) distances and the characteristic angles O=C=O...M^{n+}) are reported in nm/10 and degree, respectively (Plant et al., 2006). (For colour version of this figure, the reader is referred to the online version of this chapter.)

Beyond these structural findings, the binding energies and the resulting adsorption enthalpies were calculated for each geometry optimized cluster. It was evidenced that the simulated adsorption enthalpies decrease from Li^+ (−33.20 kJ/mol) to Cs^+(−17.17 kJ/mol), which correlate well with the increasing trend for the M^{n+}—O(CO_2) distances (Li^+: 0.205 nm; Cs^+: 0.335 nm). The closer the oxygen of carbon dioxide approaches the extra-framework cation, the stronger the interaction becomes. The energetic values have been further favourably compared for both LiY (−33.2 kJ/mol) and NaY (−28.7 kJ/mol) to those obtained by microcalorimetry measurements (−38.0 kJ/mol and −29.5 kJ/mol, respectively) realized on the corresponding faujasite systems.

Density functional theory calculations performed using periodic boundary conditions are also employed to gain insight into the adsorption of small gas molecules in crystalline porous solids. The hybrid B3LYP+D* (Beck, three parameters exchange,

Lee, Yang and Parr) functional empirically corrected to include the long-range dispersion term has been coupled with a triple-zeta valence basis set to deal with the adsorption of CO_2, CO and N_2 in the MOF-type CPO-27 material containing Mg^{2+} as cus (Valenzano et al., 2010). A full relaxation of the structure containing one adsorbed molecule per cus was performed to determine the most preferential arrangements of each adsorbate around the Mg^{2+}. It was thus evidenced that while CO and N_2 lead to quasi-linear Mg^{2+}–CO and Mg^{2+}–N_2 adducts, the Mg^{2+}–OCO complex formed by CO_2 is significantly angular (Figure 6.20). This deviation was explained by the additional lateral interactions between the oxygen atoms surrounding the Mg^{2+} and the carbon atom of CO_2. It was further shown that, in the three cases, the adsorption induces only a small volume change of the unit cell (~1% increase) associated to a tiny displacement of Mg^{2+} away from the framework when coordinated by the adsorbates. Such a conclusion was validated by *in situ* neutron diffraction powder measurements for the CO_2/CPO-27(Mg) system. The resulting simulated integral molar enthalpies of adsorption enthalpies of −30.0, −25.2 and −47 kJ/mol for CO and CO_2 at 298 K and for N_2 at 100 K compare reasonably well with the experimental values (−29.0, −21.0 and −47 kJ/mol) deduced from variable-temperature IR spectroscopy measurements using a Van't Hoff plot.

This investigation was extended to the adsorption of both CO_2 and CO in the same CPO-27 MOF but this time containing Ni^{2+} or Zn^{2+} as cus (Valenzano et al., 2011). An interesting hybrid approach consisted of adding to the periodic density functional theory (B3LYP+D*) mentioned above a high-level *ab initio* post-Hartree Fock correction for the adsorption site issued from a cluster calculation. It was emphasized that this more elaborated model leads to a better reproduction of the experimental adsorption enthalpy extracted from the isosteric method and the microcalorimetry measurements for all CPO-27/adsorbates with an average maximum deviation of 2 versus 5 kJ/mol considering only the periodic B3LYP+D* method.

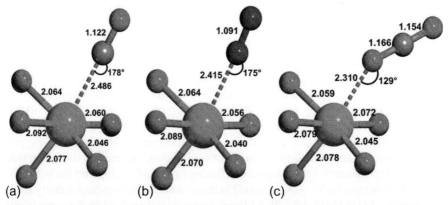

Figure 6.20 Optimized geometry of the (a) CO, (b) N_2 and (c) CO_2 adsorption adducts in CPO-27(Mg) issued from the periodic density functional theory calculations. The characteristic distances and angles are reported in nm/10 and degree, respectively. (For colour version of this figure, the reader is referred to the online version of this chapter.)
Courtesy: Valenzano et al. (2010).

6.6 Modelling of Diffusion in Porous Solids

6.6.1 Basic Principles

A reliable quantitative description of the diffusion of guest molecules in porous solids is of crucial importance to fully understand the mechanisms that govern the adsorption and the separation in many pressure swing adsorption and membrane-based processes. In complement to several experimental approaches including the quasi-elastic neutron scattering (QENS), the pulse-field gradient NMR and the zero length column, equilibrium MD is a valuable computational tool to probe the diffusivity of various guest molecules in porous solids. Details of this simulation technique are not provided; the reader is referred to authoritative books and review articles (Kärger and Ruthven, 1992; Demontis and Suffritti, 1997; Ruthven, 2005; Helmut, 2007; Jobic and Theodorou, 2007; Freeman and Yampolskii, 2010). Basically, a distribution of the adsorbates within the porosity of the solid is usually produced by preliminary MC simulations and each atom of the system is randomly assigned an initial velocity following a Boltzmann distribution. The time-dependent trajectory of the system consisting of a sequence of atomic positions for the diffusive molecules with respect to time is then generated by integrating the Newton's equations of motion numerically over short-time steps via appropriate algorithms. Different types of molecular diffusion can be further extracted from such equilibrium MD simulations usually realized in the canonical (*NVT*) or microcanonical (*NVE*) ensembles.

Indeed, the *self-diffusivity*, labelled as D_s, is calculated from the slope of the mean square displacement (MSD) of *individual* molecules with respect to time through the following Einstein relation (Equation 6.19):

$$D_s = \lim_{t \to \infty} \left\{ \frac{1}{6t} \frac{1}{N} \left\langle \sum_{i=1}^{N} [r_i(t) - r_i(0)]^2 \right\rangle \right\} \qquad (6.19)$$

where the angular brackets denote averaging over all diffusive molecules, $r(t)$ is the position of a molecules at time t and N is the total number of molecules in the system. A consideration of several trajectories generated from different starting configurations and an average over all time origin contributes to improve the statistics of the calculations. The observation time should be long enough to get a linear evolution of the MSD with time avoiding any anomalous diffusion such as the ballistic regime or the exclusion of mutual exchange (called 'single file' diffusion behaviour) in the short-time limit.

The *transport* or Fickian *diffusivity* corresponding to the *collective dynamics*, labelled as D_t, is also of great interest in real applications. It is defined as the proportional constant relating a macroscopic gradient in the chemical potentials of the adsorbate molecules to the flux generated by this gradient. Based on the linear response theory, this diffusivity is often expressed by Equation (6.20)

$$D_t = D_0 \left(\frac{\partial \ln f}{\partial \ln c} \right) \qquad (6.20)$$

where D_0 is the *corrected diffusivity*, also called Maxwell–Stefan diffusivity, f and c are the fugacity and the concentration of the adsorbate, respectively. The term $(\partial \ln f / \partial \ln c)$ usually labelled as Γ, which provides a measure of the density fluctuations in the adsorbate at equilibrium, is referred to as the thermodynamic correction factor which is most often estimated from the slope of the adsorption isotherm.

The corrected diffusivity can be calculated from the equilibrium MD simulations using an Einstein relation similar to Equation (6.19) which measures the MSD of the centre of mass of the considered swarm of the diffusive molecules denoted R as indicated in Equation (6.21)

$$D_0 = \frac{N}{6} \lim_{t \to \infty} \left\{ \frac{1}{6t} \frac{1}{N} \left\langle \sum_{i=1}^{N} [R_i(t) - R_i(0)]^2 \right\rangle \right\}$$
(6.21)

The estimation of this D_0 usually requires longer MD runs than for D_s as the statistical error is typically much larger than that for D_s, whose calculation offers the additional advantage of averaging over all molecules.

These self-, corrected and transport diffusivities are in general concentration dependent and only strictly equal in the limit of vanishing concentration of the adsorbate. Green–Kubo expressions based on the calculations of the velocity autocorrelation functions can also be employed to extract both D_s and D_0 although they lead to a significant statistical error at long times due to limited sample size. (Allen and Tidesley, 1987).

The resulting intra-crystalline diffusivities for species confined in porous solids are directly comparable with the corresponding experimental data issued from QENS measurements (D_s and D_t for molecules with incoherent and coherent cross section, respectively) as both techniques track the same time (10^{-3}–100 ns) and length (0.1–100 nm) scales. *One should bear in mind that equilibrium MD simulations covering a finite time span can only accurately capture the diffusivity of systems faster than 10^{-11} m^2/s.* For slower processes, other computational techniques are recommended including the transition state theory or the kinetic MC. Beyond the prediction of the diffusivity values, a careful analysis of the MD trajectories shed light into the diffusion microscopic mechanism for both single components and mixtures.

The next section presents some typical examples emphasizing the complementarity of MD and QENS to tackle the self- and transport diffusivity of both single gas and binary mixture in porous solids that can be considered as rigid in a first approximation or as flexible when the pore width becomes critically similar to the kinetic diameter of the diffusive species.

6.6.2 Single Component Diffusion

MD is widely employed as a predictive tool to determine the self- (D_s) and transport (D_t) diffusivities for a large variety of adsorbates in porous solids. As an illustration, a super-mobility for H_2 in the terephtalate MIL-47(V) MOF-type solid at 77 K was suspected at low concentration with self-diffusivity values about two orders of magnitude higher than in zeolites (Salles et al., 2008b). This prediction based on the use of

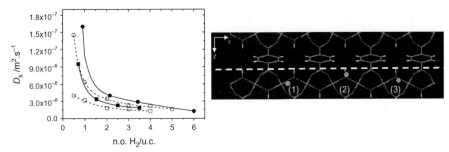

Figure 6.21 Left: self-diffusivity for H_2 in MIL-47(V) ($\circ$) and MIL-53(Cr) ($\blacksquare$) at 77 K as a function of the loading: QENS (full symbols) and MD (empty symbols). ($\square$). Right: typical illustration of the diffusion mechanism of H_2 in MIL-53(Cr); the snapshots from 1 to 3 correspond to the jump sequence of H_2 along the tunnel observed during the MD runs (Salles et al., 2008b). (For colour version of this figure, the reader is referred to the online version of this chapter.)

simple standard force fields for both the MOF (UFF force field; Rappé et al., 1992) and H_2 (single uncharged LJ site model; Frost et al., 2006) was further confirmed by QENS measurements which evidenced a similar sudden increase of D_s at low loading (Figure 6.21). Such a fast hydrogen diffusion rate also predicted in carbon nanotubes makes these porous materials very promising for the development of transport kinetics-based devices for the hydrogen fuel economy. QENS measurements evidenced that such a spectacular diffusivity is also observed in the isostructural analogue MIL-53(Cr) which again was supported by MD simulations. It was further stated that this unusual dynamic behaviour can be explained not only by a flat smoothness of the PESs but also by a necessary critical balance between the H_2/H_2 and H_2/MOF interactions leading to similar energetic profiles for both contributions within the pore of these two solids.

Beyond the simulation of the D_s values, a careful analysis of the MD trajectories provides a microscopic picture of the diffusion mechanism at play. While the H_2 molecules follow a 3D diffusion process with random motions within the pores of MIL-47 (V), the scenario significantly differs in MIL-53(Cr) where a 1D diffusion occurs along the tunnel via a jump sequence through the μ_2-OH groups present at the MOF surface (Figure 6.21). Such conclusions were supported by the use of a 1D and 3D diffusion model to fit the QENS spectra for the MIL-53(Cr) and MIL-47 (V), respectively.

MD is also a valuable tool to interpret the experimental concentration dependence of the diffusivity for a range of adsorbates confined in porous solids. This becomes even more true when this profile shows some peculiarity such as in the Zr-MOF UiO-66(Zr) for which QENS measurements evidenced a non-monotonic evolution of Ds for CH_4 at 230 K with the presence of a maximum (Figure 6.22; Yang et al., 2011a,b). The MD simulations were conducted in the *NVT* ensemble using the microscopic models and force field parameters validated on the thermodynamics properties of this system, that is, the generic DREIDING force field (Mayo et al., 1990) for the LJ parameters of each MOF atom and a standard neutral united atom model

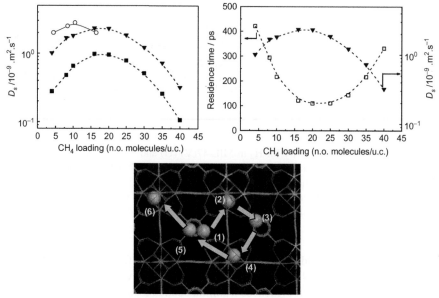

Figure 6.22 Top left: evolution of the self-diffusion coefficients as a function of the CH_4 loading in UiO-66(Zr) at 230 K: QENS (black empty circles), simulations with rigid (solid squares) and flexible (solid down-triangles) framework. Top right: residence time profiles for CH_4 in the tetrahedral cages (empty squares) and self-diffusivity (solid down-triangles) simulated using a flexible framework. Bottom: typical illustration of the global diffusion mechanism by following one targeted methane molecule, the positions from 1 to 6 correspond to jump sequences of CH_4 observed during the MD trajectories (Yang et al., 2011a,b). (For colour version of this figure, the reader is referred to the online version of this chapter.)

for CH_4 (see Section 6.3.2). Using the approximation of a rigid UiO-66(Zr) framework leads to a simulated D_s profile with a maximum at a CH_4 loading comparable to the experimental one (~15 CH_4/u.c. vs. 10.5 CH_4/u.c.); however, the absolute D_s values significantly underestimate the QENS data, especially at low loading (Figure 6.22). The implementation of a flexible force field for UiO-66(Zr) in an $N\sigma T$ ensemble, where the shape and dimensions of the unit cell can vary, leads to a much better reproduction of the experimental D_s values, while the D_s profile remains the same than those obtained using a rigid framework. This observation clearly emphasized that the flexibility of the framework plays a dominant role in the dynamics of guest molecules in narrow windows/channels solids, more particularly when the pore width of the solids is similar to the kinetic diameter of the diffusive species. Indeed, in contrast to thermodynamic studies where a flexible framework has only a negligible impact on the adsorption isotherms/enthalpies, it is crucial to consider a dynamic framework when one aims to probe the diffusivity of strongly confined molecules in porous solids. This has been demonstrated for a series of systems including methane and propane in the zeolites LTA and CHA, respectively (Combariza et al., 2013), as well as ethane in the MOF material Zn(tbip) (Seehamart et al., 2011).

In the case of UiO-66(Zr), the flexibility of the framework has been shown to strongly increase the intercage hopping rate at the origin of the 3D diffusion mechanism via jump sequences 'tetrahedral cages–octahedral cages–tetrahedral cages'.

Based on these observations, the microscopic picture of the diffusion has been rationalized by the plot of the residence time for CH_4 inside the tetrahedral cages as a function of the loading. As can be observed in Figure 6.22, the profile for the residence time is consistent with the trend observed for the self-diffusivity. At low loading, due to the confinement and energetic effect in the tetrahedral cages, the CH_4 molecules are preferentially adsorbed and mainly remain trapped inside these cages, which is consistent with long residence time and medium range values for the self-diffusivity. As the loading increases, the additional CH_4 molecules present in the tetrahedral cages involve less favourable interactions with the pore wall and thus increase the net driving force to promote the hopping of CH_4 towards the octahedral cages, which leads to a decrease in the residence time in the tetrahedral cages. Further, when the CH_4 molecules reach the octahedral cages, they are subjected to less energetic interactions with the pore wall as previously evidenced. Both factors lead to an increase in D_s in this range of loading as shown in Figure 6.22. With a further increase in the loading, the effect of the steric hindrance becomes evident, resulting in the increase in the residence time in the tetrahedral cages and a decrease in the self-diffusivity.

Beyond the self-diffusivity, MD simulations are also employed to explore the collective motion of confined molecules via the estimation of their corrected and transport diffusivities that can be compared to QENS data when the probed molecules are coherent scatterers. As an illustration, all these diffusivities were simulated for CO_2 in MIL-47(V) at 230 K using NVT MD simulations (Salles et al., 2010) based on standard force field for both MOF (UFF force field, Rappé et al., 1992) and CO_2 (EPM2 model, Harris and Yung, 1995). While the three quantities D_t, D_0 and D_s have been predicted to become almost equal at the limit of zero concentration as observed in the majority of porous solids, the simulations in tandem with QENS measurements emphasized two peculiarities at low loading with (i) D_0 values unusually higher than the D_t ones and (ii) a decrease in the D_t values which is never observed in a microporous solid. This was attributed to the unusual value of Γ which passes below 1 in this range of loading due to the presence of an inflection point in both the simulated and experimental CO_2 adsorption isotherm (Figure 6.23). Such a behaviour that significantly differs with the standard convex shape of the I-type isotherm commonly observed for microporous solids was associated with two different subregimes with a preferential location of CO_2 close to the pore wall at the initial stage of the adsorption followed by a filling of the centre of the pore at higher pressure due to strong CO_2/CO_2 interactions. It was further mentioned that the simulated D_t values underestimate the experimental ones by a factor of 2 and 7 at low and high CO_2 concentrations, respectively, although the experimental adsorption isotherm is very well reproduced using the same force field parameters. Indeed such an observation clearly emphasizes that a force field validated on the thermodynamics properties of the investigated system does not always guarantee that the dynamics features would match perfectly the experimental data.

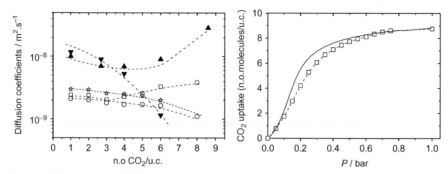

Figure 6.23 Evolution of the experimental D_0 (solid down-triangles) and D_t (solid up-triangles) and of the simulated D_s (empty circle), D_0 (empty star) and D_t (empty square) as a function of the CO_2 concentration in MIL-47(V) at 230 K. Simulated (empty squares) and experimental (black line) CO_2 adsorption isotherms for MIL-47(V) at 230 K (Salles et al., 2010).

6.6.3   Gas Mixture Diffusion

To obtain a full picture of the separation process in porous solids, besides the investigations on their equilibrium performances, probing the dynamics of the binary mixture is of great importance. A typical example is the permeation selectivity of H_2 from gas mixtures containing CO_2 and CH_4 evidenced in different porous solids, which mainly relies on the fact that H_2 diffuses much faster than the other components. While the diffusivity of single components has been widely investigated both experimentally and theoretically, the experimental measurements are much more delicate, leading to only a few multi-component diffusion data available in the literature. To overcome such a limitation, MD simulations have been employed to explore the co-diffusion behaviour of different mixtures in carbons, zeolites and MOFs. Indeed, the modeling of the diffusivity for CO_2/CH_4 gas mixture in porous solids is relatively well documented, sometimes in conjunction with QENS experiments. It has been thus predicted in various zeolites with narrow windows such as LTA, CHA and DDR that both molecules are usually diffusing independently leading to the absence of mutual speeding up or slowing down of the partner species (Krishna et al., 2006; Krishna and Van Baten, 2008a,b). In some other cases, it was also observed that the faster CO_2 molecule due to its slender shape retards the slowly diffusing CH_4 (Krishna and Van Baten, 2008a,b). An opposite trend was simulated in the narrow windows Zr-MOF UiO-66(Zr) at 230 K where the slower CO_2 unusually tends to enhance the diffusivity for the fast CH_4 thus leading to D_s values larger than those in the pure gas (Yang et al., 2011a,b). Such a prediction has been further confirmed by QENS measurements as shown in Figure 6.24.

A careful analysis of the MD trajectories for the CO_2/CH_4 mixture has evidenced that in the presence of CO_2, the CH_4 molecules are more frequently pushed into the octahedral cages than in the single component diffusion, while the CO_2 molecules spend more time in the tetrahedral cages as they are more strongly interacting with

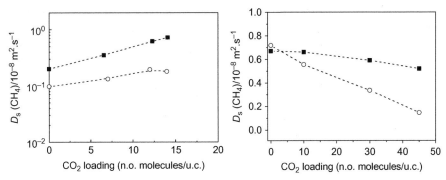

Figure 6.24 D_s for CH$_4$ as a function of the CO$_2$ concentration in UiO-66(Zr) at 230 K (left) (Yang et al., 2011a,b) and in NaY at 200 K (right): QENS (full squares), MD (empty circles) (Déroche et al., 2010).

the pore wall. Such a situation explains the faster diffusivity for CH$_4$ as these molecules spend less time in the tetrahedral cages, where the stronger interactions with the pore wall occur.

A joint QENS/MD strategy has also evidenced that the self-diffusivity for CH$_4$ significantly decreases in the presence of CO$_2$ in the relatively large channel zeolite type NaY (Figure 6.24; Déroche et al., 2010). This trend was attributed to a combination of two opposite effects. While the CO$_2$ molecules preferentially interacting with the Na$^+$ ions present in the supercage 'screen' these possible attractive sites for CH$_4$ that should lead to an enhancement of their diffusivities, they simultaneously occupy and/or crowd the supercage and consequently reduce the effective diffusing space for the CH$_4$ molecules responsible for the slight decrease in their diffusivities.

6.7   Conclusions and Future Challenges

Modelling tools have been proved to be essential for not only offering microscopic models of the existing or newly discovered crystalline porous solids and characterizing their porosities but also generating a huge collection of hypothetical crystalline porous solids since the past few years. However, even armed with appropriate computational techniques and high computing power, screening accurately the adsorption properties of such a wide range of solids is still hardly achieved in a reasonable time. One of the great challenges is to develop sophisticated automated method based on mathematical algorithms to efficiently characterize, categorize and screen such a large database of materials. A high-throughput analysis of the accessibility of thousands of porous solids to guest molecules and probes of interests computed only in a few seconds would be of considerable value. Such a modelling effort, still at its very early stage for crystalline porous solids including zeolites and MOFs, would allow to narrow down the list of materials respecting this geometric criterion which thus becomes the domain for microscopic simulations (GCMC and MD) or even less time-consuming semi-empirical approaches (macroscopic models such as the IAST) executed to screen rapidly the properties of interest and to further prioritize the most promising solids.

Nowadays, another challenge is to intelligently design/tune materials with topological, chemical and electronic features specifically targeted for adsorption/separation applications with the development of new concepts and innovative strategies capitalized on state-of-the-art advanced multi-scale approach ranging from *ab initio* calculations to mathematical tools in strong interrelation with the experimental side.

Rationalizing the adsorption/separation properties of a large series of porous materials is also of upmost importance. Top-of-the-art statistical tools including the quantitative structure property relationship (QSPR) approach, widely employed in the pharmaceutical drug design, are currently transferred to the field of porous materials. Here the objective is to define the key variables of the solids (chemical, topological, electronic, etc.) that mainly impact their performances in targeted adsorption applications. Such conclusions are a valuable complement to 'chemical intuition' and can be used to guide the synthesis effort towards tuned materials with the required features for an optimization of their properties. As a typical example, such a QSPR strategy has been recently deployed to reveal the structure property relationship of a wide variety of hundred MOFs for CO_2 capture from flue gas (Wu et al., 2012). It was evidenced that the best derived QSPR model correlates the CO_2/N_2 selectivity for a binary mixture $CO_2:N_2 = 15:85$ with the $(\Delta Q_{st}^0/\varphi)$ variable calculated for the whole series of MOFs where ΔQ_{st}^0 and φ correspond to difference in adsorption enthalpy between CO_2 and N_2 at infinite dilution and the porosity defined by the free volume to the total volume of the unit cell ratio, respectively (Figure 6.25). These findings were further used to tune a novel Zr-based MOF grafted with two carboxylic functions with a very high CO_2/N_2 selectivity.

Finally, beyond these challenging objectives, a modelling effort is still required to fully understand the microscopic mechanisms at the origin of the spectacular adsorption/separation performances of some porous solids. There is a growing need to combine standard force field-based simulations and quantum chemical calculations through QM/MM hybrid methods to get a precise microscopic description of the adsorption phenomena in large size crystalline materials.

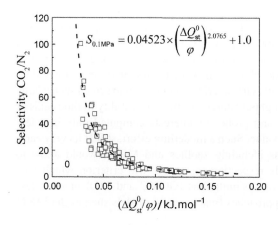

$$S_{0.1\,MPa} = 0.04523 \times \left(\frac{\Delta Q_{st}^0}{\varphi}\right)^{2.0765} + 1.0$$

Figure 6.25 Relationship between the adsorption CO_2/N_2 selectivity and the $\Delta Q_{st}^0/\varphi$ variable in 105 MOFs for a binary CO_2/N_2 mixture $(CO_2:N_2 = 15:85)$ at 0.1 MPa and 298 K (Wu et al., 2012).

References

Allen, M.P., Tildesley, D.J., 1987. Computer Simulation of Liquids. Clarendon, Oxford.

Bae, Y.S., Yazaidyn, O., Snurr, R.Q., 2010. Langmuir. 26, 5475.

Beauvais, C., Boutin, A., Fuchs, A.H., 2004. Chem. Phys. Chem. 5 (11), 1791.

Bezus, A.G., Kiselev, A.V., Lopatkin, A.A., Du, P.Q., 1978. J. Chem. Soc. Faraday Trans. 74, 367.

Bhattacharya, S., Gubbins, K.E., 2006. Langmuir. 22, 7726.

Bhattacharya, S., Coasne, B., Hung, F.R., Gubbins, K.E., 2009. Langmuir. 25, 5802.

Billemont, P., Coasne, B., De Weireld, G., 2010. Langmuir. 27, 1015.

Cavenati, S., Grande, C.A., Rodrigues, A.E., 2004. J. Chem. Eng. Data. 49, 1095.

Chen, L., Grajciar, L., Nachtigall, P., Düren, T., 2011. J. Phys. Chem. C 115, 23074.

Coasne, B., Pellenq, R.J.M., 2004. J. Chem. Phys. 121, 3767.

Coasne, B., Hung, F.R., Pellenq, R.J.M., Siperstein, F.R., Gubbins, K.E., 2006. Langmuir. 22, 194.

Combariza, A.F., Gomez, D.A., Sastre, G., 2013. Chem. Soc. Rev. 42, 114.

Demontis, P., Suffritti, G.B., 1997. Chem. Rev. 97, 2845.

Déroche, I., Maurin, G., Borah, B.J., Yashonath, S., Jobic, H., 2010. J. Phys. Chem. C 114, 5027.

Devic, T., Horcajada, P., Serre, C., Salles, F., Maurin, G., Moulin, B., Heurtaux, D., Clet, G., Vimont, A., Grenèche, J.M., Le Ouay, B., Moreau, F., Magnier, E., Filinchuk, Y., Marrot, J., Lavalley, J.C., Daturi, M., Férey, G., 2010. J. Am. Chem. Soc. 132, 1127.

Do, D.D., Nicholson, D., Fan, C., 2011. Langmuir. 27, 14290.

Dubbeldam, D., Calero, S., Vlugt, T.J.H., Krishna, R., Maesen, T.L.M., Beerdsen, E., Smit, B., 2004. Phys. Rev. Lett. 93 (8), 088302.

Düren, T., Millange, F., Férey, G., Walton, K.S., Snurr, R.Q., 2007. J. Phys. Chem. C. 111, 15350.

Férey, G., Serre, C., Devic, T., Maurin, G., Jobic, H., Llewellyn, P.L., De Weireld, G., Vimont, A., Daturi, M., Chang, J.S., 2011. Chem. Soc. Rev. 40, 550.

Fischer, M., Gomes, J.R.B., Froba, M., Jorge, M., 2012. Langmuir 28, 8537.

Freeman, B., Yampolskii, Y., 2010. Membrane Gas separation. Wiley, New York.

Frenkel, D., Smit, B., 2002. Understanding Molecular Simulations from Algorithm to Applications. Academic Press.

Frost, H., Düren, T., Snurr, R.Q., 2006. J. Phys. Chem. B 110, 9565.

Fuchs, A.H., Cheetham, A.K., 2001. J. Phys. Chem. B 105, 31.

Gelb, L.D., 2002. Mol. Phys. 100, 2049.

Ghoufi, A., Maurin, G., 2010. J. Phys. Chem. C 114, 6496.

Ghoufi, A., Subercaze, A., Ma, Q., Yot, P., Ke, Y., Puente, O.I., Devic, T., Guillerm, V., Zhong, C., Serre, C., Férey, G., Maurin, G., 2012. J. Phys. Chem. C 116, 13289.

Grosch, J.A., Paesani, F., 2012. J. Am. Chem. Soc. 134, 4207.

Guillerm, V., Ragon, F., Dan-Hardi, M., Devic, T., Vishnuvarthan, M., Campo, B., Vimont, A., Clet, G., Yang, Q., Maurin, G., Férey, G., Vittadini, A., Gross, S., Serre, C., 2012. Angew. Chem. Int. Ed. 51 (37), 9267.

Haldoupis, E., Nair, S., Sholl, D.S., 2012. J. Am. Chem. Soc. 134, 4313.

Hamon, L., Heymans, N., Llewellyn, P.L., Guillerm, V., Ghoufi, A., Vaesen, S., Maurin, G., Serre, C., De Weireld, G., Pirngruber, G., 2012. Dalton Trans. 41, 4052.

Harris, J.G., Yung, K.H., 1995. J. Phys. Chem. 99, 12021.

Heinz, H., Suter, U.W., 2004. J. Phys. Chem. B. 108, 18341.

Helmut, M., 2007. In: Diffusion in Solids. Springer Series in Solid State Science, Vol. 155. Springer, Berlin.

IZA Structure Commission, 2013. http://www.iza-structure.org/.

Jain, K., Pikunic, J., Pellenq, R.J.M., Gubbins, K.E., 2005. Adsorption 11, 355.

Jeffroy, M., Fuchs, A.H., Boutin, A., 2011. Chem. Commun. 28, 3275.

Jobic, H., Theodorou, D.N., 2007. Micropor. Mesopor. Mater. 102, 21.

Jorgensen, W.L., Maxwell, D.S., Tirado-Rives, J., 1996. J. Am. Chem. Soc. 118, 11225.

Kärger, J., Ruthven, D.M., 1992. Diffusion in Zeolites and other microporous Solids. Wiley, New York.

Koch, W., Holthausen, M.C., 2000. A Chemist's Guide to Density Functional Theory. Wiley-VCH, Weinheim.

Krishna, R., Van Baten, J.M., 2008a. Micropor. Mesopor. Mater. 109, 91.

Krishna, R., Van Baten, J.M., 2008b. Sep. Purif. Technol. 61, 414.

Krishna, R., Van Baten, J.M., García-Pérez, E., Calero, S., 2006. Chem. Phys. Lett. 429, 219.

Leach, A.R., 2001. Molecular Modelling, Principle and Applications, second ed. Prentice Hall, England.

Levitz, P., 1998. Adv. Colloid Interface Sci. 76–77, 71.

Llewellyn, P.L., Bourrelly, S., Vagner, C., Heymans, N., Leclerc, H., Ghoufi, A., Bazin, P., Vimont, A., Daturi, M., Devic, T., Serre, C., De Weireld, G., Maurin, G., 2013. J. Phys. Chem. C 117, 962.

Lowenstein, W., 1954. Am. Mineral. 39, 92.

Martin, M.G., Siepmann, J.I., 1998. J. Phys. Chem. B 102, 2569.

Maurin, G., Senet, P., Devautour, S., Gaveau, P., Henn, F., Van Doren, V.E., Giuntini, J.C., 2001. J. Phys. Chem. B 105, 9157.

Maurin, G., Llewellyn, P.L., Bell, R.G., 2005. J. Phys. Chem. B 109, 16084.

Maurin, G., Plant, D.F., Henn, F., Bell, R.G., 2006. J. Phys. Chem. B. 110, 18447.

Mayo, S.L., Olafson, B.D., Goddard III, W.A., 1990. J. Phys. Chem. 94, 8897.

Mellot-Draznieks, C., 2007. J. Mater. Chem. 17 (41), 4348.

Metropolis, N., Rosenbluth, A., Rosenbluth, M., Teller, A., Teller, E., 1953. J. Chem. Phys. 21, 1087.

Mezei, M., 1980. Mol. Phys. 40, 901.

Mulliken, R.S., 1955. J. Chem. Phys. 23 (10), 1833.

Myers, A.L., Monson, P.A., 2002. Langmuir 18, 10261.

Myers, A.L., Prausnitz, J.M., 1965. AIChE J. 11, 121.

Nicholson, D., Parsonage, N.G., 1982. Computer Simulation and the Statistical Mechanics of Adsorption. Academic Press, London.

Pellenq, R.J.M., Nicholson, D., 1994. J. Phys. Chem. 98, 13339.

Plant, D.F., Déroche, I., Gaberova, L., Llewellyn, P.L., Maurin, G., 2006. Chem. Phys. Lett. 426, 387.

Rappé, A.K., Goddard III, W.A., 1991. J. Phys. Chem. 95, 3358.

Rappé, A.K., Casewit, J., Colwell, K.S., Goddard III, W.A., Skiff, W.M., 1992. J. Am. Chem. Soc. 114, 10024.

Rideal, E.K., 1930. An Introduction to Surface Chemistry. Cambridge University Press, London, pp. 175–176.

Rosenbach, N., Ghoufi, A., Déroche, I., Llewellyn, P.L., Devic, T., Bourrelly, S., Serre, C., Férey, G., Maurin, G., 2010. Phys. Chem. Chem. Phys. 12, 6428.

Ruthven, D.M., 2005. Introduction to Zeolite Science and Practice, Studies in Surface Science and Catalysis, Vol. 168, 737 pp.

Salles, F., Ghoufi, A., Maurin, G., Bell, R.G., Mellot-Draznieks, C., Férey, G., 2008a. Angew. Chem. Int. Ed. 47, 8487.

Salles, F., Jobic, H., Maurin, G., Koza, M.M., Llewellyn, P.L., Serre, C., Devic, T., Férey, G., 2008b. Phys. Rev. Lett. 100, 245901.

Salles, F., Jobic, H., Devic, T., Llewellyn, P.L., Serre, C., Férey, G., Maurin, G., 2010. ACS Nano. 4 (1), 143.

Salles, F., Douillard, J.M., Bildstein, O., Van Damme, H., 2011. Appl. Clays Sci. 53 (3), 379.

Sauer, J., 1989. Chem. Rev. 89, 199.

Sauer, J., Ugliengo, P., Garrone, E., Saunders, V.R., 1994. Chem. Rev. 94, 2095.

Schumacher, C., Gonzalez, J., Wright, P.A., Seaton, N.A., 2006. J. Phys. Chem. B. 110, 319.

Seehamart, K., Chmelik, C., Krishna, R., Fritzche, S., 2011. Micropor. Mesopor. Mater. 143, 125.

Sholl, D.S., Steckel, J.A., 2009. Density Functional Theory: A Practical Introduction. John Wiley & Sons, Hoboken, NJ.

Siperstein, F.R., Gubbins, K.E., 2003. Langmuir 19, 2049.

Snurr, R.Q., Bell, A.T., Theodorou, D.N., 1994. J. Phys. Chem. 98, 5111.

Suwanayuen, S., Danner, R.P., 1980. AIChE J. 26, 68.

Szabo, A., Ostlund, N.S., 1996. Modern Quantum Chemistry. Dover Publications Inc, Mineola.

Tafipolsky, M., Amirjalayer, S., Schmid, R., 2007. J. Comput. Chem. 7 (28), 1169.

Talu, O., Myers, A.L., 2001. AIChE J. 47, 1160.

Valenzano, L., Civalleri, B., Chavan, S., Palomino, G.T., Arean, C.O., Bordiga, S., 2010. J. Phys. Chem. C 114, 11185.

Valenzano, L., Civalleri, B., Sillar, K., Sauer, J., 2011. J. Phys. Chem. 115, 21777.

Van Duin, A.C.T., Dasgupta, S., Lorant, F., Goddard III, W.A., 2001. J. Phys. Chem. A 105, 9396.

Vitale, G., Mellot, C.F., Bull, L.M., Cheetham, A.K., 1997. J. Phys. Chem. 101, 4559.

Vlugt, T.J.H., García-Pérez, E., Dubbeldam, D., Ban, S., Calero, S., 2008. J. Chem. Theory Comput. 4, 1107.

Walton, K.S., Snurr, R.Q., 2007. J. Am. Chem. Soc. 129, 8552.

Wilmer, C.E., Snurr, R.Q., 2011. Chem. Eng. J. 171 (3), 775.

Wilmer, C.E., Leaf, M., Lee, C.Y., Farha, O.M., Hauser, B.G., Hupp, J.T., Snurr, R.Q., 2011. Nat. Chem. 4, 83.

Wu, D., Yang, Q., Zhong, C., Liu, D., Huang, H., Zhang, W., Maurin, G., 2012. Langmuir 28 (33), 12094.

Yang, Q., Jobic, H., Salles, F., Kolokolov, D., Guillerm, V., Serre, C., Maurin, G., 2011a. Chem. Eur. J. 17, 8882.

Yang, Q., Wiersum, A., Jobic, H., Guillerm, V., Serre, C., Llewellyn, P.L., Maurin, G., 2011b. J. Phys. Chem. C 115, 13768.

Yang, Q., Wiersum, A., Llewellyn, P.L., Guillerm, V., Serre, C., Maurin, G., 2011c. Chem. Commun. 47, 9603.

Yang, Q., Guillerm, V., Ragon, F., Wiersum, A., Llewellyn, P.L., Zhong, C., Devic, T., Serre, C., Maurin, G., 2012a. Chem. Commun. 48, 9831.

Yang, Q., Vaesen, S., Vishnuvarthan, M., Ragon, F., Serre, C., Vimont, A., Daturi, M., De Weireld, G., Maurin, G., 2012b. J. Mater. Chem. 22, 10210.

Yun, J.H., Park, H.C., Moon, H., 1996. Kor. J. Chem. Eng. 13, 246.

7 Assessment of Surface Area by Gas Adsorption

Kenneth S.W. Sing

Aix Marseille University-CNRS, MADIREL Laboratory, Marseille, France

---

Chapter Contents

7.1   Introduction

Adsorption methods are universally employed for the determination of the specific surface area of fine powders and porous solids. It must be kept in mind, however, that unless a solid material is atomically flat its effective 'surface area' is not a simple property. In the context of adsorption, there are two main reasons for this complexity: (1) at the molecular level, the dimensions of the solid are dependent on the convention chosen to locate the surface – as explained in Chapter 6 and (2) the available area is dependent on the extent of surface roughness and porosity and the dimensions of the adsorbed molecules. In addition, most adsorbents of technological importance are in some respects heterogeneous and all experimental methods or theoretical treatments have limitations of one sort or another. For all these reasons, when reporting derived quantities such as surface area or pore size distribution it is essential to provide a detailed account of the chosen experimental technique including the computational procedure used for data processing.

Adsorption by Powders and Porous Solids. http://dx.doi.org/10.1016/B978-0-08-097035-6.00007-3

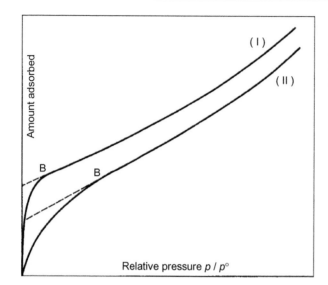

Figure 7.1 Typical Type II isotherms: (I) with sharp knee and (II) with rounded knee.
After Gregg and Sing (1982).

The appearance of Langmuir's comprehensive treatment of monolayer adsorption (Langmuir, 1916, 1918) prompted several investigators to consider the possibility of using gas adsorption for surface area determination. Early attempts were made by Williams (1919) and Benton (1926), but these led to inconclusive findings. The first significant advances were made by Brunauer and Emmett (1935, 1937) and their work prepared the way for the development of the Brunauer–Emmett–Teller (BET) theory in 1938.

As described in Chapter 5, Type II isotherms (see Figure 7.1) often display a rather long, almost linear middle range (i.e. BCD in Figure 5.2). The point at which this linear part begins was designated 'Point B' by Brunauer and Emmett (1937) and was shown empirically to indicate the completion of monolayer coverage and the beginning of multilayer adsorption. From the amount adsorbed at Point B, Emmett and Brunauer (1937) went on to calculate the surface area by assuming the monolayer to be molecularly close packed.

The BET extension of the Langmuir monolayer model to multilayer adsorption was of great historical importance and appeared to justify the interpretation of Point B. Over the past 70 years, the BET equation has continued to be extremely popular for assessing the specific surface areas of adsorbents, catalysts, pigments and many other finely divided and porous materials. For this purpose, nitrogen is still generally used as the adsorptive at 77 K (Gregg and Sing, 1982; Lowell et al., 2004; ISO, 2010; ASTM, 2012). However, the apparent success of the BET nitrogen method has tended to overshadow the fundamental weakness of the underlying theory (see Chapter 5). In fact, the deficiencies of the BET model were pointed out at a very early stage (Cassel, 1944; Hill, 1946; Gregg and Jacobs, 1948; Halsey, 1948). As will be explained in this chapter, the use of the BET *method* is essentially an *empirical* procedure since it is not dependent on the BET *theoretical model*. This necessarily reinforces the need to examine the limitations of the BET method and in particular to attempt to define the conditions which govern its application.

Various empirical methods for the analysis of physisorption isotherms were developed in the 1960s by Kiselev, Pierce, de Boer, Sing and others (see Gregg and Sing, 1982). This approach involves the measurement of standard isotherm data on well-defined non-porous adsorbents and the application of certain theoretical principles – notably those proposed by Kiselev (1965, 1972). In this manner, it is possible to check the validity of BET areas and extract useful information from composite isotherms (e.g. the evaluation of external areas).

7.2   The BET Method

7.2.1   Introduction

Two stages are involved in the evaluation of the surface area from physisorption isotherm data by the BET method. First, it is necessary to construct the BET plot and from it to derive the value of the *monolayer capacity*, n_m. The second stage is the calculation of the specific surface area, a(BET), from n_m and this requires a knowledge of the average area, σ, occupied by each molecule in the completed monolayer (i.e. *the molecular cross-sectional area*). Questionable assumptions are introduced at each stage and these therefore require careful consideration.

7.2.2   The BET Plot

As explained in Chapter 5, the BET equation is conveniently expressed in the linear form:

$$\frac{p/p^o}{n(1 - p/p^o)} = \frac{1}{n_m C} + \frac{C - 1}{n_m C}\left(\frac{p}{p^o}\right) \tag{7.1}$$

where $n(=n^a/m^s)$ is the amount adsorbed at relative pressure p/p^o and $n_m(=n_m^a/m^s)$ is the monolayer capacity. In the BET theory, the parameter C is exponentially related to E_1 (the first-layer adsorption energy). Although the adsorbent–adsorbate energy before the completion of the monolayer does generally vary in the same direction as C (Gregg and Sing, 1982), their exact relationship remains unknown (see Section 5.2.3).

According to Equation (7.1), the BET plot of $(p/p^o)/[n(1 - p/p^o)]$ versus p/p^o should be a straight line with slope $s = (C - 1)/n_m C$ and intercept $i = 1/n_m C$. By solving these two simultaneous equations, we obtain:

$$n_m = \frac{1}{s + i} \tag{7.2}$$

and

$$C = (s + i) + 1 \tag{7.3}$$

As is pointed out in Chapter 5, Equation (7.1) is obeyed only over limited regions of all known physisorption isotherms (see Figure 5.4). In their original work, Brunauer et al.

(1938) found that their Type II isotherms on certain adsorbents (a silica gel, a chromia gel and two iron catalysts) gave linear BET plots over the approximate range of $p/p^o = 0.05$–0.35 with n_m located at $p/p^o \sim 0.1$. Subsequent studies have revealed that BET plots for nitrogen at 77 K generally exhibit a somewhat shorter linear region, so that $p/p^o = 0.05$–0.35 should not be taken as a 'standard BET range' (Rouquerol et al., 1964; Sing, 1964; Everett et al., 1974; Gregg and Sing, 1982; Badalyan and Pendleton, 2003; Lowell et al., 2004). BET nitrogen plots on non-porous and meso-porous silicas begin to deviate from linearity at $p/p^o \sim 0.25$ (Everett et al., 1974) with $C \sim 80$–150. There are numerous other examples (not only with nitrogen) where the departure from linearity starts at relative pressures below ~ 0.2 (Gregg and Sing, 1982). A similar upper limit was reported (Choma and Jaroniec, 2001) for nitrogen on apparently non-porous carbon blacks, but the values of C were in the much larger range of 200–900.

According to Equation (7.1), the relative pressure corresponding to monolayer completion is inversely dependent on the C value:

$$\left(\frac{p}{p^o}\right)_{n_m} = \frac{1}{\sqrt{C} + 1} \tag{7.4}$$

For example, if $C > 350$, the BET monolayer capacity is located at $p/p^o < 0.05$ and if $C < 20$, n_m is at $p/p^o > 0.18$. As pointed out in Chapter 5, the knee of the isotherm also becomes less sharp as the value of C is reduced (see Figure 5.3). Although the Type II character is not lost until $C < 2$, the curvature of the isotherm becomes more gradual with decrease in C; especially so if $C < \sim 50$.

An interesting feature of the BET equation is that it appears to allow for the possibility that when the surface is covered with a 'statistical monolayer', a fraction remains uncovered. We may note, with Hill (1946), that this fraction $(\theta_0)n_m$ is directly dependent on the value of C:

$$(\theta_0)_{n_m} = \frac{1}{\sqrt{C} + 1} \tag{7.5}$$

so that the higher the C value, the smaller the uncovered fraction of the surface when the statistical monolayer is formed. One immediately sees that for C values of 1, 9 and 100, the corresponding fractions of uncovered surface are, respectively, 50%, 25% and ca. 10%.

Comparison of Equations (7.4) and (7.5) leads to:

$$(\theta_0)_{n_m} = \frac{p}{p^o} n_m \tag{7.6}$$

which, according to the BET model, provides a simple way of evaluating the fraction of uncovered surface.

A reliable analysis of the BET plot requires a certain number of experimental points: 10 is, we consider, a minimum in the exploratory range of relative pressures from 0.01 to 0.30. However, it is inadvisable to fit the 'best' straight line over any predetermined p/p^o range. The location and extent of the linear region of a BET plot are dependent on the system and operational temperature, and if the isotherm is Type II or Type IV, the BET plot should always be located around the knee of the isotherm (i.e. to straddle Point B). Nevertheless, the selection of the appropriate pressure range often entails some degree of qualitative judgement and several narrow, adjacent, pressure ranges may seem to offer possible ranges of linearity. To overcome this uncertainty, the following simple criteria have been proposed (Rouquerol et al., 1964, 2007) and adopted (e.g. in ISO 9277:2010, E, 2010; ASTM C1274-12, 2012):

a. the quantity C should be positive (i.e. any negative intercept on the ordinate of the BET plot is an indication that one is outside the valid range of the BET equation);
b. the application of the BET equation should be limited to the pressure range where the term $n(p^o - p)$ or alternatively $n(1 - p/p^o)$ continuously increases with p/p^o; all data points above the maximum in the plot should be discarded;
c. the pressure corresponding to n_m should be within the pressure range selected for the calculation and
d. the calculated value of $(p/p^o)n_m$ (given by Equation 7.4) should not differ, say, by more than 10%, from the value of p/p^o corresponding to the BET n_m value obtained by application of Equation (7.1). Otherwise, it is necessary to change the chosen range of relative pressures.

Criterion (a) follows from Equation (5.26), which defines C as an exponential, whereas criteria (c) and (d) can be considered as simple criteria of self-consistency, but of course within the confines of the BET theory.

Justification of criterion (b) can be seen by applying an alternative form of the BET equation proposed by Keii et al. (1961):

$$\frac{1}{n(1 - p/p^o)} = \frac{1}{n_m} + \frac{1}{n_m C} \cdot \frac{1 - p/p^o}{p/p^o} \tag{7.7}$$

This equation was applied by Parra et al. (1994) for the processing of experimental data for the adsorption of nitrogen by microporous activated carbons. It is precisely when the term $n(1 - p/p^o)$ starts to decrease (as p/p^o is increased) that the new plot clearly begins to deviate from linearity. Equation (7.7) is specially suited for the evaluation of a high C value. The intercept on the ordinate gives n_m, and the slope of the plot gives the product $n_m C$. However, for general use, we consider that the standard BET equation is more convenient.

As was noted in Chapter 5, the BET treatment leads to a modified equation if the amount adsorbed at the saturation pressure is restricted to a finite number of layers, N. Thus, in place of Equation (7.1), Brunauer et al. (1938) obtained:

$$\frac{n}{n_m} = \frac{C(p/p^o)}{1 - (p/p^o)} \cdot \frac{1 - (N+1)(p/p^o)^N + N(p/p^o)^{N+1}}{1 + (C-1)(p/p^o) - C(p/p^o)^{N+1}} \tag{7.8}$$

In practice, Equation (7.8) must be regarded as another empirical relation, since a value of N is chosen to give the best fit in the multilayer range. For example, it was found by Brunauer (1945) that for the adsorption of nitrogen on an iron catalyst at 77 K, by taking $N = 6$, the upper limit of the range of fit could be extended from p/p^o of 0.35 to 0.7. In view of the improved range of applicability of Equation (7.8), it might be expected that this equation would be capable of yielding a more reliable evaluation of n_m. It turns out, however, that the difference in the location of n_m when comparing Equations (7.1) and (7.8) is not more than a few percent provided that $N > 4$. In view of the underlying weakness of the BET theory, we conclude that there is no advantage to be gained by using Equation (7.8) to obtain n_m.

7.2.2.1 The Single-Point Method

For some routine or exploratory work, it is expedient to adopt a simplified experimental procedure, which involves the determination of a single point on the isotherm – preferably within the BET range. The assumption is then made that C is sufficiently large to give an almost zero intercept. If we assume that the intercept is zero and that $C - 1 \approx C$ we may write:

$$n_m = n\left(1 - \frac{p}{p^o}\right) \tag{7.9}$$

The acceptability of this simplifying assumption is evidently dependent on the isotherm shape: the error is likely to be within a few percent provided that $C \approx 100$. Experience confirms that the errors in the estimation of n_m by the single-point method become appreciable when $C \leq 80$.

7.2.3 Validity of the BET Monolayer Capacity

As outlined in Chapter 5, in their systematic study of the adsorption of nitrogen and other gases on a number of different adsorbents, Emmett and Brunauer (1937) came to the conclusion that Point B marked the boundary between monolayer and multilayer adsorption. It will be recalled that Point B was defined as the beginning of the middle, nearly linear, region of an adsorption isotherm (see Figures 5.2 and 7.1). On an empirical basis, a number of other characteristic features (designated A, C, D and E) were rejected in favour of Point B. The main reasons for the selection of Point B were because: (1) it gave fairly consistent values of surface area and (2) at this point, the isosteric enthalpy of adsorption appeared to undergo an appreciable decrease.

The theoretical significance of Point B was explained in terms of the BET theory by Brunauer et al. (1938). Thus, for nitrogen adsorption on 12 different adsorbents, the values of n_m and the corresponding uptakes (n_B) at Point B were generally found to be in fairly close agreement (to within ca. 10%). In fact, the agreement is within a few percent if the C value is sufficiently high to allow a well-defined Point B to be identified (Gregg and Sing, 1982). Brunauer et al. concluded that any evidence supporting the validity of Point B can be regarded as also supporting the validity of n_m. It is perhaps surprising to find such close agreement because the experimental location of Point B is not strictly compatible with the properties of the BET equation, which, as indicated in Figure 5.3, has a point of inflection and not a middle linear section.

Further confirmation for the validity of n_B, and hence of n_m, is provided *inter alia* by experimental energy of adsorption data. For the reasons given in Chapter 3, it is preferable to use calorimetry rather than the indirect isosteric method for the evaluation of energies of adsorption. Calorimetric measurements made over the past 60 years have confirmed that with certain non-porous and mesoporous solids, there is an appreciable decrease in the differential energy of adsorption at, or in the region of, Point B (Beebe and Young, 1954; Avgul et al., 1962; Holmes, 1967; Berezin et al., 1969; Berezin and Sagatelyan, 1972; Rouquerol et al., 1979). All these results are, of course, consistent with the interpretation that at Point B, monolayer adsorption is very close to completion and multilayer adsorption has just begun. It follows that if the isotherm has a sharp knee – see (I) in Figure 7.1 – Point B can be identified and n_m can be provisionally accepted as the *effective* monolayer capacity. The physical interpretation of n_m is more difficult, however, if the knee is absent or rounded, as in isotherm (II) in Figure 7.1.

We must emphasize that the calorimetric measurements discussed in later chapters reveal that monolayer–multilayer physisorption does not occur in accordance with the BET model. Thus, pronounced energetic heterogeneity is shown by some physisorption systems over the complete range of monolayer coverage. With highly homogeneous surfaces such as graphitized carbons there is a progressive *increase* in the adsorption energy of certain adsorptives as the monolayer coverage is increased (as in Figure 10.10). This increase is undoubtedly due to adsorbate–adsorbate interactions, which can begin at quite low surface coverage. As we have seen, neither of these effects is compatible with the BET theory.

It is generally agreed that a low C value and indistinct Point B are associated with a significant overlap of monolayer coverage and multilayer development (as to be expected from Equation 7.5), which is due to relatively weak adsorbent–adsorbate interactions (e.g. on a low-energy polymer surface) and/or configurational entropy effects (e.g. water on carbons or n-hydrocarbons on silica). Thus, if $C < \sim 50$, the transition from the first to the next layer occurs over a fairly broad range of coverage. Isotherms of n-alkanes and benzene on silica are typical examples of this shape (Kiselev, 1958). For example, the isotherms of n-pentane and n-hexane at 293 K on non-porous and mesoporous forms of silica are of gradual curvature with no sign of a Point B, with values of $C \sim 10$. Similarly, with nitrogen isotherms at 77 K surface modification of mesoporous silica brought about an appreciable change in isotherm shape and a reduction in C from ~ 110 to ~ 20 (Choma et al., 2003).

The BET theory is not able to account for the Type VI isotherm, which is the simplest type of layer-by-layer multilayer isotherm. The steps of a Type VI isotherm tend to lose their sharpness as the temperature is raised. However, the results of Prenzlow and Halsey (1957) showed that the midpoint (inflection) of the tread is rather insensitive to the change in temperature. This is an indication that in this case, it is the step height, and not Point B, which corresponds to the monolayer completion.

The existence of sub-steps poses another problem in the application of the BET method. As will be seen in Chapter 10, the presence of a sub-step is a sign that the monolayer can undergo a phase change, which must result in an increase in the monolayer density. Under these conditions, it is unlikely that the BET method is capable of providing an accurate evaluation of the monolayer capacity.

Before we discuss the evaluation of the BET area, it is appropriate to summarize the status of the BET equation and the derived monolayer capacity. It is evident that the BET model gives an oversimplified picture of monolayer–multilayer physisorption. The artificial extension of the simple Langmuir mechanism of localized monolayer adsorption on a uniform surface does not allow for the effects produced by adsorbate–adsorbate interactions or different surface structures (e.g. degrees of heterogeneity). Various theoretical refinements have been introduced and with the aid of the additional empirical parameters it is not difficult to extend the range of linearity of a modified BET plot. However, the increased complexity of the model does not compensate for the loss of simplicity of the BET method.

In the absence of other complicating factors (e.g. microporosity or highly active sites), the BET plot of a Type II or Type IV isotherm does appear to provide a fairly reliable assessment of n_m, provided that the knee of the isotherm is sharp with a clearly identifiable Point B. Evidently, it is not appropriate to apply the BET equation to Type III or Type V isotherms. Furthermore, caution should be exercised in interpreting the BET monolayer capacities derived from Type II or Type IV isotherms giving low C values (say, lower than $\sim$50) or abnormally high C values (say, $\sim$200 or above). We return to this problem in Section 7.2.4 and the particular difficulties involved in the interpretation of Type I isotherms are discussed in Section 7.2.5.

7.2.4 The BET Area of Non-porous and Mesoporous Adsorbents

The specific surface area, a(BET), is obtained from the BET monolayer capacity, n_m, by the application of the simple relation:

$$a(\text{BET}) = n_m \cdot L \cdot \sigma \tag{7.10}$$

where L is the Avogrado constant and σ is the average area occupied by each molecule in the completed monolayer.

The successful application of Equation (7.10) is, of course, dependent first on the validity of n_m and second on whether σ is already known or is determinable by some other method. It follows that it may have to be accepted that because of the uncertain interpretation of n_m, the search for an exact value of σ may be fruitless.

Emmett and Brunauer (1937) proposed that the molecular cross-sectional area, σ, can be calculated from the density of the liquid adsorptive in the bulk liquid state. Thus,

$$\sigma = f\left(\frac{M}{\rho L}\right)^{2/3} \tag{7.11}$$

where f is a packing factor, which for hexagonal close-packing becomes 1.091, ρ is the absolute density of the liquid adsorptive at the operational temperature and M is the molar mass of the adsorptive.

For the important case of nitrogen adsorption at 77 K, the value of $\sigma(N_2)$ is usually taken by convention, as 0.162 nm^2, which was the molecular area originally used by

Emmett and Brunauer (1937). Actually, if we insert in Equation (7.11) the most recent data available for the liquid density of nitrogen at 77 K, we obtain a molecular cross sectional of 0.163 nm^2, but the difference is so small that it is insignificant in comparison with the range of uncertainty of the effective molecular area.

When the value $\sigma(N_2) = 0.162 \text{ nm}^2$ is inserted into Equation (7.10), we obtain

$$\frac{a(\text{BET})}{\text{m}^2 \text{g}^{-1}} = 0.097 \frac{n_m}{\mu\text{mol g}^{-1}} \tag{7.12}$$

or when $v^{\sigma}(\text{STP})$ is used, we have

$$\frac{a(\text{BET})}{\text{m}^2 \text{g}^{-1}} = 4.35 \frac{v_m^{\sigma}(\text{STP})}{\text{cm}^3 \text{g}^{-1}} \tag{7.13}$$

Unfortunately, the various attempts made to compare BET nitrogen areas with independent values of surface area have not helped to clarify the situation (see Gregg and Sing, 1982). The most direct approach would be to compare a BET area with the simple geometrical area of a solid mass with no surface roughness, but in practice, this is extremely difficult because of the very small uptake of gas and the possible presence of molecular-scale surface roughness. It is not surprising that very few studies of this type have ever been attempted (see Gregg and Sing, 1982). Most tests of the validity of the BET area have been carried out with finely divided solids, and generally electron microscopy has been used to determine the particle size distribution. There are many sources of error (e.g. a wide particle size distribution and particle shape factors), and we are left with the impression that the agreement between the corresponding values of specific surface area is no better than within about ±20%.

The fact that nitrogen has a permanent quadrupole moment is considered to be responsible for the formation of an almost 'standard' nitrogen monolayer on many surfaces. Generally, the level of specificity is not high enough to give strong localization at 77 K. An exception is the exposure of small cationic sites on the surface of some oxides (see Chapter 11). In this case, at low coverage the differential enthalpy of adsorption is abnormally high, indicating strong preferential adsorption on a small fraction of the surface. The knee of the nitrogen isotherm is then extremely sharp at very low p/p^o and the C value is correspondingly high. As already mentioned, a change in shape of the nitrogen isotherm in the other direction is brought about by the surface modification of silica or carbon, which can lead to a reduction in the C value to $\sim$20–30 (Choma et al., 2003; Trens et al., 2004) and hence an increase in the degree of uncertainty in the estimation of n_m.

From an experimental standpoint, the availability of liquid nitrogen and the present range of commercial equipment make it relatively easy to determine full nitrogen adsorption–desorption isotherms at 77 K. This is an additional reason why nitrogen is internationally accepted as the standard BET adsorptive (IUPAC, Sing et al., 1985), with the convention that $\sigma(N_2) = 0.162 \text{ nm}^2$. Thus, for routine work, it is assumed that the nitrogen monolayer is in a close-packed 'liquid' state at 77 K, irrespective of the actual structure of the BET monolayer.

Argon is a possible alternative adsorptive for surface area determination: it is chemically inert and its molecule is symmetrical and monoatomic. Although the polarizabilities of argon and nitrogen are remarkably similar, their electronic structures are quite different. When the molecular area of argon is calculated from extrapolated liquid density at 77 K (Brunauer and Emmett, 1937), we obtain $\sigma(Ar) = 0.138$ nm^2, which was the value recommended by Mc Clellan and Harnsberger (1967). Comparisons between BET argon and BET nitrogen areas have revealed that significant adjustments have to be made in the respective molecular areas, which depend on the adsorbent (see Gregg and Sing, 1982). Thus, for graphitized carbon blacks, the use of $\sigma(Ar) = 0.138$ nm^2 and $\sigma(N_2) = 0.162$ nm^2 generally gives excellent agreement, but this is not the case for oxides. To obtain agreement, we now find that either the argon molecular area must be increased (to ca. 0.166 nm^2) or the nitrogen molecular area must be decreased (to ca. 0.13 nm^2). So far, this problem has not been completely resolved, although neutron diffraction and adsorption microcalorimetry studies do indicate that the orientation of the nitrogen quadrupole is dependent on the surface structure. It appears likely that this changes with the progressive dehydroxylation of a silica surface. If the nitrogen molecules are able to interact 'vertically' with surface hydroxyl groups, then $\sigma(N_2)$ could be reduced to ca. 0.11 nm^2 in the completed monolayer (Rouquerol et al., 1979, 1984).

As will be seen in the later chapters of this book, there is much to be said in favour of using *both* argon and nitrogen for the characterization of solid surfaces. However, for convenience and cost reasons, it is at 77 K that argon adsorption was mainly used until now and, at this temperature, there are several reasons why argon adsorption could not be adopted in place of nitrogen. First, 77 K is below the triple point of argon (83.8 K) and therefore since its bulk reference state is in doubt, it is not possible to use argon at this temperature for the determination of the mesopore size distribution. Second, the overall character of the argon isotherm at 77 K is generally more sensitive than that of nitrogen to any change in the surface structure. Nitrogen does not give any well-defined Type VI isotherms at 77 K and its multilayer development appears to remain remarkably constant from one surface to another (Carrott and Sing, 1989).

The state of affairs for argon is different at 87 K since at that temperature, argon is above its triple point so that (i) capillary condensation can normally take place; (ii) the saturation vapour pressure p^o can be measured and (iii) in principle, the mesopore size distribution can be determined. In the presence of micropores, adsorption equilibrium is appreciably quicker to reach at 87 K than at 77 K, due to the thermal activation of gas diffusion. For these reasons (see Thommes, 2004), we can expect argon adsorption to be used more in the future, when new cryostats will make experiments at 87 K relatively easy and cheap to carry out.

For operational reasons, it becomes more difficult to measure nitrogen or argon isotherms on low-area adsorbents (if $a < 1$ m^2 g^{-1}). To overcome this problem, *krypton* is widely used. As a consequence of its low p^o at 77 K (≈ 2 mbar), the 'dead space' correction for unadsorbed gas is relatively small and it becomes possible to measure small uptakes of gas with acceptable accuracy.

Unfortunately, as with argon at 77 K, the interpretation of a krypton isotherm is not always straightforward. At 77 K, krypton is well below its triple point and it would seem necessary to take the solid as the reference state to calculate p/p^o. Nevertheless, as for argon, there is some evidence from microcalorimetry and neutron diffraction studies (Grillet et al., 1985) that in the BET region, the adsorbate may well be in a liquid-like state. It has therefore become customary, for the construction of the BET plot, to adopt the saturation pressure, p^o (liquid), of the supercooled liquid as the effective p^o at 77 K. By applying Equation (7.11) and taking the absolute density of the supercooled liquid at 77 K, we obtain a value of 0.152 nm^2 for the molecular area of krypton. However, Beebe et al. (1945) found it necessary to adopt $\sigma(Kr) =$ 0.195 nm^2 and this still remains a useful empirical value. We conclude that krypton adsorption is useful for routine work on low-area powders, but that the results are not necessarily consistent with those one would obtain from nitrogen adsorption measurements. The values used for p^o and σ should always be recorded.

Over the past 70 years, many different gases and vapours have been proposed as suitable adsorptives for surface area determination (see Young and Crowell, 1962; Mc Clellan and Harnsberger, 1967; Gregg and Sing, 1982; Lowell et al., 2004). As indicated in Table 7.1, the adsorptives have included oxygen and xenon (at low temperature) and organic vapours such as butane and benzene (at or near 'ambient' temperatures). A very confused picture seems to emerge from the many attempts made to compare the estimated and derived values of σ for these and other adsorptives. How can we explain the wide range of derived values of σ for a given adsorptive?

To throw some light on this question, it is instructive to consider the adsorption of *n*-alkanes and benzene on various surfaces and to revisit the pioneering work of Kiselev (1957, 1958) (see Gregg and Sing, 1982). It was found *inter alia* that physisorbed *n*-alkane molecules interact strongly with the surface of graphitized carbon and

Table 7.1 Molecular Areas of Some Adsorptives

Adsorptive	T (K)	Cross-Sectional Area σ (nm^2)		
		Literature Range[a]	In Close-Packed Liquid Monolayer[b]	Customary Value
Nitrogen	77	0.13–0.20	0.162	0.162
Argon	77	0.10–0.19	0.138	0.138
Krypton	77	0.14–0.24	0.152	0.202
Xenon	77	0.16–0.25	0.168[c]	0.170
Oxygen	77	0.13–0.20	0.141	0.141
Carbon dioxide	195	0.14–0.22	0.163	0.210
n-Butane	273	0.32–0.53	0.321	0.440
Benzene	293	0.25–0.51	0.307	0.430

[a]Values quoted by Mc Clellan and Harnsberger (1967) and Gregg and Sing (1982).
[b]By application of Equation (7.11).
[c]Taking the density of the solid.

that there is a regular increase in the differential energy of adsorption with the carbon number (see Figure 1.3). Moreover, on a uniform graphitic surface, each n-alkane isotherm has a well-defined Point B. The proposed flat molecular orientation was confirmed by the steady increase in the derived values of molecular area $\sigma(n\text{-alkane})$, which are calculated from $a(\text{BET}, N_2)$ and $\boldsymbol{n}_m(n\text{-alkane})$ – the BET n-alkane monolayer capacity. In the case of n-hexane, the BET derived value of $\sigma(C_6H_{14})$ turned out to be 0.51 nm^2, which is close to $\sigma(C_6H_{14}) = 0.54 \text{ nm}^2$ for a flat molecular orientation (Isirikyan and Kiselev, 1961).

The physisorption behaviour of the n-alkanes on silica was quite different. In the case of n-pentane adsorption on silica, the values of $\sigma(C_5H_{12})$ and C are, respectively, 0.7 nm^2 and 10, in poor agreement with the corresponding values of 0.5 nm^2 and 60 on graphitized carbons (Kiselev and Eltekov, 1957). As mentioned already, the isotherms on silica had no distinctive knee or Point B, and the derived values of σ could not be explained in terms of a flat or a vertical molecular orientation (Gregg and Sing, 1967).

The adsorption of benzene was also studied in some detail by Kiselev and also by Dubinin and others (see Chapter 10). Again, on graphitized carbons, there is strong evidence from the isotherm shape and adsorption calorimetry (see Figure 10.10) to support a flat molecular orientation in a well-defined monolayer. Thus, with a number of graphitized carbon blacks, excellent agreement was found in each case between $a(\text{BET}, N_2)$ and $a(\text{BET}, C_6H_6)$, by taking $\sigma(C_6H_6) = 0.40 \text{ nm}^2$.

Benzene adsorption on silica is much more difficult to interpret (see Gregg and Sing, 1982). At temperatures $\sim 293 \text{ K}$, benzene isotherms have gradual curvature with no discernible Point B. On a series of silica gels characterized by nitrogen adsorption, the derived values of $\sigma(C_6H_6)$ were in the range of 0.3–0.5 nm^2 (Horvat and Sing, 1961). Another complication is the effect produced by surface dehydroxylation in removing the specific interactions associated with the aromatic ring (see Table 1.4). After dehydroxylation, Point B is lost entirely and C is reduced to below 10.

In comparison with the behaviour of the hydrocarbons, ethanol adsorption on hydroxylated silica is rather more straightforward (Madeley and Sing, 1959; Horvat and Sing, 1961; Branton et al., 1995). Hydrogen bonding is now responsible for the formation of a localized monolayer with little multilayer overlap and the presence of a fairly well-defined Point B. With the series of silica gels referred to above, the derived $\sigma(\text{EtOH})$ values were mostly in the range of 0.28–0.31 nm^2 for ethanol adsorption at 293 K.

One could try to explain the above findings in a number of different ways. It could be argued that the appreciable variance in the derived values of σ is simply due to the different monolayer structures. Indeed, on some surfaces, the molecular orientation does appear to be responsible for differences in the monolayer structure (Karnaukhov, 1985). Curvature of the adsorbent surface can also change the adsorbate molecular packing (Ohba et al., 2007). However, in our view, any difference in monolayer structure cannot by itself explain the large ranges of apparent molecular area shown in Table 7.1. For this reason and keeping in mind the shortcomings of the BET model, we must return to the question of the validity of $\boldsymbol{n}_m$.

As already indicated, if the knee of an isotherm on a non-porous or mesoporous adsorbent is sharp, it is possible, in principle, to estimate the 'true' monolayer capacity from the BET plot. In this case, Point B is usually located at $p/p^o < \sim 0.15$. In contrast, the BET plot cannot provide a reliable analysis of an isotherm of Type III or Type V since monolayer coverage does not take place evenly over the entire surface. Instead, the adsorbate molecules are initially located on particular surface sites and a patch-wise multilayer development then follows (e.g. water on carbons – see Chapter 10).

Now, we must return to the intermediate case and pose the question: can the BET plot be used to evaluate the 'statistical monolayer capacity' from a Type II or Type IV isotherm with no clearly identifiable Point B (see Figure 7.1) and a value of $C < \sim 50$? The surface coverage may be fairly uniform, but – as we have already seen – there is an appreciable overlap of monolayer and multilayer development. In principle, even if the molecular packing is unknown, it should be possible to define the statistical amount of adsorbate in the complete monolayer under controlled conditions. But, bearing in mind the artificial nature of the BET model and the empirical application of the BET plot, if $C < \sim 50$, the validity of n_m(BET) must be highly questionable. Indeed, it seems likely that it is the incorrect BET assessment of the *true* monolayer capacity, which is largely responsible for the wide variance in the reported values of molecular area.

Should we attempt to specify an upper limit for C? Again, we must not be too dog-matic in trying to answer this second question. However, a very high initial adsorption energy accompanying a sharp isotherm knee at very low p/p^o is a clear indication of either the presence of high energy sites on a small fraction of the surface or enhanced adsorbent–adsorbate interactions in narrow micropores. In either case, this can result in an erroneous assessment of n_m.

We conclude that the most reliable BET evaluation of n_m is obtained from a Type II or Type IV isotherm having a well-defined Point B located within a linear section of the BET plot and a value of C in the range of 50–150. We can then also regard the derived BET nitrogen area – obtained by taking $\sigma(N_2) = 0.162$ nm^2 – as a reasonable estimate of the *effective surface area* available for nitrogen adsorption at 77 K.

7.2.5 The BET Area of Microporous Solids

Considerable caution must be exercised in the interpretation of Type I isotherms or composite isotherms (i.e. combinations of I and II or I and IV). It is not surprising to find that the BET plots given by microporous adsorbents are generally of very lim-ited linearity. In fact, the BET plots of Type I nitrogen isotherms on some microporous solids (e.g. activated carbons and zeolites) become non-linear at $p/p^o < 0.1$. This is in contrast to the corresponding Langmuir plots, which are often linear up to high p/p^o (Gregg and Sing, 1967; Alaya et al., 2001). It might be thought that the Langmuir plot therefore provides a more reliable way of evaluating the monolayer capacity, but of course this does not allow for the fact that the Langmuir monolayer-coverage mech-anism cannot be applied to micropore filling. We therefore recommend that Langmuir plots should *never* be used for the evaluation of monolayer capacities (or surface areas) from Type I *physisorption* isotherms.

For rather similar reasons, when a Type I(a) isotherm (see Figures 1.4 and 7.5) is plotted in the BET coordinates, the derived value of n_m cannot be accepted as a true monolayer capacity. The difference between the mechanisms of physisorption on an open surface and in narrow micropores has been clearly revealed by adsorption calorimetry and α_s-plots (see Chapters 9–11). Thus, the enhanced adsorbent–adsorbate interactions in pores of molecular dimensions are responsible for a pronounced distortion of isotherm shape, as is clearly evident in Figures 10.19 and 10.20. As already indicated, as a result of primary micropore filling, the value of C is correspondingly increased. For example, the following results were obtained (Rouquerol et al., 1979) for the adsorption of nitrogen and argon at 77 K by two hydroxylated silicas: the C values were, respectively, 133 and 32 for the mesoporous silica and 238 and 40 for the microporous silica. A more gradual approach to the plateau is evident in the case of isotherms of Type I(b) in Figure 7.5. In wider micropores (up to at least ~5 molecular diameters), cooperative adsorbate–adsorbate interactions come into play and quasi-multilayer development can then take place. These effects are discussed in Chapter 9.

We conclude that although the BET nitrogen area (or BET argon area) of a microporous material is a useful indication of its adsorbent 'activity' (see Rouquerol et al., 2007), it should not be regarded as an *effective area* – or a fundamental property of the adsorbent – and is only the first step in the characterization of a microporous material.

The criteria listed in Section 7.2.2 are especially useful in the case of adsorbents containing micropores, as illustrated by Figures 7.2–7.4 (after Rouquerol et al., 2007) for the system argon – Zeolite 13X at 87 K. In Figure 7.2, the BET plot appears to be reasonably linear in several relative pressure ranges (0.01–0.2, 0.02–0.05 and 0.05–0.15), which are all consistent with a positive value of C. The resulting monolayer capacities range from 40 to 52 μmol g^{-1} (i.e. a variation of 30%), which illustrates the need for other criteria in addition to the linearity and a positive C.

Figure 7.3 shows the application of the criterion requiring a continuous increase in $n(p^\circ - p)$ versus p/p°: this provides a 'cut-off' $p/p^\circ \sim 0.04$, above which the BET equation should not be applied. Finally, Figure 7.4 shows the linearity of the BET plot corresponding to the selected pressure range (regression factor $R^2 = 1.000$) and the location of the calculated n_m, which is well within the pressure range selected

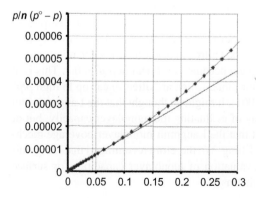

Figure 7.2 BET plot for Ar on zeolite 13X at 87 K. (For colour version of this figure, the reader is referred to the online version of this chapter.)

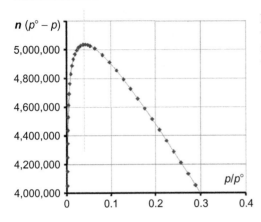

Figure 7.3 Plot of the term $n^a(p^o - p)$ versus p/p^o. (For colour version of this figure, the reader is referred to the online version of this chapter.)

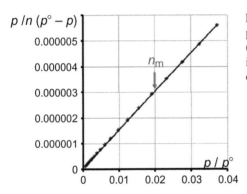

Figure 7.4 BET plot in the finally selected pressure range for Ar on zeolite 13X at 87 K. (For colour version of this figure, the reader is referred to the online version of this chapter.)

(i.e. another essential criterion). These criteria are particularly useful when applied to Type I isotherms, but it is advisable to apply them systematically since one never knows in advance whether a new sample is to some extent microporous. It is necessary to stress again that this procedure is designed to ensure that compatibility is achieved when sets of BET areas are quoted and compared, but it does not allow one to establish the validity of experimental or computed values of BET area.

7.2.6 BET Areas – Some Applications

In spite of its limitations, the BET method is extensively applied in research, process control and product development. It is therefore customary to specify the BET nitrogen areas of materials such as industrial adsorbents, catalysts, pigments, cements and polymers. An indication of the way in which BET areas can be exploited is illustrated by a brief description of some typical applications. The importance of a combination of adsorption and other measurements is clearly evident, as stressed in various chapters of this book.

Of course, the adsorbent must be outgassed in some way before any gas adsorption measurements can be undertaken (as described in Chapter 3). Hard granular or porous adsorbents do not present any special problems. Soft and flexible materials are assumed to become rigid at 77 K, but ageing often occurs when the wet material is stored or during the initial stages of sample preparation and outgassing. By following the changes in BET area, one can study the ageing of gels, pastes and precipitates (Baker et al., 1971; Bye and Sing, 1973) and establish the conditions required to minimize the collapse of a hydrogel (Kenny and Sing, 1994) or some other open structure such as in pulpwood fibres (Swanson, 1979).

It is well known that surface modification (e.g. of silica) is extremely important in the preparation of a chromatographic material and for other applications (Unger, 1979; Choma et al., 2003). In the work of Trens et al. (2004), argon and nitrogen adsorption isotherms were determined on modified macroporous silica, its surface becoming increasingly hydrophobic after treatment with various alkylchlorosilanes. Molecular areas of $\sigma(N_2) = 0.162$ nm^2 and $\sigma(Ar) = 0.138$ nm^2 were assumed for the calculation of the BET areas, which for the unmodified silica were found to be $a(BET, N_2) = 17.4$ m^2 g^{-1} and $a(BET, Ar) = 11.7$ m^2 g^{-1}. After surface modification, there was a progressive reduction in the BET nitrogen area to a final value ~ 14 m^2 g^{-1}, while the corresponding BET argon area underwent a small increase to become ~ 13.5 m^2 g^{-1}. It would at first seem that these differences were due to a change in the orientation of the adsorbed N_2 molecules. However, an alternative explanation is that the pronounced difference in shape of the nitrogen isotherm is largely responsible for the variation in the BET monolayer capacity. This interpretation is supported by the fact that the nitrogen C value was reduced from the initially high level ~ 200 to ~ 30 on the modified surface, whereas the argon value of C underwent a relatively small decrease (from ~ 30 to ~ 15). Whichever explanation is correct, the actual change in surface area was probably quite small. We must again stress the importance of recording the extent of the linear range of the BET plot along with the values of C and σ.

Physisorption studies have been made on various organic pigments (Sappok and Honigmann, 1976). In the work of Mather and Sing (1977) on copper phthalocyanine pigments, small samples were outgassed for ~ 20 h at room temperature prior to the measurement of nitrogen and argon isotherms at 77 K. A small amount of low-pressure hysteresis was observed with some samples, while others gave stable and completely reversible nitrogen and argon isotherms. The BET nitrogen areas were in the range of 43–88 m^2 g^{-1} with C values of ~ 30–70. The corresponding C values for Ar were ~ 20–55 and the derived values of $\sigma(Ar)$ were 0.15–0.16 nm^2. Toluene and n-propanol isotherm measurements at 298 K (Dean et al., 1978, 1979) revealed much larger differences in the behaviour of the various pigments. The β form of copper phthalocyanine gave almost reversible toluene isotherms in contrast to the pronounced hysteresis and ageing exhibited by α-copper phthalocyanine. These and other results were thought to be due to differences in particle morphology and aggregate structure (Dean et al., 1979). It is noteworthy that with the aid of the BET nitrogen and BET argon areas, it was possible to interpret the organic vapour sorption data and elucidate the ageing mechanism.

The BET areas of inorganic pigments are often recorded in the scientific and technical literature, although because of the complexity of their surface structures, the interpretation of the physisorption data may not be entirely straightforward. In the case of white titania pigments, the dispersibility and durability are considerably improved by surface treatment which generally involves the deposition of silica (and possibly also alumina). An understanding of the nature of the 'dense silica' coating of rutile particles was gained by studying the adsorption of nitrogen and water vapour along with electrophoretic mobilities (see Furlong, 1994). This work revealed that the high-energy cationic sites could be effectively screened by the deposition of a small amount of silica without any loss of optical quality.

7.3 Empirical Methods for Isotherm Analysis

7.3.1 Standard Adsorption Isotherms

As already noted, the detailed course of a physisorption isotherm is dependent on the nature of the gas–solid system and the operational temperature. In view of the wide variation in adsorbent–adsorbate interaction energies discussed in Chapter 1, it is not surprising to find that the shape of the isotherm in the *monolayer* region is especially sensitive to any variation in the surface structure of the adsorbent. However, as already indicated in Chapter 5, for some adsorptives (including nitrogen at 77 K), the shape of the corresponding *multilayer* isotherm is much less dependent on the adsorbent structure.

In this case, the thickness t of an adsorbed multilayer depends essentially on the equilibrium pressure and temperature and much less on the nature of the adsorbent. The multilayer thickness curve can be evaluated after normalizing an adsorption isotherm of Type II, that is, obtained in the absence of any micropore or mesopore filling. It is indeed possible to superpose, over an appreciable multilayer range, a series of normalized nitrogen adsorption isotherms determined at 77 K on non-porous oxides and carbons (Harris and Sing, 1959; de Boer et al., 1966; Gregg and Sing, 1982; Carrott and Sing, 1989; Jaroniec et al., 1999).

An early normalizing procedure, proposed by Kiselev (1957) to compare adsorption isotherms of hydrocarbons, water vapour, etc. on a series of different adsorbents, was simply to plot versus $p/p°$ the surface excess concentration $\Gamma(=n/a)$, obtained from the BET nitrogen area, a(BET). It is also possible to plot, instead of Γ, the 'reduced adsorption', n/n_m, which still relies on the BET method to determine the monolayer capacity n_m but does not require knowledge of the molecular cross-sectional area σ.

If the multilayer thickness is required, the next step is to convert the reduced adsorption (i.e. the statistical number of adsorbed layers) into t, by the relation

$$t = \frac{n}{n_m}d' \tag{7.14}$$

where d' is the effective thickness of a monolayer. The assumption is generally made that the absolute density of the adsorbed layer is identical to that of the bulk liquid adsorptive at the operational temperature, so that:

$$d' = \frac{M}{\sigma \times L \times \rho^\ell} \tag{7.15}$$

where the terms are all as previously defined.

For nitrogen adsorbed at 77 K, from Equations (7.11) and (7.15) and by taking $\rho = 0.809$ g cm^{-3}, $\sigma = 0.162$ nm^2 and $M = 28.01$ g mol^{-1}, one finds $d' = 0.354$ nm. This value, which is an average molecular thickness, compares well with the kinetic diameter of nitrogen (0.364 nm). The notional thickness t of a close-packed nitrogen multilayer then becomes:

$$t = 0.354 \frac{n}{n_m} \tag{7.16}$$

This is the equation used by Lippens and de Boer (1964) to plot a t-curve, that is, the multilayer thickness versus p/p^o, for nitrogen adsorption on various non-porous oxides at 77 K.

The observation that the t-curves for nitrogen on some apparently non-porous oxides (and carbons) were in fairly good agreement appeared to support the concept of a 'universal multilayer thickness curve' for nitrogen adsorption and appeared to offer the possibility of using this t-curve as a standard isotherm for nitrogen adsorption at 77 K (de Boer et al., 1965). Although this claim is now regarded as an oversimplification, the fact that the multilayer sections of many *nitrogen* adsorption isotherms can be superposed (see Chapters 10 and 11) has important implications for both surface area determination and pore size analysis.

The empirical methods described hereafter make use of the concept of the standard isotherm to provide an independent approach to isotherm analysis and surface area determination and so avoid complete reliance on the BET method.

7.3.2   The t-Method

The way in which Lippens and de Boer (1965) made use of the universal t-curve is simple. The experimental isotherm is transformed into a t-plot in the following manner: the amount adsorbed, n, is replotted against t, the standard multilayer thickness on the reference non-porous material at the corresponding p/p^o. Any difference in shape between the experimental isotherm and the standard t-curve is thus revealed as a non-linear region of the t-plot and/or a finite (positive or negative) intercept of the extrapolated t-plot (i.e. at $t = 0$). By this method, a specific surface area, denoted $a(t)$, can be calculated from the slope, $s_t = n/t$, of a linear section. From Equations (7.10), (7.11) and (7.15), we then get:

$$a(t) = \frac{M}{\rho} \cdot \left(\frac{n}{t}\right) \tag{7.17}$$

and, taking $\rho = 0.809$ g cm^{-3} for nitrogen at 77 K,

$$\frac{a(t)}{m^2 g^{-1}} = 0.0346 \frac{s_t}{\mu mol\, nm^{-1}} \tag{7.18}$$

In their original work, Lippens and de Boer (1965) had assumed that monolayer adsorption could occur on the micropore walls in the same manner as on the open surface and on the walls of mesopores. So the t-method did not allow for the special nature of micropore filling. A little later, it was pointed out (Sing, 1967) that, under favourable conditions, a t-plot can provide a means of assessing the micropore volume and the external area. These aspects are more fully discussed in relation to the α_s-method in the next section and in Chapters 10 and 11.

According to Brunauer et al. (1969), the correct t-curve should be the one having the same C value as the adsorption isotherm under investigation. However, this does not take the essential nature of micropore filling into account and therefore is not considered to be an acceptable approach. It is now generally agreed (IUPAC, Sing et al., 1985) that the appropriate standard isotherm must be determined on a non-porous solid having the same surface structure (i.e. surface chemistry) as that of the test adsorbent (Sing, 1970). The most serious limitation of the t-method, however, is that it is necessarily dependent on the BET evaluation of the monolayer capacity of the reference material, since t is derived from n/n_m: this presents a special problem if the C value is relatively low.

7.3.3   The α_s-Method

By a simple modification of the t-method, it is possible to avoid the prior evaluation of n_m and thereby extend the analysis to virtually any type of physisorption system (Sing, 1968, 1970; Sing and Williams, 2005; Badalyan and Pendleton, 2007).

To convert the standard adsorption data into an alternative dimensionless form, n_m is replaced by n_s the amount adsorbed at a pre-selected relative pressure $(p/p^o)_s$. In practice, it is usually convenient to take $(p/p^o)_s = 0.4$. The corresponding reduced adsorption is then $n/n_{0.4}$ and is called α_s. The reduced isotherm for the non-porous reference adsorbent (or 'α_s-curve', i.e. α_s vs. p/p^o) is therefore arrived at empirically without any need to determine the BET monolayer capacity.

Standard α_s-curves were proposed for nitrogen and argon at 77 K on non-porous silicas (Bhambhani et al., 1972; Payne et al., 1973; Kruk et al., 1999) and carbons (Carrott et al., 1987; Carrott and Sing, 1989; Gardner et al., 2001). In addition, standard physisorption data have been reported for organic vapours, including carbon tetrachloride (Cutting and Sing, 1969), benzene (Isirikyan and Kiselev, 1961; Carrott et al., 2000), methanol (Carrott et al., 2001a) and dichloromethane (Carrott et al., 2001b).

The α_s-plot is constructed in an analogous manner to the t-plot, with the α_s data for the particular gas–solid system (at the operational temperature) used in place of the t data. In place of Equation (7.17), we now have:

$$\frac{a_{\text{test}}}{a_{\text{ref}}} = \frac{n_{\text{test}}/(n_{\text{ref}})_{0.4}}{n_{\text{ref}}/(n_{\text{ref}})_{0.4}} \tag{7.19}$$

where $n_{\text{ref}}/(n_{\text{ref}})_{0.4} = \alpha_s$.

So by writing $n_{test}/\alpha_s = s_s$ we obtain:

$$a_{test} = \frac{a_{ref}}{(n_{ref})_{0.4}} \times s_s \qquad (7.20)$$

where a_{ref} is the specific BET surface area of the reference material.

In principle, the α_s-method can be applied to any gas–solid physisorption system irrespective of the shape of its isotherm. The method can be used to check the validity of the BET area and also to identify the adsorption and pore filling mechanisms (monolayer–multilayer adsorption, primary and secondary micropore filling or capillary condensation).

Numerous examples of different α_s-plots are to be found in subsequent chapters. Accurate adsorption measurements at low p/p^o have made it possible to construct high-resolution α_s-plots for N_2 and Ar on oxides and carbons (e.g. by Carrott et al., 1989; Carrott et al., 1991; Kaneko, 1996; Branton et al., 1997; Ribeiro Carrott et al., 1997; Llewellyn et al., 2000; Choma et al., 2003; Fukasawa et al., 2004; Arai et al., 2007). In addition to low-temperature nitrogen, argon and oxygen isotherms, various organic vapour isotherms have been analysed in this manner, including various alkanes (Carrott and Sing, 1988; Carrott et al., 1991), benzene (Carrott et al., 2000), methanol (Carrott et al., 2001a) and dichloromethane Carrott et al. (2001b).

Here, we are concerned with the general principles of the α_s-method of isotherm analysis with particular reference to the evaluation of surface area. The distinctive features of various hypothetical α_s-plots are revealed in Figure 7.5, where the α_s-plots are classified in relation to isotherms of Types II, IV and I, respectively.

The most straightforward form of α_s-plot is Type II(a) in Figure 7.5, which is for a typical Type II isotherm with a moderate value of C (~ 100). The extensive range of linearity and the zero intercept are the result of unrestricted monolayer–multilayer adsorption on a non-porous solid of very similar surface structure to that of the reference material. Obviously, in this case, the shapes of the experimental and standard isotherms are virtually identical and therefore the slope of the α_s-plot is directly proportional to the ratio of the surface areas, $a(S)/a_{ref}$. Thus, if the value of a_{ref} is already known, it is a simple matter to calculate a_{test}, which we denote here as $a(S)$ to indicate it is calculated by the α_s-method.

An α_s-plot of Type II(b) or II(c) in Figure 7.5 is obtained when the multilayer section of the experimental isotherm conforms to that of the standard, but the monolayer section is different. The form of plot II(b) is indicative of relatively strong adsorbent–adsorbate interactions, whereas plot II(c) is associated with significantly weaker interactions. With either II(b) or II(c), $a(S)$ may be calculated from the slope of the linear multilayer region.

Type IV isotherms, which are normally associated with capillary condensation within the mesopore structure, generally give α_s-plots of Type IV in Figure 7.5. In this case, it is the initial part of the isotherm which corresponds to monolayer–multilayer adsorption on the mesopore walls. If the corresponding section of the α_s-plot is linear and back-extrapolates to the origin, the slope provides a measure of $a(S)$ which is now the total available surface area. We may also conclude that there are no additional

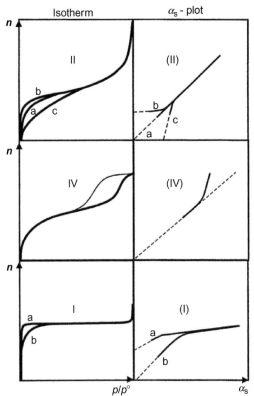

Figure 7.5 Hypothetical α_s-plots (right) and corresponding adsorption isotherms (left): the adsorbent is non-porous (top), mesoporous (middle) and microporous (bottom).

complicating factors such as a micropore filling contribution. The upward deviation is due to the onset of capillary condensation.

The α_s-plots of Type I(a) and I(b) are typical of those given by microporous adsorbents. A long linear multilayer section allows an assessment to be made of the external surface area. Any upward deviation (not represented here) is likely to be due to capillary condensation in a separate mesopore structure. If the intermediate linear region is not too short, its slope can provide a useful indication of the mesopore area. If we are confident that there are no complications associated with different surface chemical structures, we may safely attribute the difference between the shapes of the two α_s-plots I(a) and I(b) to the effects of primary and secondary micropore filling (see Chapters 5 and 9). Thus, the form of plot I(a) is due to distortion of the isotherm in the monolayer region, which is itself associated with the enhanced adsorbent–adsorbate interactions in the pores of molecular dimensions (i.e. primary micropore filling). On the other hand, the initial linear section of I(b), which can be back-extrapolated to the origin, is evidently due to a range of undistorted monolayer adsorption on the walls of wider micropores. It is only in the latter case that we can use the α_s-plot to derive a reliable value of the *internal* surface area.

The α_s-method for surface area determination has been calibrated against certain non-porous reference materials. For example, for N_2 adsorption on a non-porous hydroxylated silica, at 77 K, with $a_{ref} = 154 \text{ m}^2 \text{ g}^{-1}$ and $(n_{ref})_{0.4} = 2387 \text{ }\mu\text{mol g}^{-1}$, Equation (7.20) may be written as:

$$\frac{a(S - N_2)}{m^2 g^{-1}} = 0.0645 \frac{s_s}{\mu\text{mol g}^{-1}} \qquad (7.21)$$

and for the adsorption of argon on the same silica at 77 K, we have:

$$\frac{a(S - Ar)}{m^2 g^{-1}} = 0.074 \frac{s_s}{\mu\text{mol g}^{-1}} \qquad (7.22)$$

where $s_s = n/\alpha_s$ is the slope of a linear α_s-plot.

If we wish to use adsorption data where the specific gas volume corresponding to n is expressed in the traditional volume units of $\text{cm}^3(STP) \text{ g}^{-1}$ (Carruthers et al., 1971), it is necessary to multiply the coefficients of Equations (6.21) and (6.22) by the molar volume of the ideal gas, $0.022414 \text{ cm}^3 \text{ }\mu\text{mol}^{-1}$ (for $\theta = 0 \text{ }^\circ C$ and $p^\circ = 1.01325 \text{ bar}$). Thus:

$$\frac{a(S - N_2)}{m^2 g^{-1}} = 2.88 \frac{v^\sigma / \text{cm}^3(STP) \text{ g}^{-1}}{\alpha_s} \qquad (7.23)$$

and

$$\frac{a(S - Ar)}{m^2 \text{ g}^{-1}} = 3.29 \frac{v^\sigma / \text{cm}^3(STP) \text{ g}^{-1}}{\alpha_s} \qquad (7.24)$$

7.3.4 Comparison Plots

A simple way of comparing the shapes of two isotherms of a given adsorptive is to plot the amount adsorbed by one against the amount adsorbed by the other at the same relative pressure (Gregg and Sing, 1982; Karnaukhov, 1985; Ribeiro Carrott et al., 1991; Branton et al., 1993). Of course, if the isotherms are identical in shape, the comparison plot will be a straight line passing through the origin, its slope being equal to the ratio of the surface areas of the two adsorbents. As with an α_s-plot, any deviation from linearity can be explained in terms of micropore filling, capillary condensation or differences in surface chemistry. Comparison plots are useful for following the changes in porosity which accompany progressive activation or ageing of a material or in an exploratory investigation in which no reference data are available for a particular adsorption system (Alario Franco and Sing, 1974; Mather and Sing, 1977; Branton et al., 1993; Fernandez et al., 2001).

7.4 The Fractal Approach

It is often extremely difficult to arrive at exactly the same result when a number of different experimental methods are used for the determination of the specific surface area. Indeed, as Adamson (1990) has pointed out, one should in general *expect* such results to differ!

To illustrate this problem, we note a possible difference between the surface area of a porous solid which is available for adsorption and the area (including that of closed pores) which can scatter low angle X-rays. Even in the former case, the extent of the adsorption is likely to be dependent on the size, shape and electronic nature of the adsorptive molecules in relation to the surface chemistry, roughness and porosity of the adsorbent.

In principle, the fractal approach provides a way of circumventing the intractable problem of evaluating the *absolute* area of a finely divided or porous solid. The aim of fractal analysis is to characterize the *effective* geometry of an adsorbent and hence arrive at a clearer understanding of its behaviour.

The application of fractal geometry may be regarded as a form of resolution analysis: it is a systematic way of studying how the magnitude of a given property (e.g. surface area or pore volume) is altered by a change in the resolution of its measurement (Pfeifer and Obert, 1989; Avnir, 1991, 1997; Sonwane et al., 1999; Neimark, 2002). In general, one makes use of a simple scaling power law of the type

$$\text{measured property} = k(\text{resolution})^D \tag{7.25}$$

where k and D are constants, which define the degree of resolution.

In the present context, it is useful to express Equation (7.25) in an explicit form, such as

$$N_m = k\sigma^{-D_a/2} \tag{7.26}$$

where N_m is the number of molecules in the completed monolayer, σ is the adsorptive molecular area and D_a is now the fractal dimension of the accessible surface (Farin and Avnir, 1989).

The magnitude of D_a is determined *inter alia* by the degree of surface roughness or porosity. In principle, a lower limit of $D_a = 2$ is obtained with a perfectly smooth surface on the molecular scale. Most non-porous materials would be expected to exhibit some surface roughness. With such a material, a constant value of D_a between 2 and 3 implies that there is a degree of self-similarity: the shape of the surface irregularities thus remaining invariant over a certain range of resolution. Thus, the physical structure of a fractal surface will appear similar when viewed at different magnifications.

Fractal plots of $\log n_m$ versus $\log \sigma$ for two porous silicas are shown in Figure 7.6 (here, n_m is the BET monolayer capacity). Both plots are linear, giving $D_a = 2.98$ for the silica gel and $D_a = 2.09$ for the controlled pore glass. These values reflect the extremes of the fractal scale, the latter being close to the ideal value for a flat surface.

Some of the many values of D_a compiled by Farin and Avnir (1989) and Avnir et al. (1992) are given in Table 7.2. Most values are within the theoretical fractal range, $D_a = 2-3$ and it is noteworthy that $D_a \approx 2$ for the graphitized carbon blacks and the pillared clays.

The values of D_a of (1.89 ± 0.09) and (1.94 ± 0.10) reported by Van Damme and Fripiat (1985) for pillared clays were derived from the multilayer capacities of nitrogen and various organic adsorptives. The fact that $D_a \approx 2$ appeared to confirm that the basal smectite surface was smooth and that the pillars were regularly distributed.

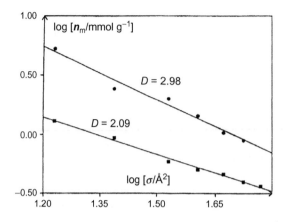

Figure 7.6 Fractal plots for the adsorption of linear alkanes on a silica gel (circles) and a controlled pore glass (squares).
Courtesy: Avnir et al. (1992) and Christensen and Topsoe (1987).

Table 7.2 Values of Accessible Fractal Dimension Evaluated from Adsorption Data

Adsorbent	Adsorptives	Fractal Dimension, D_a
Graphitized C blacks	N_2 and alkanes	1.9–2.1
Activated carbons	N_2 and organic molecules	2.3–3.0
Pillared clays	N_2 and organic molecules	1.8–2.0
Silica gels	Alkanes	2.9–3.4
Other oxides	N_2 and alkanes	2.4–2.7

Source: Taken from Farin and Avnir (1989).

It was argued by Van Damme and Fripiat that a random distribution of the pillars would necessarily lead to some localized molecular sieving and that this in turn would result in $D_a > 2$.

Various other aspects of fractal analysis have been discussed by Van Damme, Fripiat and their co-workers. For example, by extending the BET model to fractal surfaces, Fripiat et al. (1986) were able to show that the apparent fractal dimension is reduced by the progressive smoothing of a molecularly rough surface. Alternatively, the effect of a micropore filling contribution is to enhance the fractal dimension.

Other investigators, including Pfeifer and Obert (1989), Pfeifer et al. (1990), Krim and Panella (1991, 1994), Neimark and Unger (1993) and Ahmad and Mustafa (2006), have also studied multilayer adsorption on fractally rough surfaces. In particular, Pfeifer and his co-workers point out that the interpretation of a fractal dimension of a porous surface presents problems when it is associated with the BET theory. In contrast, an FHH fractal treatment appears to be more promising, provided that wetting effects are taken into account. This was exactly the approach adopted by Krim and her co-workers.

In their work on carefully prepared homogeneous silver films, Krim and Panella (1991) and Panella and Krim (1994) applied the FHH equation in the following form

$$\ln\left(\frac{p}{p^{\mathrm{o}}}\right) = -\frac{\alpha}{(kT\theta^n)} \tag{7.27}$$

where the coefficient α is dependent on the nature of the adsorbent–adsorbate and adsorbate–adsorbate interactions and the corresponding fractal dimension is determined by the magnitude of the exponent n (de Gennes, 1985). Nitrogen and oxygen isotherms at 77.4 K gave linear FHH plots over a wide range of multilayer coverage. The values of n for nitrogen on smooth and rough Ag substrates were 3.0 and 4.7, respectively, the former giving the expected fractal dimension, $D_a = 2$. However, the value $n \approx 3$ was also obtained for oxygen on the *rough* surface, which was attributed to surface tension effects (since the surface tension of liquid oxygen is much higher than that of liquid nitrogen).

Panella and Krim suggested that the high value $n = 4.7$ was more consistent with the properties of a self-affine surface rather than a self-similar one. Self-affine fractals are associated with asymmetric scaling; that is different scaling relations in different directions (Avnir, 1997).

The FHH fractal approach was also adopted by Ehrburger-Dolle et al. (1994) as part of a systematic study of the properties of silica aerogels. The value $D_a = 2.1$ was evaluated from the linear FHH plot for nitrogen on Aerosil 200, whereas values of 2.64 and 2.95 were obtained from the nitrogen multilayer isotherms on various aerogels. The latter value, which is obviously close to 3, appeared to be consistent with volume filling by capillary condensation as the predominant process at high p/p^{o}. Evidently, the low fractal dimension given by Aerosil 200 is associated with its molecularly smooth surface.

Another aspect of fractal geometry which has been discussed in some detail by Avnir and his co-workers is the relationship between surface area and particle size. If the particles of a powder are smooth and non-porous, we would expect to find a simple form of inverse relation

$$a = \frac{k}{d} \tag{7.28}$$

where the constant k is dependent on the density and the particle shape.

In practice, Equation (7.28) will not hold if the particles are either rough or porous. If the particles are highly porous, we have the extreme case of a negligible external area in comparison with the internal area.

For the intermediate case between the extremes of smooth and highly porous particles, Farin and Avnir (1988) applied an expression of the form

$$\frac{a}{a_{\mathrm{g}}} = R^{D_r - 2} \tag{7.29}$$

where a is the total apparent specific surface area, a_{g} is the 'geometric' area, which is based on the average particle radius, R, and D_r is the roughness fractal dimension.

Table 7.3 Value of the Roughness Fractal Dimension Derived from Fractal Plots of log a versus log R (Farin and Avnir, 1989)

Adsorbent	Adsorptive	Fractal Dimension, D_r
Aerosil	Nitrogen	2.0
Vitreous silica	Nitrogen	2.0
Montmorillonite	Nitrogen	2.0
Quartz	Nitrogen	2.1
Snowit	Nitrogen	2.2
Iceland spar	Krypton	2.2
Coke	Nitrogen	2.5
	Carbon dioxide	2.5
Dolomite	Krypton	2.6–2.9
Bio-carbonates	Krypton	2.7–3.0
Porous silica	Nitrogen	3.0
	Ethanol	3.0

Linear fractal plots of log a versus log R have been reported for a number of different porous and non-porous adsorbents (Farin and Avnir, 1989). A selection of the derived values of D_r is given in Table 7.3. Most of adsorption measurements were made with nitrogen or krypton at 77 K and it would now be of interest to extend the comparisons to include other adsorptive molecules.

On the whole, the overall pattern of the results in Table 7.3 is as expected. It is not surprising to find that Aerosil and vitreous silica gave low values of D_r which correspond to minimal surface roughness. The high D_r values obtained with the skeletal carbonates and porous silica are also to be expected. Although it is more difficult to understand the significance of some of the intermediate values (e.g. of coke), the linearity of the fractal plots is of practical value.

On the basis of all these findings, it would seem that a simple fractal approach offers a useful way of detecting and characterizing the effects of surface roughness and porosity. At first sight, fractal analysis appears to be more sophisticated than the traditional assumption that the same area is available to all adsorptives. However, it must be kept in mind that, because of experimental constraints, the application of the power law is limited to a fairly short range of molecular size.

Furthermore, as Drake et al. (1990) pointed out and as we have already seen in this chapter, molecular cross-sectional areas are difficult to establish unambiguously. Avnir et al. (1992) also drew attention to the need for caution in accepting molecular areas which are based on a standardized method (usually BET nitrogen areas). In a critical appraisal of the different methods for determining surface fractal dimensions, Neimark (1990) stressed the importance of taking account of the different mechanisms of physisorption (e.g. at high p/p°, the combination of multilayer adsorption and capillary condensation). Conner and Bennett (1993) also warned of the risk of an over-simplistic interpretation of a linear log–log fractal plot. In our view, an over-simplified application of fractal analysis may tend to obscure rather than clarify the interpretation of adsorption data. In practice, there are three complicating factors: (1) derived values of n_m are not always reliable, (2) different mechanisms of adsorption and pore filling and (3) non-uniformity of the surface and pore structure.

A serious limitation of fractal analysis is that the precise nature of the surface roughness or porosity cannot be deduced from a single value of D_a. In an extreme case of molecular sieving, differences in accessibility are obvious; but with many systems (especially heterogeneous substrates), it is much more difficult to distinguish between the effects of surface roughness and surface structure. We are drawn to the conclusion that log–log fractal plots are useful for the correlation of adsorption data – especially on well-defined porous or finely divided materials. A derived fractal dimension can also serve as a characteristic empirical parameter, provided that the system and operational conditions are clearly recorded. In some cases, the fractal self-similarity (or self-affine) interpretation appears to be straightforward, but this is not so with many adsorption systems which are probably too complex to be amenable to fractal analysis.

7.5 Conclusions and Recommendations

Although real gas–solid physisorption systems do not conform to the BET *model*, the BET *method* is widely employed for determining the BET area, a(BET), of many kinds of porous and non-porous materials. For this purpose, nitrogen at 77 K is the most popular adsorptive, although argon at 87 K is beginning to be favoured by some investigators.

Two stages are involved in the evaluation of a(BET) from a physisorption isotherm. The BET monolayer capacity, n_m, must first be derived from the linear part of the BET plot. If possible, this should straddle the knee of a Type II or Type IV isotherm (i.e. be located around Point B). The application of a few criteria is recommended to ensure that the BET range is selected objectively. If the BET C value lies outside the range of $\sim$50–150, the validity of n_m is questionable: thus, if $C < 50$, there is an appreciable overlap of monolayer and multilayer adsorption; if $C > 150$, there is either strong localized adsorption or a micropore filling contribution. Micropore filling is clearly evident in the case of a Type I isotherm so the physical validity of n_m is doubtful and should not be regarded as a true monolayer capacity.

The BET method is strictly not applicable to Type III or Type V isotherms.

In the second stage of the BET method, the calculation of a(BET) from n_m requires the effective molecular area, σ, to be known or assumed. In applying the 'standard' BET method, the completed nitrogen monolayer is assumed to be close-packed at 77 K so that $\sigma(N_2) = 0.162$ nm^2. This is unlikely to be always true, but as most surfaces are to some extent heterogeneous little is gained by adjusting the value of $\sigma(N_2)$, unless independent evidence can be obtained. Provided that certain conditions are fulfilled, the BET area derived from a nitrogen isotherm of Type II or Type IV can be regarded as an *effective* surface area available for nitrogen adsorption at 77 K. Although the BET area derived from a Type I isotherm should not be treated as a 'real' surface area, it is useful as an indication of adsorbent 'activity'.

It is recommended that when a BET area is reported, the selected range of linearity of the BET plot and the particular values of σ and C should always be recorded.

Another method is required to throw more light on the adsorption mechanism. A fundamentally rewarding approach is to study the energetics of adsorption – preferably by adsorption calorimetry – alongside the isotherm measurement. When enhanced adsorbent–adsorbate interactions are revealed, they can be attributed to micropore filling or adsorption on high energy sites. However, adsorption calorimetry is experimentally and thermodynamically demanding and not to be undertaken lightly!

A simpler approach is to use an empirical procedure for the analysis of a physisorption isotherm. To do this, it is first necessary to determine standard isotherm data for monolayer–multilayer adsorption on a well-characterized non-porous reference adsorbent. It is then possible to transform an isotherm for the same gas–solid system into the α_s-plot. Deviations from linearity are produced by different stages of surface coverage or by pore filling. The α_s-method can provide an independent assessment of the total available surface area of a mesoporous adsorbent or the external area of a microporous adsorbent and can be applied to any physisorption system.

References

Adamson, A.W., 1990. Physical Chemistry of Surfaces. Wiley, New York, p. 561.
Ahmad, A.L., Mustafa, N.N.N., 2006. J. Colloid Interface Sci. 301, 575.
Alario Franco, M.A., Sing, K.S.W., 1974. An. Quim. 70, 41.
Alaya, M.N., Hourieh, M.A., El-Sejariah, F., Youssef, A.M., 2001. Adsorpt. Sci. Technol. 19, 321.
Arai, M., Kanamaru, M., Matsumura, T., Hattori, Y., Utsumi, S., Ohba, T., Tanaka, H., Yang, C.M., Kanoh, H., Okino, E., Touhara, H., Kaneko, K., 2007. Adsorption. 13, 509.
ASTM C1274-12, 2012. Standard test method for advanced ceramic specific surface area by physical adsorption.
Avgul, N.N., Kiselev, A.V., Lygina, I.A., Mikailova, E.A., 1962. Izv. an SSSR, Otd Khim Nauka. 769.
Avnir, D., 1991. Chem. Ind. 912.
Avnir, D., 1997. In: Ertl, G., Knozinger, H., Weitkamp, J. (Eds.), Handbook of Heterogeneous Catalysis, Vol. 2. Wiley-VCH, Weinheim, p. 598.
Avnir, D., Farin, D., Pfeifer, P., 1992. New J. Chem. 16, 439.
Badalyan, A., Pendleton, P., 2003. Langmuir 19, 7919.
Badalyan, A., Pendleton, P., 2007. In: Lewellyn, P.L., Rodriguez-Reinoso, F., Rouquerol, J., Seaton, N. (Eds.), Characterization of Porous Solids VII. Elsevier, Amsterdam, p. 383.
Baker, F.S., Carruthers, J.D., Day, R.E., Sing, K.S.W., Stryker, L.J., 1971. Disc. Faraday Soc. 52, 173.
Beebe, R.A., Young, D.M., 1954. J. Phys. Chem. 58, 93.
Beebe, R.A., Beckwith, J.B., Honig, J.M., 1945. J. Am. Chem. Soc. 67, 1554.

Benton, A.F., 1926. J. Am. Chem. Soc. 48, 1850.

Berezin, G.I., Sagatelyan, R.T., 1972. In: Thermochimie. Coll. Internationaux CNRS, Vol. 201. Editions du CNRS, Paris, p. 561.

Berezin, G.I., Kiselev, A.V., Sagatelyan, R.T., Serdobov, M.V., 1969. Zh. Fiz. Khim. 43, 224.

Bhambhani, M.R., Cutting, P.A., Sing, K.S.W., Turk, D.H., 1972. J. Colloid Interface Sci. 38, 109.

Branton, P.J., Hall, P.G., Sing, K.S.W., 1993. J. Chem. Soc., Chem. Commun, 1257.

Branton, P.J., Hall, P.G., Sing, K.S.W., 1995. Adsorption 1, 77.

Branton, P.J., Sing, K.S.W., White, J.W., 1997. J. Chem. Soc., Faraday Trans. 93, 2337.

Brunauer, S., 1945. The Adsorption of Gases and Vapors. Princeton University Press, Princeton.

Brunauer, S., Emmett, P.H., 1935. J. Am. Chem. Soc. 57, 1754.

Brunauer, S., Emmett, P.H., 1937. J. Am. Chem. Soc. 59, 2682.

Brunauer, S., Emmett, P.H., Teller, E., 1938. J. Am. Chem. Soc. 60, 309.

Brunauer, S., Skalny, J., Bodor, E.E., 1969. J. Colloid Interface Sci. 30, 546.

Bye, G.C., Sing, K.S.W., 1973. In: Smith, A.L. (Ed.), Particle Growth in Suspensions. Academic, London, p. 29.

Carrott, P.J.M., Sing, K.S.W., 1988. Langmuir 4, 740.

Carrott, P.J.M., Sing, K.S.W., 1989. Pure Appl. Chem 61, 1835.

Carrott, P.J.M., Roberts, R.A., Sing, K.S.W., 1987. Carbon 25, 59.

Carrott, P.J.M., Drummond, F.C., Kenny, M.B., Roberts, R.A., Sing, K.S.W., 1989. Colloids Surf. 37, 1.

Carrott, P.J.M., Ribeiro Carrott, M.M.I., Roberts, R.A., 1991. Colloids Surf. 58, 385.

Carrott, P.J.M., Ribeiro Carrott, M.M.L., Cansado, I.P.P., Nabais, J.M.V., 2000. Carbon 38, 465.

Carrott, P.J.M., Ribeiro Carrott, M.M.L., Cansado, I.P.P., 2001a. Carbon 39, 193.

Carrott, P.J.M., Ribeiro Carrott, M.M.L., Cansado, I.P.P., 2001b. Carbon 39, 465.

Carruthers, J.D., Payne, D.A., Sing, K.S.W., Stryker, L.G., 1971. J. Colloid Interface Sci. 36, 205.

Cassel, H., 1944. J. Phys. Chem. 48, 195.

Choma, J., Jaroniec, M., 2001. Adsorpt. Sci. Technol. 19, 765.

Choma, J., Kloske, M., Jaroniec, M., 2003. J. Colloid Interface Sci. 266, 168.

Christensen, S.V., Topsoe, H., 1987. Haldor Topsoe Co, Denmark, private communication quoted by Avnir et al. (1992).

Conner, W.M., Bennett, C.O., 1993. J. Chem. Soc. Faraday Trans. 89, 4109.

Cutting, P.A., Sing, K.S.W., 1969. Chem. Ind, 268.

de Boer, J.H., Linsen, B.G., Osinga, Th.J., 1965. J. Catalysis. 4, 643.

de Boer, J.H., Lippens, B.C., Linsen, B.G., Broekhoff, J.C.P., van den Heuvel, A., Osinga, Th.J., 1966. J. Colloid Interface Sci. 21, 405.

Dean, C.R.S., Mather, R.R., Sing, K.S.W., 1978. Thermochimica Acta. 24, 399.

Dean, C.R.S., Mather, R.R., Segal, D.L., Sing, K.S.W., 1979. In: Gregg, S.J., Sing, K.S.W., Stoeckli, H.F. (Eds.), Characterisation of Porous Solids. Society of Chemical Industry, London, p. 359.

de Gennes, P.G., 1985. In: Adler, D., Fritzsche, H., Ovshinsky, S.R. (Eds.), Physics of Disordered Materials. Plenum, New York.

Drake, J.M., Levitz, P., Klafter, J., 1990. New. J. Chem. 14, 77.

Ehrburger-Dolle, F., Dallamano, J., Pajonk, G.M., Elaloui, E., 1994. In: Rouquerol, J., Rodriguez-Reinoso, F., Sing, K.S.W., Unger, K.K. (Eds.), Characterization of Porous Solids III. Elsevier, Amsterdam, p. 715.

Emmett, P.H., Brunauer, S., 1937. J. Am. Chem. Soc. 59, 1553.

Everett, D.H., Parfitt, G.D., Sing, K.S.W., Wilson, R., 1974. J. Appl. Chem. Biotechnol. 24, 199.

Farin, D., Avnir, D., 1988. In: Unger, K.K., Rouquerol, J., Sing, K.S.W., Kral, H. (Eds.), Characterization of Porous Solids I. Elsevier, Amsterdam, p. 421.

Farin, D., Avnir, D., 1989. In: Avnir, D. (Ed.), The Fractal Approach to Heterogeneous Chemistry. John Wiley, Chichester, p. 271.

Fernandez, E., Centeno, T.A., Stoeckli, F., 2001. Adsorpt. Sci. Technol. 19, 645.

Fripiat, J.J., Gatineau, L., Van Damme, H., 1986. Langmuir 2, 562.

Fukasawa, K., Ohba, T., Kanoh, H., Toyoda, T., Kaneko, K., 2004. Adsorpt. Sci. Technol. 22, 595.

Furlong, D.N., 1994. In: Bergna, H.E. (Ed.), The Colloid Chemistry of Silica. American Chemical Society, Washington, p. 535.

Gardner, L., Kruk, M., Jaroniec, M., 2001. J. Phys. Chem. B. 105, 12516.

Gregg, S.J., Jacobs, J., 1948. Trans. Faraday Soc. 44, 574.

Gregg, S.J., Sing, K.S.W., 1967. Adsorption, Surface Area and Porosity, first ed. Academic Press, London.

Gregg, S.J., Sing, K.S.W., 1982. Adsorption, Surface Area and Porosity, second ed. Academic Press, London.

Grillet, Y., Rouquerol, F., Rouquerol, J., 1985. Surface Sci. 162, 478.

Halsey, G.D., 1948. J. Chem. Phys. 16, 931.

Harris, M.R., Sing, K.S.W., 1959. Chem. Ind, 487.

Hill, T.L., 1946. J. Chem. Phys. 14, 268.

Holmes, J.M., 1967. In: Flood, E.A. (Ed.), The Solid-Gas Interface. Marcel Dekker, New York, p. 127.

Horvat, D.M., Sing, K.S.W., 1961. J. Appl. Chem. 11, 313.

Isirikyan, A.A., Kiselev, A.V., 1961. J. Phys. Chem. 65, 601.

ISO 9277:2010 (E), 2010. Determination of the specific surface area of solids by gas adsorption – BET method.

Jaroniec, M., Kruk, M., Olivier, J.P., 1999. Langmuir 15, 5410.

Kaneko, K., 1996. In: Dabrowski, A., Tertykh, V.A. (Eds.), Adsorption on New and Modified Inorganic Sorbents. Elsevier, Amsterdam, p. 573.

Karnaukhov, A.P., 1985. J. Colloid Interface Chem. 103, 311.

Keii, T., Takagi, T., Kanataka, S., 1961. Anal. Chem. 33, 1965.

Kenny, M.B., Sing, K.S.W., 1994. In: Bergna, H.E. (Ed.), The Colloid Chemistry of Silica. American Chemical Society, Washington, p. 505.

Kiselev, A.V., 1957. In: Schulman, J.M. (Ed.), Second International Congress of Surface Activity II. Butterworths, London, p. 168.

Kiselev, A.V., 1958. In: Everett, D.H., Stone, F.S. (Eds.), The Structure and Properties of Porous Materials. Butterworths, London, p. 195.

Kiselev, A.V., 1965. Disc. Faraday Soc. 40, 205.

Kiselev, A.V., 1972. Disc. Faraday Soc. 52, 14.

Kiselev, A.V., Eltekov, Y.A., 1957. In: Schulman, J.H. (Ed.), Second International Congress of Surface Activity II. Butterworths, London, p. 228.

Krim, J., Panella, V., 1991. In: Rodriguez-Reinoso, F., Rouquerol, J., Sing, K.S.W., Unger, K.K. (Eds.), Characterization of Porous Solids II. Elsevier, Amsterdam, p. 217.

Kruk, M., Jaroniec, M., Sayari, A., 1999. Langmuir 15, 5683.

Langmuir, I., 1916. J. Am. Chem. Soc. 38, 2221.

Langmuir, I., 1918. J. Am. Chem. Soc. 40, 1361.

Lippens, B.C., de Boer, J.H., 1964. J. Catal. 3, 44.

Lippens, B.C., de Boer, J.H., 1965. J. Catal. 4, 319.

Llewellyn, P.L., Rouquerol, F., Rouquerol, J., Sing, K.S.W., 2000. In: Unger, K.K., Kreysa, G., Basel, J.P. (Eds.), Characterization of Porous Solids V. Elsevier, Amsterdam, p. 421.

Lowell, S., Shields, J.E., Thomas, M.A., Thommes, M., 2004. Characterization of Porous Solids and Powders: Surface Area, Pore Size and Density. Kluwer, Dordrecht.

Madeley, J.D., Sing, K.S.W., 1959. Chem. Ind, 289.

Mather, R.R., Sing, K.S.W., 1977. J. Colloid Interface Sci. 60, 60.

Mc Clellan, A.L.H., Harnsberger, H.F., 1967. J. Colloid Interface Sci. 23, 577.

Neimark, A.V., 1990. Adsorpt. Sci. Technol. 7, 210.

Neimark, A.V., 2002. In: Schüth, F., Sing, K.S.W., Weitkamp, J. (Eds.), Handbook of Porous Solids I. Wiley-VCH, Weinheim, p. 81.

Neimark, A.V., Unger, K.K., 1993. J. Colloid Interface Sci. 158, 412.

Ohba, T., Matsumura, T., Hata, K., Yumura, M., Iijima, S., Kanoh, H., Kaneko, K., 2007. J. Phys. Chem. C 111, 15660.

Panella, V., Krim, J., 1994. In: Rouquerol, J., Rodriguez-Reinoso, F., Sing, K.S.W., Unger, K.K. (Eds.), Characterization of Porous Solids III. Elsevier, Amsterdam, p. 91.

Parra, J.B., de Sousa, J.C., Bansal, R.C., Pis, J.J., Pajares, J.A., 1994. Adsorpt. Sci. Technol. 11, 51.

Payne, D.A., Sing, K.S.W., Turk, D.H., 1973. J. Colloid Interface Sci. 43, 287.

Pfeifer, P., Obert, M., 1989. In: Avnir, D. (Ed.), The Fractal Approach to Heterogeneous Chemistry. John Wiley, Chichester, p. 11.

Pfeifer, P., Obert, M., Cole, M.W., 1990. In: Fleischmann, M., Tildesly, D.J., Ball, R.C. (Eds.), Fractals in the Natural Sciences. Princeton University Press, Princeton, NJ, p. 169.

Prenzlow, C.F., Halsey, G.D., 1957. J. Phys. Chem. 61, 1158.

Ribeiro Carrott, M., Carrott, P., Brotas de Carvalho, M., Sing, K.S.W., 1991. J. Chem. Soc. Faraday Trans. 87, 185.

Ribeiro Carrott, M.M.I., Carrott, P.J.M., Candeias, A.J.E.G., 1997. In: McEnaney, B., Mays, T.J., Rouquerol, J., Rodriguez-Reinoso, F., Sing, K.S.W., Unger, K.K. (Eds.), Characterisation of Porous Solids IV. Royal Society of Chemistry, Cambridge, p. 103.

Rouquerol, F., Rouquerol, J., Imelik, B., 1964. Bull. Soc. Chim. Fr, 635.

Rouquerol, J., Rouquerol, F., Pérès, C., Grillet, Y., Boudellal, M., 1979. In: Gregg, S.J., Sing, K.S.W., Stoeckli, H.F. (Eds.), Characterization of Porous Solids. Society of Chemical Industry, London, p. 107.

Rouquerol, J., Rouquerol, F., Grillet, Y., Torralvo, M.J., 1984. In: Meyer, A.L., Belfort, G. (Eds.), Fundamentals of Adsorption. Engineering Foundation, New York, p. 501.

Rouquerol, J., Llewellyn, P., Rouquerol, F., 2007. In: Llewellyn, P. et al., (Ed.), Characterization of Porous Solids VII. Studies in Surface Science and Catalysis, Vol. 160. Elsevier, Amsterdam, p. 49.

Sappok, R., Honigmann, B., 1976. In: Parfitt, G.D., Sing, K.S.W. (Eds.), Characterization of Powder Surfaces. Academic Press, London, p. 231.

Sing, K.S.W., 1964. Chem. Ind. 321.

Sing, K.S.W., 1967. Chem. Ind. 829.

Sing, K.S.W., 1968. Chem. Ind. 1520.

Sing, K.S.W., 1970. In: Everett, D.H., Ottewill, R.H. (Eds.), Surface Area Determination. Butterworths, London, p. 15.

Sing, K.S.W., Williams, R.T., 2005. Adsorpt. Sci. Technol. 23, 839.

Sing, K.S.W., Everett, D.H., Haul, R.A.W., Moscou, L., Pierotti, R.A., Rouquerol, J., Siemieniewska, T., 1985. Pure Appl. Chem. 57, 603.

Sonwane, C.G., Bhatia, S.K., Calos, N.J., 1999. Langmuir 15, 4603.

Swanson, J.W., 1979. In: Gregg, S.J., Sing, K.S.W., Stoeckli, H.F. (Eds.), Characterisation of Porous Solids. Society of Chemical Industry, London, p. 339.

Thommes, M., 2004. In: Lu, G.Q., Zhao, X.S. (Eds.), Nanoporous Materials: Science and Engineering. Imperial College Press, London, p. 317.

Trens, P., Denoyel, R., Glez, J.C., 2004. Colloids Surf. A 245, 93.

Unger, K.K., 1979. Porous Silica. Elsevier, Amsterdam.
Van Damme, H., Fripiat, J.J., 1985. J. Chem. Phys. 82, 2785.
Williams, A.M., 1919. Proc. R. Soc. A96, 298.
Young, D.M., Crowell, A.D., 1962. Physical Adsorption of Gases. Butterworths, London.

8 Assessment of Mesoporosity

Kenneth S.W. Sing, Françoise Rouquerol, Jean Rouquerol, Philip Llewellyn

Aix Marseille University-CNRS, MADIREL Laboratory, Marseille, France

Chapter Contents

8.1   Introduction

The aims of this chapter are to discuss in general terms the use of physisorption for mesopore size analysis (i.e. for adsorbents having effective pore widths in the approximate range of 2–50 nm) and hence also the interpretation of the Type IV isotherm in the IUPAC classification (see Figure 1.4). Part of the chapter is along 'classical' lines: it is based on the concept of capillary condensation and the use of the Kelvin equation in evaluating the mesopore size distribution. There follows an account of the advantages and limitations of using density functional theory (DFT) for pore size

Adsorption by Powders and Porous Solids. http://dx.doi.org/10.1016/B978-0-08-097035-6.00008-5

analysis. Finally, the interpretation of adsorption hysteresis is briefly discussed in terms of capillary condensation, network–percolation, cavitation and molecular modelling.

Virtually all the computational procedures used for pore size analysis from physisorption data start from the assumption that the pores are rigid and of well-defined shape. Many investigators have assumed the mesopores to be either regular cylinders or parallel-sided slits, but in fact there are relatively few real adsorbents which conform exactly to either of these shapes. We should keep in mind that most porous adsorbents of technological importance are composed of complex interconnected pores of irregular shape.

Nitrogen adsorption at 77 K has become generally accepted as the standard method for mesopore size analysis (Gregg and Sing, 1982; Lowell et al., 2004). The recommended experimental and computational procedures involved have been described in various official publications (e.g. IUPAC: Sing et al., 1985 and Rouquerol et al., 1994; British Standard, 1992; International Organization for Standardization, ISO 15901-3). However, over the past few years, attention has been drawn to the advantages of using argon adsorption at 87 K as an alternative method.

A long-standing problem is the interpretation of physisorption hysteresis, when it appears in the form of an adsorption–desorption loop. Adsorption hysteresis may originate in various ways (e.g. by delayed condensation or network–percolation effects), but the shape of the loop can sometimes provide useful qualitative information about the mesopore structure. At one time, it was thought that hysteresis was associated with all Type IV isotherms, but a few completely reversible Type IV isotherms are now well documented, particularly on certain forms of MCM-41; these include isotherms of nitrogen at 77 K (Branton et al., 1993), neopentane at 273 K and toluene at 298 K (Russo et al., 2009).

8.2 Mesopore Volume, Porosity and Mean Pore Size

8.2.1 Mesopore Volume

It is customary to take the total specific pore volume, v_p, of an adsorbent as the *liquid* volume adsorbed at a predetermined high p/p^o (e.g. at $p/p^o = 0.95$). This procedure is not always satisfactory, however, because the total adsorption capacity (i.e. the amount adsorbed as $p/p^o \to 1$) is dependent on the magnitude of any non-porous area and also on the upper limit of the pore size distribution.

If a Type IV isotherm has a distinctive plateau, which cuts the p^o axis at an angle $\sim 90°$, we may generally arrive at an acceptable assessment of the total mesopore volume, v_p. The amount adsorbed, n_{sat}, at the plateau is a measure of the total adsorption capacity, and to obtain v_p, it is assumed that the adsorbate has the normal molar volume, V_m^ℓ, of the liquid at the operational temperature. This simple method for the determination of the pore volume is based on a general principle, which was put forward 80 years ago by Gurvich (1915) and is still known as the Gurvich rule.

Confirmation of the validity of the Gurvich rule has been obtained with a number of mesoporous adsorbents for a wide range of adsorptives. The results in Table 8.1 are typical of the level of agreement to be expected between the values of v_p derived from a number of isotherms determined on a given mesoporous adsorbent. Although the

Table 8.1 Uptake at Saturation on a Silica Gel at 25 °C, Calculated as a Volume of Liquid and as an Amount

Adsorbate	$\dfrac{v_{\text{sat}}}{\text{cm}^3\,\text{g}^{-1}}$	$\dfrac{n_{\text{sat}}}{\text{cm}^3\,\text{g}^{-1}}$
n-Hexane	0.431	3.28
2,3-Dimethylbutane	0.429	3.28
2-Methylpentane	0.431	3.28
n-Heptane	0.431	2.91
2,2,3-Trimethylbutane	0.420	2.88
n-Octane	0.434	2.66
2,2,4-Trimethylpentane	0.439	2.63
2,3,4-Trimethylpentane	0.425	2.66
Cyclohexane	0.421	3.88
Methyl cyclohexane	0.425	3.32
Ethyl cyclohexane	0.426	2.99
Benzene	0.440	4.92
Nitromethane	0.449	8.33
Nitroethane	0.434	6.03
Carbon tetrachloride	0.421	4.30

Source: From Mc Kee (1959).

adsorptives differ widely in their chemical and physical properties, the deviation from the mean value of v_p is within ca. 5%.

Some Type IV isotherms do not obey the Gurvich rule. A possible explanation for this failure is that the particular adsorbent has a range of narrow micropores, giving molecular sieving, in addition to its mesopore structure. The overall Type IV shape would then be misleading and it would be necessary to undertake a more detailed analysis (e.g. by applying the α_s-method) in order to evaluate the micropore filling contribution with different adsorptives. Another reason for failure of the Gurvich rule may be the lack of rigidity of the mesopore structure.

If the Type IV isotherm has a short plateau, which is followed by an upward swing, the amount adsorbed at the plateau can be regarded as the capacity of the particular range of mesopores. A useful assessment of the upper limit of mesopore size can then be obtained with the aid of the Kelvin equation corrected for multilayer adsorption.

In converting n_p to v_p, we have assumed that the condensate and the adsorbed layer together have an average density close to that of the bulk liquid adsorptive. Strictly, an allowance should be made for differences in density of the various parts of the adsorbate, but these corrections are probably small in comparison with the other uncertainties in the evaluation of v_p.

Table 8.2 Properties of Pore Structures Obtained by Three Types of Packing of Spherical Particles of Radius r

Type of Packing	N	ε	Radius of Spheres Inscribed in Cavities	Radius of Circle Inscribed in Connecting Throats
Hexagonal close packed	12	0.260	0.225 r-octahedral 0.414 r-tetrahedral	0.155 r
Primitive hexagonal	8	0.395	0.527 r	0.414 r 0.155 r
Tetrahedral	4	0.660	1.00 r	0.732 r

Source: After Avery and Ramsay (1973).

8.2.2 Porosity

For the present purpose, we define the porosity, ε, as the ratio of the total accessible pore volume V_p to the apparent volume of the adsorbent. Thus,

$$\varepsilon = \frac{V_p}{V_p + V^s} \tag{8.1}$$

where V^s is the inaccessible volume of the solid. In general, we should not expect to find a simple relationship between the porosity and the coordination number of packed spheres (see Table 8.2). However, Karnaukhov (1971, 1979) has shown that for a wide range of regular packings ($N = 3\text{--}12$), an 'ideal' porosity can be specified ($\varepsilon = 0.815\text{--}0.260$). In practice, it is not easy to relate the properties of random sphere packings to those of regular packings (Haynes, 1975) as the overall pore space may be defined by an infinite variety of local packings.

8.2.3 Hydraulic Radius and Mean Pore Size

The ratio of specific pore volume to specific surface area, v_p/a, has been used for many years as a simple means of characterising the pore size. This volume-to-surface ratio, when applied to a group of pores, is known as the hydraulic radius, r_h, and has an unambiguous physical significance, provided that the pore geometry can be specified by a single parameter (Everett, 1958).

For example, if the mesopore structure consists of a set of open-ended, non-intersecting cylinders, the mean pore radius, $\bar{r}_p$, is given by

$$\bar{r}_p = 2v_p/a \tag{8.2}$$

which is, of course, based on the assumption that the specific surface area, a, is confined to the cylindrical pore walls. This condition is fulfilled in the case of MCM-41 (see Chapter 13).

Other pore structures amenable to this simple treatment are parallel-sided slit-shaped pores or aggregates of parallel plates. The appropriate volume/surface relation is similar to Equation (8.2), but now the mean pore width, $\bar{w}_p$, replaces $\bar{r}_p$. Thus,

$$\bar{w}_p = 2v_p/a \qquad (8.3)$$

The mathematical significance of r_h has been established for a number of other pore geometries (Everett, 1958), but with most real systems, it is not possible to arrive at an unambiguous evaluation of $\bar{r}_p$ or $\bar{w}_p$ or a useful interpretation of r_K. For example, with an assemblage of packed spheres, as already noted, the porosity is dependent on the packing density as well as the particle size. Similarly, in the case of a network of intersecting pores, the value of r_h is dependent on both the pore radius and the lattice spacing of the intersections.

8.3   Capillary Condensation and the Kelvin Equation

8.3.1   Derivation of the Kelvin Equation

It is well known (Defay and Prigogine, 1951) that a spherical interface of radius of curvature r and surface tension γ can maintain mechanical equilibrium between two fluids at different pressures p'' and p'. The phase on the concave side of the interface experiences a pressure p'' which is greater than that on the convex side. The mechanical equilibrium condition is given by the Laplace equation:

$$p'' - p' = \frac{2\gamma}{r} \qquad (8.4)$$

Indeed, in the particular case when $r \rightarrow \infty$, $p'' = p'$, mechanical equilibrium between two phases separated by a plane interface can only be obtained if their pressures are equal.

If, instead of a spherical surface, one considers any interface having two principal radii of curvature of the surfaces r_1 and r_2, the condition of mechanical equilibrium at each point on the interface is:

$$p'' - p' = \gamma \left[\frac{1}{r_1} + \frac{1}{r_2} \right] \qquad (8.5)$$

Or introducing the radius r_m of mean curvature defined by:

$$\frac{1}{r_m} = \frac{1}{2} \left[\frac{1}{r_1} + \frac{1}{r_2} \right] \qquad (8.6)$$

the equation becomes:

$$p'' - p' = \frac{2\gamma}{r_m} \qquad (8.7)$$

which is a generalisation of Equation (8.4) and is sometimes referred to as the Young–Laplace equation (see Sing and Williams, 2012).

A related phenomenon is the difference in vapour pressure between the flat and curved surfaces of a given liquid. The application of classical thermodynamics (cf. Defay and Prigogine, 1951) allows us to replace the difference in mechanical pressure, $\Delta p = p^g - p^\ell$, by a function of the relative vapour pressure, p/p^o. The condition for physico-chemical equilibrium is

$$\mu^\ell = \mu^g \tag{8.8}$$

If we now move from one equilibrium state to another neighbouring equilibrium state, at constant temperature, then

$$dp^g - dp^\ell = d\left(\frac{2\gamma}{r_m}\right) \tag{8.9}$$

and

$$d\mu^\ell = v^\ell dp^\ell = d\mu^g = v^g dp^g \tag{8.10}$$

which allows Equation (8.9) to be written in the form:

$$d\left(\frac{2\gamma}{r_m}\right) = \frac{v^\ell - v^g}{v^\ell} dp^g \tag{8.11}$$

If we neglect the molar volume of the liquid v^ℓ in comparison with the molar volume of the vapour, v^g, and assume the vapour to be ideal, Equation (8.11) becomes

$$d\left(\frac{2\gamma}{r_m}\right) = -\frac{RT}{v^\ell} \frac{dp^g}{p^g} \tag{8.12}$$

If we now integrate this equation from zero curvature ($1/r_m = 0, p^g = p^o$) to some other state ($1/r_m, p$) and assume that v^ℓ is nearly constant, we have

$$\ln \frac{p}{p^o} = -\frac{2\gamma v^\ell}{r_K RT} \tag{8.13}$$

which gives the dependence of p/p^o on the mean radius of curvature of the meniscus r_m, now replaced by r_K. Equation (8.13) is often referred to as the Kelvin Equation (see Sing and Williams, 2012) as it is closely related to an equation originally proposed by Lord Kelvin (Thomson, 1871). Note that in this equation, γ and v^ℓ are assumed to be independent of r_K.

For nitrogen adsorbed at 77.35 K, we can use the following values: $\gamma = 8.85$ m Nm^{-1}; $\rho^\ell = 0.807\,g\,cm^{-3}$; $M = 28.01$ g mol^{-1}; $V_m^\ell = 34.71\,cm^3\,mol^{-1}$. Then:

$$\frac{r_K}{nm} = -\frac{0.415}{\log_{10}(p/p^o)}$$

A more general equation, derived by Everett (1979), is appropriate if the vapour pressure is high or the liquid–vapour adsorptive is close to its critical state.

8.3.2 Application of the Kelvin Equation

We are now in a position to consider the significance of Equation (8.13) in relation to physisorption. First, let us consider an assemblage of cylindrical mesopores in which all the pores have exactly the same radius, r_p. In the 'ideal' (and simplified) situation of strict thermodynamic reversibility, we would expect the pore filling to be indicated by a vertical riser in accordance with Equation (8.13).

In this particular case of a cylindrical pore shape, it seems reasonable to assume that the *condensate* has a meniscus of spherical form and radius r_K. However, as some physisorption has already occurred on the mesopore walls, it is evident that r_K and r_p are not equal. If the thickness of the adsorbed multilayer is t, and the contact angle is assumed to be zero, the *radius of the cylindrical pore* is simply

$$r_p = r_K + t \tag{8.14}$$

as in Figure 8.1.

However, if there is a finite contact angle, θ, between the adsorbed film and the capillary condensate, the relation between r_p and r_K becomes

$$r_p = r_K \cos \theta + t \tag{8.15}$$

In the application of Equation (8.15), it is usually assumed that $\theta = 0$ and therefore $\cos \theta = 1$. The link between r_K and r_p is somewhat more complex when we consider other

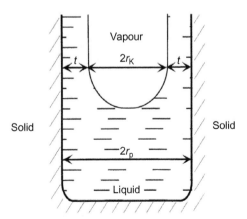

Figure 8.1 Relation between the Kelvin radius r_K and the pore radius r_p in a cylindrical mesopore.

pore shapes. Two types of particular importance are parallel-sided slits and the interstices between spheroidal particles. At first sight, the former type would appear to be more amenable to analysis.

The form of a meniscus formed in a *slit-shaped pore* cannot be spherical and therefore r_1 and r_2 in Equation (8.5) are no longer equal. In the simplest case, its form is hemicylindrical. The curvature is confined to the one axis across the pore, the curvature in the other principal direction being infinite so that $1/r_2 = 0$. We now have $1/r_K = 1/2r_1$ and therefore r_K is directly related to the effective pore width, w_p. It follows that in place of Equation (8.14), we have

$$w_p = r_K + 2t \tag{8.16}$$

We consider next the case of capillary condensation within a *system of packed spheres*: this normally occurs in three stages. The initial adsorption on the overall surface is now accompanied by the first stage of condensation around the points of contact of the particles. The initial interparticle condensation is reversible, but as the advancing menisci meet the narrow openings between the particles (i.e. the 'windows' or 'throats') are spontaneously filled. The third stage involves the filling of the larger voids (or cavities) within the packed particles.

The initial process of monolayer–multilayer adsorption on the available surface of the packed spherical particles is complicated by two opposing effects. First, there is always a significant reduction in adsorption due to the loss of surface area between adjacent particles (see Figure 8.2a). On the other hand, because of the close proximity of the two surfaces, the adsorption is enhanced in the inner part of the annular space (i.e. similar to small-scale micropore filling).

A saddle-shaped meniscus (or pendular ring) is developed in the first stage of condensation. Application of Equation (8.13) now requires the designation of two radii

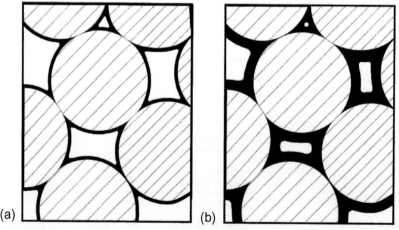

(a)   (b)

Figure 8.2 Sections of the pore space between solid spheres in irregular packing. First (a) and second (b) stage of condensation.

of curvature of opposite sign, one being concave and the other convex. The Kelvin equation therefore takes the form

$$\ln(p/p^{\circ}) = -\frac{\gamma v^{\ell}}{RT}\left[\frac{1}{r_1} - \frac{1}{r_2}\right] \tag{8.17}$$

where r_1 is the concave radius and r_2 is the convex radius, which is directly related to the particle radius. If we assume that the spherical particles all have the same radius, R, the radius of curvature r_2 as given by Wade (1964, 1965) is $\left[\sqrt{(R+t+r_1)^2 - R^2} - r_1 + R + t\right]/2.$

The second and third stages of capillary condensation are evidently dependent on both the particle size and the degree of packing (i.e. the coordination number) of the spherical particles. The windows may be pictured as the space between three or four neighbouring particles as in Figure 8.2b. If the particles are in a triangular array, the second stage of condensation is controlled by the radius of the inscribed circle. Although the amount taken up may be relatively small, this stage is important in relation to hysteresis (see Section 8.6).

The overall pore volume of the bed of packed spheres is mainly determined by the size of the inner cavities, which is in turn dependent on the particle radius, R, and the coordination number, N. The effective Kelvin radius corresponding to this major stage of capillary condensation is approximately given by the radius, r_s, of the inscribed sphere within the cavity (Karnaukhov, 1971).

A number of theoretical and experimental studies have been made of physisorption by compacted non-porous particles (see Haynes, 1975). Pioneering work in this field was carried out by Wade (1964), Broekhoff and Linsen (1970), Karnaukhov (1971) and Kanellopoulos et al. (1983). In his treatment of the random packing of equal spheres, Mason (1971) showed that tetrahedral sub-units could be constructed by taking the edges at random from a distribution function, which was considered to be a characteristic property of the packing. This model was applied by Kanellopoulos et al. (1983) to a system of densely packed spheres. A two-dimensional network of irregular tetrahedra was established, and cavities within the tetrahedra were assumed to be interconnected through the irregular triangular windows.

A systematic study of the adsorption of nitrogen by packed assemblages of spheroidal particles was undertaken by Adkins and Davis (1986, 1987). After the consideration of various pore filling models, it was concluded that the desorption process can be adequately described by the instability of a Kelvin, hemispherical meniscus in the neck (i.e. the window) of the structure, and the adsorption process can be viewed as a delayed Kelvin condensation in the largest dimension of the void structure. This reasoning is consistent with the network–percolation theory of hysteresis, which is discussed briefly in Section 8.6.

In the experimental investigations of Avery and Ramsay (1973), it was found that progressive compaction of very small spheroidal oxide particles led to the sequential change in nitrogen isotherm type from II to IV and finally to I. This was a clear demonstration of a decrease in pore width as a result of increase in particle coordination number.

Table 8.2 gives some of the properties of the pore structures produced by the different degrees of packing of spherical particles. It can be seen that it is only in the case of tetrahedral packing ($N = 4$) that the effective size of the cavities is approximately equal to the particle size.

Simulation studies have been made also on packed beds of other corpuscular systems, such as discs and rods (Karnaukhov 1979). Although this approach is based on an over-simplified picture of real powders, it does serve a useful purpose in providing a semi-quantitative analysis of the stages of capillary condensation. These mechanisms have been broadly confirmed (e.g. by Avery and Ramsay (1973) and Unger and his co-workers (Bukowiecki et al., 1985)).

Scherer (1998) has drawn our attention to a problem which must be faced in the characterisation of many aerogels. Even when an aerogel is rigid enough to withstand the capillary pressure which is exercised during the condensation of a vapour, the usual values of Kelvin radius and pore volume may be seriously in error. For example, in the case of a particular silica aerogel, it was found that the pore volume registered by nitrogen adsorption amounted to less than 60% of the total pore volume evaluated by other methods. Scherer suggests that the failure of the nitrogen to condense was not due to the pore *size*, but was instead the result of the peculiar curvature of the solid surface. Scherer pictures the gel as a network of intersecting solid cylinders joined at nodes. The adsorption/condensation will begin around the nodes, but will soon be arrested if the distance between the nodes is large in comparison with the radius of the cylinders joining the nodes. Thus, at an early stage, the meniscus adopts zero curvature and the overall pore space thus remains partially empty, regardless of the effective pore size, until $p/p^o \to 1$. A quantitative treatment of the condensation process is proposed by Scherer, which involves the evaluation of meniscus shape by minimising its surface area.

8.4   Classical Computation of the Mesopore Size Distribution

8.4.1   General Principles

Over the period 1945–1970, many different mathematical procedures were proposed for the derivation of the pore size distribution from nitrogen adsorption isotherms. It is appropriate to refer to these computational methods as 'classical' as they were all based on the application of the Kelvin equation for the estimation of pore size. Among the methods which remain in current use were those proposed by Barrett et al. (1951), which is still the most popular, Cranston and Inkley (1957), Dollimore and Heal (1964), Roberts (1967) and Kruk et al. (1997). In the early work, it was customary to assume the pore shape to be cylindrical, but now the slit-shaped and packed sphere models are considered to be more suitable for some systems.

There are several reasons why nitrogen (at 77 K) is generally accepted as the most suitable adsorptive for mesopore size analysis. First, the thickness of the N_2 multilayer is largely insensitive to differences in adsorbent particle size or surface structure (Carrott and Sing, 1989). Second, the same isotherm can be used for the evaluation of both the surface area and the mesopore size distribution (Sing et al., 1985).

However, in spite of these considerations, there is an emerging view that there are advantages in using argon adsorption at 87 K for pore size analysis by the application of newer 'microscopic' methods based on the DFT and molecular simulation (Thommes et al., 2012). Indeed, ideally more than one adsorptive should be used for the characterisation of mesoporous solids (e.g. see Llewellyn et al., 1997; Machin and Murdey, 1997).

In adopting the classical approach, one must necessarily assume that:

1. the Kelvin equation is applicable over the complete mesopore range;
2. the meniscus curvature is controlled by the pore size and shape and that $\theta = 0$;
3. the pores are rigid and of well-defined shape;
4. the distribution is confined to the mesopore range;
5. the filling (or emptying) of each pore does not depend on its location within the network;
6. the adsorption on the pore walls proceeds in exactly the same way as on the corresponding open surface.

The relation between pore volume and pore dimensions is presented graphically either in the form of a cumulative plot of pore volume against mean pore size (i.e. v_p vs. $\bar{r}_p$) or often more usefully as a distribution (or frequency curve), dv_p/dr_p versus $\bar{r}_p$ (or $\bar{w}_p$). As the computation is usually based on the notional removal of the condensate by a stepwise lowering of p/p^o, in practice the pore size distribution is expressed in the form of $\delta v_p/\delta r_p$ versus $\bar{r}_p$. The computation is somewhat complicated because allowance must be made at each desorption step for the thinning of the adsorbed multilayer in pores from which the capillary condensate has already been removed.

Before any attempt is made to evaluate the pore size distribution, it is necessary to decide which branch of the hysteresis loop of a Type IV isotherm is to be used for the analysis. It will be evident from the discussion of hysteresis in Section 8.6 that this choice is not easy and at this point it may be helpful to summarise the main implications.

A relatively simple pore structure of fairly uniform tubular pores would be expected to give a narrow Type H1 hysteresis loop (see Figure 8.3), and in this case, the desorption branch is generally used for the analysis. On the other hand, if there is a broad distribution of interconnected pores, it would seem safer to adopt the adsorption branch as the location of the desorption branch is largely controlled by network–percolation effects. If a Type H2 loop is very broad, neither branch can be used with complete confidence because of the possibility of a combination of effects (i.e. both delayed condensation and network–percolation). Furthermore, the condensate becomes unstable and pore emptying occurs when the steep desorption branch is located at a critical p/p^o (i.e. at ca. 0.42 for N_2 adsorption at 77 K).

8.4.2 Computation Procedure

The plateau of a Type IV isotherm is normally taken as the starting point for the computation of the mesopore size distribution. If all the pores are full, the first step in the notional desorption process (e.g. from p/p^o of 0.95 to 0.90) involves only the removal of the capillary condensate. Each subsequent step involves both the removal of the condensate from the *cores* of a group of pores and the thinning of the multilayer in the larger pores (i.e. those pores already emptied of the condensate). In the following treatment, the symbol v_K is used to represent the inner core volume and, as before, v_p is the pore volume. The corresponding radii are r_K and r_p.

Let us suppose that the amount of nitrogen removed in each desorption step 'j' is $\delta n(j)$, for the purpose of the pore size calculations, this amount is expressed as the volume, $\delta v^\ell(j)$, of *liquid* nitrogen. In the first desorption step ($j=1$), the initial removal is the result of capillary evaporation alone and therefore the volume of core space released is equal to the volume of nitrogen removed, that is, $\delta v_K(1)=\delta v^\ell(1)$.

If the pores are cylindrical, we have a simple relation between the core volume $v_K(1)$ and the pore volume, $v_p(1)$, of the first, and largest, group of mesopores. Thus,

$$v_p(1) = \frac{\bar{r}^2_p(1)}{\bar{r}^2_K(1)} v_K(1) \tag{8.18}$$

where $\bar{r}_p(1)$ and $\bar{r}_K(1)$ are the mean pore and core radii of the first step.

As the stepwise removal proceeds ($j\neq1$), we must allow for the contribution from the thinning of the multilayer thickness, $\delta t(j)$. For the step j:

$$\delta v(j) = \delta v_K(j) + \delta v_t(j) \tag{8.19}$$

where $\delta v_K(j)$ is the core volume emptied in step j and $\delta v_t(j)$ is the equivalent liquid volume removed from the multilayer.

The volume, $\delta v_p(j)$, of the group of pores emptied of the condensate in step j is given by

$$\delta v_p(j) = \frac{\bar{r}_p^2(j)}{[\bar{r}_K(j) + \delta t(j)]^2} \cdot \delta v_K(j) \tag{8.20}$$

where $\bar{r}_p(j)$ and $\bar{r}_K(j)$ are now the mean pore and core radii for the step j.

We can make use of Equations (8.18)–(8.20) to obtain all the successive contributions to the total pore volume, that is, $\delta v_p(1)$, $\delta v_p(2)$, $\delta v_p(j)$; but to do this we need to know the individual values of $\delta v_t(j)$ (the changes in multilayer volumes) for each stage of the stepwise procedure.

Various methods have been devised for calculating the values of δv_t (j) the simplest approach being to adapt Equation (8.2) for the individual groups of cylindrical pores and cores. For example, the core area is

$$\delta a_K(j) = 2\delta v_K(j)/\bar{r}_K(j) \tag{8.21}$$

and similarly, the pore area is

$$\delta a_p(j) = 2\delta v_p(j)/\bar{r}_p(j)$$

The corresponding core and pore areas are related by the equation

$$\delta a_K(j) = \delta a_p(j) \frac{\bar{r}_p(j) - \bar{t}(j)}{\bar{r}_p(j)}$$

$$= \delta a_p(j) \cdot \rho(j)$$

(8.22)

In the original Barrett, Joyner and Halenda (BJH) method, $\rho(j)$ was given a single value, which corresponded to the most frequent pore size, whereas in the procedure adopted by Pierce (1953), it was considered enough to take $\rho(j) = 1$, that is, to omit the correction. Of course, with the aid of a modern computer, it is not difficult to apply a separate correction factor for each individual step (Montarnal, 1953).

The cumulative pore volumes and pore areas are obtained by the summation of all the respective contributions, $\delta v_p(j)$ and $\delta a_t(j)$. As a check on the overall consistency of the pore volume-area analysis, it is useful to compare the cumulative values of $v_p(j_{max})$ and $a_p(j_{max})$ with the corresponding Gurvich volumes and BET areas. In view of all the uncertainties and approximations, one should not expect to obtain perfect agreement (Gregg and Sing, 1982). Indeed, agreement to within, say, 5% may be fortuitous!

Although many adsorbents possess exceedingly complex pore structures, the meso-pore size analysis carried out by several groups (e.g. Dollimore and Heal, 1973; Havard and Wilson, 1976) appears to indicate that the calculated pore size distribution is rather insensitive to the model. However, as was pointed out by Haynes (1975) such results may be due in part to the over-simplification of the computations.

Figure 8.3 shows a typical example of the BJH pore size analysis. The adsorbent was a mesoporous alumina (Rouquerol et al. 2003). The Type IV nitrogen isotherm at 77 K in Figure 8.3a has a fairly clear plateau. The narrow hysteresis loop is Type H1, which is consistent with a rigid mesopore structure. Because any percolation effects are likely to be minimal, the desorption branch was used for the mesopore size analysis. The Harkins and Jura (1944) multilayer equation and the Montarnal (1953) correction for the pore diameter were applied. The pore size distribution is displayed as the *cumulative* mesopore volume versus pore diameter in Figure 8.3b and in the *differential* form in Figure 8.3c. Although overall there is an extensive range of

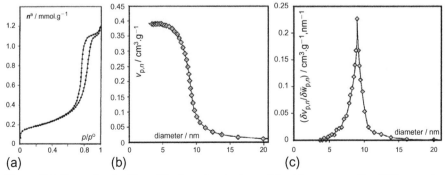

Figure 8.3 Example of a Type IV adsorption isotherm (a) with corresponding BJH pore size distribution expressed in cumulative form (b) and differential form (c) of mesopore volume versus pore diameter.
After Rouquerol et al. (2003).

pore size (from ~5 to ~15 nm), the sharpness of the differential peak indicates that the distribution was relatively narrow with more than 50% of the mesopore volume located within 8–10 nm pores.

In view of the limitations of the Kelvin equation, it is now generally accepted that the BJH method cannot be expected to provide more than rough – but possibly quite useful – assessment of the mesopore size distribution. Various changes have been introduced, which have undoubtedly improved this situation (see Thommes, 2004). One such empirical procedure was proposed by Kruk et al. (1997), the KJS method. In the KJS method, the relation between the pore diameter and the relative equilibrium pressure was empirically derived from adsorption data on ordered mesoporous materials such as MCM-41, which possessed open-ended cylindrical pores of controlled size. The multilayer thickness curves were also obtained from experimental data on macroporous silicas. This has the advantage that the method can be applied to modified mesoporous materials by determining new standard multilayer isotherms (Gierszal et al., 2013).

8.4.3   The Multilayer Thickness

It is evident that the corrections for multilayer thickness and thinning effects are of great importance. For example, for nitrogen adsorption at 77 K, $t \sim 0.6$ nm at $p/p^o = 0.5$, just before the condensation occurs in a cylindrical pore of $r_p \sim 2$ nm. The corresponding values at $p/p^o = 0.8$ are $t \sim 0.9$ nm and $r_p \sim 5$ nm.

Values of t are obtained from n/n_m by assuming an average molecular layer thickness. Following the recommendation of Lippens et al. (1964), the *average* adsorbed nitrogen layer thickness is usually taken as 0.35 nm. This value was based on the assumption of a hexagonally close-packed multilayer and molecular diameter of 0.43 nm.

The assumption that multilayer coverage of the mesopore walls proceeds in exactly the same manner as on the open surface has been questioned by several investigators (see Everett, 1988). Evans and Marconi (1985) first suggested on theoretical grounds that in a narrow mesopore the thickness of a multilayer is likely to be somewhat greater than on the non-porous surface, and similar views have been expressed by others (e.g. Thommes et al., 2012). On the other hand, there is a good deal of experimental evidence in support of the long-standing assumption that in the pre-capillary condensation region, the corresponding isotherms on mesoporous and non-porous adsorbents follow a very similar path (Gregg and Sing, 1982; Milburn and Davis, 1997). The equations proposed by Harkins and Jura (1944), Halsey (1948) and Shull (1948) to describe the thickness of the nitrogen multilayer are still generally used, but more work is clearly required. With the aid of small-angle neutron and X-ray scattering (see Schreiber et al., 2007), it is now possible to obtain an independent assessment of the thickness of some adsorbed layers.

8.4.4   Validity of the Kelvin Equation

We turn now to the question of validity of the Kelvin equation. Although the thermodynamic basis of the Kelvin equation is well established (Defay and Prigogine, 1966), its reliability for pore size analysis is questionable. In this context, there are three related questions: (1) What is the exact relation between the meniscus curvature and the pore size and shape? (2) Is the Kelvin equation applicable in the range of narrow mesopores (say, $w_p < 5$ nm)? (3) Does the surface tension vary with pore width? The answers to these questions are still elusive, but recent theoretical work has improved our understanding of mesopore filling and the nature of the condensate.

It was recognised many years ago (Foster, 1932) that the Kelvin equation is likely to break down as the meniscus curvature approaches a limiting value. Many early attempts were made to correct the Kelvin equation (see Brunauer, 1945; Sing and Williams, 2012). As already indicated, when the Kelvin equation is applied to capillary condensation, it is normally assumed that the reduction in chemical potential is entirely dependent on the curvature of the meniscus. This assumption implies a sharp discontinuity between the state of the adsorbed layer and the condensate. However, as Derjaguin (1957) first suggested, the transition is more likely to be a gradual one. This problem was also discussed by Everett and Haynes (1972,1973).

Evans et al. (1986) pointed out that a statistical mechanical treatment may be used to derive the Kelvin equation. This approach, which was designed to avoid the difficulties associated with the exact form of the meniscus, led to a new mathematical description of the effect of confining a fluid in pores of different size and shape on its liquid–gas co-existence curve. An equation of the same mathematical form as Equation (8.13) was obtained, provided that the 'undersaturation' was not too great, that is, that p/p^o was not too low. It was shown that this simple equation becomes less accurate as r_K is reduced and is no longer applicable beyond a 'capillary critical point'. At a lower r_K or higher T, the two-phase relation fails because of the existence of only one stable fluid configuration in the pore.

Since the capillary condensate in a particular mesopore is in thermodynamic equilibrium with the vapour, its chemical potential, μ^σ, must be equal to that of the gas (under the given conditions of T and p). As we have seen, the difference between μ^σ and μ^l (the chemical potential of the free liquid) is normally assumed to be entirely due to the Laplace pressure drop, Δp, across the meniscus. However, in the vicinity of the pore wall, a contribution from the adsorption potential, $\phi(z)$, should be taken into account. Thus, if the chemical potential is to be maintained constant throughout the adsorbed phase, the capillary condensation contribution must be reduced.

Molecular simulation studies (e.g. Jessop et al., 1991; Lastoskie et al., 1993; Lastoskie and Gubbins, 2000) have also indicated that the Kelvin equation fails to account for the effects of the fluid–wall interactions and the associated inhomogeneity of the pore fluid, and it appears that the Kelvin equation underestimates the pore size and that its reliability may not extend below a pore size of, say, $\sim$7.5 nm. Some other strong evidence reveals that the application of the corrected Kelvin equation underestimates the pore size by about 25% for pores smaller than 10 nm (see Thommes, 2004). Indeed, it seems likely that within a particular mesopore, the curvature of the meniscus is not constant. In the middle of a cylindrical pore, where the wall effect is negligible, the radius of curvature is r_K; whereas as the wall is approached, the radius of curvature is progressively increased. Of course, such findings serve to justify the DFT approach, which is dealt with in Section 8.5.

8.5 DFT Computation of the Mesopore Size Distribution

8.5.1 General Principles

The pioneering work in this area was carried out by Seaton et al. (1989), who adapted a statistical mechanical approach originally known as mean field theory (Ball and Evans, 1989). At the time of their early work (Jessop et al., 1991), the mean field theory was already known to become less accurate as the pore size was reduced, but even so it was claimed to offer a more realistic way of determining the pore size distribution than the classical methods based on the Kelvin equation.

More recently, a revised form of DFT has become a powerful tool for the interpretation of physisorption data (Balbuena and Gubbins, 1992,1993; Lastoskie et al., 1993; Cracknell et al., 1995; Maddox and Gubbins, 1995; Olivier, 1995; Ravikovitch and Neimark, 2001; Neimark and Ravikovitch, 2001; Neimark et al., 2003). In particular, the approach must now be regarded as the recommended procedure for evaluating the pore size distribution (Lastoskie et al., 1994; Olivier et al., 1994; Thommes, 2004; Gor et al., 2012).

One can be somewhat overwhelmed by the different terms used (e.g. DFT, local DFT, non-local DFT, quenched solid DFT), and the following provides a simple overview of density functional theory as applied to physisorption and pore filling. We begin by considering a one-component fluid confined in a pore of given size and shape, which is itself located within a well-defined solid structure. We suppose that the pore is open and that the confined fluid is in thermodynamic equilibrium with the same fluid (gas or liquid) in the bulk state and held at the same temperature. As indicated in Chapter 2, under conditions of equilibrium, a uniform chemical potential is established throughout the system. As the bulk fluid is homogeneous, its chemical potential is simply determined by the pressure and temperature. The fluid in the pore is not of constant density, however, as it is subjected to adsorption forces in the vicinity of the pore walls. This *inhomogeneous* fluid, which is stable only under the influence of the external field, is in effect a layerwise distribution of the adsorbate. The density distribution can be characterised in terms of a *density profile*, $\rho(r)$, expressed as a function of distance, r, from the wall across the pore. More precisely, r is the generalised coordinate vector.

In the DFT treatment, the statistical mechanical grand canonical ensemble is considered. The appropriate free energy quantity is the grand Helmholtz free energy, or grand potential functional, $\Omega[\rho(r)]$. This free energy functional is expressed in terms of the density profile, $\rho(r)$: then, by minimising the free energy (at constant μ, V, T), it is possible in principle to obtain the equilibrium density profile.

For a one-component fluid, which is under the influence of a spatially varying external potential, the grand potential functional becomes

$$\Omega[\rho(r)] = F[\rho(r)] + \int \mathrm{d}r \rho(r)[U_{\mathrm{ext}}(r) - \mu] \tag{8.23}$$

where $F[\rho(r)]$ is the intrinsic Helmholtz free energy functional, $U_{\mathrm{ext}}(r)$ is the external interaction energy and the integration is performed over the pore volume, V.

The $F[\rho(r)]$ functional can be separated into an ideal gas term and contributions from the repulsive and attractive forces between the adsorbed molecules (i.e. the fluid–fluid interactions). Hard-sphere repulsion and pairwise Lennard-Jones 12-6 potential are usually assumed, and a mean field treatment is generally applied to the long-range attraction. However, the evaluation of the density profile of an inhomogeneous hard sphere fluid near a solid surface presents a special problem. The mean field approach therefore provides an unrealistic picture of the density profile in this region.

It was this '*local*' DFT approach which was developed for the analysis of nitrogen isotherms by Seaton et al. (1989). However, although the local density approximation is acceptable for fluid–fluid interactions, it breaks down for fluids close to solid surfaces. This is because short-range correlations near such walls are not allowed for in the local density approximation. For this reason, non-local density functions were developed.

The *nonlocal density functional theory* (NLDFT) approach involves the incorporation of short-range smoothing approximation of the fluid density and weighting functions. Various procedures have been adopted, such as the smoothed density approximation (Tarazona, 1985), which are discussed by Ravikovitch and Neimark (2001). An improved description is obtained of the uniform fluid over a wide range of densities in the confined state. In this manner, it has been possible to obtain a good agreement with the density profiles determined by Monte Carlo molecular simulation (Lastoskie et al., 1993; Cracknell et al., 1995; Neimark et al., 2003).

In the special case of the interaction of nitrogen with two graphitic slabs (i.e. within a carbon slit), the Steele 10-4-3 potential has been used to calculate the effective external interaction energy (Lastoskie et al., 1993). It is this system which has so far received most attention and generally the surface is assumed homogeneous so that the interaction energy is constant in the xy plane.

NLDFT appears to give a satisfactory description of adsorption and phase transitions in slit-shaped and cylindrical pores mesopores. It can also describe different types of isotherm, including the layer-by-layer Type VI isotherm. However, this approach does have limitations. It fails for very narrow pores and also cannot predict the solid–liquid adsorbate transitions. Normally, the adsorbent surface is assumed to be molecularly smooth. Experimentally, this is only really observed for graphitic carbon (or some types of boron nitride) and we know that such surfaces give stepwise Type VI isotherms. If layering is also predicted for other systems, anomalies are obtained in the derived pore size distribution.

Neimark and his co-workers have addressed this problem and have proposed *Quenched Solid DFT*, which takes into account the 'roughness' of the adsorbent surface (Gor et al. 2012). The density profile of the solid component is included in the description of the grand potential, and a two component density functional thus models the overall solid–fluid interaction as hardcore spheres interacting with the fluid molecules via a pairwise attractive potential. A key term in this approach is the excess free energy of the solid–fluid hard sphere mixture. The solid is 'quenched' in that it is considered completely rigid for the calculations, via a roughness parameter, and the density of this quenched solid is used in the optimisation of the grand potential.

In each of these methods, once the density profile is defined for a given pressure, the amount adsorbed by a particular pore can be obtained from the area under the curve. The surface excess number of molecules adsorbed, which as we have seen corresponds to the experimentally determined adsorption, is then given by $[\rho(r) - \rho_B]dr$, where ρ_B is the density of the bulk phase at (μ, T).

Thus, NLDFT or QSDFT can be used to generate a series (or *kernel*) of hypothetical 'individual pore isotherms' for a range of pore sizes and wall potentials. It is also possible to construct such a kernel by another theoretical approach or by molecular simulation (e.g. GCMC). The main difference between NLDFT and QSDFT is that the former assumes the adsorbent surface to be homogeneous, whereas the latter takes into account the surface roughness. In contrast to the NLDFT kernels, QSDFT and GCMC isotherms are smooth prior to capillary condensation, that is, they do not exhibit stepwise inflections caused by artificial layering transitions (Gor et al., 2012).

The kernel can be regarded as a theoretical reference for a given class of adsorbate/adsorbent system. This theoretical treatment provides a basis for a new classification based on three regimes: continuous pore filling, capillary condensation and layering transitions. It was therefore proposed (Lastoskie et al., 1993) that the filling behaviour rather than the pore width should be adopted for the classification of physisorption systems. Thus, according to the DFT model, the critical pore widths which correspond to the boundaries between these regimes are strongly dependent on temperature. Furthermore, it is the magnitude of the solid–fluid interactions which govern the pressure at which pore filling occurs, the type of filling being dependent on the ratio of pore width to adsorbate molecular diameter.

For a given local isotherm in a kernel, it is normally assumed that all the pores are of the same size and shape and that the pore size has a minimum and a maximum value. In addition to pores of a given geometry (slit shape, cylinder, spherical), it is also possible to set up more complex kernels including different pore geometries. To arrive at the pore size distribution, it is assumed that a porous adsorbent has an array of noninteracting pores (i.e. there are no network or percolation effects) and that the distribution of pore widths can be described by a continuous function $f(w)$. The experimental isotherm can then be regarded as a composite of isotherms for each group of pores.

The pore size distribution is obtained by solving the generalised adsorption isotherm (GAI) integral equation, which correlates the kernel of the theoretical isotherms with the experimental isotherm and which is now expressed in the form:

$$N_{\exp}\left(\frac{p}{p^o}\right) = \int_{W_{\min}}^{W_{\max}} N_{\text{theo}}\left(\frac{p}{p^o}, w\right) f(w)\mathrm{d}w \qquad (8.24)$$

The GAI equation is based on the assumption that the measured isotherm is composed of a number of individual single pore isotherms and is dependent on their distribution, $f(w)$ over a finite range of pore size. The formal similarity of Equations (8.22) and (5.55) is evident and the solution of each equation is strictly an ill-posed problem and requires some degree of regularisation. There are several regularisation

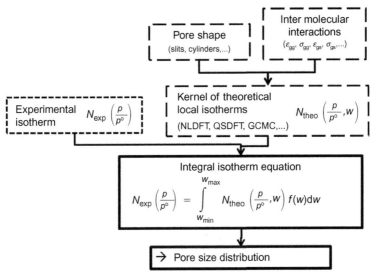

Figure 8.4 Flow diagram of how the pore size distribution is calculated by isotherm reconstruction methods.

algorithms, which allow meaningful solutions to be obtained either by using summation and numerical deconvolution (Olivier et al., 1994) or by a non-negative least squares method (Ravikovitch and Neimark, 2001).

The computational approach is shown in the flow diagram in Figure 8.4.

In applying DFT for pore size analysis, one must first choose the kernel that corresponds most closely to the adsorbent sample (i.e. between pore shape, pore chemistry, temperature and probe molecule). A comparison can then be made between the regression fit and the experimental isotherm. If the agreement is satisfactory, it is then possible to extract the pore size distribution.

8.5.2   Nitrogen Adsorption at 77 K

A typical example of mesopore size analysis by NLDFT and BJH is given in Figure 8.5. The nitrogen isotherm at 77 K on a sample of mesoporous silica is clearly Type IV with a well-defined plateau and a narrow H1 hysteresis loop. The kernel for the NLDFT analysis was for nitrogen adsorption in cylindrical pores of silica, and the desorption isotherm was assumed to be the equilibrium curve.

It is evident that there are significant differences in the pore size distribution curves assessed by the DFT and BJH methods. Although the two 'frequency' curves in Figure 8.5 are of similar width, their locations are quite different: DFT has given a distribution of pore width of ca. 7–10 nm, whereas the BJH distribution appears to extend from ca. 5 to 8 nm. Similar results have been reported by others (Neimark and Ravikovitch, 2001; Lowell et al., 2004; Thommes, 2004; Thommes et al., 2012).

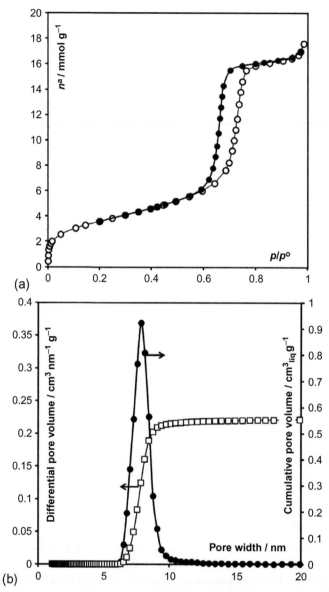

Figure 8.5 Comparison of the pore size distributions obtained for the nitrogen adsorption at 77 K on a mesoporous silica (a) from the DFT (b) and BJH

Continued

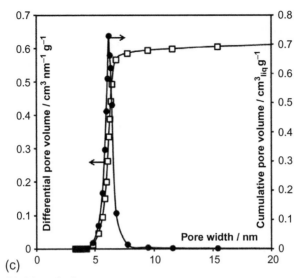

Figure 8.5, cont'd (c) methods.

The soundness of the DFT method for the assessment of mesopore size distribution has been confirmed by its application to a few ordered pore structures such as MCM-41 and SBA-15. Thus, good quantitative agreement was obtained between parts of the adsorption–desorption isotherms predicted by NLDFT and the corresponding exper-imental data (Neimark and Ravikovitch, 2001). This work also confirmed that meta-stable states on the adsorption branch of the hysteresis loop must be taken into account, while the desorption process occurs under conditions of liquid–vapour co-existence (i.e. in thermodynamic equilibrium). The situation is more complex for ordered three-dimensional networks (e.g. MCM-48) and especially so with highly dis-ordered materials such as silica gels or porous vycor glass. There are two separate problems: (a) in having the correct kernel for the set of single pore isotherms and (b) in allowing for network blocking and cavitation on the desorption path and delayed condensation on the adsorption path. The great advantage of DFT over the classical approach is that, in principle, it can be applied to both capillary condensation and micropore filling. This latter application is discussed in Chapter 9.

8.5.3 Argon Adsorption at 87 K

Most of the many argon isotherms in the literature were determined at 77 K. At this tem-perature, argon is well below its triple point temperature, and therefore, mesopore size analysis is limited to pore diameters smaller than ~15 nm (Thommes, 2004). However, by increasing the operational temperature to ca. 87 K, it is possible to achieve capillary condensation up to the saturation pressure. As discussed in Chapter 9, the use of argon adsorption at 87 K is of particular interest in the characterisation of microporous adsor-bents. But, even with mesoporous adsorbents, the absence of a permanent quadrupole moment makes it easier to calculate the kernel because it is less sensitive to the surface chemistry. Although this approach is still mainly research based, it has been found fruitful already, especially in comparing the DFT analysis carried out on adsorption isotherms obtained with nitrogen at 77 K and argon at 87 K. For example, a comparison was made for a lignocellulosic-derived carbon, which was eventually shown to have large mesoporous cavities connected to the exterior through narrower mesopore necks (Silvestre-Albero et al., 2012). In the 4–7 nm range, there was a discrepancy between the PSD curves obtained by NLDFT from the desorption branches of the isotherms of the two adsorptives. This was due to cavitation occurring on desorption from pores in the region of 4 nm, which made the NLDFT analysis of the desorption branch unreliable. Of course, the BJH analysis would have suffered from the same limitation. This led the authors to use QDFT for the analysis of the metastable *adsorption* branch of the nitrogen isotherm, where cavitation does not take place. In this way, it was possible to obtain a more reliable assessment of the pore size distribution.

8.6   Hysteresis Loops

Hysteresis loops, which appear in the multilayer range of physisorption isotherms, are generally associated with capillary condensation. It is well known that most mesoporous adsorbents give distinctive and reproducible hysteresis loops (de Boer, 1958; Sing et al., 1985). But, according to the laws of classical thermodynamics, the amount adsorbed is controlled by the chemical potential of the adsorptive (see Section 2.3). It follows that the two branches of a loop cannot satisfy both the requirements of thermodynamic reversibility. The appearance of reproducible and stable hysteresis therefore implies the existence of certain well-defined metastable states. Over the past 100 years, the interpretation of adsorption hysteresis has attracted much attention (see Sing and Williams, 2004).

Many different forms of loop have been reported in the literature, but the major types are represented in a classification proposed by the IUPAC (Sing et al., 1985), which is given in Figure 8.6. Types H1, H2 and H3 were included in the first classification of hysteresis loops put forward by de Boer (1958). Type H1 (originally known as Type A) is a fairly narrow loop with very steep and nearly parallel adsorption and desorption branches. In contrast, the Type H2 loop (formerly Type E) is broad with a long and almost flat plateau and a steep desorption branch. Types H3 (formerly Type B) and H4 do not terminate in a plateau at high p/p^o, and the limiting desorption boundary curve is therefore more difficult to establish.

The characteristic features of some types of loop are associated with certain well-defined pore structures. Thus, Type H1 loops are given by adsorbents with a narrow distribution of uniform pores (e.g. open-ended tubular pores as in MCM-41 and SBA-15 – see Chapter 13). Many inorganic oxide gels give the more common Type H2 loops. The pore structures in these materials are complex and tend to be made up of interconnected networks of pores of different size and shape. Type H3 loops are often given by the aggregates of platy particles (Rouquerol et al., 1970), while Type H4 loops are obtained with activated carbons and other adsorbents having slit-shaped pores – mainly in the micropore range.

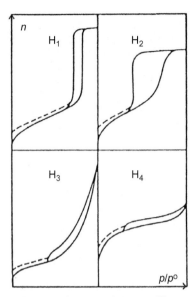

Figure 8.6 The IUPAC classification of hysteresis loops (Sing et al., 1985).

A feature common to many H2 and H4 loops is that for a given adsorptive (at a particular temperature), the very steep region of the desorption branch joins the adsorption branch at a limiting relative pressure. It was once thought that this lower limit of hysteresis is dependent on the adsorptive and the operational temperature but not at all on the adsorbent (see Gregg and Sing, 1982). It is now known that this is not strictly true (Rasmussen et al., 2010). However, in the case of nitrogen adsorption at 77 K, the lower closure point is often located at the limiting $p/p^o \sim 0.42$. Any hysteresis recorded below this p/p^o is therefore due to an irreversible change such as swelling of the adsorbent or chemisorption (see Chapters 10 and 11).

In principle, the processes of capillary condensation and evaporation should occur reversibly in a closed tapering pore (Everett, 1967). At low relative pressures, there is an enhanced concentration of adsorbed molecules in the narrow end of the pore (i.e. a micropore filling effect) as in Figure 8.7. At a certain p/p^o, a meniscus begins to form which, with the increase in p/p^o, then moves steadily up towards the pore entrance. Evaporation proceeds in the reverse direction, but involves the same elemental steps (i.e. the meniscus configurations), and therefore, the entire isotherm is reversible.

The systematic studies of adsorption hysteresis made over a period of many years by Everett and his co-workers revealed that the phenomena are temperature dependent (Amberg et al., 1957; Everett, 1967; Burgess et al., 1989). Thus, with increase in temperature, most hysteresis loops undergo a reduction in size. These findings were extended by adsorption measurements on mesoporous controlled-pore glass (Findenegg et al., 1994; Machin, 1994; Brown et al., 1997; Pellenq et al., 2007), MCM-41 (Branton et al., 1997; Morishige and Nakamura, 2004) and graphitic carbons (Lewandowski et al., 1991).

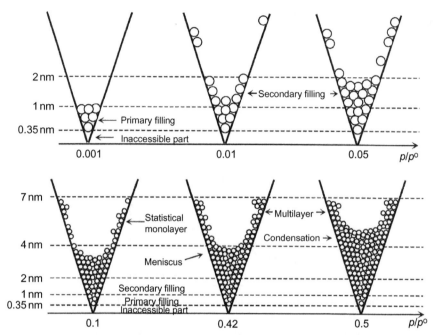

Figure 8.7 Main steps of the successive filling of micropores and mesopores during adsorption. The pressure scale gives orders of magnitudes for the case of N_2 at 77 K. The size scale is twice larger in the upper row.

As with bulk liquid–vapour equilibria, hysteresis (or capillary) phase diagrams can be presented in different ways. By taking the lower and upper hysteresis closure points to represent the limits of the region of co-existence of the vapour and capillary condensed liquid, one can obtain, for example, a smooth curve relating the two densities of the condensed phase to the temperature. The two curves come together at the hysteresis critical temperature, $T_c(h)$, for the given system. Such hysteresis phase diagrams are similar in shape to the phase diagrams for the corresponding bulk liquids (e.g. CO_2/Vycor), but with the important difference that $T_c(h)$ is consistently much lower than the corresponding bulk critical temperature, T_c.

The fact that the isotherms become reversible at temperatures above $T_c(h)$ may seem to indicate that the distinction between the pore condensate and vapour must disappear and that the surface tension becomes zero: that the adsorbate loses its liquid-like properties at $T_c(h)$. However, this explanation is probably over-simplified in the light of work on highly uniform pore structures such as MCM-41 (see Chapter 13). As already indicated, and will be discussed in more detail later, it is evident that adsorption hysteresis can originate in different ways and also that *reversible*, stepwise condensation can occur under certain special conditions.

Most of the older theories of adsorption hysteresis made explicit use of the Kelvin equation. Zsigmondy (1911) was the first to suggest that the phenomenon was due to a difference in the contact angles of the condensing and evaporating liquid. This explanation may account for the some of the anomalous effects produced by the presence of surface impurities, but in its original form it cannot explain the permanence and reproducibility of the majority of recorded loops. However, as it will become evident, the notion of delayed condensation is still of great importance.

Another early theory, which also attracted a great deal of attention, was the 'ink-bottle theory': this was originally put forward by Kraemer (1931) and subsequently developed by McBain (1935). Kraemer pointed out that the rate of evaporation of a liquid in a relatively large pore is likely to be retarded if the only exit is through a narrow channel. This argument led Brunauer (1945) to conclude that the liquid in the pore cannot be in true equilibrium with its vapour during the desorption process, and therefore, it is the *adsorption* branch of the loop which represents thermodynamic reversibility.

The mechanisms of meniscus formation, modification and removal were discussed by many investigators including Foster (1932), Cohan (1944), de Boer (1958), Everett (1967, 1979), Broekhoff and Linsen (1970), Haynes (1975), Lewandowski et al. (1991); Machin (1992) and Findenegg et al. (1994). A substantial amount of evidence is now available in support of Foster's view that condensation does not always occur in accordance with the simple Kelvin equation. The most striking examples of delayed meniscus formation are afforded by many clays and oxides, which give Type H3 loops, and by some activated carbons, which give Type H4 loops. In a slit-shaped pore, the meniscus is formed only at high p/p^{o}; consequently over a wide range, the *adsorption* isotherm is of Type II (but is not reversible) and may appear to approach the p^{o} axis asymptotically. Such an isotherm could be regarded as a pseudo-Type II isotherm, but we prefer to adopt the designation Type IIb as there may be virtually no detectable capillary condensation over much of the adsorption branch.

Cohan (1938) suggested that in an open-ended pore, a difference in meniscus shape is involved in the processes of condensation and evaporation. According to this view, condensation is governed by the initial formation of a cylindrical meniscus, whereas two hemispherical menisci are developed once the pore has been filled. By applying Equations (8.11) and (8.13), we find that a given value of r_K represents the corrected *width*, $2(r_p - t)$, for pore filling and the corrected *radius*, $r_p - t$, for pore emptying. Thus, the hysteresis in the filling and emptying of a single open-ended tubular pore is apparently simply due to the processes of condensation and evaporation occurring at different relative pressures.

Everett (1979) pointed out that several irreversible steps are involved when condensation occurs in a single open-ended cylinder. Since a cylindrical meniscus is unstable, a spontaneous change leads to the development of an unduloid. In the next stage, the pore becomes blocked by the formation of a biconcave lens of liquid. The evaporation process proceeds in a thermodynamically reversible manner in accordance with the Kelvin equation, the relative pressure now being dependent on the radius of curvature of the hemispherical menisci. Thus, in the case of this simple system, it is the location of the desorption branch which should be used for the calculation of r_p.

A different approach was adopted by Broekhoff and de Boer (1967) and Saam and Cole (1975), who explained the hysteresis exhibited by a cylindrical pore in terms of the regimes of stability, metastability and instability of the multilayer film. The applicability of the Saam–Cole theory was explored in some detail by

Findenegg et al. (1994) and Lewandowski et al. (1991) and Michalski et al. (1991). Thus, there are two opposing effects which govern the range of metastability of a multilayer film in a cylindrical mesopore. The long-range adsorption forces help to stabilise the film, while capillary forces are responsible for the condensation of the liquid. At a critical film thickness, t_c, the curved film becomes unstable and condensation occurs. Evaporation of the liquid condensate requires a lower p/p^o and now the residual film thickness is t_m. We may therefore regard the difference $(t_c - t_m)$ as the metastability thickness range of the multilayer film.

On the basis of the Saam–Cole–Findenegg approach, we can envisage an 'ideal isotherm' for the physisorption of a vapour in an assemblage of uniform open-ended cylindrical mesopores, as shown in Figure 8.8. Here, C represents the limit of metastability of the multilayer (of thickness t_c) and M is the point at which the three phases (multilayer, condensate and gas) all co-exist. Along MC, the multilayer and gas are in metastable equilibrium.

The chemical potential of an adsorbed multilayer film in cylindrical pore may be expressed as

$$\mu_m = \mu_0 + U(r_p) - \gamma/r_K \Delta\rho \qquad (8.25)$$

where $U(r_p)$ is the interaction energy between the film and the adsorbent, r_K is the inner radius $(r_K = r_p - t)$ and $\Delta\rho$ is the difference in density between the liquid and vapour. The term $\gamma/r_K \Delta\rho$ represents the energy associated with the curved film/vapour interface.

If the multilayer is sufficiently thick, the interaction energy, $U(r_p)$, is obtained by integration of the C/r^6 pair potentials. From the Saam–Cole theory, it then becomes possible to arrive at the critical film thickness, t_c, and metastable limit, t_m, in terms of t/r_p. The following qualitative behaviour was predicted (Findenegg et al., 1994): (1) as

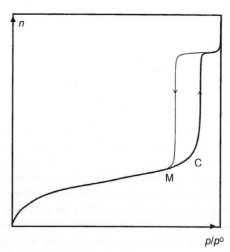

Figure 8.8 An adsorption–desorption isotherm for a mesoporous solid with cylindrical pores, all having the same radius.

the pore radius is decreased, the stability limits are shifted to larger relative film thickness, t/r_p; (2) for a given pore radius, the stability limits are temperature dependent. The latter effect is dominated by the negative temperature coefficient of the surface tension, and therefore the stability limits are shifted to larger reduced film thickness as $T \to T_c$.

The applicability of the Saam–Cole theory was tested by Findenegg et al. (1993, 1994). Their adsorption measurements of certain organic vapours on carefully selected grades of controlled-pore glass provided semi-quantitative confirmation of the theoretical treatment adopted so far. However, it is evident that some refinement is required in assessing $U(r_p)$ for a selection of well-defined adsorbents having narrow size and shape distributions of easily accessible mesopores.

Early attempts were made by Everett (1955, 1967) to formulate the behaviour of groups of pores in terms of the domain theory. In the independent domain theory, it was assumed that each pore behaves as though it were interacting in isolation with the vapour, and therefore, the overall behaviour was considered to depend on the statistical distribution of the individual pore properties. The pathways of crossing the loop (i.e. scanning behaviour) were predicted. For example, according to the independent domain model, if scanning loops are traced within given limits, the areas of these subsidiary loops should be constant and independent of their location within the main loop. Generally, this prediction is not fulfilled and it must be concluded that the domains do not behave independently (Everett, 1979).

In fact, most mesoporous adsorbents possess complex networks of pores of different size. It is therefore unlikely that the condensation–evaporation processes can occur independently in each pore. Let us postulate that the condensate cannot leave a pore (or segment of a pore) at the 'Kelvin relative pressure' unless there is a continuous channel of *vapour* leading to the adsorbent surface. It follows that the probability that a portion of the condensate will be trapped will depend on the nature of the pore network and the numerical and spatial distribution of pore size. Thus, a large amount of entrapment – and hence pronounced hysteresis – will be observed if a high proportion of the pore volume is only accessible through narrow channels with a relatively low level of pore connectivity. Many different model structures have been proposed for such ordered and disordered nanoporous networks (see Ravikovitch and Neimark, 2002; Bandosz et al. 2003; Palmer and Gubbins, 2012).

Computer modelling of network effects is simplified if it is assumed that pore filling occurs 'reversibly' (i.e. in accordance with the Kelvin equation) along the adsorption branch of the loop. Percolation theory has been applied by Mason (1988), Liu et al. (1993, 1994), Lopez-Ramon et al. (1997) and others (Zhdanov et al., 1987; Neimark, 1991). One approach is to picture the pore space as a three-dimensional network (or lattice) of cavities and necks. If the total neck volume is relatively small, the location of the adsorption branch should be mainly determined by the cavity size distribution. On the other hand, if the evaporation process is controlled by percolation, the location of the desorption branch is determined by the network coordination number and neck size distribution.

An alternative structural model, which was favoured by Seaton (1991), is one made up of an aggregate of small particles, which contain a network of mesopores (and possibly also micropores). The interstices between the particles form a continuous network of macropores. In Seaton's model, the mesopore network is represented by a simple cubic lattice (Liu et al., 1992). In percolation terminology, each pore is regarded as a bond in the lattice, and each pore junction is a node. During the desorption process, the liquid condensate leaves the pore when two conditions are satisfied: (1) the relative pressure must be low enough for vaporisation to be thermodynamically favourable; (2) the pore must have direct access to the vapour phase, either because it is at the surface or because it is adjacent to another pore from which the condensate has already been vaporised. At the beginning of the desorption process, some condensate is removed from the wider pores (i.e. unoccupied bonds) near the surface. As the pressure is reduced, the vapour-filled pores (occupied bonds) form clusters, which eventually extend across the particle. The stage at which a spanning cluster is formed across the particle corresponds to the percolation threshold, when the pore emptying becomes rapid. This stage corresponds to the knee of an H2 hysteresis loop. As one would expect, the location of the 'accessibility curve' is highly dependent on the lattice size and coordination number. Although these basic principles appear to be sound, the physical significance of the unknown lattice parameters are questionable. The application of the percolation theory is likely to be more useful when it can be clearly related to a characteristic isotherm type and range of pore size, shape and disposition.

The importance of molecular dynamics in controlling capillary condensation and evaporation has been argued by Monson and his co-workers (Woo and Monson, 2003; Monson, 2008, 2009). Their recent approach (Edison and Monson, 2012) is based on the application of dynamic mean field theory (DMFT) to three networks in the form of the cubic two-dimensional lattice gas model. This involves the numerical solution of molecular transition probabilities from one site to another in order to compute the change of occupational density with time. In the simplest case of a single slit width, the dependence of adsorption hysteresis on the segment length is evident, whereas the behaviour becomes more complex with the introduction of arrays of two pore widths and inkbottles. The equilibration is then critically dependent on the ease of transport of fluid to and from the larger pores. These first results indicate that the adsorption dynamics may affect both adsorption and desorption equilibration; the proposed extension of the application of the DMFT to networks with the three-dimensional connectivity will be of considerable interest.

NLDFT and molecular simulation have been used by Neimark, Ravikovitch and Thommes and their co-workers (e.g. Neimark and Ravikovitch, 2001; Ravikovitch and Neimark, 2001, 2002; Thommes et al., 2006, 2012; Rasmussen et al., 2010) in their extensive studies of the mechanisms of pore filling in ordered nanoporous structures, which were characterised by X-ray diffraction, electron microscopy and other techniques. Their work on MCM-41 and SBA-15 confirmed that delayed condensation on the *adsorption* branch is responsible for the appearance of a Type H1 hysteresis loop. Thus, in relatively large mesopores, spontaneous spinodal condensation occurs when the metastable multilayer becomes unstable, whereas on the desorption path there are no metastable states since evaporation takes place under the conditions of liquid–vapour co-existence.

As was found in earlier studies (e.g. Branton et al., 1983), the Type IV nitrogen isotherms at 77 K on narrow mesoporous MCM-41 were completely reversible. Similarly, MCM-48 – having a regular network of relatively narrow mesopores – also gave reversible nitrogen and argon isotherms at 77 K (Schumacher et al., 2000). It is therefore apparent that in these relatively narrow mesopores the interactions with the pore walls are strong enough to initiate a form of pore filling, which does not involve any delayed condensation.

In contrast, large hysteresis loops of Type H2 are given by networks of pores of different widths or containing spheroidal cavities (ink bottles) such as in SBA-16 (Ravikovitch and Neimark, 2002; Rasmussen et al., 2010; Edison and Monson, 2012; Monson, 2012). Two basic mechanisms of desorption from such pore networks were identified as percolation pore blocking and cavitation. As already indicated, in the former case, the onset of evaporation from the network is associated with a percolation threshold: the unblocking of trapped condensate then allows free access to the multilayer and any remaining condensate. However, if the neck width is less than a certain critical size (estimated to be ca. 5 nm for nitrogen adsorption at 77 K), the mechanism of desorption from the pore body involves cavitation: this is the spontaneous nucleation and growth of gas bubbles in the metastable condensed fluid while the narrow necks may remain filled. In this case, the desorption pressure does not depend on the neck size, but is mainly dependent on the properties of the condensed fluid, so that the corresponding portion of the desorption branch does not provide any reliable information about the width of the necks or of the pore body.

We conclude that the adsorption hysteresis may be generated in a number of different ways. In the context of the assessment of mesoporosity, we have seen that there are two major contributing factors: (a) on the adsorption branch, the development of a metastable multilayer and the associated delay in capillary condensation; (b) on the desorption branch, the entrapment of the condensate through the effect of network–percolation and/or cavitation. Many porous materials are composed of complex networks so that a combination of different pore filling and emptying mechanisms may have to be accepted; but in order to interpret a Type IV physisorption isotherm, it is always helpful to record the size (or absence) and shape of the hysteresis loop.

8.7   Conclusions and Recommendations

1. If a physisorption isotherm is Type IV with a well-defined plateau at high relative pressure and there is no detectable microporosity, the effective mesopore volume can be evaluated from the amount adsorbed at, say, $p/p^\circ = 0.95$.
2. If the hysteresis loop is H1 type, because of metastability of the multilayer and delayed capillary condensation, the *desorption* branch should be used to compute the mesopore size distribution.
3. If the hysteresis loop is H2 type, the location of the steep desorption branch is either dependent on network pore-blocking effects or on cavitation of the condensate. The limiting p/p° of stability of the condensate is not directly related to the pore size distribution, but is instead mainly governed by the properties of the adsorptive and operational temperature. Useful

qualitative information is obtained from the shape of the loop and semi-quantitative pore size analysis may be possible from the *adsorption* branch.

4. In the case of an H3 loop, the isotherm is *not* Type IV (there is no plateau) and mesopore size analysis should not be undertaken.

5. Provided the appropriate kernel is available for the type of adsorbent studied, it is generally preferable to use a DFT-based method for pore size analysis, rather than rely on the 'classical' BJH method. However, in the application of the readily available DFT software, there are various assumptions and limitations, which should be taken into account. Unless the surface composition and adsorbent texture are uniform, the derived pore size distribution is unlikely to provide a completely true portrayal of the real pore structure.

6. One advantage of the BJH method is that its application is less 'software dependent' than the DFT method as there are variations in the operational kernels provided by the different commercial software suppliers. Therefore, the BJH method allows the user to have more direct control in data processing. Also, for many purposes, it may be acceptable to use an empirical procedure, such as that proposed by Kruk et al. (1997) for mesopore size analysis.

7. In future, argon adsorption at 87 K may become more attractive than nitrogen at 77 K for pore size analysis.

References

Adkins, B.D., Davis, B.H., 1986. J. Phys. Chem. 90, 4866.

Adkins, B.D., Davis, B.H., 1987. Langmuir. 3, 722.

Amberg, C.H., Everett, D.H., Ruiter, L.H., Smith, E.W., 1957. In: Solid/Gas Interface. Proceedings of the Second International Congress of Surface Activity. Butterworths Scientific Publications, London, p. 3.

Avery, R.G., Ramsay, J.D.F., 1973. J. Colloid Interface Sci. 42, 597.

Balbuena, P.B., Gubbins, K.E., 1992. Fluid Phase Equilib. 76, 21.

Balbuena, P.B., Gubbins, K.E., 1993. Langmuir. 9, 1801.

Ball, P.C., Evans, R., 1989. Langmuir. 5, 714.

Bandosz, T.J., Biggs, M.J., Gubbins, K.E., Hatton, Y., Iiyama, T., Kaneko, K., Pikunic, J., Thomson, K.T., 2003. In: Radovic, I.R. (Ed.), Chemistry and Physics of Carbon, Vol. 28. Marcel Dekker, NewYork.

Barrett, E.P., Joyner, L.G., Halenda, P.H., 1951. J. Am. Chem. Soc. 73, 373.

Branton, P.J., Hall, P.G., Sing, K.S.W., 1983. J. Chem. Soc. Chem. Commun. 16, 1257.

Branton, P.J., Hall, P.G., Sing, K.S.W., 1993. J. Chem. Soc. Chem. Commun. 1257.

Branton, P.J., Sing, K.S.W., White, J.W., 1997. J. Chem. Soc. Faraday Trans. 93, 2337.

British Standard, 1992. BS 7591 – Part 2, British Standards Institution, London.

Broekhoff, J.C.P., de Boer, J.H., 1967. J. Catal. 9, 9.

Broekhoff, J.C.P., Linsen, B.G., 1970. In: Linsen, B.G. (Ed.), Physical and Chemical Aspects of Adsorbents and Catalysts. Academic Press, London.

Brown, A.J., Burgess, C.G.V., Everett, D.H., Nuttal, S., 1997. In: McEnaney, B., Mays, T.J., Rouquerol, J., Rodriguez-Reinoso, F., Sing, K.S.W., Unger, K.K. (Eds.), Characterization of Porous Solids IV. The Royal Society of Chemistry, Cambridge, p. 1.

Brunauer, S., 1945. The Adsorption of Gases and Vapours. Oxford University Press, London p. 126.

Bukowiecki, S.T., Straube, B., Unger, K.K., 1985. In: Haynes, J.M., Rossi-Doria, P. (Eds.), Principles and Applications of Pore Structural Characterization. Arrowsmith, Bristol, p. 43.

Burgess, C.G.V., Everett, D.H., Nutall, S., 1989. Pure Appl. Chem. 61, 1845.

Carrott, P.J.M., Sing, K.S.W., 1989. Pure Appl. Chem. 61, 1835.

Cohan, L.H., 1938. J. Am. Chem. Soc. 60, 433.

Cohan, L.H., 1944. J. Am. Chem. Soc. 66, 98.

Cracknell, R.F., Gubbins, K.E., Maddox, M., Nicholson, D., 1995. Acc. Chem. Res. 28, 281.

Cranston, R.W., Inkley, F.A., 1957. Advances in Catalysis, Vol. 9. Academic Press, New York, p. 143.

de Boer, J.H., 1958. In: Everett, D.H., Stone, F.S. (Eds.), The Structure and Properties of Porous Materials. Butterworths, London, p. 68.

Defay, R., Prigogine, I., 1951. Tension Superficielle et Adsorption. Dunod, Paris.

Defay, R., Prigogine, I., 1966. Surface Tension and Adsorption. Longmans, Green and Co., Bristol.

Derjaguin, B.V., 1957. Proceedings of the Second International Congress on Surface Activity, Vol. II. Butterworths, London, p. 154.

Dollimore, D., Heal, G.R., 1964. J. Appl. Chem. 14, 109.

Dollimore, D., Heal, G.R., 1973. J. Colloid Interface Sci. 42, 233.

Edison, J.R., Monson, P.A., 2012. Micropor. Mesopor. Mater. 154, 7.

Evans, R., Marconi, U.M.B., 1985. Chem. Phys. Lett. 114, 415.

Evans, R., Marconi, U.M.B., Tarazona, P., 1986. J. Chem. Phys. 84, 2376.

Everett, D.H., 1955. Trans. Faraday Soc. 51, 1551.

Everett, D.H., 1958. In: Everett, D.H., Stone, F.S. (Eds.), Structure and Properties of Porous Materials. Butterworths, London, p. 95.

Everett, D.H., 1967. In: Flood, E.A. (Ed.), The Solid-Gas Interface. Edward Arnold, London, p. 1055.

Everett, D.H., 1979. In: Gregg, S.J., Sing, K.S.W., Stoeckli, H.F. (Eds.), Characterization of Porous Solids. Society of Chemical Industry, London, p. 229.

Everett, D.H., 1988. In: Unger, K.K., Rouquerol, J., Sing, K.S.W., Kral, H. (Eds.), Characterization of Porous Solids I. Elsevier, Amsterdam, p. 1.

Everett, D.H., Haynes, J.M., 1972. J. Colloid Interface Chem. 38, 125.

Everett, D.H., Haynes, J.M., 1973. In: Colloid Science. Chemical Society, London, p. 123.

Findenegg, G.H., Groß, S., Th, Michalski, 1993. In: Suzuki, M. (Ed.), Fundamentals of Adsorption. Kodanska, Tokyo, p. 161.

Findenegg, G.H., Groß, S., Th, Michalski, 1994. In: Rouquerol, J., Rodriguez-Reinoso, F., Sing, K.S.W., Unger, K.K. (Eds.), Characterization of Porous Solids III. Elsevier Science BV, Amsterdam, p. 71.

Foster, A.G., 1932. Trans. Faraday Soc. 28, 645.

Gierszal, K., Kruk, M., Jaroniec, M., 2013. Adsorpt. Sci. Technol. 31, 153.

Gor, G.Y., Thommes, M., Cychosz, K.A., Neimark, A.V., 2012. Carbon. 50, 1583.

Gregg, S.J., Sing, K.S.W., 1982. Adsorption, Surface area and Porosity, 2nd edn, Academic Press, London.

Gurvich, L., 1915. J. Phys. Chem. Soc. Russ. 47, 805.

Halsey, G.D., 1948. J. Chem. Phys. 16, 93.

Harkins, W.D., Jura, G., 1944. J. Am. Chem. Soc. 66, 1366.

Havard, D.C., Wilson, R., 1976. J. Colloid Interface Sci. 57, 276.

Haynes, J.M., 1975. Colloid Science, Vol. 2. Chemical Society, London, p. 101.

Jessop, C.A., Riddiford, S.M., Seaton, N.A., Walton, J.R.P.B., Quirke, N., 1991. In: Rodriguez-Reinoso, F., Rouquerol, J., Sing, K.S.W., Unger, K.K. (Eds.), Characterization of Porous Solids II. Elsevier, Amsterdam, p. 123.

Kanellopoulos, N.K., Petrou, J.K., Petropoulos, J.H., 1983. J. Colloid Interface Sci. 96, 90.

Karnaukhov, A.P., 1971. Kinet. Catal. 12 (908), 1096.

Karnaukhov, A.P., 1979. In: Gregg, S.J., Sing, K.S.W., Stoeckli, H.F. (Eds.), Characterization of Porous Solids. Society of Chemical Industry, London, p. 301.

Kraemer, E.O., 1931. In: Taylor, H.S. (Ed.), A Treatise of Physical Chemistry. Macmillan, New York, p. 1661.

Kruk, M., Jaroniec, M., Sayari, A., 1997. Langmuir. 13, 6267.

Lastoskie, C., Gubbins, K.E., 2000. In: Unger, K.K., Kreysa, G., Baselt, J.P. (Eds.), Characterization of Porous Solids V. Elsevier, Amsterdam, p. 41.

Lastoskie, C., Gubbins, K.E., Quirke, N., 1993. J. Phys. Chem. 97, 4786.

Lastoskie, C., Gubbins, K.E., Quirke, N., 1994. In: Rouquerol, J., Rodriguez-Reinoso, F., Sing, K.S.W., Unger, K.K. (Eds.), Characterization of Porous Solids III. Elsevier, Amsterdam, p. 51.

Lewandowski, H., Michalski, T., Findenegg, G.H., 1991. In: Mersmann, A.B., Scholl, S.E. (Eds.), Fundamentals of Adsorption, III. Technical University of Munich, Munich, Germany, p. 497.

Lippens, B.C., Linsen, B.G., de Boer, J.H., 1964. J. Catal. 3, 32.

Liu, H., Zhang, L., Seaton, N.A., 1992. Chem. Eng. Sci. 47, 4393.

Liu, H., Zhang, L., Seaton, N.A., 1993. Langmuir. 9, 2576.

Liu, H., Zhang, L., Seaton, N.A., 1994. In: Rouquerol, J., Rodriguez-Reinoso, F., Sing, K.S.W., Unger, K.K. (Eds.), Characterization of Porous Solids III. Elsevier Science BV, Amsterdam, p. 129.

Llewellyn, P.L., Sauerland, C., Martin, C., Grillet, Y., Coulomb, J.P., Rouquerol, F., Rouquerol, J., 1997. In: McEnaney, B., Mays, T.J., Rouquerol, J., Rodriguez-Reinoso, F., Sing, K.S.W., Unger, K.K. (Eds.), Characterization of Porous Solids IV. The Royal Society Chemistry, Cambridge, p. 111.

Lopez-Ramon, M.V., Jagiello, J., Bandosz, T.J., Seaton, N.A., 1997. In: McEnaney, B., Mays, T.J., Rouquerol, J., Rodriguez-Reinoso, F., Sing, K.S.W., Unger, K.K. (Eds.), Characterization of Porous Solids IV. The Royal Society Chemistry, Cambridge, p. 73.

Lowell, S., Shields, J.E., Thomas, M.A., Thommes, M., 2004. Characterization of Porous Solids and Powders: Surface Area, Pore Size and Density. Kluwer, Dordrecht.

Machin, W.D., 1992. J. Chem. Soc. Faraday Trans. 88, 729.

Machin, W.D., 1994. Langmuir. 10, 1235.

Machin, W.D., Murdey, R.J., 1997. In: McEnaney, B., Mays, T.J., Rouquerol, J., Rodriguez-Reinoso, F., Sing, K.S.W., Unger, K.K. (Eds.), Characterization of Porous Solids IV. The Royal Society Chemistry, Cambridge, p. 221.

Maddox, M.W., Gubbins, K.E., 1995. Langmuir. 11, 3988.

Mason, G., 1971. J. Colloid Interface Sci. 35, 279.

Mason, G., 1988. In: Unger, K.K., Rouquerol, J., Sing, K.S.W., Kral, H. (Eds.), Characterization of Porous Solids I. Elsevier, Amsterdam, p. 323.

McBain, J.W., 1935. J. Am. Chem. Soc. 57, 699.

Mc Kee, D.W., 1959. J. Phys. Chem. 63, 1256.

Michalski, T., Benini, A., Findenegg, G.H., 1991. Langmuir. 7, 185.

Milburn, D.R., Davis, B.H., 1997. In: McEnaney, B., Mays, T.J., Rouquerol, J., Rodriguez-Reinoso, F., Sing, K.S.W., Unger, K. (Eds.), Characterization of Porous Solids IV. The Royal Society Chemistry, Cambridge, p. 274.

Monson, P.A., 2008. J. Chem. Phys. 128, 084701.

Monson, P.A., 2009. In: Kaskel, S., Llewellyn, P., Rodriguez-Reinoso, F., Seaton, N.A. (Eds.), Characterisation of Porous Solids VIII. Royal Society of Chemistry, p. 103.

Monson, P.A., 2012. Micropor. Mesopor. Mater. 160, 47.

Montarnal, R., 1953. J. Phys. et Rad. 12, 732.

Morishige, K., Nakamura, Y., 2004. Langmuir. 20, 4503.

Neimark, A.V., 1991. In: Rodriguez-Reinoso, F., Rouquerol, J., Sing, K.S.W., Unger, K.K. (Eds.), Characterization of Porous Solids II. Elsevier, Amsterdam, p. 67.

Neimark, A.V., Ravikovitch, P.I., 2001. Micropor. Mesopor. Mater. 44, 697.

Neimark, A.V., Ravikovitch, P.I., Vishnyakov, A., 2003. J Phys-Condens Mater. 15 (3), 347.

Olivier, J.P., 1995. J. Porous Mater. 2, 9.

Olivier, J.P., Conklin, W.B., Szombathely, M.V., 1994. In: Rouquerol, J., Rodriguez-Reinoso, F., Sing, K.S.W., Unger, K.K. (Eds.), Characterization of Porous Solids III. Elsevier, Amsterdam, p. 81.

Palmer, J.C., Gubbins, K.E., 2012. Micropor. Mesopor. Mater. 154, 24.

Pellenq, R.J.-M., Coasne, B., Denoyel, R.O., Puibasset, J., 2007. In: Llewellyn, P.L., Rodriguez-Reinoso, F., Rouquerol, J., Seaton, N. (Eds.), Characterization of Porous Solids VII. Elsevier, Amsterdam, p. 1.

Pierce, C., 1953. J. Phys. Chem. 57, 149.

Rasmussen, C.J., Vishnyakov, A., Thommes, M., Smarsly, B.M., Kleitz, F., Neimark, A.V., 2010. Langmuir. 26, 10147.

Ravikovitch, P.I., Neimark, A.V., 2001. J. Phys. Chem. B. 105, 6817.

Ravikovitch, P.I., Neimark, A.V., 2002. Langmuir. 18, 1550.

Roberts, B.F., 1967. J. Colloid Interface Sci. 23, 266.

Rouquerol, F., Rouquerol, J., Imelik, B., 1970. Bull. Soc. Chim. Fr. 10, 3816.

Rouquerol, J., Avnir, D., Fairbridge, C.W., Everett, D.H., Haynes, J.M., Pernicone, N., Ramsay, J.D.F., Sing, K.S.W., Unger, K.K., 1994. Pure Appl. Chem. 66, 1739.

Rouquerol, F., Luciani, L., Llewellyn, P., Denoyel, R., Rouquerol, J., 2003. Techniques de l'Ingénieur, Traité Analyse et Caractérisation, Paris, p. 1050.

Russo, P.A., Ribeiro Carrott, M.M.L., Conceicao, F.M.L., Carrott, P.J.M., 2009. In: Kaskel, S., Llewellyn, P., Rodriguez-Reinoso, F., Seaton, N.A. (Eds.), Characterisation of Porous Solids VIII. Royal Society of Chemistry, Amsterdam, p. 295.

Saam, W.F., Cole, M.W., 1975. Phys. Rev. B. 11, 1086.

Schreiber, A., Ketelsen, I., Findenegg, G.H., Hoinkis, E., 2007. In: Llewellyn, P.L., Rodriguez-Reinoso, F., Rouquerol, J., Seaton, N. (Eds.), Characterization of Porous Solids VII. Elsevier, Amsterdam, p. 17.

Schumacher, K., Ravikovitch, P.I., Du Chesne, A., Neimark, A.V., Unger, K.K., 2000. Langmuir. 16, 4648.

Seaton, N.A., 1991. Chem. Eng. Sci. 46, 1895.

Seaton, N.A., Walton, J.P.R.B., Quirke, N., 1989. Carbon. 27, 853.

Shull, C.G., 1948. J. Am. Chem. Soc. 70, 1410.

Silvestre-Albero, A., Gonçalvez, M., Itoh, T., Kaneko, K., Endo, M., Thommes, M., Rodriguez-Reinosos, F., Silvestre-Albero, J., 2012. Carbon. 50, 66.

Sing, K.S.W., Williams, R.T., 2004. Adsorpt. Sci. Technol. 22, 773.

Sing, K.S.W., Williams, R.T., 2012. Micropor. Mesopor. Mater. 154, 16.

Sing, K.S.W., Everett, D.H., Raul, R.A.W., Moscou, L., Pierotti, R.A., Rouquerol, J., Siemieniewska, T., 1985. Pure Appl. Chem. 57 (4), 603.

Tarazona, P., 1985. Phys. Rev. A. 31 (4), 2672.

Thommes, M., 2004. In: Lu, G.Q., Zhao, X.S. (Eds.), Nanoporous Materials: Science and Engineering. Imperial College Press, London, p. 317.

Thommes, M., Smarsly, B., Groenewolt, M., Ravikovitch, P.I., Neimark, A.V., 2006. Langmuir. 22, 756.

Thommes, M., Cychosz, K.A., Neimark, A.V., 2012. In: Tascon, J.M.D. (Ed.), Novel Carbon
    Adsorbents. Elsevier, Amsterdam, p. 107.
Thomson, W.T., 1871. Philos. Mag. 42, 448.
Wade, W.H., 1964. J. Phys. Chem. 68, 1029.
Wade, W.H., 1965. J. Phys. Chem. 69, 322.
Woo, H.J., Monson, P.A., 2003. Phys. Rev. E67: 041207, 109.
Zhdanov, V.P., Fenelonov, V.B., Efremov, D.K., 1987. J. Colloid Interface Sci. 120, 218.
Zsigmondy, A., 1911. Z. Anorg. Chem. 71, 356.

9 Assessment of Microporosity

Kenneth S.W. Sing, Françoise Rouquerol, Philip Llewellyn, Jean Rouquerol

Aix Marseille University-CNRS, MADIREL Laboratory, Marseille, France

---

Chapter Contents

9.1 Introduction

We recall that the filling of micropores (of width <2 nm) takes place in the pre-capillary condensation range of a physisorption isotherm (see Figure 8.7). If the pore width, w, is no greater than a few molecular diameters (i.e. a 'narrow micropore' or an 'ultramicropore'), pore filling occurs at a very low p/p^o. This process, which we refer to as 'primary micropore filling', is associated with enhanced adsorbent–adsorbate interactions and always involves some distortion of the sub-monolayer isotherm. Wider micropores (the 'wide micropores' or 'supermicropores') are filled by a secondary, or cooperative, process over a somewhat higher range of p/p^o, which may extend into the multilayer region. Although these mechanisms have been known for many years (see Gregg and Sing, 1982), they can now be studied more rigorously by the application of molecular simulation, density functional theory (DFT) and

Adsorption by Powders and Porous Solids. http://dx.doi.org/10.1016/B978-0-08-097035-6.00009-7

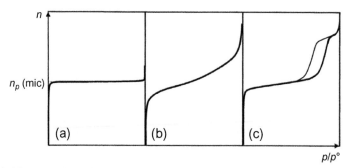

Figure 9.1 Nitrogen adsorption isotherms corresponding to adsorption (a) in narrow micropores (or 'ultramicropores'), (b) in wide micropores (or 'supermicropores') and on external surfaces, and (c) in micropores and mesopores.

high-resolution adsorption measurements on ordered pore structures. In this chapter, we introduce the currently most popular adsorption methods used for micropore size analysis: our aim is to outline in general terms the relative merits and limitations of these procedures. Their application is discussed more fully in later chapters in relation to the characterization of particular adsorbents.

As we have already seen, an ideal Type I isotherm has a long, almost horizontal plateau, which extends up to $p/p^o \rightarrow 1$, as in Figure 9.1a. In this case, the micropore capacity, $n_p(\text{mic})$, is registered directly as the amount adsorbed at the plateau. Such well-defined Type I isotherms are given by large crystals of molecular sieve zeolites and some microporous carbons. Many other porous adsorbents contain wide ranges of micropores and mesopores. Also, some microporous adsorbents are composed of very small agglomerated particles, which exhibit a significant external area. Such materials give composite isotherms with no distinctive plateau, as in Figure 9.1b. The presence of mesopores can often be detected by the appearance of a hysteresis loop with a final saturation plateau – as in Figure 9.1c.

Here, multilayer adsorption on the mesopore walls is followed by an upward deviation of the isotherm due to capillary condensation at high p/p^o. Of course, no intermediate multilayer adsorption would be detectable, if there is a continuous pore size distribution extending from wide micropores into the narrow mesopore range.

Many attempts have been made to obtain the micropore capacity by the analysis of composite isotherms. The calculation of the micropore volume, $v_p(\text{mic})$, from $n_p(\text{mic})$ is almost invariably based on the assumption that the adsorbate in the micropores has the same density as the adsorptive in the liquid state at the operational temperature. We saw in Chapter 8 that for the condensate in a mesoporous adsorbent this assumption (i.e. the Gurvich rule) appears to be justified. The situation is quite different, however, with a microporous material – particularly, when the pore dimensions are in the ultramicropore range.

Studies of the entry and packing of molecules in cylindrical and slit-shaped pores have revealed the importance of the molecular dimensions and of both the width and shape of the micropores (Carrott et al., 1987; Carrott and Sing, 1988; Balbuena and Gubbins, 1994; Sing and Williams, 2004). An indication of the effect of pore size on the packing density of spherical particles is given in Figure 9.2. Here, the degree of

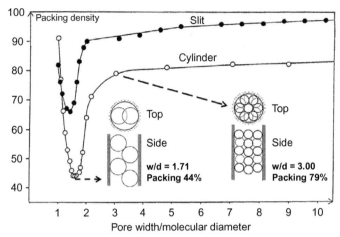

Figure 9.2 Packing of spherical particles in narrow slits or cylinders, with illustration for cylinders of two sizes.
After Carrott et al. (1987).

packing in cylinders and slits is expressed as a percentage of the packing density in the corresponding close-packed state. Although this is an oversimplified picture since it does not allow for the adsorption forces, it does illustrate the difficulty of arriving at an unambiguous assessment of the accessible pore volume. Inspection of Figure 9.2 reveals that the Gurvich rule is more likely be obeyed if $w/d > 4$. Of course, conformity to the Gurvich rule does not by itself guarantee that the packing densities are the same as in the bulk liquid. The most striking feature of Figure 9.2 is the effect of the additional degree of freedom provided by a parallel-sided slit, which allows a substantially higher packing density. Indeed, this difference in the packing density in slits and cylinders will be seen to be of great importance when we consider the adsorptive properties of molecular sieve carbons and certain zeolites.

In assessing the packing densities within narrow slits and cylinders, we have assumed that the same width is available for all molecules within a given slit or cylinder. This cannot be strictly true, however, as the effective pore width is to some extent dependent on the molecular interactions (Everett and Powl, 1976) and the density of a confined inhomogeneous fluid is not uniform within a pore structure (Gubbins, 1997).

Of course, one possible reason for an observed large departure from the Gurvich rule is as a direct result of molecular sieving: that is the inability of larger adsorptive molecules to enter a certain range of narrow pores (see Sing and Williams, 2004). Behaviour of this type is well documented for many zeolites and some activated carbons. Size exclusion measurements (gas adsorption or immersion calorimetry) provide an obvious way of determining the micropore size distribution (Denoyel et al., 1993; Gonzalez et al., 1997; Lopez-Ramon et al., 1997). Even when exclusion effects are small, or even absent, the use of molecules of different size and polarity is strongly recommended for the characterization of microporous adsorbents (see Lowell et al., 2004).

9.2   Gas Physisorption Isotherm Analysis

9.2.1   Empirical Methods

Several early attempts were made to adapt the t method for micropore analysis. In their original work, Lippens and de Boer (1965) had proposed that the *total* surface area of a porous solid was directly proportional to the slope of the initial linear section of their t-plot (see Chapter 7). Also, in the MP method of Mikhail et al. (1968), tangents to the t-plot were taken to represent the surface areas of different groups of micropores. The pore volume distribution was then determined for a given pore shape (e.g. parallel-sided slits). Although this apparently simple procedure initially attracted a good deal of interest, Dubinin (1970) and others (e.g. Gregg and Sing, 1976) pointed out that the MP method was fundamentally unsound. Thus, it is not safe to select a standard isotherm with the same BET C value (i.e. the procedure recommended by Brunauer, 1970 and Lecloux and Pirard, 1979), as this does not allow for the fact that the sub-monolayer isotherm shape is dependent on both the surface chemistry and the micropore structure. However, a t-plot can be used to assess the micropore capacity provided that the standard multilayer thickness curve has been determined on a non-porous reference material with a similar surface structure to that of the microporous sample (Sing, 1967, 1970).

   An important advantage of the α_s method (see Chapter 7) is that it does not depend on any *a priori* assumptions concerning the mechanism of adsorption by the reference material, and therefore its application is not restricted to nitrogen adsorption. For this reason, it can be used to explore the various stages of micropore filling with a number of different adsorptives (Sing and Williams, 2005). As explained in Chapter 7, the standard isotherm is plotted in the reduced form, $(n/n_x)_s$ versus p/p^o, the normalizing factor, n_x, being taken as the specific amount adsorbed at a pre-selected p/p^o (generally, it is convenient to take $p/p^o = 0.4$).

   To construct an α_s plot for a given microporous adsorbent, the amount adsorbed, n, is plotted against the reduced standard adsorption, $\alpha_s = (n/n_x)_s$. Hypothetical α_s plots for two different microporous adsorbents are shown in Figure 9.3.

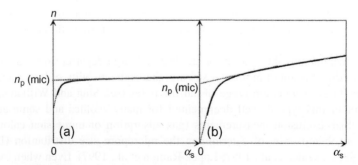

Figure 9.3 Hypothetical α_s plots for a sample (a) with narrow micropores (or 'ultramicropores') and (b) with only wide micropores (or 'supermicropores').

In the case of (a), there is no initial linear region since the isotherm is distorted at low p/p^o as a result of the enhanced adsorbent–adsorbate interactions in ultramicropores. On the other hand, the initial linear section of the α_s plot in (b) corresponds to monolayer adsorption on the walls of supermicropores. Cooperative filling of the supermicropores occurs progressively over a range of higher p/p^o and in some cases is manifested by an upward deviation of the α_s plot. Once the micropores have been filled, both plots in Figure 9.3 become linear, provided that capillary condensation is absent (or only detectable at high p/p^o). The low slope signifies that multilayer adsorption has occurred on a relatively small external surface. Back-extrapolation of the linear multilayer section gives the specific micropore capacity, $n_p(\text{mic})$, as the intercept on the n axis.

As already noted, the effective micropore volume, $v_p(\text{mic})$ is as follows:

$$v_p(\text{mic}) = n_p(\text{mic})\, M/\rho \tag{9.1}$$

where M is the molar mass of the adsorptive and ρ is the average absolute density of the adsorbate, which is generally assumed to be the same as the absolute density of the liquid adsorptive.

The α_s method has been used by a number of investigators to study the various stages of micropore filling (Carrott et al., 1989; Fernandez-Colinas et al., 1989; Kenny et al., 1993; Kaneko, 1996; Sing and Williams, 2005; Almazan-Almazan et al., 2009; Villarroel-Rocha et al., 2013). For example, high-resolution nitrogen measurements have confirmed the difference in the adsorptive behaviour of ultramicroporous and supermicroporous carbons (Kenny et al., 1993). Also, by constructing high-resolution α_s plots for probe molecules of different size, one can study the effect of changing the ratio of pore width/molecular size, w/d, on the mechanism of pore filling (Carrott et al., 1987, 1988a,b; Sing and Williams, 2004, 2005). In this manner, it is possible to gain a semi-quantitative estimate of the pore size distribution as described in Chapter 10.

9.2.2   Dubinin–Radushkevich–Stoeckli Methods

The Dubinin–Radushkevich (DR) equation, which in Chapter 5 was given as Equation (5.44), can also be expressed in the form

$$\log_{10}(n) = \log_{10}\big(n_p(\text{mic})\big) - D \log_{10}^2(p^o/p) \tag{9.2}$$

where D is an empirical constant – as defined in Equation (5.45). Thus, according to the DR theory, a plot of $\log_{10}(n)$ against $\log_{10}^2(p^o/p)$ should be linear with slope D and intercept $\log_{10}(n_p(\text{mic}))$. As before, Equation (9.1) is used to obtain $v_p(\text{mic})$ from $n_p(\text{mic})$.

Various DR plots for activated carbons are given in Chapter 10. For our present purpose, the two hypothetical examples in Figure 9.4 will suffice to illustrate some important features.

As a general rule, an ultramicroporous carbon gives a linear DR plot over a wide range of p/p^o (see Figure 9.4a) provided that the isotherm is reversible (i.e. there is no

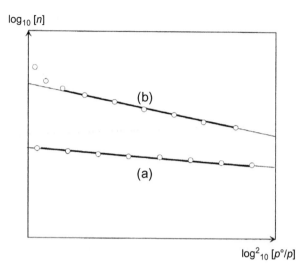

Figure 9.4 Hypothetical DR plots for a sample (a) with only narrow micropores (or 'ultramicropores') and (b) also including wide pores.

low-pressure hysteresis). Generally, any significant adsorption in wider pores or on an external surface leads to a departure from linearity (see Figure 9.4b). But, some DR plots on non-porous or mesoporous solids do exhibit limited ranges of linearity. In fact, it was this behaviour which led Kaganer (1959) to propose an analogous form of Equation (9.2) in which $v_m(mic)$ is replaced by n_m, the monolayer capacity.

It is apparent that any limited range of linearity of a simple DR plot cannot be used to give a reliable evaluation of the pore size distribution. In order to describe the bimodal micropore size distribution in certain active carbons, Dubinin (1975) applied a two-term equation, which we now write in the form

$$n = n_{p,1} \exp\left[-(A/E_1)^2\right] + n_{p,2} \exp\left[-(A/E_2)^2\right] \tag{9.3}$$

where $n_{p,1}$ and $n_{p,2}$ are the micropore capacities of the two groups of pores, E_1 and E_2 are now the corresponding characteristic energies and the adsorption potential, $A = RT\ln(p^o/p)$.

The application of Equation (9.3) led Dubinin (1975) to refer to the group of wider pores as 'supermicropores' (~ 1.5 nm), which are filled at higher p/p^o.

Equation (9.3) can be extended to include a much larger number of different groups of pores as envisaged by Stoeckli (1977). By assuming a Gaussian distribution, Stoeckli (1977) and Stoeckli et al. (1979) attempted to allow for a continuous distribution of pore size by replacing the summation of the individual DR contributions by integration (see Chapter 5). The integral transform was then solved by the use of a mathematical device equivalent to an 'error function' to give an exponential relation between the micropore size distribution and the Gaussian half-width, Δ. In other words, Δ was taken to be a measure of the dispersion of B around a mean value of B in Equation (5.48).

To explain the significance of the various terms adopted by Dubinin and Stoeckli, it is convenient to write the Dubinin–Astakov (DA) equation – Equation (5.46) – in the form

$$n = n_p \exp\left[-(A/\beta E_0)^N\right] \tag{9.4}$$

where β and N are as defined in Chapter 5 and E in Equation (5.46) is replaced by βE_0. On the basis of a limited amount of experimental evidence, an inverse relation was proposed (Dubinin, 1979) between E_0 and the pore width, w:

$$w = K/E_0 \tag{9.5}$$

where K is an empirical constant, which was reported to be ~ 17.5 nm kJ mol^{-1} for a series of activated carbons (see Bansal et al., 1988). The micropore size became somewhat smaller and the distribution more homogeneous with increase in N over the range 1.5–3.0. The extreme value of $N=3$ was found with a carbon molecular sieve, which had a particularly narrow distribution of pore size (Dubinin and Stoeckli, 1980).

The mathematical elegance of the Dubinin–Stoeckli approach is impressive, but the basic DR equation is unlikely to be strictly applicable to secondary micropore filling of supermicropores (see Chapter 10). It is noteworthy that the method appears to be most successful when applied to a relatively narrow range of ultramicropores in molecular sieve carbons, that is, when only primary micropore filling is involved (Martin-Martinez et al., 1986). Although the DA equation can be applied to physisorption isotherms on molecular sieve zeolites, the results are much more difficult to interpret. This is an indication that consideration must be given to both the pore structure and the surface chemistry.

Although the DR equation continues to be extensively applied to physisorption isotherms on microporous carbons (e.g. Almazan-Almazan et al., 2009; Thommes et al. 2012; Villarroel-Rocha et al., 2013), it is of questionable value for assessment of micropore capacities (see Figure 10.11 and Table 10.1). However, as an empirical relation within prescribed limits, it can be used to interpolate or extrapolate experimental data.

9.2.3   The Horvath–Kawazoe (HK) Method

A method for determining the micropore size distribution was introduced by Horvath and Kawazoe in 1983. In its original form, the HK analysis was applied to nitrogen isotherms determined on molecular sieve carbons, the assumption being made that these adsorbents contained slit-shaped graphitic pores. The HK treatment was also extended to include argon and nitrogen adsorption in cylindrical and spherical pores of zeolites and aluminophosphates (Venero and Chiou, 1988; Davis et al, 1989; Saito and Foley, 1991; Cheng and Yang, 1995; Horvath and Suzuki, 1997).

The HK method is based on the general idea that the relative pressure required for the filling of micropores of a given size and shape is directly related to the adsorbent–adsorbate interaction energy. As in the FHH theory (see Chapter 5), the assumption is made that the entropy contribution to the free energy of adsorption is small in comparison with the large change in internal energy, which is itself largely dependent on the depth of the potential energy well. Expressed in another way, it is assumed that the molar entropy of the adsorbed phase is not very sensitive to a change in the pore dimensions. In this manner, it was possible to evaluate the required relative pressure, $(p/p^o)_{pore}$ to fill pores of a given size since it was assumed that this depended only on the interaction energy, $\phi(z)_{\textbf{pore}}$. Thus

$$RT \ln(p/p^o)_{pore} = f\left[\phi(z)_{\textbf{pore}}\right] \tag{9.6}$$

By making use of the Kirkwood–Müller equation and substituting the available experimental data for the physical properties of nitrogen and carbon, Horvath and Kawazoe (1983) arrived at an equation for the adsorption of nitrogen by molecular sieve carbons at 77 K. A few representative values of $(p/p^o)_{pore}$ obtained by the application this equation are given in Table 9.1. Although the HK method is based on questionable principles and is now overshadowed by the application of DFT, it merits recognition as the first stage in the development of a unified theoretical treatment of micropore filling.

Although the estimated values of $(p/p^o)_{pore}$ in Table 9.1 are unlikely to agree exactly with the equivalent experimental values, the predicted range is of interest. Thus, the filling of ultramicropores is expected to occur at very low relative pressures. In fact, high-resolution adsorption measurements (e.g. Conner, 1997; Maglara et al., 1997; Ravikovitch et al., 2000; Thommes et al., 2006) have confirmed that to investigate the first stages of micropore filling, it is necessary to work at $p/p^o < 10^{-6}$.

9.2.4   Density Functional Theory

As indicated in Chapters 6 and 8, there is now considerable interest in the application of computer simulation (e.g. molecular dynamics and Monte Carlo simulation) and DFT for investigating the behaviour of fluids confined in nanopores (Olivier, 1995; Gubbins, 1997; Jagiello et al., 2011; Thommes et al., 2012; Landers et al., 2013).

Here, we discuss the application of non-local DFT (NLDFT) and quenched-solid DFT (QSDFT) for nanopore size analysis. Unlike the older methods, such as the Barrett, Joyner and Halenda (BJH) method, DFT is independent of capillary

Table 9.1 HK-Predicted Values of p/p^o of N_2 Required to fill Carbon Slit-Shaped Pores

Effective Pore Width, w/nm	$(p/p^o)_{\textbf{pore}}$
0.4	1.46×10^{-7}
0.5	1.05×10^{-5}
0.6	1.54×10^{-4}
0.8	2.95×10^{-3}
1.5	7.59×10^{-2}

condensation and can be applied across the complete micropore–mesopore range. Furthermore, easily accessible NLDFT and QSDFT software are now commercially available for a number of adsorbent–adsorptive systems.

An outline of DFT will suffice since a general description is given in Chapter 8. NLDFT is based on the well-established principles of classical and statistical thermodynamics and it is assumed that, under certain controlled conditions, the adsorbate is in thermodynamic equilibrium with the adsorptive in the gas phase. The distribution in density, $\rho(r)$, of the pore fluid is determined by minimizing the grand thermodynamic potential (see Equation 8.26), which is dependent on the fluid–fluid and fluid–solid interactions. Once $\rho(r)$ is known, the adsorption isotherm and other thermodynamic properties can be calculated. The parameters of the fluid–fluid interactions are cross-checked to ensure consistency with the bulk properties, and the fluid–solid interactions are obtained by adjusting the calculated isotherms to fit experimental data on non-porous reference adsorbents. It is generally assumed that the pores are of simple geometry (usually slits, cylinders or spheres). NLDFT considers the pore wall as smooth which can lead to anomalies in the analysis at low pressure. To overcome this drawback, QSDFT adds a certain degree of roughness to the pore surface (Gor et al., 2012). In a similar way, a modified form of NLDFT was proposed (Jagiello and Olivier, 2013a,b), which takes into account the energetical heterogeneity and the geometrical corrugation of the surface.

Thus, a series of theoretical isotherms can be calculated for a particular gas–solid system and pores of a given shape. This series of theoretical isotherms is called the *kernel*. Calculation of the pore size distribution function, $f(w)$, is based on a solution of the generalized adsorption isotherm (GAI) equation as follows:

$$N_{\text{exp}}\left(\frac{p}{p^{\text{o}}}\right) = \int_{w_{\text{min}}}^{w_{\text{max}}} N_{\text{theo}}\left(\frac{p}{p^{\text{o}}}, w\right) f(w)\,\mathrm{d}w \tag{9.7}$$

where $N_{\text{exp}}(p/p^{\text{o}})$ is the measured number of adsorbed molecules and $N_{\text{theo}}(p/p^{\text{o}}, w)$ is the kernel of theoretical isotherms in model pores. The integration is over a finite range of pore size. Although the solution of the GAI equation is strictly an ill-posed problem, meaningful solutions can be obtained by using regularization algorithms (Dombrowski et al., 2000; Ravikovitch et al., 2000).

The application of QSDFT is illustrated in Figure 9.5, in which the pore size distribution plots are given for two samples of carbon cloth (Llewellyn et al., 2000). Commercial software was used for the computation, with a kernel based on the assumption of slit-shaped pores in graphitic carbon. The shape of the nitrogen isotherms (at 77 K) and the corresponding high-resolution α_s plots indicated that while samples A and B were both microporous, sample B was to some extent also mesoporous. More information is clearly provided by the DFT analysis in Figure 9.5, which reveals the presence of a distinctive bimodal micropore structure in each carbon cloth. It is evident that similar ranges of ultramicropores and supermicropores were present in both adsorbents, but that the scale of the supermicropore structure was greater in the case of sample A. The QSDFT kernel was used here because in the 'basic' NLDFT of

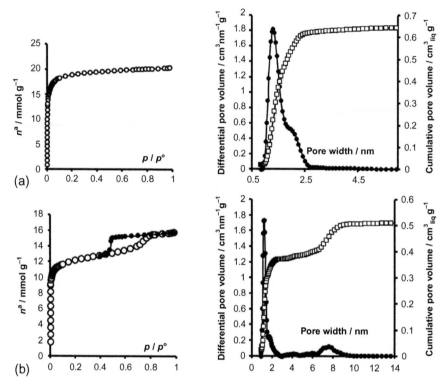

Figure 9.5 N_2 adsorption isotherms at 77 K (left) and corresponding pore size distributions (right) obtained by QSDFT on two carbon cloths: only microporous (a) or both microporous and mesoporous (b).

layer-by-layer adsorption no allowance is made for any surface heterogeneity, roughness or disordered pore structure (see Figure 10.14). Nevertheless, the degree of disorder in the QSDFT model may not completely reflect that of the actual solid. As such, the pore size distributions in Figure 9.5 should therefore be regarded as giving only a semi-quantitative assessment of the effective pore size distribution. Considerable progress has been made, however, by studying ordered porous materials with well-characterized properties and it has been possible to obtain remarkably good agreement between the pore sizes derived from the isotherms of nitrogen and argon (Ravikovitch et al., 2000; Thommes et al., 2006, 2012).

9.2.5 Nonane Pre-adsorption

An interesting method for the evaluation of microporosity was introduced by Gregg and Langford (1969). The aim was to fill the micropores of an adsorbent with *n*-nonane, while leaving the wider pores and open surface still available for the adsorption of nitrogen at 77 K. Nonane was chosen as the pre-adsorptive because of its relatively large physisorption energy, which in turn results in a high energy barrier (i.e. activation energy) for desorption. As a consequence, elevated temperatures are required to remove the nonane molecules from the narrow pores at a measurable rate. Moreover, the long-chain molecules of n-nonane are able to enter narrow micropores of width as small as 0.4 nm, which is not the case with more bulky hydrocarbon molecules.

The following experimental procedure was used by Gregg and Langford. The outgassed adsorbent was first exposed to n-nonane vapour at 77 K, re-outgassed at room temperature and the first nitrogen isotherm determined at 77 K. The sample was then outgassed at a number of increasingly higher temperatures, and nitrogen isotherms were successively determined after each stage of outgassing until the nonane had been completely removed.

The first material to be studied in this way was an activated sample of carbon black and it was found that prolonged outgassing at a temperature of 350 °C was required to achieve complete removal of the pre-adsorbed nonane. It was reported that all the intermediate nitrogen isotherms were parallel in the multilayer range and that the vertical separation between the isotherms obtained after outgassing at 20 and 350 °C provided a satisfactory measure of the micropore capacity. Convincing evidence was also obtained that the nonane was removed only from the external surface at 20 °C.

Some more recent investigations of the pre-adsorption method have shown, however, that the results are not always so easy to interpret (Martin-Martinez et al., 1986; Carrott et al., 1989). As would be expected, the nonane molecules are more strongly trapped in narrow ultramicropores than in the wider supermicropores, but since many microporous adsorbents have complex networks of pores of different size, the retention of the nonane molecules in narrow pores also leads to blocking of some wider pores.

9.2.6   Choice of Adsorptive and Temperature

For many years, it has been customary to use *nitrogen* (at 77 K) as the adsorptive for the assessment of both the surface area and the pore size distribution (Gregg and Sing, 1982; Sing et al., 1985; Lowell et al., 2004). Because of the ready availability of commercial equipment and user-friendly software and a good supply of inexpensive liquid nitrogen, until recently, there seemed little tendency to favour the use of another gas or vapour. Although nitrogen at 77 K is still the generally preferred adsorptive for mesopore size analysis, it is now apparent that there may be advantages in using argon at 87 K – especially for micropore size analysis.

As already mentioned, at 77 K the adsorption of nitrogen by certain microporous adsorbents (e.g. zeolites) is detectable at $p/p^{\circ} < 10^{-6}$. At these very low pressures, the rates of diffusion and adsorption are exceedingly slow. Furthermore, strong physisorption in narrow pore entrances can block the passage of molecules into larger pores or cavities. These effects are minimized if the adsorption measurements are made at higher temperature, but obviously with nitrogen, a temperature above the normal boiling point is inconvenient if the complete isotherm is required.

It turns out that there are several advantages in using *argon at* 87 K. At this temperature, argon starts to fill ultramicropores at an appreciably higher p/p^o than nitrogen does at 77 K. This leads to accelerated diffusion and faster equilibration times (Thommes et al., 2012) and also facilitates the measurement of the equilibrium pressures. Because the argon molecule is nonpolar, the specific interactions associated with the quadrupolar nitrogen molecule are avoided and therefore argon standard isotherms for non-porous reference materials are not highly dependent on the surface chemistry. This makes it easier to apply the α_s method and also to calculate the DFT kernel.

As explained in Chapter 10, *carbon dioxide* adsorption at 273 K is widely used for investigating the ultramicropore structure in the 'low burn-off' activated carbons (Thommes et al., 2012). The uptake of carbon dioxide is often found to be much larger than that of nitrogen, although the molecular sizes of CO_2 and N_2 are not very different. It is evidently the higher operational temperature which is the most important factor in producing the enhanced uptake of CO_2. However, because of the quadrupolar nature of CO_2, strong specific interactions are involved with zeolites and some microporous oxides and such differences in adsorptive behaviour are more difficult to interpret.

For some purposes, there are good reasons for using a range of adsorptives of different molecular size and polarity (Sing and Williams, 2004; Thommes et al, 2012). As pointed out in Chapter 10, by selecting a series of 'globular' alkane molecules one can investigate differences in the Type I isotherm character in relation to the molecular dimensions of the adsorptive. Gas chromatographic retention measurements are also useful for studying the interaction of organic probe molecules with microporous adsorbents. The use of water as a probe molecule has attracted a considerable amount of interest. As is well known, the water molecule is relatively small, but it is highly polar and reactive and has a propensity for strong hydrogen bonding. It is hardly surprising that water has a unique role as a molecular probe!

9.3   Microcalorimetric Methods

Because of the appreciable enhancement of the adsorption energy in micropores of a few molecular diameters in width, microcalorimetry can provide a useful means of assessing microporosity. The available experimental procedures are outlined in the following sections.

9.3.1   Immersion Microcalorimetry

9.3.1.1   Immersion of Various Dry Samples in the Same Liquid

For this approach, the liquid is usually selected to fill the micropores as completely as possible. The most popular liquids are water, methanol, benzene, cyclohexane and *n*-hexane; the choice depends on the expected polarity of the surface and shape of the micropores (cylindrical, slit-shaped, etc.).

The first stage is usually a simple comparison of the specific energies of immersion. In the absence of a relatively large proportional of external surface area, these energies are controlled by the nature of the microporous network (see, for instance, Stoeckli and Ballerini, 1991 or Rodriguez-Reinoso et al., 1997).

For a more detailed evaluation of the microporosity, one must be able to interpret the energy of immersion data. At present, there are two procedures favoured by different investigators.

The first procedure is based on the Dubinin–Stoeckli principles of volume filling (see Section 5.2.4). The energy of immersion $\Delta_{\mathrm{imm}}U$ is related to the *micropore volume* $W_{\mathrm{o}}(d)$ and the 'characteristic energy' E for a given micropore size and immersion liquid (Bansal et al., 1988) by the expression:

$$\frac{\Delta_{\mathrm{imm}}U}{m_{\mathrm{s}}} = \left[-\beta EW_0(d)(1 + \alpha T)\sqrt{\Pi}/2V_{\mathrm{m}}^{\mathrm{l}}\right] + u^{\mathrm{i}}a(\mathrm{ext}) \tag{9.8}$$

where m_{s} is the mass of adsorbent, d is the molecular diameter, $V_{\mathrm{m}}^{\mathrm{l}}$ is the molar volume of the liquid, u^{i} is the energy of immersion of 1 m^2 of external surface area $a(\mathrm{ext})$, β is a scaling factor and α is the coefficient of thermal expansion of the liquid. Provided the external surface area is negligible, the energy of immersion appears to provide an approximate comparative assessment of the microporous volume $W_{\mathrm{o}}(d)$ accessible to the immersion liquid. However, this only holds if the characteristic energy, E, remains the same for all samples or if allowance can be made for the likely variation in E. Generally, this poses a serious problem.

The second procedure directly relates the energy of immersion $\Delta_{\mathrm{imm}}U$ to the *micropore surface area* $a(\mathrm{mic})$ as described in Section 4.2.3. As we saw, very simply:

$$\Delta_{\mathrm{imm}}U = u_{\mathrm{i}}a(\mathrm{mic}) + u_{\mathrm{i}}a(\mathrm{ext}) \tag{9.9}$$

It is of interest that, in spite of their great difference in complexity, Equations (9.8) and (9.9) are at least mutually consistent: since E is an inverse function of the pore width w, EW_{o} may be approximately proportional to $a(\mathrm{mic})$.

9.3.1.2 Immersion of Dry Samples in Liquids of Different Molecular Size

This method is designed to take advantage of molecular sieving. The basic data are simply in the form of a curve of the specific energy of immersion versus the molecular size of the immersion liquid. This provides immediate information on the micropore size distribution as directly assessed by probe molecules, that is, in a realistic and practical manner which takes into account any imperfection of the sample and any actual hindrance to the diffusion of the probe (e.g. due to tortuosity or constrictions). In this respect, this approach offers a useful complement to the methods based on the adsorption of a single adsorptive with small molecular size, which were examined in Section 9.2. For room-temperature experiments one can use the liquids listed in Table 9.2, which are well suited for the study of carbons. Because of the various ways of expressing 'critical dimension' of a molecular probe or its 'molecular size', one must be careful to use a consistent set of data (hence the two separate lists in Table 9.2).

Table 9.2 Immersion Liquids Used to Probe Micropore Size

Liquid	Critical Dimension/nm (after Stoeckli, 1995)	Liquid	Molecular Size/nm (after Denoyel et al., 1993)
Dichloromethane	0.33	Benzene	0.37
Benzene	0.41	Methanol	0.43
Cyclohexane	0.54	Isopropanol	0.47
Carbon tetrachloride	0.63	Cyclohexane	0.48
1,5,9-Cyclododecatriene	0.76	*ter*-Butanol	0.6
Tri 2,4-xylylphosphate	1.5	α-Pinene	0.7

Again, one can process the microcalorimetric data to compare either the micropore volumes accessible to the various molecules (see Stoeckli et al., 1996), or the micropore surface areas, as illustrated in Figure 9.6 (see also Rodriguez-Reinoso et al., 1997, Villar-Rodil et al., 2004).

A particular advantage of using immersion microcalorimetry for the study of ultra-microporous materials is that the molecular entry into very fine pores takes place

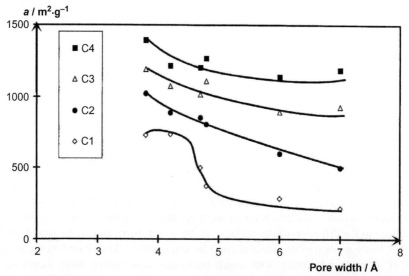

Figure 9.6 Accessible surface area versus pore width from immersion microcalorimetry in the liquids listed in Table 9.2 (right-hand side) for a set of charcoals (activation increasing from C1 to C4).
After Denoyel et al. (1993).

much more rapidly from the liquid phase than from a gas. There are two reasons for this difference: gas diffusion may be slow (thermally activated) – especially at 77 K – and secondly the higher liquid density also favours a more rapid molecular penetration.

9.3.1.3 Immersion of Samples Partially Pre-covered by Vapour Adsorption

This is an indirect way of assessing the energetics of gas adsorption in micropores. The pre-adsorbed vapour can be that of the immersion liquid or it can be another adsorptive: for instance, Stoeckli and Huguenin (1992) devised an experiment with water pre-adsorption prior to immersion calorimetry (in water or in benzene), in order to study the water filling mechanism in microporous carbons.

9.3.2 Gas Adsorption Microcalorimetry

The enhancement of the adsorption energy in micropores is best assessed by direct microcalorimetric measurements of $\Delta_{ads}\dot{u}$. Moreover, a comparison between N_2 and Ar adsorption allows one to distinguish between the enhancement due to the confinement in micropores and that due to specific adsorbent–adsorbate interactions. Both effects are manifested in the low-pressure range of the nitrogen isotherm, but the specific interactions are virtually absent with argon.

An important feature of gas adsorption microcalorimetry for the study of microporosity is that *its highest sensitivity is in the low-pressure region* (say, $p/p^{\circ} < 0.05$), which is precisely the region of micropore filling and where the adsorption isotherm measurement often lacks accuracy with respect to the equilibrium pressure. It follows that it is usually advisable to plot $\Delta_{ads}\dot{u}$ versus $\boldsymbol{n}$ rather than versus p/p°. It is then possible to distinguish various types of micropores from the value of $\Delta_{ads}\dot{u}$ and to obtain information on their effective volumes. This is illustrated by the microcalorimetric curves of Figure 10.15, obtained for nitrogen and argon adsorption on two activated carbons. In this manner, microcalorimetry is an independent way of checking the analysis of the isotherm. Values of $\Delta_{ads}\dot{u}$ can be meaningful. For instance, with molecular sieve carbons, the initial values of $\Delta_{ads}\dot{u}$ for nitrogen and argon are approximately twofold higher than those obtained on non-porous carbon black: this is the maximum enhancement predicted for the adsorption in slit-shaped micropores (Everett and Powl, 1976).

9.4 Conclusions and Recommendations

1. The physisorption isotherm given by a purely microporous adsorbent is Type I in the IUPAC classification. Micropore filling in pores of molecular dimensions occurs at very low p/p°, whereas wider micropores are filled over a range of higher p/p°.

2. The α_s-method provides a useful way of analysing a composite isotherm to obtain an indication of the mechanisms of pore filling and assessment of the micropore capacity and the effective external surface area.

3. The Dubinin–Raduskevich (DR) plot can be used for the interpolation or extrapolation of adsorption data, but cannot provide an entirely reliable estimate of the micropore volume.

4. The application of DFT is generally recommended for pore size analysis, with a version of NLDFT taking into account the surface heterogeneity and roughness. A semi-quantitative evaluation of the effective pore size distribution is then obtained.

5. In future, argon adsorption at 87 K may replace nitrogen at 77 K for nanopore size analysis.

6. A simple check of the micropore size and accessibility is possible by immersion microcalorimetry in liquids of various molecular sizes. The same molecules can also be adsorbed from the gas phase, though more attention must be paid to the problem of diffusion in narrow micropores.

References

Almazan-Almazan, M.C., Perez-Mendoza, M., Fernandez-Morales, I., Domingo-Garcia, M., Lopez-Garzon, F.J., Martinez-Alonso, A., Suarez-Garcia, F., Tascon, J.M.D., 2009. In: Kaskel, S., Llewellyn, P., Rodriguez-Reinoso, F., Seaton, N.A. (Eds.), Characterization of Porous Solids VIII. Royal Society of Chemistry, Cambridge, p. 159.

Balbuena, P.B., Gubbins, K.E., 1994. In: Rouquerol, J., Rodriguez-Reinoso, F., Sing, K.S.W., Unger, K.K. (Eds.), Characterization of Porous Solids III. Elsevier, Amsterdam, p. 41.

Bansal, R.C., Donnet, J.B., Stoeckli, H.F., 1988. Active Carbon. Marcel Dekker, New York, p. 139.

Brunauer, S., 1970. In: Everett, D.H., Ottewill, R.H. (Eds.), Surface Area Determination. Butterworths, London, p. 63.

Carrott, P.J.M., Sing, K.S.W., 1988. In: Unger, K.K., Rouquerol, J., Sing, K.S.W., Kral, H. (Eds.), Characterization of Porous Solids I. Elsevier, Amsterdam, p. 77.

Carrott, P.J.M., Roberts, R.A., Sing, K.S.W., 1987. Chem. Ind. 855.

Carrott, P.J.M., Roberts, R.A., Sing, K.S.W., 1988a. In: Unger, K.K., Rouquerol, J., Sing, K.S.W., Kral, H. (Eds.), Characterization of Porous Solids I. Elsevier, Amsterdam, p. 89.

Carrott, P.J.M., Roberts, R.A., Sing, K.S.W., 1988b. Langmuir. 4, 740.

Carrott, P.J.M., Drummond, F.C., Kenny, M.B., Roberts, R.A., Sing, K.S.W., 1989. Colloids Surf. 37, 1.

Cheng, L.S., Yang, R.T., 1995. Adsorption. 1, 187.

Conner, W.C., 1997. In: Fraissard, J., Conner, W.C. (Eds.), Physical Adsorption: Experiment, Theory and Applications. Kluwer, Dordrecht, p. 33.

Davis, M.E., Montes, C., Hathaway, P.E., Arhancet, J.P., Hasha, D.L., Garces, J.E., 1989. J. Am. Chem. Soc. 111, 3919.

Denoyel, R., Fernandez-Colinas, F., Grillet, Y., Rouquerol, J., 1993. Langmuir. 9, 515.

Dombrowski, R.J., Hyduke, D.R., Lastoskie, C.M., 2000. Langmuir. 16, 5041.

Dubinin, M.M., 1970. In: Everett, D.H., Ottewill, R.H. (Eds.), Surface Area Determination. Butterworths, London, p. 123.

Dubinin, M.M., 1975. In: In: Cadenhead, D.A. (Ed.), Progress in Surface and Membrane Science, Vol. 9. Academic Press, New York, p. 1.

Dubinin, M.M., 1979. In: Gregg, S.J., Sing, K.S.W., Stoeckli, H.F. (Eds.), Characterisation of Porous Solids. Society of Chemical Industry, London, p. 1.

Dubinin, M.M., Stoeckli, H.F., 1980. J. Colloid Interface Sci. 75, 34.

Everett, D.H., Powl, J.C., 1976. J. Chem. Soc. Faraday Trans. I. 72, 619.

Fernandez-Colinas, J., Denoyel, R., Grillet, Y., Rouquerol, F., Rouquerol, J., 1989. Langmuir. 5, 1205.

Gonzalez, M.T., Rodriguez-Reinoso, F., Garcia, A.N., Marcilla, A., 1997. Carbon. 35, 8.

Gor, G.Y., Thommes, M., Cychosz, K.A., Neimark, A.V., 2012. Carbon. 50, 1583.

Gregg, S.J., Sing, K.S.W., 1982. Adsorption, Surface Area and Porosity, 2nd edn Academic Press, London.

Gregg, S.J., Langford, J.F., 1969. Trans. Faraday Soc. 65, 1394.

Gregg, S.J., Sing, K.S.W., 1976. In: In: Matijevic, E. (Ed.), Surface and Colloid Science, Vol. 9. Wiley, New York, p. 336.

Gubbins, K.E., 1997. In: Fraissard, J., Conner, C.W. (Eds.), Physical Adsorption: Experiment, Theory and Applications. Kluwer, Dordrecht, p. 65.

Horvath, G., Kawazoe, K., 1983. J. Chem. Eng. Jpn. 16, 470.

Horvath, G., Suzuki, M., 1997. In: Fraissard, J., Conner, C.W. (Eds.), Physical Adsorption: Experiment, Theory and Applications. Kluwer, Dordrecht, p. 133.

Jagiello, J., Kenvin, J., Olivier, J., Contescu, C., 2011. Adsorpt. Sci. Technol. 29, 769.

Jagiello, J., Olivier, J.P., 2013a. Carbon. 55, 70.

Jagiello, J., Olivier, J.P., 2013b. Adsorption. 19, 777.

Kaganer, M.G., 1959. Zhur. Fiz. Khim. 33, 2202.

Kaneko, K., 1996. In: Dabrowski, A., Tertykh, V.A. (Eds.), Adsorption on New and Modified Inorganic Sorbents. Elsevier, Amsterdam, p. 573.

Kenny, M.B., Sing, K.S.W., Theocharis, C., 1993. In: Suzuki, M. (Ed.), Proceedings of the 4th International Conference on Fundamentals of Adsorption. Kodansha, Tokyo, p. 323.

Landers, J., Yu, G., Neimark, A.V., 2013, Colloids and Surfaces A: Physicochemical and Engineering Aspects, in press.

Lecloux, A., Pirard, J.P., 1979. J. Colloid Interface Sci. 70, 265.

Lippens, B.C., de Boer, J.H., 1965. J. Catal. 4, 319.

Llewellyn, P.L., Rouquerol, F., Rouquerol, J., Sing, K.S.W., 2000. Studies in Surface Science and Catalysis, Vol 128. Elsevier, Amsterdam p. 421.

Lopez-Ramon, M.V., Jagiello, J., Bandosz, T.J., Seaton, N.A., 1997. Langmuir. 13, 4435.

Lowell, S., Shields, J.E., Thomas, M.A., Thommes, M., 2004. Characterization of Porous Solids and Powders: Surface Area, Pore Size and Density. Kluwer, Dordrecht.

Maglara, E., Pullen, A., Sullivan, D., Conner, W.C., 1997. Langmuir. 10, 11.

Martin-Martinez, J.M., Rodriguez-Reinoso, F., Molina-Sabio, M., McEnaney, B., 1986. Carbon. 24, 255.

Mikhail, R.Sh., Brunauer, S., Bodor, E.E., 1968. J. Colloid Interface Sci. 26, 45.

Olivier, J.P., 1995. J. Porous Mat. 2, 9.

Ravikovitch, P.I., Vishnyakov, A., Russo, R., Neimark, A.V., 2000. Langmuir. 16, 2311.

Rodriguez-Reinoso, F., Molina-Sabio, M., Gonzalez, M.T., 1997. Langmuir. 13, 8.

Saito, A., Foley, H.C., 1991. AIChE J. 37, 429.

Sing, K.S.W., 1967. Chem. Ind. 829.

Sing K.S.W. (1970) In: Surface Area Determination (D.H. Everett and R.H. Ottewill, eds), Butterworths, London, p.15.

Sing, K.S.W., Everett, D.H., Haul, R.A.W., Moscou, L., Pierotti, R.A., Rouquerol, J., Siemieniewska, T., 1985. Pure Appl. Chem. 57, 603.

Sing, K.S.W., Williams, R.T., 2004. Part. Part. Syst. Charact. 20, 1.

Sing, K.S.W., Williams, R.T., 2005. Adsorpt. Sci. Technol. 23, 839.

Stoeckli, H.F., 1977. J. Colloid Interface Sci. 59, 184.

Stoeckli, H.F., Ballerini, L., 1991. Fuel. 70, 557.

Stoeckli, H.F., Huguenin, D., 1992. J. Chem. Soc. Faraday Trans. 88 (5), 737.

Stoeckli, H.F., Houriet, J.Ph., Perret, A., Huber, U., 1979. In: Gregg, S.J., Sing, K.S.W., Stoeckli, H.F. (Eds.), Characterisation of Porous Solids. Society of Chemistry Industry, London, p. 31.

Stoeckli, H.F., 1995. In: Patrick, J.W. (Ed), Porosity in Carbons. E.Arnold. London, Chapter. 3, 66–97.

Stoeckli, H.F., Centeno, T.A., Fuertes, A.B., Muniz, J., 1996. Carbon. 34, 10.

Thommes, M., Cychosz, K.A., Neimark, A.V., 2012. In: Tascon, J.M.D. (Ed.), Novel Carbon Adsorbents. Elsevier, Amsterdam, p. 107.

Thommes, M., Smarsly, B., Groenewolt, M., Ravikovitch, P.I., Neimark, A.V., 2006. Langmuir. 22, 756.

Venero, A.F., Chiou, J.N., 1988. Mater. Res. Soc. Symp. Proc. 111, 235.

Villar-Rodil, S., Denoyel, R., Rouquerol, J., Martinez-Alonso, A., Tascon, J.M.D., 2004. Thermochim. Acta. 420, 141.

Villarroel-Rocha, J., Barrera, D., Garcia Blanco, A.A., Jalil, M.E.R., Sapag, K., 2013. Adsorpt. Sci. Technol. 31, 165.

10 Adsorption by Active Carbons

Kenneth S.W. Sing

Aix Marseille University-CNRS, MADIREL Laboratory, Marseille, France

---

Chapter Contents

Adsorption by Powders and Porous Solids. http://dx.doi.org/10.1016/B978-0-08-097035-6.00010-3

10.1   Introduction

In this chapter, the term *active carbon* is applied to a porous or finely divided carbon of appreciable internal and/or external surface area. Numerous investigations have been undertaken of the adsorptive properties of a wide variety of such carbons with Brunauer-Emmett-Teller (BET) areas between, say, $\sim 5 \, \mathrm{m^2 \, g^{-1}}$ and $\sim 3000 \, \mathrm{m^2 \, g^{-1}}$. Carbons at the lower end of this wide range are usually regarded as non-porous powders, but this is not always true. As with other materials, an active carbon with a specific surface area of $> \sim 100 \, \mathrm{m^2 \, g^{-1}}$ is almost invariably to some extent porous.

The term *activated carbon* is used more specifically to indicate a highly porous material (generally microporous) produced from a carbon-rich precursor by some form of chemical or physical activation. Most commercial grades of activated carbon possess large internal areas located within disordered micropore structures. However, carbon adsorbents and catalyst supports are now available in the form of activated carbon fibres (ACFs) and cloth, monoliths, gels and membranes. Furthermore, the pore size can be 'finely tuned' within narrow micropore, mesopore or macropore limits.

Carbon blacks are composed of small globular particles and are important industrial products. Although inevitably of much smaller specific surface than activated carbons, as adsorbents they are important research and reference materials. After graphitization at high temperature, the spheroidal particles become polyhedral with the development of a remarkably uniform surface structure. The *graphitized blacks* are of particular value as model adsorbents.

In addition to the above poorly ordered carbons, there are some well-defined carbon structures, which are now attracting much attention. These include the *fullerenes*, *carbon nanotubes* (CNTs) and *ordered mesoporous carbons* (OMCs). The long-range order and unique properties of these carbons offer new opportunities in both fundamental and applied research.

Low-temperature nitrogen adsorption at 77 K is generally used as the first stage in the characterization of the surface area and porosity of all active carbons. Progress has been made in the interpretation of the adsorption data by the application of molecular simulation (MS) and density functional theory (DFT), but in routine work (e.g. for quality control or in a patent specification), reliance is still placed on the use of 'classical' methods of isotherm analysis – particularly, the BET-method for surface area determination and the Dubinin-Radushkevich (DR) and Barrett–Joyner–Halenda (BJH) methods for micropore and mesopore size analysis. As explained in other chapters, these procedures all have their limitations. Indeed, even the application of DFT and MS involves various assumptions, which should be taken into account. By applying an empirical method (e.g. an α_s or comparison plot), one can assess the effective area of the external surface (i.e. outside the micropores) and the micropore capacity and also check the validity of the BET-area.

To obtain an understanding of the behaviour of an active carbon, it is generally necessary to employ other measurements in addition to low-temperature nitrogen adsorption. For some purposes, probe molecules of different size, shape and polarity are useful. Also, it may be necessary to undertake high pressure or dynamic measurements, or to investigate adsorption at the carbon/liquid interface. Selection of the most appropriate characterization procedures is thus dependent on the purpose of the work or the application.

As in other chapters of this book, our approach here is necessarily selective. In a single chapter, it would be impossible to deal adequately with all aspects of the published research on active carbons. Our main aim is to discuss the interpretation of physisorption isotherms and other adsorption data on some typical active carbons. A short account is first given of the preparation, properties and applications of the principal types of active carbons.

10.2 Active Carbons: Preparation, Properties and Applications

The preparation and properties of various active carbons are described briefly in this section. More detailed information is to be found in a number of excellent books and reviews (see Bansal et al., 1988; Patrick, 1995; McEnaney, 2002; Rodriguez-Reinoso, 2002; Bandosz et al., 2003; Marsh and Rodriguez-Reinoso, 2006; Bottani and Tascon, 2008; Palmer and Gubbins, 2012; Tascon, 2012).

10.2.1 Graphite

As is well known, graphite is the stable form of elemental carbon at ambient temperature and pressure. Its anisotropic atomic structure is made up of planar graphene layers, composed of interlinked six-membered rings of carbon atoms (see Figure 10.1). There are two forms: in the normal α (or hexagonal) form, the graphene layers alternate, ABAB$\cdots$, whereas in the β (rhombohedral) form, the stacking sequence is ABCABC$\cdots$. In each case, the carbon–carbon distance within the graphene layer is 0.142 nm and the interplanar spacing is normally ca. 0.335 nm. This highly characteristic layer structure is associated with the sp^2 electronic configuration. Thus, three of the outer electrons of the C atom hybridize to form localized sp^2 bonds (i.e. three trigonal σ-sp^2 planar orbitals), while the remaining electrons are delocalized in the form of π molecular orbitals.

The graphene layer structure accounts for the strong directional dependence of the electrical and thermal conductivity of graphite and its high thermal stability. The easy cleavage along the basal plane and its readiness to form intercalation compounds are also characteristic properties. Unlike diamond, graphite is a relatively soft material, which can easily be broken down into small flaky particles.

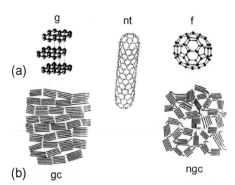

Figure 10.1 (a) Structures of graphite (g), carbon nanotube (nt) and fullerene (f).
(b) Graphitizing carbon (gc) and non-graphitizing carbon (ngc).
After Franklin (1951).

There is still some confusion about the use of the term 'graphitic carbon'. According to the IUPAC criterion (see Fitzer et al., 1995), the graphene layers must be arranged parallel to each other in a three-dimensional crystalline network. It follows that the term graphitic carbon should not be applied to any carbon, which does not possess a long-range graphitic structure. In her pioneering work on the structure of carbons, Franklin (1951) introduced the terms 'graphitizing' and 'non-graphitizing' carbons (see Figure 10.1b), the former being capable of transformation into graphite at high temperature. Although now recognized as an oversimplification (see Bandosz et al., 2003), this model is still widely accepted as providing a useful distinction between two extreme types of poorly ordered carbons. In the case of a graphitizing carbon, the small structural units containing graphitic crystallites are roughly parallel to each other allowing a relatively facile cross-linking to occur. This transformation is not possible in the case of a non-graphitizing carbon because the random orientation of the crystallites inhibits the development of long-range order.

As explained in Chapters 5 and 6, the graphitic structure has been extensively used in the development of physisorption theory, modelling and MS (see Steele, 1974; Ravikovitch et al., 2000; Do and Do, 2002, 2003; Bandosz et al., 2003; Bock et al., 2008; Bojan and Steele, 2008; Do et al., 2008; Olivier, 2008; Neimark et al., 2009). The exposed graphitic surface is largely made up of the graphene basal plane. The absence of energetic heterogeneity therefore simplifies the calculation of the gas–solid interaction energy.

Natural and synthetic graphites generally contain defects within their structure. Porosity is dependent on the manufacturing process and may be as high as 30%. It can be in the form of slit-shaped cracks or roughly spheroidal pores and these two forms may be connected so that the overall pore structure is an irregular network (Patrick and Hanson, 2002).

Graphite was one of the first adsorbents to be studied by Brunauer and his co-workers (see Brunauer, 1945) in their early work on the application of the BET-method. The graphite powder used by Harkins and Jura (1944a,b) in the development of their 'absolute method' for surface area determination had a BET-nitrogen area of $4.3 \, \text{m}^2 \, \text{g}^{-1}$ in good agreement with $4.4 \, \text{m}^2 \, \text{g}^{-1}$, the value obtained from the immersion method. This agreement was taken as confirmation that their sample of graphite was non-porous.

The surface area of graphite can be considerably enlarged in two ways: either by milling or by exfoliation. Fine grinding inevitably results in an increase in surface area, but it also leads to an increase in the degree of surface heterogeneity (Olivier, 2008). The effect of prolonged grinding was investigated by Gregg and Hickman (see Gregg, 1961). After prolonged ball milling, the BET-nitrogen area was found to increase enormously, finally becoming $\sim$600 m^2 g^{-1}, and accompanied by the development of a large pore volume.

In an investigation by de Boer and his co-workers (see van der Plas, 1970), two samples of milled graphite, which had BET-areas of 75 and 218 m^2 g^{-1}, were reported to be microporous. Special techniques were used by Thomy and Duval (1969) to prepare exfoliated graphites with a surface area of $\sim$20 m^2 g^{-1}. This form of exfoliated graphite was evidently non-porous with a remarkably homogeneous basal plane surface (Thomy et al., 1972) as indicated by the stepwise character of the krypton isotherm in Figure 5.5.

Porosity must be taken into account in the many applications of graphite (Patrick and Hanson, 2002). The presence of slit-shaped cracks adversely affects its mechanical performance and conductivity. The pore structure also has an important role in gasification reactions and must be controlled in the development of graphite as a nuclear moderator. For some other applications, the nature of the surface is of primary importance. The performance of the graphite anode in lithium ion batteries is known to be dependent on a reduction in the fractional area of the basal plane (see Olivier, 2008).

10.2.2   Fullerenes and Nanotubes

Other well-defined polymorphic forms of carbon are the fullerenes and CNTs (see Minett et al., 2002; Bottani and Tascon, 2008). Their identification and characterization followed the major discovery of the buckminsterfullerene molecule, C_{60}, by Kroto et al. (1985). As is well known, this spheroidal molecule (diameter $\sim$1 nm) has a soccerball-like structure, being composed of a stable cage of carbon atoms arranged in the form of six- and five-membered rings (see Figure 10.1). The name 'buckminster fullerene' (now popularly reduced to 'buckyball') was inspired by the geodesic dome designed by the architect R. Buckminster Fuller. Although at first regarded as a rarity, it was soon discovered that C_{60} could be readily synthesized (along with C_{70}). The availability of gram quantities has allowed the chemical and physical properties of the C_{60} fullerene to be studied in some detail.

The fullerenes are less stable than graphite and their structures and electronic properties are quite distinctive. Thus, the aromatic character of the graphitic basal plane is modified by a reduction in the degree of π bonding and an increase in the σ character. The resulting curvature of the carbon layers leads to the formation of C_{60} and other giant molecules, including nanotubes. An extensive range of multilayered fullerenes include 'carbon onions' or 'bucky onions', which have been isolated from soot (Ugarte, 1995). At moderate temperatures and pressures, molecular solids termed fullerites are formed by the aggregation of fullerene molecules.

Many experimental investigations have been undertaken of the physisorption of gases by fullerenes of questionable purity (see Suarez-Garcia et al., 2008). In the work of Ismail and Rodgers (1992) on C_{60} crystals, a very large difference was reported between the BET-areas available at 77 K for nitrogen and krypton adsorption (i.e. ~ 5 m^2 g^{-1}) and at 298 K for carbon dioxide adsorption (i.e. ~ 130 m^2 g^{-1}). Obviously, this discrepancy cannot be explained in terms of molecular packing within the adsorbed monolayer. Instead, it was much more likely to be due to the greater translational energy of CO_2 at the higher temperature, which allowed it to penetrate into regions which were inaccessible to N_2 and Kr. At first, it was not clear whether these regions were well-defined micropores or simply vacancies (defects) between the individual C_{60} molecules.

The microporous character of a particular batch of C_{60} powder was confirmed by Kaneko et al. (1993). Analysis of the Type I nitrogen isotherms by the α_s-method gave an external area of ca. 1 m^2 g^{-1} (the BET-area being 24 m^2 g^{-1}). This value of the external area was consistent with the crystal size determined by electron microscopy. The porous nature of the C_{60} crystals was examined in more detail by Kaneko and others (Setoyama et al., 1996; Thess et al., 1996). The C_{60} powder was recrystallized from carbon disulfide and then studied before and after annealing by heat treatment. Nitrogen adsorption measurements revealed that the recrystallized material was both microporous and mesoporous; the mesoporosity was completely removed by heat treatment, but the sample remained microporous. Since the micropores and the C_{60} molecules appeared to be of similar size, it was concluded that the micropores were largely in the form of molecular defects and lattice vacancies.

A few comparisons have been made between physisorption isotherms determined experimentally and by GCMC simulation. For example, in the work of Martinez-Alonso et al. (2000, 2001) on high-purity C_{60}, excellent agreement was obtained between the corresponding N_2, Ar and CO_2 isotherms. Further work (see Suarez-Garcia et al., 2008) was undertaken on the effect of hypothetical defect structures on the degree of energetic heterogeneity. Other theoretical studies have explored the dependence of adsorption energetics on the topography of C_{60} and C_{70} crystals (e.g. Arora and Sandler, 2005). The difference in the surface curvatures and electronic configurations of the two crystal structures and of graphite would appear to appreciably affect the nitrogen adsorption energies at 77 K and low surface coverage.

The interactions between organic molecules and fullerenes have been investigated with the aid of inverse gas chromatography (IGC) by several groups (Papirer et al., 1999; Bottani and Tascon, 2004). As might be expected, the dispersion adsorption potential of fullerene crystals is much lower than that of graphitized carbon black – as indicated by the smaller differential energies of adsorption for the alkanes. However, the stronger specific interactions shown by polar molecules appeared to be associated with the electron-donor character of the C_{60}.

The feasibility of using fullerenes (or fullerites) for hydrogen storage has been examined by several groups (see Suarez-Garcia et al., 2008). Of course, at ambient temperatures, the amount of hydrogen physisorbed on the external surface of C_{60} crystals is likely to be much too small to be useful. However, hydrogen sorption could take place in another way. Sorbed hydrogen could be located in interstitial lattice sites in the form of a 'solid solution' or the sorbed state might involve the chemical bonding of hydrogen atoms to carbon atoms in the fullerene cage. The interstitial uptake is likely to be limited, but a fully hydrogenated state of $C_{60}H_{60}$ would, in principle, offer a possible way of meeting the goal of an effective uptake of ca. 7 wt% H_2. For a number of reasons, however, it is doubtful whether fullerenes could ever be commercially viable for hydrogen storage, but the scientific interest in hydrogen sorption is likely to continue.

CNTs were first identified by Iijima (1991). They are found in various kinds of soot and are formed by the incomplete combustion of diesel oil, etc. Various methods are now available for the preparation of CNTs (see Bandosz et al., 2003), but the most common synthetic route is probably still the carbon-arc method (Ebbesen and Ajayan, 1992). Purification is required since the raw material is usually contaminated with amorphous carbon and graphite. Separation can be achieved by the application of size-exclusion chromatography (Minett et al., 2002).

Multiwall carbon nanotubes (MWCNTs) are made up of a number of concentric C nanotubes (up to ~80), with external diameters of up to ~55 nm and correspondingly inner diameters of ~2–50 nm. Defects are usually present and MWCNTs are rarely perfect straight cylinders. Thus, tangled aggregates are often mixed with nano-sized particles. Remarkably, strong MWMTs of high quality can now be produced by CVD methods (see Minett et al., 2002). However, because of their hydrophobic nature, they are not easily dispersible. In a recent investigation by Bradley et al. (2012), a significant increase in the surface polarity, as indicated by the enhanced sorption of water vapour, has been achieved by carefully controlled hydroxylation with little change in the texture of the material.

Single-wall carbon nanotubes (SWCNTs) are normally produced in the form of tangled mats or bundles – 'ropes' (Journet et al., 1997; Lambin et al., 2000). A particle-coagulation spinning process was developed by Poulin et al. (2002) to produce fibres, which consist of interconnected networks of polymer chains and SWNTs. Bundles of SWNTs are 10–30 nm diameter and are grouped together within filaments of ~0.2–2 μm diameter. The BET-nitrogen area of this type of material was found to be 160 $m^2\ g^{-1}$ and DFT and comparison plot analysis revealed the presence of a wide range of mesopores and the absence of microporosity (Neimark et al., 2003).

The selective opening of CNTs was first accomplished by Ajayan and Iijima (1993). This method, which involved the use of molten lead, has been superseded by other procedures including refluxing with nitric acid. Open SWCNTs have extremely high BET surface areas, which in principle could amount to 2630 $m^2\ g^{-1}$ (taking account of the exposure of all the component C atoms). Once opened, the CNT can be filled with nanoparticles of various kinds. This offers the possible development of novel adsorbents, catalysts and composites.

Problems involved in determining the internal and external surface areas of the nanotube bundles were discussed by Ohba and Kaneko (2002), Esteves et al. (2009) and Furmaniak et al. (2010). It was pointed out that the BET-area is likely to be misleading because of the presence of microporosity. In addition, the usual assumption of a close-packed monolayer is strictly only applicable to a 'flat' surface. The pronounced surface curvature of narrow nanotubes must to some extent affect the monolayer structure. A related problem is the degree of energetic heterogeneity associated with the different accessible areas and pores in the CNT microstructures.

In another recent investigation, Yamamoto et al. (2011) were able to prepare several different bundle structures under carefully controlled conditions. Nitrogen comparison plots provided an approximate evaluation of the micropore volume and external area of each bundle.

As already indicated, in the context of clean energy requirements, the storage of hydrogen has received a great deal of attention. Although the situation is not at all clear, the prospects for the application of CNTs seem rather brighter than for fullerenes. The unexpected results reported by Chambers et al. (1998) stimulated much interest and appeared to offer the possibility of achieving a storage capacity of over 20% in 'carbon nanofibres' at room temperature and pressures ~120 bar. This remarkable level of hydrogen uptake has not been confirmed by any other group and it now seems that these findings are of questionable reliability. Indeed, much lower uptakes by SWNTs have been reported, but the amount appears to be highly dependent on the conditions of preparation and pre-treatment (Lee et al., 2001; Johnson and Cole, 2008).

The work of Yamamoto et al. (2011) on the effect of nanoscale curvature and bundle structure may help to explain why it has been difficult to reproduce the adsorptive properties of SWCNTs. It appears that the internal tube walls have a higher adsorption affinity for supercritical H_2 and CH_4 than the external walls. Furthermore, in the close-packed bundle, the highest interaction energy is located in the interstitial pore space. Of course, this analysis assumes a rigid assemblage of equal-sized CNTs. However, it is known (Futaba et al., 2006) that there is normally a wide range of SWCNT tube diameters (e.g. 1–5 nm), that the packing within a bundle is not uniform and that the bundle is not really a rigid structure (Wesolowski et al., 2011).

10.2.3 Carbon Blacks

Carbon blacks are of great industrial importance. They are the preferred black pigments and fillers in printing inks, paints, plastics and pneumatic tyres. These and other applications are dependent on the control of the surface chemistry together with the particle size and aggregate structure. Interaction of the particle surface with a dispersion medium involves adsorption at the solid/liquid interface, which in turn affects the quality and stability of the dispersion. In this respect, changes in the surface chemistry can have a major effect on the performance of a carbon black. For example, a high concentration of chemisorbed oxygen renders the surface more hydrophilic and leads to improved dispersion in an aqueous medium.

Carbon blacks are produced by the controlled pyrolysis or combustion of a variety of organic precursors (solids, liquids or gases). They are composed of discrete or aggregated spheroidal particles, generally within the size range of 10–1000 nm (see Figure 10.2). The carbon atoms are arranged in localized and distorted graphene layers. These layers are disordered in the centre of the particles but may tend to lie parallel to the particle surface (Medalia and Rivin, 1976). The surface is not atomically smooth and is pictured as overlapping 'quasi-graphitic scales' (Donnet, 1994).

Various manufacturing processes are employed in the production of different grades of carbon black (Medalia and Rivin, 1976). In the oil furnace process,

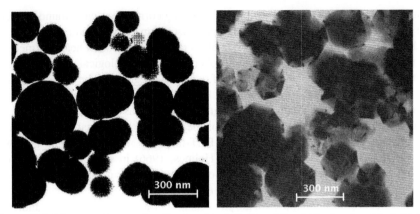

Figure 10.2 Electron micrographs of carbon black (left) and Graphitized carbon black (right). After Medalia and Rivin (1976).

petroleum residue is subjected to partial combustion and cracking in a refractory furnace. Colloidal carbon is produced in the form of a hot smoke of small globular particles (typically ~30 nm), which on cooling undergo aggregation to give branched chain-like structures. The loose aggregates can be converted into well-defined mesoporous materials by compaction (Kiselev, 1968).

Channel blacks are composed of very small particles, with BET-areas ~100 m² g⁻¹. Many early physisorption studies (e.g. by Joyner and Emmett, 1948) were undertaken on Spheron 6, a channel black having a BET-nitrogen area of ca. 120 m² g⁻¹. At that time, it was assumed that the carbon particles (of ca. 25 nm diameter) were non-porous, but later work (Carrott et al., 1987; Sing, 1994) revealed that about 20% of the BET-area could be attributed to micropore filling. Stoeckli et al. (1994a,b) used a combination of techniques, including the physisorption of different vapours and immersion calorimetry, to study the surface area and micropore structure of a series of carbon blacks. It was concluded that within certain limits it is possible to make use of the same standard techniques as those employed for characterizing activated carbons (Stoeckli et al., 1994a,b).

Globular particles of larger size (~200 nm) and lower BET-area (~10 m² g⁻¹) are produced by the thermal process, which involves the thermal cracking of natural gas. The thermal blacks (e.g. Sterling FT) have lower ash content than the furnace blacks: for this reason and because of the discrete nature of their primary particles, they are now often favoured for fundamental adsorption studies (Olivier, 2008).

The surface properties of ungraphitized carbon blacks are to a great extent influenced by the presence of surface oxides and oxygen-containing groups (Medalia and Rivin, 1976; Boehm, 2008). Thus, the acidity of aqueous dispersions of carbon blacks is dependent on their oxygen content and this has been attributed to the presence of carboxyl and other functional groups such as lactols and phenols (Boehm, 1966, 1994, 2008). Carbon blacks with low oxygen contents generally exhibit basic properties. This behaviour is to some extent associated with the basic character of the aromatic ring structure. The physisorption specificity associated with the surface oxygen content was studied by Andreu et al. (2007). Strong correlations were found between the surface oxygen content and the adsorption isotherms of the lower alcohols.

Progressive graphitization of a carbon black occurs as a result of heat treatment (at temperatures $>1000\,°C$) in an inert atmosphere. Some technological grades of ungraphitized carbon blacks have BET-areas of over $200\,m^2\,g^{-1}$, but the surface areas of the graphitized materials are smaller ($<100\,m^2\,g^{-1}$). It was originally shown by Schaeffer et al. (1953) that the degree of graphitization was a function of the temperature of heat treatment and also depended on the nature of the original black. Graphon, the graphitized form of Spheron with a BET-nitrogen area of ca. $90\,m^2\,g^{-1}$, was used by Kiselev (1968) and others (see Holmes, 1966) in some of the first well-documented calorimetric measurements of the energetics of physisorption. In its uncompacted state, Graphon does appear to be non-porous.

The effects of heat treatment on the physical properties of Vulcan 3 were studied in some detail by Everett and Ward (1986). An unexpected finding was that the BET-nitrogen area exhibited a sharp peak in the range of $1035-1050\,°C$. The steady decrease in area between 1100 and $2700\,°C$ was in accordance with earlier work, but the peak appeared to be due to the removal of volatile material and the development of some surface porosity, which is removed at higher temperature.

On graphitization, the relatively large particles in the thermal blacks are converted into well-defined polyhedra with a high proportion of their surface in the form of graphitic basal planes (see Figure 10.2). Because of the uniformity of their surface, graphitized blacks have been adopted as model adsorbents in many fundamental investigations of adsorption from both the gas phase and the liquid phase (see Parkyns and Sing, 1975; Gregg and Sing, 1982; Bottani and Tascon, 2008; Denoyel et al., 2008).

Various attempts have been made to use carbon blacks as non-porous reference materials for the determination of standard adsorption data (see Sing, 1994; Carrott et al., 2001; Gardner et al., 2001; Guillot and Stoeckli, 2001). The globular shape of carbon particles in ungraphitized blacks would seem to make them attractive reference materials. In practice, however, a number of difficulties are encountered. The fact that a particular carbon black gives a reversible Type II isotherm may be deceptive: more detailed analysis is required to establish the absence of microporosity. Also, aggregation of the small particles can lead to inter-particle capillary condensation at high p/p^o, which in turn can result in irreversible shrinkage (compaction) of the aggregate. Another difficulty is the heterogeneous nature of the particle surface, which must vary from one type of black to another.

In spite of these problems, useful progress has been made in the determination of standard physisorption isotherms of a number of gases and vapours on several commercial grades of ungraphitized carbon black, including Vulcan, Elftex and Sterling (Carrott et al., 1987, 2000, 2001).

10.2.4 Activated Carbons

The recorded history of the medicinal and metallurgical uses of charcoal goes back to the ancient Egyptians, Greeks and Romans, but it was not until the latter part of the eighteenth century that the adsorptive properties of charcoal were first noted (see

Forrester and Giles, 1971; Derbyshire et al., 1995). The industrial production of activated charcoal was started in the early years of the twentieth century and the scale was considerably increased during World War I. The activated material was produced in a granulated form from wood chips treated with zinc chloride. Cheap and readily available materials such as wood, peat, coal and nut shells are still widely used for the large-scale production of activated carbons (Baker, 1992; Derbyshire et al., 1995; Rodriguez-Reinoso, 2002; Marsh & Rodriguez-Reinoso, 2006).

The molecular structure of activated carbons has been the subject of many investigations (see Bandosz et al., 2003; Bottani and Tascon, 2008; Palmer and Gubbins, 2012) and numerous structural models have been proposed (see Bandosz et al., 2003; Palmer and Gubbins, 2012). For the computation of fluid–carbon interaction energies and in the application of MS or DFT, it is usually assumed that the pores are slit-shaped with walls composed of stacked parallel graphene layers. Each graphene layer is generally assumed to be a homogeneous hexagonal arrangement of carbon atoms – without any defects or chemical impurities (see Bandosz et al., 2003; Do and Do, 2003; Do et al., 2008).

10.2.4.1 Carbonization

Organic materials undergo pyrolytic decomposition when heated in an inert atmosphere. Polyaromatic ring structures are developed in the early stages of carbonization. As the heat-treatment temperature (HTT) is increased, the solid char or coke begins to acquire short-range order with the formation of distorted graphitic lamellae. In addition, localized and anisotropic densification leads to the development of free space between the lamellae.

Wood and other naturally occurring precursors are composed of three-dimensional polymeric networks of cellulose (CEL) and lignin. Pyrolysis at temperatures $<700\ ^\circ C$ results in the loss of water, carbon dioxide and a wide range of organic molecules (e.g. alcohols, ketones, acids). There follows a progressive increase in the C/H and C/O ratios, but hetero atoms (O, H, Cl, N, S, etc.) remain chemically bonded at the edges of the aromatic macromolecules and these are finally transformed into surface complexes. Some of the purest chars are derived from sugars, regenerated CEL and certain synthetic polymers, but even these do not possess completely clean carbon surfaces unless they have been subjected to special treatment.

With the aid of high-resolution electron microscopy and other techniques, Oberlin and her co-workers followed the changes in microstructure which accompany carbonization (Bonijoly et al., 1982; Rouzard and Oberlin, 1989; Oberlin et al., 1999). The basic structural units (BSU) were identified as small assemblages of the polyaromatic rings. Increase in HTT caused the BSU to form distorted stacks (or columns), but crystal growth of the long-range graphitic structure required the removal of various defects. The 'non-graphitizable' chars produced from sugar contained highly disordered stacks with defects in the form of functional groups firmly attached to the BSU. In contrast, 'graphitizable coke' derived from anthracene appeared to undergo a progressive ordering of the BSU stacks with the formation of the long-range graphitic structure.

In a char, which has been subjected to HTT of, say, 800–900 °C, the interstices between the BSU have the dimensions of narrow micropores, but generally they are not easily accessible to adsorptive molecules.

10.2.4.2 Activation

Activation of the char is undertaken to improve the accessibility of the pore structure and also, if required, to increase the pore width and pore volume. Changes in the BET-area, pore dimensions and cell wall thickness as a function of HTT of thermal and acid-treated carbons were studied by Jagtoyen and Derbyshire (1993). In this work, an initial contraction was followed by a considerable expansion, which accompanied the development of the internal surface area. At HTT >ca. 450 °C, there was a reduction in surface area and pore volume along with a small contraction.

Activation always involves some form of chemical attack. However, *chemical activation* is a term often used to indicate some form of chemical impregnation before heat treatment, whereas *thermal activation* (also called *physical activation*) implies heat treatment of the char in a mildly reactive atmosphere (e.g. CO_2). Micropores are opened and widened as a result of progressive activation. The complex reactions involved are catalysed by various inorganic compounds in the precursor (see Rodriguez-Reinoso, 2002).

Chemical activation entails pre-treatment of the precursor, which is often a ligno-cellulosic material, with a chemical agent such as zinc chloride or phosphoric acid. Although $ZnCl_2$ was once the preferred reagent, its use is now in decline in Europe and North America (Rodriguez-Reinoso, 2002). For various reasons, the most popular dehydrating agent is phosphoric acid, but KOH and other reagents can be used to pro-duce special grades of superactive and mesoporous carbons (Freeman and Sing, 1991; Silvestre-Albero and Rodriguez-Reinoso, 2012).

The extent of the activation reaction is characterized by the *burn off* as deter-mined by the change in mass of the char, expressed as the percentage weight loss of the carbonized material as a result of HTT under controlled conditions. With some chars, the burn off increases linearly with the time of HTT at a constant temperature. This form of linear dependence has been reported by Rodriguez-Reinoso (1986) for the activation of carbonized olive stones and almond shells in CO_2 at temp-eratures around 850 °C. The extensive linear relationship was a clear indication that the reaction rate was almost constant over a very wide range of burn off (i.e. 8–80%).

Activation in CO_2 is often used on a laboratory scale, but steam activation is generally favoured for the large-scale production of most activated carbons of indus-trial importance (Baker, 1992). The steam reaction is considerably faster than the carbon dioxide reaction (Wigmans, 1989). Steam activation is normally carried out at temperatures of 750–950 °C. Direct contact between oxygen and carbon must be avoided, as at these temperatures, oxygen would aggressively attack the carbonized material.

10.2.4.3   Structure and Properties

Most activated carbons are to a large extent microporous, but for some purposes, it is desirable to extend the range of pore size into the mesopore or macropore range. Progress in this direction has been made by the use of special pre-treatment procedures and careful control of the conditions of carbonization and activation (Freeman and Sing, 1991; Jagtoyen and Derbyshire, 1993; Rodriguez-Reinoso, 2002; Silvestre-Albero and Rodriguez-Reinoso, 2012). It is not difficult to prepare a mixed system of micropores and mesopores, but the production of a well-defined mesopore structure is more demanding.

Many investigations have been made of the surface composition and structure of active carbons (Boehm, 1966, 1994; Bansal et al., 1988; Schlögl, 2002). It has been known for many years that activated carbons appear to exhibit both acidic and basic properties (Boehm, 1966, 2008; Puri, 1970; Bansal et al., 1978a,b). The acidic properties have been attributed to the presence of phenolic and carboxylic acid groups, whereas the basic character has been more difficult to explain (see Bansal et al., 1988). Aldehydes, ketones, lactones, and quinones have been identified, but the overall surface composition is likely to be complex and not easy to establish.

Some of the sp^2 and sp^3 defect sites may be occupied by H atoms and are thus rendered hydrophobic – as is the basal graphene surface. The remaining smaller part of the surface is hydrophilic, and it is generally agreed that this is due to the presence of carbon–oxygen complexes, which have a great influence on the chemical reactivity of the carbon (Schlögl, 2002).

In spite of the experimental difficulty of using infrared spectroscopy to study the surface structure of carbon, IR measurements (especially in the form of FTIR) have revealed important information on the changes in surface chemistry produced by oxidation and substitution reactions. As a result of the systematic FTIR studies of Starsinic et al. (1983), van Driel (1983) and others (see Zawadzky, 1989; Boehm, 1994), considerable progress has been made in the assignment of the IR bands.

10.2.4.4   Applications of Activated Carbons

Activated carbons are the most widely used of all the relatively cheap general-purpose adsorbents. The obvious reason for this universal importance is their large internal surface, which is generally located within a wide range of micropores. A network of macropores provides the pathway for molecules to be transported to the micropores. For many applications, such as the removal of impurities from gases and liquids, the micropore structure provides both a large adsorption capacity and a high adsorption affinity. However, it is well known that the surface chemistry also plays an important role in controlling the properties, and therefore the applications, of activated carbons (see Rodriguez-Reinoso, 2002; Boehm, 2008).

Certain commercial grades, in powder or granular form, are extensively employed for the treatment of potable water (see Newcombe, 2008). For the optimum purification of drinking water, it is clearly necessary to use an activated carbon of adequate pore size to achieve the removal of fairly large solute molecules such as pesticides and toxins. For some other applications, including solvent recovery and gas separations, microporous carbons are required.

Organic solvent recovery is probably the most important gas-phase application since it is used in the production of plastics, rubber, synthetic fibres, printing inks, etc. For this purpose, the adsorbent is usually in the form of hard granules or pellets (i.e. dust-free). Other gas-phase applications of microporous carbons include respiratory protection, gas separation, gas storage, the removal of toxic components from exhaust gases and the recovery of H_2S and CS_2. Odour control is another application of increasing importance.

Miscellaneous applications of activated carbons in the food and beverage industries include the decolourization of sugar and the removal of caffeine and chlorine. The use of activated carbons in the recovery of gold is also of long-standing importance (see Bansal et al., 1988). Numerous small-volume applications include the medical treatment of blood and electroplating and dry cleaning.

10.2.5 Superactive Carbons

It is not difficult to produce activated carbons with BET-areas of at least $1000 \text{ m}^2 \text{ g}^{-1}$. Indeed, the adsorbent activity can be increased well beyond this level by chemical activation with phosphoric acid (Baker, 1992; Rodriguez-Reinoso, 2002) followed by the controlled heat treatment of certain impregnated precursors such as hardwoods (Jagtoyen and Derbyshire, 1993) or olive stones (Molina-Sabio et al., 1995; Rodriguez-Reinoso et al., 1995). Typical commercial products have BET-areas in the range of $500-2000 \text{ m}^2 \text{ g}^{-1}$(Baker, 1992), but it is evidently more technologically demanding to manufacture 'superactive carbons' with significantly larger BET-areas.

A novel method of chemical activation for producing some of the most active carbons so far developed was disclosed by the AMOCO (Standard Oil Company) scientists Wennerberg and O'Grady in 1978. These superactive carbons were prepared on a small scale by the high-temperature heat treatment of petroleum coke or coal mixed with excess amounts of potassium hydroxide. A typical research sample, PX21, had a BET-nitrogen area of $\sim 3700 \text{ m}^2 \text{ g}^{-1}$ and a total micropore volume of $\sim 1.75 \text{ cm}^3 \text{ g}^{-1}$ (see Atkinson et al., 1982).

The first commercial plant for the production of the AMOCO superactive carbons was constructed by the Anderson Development Company, Michigan. Their products (e.g. AX21) had BET-areas in the range of $2800-3500 \text{ m}^2 \text{ g}^{-1}$ and total pore volumes of $1.4-2.0 \text{ cm}^3 \text{ g}^{-1}$ (see Carrott et al., 1987, 1988a,b, 1989). The AMOCO process was further developed and extended by the Kansai Coke and Chemicals Company of Japan. Their extremely active carbons, marketed as powdered and granular forms of MAXSORB, are produced by the KOH activation of petroleum coke, coal or coconut shell char.

In the preparative procedure described by Otowa et al. (1993), mixtures of petroleum coke and excess amounts of KOH were dehydrated at $400 \text{ }^\circ\text{C}$ and then heated under nitrogen flow at $600-900 \text{ }^\circ\text{C}$. The activated materials were washed thoroughly with water to remove KOH and K_2CO_3. Certain batches of MAXSORB prepared in this manner were reported to have BET-areas >3100 m^2 g^{-1} and total pore volumes >2.5 cm^3 g^{-1} (Otowa, 1991; Otowa et al., 1993, 1996). In a systematic investigation of the activation process, Otowa et al. (1993, 1996) found that the K$_2$O, formed by the dehydration of KOH, reacted with CO$_2$ (produced by the water gas shift reaction) to give K$_2$CO$_3$. Intercalation of metallic potassium, which was also formed at temperatures >700 °C, appeared to be responsible for the drastic expansion of the carbonized material and hence the creation of large internal surface areas and pore volumes (McEnaney, 2002).

In view of the complexity of the system, it is not surprising that the untreated MAXSORB had a high concentration of surface functional groups (—COOH, —OCO and —OH, in similar proportions). However, it appears that these could be largely removed by further heat treatment in an inert atmosphere at 700 °C (Otowa et al., 1996). This form of after-treatment did not affect the excellent correlation obtained between methylene blue adsorption and the BET-area, but it did lead to a considerable improvement in the breakthrough performance for the removal of CHCl$_3$ from water. However, the presence of the surface functional groups was found to be beneficial for some applications (e.g. for the double layer capacitance storage of electricity).

10.2.6   Carbon Molecular Sieves

There is some confusion in the literature over the use of the term 'carbon molecular sieve' (CMS), which has been applied to two different types of pore structure: (1) an assembly of inter-connected ultramicropores ($w <$ ca. 7 nm) and (2) an assortment of pores of different sizes with molecular-sized entrances.

The synthesis of a CMS was first reported by Walker and his co-workers (Lamond et al., 1965; Walker and Janov, 1968; Walker and Patel, 1970; Nandy and Walker, 1975) and involved the use of polymers as precursors (e.g. polyacrylonitrile (PAN), polyvinylidene chloride and co-polymers). Such activated chars are almost pure with a well-defined ultramicropore structure, but they are expensive. Other less-expensive precursors such as anthracite, synthetic polymers and lignocellulosic materials can be used, provided the carbonization and activation stages are very well controlled to give a low burn off (Freeman and Sing, 1991; Rodriguez-Reinoso, 2002).

In the case of the second group of CMS, the pore entrance size is controlled by the deposition of carbon by the cracking or pyrolysis of organic vapours. The application of this type of material for gas separation is then dependent on molecular sieving or kinetic selectivity (i.e. diffusion control) together with equilibrium capacity (Sircar, 2008). The commercial recovery of nitrogen from air operated by Bergbau-Forschung involved the development of this type of CMS (Schröter and Jüntgen, 1988).

Superactive grades of CMS (with high BET-areas ~3000 m^2 g^{-1} and narrow distributions of ultramicropores) can be prepared by the pyrolysis of mesophase pitch obtained from petroleum residues (Wahby et al., 2010). The possible use of this type of CMS for CO$_2$ capture and storage is of current interest (see Wahby et al., 2010; Silvestre-Albero and Rodriguez-Reinoso, 2012). It is evident that high CO$_2$ adsorption capacity is dependent on both the pore size and the surface chemistry. Tailored ultramicroporous carbons appear to offer a number of advantages over other materials (e.g. MOFs and zeolites): the CO_2 adsorption is rapid and reversible over wide ranges of pressure.

As Carruthers et al. (2012) have pointed out, it is important to keep in mind that the effectiveness of a CMS for CO_2 capture is governed by the affinity of adsorption which in turn is critically dependent on the pore width in relation to the dimensions of the CO_2 molecules. Other factors to be taken into account include the effective dimensions of the pore entrances, which control the ease of entry and exit and the extent and spatial distribution of larger pores.

10.2.7   ACFs and Cloth

The first high-strength carbon fibres were produced in the 1950s (see Donnet and Bansal, 1990). The early carbonized products were rayon based, but it was soon found that the mechanical properties and the carbon yield could be improved by using PAN as the precursor. Also, less-expensive fibres of somewhat lower strength and modulus could be made from various other precursors including petroleum pitch and lignin. However, cotton and other forms of natural CEL fibres possess discontinuous filaments and the resulting mechanical properties are consequently found to be inferior to those of the rayon-based fibres.

It was not long before the first ACFs were developed (see Mays, 1999; Rodriguez-Reinoso, 2002). In the work of Economy and Lin (1971, 1976), highly porous carbon fibres were prepared from Kynol, a fibrous phenolic precursor. Carbonization was carried out in nitrogen at 800 °C and activation in steam at 750–1000 °C. The products appeared to be predominantly microporous and were found to be effective for the removal of low levels of certain pollutants (e.g. phenol and pesticides) from air or aqueous solutions.

Extensive studies have been undertaken by Kaneko and his co-workers of the properties of ACFs produced from CEL, PAN and pitch. X-ray diffraction and electron microscopy revealed that the PAN- and pitch-based fibres had more uniform pore structures than that of the CEL-based material, although the latter had the largest BET-area and pore volume (Kakei et al., 1991). Setoyama et al. (1996) investigated the properties of fluorinated activated carbons. Analysis of their nitrogen isotherms determined before and after the fluorination of CEL-based ACF indicated that although the micropore capacity and pore width were reduced, the micropore structure appeared to become more homogeneous as a result of fluorination. In another investigation, Wang and Kaneko (1995) determined SO_2 isotherms and energies of adsorption on a pitch-based ACF. The relatively strong adsorption at low p/p^o was ascribed to a combination of enhanced non-specific dispersion interactions and specific permanent dipole–induced dipole interactions.

An important development was the disclosure by Bailey and Maggs (1972, 1973) of a novel procedure for the manufacture of 'charcoal cloth'. The continuous process developed in the laboratories of the Chemical Defence Establishment, Porton Down, England, involved three main stages: (i) immersion of the roll of viscose rayon cloth in an aqueous solution of inorganic chlorides (e.g. $ZnCl_2$, $AlCl_3$ and NH_4Cl); (ii) oven drying in nitrogen; (iii) carbonization and activation in carbon dioxide.

The original Porton material was fairly strong and flexible with a BET-area $\sim 1200 \, m^2 \, g^{-1}$, a wide distribution of micropores and no detectable mesoporosity or macroporosity (see Atkinson et al., 1982; Hall and Williams, 1986). In view of its early promise, it was logical to attempt to control the pore structure of charcoal cloth and therefore a systematic study of the development of porosity was undertaken by Atkinson et al. (1982, 1984), Freeman and Sing (1991) and Freeman et al. (1987, 1988, 1993).

Pre-treatment of the rayon cloth with sodium dihydrogen phosphate, in addition to the mixed chlorides, provided a way of generating a well-defined mesopore structure at high burn off (Freeman et al., 1988). Electron microscopy confirmed that there was an important difference between this mesopore structure and the pore widening produced by steam activation. The effect of impregnating the viscose rayon with transition metal salts and oxo-complexes was also explored together with the use of more highly ordered precursors in place of viscose rayon (Freeman et al., 1993). It was found that activated carbons prepared from Kevlar interacted much more strongly with CO_2 than the rayon-based products.

Various commercial grades of activated carbon cloth are now available. These products are designed to provide protection against noxious gases (Hall and Sing, 1988), to remove unpleasant odours in medical dressings (Wright et al., 1988) or for water treatment. Specific adsorptive properties and catalytic activity can be produced by the impregnation of the activated material with metals (e.g. silver, copper or platinum).

In separational and catalytic technology, there are several advantages of using fibrous materials in place of powders, pellets or granules. An important operational advantage is the low resistance to fluid flow which, together with the relatively short molecular diffusion path within the fibre, can result in remarkably fast adsorption kinetics (Gimblett et al., 1989; Suzuki, 1994). The unique combination of texture and adsorbent properties provided by carbon cloth has been found to be of particular significance for gas chromatography. Thus, an efficient chromatographic 'column' could be constructed by clamping together 15–20 discs of microporous carbon cloth (Carrott and Sing, 1987). This arrangement was used for studying the low-coverage energetics of alkanes, alkenes and other organic molecules (Carrott and Sing, 1988).

10.2.8 Monoliths

It is now possible to prepare a range of carbon monoliths of different shape and internal porosity (e.g. honeycomb type). The conventional approach is to use a binder to achieve consolidation of the pulverized material. This generally leads to some loss of porosity and surface area by pore blocking, the extent of which depends on the type of binder. For example, the reduction in micropore volume can be minimized by using sepiolite.

Binderless activated carbon monoliths (ACMs) have been prepared in various ways. Rodrigues-Reinoso and his co-workers (e.g. Ramos-Fernandez et al., 2008) made use of a mesophase-based material as the precursor. A mesophase pitch, which was obtained by pyrolysis of a petroleum residue, was ball-milled with a controlled amount of KOH before heat treatment in nitrogen. By adapting the superactive carbon preparative technique, it was possible to produce microporous ACMs of high internal surface area and acceptable mechanical strength. The BET-area ($\sim$1500–3000 m^2 g^{-1}), micropore width and volume (up to 1 cm^3 g^{-1}) were all dependent on the KOH/C ratio. Another approach is via a carbon gel, as described in the following section.

ACMs are now of technological interest as advanced adsorbents, catalyst supports and for energy storage. They offer the advantages of large adsorbent capacity together with minimum void volume for the storage of natural gas (Sircar, 2008).

10.2.9 Carbon Aerogels and OMCs

As already indicated (see Section 10.2.4), activated carbons normally exhibit some short-range order, which is brought about by the localized development of distorted graphitic lamellae. The associated micropore structure is predominantly made up of non-rigid, irregular slits. Pore widening occurs with increased burn off, but this does not generally result in the formation of regular mesopore structures. Special techniques are required, therefore, for the preparation of carbons with well-defined mesopores of controlled size and shape.

One approach is to prepare an organic gel as a precursor. An aerogel in the form of a network of polydisperse polymeric beads is obtained by a sol–gel polycondensation process followed by supercritical drying (e.g. with CO_2) to minimize the collapse of product. The mesopore–macropore inter-particle network is then preserved after pyrolysis. Carbon aerogels were first prepared in this manner by Pekala (1989). The organic aerogel precursors were synthesized by a sol–gel polycondensation reaction of resorcinol with formaldehyde and subjected to supercritical drying and pyrolysis in an inert atmosphere to produce C aerogels with BET-areas of 500–1200 m^2 g^{-1} and pore volumes $\sim$0.9 cm^3 g^{-1}. More recent work (see Bandosz et al., 2003) reveals that the pore structures can be changed by varying the reaction conditions. Kaneko et al. (1999) investigated the texture of three aged carbon aerogels prepared in a similar way. Transmission electron microscopy revealed an interconnecting network of carbon particles of diameter of $\sim$5 nm with different degrees of agglomeration. Nitrogen high-resolution α_s-plots indicated that the gels were mesoporous and to a small extent also microporous.

In the work of Yamamoto et al. (2011), the preparative procedure was extended to give carbon cryogel microspheres. In this case, emulsion polymerization was followed by freeze-drying and pyrolysis, and it was possible to prepare both mesoporous microspheres and also microspheres coated with surface molecular sieve layers. Water sorption measurements showed that the degree of hydrophobicity increased with increase in pyrolysis temperature. BET-nitrogen areas were in two distinct ranges: 525–750 and 4–9 m^2 g^{-1}, depending on the reaction conditions.

Carrott et al. (2007) prepared a range of pore structures by the chemical activation of resorcinol-formaldehyde aerogels with phosphoric acid. The microporous and mesoporous products were characterized by the construction of nitrogen α_s-plots before and after the pre-adsorption of n-nonane. Remarkably, large and easily accessible pore volumes (e.g. $\sim$1.4 cm^3 g^{-1}) were obtained by the careful control of the acid/aerogel ratio.

For particular applications, carbon aerogels can be produced in microspheroidal or monolithic form or as thin films. Interest has been shown in their special electrical properties in the context of fuel cell and supercapacitor technology. Because of their low thermal conductivity, they are of potential value as thermal insulators. The small spheroidal particles (e.g. of 5–10 nm diameter) in carbon aerogels are interlinked in random networks and the inter-particle pores are therefore distributed in disordered arrays (Bandosz et al., 2003).

OMCs are synthesized in a completely different manner, with the aid of a matrix (Schüth, 2003; Darmstadt and Ryoo, 2008). Two different procedures have been developed, which have been termed 'soft' templating and 'hard' templating (see Cao and Kruk, 2010). The first approach involves the use of self-assembled surfactant molecules or blockcopolymers as a template to allow the carbon precursor to form a regular cross-linked polymeric network. Removal of the template is followed by pyrolysis of the precursor (Li and Jaroniec, 2004). In hard templating, a solid nano-structure is used as the template so that after removal of the template, its pore morphology is retained. The first template to be used was MCM-48 silica (Lee et al., 2001; Ryoo et al., 2001), but now SBA-15 is often favoured along with meso-phase pitch (Cao and Kruk, 2010).

OMCs are unique in having ordered nanoscale structures, narrow controllable mesopore size distributions, large pore volumes (up to 2 $cm^3 g^{-1}$) and appreciable specific surface areas (up to 500 $m^2 g^{-1}$). As one would expect, their nitrogen iso-therms exhibit the typical Type IV character. A number of potential applications are now under investigation including the removal of large molecules from aqueous solution (Darmstadt and Ryoo, 2008; Figueiredo et al., 2011).

10.3 Physisorption of Gases by Non-Porous Carbons

10.3.1 Adsorption of Nitrogen and Carbon Dioxide on Carbon Blacks

Nitrogen adsorption (at ca. 77 K) is used routinely for the characterization of both porous and non-porous carbons. In this section, we examine the distinguishing fea-tures of nitrogen isotherms on various graphitized and ungraphitized carbon blacks and note the changes produced by modifying the surface structure. Carbon dioxide is becoming a popular adsorptive for investigating the micropore structure of active carbons, and therefore it is of interest to review the few CO_2 adsorption measurements on carbon blacks.

10.3.1.1 Nitrogen Adsorption

Because of their globular shape, carbon black particles were used in some of the earliest attempts to validate the BET-method (Emmett and DeWitt, 1941). Com-parison of the values of surface area obtained by nitrogen adsorption and electron microscopy appeared to provide sound evidence for the validity of the BET-areas (Brunauer, 1945; Gregg and Sing, 1982, p. 64). It now seems likely that the good agreement obtained in the early measurements may have been to some extent fortuitous (Sing, 1994). However, as previously noted, the thermal blacks must be regarded as useful non-porous reference adsorbents. Also, although the graphitized thermal blacks are in short supply, they are still particularly important as research materials.

In the work of Isirikyan and Kiselev (1961), adsorption isotherms of nitrogen were determined at 77 K in considerable detail on four different graphitized thermal blacks (with BET-areas in the range of 6.5–29.1 m^2 g^{-1}). The isotherms are plotted in Figure 10.3 in a normalized form, as the amount adsorbed per unit area (in μmol m^{-2}) against the relative pressure, p/p^o. Kiselev and his co-workers referred to such isotherm plots as 'absolute adsorption isotherms', but of course they are not strictly absolute since they are dependent on the validity of the BET-nitrogen areas – with the usual assumption that $\sigma(N_2) = 0.162$ nm^2.

Inspection of the common normalized isotherm in Figure 10.3 reveals a number of distinctive features. At very low p/p^o, the isotherm is slightly convex with respect to the p/p^o axis and it is evident that the linear, Henry's law region does not extend above $p/p^o \sim 5 \times 10^{-4}$. Although the isotherm is not truly stepwise (i.e. it is not a true Type VI isotherm), it does exhibit a characteristic monolayer step. This is followed by a wavy second-layer region and then a smooth multilayer curve. Thus, as the multilayer coverage is increased, the isotherm appears to conform to the normal Type II shape.

The effect of the graphitization process on the shape of the nitrogen isotherm is revealed more clearly in Figure 10.4a. Here, isotherms on a number of graphitized and ungraphitized carbon blacks are plotted together in the reduced form of $n/n_{0.4}$

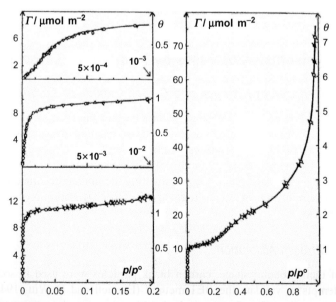

Figure 10.3 Adsorption isotherm of N$_2$ at 77 K on six different graphitized thermal blacks, at various scales of p/p^o.
After Isirikyan and Kiselev (1961).

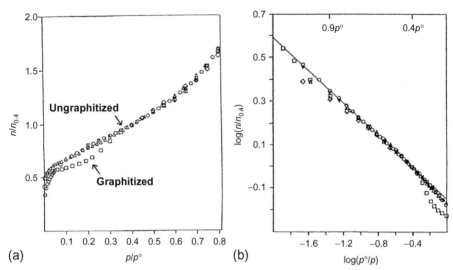

Figure 10.4 Standard nitrogen isotherms for carbons (a) and corresponding FHH plots (b) (circles, Carrott et al., 1988a,b; diamonds, squares, Isirikyan and Kiselev, 1961; triangles down, Pierce, 1969; triangles up, Rodriguez-Reinoso et al., 1987) (Carrott and Sing, 1989).

against p/p^o, where $n_{0.4}$ is the amount adsorbed at $p/p^o = 0.4$. It is apparent that all the isotherms follow a nearly common path over the multilayer range of $p/p^o = 0.3–0.8$, but that only the ungraphitized blacks give completely smooth Type II isotherms.

A more rigorous test of the multilayer conformity is provided by linearizing the multilayer isotherms in the Frenkel–Halsey–Hill (FHH) coordinates. The FHH plots of the adsorption isotherms in Figure 10.4a are shown in Figure 10.4b. Here, several of the plots of $\log_{10}(n/n_{0.4})$ versus $\log_{10}[\log_{10}(p^o/p)]$ are linear over a wide multilayer range, in accordance with the FHH equation,

$$\ln(p^o/p) = k(n/n_{0.4})^s \tag{10.1}$$

where s is an empirical constant. From the slope of the linear section of the FHH plots, we obtain $s = 2.70 \pm 0.05$.

Many attempts have been made to determine standard nitrogen isotherms on well-defined non-porous carbons (see Sing, 1994; Kruk et al., 1996; Choma et al., 2002), but it is evident that there is no single isotherm which could serve as standard for both graphitized and non-graphitized carbons. There are three main reasons why there are significant discrepancies between proposed standard data to be found in the literature. First, any significant differences in the carbon surface structure or polar groups will have some effect on the isotherm shape – especially at low surface coverage. Second, any inter-particle capillary condensation will produce an upward deviation in the multilayer/capillary condensation region. Third, any microporosity will enhance the adsorption in the sub-monolayer region and will tend also to reduce the isotherm slope in the multilayer region.

As explained in Chapters 7 and 9, by applying the α_s-method, we have a simple way of checking the validity of the BET-area and detecting the presence of microporosity. Many carbon blacks are essentially non-microporous (Carrott et al., 1987, 2001; Carrott and Sing, 1989; Bradley et al., 1995), and the corresponding values of BET and α_s areas are then in good agreement. However, in a few cases, back-extrapolation of the α_s-plot has given a small positive intercept on the adsorption axis, which is an indication of some microporosity. The microporous nature of some carbon blacks has been confirmed in several investigations (Stoeckli et al., 1994a,b; Kruk et al., 1996). As one might expect, oxidation generally leads to a considerable increase in the level of the microporosity (Bradley et al., 1995).

In the work of Andreu et al. (2007), carefully controlled ozone treatment provided a means of increasing the level of surface oxygen (as determined by X-ray photoelectron spectroscopy, XPS) with very little development of porosity. In this case, the Type II nitrogen isotherms (at 77 K) and the corresponding α_s-plots underwent a small change (giving an increase of ca. 8% in the BET-area), which was probably associated with the enhanced field gradient–quadrupole interaction. However, this contribution was very small in comparison with the strong specific interaction between methanol and the oxygenated surface.

The strong energetic heterogeneity exhibited by Spheron 6 was first shown calorimetrically by Beebe and his co-workers (Beebe et al., 1947; Kington et al., 1950). This work also revealed that the surface of Graphon was much less heterogeneous than that of the original carbon black. The results of a more detailed investigation of the effect of thermal treatment of carbon black on the energetics of nitrogen adsorption (i.e. variation of $\Delta_{ads}h$ with coverage θ) are shown in Figure 10.5. Microcalorimetric measurements were undertaken on a sample of heat-treated Sterling FT-FF (i.e. a thermal black).

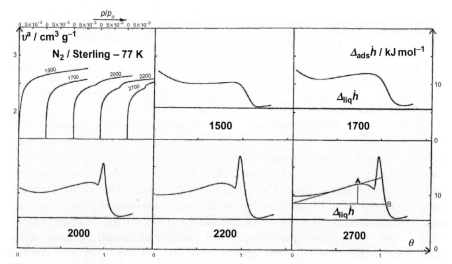

Figure 10.5 Differential enthalpies of N_2 adsorption at 77 K on heat-treated blacks (at temperatures from 1500 to 2700 °C), as a function of coverage (Grillet et al., 1979).

The interpretation of the microcalorimetric data in Figure 10.5 seems at first sight to be quite straightforward. Thus, heat treatment at temperatures >2000 °C evidently removed much of the adsorbent–adsorbate heterogeneity since in the region of low θ the differential enthalpy underwent very little variation. This is, of course, consistent with the picture of the development of a uniform surface composed largely of the graphitic basal planes.

However, it is possible that this interpretation is somewhat over-simplified. The simulation study of the energetics of nitrogen adsorption on graphite by Steele and Bojan (1989) indicated that there is a steady decrease in the adsorbent–adsorbate interaction with increase in surface coverage. This unexpected finding was attributed to a change in the orientation of the nitrogen molecules as a result of molecule–molecule interactions. Thus, as the surface population is increased, there is a greater probability that the adsorbed molecules will adopt a vertical orientation with respect to the basal plane. It follows that the almost constant adsorption energy observed experimentally may be the result of compensation between an increase in adsorbate–adsorbate interactions and a decrease in the adsorbent–adsorbate interactions.

Turning now to the regions of high surface coverage in Figure 10.5, we note the appearance of two maxima in the differential enthalpy curves for the high-temperature graphitized Sterling. The first is a broad peak with a maximum attained at $\theta \sim 0.8$ and the second is a sharp peak in the vicinity of $\theta \sim 0.1$. The initial increase in $\Delta_{ads} h$ can be attributed to the normal attractive interactions between neighbouring adsorbate molecules, whereas the second peak is the result of a two-dimensional (2D) phase transformation (Rouquerol et al., 1977).

The use of the 'continuous quasi-equilibrium technique' (see Chapter 3) made it possible to determine the corresponding adsorption isotherm in sufficient detail to reveal the sub-step shown in Figure 10.5 located at the same θ as the sharp calorimetric peak. The isotherm sub-step and calorimetric peak are evidently associated with an increase in packing density of the adsorbate. Rouquerol et al. (1977) and Grillet et al. (1979) concluded that these changes were due to a degenerated first-order transition from a hypercritical 2D fluid state to a 2D localized state.

Many years ago, Isirikyan and Kiselev (1961) and Pierce and Ewing (1962, 1967) came to the conclusion that the nitrogen monolayer on graphite was localized at 77 K, the most favourable site being at the centre of the hexagon of carbons (see Figure 10.6). At that time, it was thought that because of its size and diatomic shape, each nitrogen molecule would demand the space provided by four hexagons. A simple calculation indicates that this is equivalent to a molecular area, σ (N_2), of 0.21 nm^2 for the completed monolayer. After making a detailed empirical analysis of the isotherms of nitrogen and other gases on Sterling MT(3100 °C), Pierce and Ewing (1967) concluded that 0.194 nm^2 was a more realistic value for the effective area of nitrogen in the BET-monolayer on graphitized carbons. This value was remarkably close to that (0.195 nm^2) proposed independently by Carrott et al. (1987).

However, a number of other investigators came to a different conclusion. They argued that a commensurate 'herringbone' structure could be adopted if the adsorbed nitrogen molecules occupied one in three of the neighbouring hexagonal sites (Rouquerol et al., 1977; Bojan and Steele, 1987; Ismail, 1990). This arrangement,

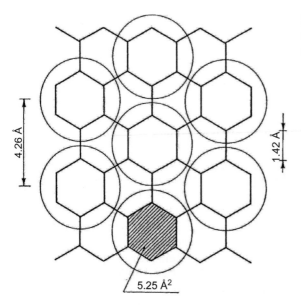

Figure 10.6 Localization of N_2 molecules on graphite (Rouquerol et al., 1977).

4.26 Å

1.42 Å

5.25 Å^2

which is pictured in Figure 10.6, would reduce the value of σ_m (N_2) to 0.157 nm^2, and therefore the commensurate monolayer would have a packing density only slightly greater than that of the hypothetical close-packed 'liquid' monolayer. Since independent support for this state has been obtained by neutron diffraction and X-ray scattering (see Ismail, 1992), this value would now seem to be the preferred one for the effective molecular area of nitrogen in the localized monolayer on the graphitic basal plane.

It is significant that the second calorimetric peak and the associated isotherm substep were detectable only if the graphitized thermal black had been heated at temperatures above 1700 °C. These results suggest that the 2D phase transformation is very sensitive to the perfection of the surface basal planes and this is a further indication that the phase change leads to the development of a commensurate structure.

10.3.1.2 Carbon Dioxide Adsorption

In view of the strong revival of interest in the adsorption of carbon dioxide by microporous carbons, it might be expected that attention would be given also to the interaction of this adsorptive with non-porous carbons. There are two related aspects of some importance: (a) the determination of reference adsorption data for CO_2 and (b) the degree of specificity associated with the CO_2 quadrupole. For characterization purposes, CO_2 isotherms are usually determined at 273 K (see Lowell et al., 2004), but at this temperature, the saturation pressure is ca. 3.5 MPa. Therefore, at pressures below atmospheric (i.e. ca. 0.1 MPa), the amounts adsorbed cannot be more than a small fraction of a possible total adsorption capacity. This is acceptable for investigating the properties of a narrow micropore structure (i.e. a pore width of $<\sim 1$ nm), but it does not allow the complete micropore range to be studied.

Guillot and Stoeckli (2001) obtained standard isotherms over a wide range of multilayer coverage by determining CO_2 isotherms on Vulcan 3G (a graphitized carbon black – reference material) at 253, 273 and 298 K at pressures extending well above atmospheric. For pressures up to 3 MPa (i.e. $p/p^o = 0.86$ at 273 K), it was found that the standard isotherm could be fitted to a simple polynomial expression over the full monolayer–multilayer range and to the DRK equation in the sub-monolayer range. Comparison plots were constructed for CO_2 adsorption isotherms on various microporous carbons and Sing's α_s-method was used to provide an approximate assessment of the micropore capacities and external areas. Although the evidence was not conclusive, it does appear that specific interactions do not complicate the interpretation of the CO_2 comparison plots at 273 K.

10.3.2 Adsorption of the Noble Gases

The stepwise character of the low-temperature isotherms of argon and krypton on graphitized carbon blacks was firmly established by the pioneering investigations of Beebe et al. (1953), Beebe and Young (1954), Beebe and Dell (1955), Polley et al. (1953) and Prenzlow and Halsey (1957). The early measurements revealed *inter alia* that argon isotherms at 77 K underwent a progressive change from Type II to VI as carbon blacks were heated at increasingly higher temperatures over the range 1000–2700 °C.

The energetic heterogeneity of the original ungraphitized surface was confirmed by the careful adsorption calorimetric measurements undertaken by Beebe et al. (1953). By comparing the changes in the differential energies of adsorption for argon on Spheron and Graphon, these investigators found that the initial steep decline in 'differential heat of adsorption' was largely removed as a result of graphitization. Instead, an increase in the differential energy was observed at a higher surface coverage: it is now generally agreed that this was due to the adsorbate–adsorbate interaction becoming more apparent as the degree of energetic heterogeneity was reduced.

A systematic study of krypton adsorption on exfoliated graphite was subsequently undertaken by Thomy and Duval (1969, 1972). Their stepwise isotherm, determined at 77.3 K, is shown in Figure 5.1. The layer-by-layer nature of the physisorption process is clearly evident – at least up to four molecular layers. This isotherm shape is remarkably similar to that of the krypton isotherm on graphitized carbon black reported by Amberg et al. (1955).

The work of Thomy and Duval (1969, 1972) provided the first well-documented evidence for the presence of a sub-step in the Kr isotherm. The effect of temperature on the shape and the location of the sub-step is shown in Figure 10.7. The fact that the riser of the sub-step remained vertical over the temperature range of 77.3–96.3 K served to confirm that the sub-step was due to a first-order 2D phase change. It was thus evident that, at a given temperature, the two sub-monolayer phases were in thermodynamic equilibrium at a characteristic pressure, p_{2D}. As discussed in Chapter 5, the 2D phase diagram is then obtained as the plot of p_{2D} against T.

As already noted, adsorption microcalorimetry is one of the most useful techniques for studying 2D phase changes. Tian-Calvet microcalorimetry was used to investigate the adsorption of argon on graphitized carbon black (Grillet et al., 1979). As with

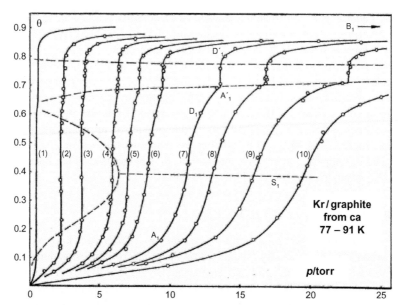

Figure 10.7 Adsorption isotherms of krypton on exfoliated graphite. Curves labelled from 1 to 10, obtained at 77.3, 82.4, 84.1, 85.7, 86.5, 87.1, 88.3, 89.0, 90.1 and 90.9 K, respectively. *Courtesy* Thomy et al. (1972).

nitrogen, the argon phase change was accompanied by a peak in $\Delta_{ads}\dot{h}$, but in this case on a smaller scale. The difference in the magnitude of the calorimetric peak was consistent with the lower height of the argon sub-step. Furthermore, the amounts adsorbed at 77 K were different, the monolayer of nitrogen being the more densely packed. It follows that, unlike nitrogen, the 2D 'solid' argon was not in registry with the graphite structure.

According to Larher (1983), the behaviour of argon is also unlike that of krypton and xenon difference in size and location of these three adsorbate molecules as illustrated in Figure 10.8 in relation to the graphitic basal plane. Here, a hypothetical commensurate hexagonal structure is shown in comparison with the dense 2D (111)

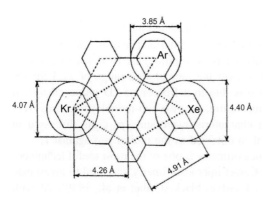

Figure 10.8 Possible structures for argon, krypton and xenon adsorbed in registry with graphite. Dotted unit cell: Xe linearly expanded by $(4.91 - 4.40)/4.40 = 11.6\%$. Dashed unit cell: Xe linearly compressed by $(4.40 - 4.26)/4.26 = 3.2\%$, Kr expanded by $(4.26 - 4.07)/4.07 = 4.7\%$ and Ar expanded by $(4.26 - 3.85)/3.85 = 10.6\%$.
Courtesy Larher (1974).

plane of the noble gases, Ar, Kr and Xe. It can be seen that the lattice mismatch is likely to be quite small for Kr and Xe, but significantly larger for Ar and evidently much larger for Ne. Therefore, we would expect Kr and Xe to undergo the commensurate–incommensurate phase changes more easily than Ar or Ne.

The 'best' value for the effective molecular cross-sectional area, $\sigma(Kr)$, of krypton in the BET-monolayer at 77 K has been under discussion for many years. In their original work on krypton adsorption, Beebe et al. (1945) recommended the value 0.195 nm^2 for σ (Kr) and this empirical value is still used by many investigators. For the adsorption of krypton on graphitized carbon, Ismail (1990, 1992) gives preference to the value $\sigma(Kr) = 0.157$ nm^2, which is fairly close to the values of molecular area calculated from the liquid density and determined by X-ray scattering. This, of course, implies that Kr and N_2 molecules undergo localized adsorption on the same sites. For ungraphitized carbons, Ismail (1992) recommends $\sigma(Kr) = 0.214$ nm^2.

It seems likely that the effective molecular area of krypton is dependent on the surface structure, but the extent of this variation is unknown. In practice, there is another problem which is encountered when the krypton-BET plot is not strictly linear (Malden and Marsh, 1959). The calculated value of n_m then varies according to where the tangent is drawn.

There is a similar difficulty in the selection of the effective molecular area with argon adsorption at 77 K on graphitized and ungraphitized carbon blacks. The results of a number of comparisons with nitrogen have led to recommended values of σ (Ar) in the range of 0.130–0.165 nm^2 (Gregg and Sing, 1982), but according to Ismail (1992), the most appropriate values are 0.138 and 0.157 nm^2 for graphitized and ungraphitized carbons, respectively.

As explained in Chapter 7, the evaluation of the BET-monolayer capacity of either Kr or Ar at 77 K is dependent on the choice of p^o for the construction of the BET plot at 77 K: that is whether the saturation pressure of the stable 3D solid, p^o (sol), or the extrapolated liquid value, p^o (liq), is used. Until recently, most workers have followed the early recommendation of Beebe et al. (1945) that the supercooled p^o (liq) value should be adopted since many Type II isotherms of argon and krypton at 77 K appear to cut the p^o (sol) axis at a sharp angle (Gregg and Sing, 1982). However, Ismail (1990) has put forward evidence in favour of p^o (sol). The case for p^o (sol) is especially strong in relation to the adsorption of Kr on graphitized carbon since a liquid type of multilayer seems unlikely to develop at 77 K after the formation of a 2-D 'solid' in registry with the basal plane. The evidence for the more general use of p^o (sol) is not convincing, however, and it appears that the effective p^o is dependent on the nature of the adsorbent.

This problem is avoided if the operational temperature is above the triple point of the adsorptive. Indeed, argon measurements at 87 K (the temperature of liquid argon) are now favoured in some laboratories (Gardner et al., 2001; Thommes et al., 2012). An advantage of adopting the higher temperature is that the stepwise isotherm is replaced by the normal Type II isotherm, which is obtained over the complete range of monolayer–multilayer coverage. Furthermore, argon does not undergo specific interactions with any surface functional groups. It is therefore possible to determine reference adsorption data for various non-porous carbons. In the work of Gardner et al. (2001), two different carbon blacks (one graphitized and the other ungraphitized) were selected as non-porous reference materials. The standard argon isotherm data at 77 and 87 K were reported in tabular form and are available for the construction of α_s-plots on activated carbons.

10.3.3 Adsorption of Organic Vapours

At appropriately low temperatures, many organic molecules undergo localized adsorption on graphite. The mode of adsorption of small molecules (e.g. methane and ethane) is highly dependent on temperature, the monolayer becoming more mobile with increase in thermal energy.

The adsorption isotherm of methane at 77 K on exfoliated graphite is shown in Figure 10.9: the stepwise character is clearly very similar to that of krypton at 77 K (see Figure 5.7). Stepwise isotherms are also given by ethane on graphite (Bienfait, 1985) and ethyl chloride on graphitized Sterling MT (Davis and Pierce, 1966). On the other hand, the isotherms of benzene and hexane at 293 K on graphitized Sterling MT are essentially Type II – although the hexane isotherm is similar to that of nitrogen in having a slight indication of a second-layer step. The isotherms of propane (at 196 K), isobutane (at 261 K) and neopentane (at 273 K) on graphitized carbon blacks are all typical Type II (Carrott and Sing, 1989).

We conclude that, provided the temperature is not too high, the simplest organic adsorptives can undergo stepwise adsorption on the graphitic basal plane to give

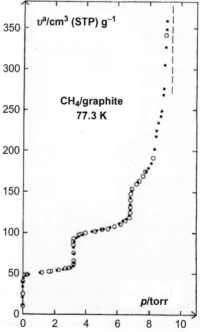

Figure 10.9 Stepwise isotherm of CH_4 on graphite foam (open circles) and exfoliated graphite (solid circles) at 77.3 K.
Courtesy Bienfait et al. (1990).

well-defined Type VI isotherms. However, most organic adsorptives give Type II adsorption isotherms with values of C (BET) tending to be high and therefore clearly marked monolayer capacities.

Sub-steps, similar to those in Figure 10.7, have been observed with both methane and ethane (Bienfait, 1980, 1985). It has been possible to construct 2D phase diagrams for several of these systems (Gay et al., 1986; Suzanne and Gay, 1996). LEED and neutron diffraction have provided information on the 2D structures. For example, seven different 2D phases have been reported for ethane on graphite over the temperature range of 64–140 K. Thus, three 'solid' commensurate phases were identified at temperatures <85 K, the S_3 phase apparently having a close-packed hexagonal structure, with σ (C_2H_6) = 0.157 nm^2.

In the case of hexane on graphite, the adsorbate molecules appear to adopt an ordered herringbone structure at temperatures below 151 K, whereas a fluid-like phase is formed at higher temperatures (Krim et al., 1985). Several investigators (e.g. Avgul and Kiselev, 1970; Clint, 1972; Gregg and Sing, 1982) have concluded that long-chain alkane molecules tend to lie flat and parallel to the basal plane, which explains the constant incremental increase in adsorption energy and molecular area with chain length.

In their comparative study of the adsorption of n-hexane and benzene on graphitized thermal blacks, Isirikyan and Kiselev (1961) found a significant difference in the behaviour of the two adsorptives. This difference is illustrated in Figure 10.10, where the differential energies of adsorption are plotted against the surface coverage. In the case of benzene, the nearly constant adsorption energy over a wide range of coverage is a clear indication of a low level of lateral interaction between neighbouring adsorbate molecules. In contrast, the pronounced increase in hexane adsorption energy provides unambiguous evidence for the steady increase in adsorbate–adsorbate interaction as the monolayer becomes more densely populated.

As a result of their study of the adsorption of CH_2Cl_2 and C_6H_6 by various carbon blacks, Stoeckli et al. (1994a,b) have demonstrated the importance of using a number of adsorption and calorimetric techniques. This work has confirmed that the high surface area of some carbon blacks is not confined to the external surface. That particular blacks were to some extent microporous was revealed by immersion microcalorimetry, gas-phase microcalorimetry and the form of their comparison and DR plots. These findings strengthen the view that the BET-method cannot always be relied upon to evaluate the effective surface area of a carbon black.

10.4 Physisorption of Gases by Porous Carbons

10.4.1 Adsorption of Argon, Nitrogen and Carbon Dioxide

10.4.1.1 Nitrogen Adsorption at 77 K

Nitrogen is still the most popular adsorptive for determining the surface area and characterizing the pore structure of activated carbons (see de Vooys, 1983; Fernandez-Colinas et al., 1989a,b; Rodriguez-Reinoso et al., 1989; Sing, 1989, 1995, 2008; Bradley and Rand, 1995; Thommes et al., 2012).

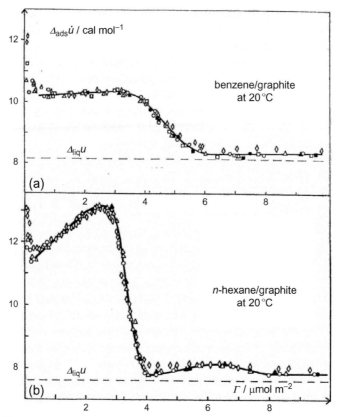

Figure 10.10 Differential energies of adsorption of benzene (top) and *n*-hexane (bottom) at 20 °C on three different graphitized thermal blacks.
After Isirikyan and Kiselev (1961).

As explained in Chapter 8, since the *multilayer* isotherm path is rather insensitive to differences in surface chemistry, for routine mesopore analysis, it is possible to make use of a 'universal' form of nitrogen isotherm. However, most activated carbons are highly microporous and the determination of the micropore size distribution remains a more difficult problem. Indeed, as discussed in Chapter 9, even the assessment of the total micropore volume presents conceptual difficulties. We should therefore regard the measurement of a nitrogen adsorption isotherm as only the first stage in the characterization of a microporous carbon.

Nitrogen isotherms on four activated carbons and the corresponding α_s-plots and Dubinin–Radushkevich (DR) plots are shown in Figure 10.11. Carbosieve and charcoal cloth JF005 are both molecular sieve carbons, while AX21 is a superactive carbon. The isotherms on Carbosieve and the carbon cloth JF005 are of well-defined Type I in the IUPAC classification, but the isotherms on the carbon cloth JF517 and the superactive carbon AX21 are evidently more complex.

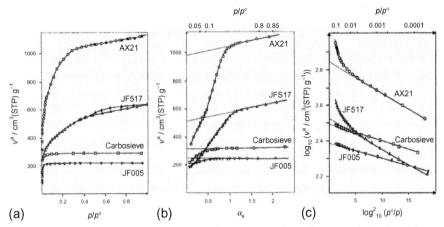

Figure 10.11 (a) Adsorption isotherms on microporous carbons (open symbols adsorption; solid symbols, desorption), (b) α_s-plots for microporous carbons and (c) DR plots for microporous carbons.

The *multilayer* section of each α_s-plot in Figure 10.11b exhibits an appreciable range of linearity. By assuming that this linear section represents unrestricted multilayer adsorption on the external surface, we can arrive at an approximate estimate of the external area, a(ext). Thus, if the amounts adsorbed, v^σ or v^a, are expressed in the units cm^3 (STP) g^{-1}, the values of a(ext) in Table 10.1 are obtained by application of the following equation:

$$a(\text{ext}) = 2.86 v^\sigma / \alpha_S \qquad (10.2)$$

where the factor 2.86 has been evaluated by calibration against the BET-areas of several non-porous carbon blacks (Carrott et al., 1987).

The values of a(BET) in Table 10.1 are the BET-nitrogen areas, which were derived from the linear regions of the BET plots, with the molecular area assumed

Table 10.1 Surface Areas and Pore Volumes of some Microporous Carbons (N_2)

Carbon	a(BET) ($m^2\,g^{-1}$)	a(ext) ($m^2\,g^{-1}$)	v_p(mic, S) ($cm^3\,g^{-1}$)	v_p(mic, D) ($cm^3\,g^{-1}$)
PX 21	3700	178	1.75	0.99
AX 21	3393	233	1.52	1.00
JF 516	2053	246	0.98	
JF 517	1657	218	0.76	0.47
JF 142	1479	54	0.55	
Carbosieve	1179	41	0.43	0.45
JF 005	882	19	0.33	0.35

to be 0.162 nm^2. The values of a(ext) in Table 10.1 are obviously much smaller than the corresponding values of a(BET). The question naturally arises: does the BET-method provide a reliable assessment of the *total area* (i.e. internal plus external area)? The non-linear character of the *low-pressure* region of each α_s-plot is a clear indication that the isotherm is distorted in the monolayer region and we may therefore conclude that a(BET) does *not* represent an *effective* surface area (see Chapter 7). Additional support for this interpretation comes from the microcalorimetric data, which are discussed later in this section.

Backward extrapolation of the linear multilayer section of the α_s-plot allows us to assess the total micropore capacity (as indicated in Chapter 9) and hence to evaluate the *effective* micropore volume, v_p(mic, S). The values of v_p(mic, S) in Table 10.1 were obtained by making the usual assumption that the pores are filled with *liquid* nitrogen (density 0.808 g cm^{-3}).

The DR plots in Figure 10.9c are all linear at very low p/p°, but those for JF517 and AX21 show strong deviations at $p/p^\circ > 0.01$. Nitrogen DR plots of similar appearance have been reported in other studies of activated carbons (e.g. Dubinin, 1966; Atkinson et al., 1984; Rodriguez-Reinoso, 1989; Linares-Solano and Cazorla-Amoros, 2008). According to the DR theory (see Section 5.2.4), the intercept of the linear plot should equal $\log_{10}v_p$(mic, D), where v_p^σ(mic) is the volume of gas required to fill the micro-pores. The apparent micropore volumes, v_p(mic, D), in Table 10.1 are obtained from the values of v^σ(mic, D) – again, by taking the liquid density.

Generally, long linear DR plots are given by carbons with narrow micropores, whereas the more restricted linearity is an indication of the presence of wider micropores and mesopores (Gregg and Sing, 1982; Atkinson et al., 1984). The wider micropores were designated 'supermicropores' by Dubinin (1975). It is evident that the corresponding values of v_p(mic, S) and v_p(mic, D) in Table 10.1 are in close agreement only if the supermicropores are absent (Carrott et al., 1987). Similar results were reported by Rodriguez-Reinoso (1989). Since the primary filling of the very narrow micropores (i.e. the ultramicropores) occurs at very low p/p°, the change in amount adsorbed at $p/p^\circ > 0.01$ is quite small (Atkinson et al., 1987; Kenny et al., 1993), and therefore it is not surprising to find that the DR plot is virtually linear over a fairly wide range of p/p°.

In a number of investigations (e.g. Fernandez-Colinas et al., 1989a,1989b; Kakei et al., 1991; Kenny et al., 1993; Kaneko, 1996; Llewellyn et al., 2000), it has been found that the shapes of high-resolution α_s-plots and DR plots provide strong evidence for a sequential filling of several groups of micropores. For example, the nitrogen iso-therms in Figure 10.12a were determined on a series of activated pine wood charcoals (Fernandez-Colinas et al., 1989a,b). The change in isotherm shape is the first indica-tion that pore widening has occurred as a result of progressive activation in steam. The α_s-plots in Figure 10.12b confirm this interpretation and indicate that this was mainly due to the development of a supermicropore structure (i.e. the wide micropores). It appears that the initial stage of primary micropore filling at very low p/p° (i.e. $p/p^\circ < 0.01$) was followed by the more gradual filling of supermicropores.

Surface coverage of the supermicropore walls is indicated by the appearance of a short linear section at $\alpha_s < 0.5$. Extrapolation of this section to $\alpha_s = 0$ provides an

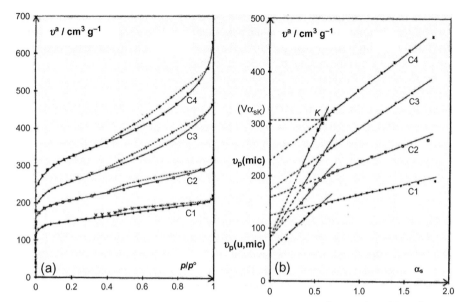

Figure 10.12 N_2 adsorption isotherms at 77 K (a) and corresponding α_s plots (b) on four activated carbons (Fernandez-Colinas et al., 1989a,b).

approximate evaluation of the *effective ultramicropore volume*, v_p(u, mic) (i.e. of the narrow micropores).

The second linear section extends over the multilayer range of each α_s – plot in Figure 10.12b. Backward extrapolation of this branch gives the total effective micropore volume, v_p(mic), from the intercept on the v^a axis. It follows that the effective supermicropore volume, v_p(sup, mic), can be regarded as the difference v_p(mic) – v_p(u, mic). It is of interest that, after an initial small change in v_p(u, mic), it has remained constant during further activation while the magnitude of v_p(sup, mic) has increased steadily.

Nitrogen isotherms, which were determined after the CO_2-activation of untreated rayon cloth chars, are similarly analysed in Table 10.2. It can be seen that the

Table 10.2 Evolution of the Micropore Structure of Untreated Carbon Cloth
(Analysis of Nitrogen Isotherms of Carrott and Freeman, 1991)

Percent burn off	20	31.2	49.7	70.1	92.0
Total micropore volume (cm^3 g^{-1})	0.35	0.47	0.67	0.85	1.11
Ultramicropore volume (cm^3 g^{-1})	(0.34)	(0.34)	0.34	0.34	0.34
Supermicropore volume (cm^3 g^{-1})	(0.01)	(0.13)	0.33	0.51	0.77
Supermicropore area (m^2 g^{-1})	–	(500)	1030	1520	1690
External area (m^2 g^{-1})	20	30	30	50	200

supermicropore volume and area increased progressively with percentage burn off, whereas the ultramicropore volume again remained virtually constant. The external area also underwent little change until the burn off exceeded 70%. The extent of the mesoporosity remained small throughout the series (Freeman et al., 1990).

As discussed in Chapter 9, the pre-adsorption of n-nonane can be used as a means of blocking narrow micropore entrances (see Section 9.2.5). Thus, in the case of an ultramicroporous adsorbent such as Carbosieve, the pre-adsorption of nonane leads to complete blockage of the pore structure. The effect of progressively removing the pre-adsorbed nonane from a supermicroporous carbon is shown in Figure 10.13. The adsorbent used in this work was a well-characterized carbon cloth with the following properties: a(BET), 1330 m^2 g^{-1}; a(ext), 25 m^2 g^{-1}; v_p(mic), 0.44 cm^3 g^{-1}; w_p, 0.6–2.0 nm (Carrott et al., 1989).

Inspection of the α_s-plots in Figure 10.13 is instructive. It can be seen that there are two linear sections: Back-extrapolation of the first giving a zero intercept when part of the pre-adsorbed nonane was removed by prolonged outgassing at 50 °C. We may conclude that the initial stage of nitrogen adsorption is then by monolayer adsorption on the walls of the supermicropores. This is followed by the formation of a quasi-multilayer until pore filling is complete at $p/p° \sim 0.4$. Increase in outgassing temperature leads to the progressive removal of nonane from the ultramicropores and narrow entrances and the original nitrogen isotherm is restored after prolonged outgassing at 200 °C. Thus, we may compare the change in the extent of total micropore capacity, v_p(mic) (obtained as described earlier by back-extrapolation of the linear multilayer branch) with the magnitude of the restoration of the ultramicropore capacity, v_p(u, mic) (as defined earlier). The fact that at each stage v_p(mic) $> v_p$(u, mic) is a clear indication that the pre-adsorption of nonane has resulted in the blockage of narrow entrances of some supermicropores.

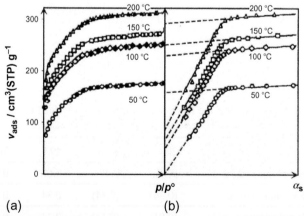

Figure 10.13 Nitrogen adsorption isotherms at 77 K (a) and corresponding α_s plots (b) for charcoal cloth JF012 after pre-adsorption of nonane followed by outgassing at indicated temperature.
After Carrott et al. (1989).

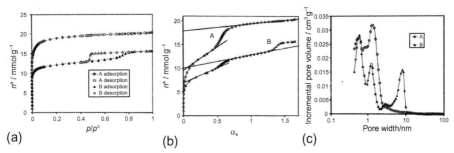

Figure 10.14 N_2 adsorption at 77 K on two microporous carbon clothes A and B. (a) Adsorption–desorption isotherms, (b) α_s plots and (c) DFT plots of pore-size distribution. After Llewellyn et al. (2000).

The application of DFT and the α_s-method is illustrated in Figure 10.14 where the nitrogen isotherms are shown for two samples of carbon cloth activated under different conditions to give different ranges of pore size. The BET-areas of samples A and B were 1666 and 1058 m^2 g^{-1}, respectively. The reversible Type I character of isotherm A is clearly associated with the microporous nature of sample A, whereas the Type IV appearance of isotherm B is indicative of capillary condensation in sample B. More qualitative and quantitative information was obtained by application of the α_s method of isotherm analysis. The α_s-plots in Figure 10.14b provide clear evidence of two stages of micropore filling in both samples A and B and also an additional mesopore-filling stage in sample B. By back-extrapolation of the linear sections, it was possible to assess the effective ultramicropore and supermicropore capacities and from the respective slopes of these linear regions, the effective external and supermicropore surface areas could be approximately evaluated (Llewellyn et al., 2000).

The pore size distribution curves in Figure 10.14c were obtained by the application of the DFT method with the aid of commercial Micromeritics software. DFT analysis has the advantage over the classical BJH method of mesopore size analysis (see Chapter 8) of being in principle applicable to both micropore filling and capillary condensation, but its limitations must be kept in mind. In particular, it is assumed that the experimental isotherm is a composite of a series of local isotherms, that the carbon surface is uniform and that the pores were all in the form of individual rigid slits.

As Do and Do (2003) have pointed out, the derived pore size distribution should be regarded as *effective* for the following reasons:

1. Pores in carbons are of irregular shape;
2. Defects are present on the graphene surface;
3. Polar groups are attached to edge sites;
4. Pores are interconnected;
5. Molecular interactions do not conform to the simple model.

The DFT-derived pore size distribution curves in Figure 10.14c are therefore effective distributions *with respect to the adsorption of nitrogen at 77 K*. In spite of these limitations, the combination of α_s-plots and DFT analysis can provide a useful approach in the interpretation of the adsorptive behaviour of porous carbons.

10.4.1.2 Argon and Nitrogen Adsorption at 77 K

A possible disadvantage of nitrogen is that, because of its diatomic molecular shape and quadrupolar nature, it is an unrepresentative adsorptive for the investigation of micropore filling. It is instructive therefore to compare the results of nitrogen and argon adsorption measurements on a series of activated carbons. For this purpose, adsorption microcalorimetry is an invaluable tool. Differential enthalpies of adsorption for argon and nitrogen are plotted in Figure 10.15 (i.e. $\Delta_{ads}\dot{h}$ vs. θ) for two of the activated charcoals, C1 and C4, featured in Figure 10.12. As expected, over most of the micropore-filling range, the nitrogen adsorption energies are appreciably above the corresponding argon energies. As discussed in Chapter 1, this difference is likely to be due to the specific field gradient–quadrupole interaction experienced by nitrogen. However, it is evident that with both C1 and C4 the corresponding adsorption energy curves have the same general appearance.

Inspection of the adsorption enthalpy curves for nitrogen in Figure 10.15 reveals that three characteristic stages of physisorption can be identified: point A is at the end of the first plateau; point B is at the beginning of a second, less well-defined, plateau; and point C is the point where the pore-filling curve crosses the corresponding curve for monolayer adsorption on ungraphitized carbon (Spheron 1500). We can attribute the high initial adsorption enthalpies ($\Delta_{ads}\dot{h} \sim 18\,\mathrm{kJ\,mol^{-1}}$ for nitrogen and $\sim 16\,\mathrm{kJ\,mol^{-1}}$ for argon) to primary micropore filling within pores of molecular dimensions. This is followed by the transitional region AB and finally the mainly cooperative filling range of BC.

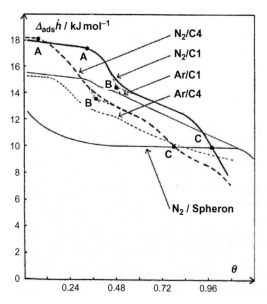

Figure 10.15 Differential enthalpies of adsorption of N_2 and Ar at 77 K on carbons C1 and C4 (same as in Figure 10.12) and on Spheron 1500 (Fernandez-Colinas et al., 1989a,b).

As pointed out in Chapters 1 and 6, high physisorption energies are produced by the overlap of the adsorbent–adsorbate interactions in pores of molecular dimensions (Everett and Powl, 1976). In the case of slit-shaped pores in carbons, a significant enhancement of adsorption energy would be expected for an effective pore width, $w < c \cdot 2d$ (d is the molecular diameter). A twofold enhancement of adsorption energy would be the maximum expected for the entry of molecules in narrow slit-shaped pores. Such values are recorded in Table 1.5, but it is evident that the initial values in Figure 10.15 are at a somewhat lower level. These findings indicate that the narrowest pore width in the series of activated charcoals is probably in the region of 0.8 nm.

A detailed microcalorimetric study of the adsorption of argon and nitrogen by a selection of microporous carbons was reported by Atkinson et al. (1987). The differential enthalpies of adsorption are presented in two ways: in Figure 10.16a and b, they are plotted in the usual manner, as a function of the fractional micropore filling, $v/v_p(\text{mic})$, and in Figure 10.17a and b, they are plotted as a function of the equilibrium pressure p/p^o. The latter presentation is of particular interest since it reveals the considerable differences in the energetics of adsorption at low p/p^o. As indicated in Chapter 3, Tian-Calvet microcalorimetry is the preferred technique for this type of investigation because the adsorption enthalpy measurements are made at constant temperature.

The nitrogen isotherms and corresponding α_s-plots for the microporous carbons JF005 (a molecular sieve carbon cloth), JF142, JF156 and PX21 (a superactive carbon) were very similar to those for carbons JF005, Carbosieve, JF517 and AX21 in Figure 10.11. Thus, JF005 and JF142 were both ultramicroporous, whereas JF516 and PX21 contained wide ranges of micropores. Values of $a(\text{BET})$, $a(\text{ext})$ and v_p, derived from the nitrogen isotherms, are included in Table 10.1.

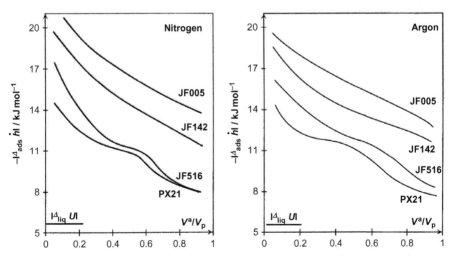

Figure 10.16 Differential enthalpies of adsorption of N_2 and Ar on microporous carbons, plotted versus fractional pore filling.
After Atkinson et al. (1987).

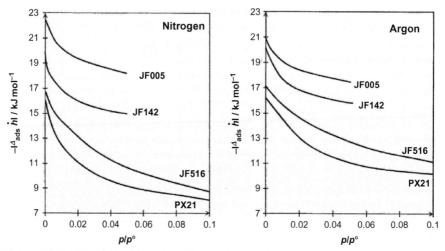

Figure 10.17 Differential enthalpies of adsorption of N_2 and Ar on microporous carbons, plotted versus relative pressure.
After Atkinson et al. (1987).

The changes in the differential enthalpies of adsorption revealed in Figures 10.16 and 10.17 illustrate the strong energetic heterogeneity associated with micropore filling by all four carbons. However, it is evident that for both nitrogen and argon adsorption, the differential enthalpy curves are of two types. Of special interest is the high level of adsorption enthalpy given by JF005 and JF142 over almost the complete range of micropore filling, which is in contrast to the more complex behaviour of JF516 and PX21. With the latter adsorbents, the two stages of micropore filling are clearly evident. It is also apparent that with the two molecular sieve carbons JF005 and JF142 over 80% of the pore filling (i.e. primary micropore filling) has taken place at $p/p^o < 0.01$.

In the light of these results, it is of interest to obtain accurate adsorption isotherm data at very low p/p^o. As explained in Chapter 3, such high-resolution adsorption measurements are not easy to make, but a preliminary investigation (Kenny et al., 1993) indicated that primary filling of the micropores in Carbosieve by nitrogen at 77 K begins at $p/p^o < 10^{-5}$ and is complete at $p/p^o \sim 10^{-2}$. These findings have been broadly confirmed by more recent work (Conner, 1997; Llewellyn et al., 2000; Thommes et al., 2011).

By applying non-local DFT, Gubbins and his co-workers (Balbuena et al., 1993) predicted that the stepwise filling of very narrow slit-shaped micropores takes place at $p/p^o < 10^{-5}$. The non-local DFT theory also predicts that there are layering transitions in the filling of supermicropores. A homogeneous graphitic-type surface was assumed in these calculations, but it was also shown (Balbuena and Gubbins, 1994) that the filling pressure is dependent on the relative strengths of the intermolecular interactions (i.e. solid–fluid and fluid–fluid) and the precise pore geometry (Bock et al., 2008).

Brauer et al. (1993) carried out GCMC simulations for argon adsorption in carbon micropores. As expected, as the pore width was increased from 0.7 to ca. 1 nm, the potential energy curves underwent a change from a single deep potential energy well to two minima, which became progressively separated into the normal wall potentials with a further increase in pore width. The simulation experiments indicated that at 87 K, stepwise filling of the 0.7-nm pore would occur at $p/p^o \sim 10^{-5}$. Pore widening caused filling to separate into two stages. The results of computer simulation are thus consistent with the DFT calculations.

10.4.1.3 Argon Adsorption at 87 K

As already indicated in Section 10.3.2, the stepwise character of argon isotherms on graphitized carbons is removed if the operational temperature is raised to 87 K, the temperature of liquid Ar. For a number of reasons, Ar adsorption at 87 K has been proposed as an alternative to N_2 at 77 K for surface area determination and pore size analysis (Thommes et al., 2012). Ar is obviously a non-polar monatomic molecule and is much less reactive than N_2. In principle, the application of DFT should be more straightforward than is the case with nitrogen. Furthermore, the higher operational temperature provides more favourable conditions for fast equilibration and for the range of micropore filling (Silvestre-Albero et al., 2012; Thommes et al., 2012). It is evident that Point B is not sharply defined at the higher temperature (as indicated by the lower C value), but this is not a serious problem if the method proposed by Rouquerol et al. (2007) is used to assess the BET-monolayer capacity. It is too early to say whether argon adsorption at 87 K will come to be accepted as a standard procedure for surface area determination and pore size analysis.

10.4.1.4 Carbon Dioxide Adsorption at 273 K

The first reliable measurements of carbon dioxide adsorption by charcoal were reported over a hundred years ago (e.g. the work of Miss I.F. Homfray in 1910 – see McBain, 1932), and many CO_2 isotherms have been subsequently determined on various forms of activated carbon. The present strong interest in CO_2 physisorption is mainly due to (a) its use as a probe for the characterization of ultramicroporous carbons and (b) the separation and capture of this troublesome 'greenhouse gas' (see Silvestre-Albero and Rodriguez-Reinoso, 2012).

When studying the properties of carbonized organic polymers, Marsh and Wynne-Jones (1964) found that the levels of uptake of carbon dioxide at 195 K were much larger than could be explained by their BET-nitrogen areas. At first sight, this seems surprising because the two molecules are not very different in size: the kinetic diameter of N_2 is 0.36 nm and that of CO_2 is 0.33 nm and the corresponding minimum dimensions are 0.30 and 0.28 nm. However, by far the most important factor in causing the greater uptake of CO_2 is the higher operational temperature (see Gregg and Sing, 1982).

This effect is shown to an even greater extent when the carbon dioxide measurements are undertaken at 273 or 298 K. At these temperatures, the carbon dioxide saturation pressure is extremely high (e.g. at 298 K $p^o = 63.4$ bar) so that the range of p/p^o is limited to ~0.02 at sub-atmospheric pressures. This has the advantage that the initial part of the isotherm can be determined with a much greater accuracy than is normally possible with nitrogen at 77 K and in addition, the DR plots are generally more linear (Rodriguez-Reinoso, 1989).

In their extensive investigations of microporous carbons, Rodriguez-Reinoso and his co-workers have used carbon dioxide as a molecular probe (generally at 273 K) alongside nitrogen at 77 K (Garrido et al., 1987; Rodriguez-Reinoso, 1989; Rodriguez-Reinoso et al., 1989; Molina-Sabio et al., 1995; Silvestre-Albero and Rodriguez-Reinoso, 2012, 2012). Three groups of porous carbons have been identified: (a) activated carbons of low burn off, giving much larger uptakes of CO_2 than N_2, because of restricted diffusion of N_2 into very narrow pores; (b) activated carbons of low-to-medium burn off, having fairly narrow micropores, giving approximately equivalent uptakes of CO_2 and N_2; and (c) activated carbons of medium-to-high burn off, having a range of wider micropores, giving larger uptakes of N_2 than CO_2.

In the work of Cazorla-Amoros et al. (1996), CO_2 isotherms were determined at 273 and 298 K over a wide range of pressure up to 4 MPa. It was confirmed that CO_2 adsorption at sub-atmospheric pressures is a useful complementary technique for the characterization of very narrow micropores, while at higher pressures, the adsorption of CO_2 and N_2 appeared to be similar. A unified theoretical approach to pore size characterization of microporous carbons was developed by Ravikovitch et al. (2000). This involved the application of NLDFT and GCMC to isotherms of CO_2 (at 273 K), N_2 and Ar (both at 77 K).

Stoeckli and his co-workers used high-pressure CO_2 adsorption (at various temperatures) alongside adsorption and immersion calorimetry and vapour adsorption measurements in their ongoing work on micropore filling (Guillot et al., 2000; Guillot and Stoeckli, 2001; Stoeckli et al., 2002). The behaviour of two well-characterized microporous carbons is of particular interest: one gave a linear DR plot over a very wide range of p/p^o and the other, a strongly activated material, gave a smooth DR plot with a much shorter linear range. The corresponding differential adsorption enthalpies for CO_2 confirmed that different micropore-filling mechanisms were involved, the linear DR region being associated with enhanced adsorbent–adsorbate interactions in ultramicropores.

With the aid of CO_2 adsorption and immersion calorimetry, Stoeckli et al. (2002) also investigated the micropore structure of several CMSs. As might be expected with such ultramicroporous carbons, the CO_2 DR plots exhibited long ranges of linearity. Analysis of the CO_2 comparison plots allowed an approximate estimate to be made of the external areas and micropore volumes.

A GCMC simulation study was made by Müller (2008) of the quadrupolar interactions between adsorbed CO_2 molecules in relation to pore size and shape. The slanted ordering found in narrow nanotubes of a certain critical size could not be detected in slit-shaped pores. Of course, this form of specific interaction might play a small part in the close-packing of CO_2 molecules in the disordered pore structure of an activated carbon.

10.4.2 Adsorption of Organic Vapours

Benzene was the most popular adsorptive in many early studies of the pore structure of activated carbons (Cadenhead and Everett, 1958; Dubinin, 1958, 1966; Smisek and Cerny, 1970). Indeed, in order to construct the characteristic curve for a given microporous carbon, Dubinin and his co-workers (Dubinin, 1966) originally adopted benzene as the standard adsorptive: thus, in the context of the Dubinin theory of the volume filling of micropores (TVFM), the scaling factor $\beta(C_6H_6) = 1$ (see Chapter 5).

Polanyi's concept of the temperature invariance of the characteristic curve became an important feature of the TVFM proposed by Dubinin (1966). The approach provided a way of bringing together a family of isotherms determined at different temperatures. The resulting common curve may be regarded as the relation between the fractional filling of the pores of a microporous carbon by a particular adsorptive and the 'adsorption potential', defined as $RT \ln (p^o/p)$. An example of a typical characteristic curve is shown in Figure 10.18, where it is evident that for the active carbon CK, the common characteristic curve is given by all the benzene isotherms determined over the temperature range of 20–140 °C.

Not all characteristic curves are temperature invariant (Aranovich, 1991; Tolmachev, 1993). Invariance over a wide temperature range has thermodynamic

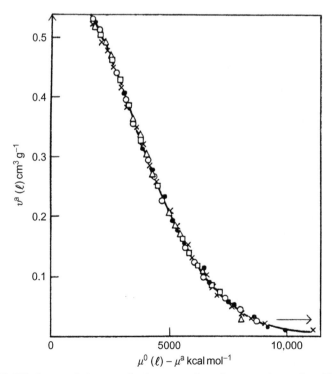

Figure 10.18 DR characteristic curve for benzene adsorption on active carbon CK, for temperatures ranging from 20 (triangles) to 140 °C (solid circles). From Dubinin (1975).

implications, which are unlikely to be consistent with the behaviour of many systems – especially when strong adsorbent–adsorbate interactions or a combination of adsorption mechanisms are involved. As might be expected, different characteristic curves are often obtained when the isotherms of various adsorptives are determined on the same microporous carbon; in fact there are relatively few well-documented examples of an identical characteristic curve being derived from the isotherms of different vapours (Bansal et al., 1988). Bradley and Rand (1995) compared the characteristic curves derived from the isotherms of a series of lower alcohols on a coal-based activated carbon. The characteristic curves for methanol and ethanol are in good agreement, but the corresponding curves for propan-2-ol and butan-1-ol deviate at low fractional loading (as does nitrogen). The authors conclude that the deviations are probably the result of molecular packing restrictions within ultramicropores.

Over the past 30 years, many organic molecules of different sizes, shapes and polarities have been used as molecular probes. A high proportion of the experimental isotherms on porous carbons have been analysed by the application of the DR equation or, in a few cases, by the Dubinin–Astakhov (DA) equation. So far, the more sophisticated Dubinin–Serpinsky treatment (Stoeckli, 1993) has been applied by very few other investigators.

As indicated earlier, an extensive range of linearity of a DR plot is usually associated with primary micropore filling. However, it must be kept in mind that the micropore-filling mechanism is dependent on the nature of the adsorption system and temperature as well as on the pore size. Since it contains an additional adjustable parameter, the DA equation is obviously more adaptable than the simple DR equation. With most activated carbons, values of the empirical exponent N in Equation (5.46) are in the range of 1.5–3.

Stoeckli (1981), McEnaney and Mays (1991), Hutson and Yang (1997) and others (see Rudzinski and Everett, 1992) have attempted to provide a theoretical basis for the DR and DA equations in terms of an integral transform or a generalized adsorption isotherm, which may be expressed in the form of Equation (5.55). However, in practice, the DR and DA equations are usually applied empirically and consequently the derived quantities (micropore volume, characteristic energy and structural constant) are not always easy to interpret.

An alternative, and perhaps more pragmatic, approach is to compare the pore-filling behaviour of a number of non-polar vapours of different molecular sizes (Carrott and Sing, 1988). In this manner, it is possible to explore changes in the isotherms as a result of increasing the size and polarizability of the adsorbate without the added complication of differences in the specific interactions. The following few examples will illustrate the applicability of this approach.

The adsorption isotherms and corresponding α_s-plots for nitrogen, propane, isobutane and neopentane on four different activated carbons are given in Figures 10.19–10.22. The isotherms in Figure 10.19 were determined on Carbosieve and those in Figures 10.20 and 10.21 on different grades of carbon cloth, while the isotherms in Figure 10.22 were on the superactive carbon AX21. To facilitate comparison of the levels of uptake, the amounts adsorbed are expressed as the equivalent volumes of *liquid* adsorptive. Of course, it must be kept in mind that the densities of the adsorbates are unlikely to be exactly the same as the corresponding bulk liquid

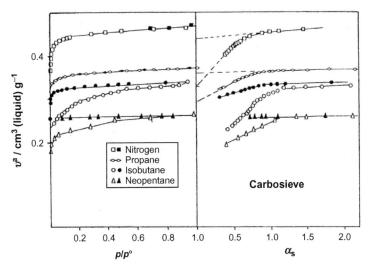

Figure 10.19 Adsorption isotherms and corresponding α_s plots for Carbosieve S (open symbols: adsorption; solid symbols: desorption) (Carrott et al., 1988a,b).

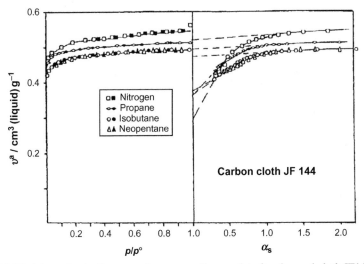

Figure 10.20 Adsorption isotherms and corresponding α_s plots for charcoal cloth JF144 (open symbols: adsorption; solid symbols: desorption) (Carrott et al., 1988a,b).

densities. Although some of the isotherms in Figures 10.19–10.22 are more complex than others, they are all essentially Type I in the IUPAC classification. Evidently therefore, the four carbons are predominantly microporous, but with different ranges of pore size.

The α_s-plots in Figures 10.19–10.22 were constructed with the aid of standard adsorption data obtained with Elftex 120 and other non-porous carbon blacks (Carrott et al., 1987; Carrott et al., 1988a,b). As explained earlier in this chapter

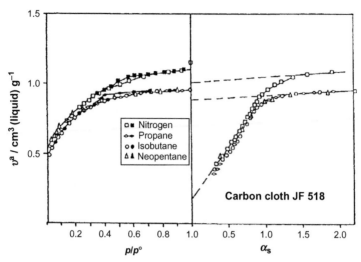

Figure 10.21 Adsorption isotherms and corresponding α_s plots for charcoal cloth JF518 (open symbols: adsorption; solid symbols: desorption) (Carrott et al., 1988a,b).

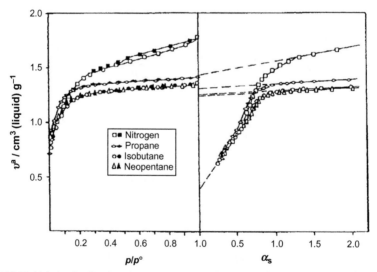

Figure 10.22 Adsorption isotherms and corresponding α_s plots for active carbon AX21 (open symbols: adsorption; solid symbols: desorption) (Carrott et al., 1988a,b).

(also see Chapter 7), an α_s-plot on a microporous adsorbent generally has two sections. If the first section is linear (at $\alpha_s < 1.0$, i.e. $p/p^o < 0.4$), it can be attributed to adsorption on the walls of the supermicropores and therefore its back-extrapolation to $\alpha_s = 0$ gives the ultramicropore capacity. The second section, at $\alpha_s > 1.0$, is associated with either capillary condensation or multilayer adsorption; if linear, the intercept corresponds to the total micropore capacity.

In attempting to make a semi-quantitative estimation of the range of pore size in the four carbons, we must assume (a) that the pores are rigid and remain so during the physisorption measurements, (b) that the micropores are all slit-shaped and (c) that primary micropore filling occurs in pores of width $<2d$ and secondary micropore filling in pores of up to $\sim 5d$ (where d is the molecular diameter). Since N_2 molecules are not able to enter 0.4-nm zeolite channels at 77 K (Breck, 1974), it is assumed that this is also the lower limit for their entry into slit-shaped pores. On the basis of these assumptions, the permitted ranges of primary and secondary micropore filling are given in Table 10.3.

The isotherms in Figure 10.19 confirm the molecular sieve and ultramicroporous character of Carbosieve. A striking feature of the isobutane and neopentane isotherms is their pronounced low-pressure hysteresis (LPH). Since the rate of uptake of the bulky isobutane and neopentane molecules was very slow, thermodynamic equilibration was not established and therefore these isotherms are not amenable to detailed analysis. In contrast, the smaller nitrogen and propane molecules are more rapidly adsorbed and their isotherms on Carbosieve are reversible although the available micropore capacities are quite different.

The α_s-plots for nitrogen and propane reveal that in each case ~ 70–80% of the total uptake is associated with primary micropore filling. If we accept the approximate limits of primary and secondary pore filling in Table 10.2 and allow for the different nitrogen and propane micropore capacities, we may tentatively conclude that $\sim 60\%$ of the nitrogen-available pore volume is in pores of effective width of ~ 0.45–0.9 nm. The entry of the more bulky isobutane and neopentane molecules is evidently more restricted and the lack of reversibility complicates the interpretation, but it seems that up to $\sim 55\%$ of the total micropore volume is available to neopentane and is located in pores of ~ 0.6 nm. It must be stressed again that the use of a few molecular probes in this way cannot be expected to provide more than a preliminary rough indication of the effective micropore size distribution. There are at least two unknown variables: the distribution of pore shape and the non-rigidity of the disordered porous carbon.

The situation is somewhat less complex in the case of the isotherms on the JF144 (Figure 10.20) since all the isotherms are now reversible, the paths of the isobutane and neopentane being virtually identical. The location of each low-coverage intercept indicates a fairly high proportion of primary micropore filling (60–70%), and it is significant that at least 90% of the micropore-nitrogen capacity is available for

Table 10.3 Primary and secondary filling of slit-shaped pores

Adsorptive	d (nm)	Effective pore widths, w (nm)	
		Primary	**Secondary**
Nitrogen	0.36	0.4–0.7	0.7–1.8
Propane	0.43	0.45–0.9	0.9–2.2
Isobutane	0.50	0.5–1.0	1.0–2.5
Neopentane	0.62	0.6–1.2	1.2–3.1

neopentane adsorption. From these results, we may conclude that the micropores in carbon cloth were mainly distributed over the effective pore width range of ~0.6–2.0 nm.

In Figure 10.21, all three hydrocarbon isotherms on JF518 are in remarkably close agreement, but the nitrogen uptake is much larger except at very low p/p^o. All the isotherms on JF518 exhibit a gradual approach to the plateau. The predominance of secondary micropore filling is confirmed by the small scale of the primary intercepts (ca. 20% of the total hydrocarbon micropore capacity). This is a strong indication that most pores are in the supermicropore range with effective widths of ~1–3 nm.

The hydrocarbon isotherms on the superactive AX21 in Figure 10.22 are also fairly close (especially at low p/p^o). In this case, a distinctive feature of the nitrogen isotherm is a narrow hysteresis loop in the capillary condensation range and the shape of the corresponding α_s-plot is indicative of a continuous range of supermicropores and narrow mesopores. Since there appears to be no sharp boundary between the completion of cooperative micropore filling and the beginning of reversible capillary condensation, we cannot at present define the upper limit of secondary micropore filling.

The fact that all the nitrogen isotherms deviate to some extent from the corresponding hydrocarbon isotherms illustrates the abnormal adsorptive behaviour of nitrogen and confirms that low-temperature nitrogen adsorption is not able to provide a complete evaluation of the adsorptive properties of a porous carbon. Furthermore, there are inescapable difficulties involved in the characterization of microporous adsorbents. As other authors have also pointed out (Gregg and Sing, 1982; Aukett et al., 1992; Bradley and Rand, 1995), adsorbate densities in micropores are unlikely to be the same as the corresponding liquid densities. Furthermore, the degree of molecular packing is to some extent dependent on both pore size and shape (Carrott and Sing, 1988). For these and other reasons, it is advisable to refer to the *effective* or *apparent* pore size or volume (Rouquerol et al., 1994) and also to specify the adsorptive and operational temperature.

10.4.3 Adsorption of Water Vapour

Isotherms of Types III and V were uncommon in 1940 when the BDDT classification was first proposed (see Brunauer, 1945). In the early 1950s, it was shown by Pierce and his co-workers that a Type III isotherm was given by the adsorption of water vapour on Graphon – a graphitized carbon black (Pierce and Smith, 1950; Pierce et al., 1951; Pierce and Smith, 1953). Kiselev and his co-workers reported that after high-temperature treatment in hydrogen, the water uptake by graphitized carbon black was so small that it was even difficult to *detect* at $p/p^o < 0.8$ (Avgul et al., 1957). Correspondingly, the energy of immersion of graphitized carbons in water was also found to be extremely low (Zettlemoyer et al., 1953; Barton and Harrison, 1975; Bansal et al., 1988).

The first well-defined Type V water isotherms on charcoal were also reported in the 1950s (Pierce et al., 1951; Arnell and McDermott, 1952; Pierce and Smith, 1953; Dacey et al., 1958). Since then, the adsorption of water vapour by microporous carbons has been studied in considerable detail (see Dubinin, 1980; Gregg and Sing, 1982; McCallum et al., 1999; Kaneko, 2000; Striolo et al., 2003; Bradley, 2011).

The reason for the low uptake of water on the graphite basal plane (i.e. on graphene) is not difficult to understand. The water molecule is small and of relatively low polarizability. In the absence of specific interactions, the adsorbent–adsorbate interactions are therefore weak, which accounts for the low energy of immersion of carbon black in water (see Bradley, 2011).

Two typical examples of water isotherms on microporous carbons are shown in Figure 10.23. Although, as we have already seen, Carbosieve and the superactive carbon AX21 were both highly microporous, their pore size distributions were quite different. It is apparent that the very narrow pores ($w \sim 0.4$–0.8 nm) in Carbosieve begin to fill with water at $p/p^{\circ} \sim 0.3$, while filling of the wider pores ($w \sim 1$–3 nm) of AX21 occurs at $p/p^{\circ} > 0.5$. Another interesting difference between the two water isotherms in Figure 10.23 is the size of their hysteresis loops: the loop given by Carbosieve is steep and narrow, while the AX21 loop is broader and more rounded. This difference is in accordance with the work of Miyawaki et al. (2001).

A qualitative explanation for isotherms of the type shown in Figure 10.23 was first put forward by Pierce and Smith (1950), who postulated that initially a few water molecules are adsorbed on polar sites (e.g. oxygen complexes) and, with increase in p/p°, clusters of molecules are then hydrogen-bonded around these favourable sites. The clusters grow with increased pressure until they merge together and the pores are filled. According to Pierce and Smith, the hysteresis is due to a difference between the steps involved in pore filling and emptying, the latter representing a more stable state. The two-stage process was the basis of the model adopted by Dubinin and Serpinski (1981) and further developed by Barton and Koresh (1983) and Talu and Meunier (1996).

Most of the Type V water isotherms in the literature exhibit pronounced hysteresis, but a few reversible – or almost reversible – water isotherms have been reported on

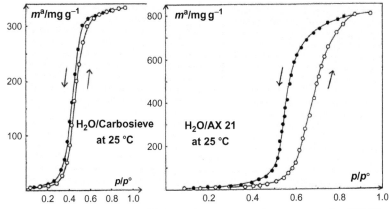

Figure 10.23 Water isotherms at 25 °C on microporous carbons outgassed at 300 °C (Sing, 1991).

certain ultramicroporous carbons. For example, the isotherm obtained by Dacey et al. (1958) on Saran charcoal had the well-defined Type V character with a steep riser at $p/p^o \sim 0.5–0.6$. There was a small amount of LPH, but the isotherm appeared to be completely reversible at $p/p^o > 0.55$. It was pointed out, however, that on the steeply rising part of the isotherm, over 24 h were required for equilibrium to be reached. Another almost reversible Type V water isotherm was reported by Kaneko (2000). These parameters were explored by Miyawaki et al. (2001), who found a semi-quantitative relation between the loop width and its p/p^o location.

Completely reversible Type V isotherms are quite rare. It is significant that the example reported by Dubinin (1980) was given by a low burn off (5.7%) carbon, which was certainly ultramicroporous. After 20% burn off, the hysteresis extended over virtually the whole range of pore filling. Similar findings were reported by Vartapetyan et al. (2005), who studied the adsorption of water on activated chars produced from a tropical wood. In the work of Barton and Koresh (1983), a reversible water isotherm was obtained after the low-temperature (i.e. 40 °C) evacuation of a carbon cloth. The molecular sieve character of this material was reduced by evacuation at 400 °C and this also led to the appearance of hysteresis in the water isotherm. Barton and Koresh (1983) came to the questionable conclusion that such a hysteresis is *mainly* due to the concentration of surface oxides 'which dictate the adsorption value at which the change from cluster adsorption to a continuous adsorbed phase takes place'.

The shape of the initial part of a Type V water isotherm is controlled by the surface chemistry (Gregg and Sing, 1982; Bansal et al., 1988; Rodriguez-Reinoso et al., 1995; Choma and Jaroniec, 1998; Kaneko, 2000; Boehm, 2008; Bradley, 2011; Thommes et al., 2011). As already indicated, the low affinity of the surface of pure carbon for water is associated with the weak non-specific interactions between the carbon surface and the adsorbate. When certain functional groups are present, specific interactions come into play and the adsorption affinity is thereby increased. The relationship between the adsorption of water and the surface concentration of chemisorbed oxygen was identified by Walker and Janov (1968). Bansal et al. (1978a,b) also studied the influence of the surface oxygen on the adsorption of water: they concluded that at $p/p^o < 0.5$, the level of water uptake is determined by the concentration of surface oxygen-containing structures.

The hydrophilic sites on the surface of ACFs were investigated by Kaneko et al. (1995) with the aid of XPS. In this work, CEL- and PAN-based ACFs were studied and samples were either chemically treated with H_2O_2 or heated in H_2 at 1000 °C. As expected, surface oxidation by the H_2O_2 treatment increased the initial uptake of water, while the H_2 reduction caused a marked decrease in the amount of water adsorbed at low p/p^o. Measurement of the peak areas of the XPS spectra provided a means of determining the fractional surface coverage by the hydrophilic sites. In this way, a linear relationship was found between the low-pressure adsorption of water vapour and the number of hydrophilic sites (mainly COOH).

The effects produced by changing the surface chemistry of porous carbons have been studied by a number of other investigators. In the systematic work of Hall and Holmes (1991, 1992, 1993), chemical modification of activated carbons was brought about by treatment with such agents as chlorine, phosgene, fluorine, dinitrogen tetroxide (NO_2) and 1,1-difluoroethene. Significant changes were observed in the $p/p°$ required for pore filling by water vapour. In some cases, the behaviour of the modified adsorbent was found to be dependent on the reaction conditions: for example, the degree of hydrophobicity produced by chlorination was affected by the reaction temperature. Although temperatures of around 180 °C were required to produce relatively hydrophobic material, it seems that chlorination at low temperature may offer a way of attaching different functional groups to the carbon surface (Hall and Holmes, 1993).

Kaneko et al. (1995) found that it was possible to produce highly hydrophobic fluorinated microporous carbon fibres. Two fluorinated carbons were reported to have BET-areas of 420 and 340 $m^2\,g^{-1}$ and micropore volumes of, respectively, 0.19 and 0.14 $cm^3\,g^{-1}$. These materials gave Type I nitrogen and methanol isotherms, but the adsorption of water vapour was too small to measure at $p/p° < 0.8$ and the uptake was very low even at $p/p° \sim 1$.

A novel method for determining the location of the primary water adsorbing sites was developed by Bailey et al. (1995). This approach involved the pre-adsorption of naphthalene, which was chosen because of its planar molecular shape and immiscibility with water. With some activated carbons, it was found that the growth of the H-bonded water clusters was inhibited by the presence of naphthalene, while in other cases, there was very little effect. It was thought that sites in larger micropores were prone to obstruction by the pre-adsorbed naphthalene. It is too early to judge the success of this interesting approach, which may turn out to be a useful alternative to pre-adsorption by n-nonane.

It is evident that the initial stage of water vapour adsorption is dependent on the presence of a number of polar sites on the carbon surface. In the theory of Dubinin and Serpinski (1981), these sites are pictured as primary adsorption centres of uniformly high energy. Water molecules are first adsorbed on the primary centres in a 1:1 ratio and these molecules then act as secondary adsorption centres for the adsorption (by H-bonding) of other molecules.

The Dubinin–Serpinsky (DS) equation, which was based on this model, can be expressed in the form

$$p/p° = \boldsymbol{n}/[k_1(n_0 + \boldsymbol{n})(1 - k_2\boldsymbol{n})] \tag{10.3}$$

where $\boldsymbol{n}$ is the specific amount of water adsorbed at $p/p°$, n_0 is the specific amount of primary centres, k_1 is the ratio of the rate constants for adsorption and desorption and k_2 is a constant related to the uptake at $p/p° = 1$.

In some studies, it has been found that Equation (10.3) is applicable to experimental water adsorption data in the region of $0.4 < p/p° < 0.8$ (Stoeckli et al., 1994a,b). In practice, the $p/p°$ range is dependent on both the nature of the carbon surface and the pore structure (Carrott et al., 1991; Carrott, 1993). The range of fit has been extended by the addition of other empirical parameters (Barton et al., 1991, 1992; Talu and Meunier, 1996; Miyawaki et al., 2001).

Stoeckli et al. (1994a,b) have shown that the adsorption branch of a water isotherm can be described by a Dubinin–Astakov (DA) type of equation,

$$n = n_p \exp\left[-(A/E)^N\right] \tag{10.4}$$

where n_p is the micropore capacity, $A = RT \ln(p^o/p)$ and N and E are temperature-invariant parameters.

As we saw in Chapter 5, Equation (10.4) has been customarily applied to Type I isotherms. However, if the value of E is sufficiently low (i.e. $E < 2$–3 kJ mol^{-1} near room temperature), it can be applied to the major part of a Type V isotherm. Furthermore, with at least some systems, it has been found that E and N do not vary appreciably with temperature and therefore the temperature-invariance condition is satisfied (Stoeckli et al., 1994a,b). On the other hand, so far it has not been possible to define the exact meaning of these terms and Equations (10.3) and (10.4) must be regarded as empirical equations.

The structure of water confined in 'carbon nanospace' was discussed *inter alia* in an important review by Kaneko (2000). The distinctive X-ray diffraction patterns obtained by Iiyama et al. (1995) had already provided new evidence for the existence of long-range order of water in microporous carbon. By examining additional X-ray diffraction and small-angle scattering data, Kaneko identified the characteristic patterns of water in micropores of different sizes. As one might expect, molecules in ultramicropores appeared to be less mobile than in supermicropores. Furthermore, this behaviour was associated with the reversibility of the ultramicropore filling in contrast to the hysteresis associated with the filling and emptying of the wider micropores. The relationship between the pore width and hysteresis of water adsorption was explored in more detail by Miyawaki et al. (2001). With a series of ACFs, increase in loop area was associated with increase in pore width.

By application of the well-known Gurvich rule, it has been customary to evaluate the apparent specific pore volume from the uptake of an adsorptive at a pressure close to saturation (say, at $p/p^o = 0.98$) and to assume the adsorbate to have liquid-like properties (the bulk liquid density of water is 0.99 g cm^{-3} at 293 K). The numerous comparisons between the micropore volumes assessed in this manner from water and other isotherms have revealed moderate agreement in some cases and also significant discrepancies between the corresponding values (e.g. in the work of Vartapetyan et al., 2005). A difference of ca. 18% was reported by Thommes et al. (2011) for the water and nitrogen micropore volumes of a coal-based AC, water vapour giving the lower value. These and other data (Iiyama et al., 2000) appeared to indicate that the effective density of water adsorbed in carbon slit-shaped micropores was between 0.81 and 0.86 g cm^{-3}. However, the interpretation of these data may not be quite as straightforward as it might appear.

Earlier results of adsorption measurements on a variety of microporous carbons (see Sing, 1991) revealed similar discrepancies between the micropore volumes derived from the corresponding water and nitrogen adsorption capacities. But, when the comparison was made between water (at 298 K) and isobutane (at 261 K) on four microporous carbons, the corresponding values of micropore volume were in remarkably good agreement. These results illustrate the difficulty in characterizing the adsorbate structure simply by comparing the corresponding values of apparent pore volume.

In spite of the theoretical difficulties of quantifying the overall water–carbon interactions, MS studies have broadly supported the original ideas of Pierce and Smith (1950, 1953). Thus, the GCMC studies of Müller, Gubbins, Seaton, Quirke and their co-workers have confirmed that the adsorption of water on carbon does not proceed in an orderly monolayer–multilayer form (see Müller et al., 1996; McCallum et al., 1999; Jorge and Seaton, 2002; Striolo et al., 2003). Instead, water adsorbs preferentially on particular sites. The adsorbed molecules then act as nucleation sites for the further adsorption of molecules, which form hydrogen-bonded three-dimensional clusters. It is evident that spontaneous pore filling occurs long before a hypothetical monolayer would be formed. The complexity of the mechanism is due to: (a) the precise location and nature of the hydrophilic sites, (b) the adsorbate–adsorbate hydrogen-bonded structure; and (c) the micropore size and shape distribution. Each activated carbon is therefore likely to behave in a unique manner.

Thommes et al. (2011) have stressed the special features of water adsorption by microporous carbons. With the aid of temperature-programmed desorption coupled with mass spectrometry together with XPS and a range of physisorption measurements, it was possible to separate the dependency of water adsorption on porosity and on surface chemistry. The comparison of the water isotherms on well-characterized ultramicroporous carbons confirmed that differences in the uptake at low p/p^o were dependent on the surface chemistry, while at higher p/p^o the location of the isotherm loop was controlled by the pore structure. A detailed study revealed that the loop shape did not depend on the degree of pore filling, but that the position of the desorption boundary curve was to some extent temperature dependent. These results were in qualitative agreement with the findings of Ohba et al. (2004) and provide additional support for the view that hysteresis is the result of delayed condensation.

All the above findings are consistent with the fact that water can be accommodated easily in narrow slit-shaped pores of carbon in contrast to the hydrophobicity of the tubular pores in Silicalite (Carrott et al., 1991). As indicated in Figure 10.24, a thin slab of hydrogen-bonded water can be placed in a slit of width $>\sim0.5$ nm with little distortion of its structure. Moreover, we can begin to see why the appearance of the hysteresis loop is dependent on pore size. It seems likely that the adsorbate in the central layer will become more liquid like as the pore width is increased. In the absence of

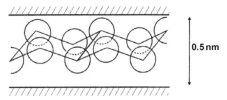

0.5 nm

Figure 10.24 H-bonded structure of water in a slit-shaped pore (Carrott et al., 1991).

pore blocking or network effects, the filling/emptying of narrow slit-shaped micropores can take place reversibly (i.e. without hysteresis). Above a limiting p/p^o, wider pores are filled in two stages, resulting in the appearance of adsorption/desorption hysteresis. For microporous carbons with uniform slit-shaped pores, the limiting p/p^o may be considered to resemble an adsorbate phase boundary and to be temperature dependent.

10.4.4   Adsorption of Helium

Helium is often used in adsorption manometry for the determination of the 'dead space' volume (see Chapter 3), but this procedure is based on the presupposition that the gas is not adsorbed at ambient temperature and that it does not penetrate into regions of the adsorbent structure which are inaccessible to the adsorptive molecules. In fact, with some microporous adsorbents, significant amounts of helium adsorption can be detected at temperatures well above the normal boiling point (4.2 K). For this reason, the apparent density (or the so-called true density) determined by helium pycnometry (Rouquerol et al., 1994) may be dependent on the operational temperature and pressure (Fulconis, 1996).

Because of its small size (collision diameter $\sim$0.20 nm), helium would appear to be a useful probe molecule for the study of ultramicroporous carbons. Obviously, the experimental difficulty of working at liquid helium temperature (4.2 K) is the main reason why helium has not been widely used for the characterization of porous adsorbents. In addition, since helium has some unusual physical properties, it is to be expected that its adsorptive behaviour will be abnormal and dependent on quantum effects.

In a series of investigations of helium adsorption by microporous carbons, Kaneko and his co-workers (Kuwabara et al., 1991; Setoyama et al., 1993; Setoyama and Kaneko, 1995; Setoyama et al., 1996) obtained strong evidence that the density of physisorbed helium is not the same as that of bulk liquid helium at 4.2 K (i.e. 0.102 g cm^{-3}). By adopting the value 0.202 g cm^{-3}, which had been proposed on theoretical grounds by Steele (1956), Kaneko and his co-workers were able to obtain fairly good agreement between the corresponding uptakes of He and N_2 by certain microporous carbons – as indicated in Figure 10.25. With some other porous carbons,

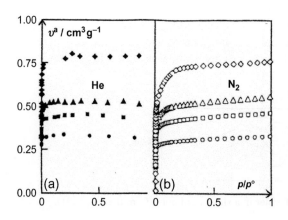

Figure 10.25 Adsorption isotherms of He at 4.2 K (a) and of N_2 at 77 K (b) on a carbon activated up to 19% (circles), 34% (squares), 52% (triangles) and 80% burn off, respectively.

Courtesy Setoyama et al. (1996).

the presence of narrow ultramicropores was demonstrated by the much larger apparent pore volumes available for helium adsorption.

The shapes of a series of helium and nitrogen isotherms are compared in Figures 10.26 and 10.27. To facilitate comparison, the amount adsorbed is expressed in the form of a liquid-like volume. As already indicated, the adjusted density of the adsorbed helium is taken as 0.202 g cm^{-3}, whereas the adsorbed nitrogen is assumed to have a normal liquid density of 0.808 g cm^{-3}. It is evident that on this basis the corresponding saturation levels are again in quite close agreement, in accordance with the Gurvich rule. A notable feature of the helium isotherms is their relative steepness at very low p/p°. The striking difference between the helium and nitrogen isotherms is revealed in Figure 10.26, where on the abscissa p/p° is replaced by $\log(p/p^\circ)$. It is now apparent that micropore filling by helium has begun at $p/p^\circ < 10^{-6}$ in comparison with the corresponding nitrogen micropore filling at $p/p^\circ \sim 10^{-4}$.

The observed difference in the low-pressure character of the helium and nitrogen isotherms is all the more remarkable in the light of the agreement already noted between the Gurvich volumes. The explanation for this difference given by Kaneko is based on the theory originally proposed by Steele (1956) that the abnormally high adsorption of helium on a graphitic surface at 4.2 K is due to 'an accelerated bilayer adsorption'. In view of this unusual behaviour, it might be expected that distinctive micropore-filling mechanisms would be exhibited by helium at 4.2 K. It turns out,

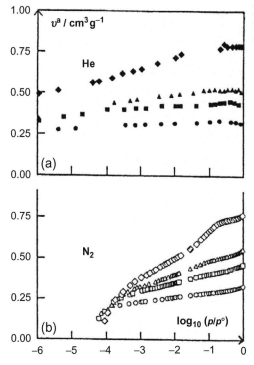

Figure 10.26 Logarithmic plot of isotherms in Figure 9.23. *Courtesy* Setoyama et al. (1996).

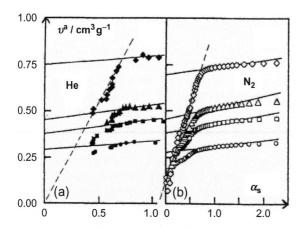

Figure 10.27 α_s plots of adsorption isotherms given in Figures 9.23 and 9.24. *Courtesy* Setoyama et al. (1996).

however, that there appears to be a close similarity between the helium and nitrogen primary and secondary stages of micropore filling.

The helium and nitrogen α_s-plots in Figure 10.27 were derived from the isotherms in Figure 10.26 with the aid of standard isotherm data determined on a non-porous carbon black (Setoyama et al., 1996). In spite of the appreciable difference in the range of p/p^o, it is apparent that the shapes of the corresponding He and N_2 α_s-plots are remarkably similar. We may conclude that, with both adsorptives, the initial stage of physisorption is either predominantly surface coverage of supermicropore walls or primary micropore filling of ultramicropores. These findings are entirely consistent with the conditions of activation of these and the other porous carbons, which had been prepared from olive stones, the levels of burn off being indicated by the sample designation (i.e. 80% and 19%, respectively).

As we have seen, the apparent density of He at 4.2 K in the slit-shaped micropores of some activated carbons appears to be ca. 0.20 g cm^{-3}. However, the true density must depend on both the nature of the adsorption system and the pore size and shape. Setoyama and Kaneko (1995) have given a possible range of 0.20–0.23 g cm^{-3} for the density of helium in the micropores of activated carbons.

10.5   Adsorption at the Carbon–Liquid Interface

10.5.1   Immersion Calorimetry

In characterizing their original batch of charcoal cloth, Bailey and Maggs (1972) used the traditional 'heat of wetting' method (see Smisek and Cerny, 1970). This form of immersion calorimetry is now known to be of limited value, but provided that it is applied under carefully controlled conditions, the technique can yield useful information on the surface area or the surface chemistry or the porosity of an active carbon (see Denoyel et al., 2008; Silvestre-Albero et al., 2012). The experimental procedure, which makes use of a sensitive Tian-Calvet microcalorimeter, is described in Section 4.4. The main differences in approach are in the choice of wetting liquids.

A selection of polar liquids is required for a systematic study of the surface chemistry. In the early work of Zettlemoyer and Chessick (1974), liquids of different dipole moment were used with the aim of obtaining a linear relation between the energy of immersion and the specific interaction of the dipole and the surface electrostatic field. Differences in the non-specific interaction energies were minimized by maintaining a constant hydrocarbon moiety (e.g. by comparing the energies of immersion of *n*-butylamine and *n*-butanol).

Rodriguez-Reinoso et al. (1997) identified the properties of two different surface oxygen groups on activated carbons by making their immersion energy measurements with benzene, methanol and water. In this manner, it was possible to distinguish between the behaviour of functional groups, which were thermally displaced in the form of either CO or CO_2. Similarly, Bradley and Stoeckli and their co-workers have employed immersion calorimetry along with other techniques in their extensive investigations of the effects of oxidation on the surface polarity of carbon blacks and CNTs (e.g. Bradley et al., 1995; Lopez-Ramon et al., 1999; Andreu et al., 2007; Bradley et al., 2012).

The enthalpies of immersion of a series of carbon blacks in water, toluene, methanol, ethanol and isopropanol together with the corresponding physisorption isotherms and XPS measurements allowed Andreu et al. (2007) to make a detailed study of the specific and non-specific interactions exhibited before and after surface oxidation. As one would expect, in the case of the alcohols, it was found that the degree of specificity was inversely related to the hydrocarbon chain length. With toluene, the enthalpy of immersion *per unit area* underwent little variation with increase in surface oxygen level and it was concluded that the interaction was essentially non-specific and furthermore there was no significant structural change as a result of the surface oxidation.

In a recent investigation of the effect of surface hydroxylation on the properties of MCNTs, Bradley et al. (2012) used enthalpy of immersion in water to follow the changes in polarity. As with carbon black, the large increase in areal immersion enthalpy (~fourfold) confirmed the highly polar (hydrophilic) nature of the treated material as did the considerably enhanced uptake of water vapour.

As already discussed in Chapter 4, the surface area of a mesoporous solid can be directly assessed by immersion calorimetry and compares well with the BET surface area. However, because of the unreliability of the BET-area, the latter is of no help for the interpretation of the immersion enthalpy data in the case of an ultramicroporous carbon. But, as indicated in Section 9.3, this does not invalidate the use of immersion calorimetry for characterizing microporous materials. For the pore size analysis of microporous carbons, a series of probe liquids of different molecular sizes should be employed. The energy of immersion can be converted into an effective area, which is accessible to each liquid, and the micropore size distribution is then obtained from the plot of surface area versus molecular size (see Figure 9.5).

The Tian-Calvet microcalorimetric technique was used in a comparative study of the properties of carbon cloth, superactive carbon and graphitized carbon black

Table 10.4 Energies of Immersion for Charcoal Cloth AM4, AMOCO Carbon PX21 and Graphitized Carbon Black Vulcan 3G in Organic Liquids at 300 K (Atkinson et al., 1982)

	Energies of immersion					
Immersion liquid	AMOCO PX21		C Cloth AM4		Vulcan 3G	
	$J\,g^{-1}$	$mJ\,m^{-2}$	$J\,g^{-1}$	$mJ\,m^{-2}$	$J\,g^{-1}$	$mJ\,m^{-2}$
n-Hexane	245	66	94	74	6.1	86
Cyclohexane	190	51	97	77	5.8	82
Neohexane	190	52	72	58	5.3	75
Toluene	271	73	155	123	8.2	115
Mesitylene	300	81	161	128	9.8	138
Isodurene	305	82	154	122	10.7	150

Areal values in $mJ\,m^{-2}$ are based on BET-nitrogen areas.

(Atkinson et al., 1982). The results in Table 10.4 were obtained on a standard batch of Porton charcoal cloth (AM4), an AMOCO superactive carbon (PX21) and a graphitized carbon black (Vulcan 3G) immersed in a series of organic liquids.

The results in Table 10.4 confirm that the areal energies of immersion (in column 7) of graphitized carbon black increase progressively with increase in carbon number, N_C. This behaviour is consistent with the dependence of E_0 on N_C in Figure 1.5 and is also broadly in agreement with the work of Bradley and his co-workers. As already indicated, because of the unreliability of the BET-areas, it is not possible to arrive at an unambiguous interpretation of the apparent areal immersion energies for the two microporous carbons.

We are able to make further progress in the following manner: by adopting a more thermodynamically rigorous treatment, we can transform the energies of immersion into integral energies of adsorption (see Chapter 5). In Figure 10.28, the derived values of $\Delta_{ads}u$ for the charcoal cloth and superactive carbon are plotted against N_C. The difference in the overall energies of adsorption by activated carbon cloth and superactive carbon is now clearly evident as is the N_C dependency. However, these results are still of limited value unless we have some additional information. In this particular case, the difference between the corresponding values of $\Delta_{ads}u$ can be attributed to the higher proportion of ultramicropores in the carbon cloth.

10.5.2 Adsorption from Solution

Adsorption at the solution/carbon interface is widely used on a large scale for water treatment, decolourizing sugars, gold recovery, etc. (Derbyshire et al., 1995; Rodriguez-Reinoso, 2002; Bottani and Tascon, 2008). In addition to these well-established applications, considerable interest is now being shown in the potential

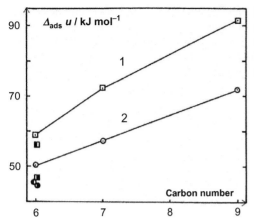

Figure 10.28 Integral molar energies of adsorption given by the immersion of carbon cloth AM4 (1) and superactive PX21 (2) in organic liquids plotted against carbon number (Atkinson et al., 1984).

use of activated carbons for the removal of a wide range of pollutants such as aromatic hydrocarbons, humic acids and heavy metal ions (Costa et al., 1988; Youssef et al., 1996; Moreno-Castilla, 2008) and for the treatment of radioactive waste (Qadeer and Saleem, 1997). In recent years, considerable progress has been made in optimizing the performance of commercial grades of activated carbon.

As explained in Chapter 4, adsorption from solution is generally more complex than physisorption at the gas–solid interface. In spite of this complexity, several attempts have been made to classify adsorption isotherms from *dilute* solutions (see Denoyel et al., 2008). Two hypothetical ideal types are shown in Figure 10.29 where the 'apparent adsorption' (i.e. the reduced surface excess amount, based on the change of solute concentration) is plotted against the equilibrium concentration. Type L is the 'classical' Langmuir form of isotherm with a plateau, which corresponds to monolayer completion, whereas in the case of Type S, the surface coverage is preceded by interactions between the solute molecules.

Many activated carbons give solute isotherms of L type, although a well-defined plateau (corresponding to the monolayer completion or the micropore capacity) is not usually attained. It is then customary to apply a simple or binary form of the Langmuir

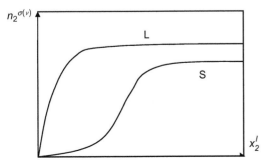

Figure 10.29 Two basic shapes of adsorption isotherms from dilute solutions (Denoyel et al., 2008).

equation or a Langmuir–Freundlich relation, which supposedly takes energetic heterogeneity into account (Jaroniec and Madey, 1988; Derylo-Marczewska, 2010).

In the past, much attention was given to the study of dye and iodine adsorption by active carbons (see Brunauer, 1945; Orr and Dalla Valle, 1959) and this continues to be of some interest. Although dye molecules of known size, shape and chemical properties have been used, the results are not easy to interpret (Giles et al., 1970; Mc Kay, 1982, 1984; Figueiredo et al., 2011). The adsorption of iodine (from aqueous solution) was investigated by Ziolkowska and Garbacz (1997) and also, in a systematic study of the porosity of four activated carbons, by Fernandez-Colinas et al. (1989a,b). In this work, the iodine isotherms were analysed by an adapted form of the α_s-method, with non-porous carbon black taken as a reference, in order to assess the available volumes of pores of effective width in the range 0.5–1.5 nm. Differential enthalpy measurements provided independent evidence for the validity of the method.

As is well known, phenolic compounds are industrial pollutants, and for this reason, their adsorptive behaviour has received attention (Moreno-Castilla, 2008). The role of microporosity in the adsorption of phenol and 4-chlorophenol from aqueous solution was identified by Juang et al. (2001) and more recently this was confirmed by Velasco and Ania (2011), who studied the mechanisms of phenol adsorption by activated carbons before and after surface oxidation. Somewhat surprisingly, these results indicated that the micropore structure was of much greater importance than the polar nature of the carbon surface. The work of Derylo-Marczewska et al. (2010) on the equilibria and kinetics adsorption of nitrophenols and chlorophenols has revealed a more complex pattern of behaviour, which appears to depend on the adsorptive, the adsorbent and the temperature. The removal of amitrole (a non-selective herbicide) from aqueous solution by activated carbon cloth was found to be highly dependent on the surface chemistry and also on the pH of the solution (Moreno-Castillo et al., 2011). The importance of pH was also encountered in the work of Ania et al. (2011) on the 'reactive adsorption' of penicillin by activated carbons.

The above findings confirm the complexity of adsorption at the carbon/ solution interface (particularly with activated carbons) and the need for caution in interpreting the apparent adsorption data. It is now evident that in addition to the nature and magnitude of the specific and non-specific solute–carbon interactions, allowance must be made for the role of the solvent (in its interactions with both the carbon and the solute) and also any temperature dependency. Furthermore, thermodynamic equilibration may not be attained if the activated carbon is ultramicroporous.

10.6   LPH and Adsorbent Deformation

10.6.1   Background

Two important assumptions are usually made in the interpretation of physisorption isotherms. The first is that, if adsorption hysteresis is present, it is always located only within the capillary condensation range of the isotherm. It follows that monolayer–multilayer coverage and micropore filling are both regarded as reversible phenomena and therefore are amenable to classical thermodynamic analysis. The second important assumption is that under the operational conditions no gas molecules are able to penetrate into the adsorbent, which remains rigid with no structural deformation.

In fact, many examples of hysteresis extending down to very low pressures have been reported (see Bailey et al., 1971; Gregg and Sing, 1976, 1982): this type of behaviour is usually referred to as LPH. Typical LPH phenomena obtained with ultramicroporous carbons are illustrated in Figure 10.19. It might seem that we should simply regard LPH as an artefact associated with the lack of thermodynamic equilibrium (see Silvestre-Albero et al., 2012). Alternatively, the appearance of LPH might be expected to tell us something useful about the properties of the adsorbent. Here, we discuss only the behaviour of porous carbons; different types of LPH are given by certain clays, oxides, zeolites and MOFs and each system requires individual consideration.

It is well known that ill-defined adsorption–desorption hysteresis may be the result of a faulty experimental technique (see Chapter 3 and Lowell et al., 2004). For example, spurious hysteresis may result if insufficient time is allowed for adsorption equilibration. Another possible source of error is the presence of impurities either in the gas phase or on the surface. These complications must be avoided, or removed, before the evidence for reproducible LPH can be accepted.

Physisorption is perhaps always accompanied by some deformation of the adsorbent, but with many systems, this appears to be very small and reversible and is usually regarded as negligible. Although the effects have been generally ignored, it has long been known that significant dimensional changes are associated with the physisorption of vapours by certain microporous carbons (see Gregg, 1961). In the light of recent work (e.g. by Balzer et al., 2011), it is now timely to discuss the implications of these phenomena.

10.6.2 Activated Entry

Since physisorption is an exothermic process, the amount adsorbed at a given equilibrium pressure should always diminish as the temperature is increased. This is, of course, found in practice, provided thermodynamic equilibrium is established at each temperature. However, some vapour isotherms on microporous carbons behave anomalously (see Gregg and Sing, 1976, 1982). Thus, if an adsorption isotherm at one temperature apparently lies *above* the corresponding isotherm at a lower temperature, the obvious inference is that thermodynamic equilibrium is not established over one, or both, isotherms. According to the 'activated entry' hypothesis put forward by Maggs in 1953 and independently by Zwietering (see Gregg and Sing, 1982), this is due to very slow diffusion of adsorptive molecules through narrow pore entrances. At sufficiently high temperatures, the rate of adsorption becomes fast enough to allow equilibrium to be reached within the time of measurement.

The experimental data recorded by Marsh and Wynne-Jones (1964) and others (see Gregg and Sing, 1976, 1982) for CO_2 and N_2 adsorption by ultramicroporous carbons appear to support the notion of activated entry. As discussed in Section 10.4.1, the fact that the uptakes of CO_2 at 195 and 273 K were much larger than the corresponding N_2 uptakes at 77 K cannot be explained in terms of simple molecular sieving. Similarly, the adsorption of butane was found to increase considerably over the temperature range of 196–273 K (Gregg and Sing, 1976).

Although all these results are consistent with the idea of activated entry, they should not be regarded as conclusive evidence for the hypothesis in its original form. As will be seen in the following sections, we must also take account of the dimensional changes which accompany the physisorption of vapours by some activated carbons.

10.6.3 Low-Pressure Hysteresis

In their work on the sorption of gases by synthetic graphites, McDermott and Arnell (1955) found two types of reproducible hysteresis. The one at higher relative pressures could be attributed to capillary condensation–evaporation, while the other, which extended down to much lower pressure, was thought to be associated with swelling of the adsorbent. Although at that time it was not possible to confirm this hypothesis, it was already known that some sorption systems do undergo swelling (see Gregg, 1961).

A major advance in the interpretation of LPH was made by Everett and his co-workers in 1971. Their comprehensive paper (Bailey et al., 1971) was based on extensive work at Bristol on the adsorption of organic vapours by porous carbons. It was found *inter alia* that the phenomenon was extremely dependent on the nature of the adsorptive and the conditions of outgassing, and in a series of experiments lasting several months, clear evidence was obtained that LPH was associated with some distortion of the carbon structure – as indicated by changes in the pore structure and the adsorptive behaviour. The proposed theory involved irreversible intercalation of adsorbate molecules in pores of molecular dimensions leading to inelastic distortion of the adsorbent. Furthermore, in some cases, there was very slow relaxation at the operational temperature, even when all the intercalating molecules had been removed.

A general thermodynamic description of LPH was put forward along the following lines. The adsorbent (1) and the adsorptive (2) together are regarded as a two-component system. At constant temperature and external pressure, the Gibbs–Duhem equation can be expressed in the following form:

$$d\mu_1 = -(n_2/n_1)d\mu_2 \qquad (10.5)$$

where μ_1 and μ_2 are the chemical potentials and n_1 and n_2 are the amounts of the respective components. If the adsorbate is in equilibrium with its vapour,

$$d\mu_2 = RTd \ln p_2 \qquad (10.6)$$

so that

$$d\mu_1 = -RT(n_2/n_1)d \ln p_2 \qquad (10.7)$$

or

$$\mu_1 = \mu_1^o - RT \int_0^P (n_2/n_1)d \ln p_2 \qquad (10.8)$$

where μ_1^o is the chemical potential of the empty solid structure.

If the solid structure is either unchanged or changes elastically (reversibly) over the course of the physisorption isotherm, μ_1^o remains constant and there is no hysteresis, but LPH occurs if at some point on the isotherm the structure jumps irreversibly to a new configuration, which when empty has a different value of μ_1^o. For this reason and because of the complex nature of an ultramicroporous carbon, the adsorption and desorption isotherms follow different smooth paths.

In the case of an activated carbon, we may picture intercalation as a form of localized interlamellar penetration and express the process in terms of a 'spreading pressure', which enables the adsorbed film to 'prise open' the lamellae. Thus, to allow LPH to occur, parts of the structure must be strained beyond their elastic limit. Consequently, LPH cannot be expected to occur if the solid structure is strong and rigid (i.e. cross-linked). The Everett model is able to explain molecular size and shape effects: LPH is absent if the adsorbate molecules are either small enough to easily penetrate into narrow slit-shaped pores or too large to penetrate into the expanded state.

10.6.4 Expansion and Contraction

The first measurements of the expansion of charcoal brought about by physisorption were undertaken in the 1920s and 1930s (see McBain, 1932; Brunauer, 1945; Gregg, 1961). The small expansion of wood charcoal, which accompanied the adsorption of CO_2, appeared to be isotropic, as first reported by Meehan in 1927. It was pointed out by McBain that any expansion of the adsorbent could not be due to capillary condensation.

By using a specially designed extensometer, Bangham and his co-workers measured the fractional change in length of rods of compacted charcoal (Bangham and Fakhoury, 1930; Bangham and Razouk, 1938; Bangham and Franklin, 1946). The expansion isotherms determined in this manner were somewhat similar in shape to the corresponding adsorption isotherms, but the expansion was found to be generally proportional to the reduction in free surface energy (*not* to the amount adsorbed), as expressed in terms of the spreading pressure. The thermodynamic treatment developed by Bangham (1937) involved the application of the Gibbs adsorption equation to gas–solid systems.

In the work of Haines and McIntosh (1947), a small contraction of the carbon rod was registered at low p/p^o with certain adsorptives (e.g. butane). With various organic vapours, over a limited range of higher p/p^o, the extension appeared to be reversible and in accordance with Bangham's linear relationship. In contrast, the desorption of water vapour gave a marked contraction of the rod to below its original length.

There is now a strong revival of interest in adsorption-induced deformation of porous carbons and other materials. In their *in situ* experimental measurements of length changes by different monolithic carbon xerogels on the adsorption of nitrogen and carbon dioxide, Reichenauer and her co-workers have explored different stages of expansion and contraction in relation to the pore structure (Balzer et al. 2011). Several mechanisms appeared to contribute to the overall changes in sample length. In the mesopore range, there is a fairly good correlation between the change in length on desorption and the change in surface free energy (Bangham's model).

Micropore filling is evidently more complex: with N_2 at 77 K, an initial small contraction was followed by a steep increase in expansion up to $p/p^\circ \sim 0.1$. It is suggested that the magnitude of these changes was probably to some extent dependent on the micropore size distribution. There is clearly considerable scope for further work, particularly in the context of gas sequestration and storage.

10.7   Characterization of Active Carbons: Conclusions and Recommendations

Because of the complexity of their surface chemistry and pore structures, the characterization of active carbons is not always straightforward. Adsorption techniques have an indispensable role, but it is evident that they all have limitations of one sort or another. To understand the behaviour of an active carbon, it is necessary therefore to apply a combination of methods. The most appropriate selection of characterization techniques is still under discussion, but guidance on the methodology can be given in the light of progress made over the past few years.

A physisorption isotherm of nitrogen at 77 K is a useful start. The overall shape of the isotherm provides the first qualitative impression of the pore structure of the carbon. Most activated carbons are microporous, as revealed by their typical Type I isotherm shape. Progressive burn off is accompanied by a change in the shape of isotherm with the loss of the flat plateau and often the development of adsorption hysteresis. This form of pore widening is not always associated with the development of a true Type IV isotherm (i.e. having a distinct plateau at high p/p°), and therefore the absence of a well-defined mesopore structure. However, by modifying the conditions of pretreatment and activation, it is possible to obtain a more regular mesoporous product.

The BET-nitrogen area is a useful characteristic quantity, but it does not have the physical significance of an effective area if the pores are no wider than ~ 0.7 nm (i.e. if the carbon is ultramicroporous). The properties of narrow micropores can be investigated in three ways: (a) by constructing α_s-plots, (b) by using various molecular probes at different temperatures (e.g. CO_2 at 273 K) and (c) by adsorption calorimetry. The DR method is a popular method for evaluating the micropore volume of activated carbons, but its empirical nature and other limitations should be kept in mind.

Although the BJH method for mesopore size analysis is still widely used for the routine analysis of nitrogen isotherms on activated carbons, it is now in competition with another more sophisticated approach. In principle, the computational procedures based on DFT (e.g. NLDFT) can be applied to microporous and mesoporous carbons, but questionable assumptions are necessarily involved so that the derived quantities are still of uncertain validity.

There is now a strong case for adopting argon at 87 K in place of nitrogen at 77 K as the preferred adsorptive for surface area and pore size analysis. More work on porous and non-porous carbons is required to establish whether the advantages of using Ar (e.g. its non-specific interactions) outweigh the higher operational cost at the temperature of liquid argon.

There are several different ways in which the surface chemistry can be investigated. One approach is to use water vapour measurements to detect H-bond interaction with exposed polar groups. Another procedure is to apply immersion calorimetry with various polar liquids to determine specific adsorbent–adsorbate interaction energies. In this connection, IGC can be used to obtain differential enthalpies of adsorption at very low surface coverage.

The precise purpose of adopting a particular technique should be critically examined. For example, BET-areas and pore sizes are often quoted in patent specifications or scientific publications, but these quantities are rarely directly related to the performance of active carbons in such applications as catalysis or water purification. To optimize the adsorbent activity, it is therefore necessary to undertake other measurements (e.g. chemisorption or adsorption from solution), which are more closely related to a given application.

References

Ajayan, P.M., Iijima, S., 1993. Nature. 361, 333.

Amberg, C.H., Spencer, W.B., Beebe, R.A., 1955. Canad. J. Chem. 33, 305.

Andreu, A., Stoeckli, H.F., Bradley, R.H., 2007. Carbon. 45, 1854.

Ania, C.O., Pelayo, J.G., Bandosz, T.J., 2011. Adsorption. 17, 421.

Aranovich, G.L., 1991. J. Colloid Interface Sci. 141, 30.

Arnell, J.C., McDermott, H.L., 1952. Canad. J. Chem. 30, 177.

Arora, G., Sandler, S.I., 2005. J. Chem. Phys. 123, 044705.

Atkinson, D., McLeod, A.I., Sing, K.S.W., Capon, A., 1982. Carbon. 20, 339.

Atkinson, D., McLeod, A.I., Sing, K.S.W., 1984. J. Chim. Phys. 81, 791.

Atkinson, D., Carrott, P.J.M., Grillet, Y., Rouquerol, J., Sing, K.S.W., 1987. In: Liapis, A.I. (Ed.), Proceedings of the 2nd International Conference on Fundamentals of Adsorption. Engineering Foundation, New York, p. 89.

Aukett, P.N., Quirke, N., Riddiford, S., Tennison, S.R., 1992. Carbon. 30, 913.

Avgul, N.N., Kiselev, A.V., 1970. Chem. Phys. Carbon. 1, p. 1.

Avgul, N.N., Kiselev, A.V., Kovalyova, N.V., Khrapova, E.V., 1957. In: Schulman, J.H. (Ed.), Proceedings of the 2nd International Congress on Surface Activity, Vol. II. Butterworths, London, p. 218.

Bailey, A., Maggs, F.A.P., 1972. British Patent 1 301 101.

Bailey, A., Cadenhead, D.A., Davies, D.H., Everett, D.H., Miles, A.J., 1971. Trans. Faraday Soc. 67, 231.

Bailey, A., Maggs, F.A.P., Williams, J.H., 1973. British Patent 1 310 011.

Bailey, A., Lawrie, G.A., Williams, M.R., 1995. Adsorption Sci. Technol. 12, 193.

Baker, F.S., 1992. Kirk-Othmer Encyclopedia of Chemical Technology, Vol. 4. John Wiley, New York, p. 1015.

Balbuena, P.B., Gubbins, K.E., 1994. In: Rouquerol, J., Rodriguez-Reinoso, F., Sing, K.S.W., Unger, K.K. (Eds.), Characterization of Porous Solids III. Elsevier, Amsterdam, p. 41.

Balbuena, P.B., Lastoskie, C., Gubbins, K.E., Quirke, N., 1993. In: Suzuki, M. (Ed.), Proceedings of the 4th International Conference on Fundamentals of Adsorption. Kodansha, Tokyo, p. 27.

Balzer, C., Wildhage, T., Braxmeier, S., Reichenauer, G., Olivier, J.P., 2011. Langmuir. 27, 2553.

Bandosz, T.J., Biggs, M.J., Gubbins, K.E., Hattori, Y., Iiyama, T., Kaneko, K., Pikunic, J., Thomson, K.T., 2003. In: Radovic, L.R. (Ed.), Chemistry and Physics of Carbon. Marcel Dekker, New York, p. 41.

Bangham, D.H., 1937. Trans. Faraday Soc. 33, 805.

Bangham, D.H., Fakhoury, N., 1930. Proc. R. Soc. A130, 81.

Bangham, D.H., Franklin, R., 1946. Trans. Faraday Soc. 42, 289.

Bangham, D.H., Razouk, R.I., 1938. Proc. R. Soc. A166, 572.

Bansal, R.C., Bhatia, N., Dhami, T.L., 1978a. Carbon. 16, 65.

Bansal, R.C., Dhami, T.L., Parkash, S., 1978b. Carbon. 16, 389.

Bansal, R.C., Donnet, J.-B., Stoeckli, F., 1988. Active Carbon. Marcel Dekker, New York.

Barton, S.S., Harrison, B.H., 1975. Carbon. 13, 47.

Barton, S.S., Koresh, J.E., 1983. J. Chem. Soc. Faraday Trans. I79, 1147–1165.

Barton, S.S., Evans, M.J.B., MacDonald, J.A.F., 1991. Carbon. 29 (8), 1099.

Barton, S.S., Evans, M.J.B., MacDonald, J.A.F., 1992. Carbon. 30 (1), 123.

Beebe, R.A., Dell, R.M., 1955. J. Phys. Chem. 59, 746.

Beebe, R.A., Young, D.M., 1954. J. Phys. Chem. 58, 93.

Beebe, R.A., Beckwith, J.B., Honig, J.M., 1945. J. Am. Chem. Soc. 67, 1554.

Beebe, R.A., Biscoe, J., Smith, W.R., Wendell, C.B., 1947. J. Am. Chem. Soc. 69, 95.

Beebe, R.A., Millard, B., Cynarski, J., 1953. J. Am. Chem. Soc. 75, 839.

Bienfait, M., 1980. In: Dash, J.G., Ruvalds, J. (Eds.), Phase Transitions in Surface Films. NATO ASI, Vol. B51. Plenum, New York, p. 29.

Bienfait, M., 1985. Surf. Sci. 162, 411.

Bienfait, M., Zeppenfeld, P., Gay, J.M., Palmari, J.P., 1990. Surf. Sci. 226 (3), 327.

Bock, H., Gubbins, K.E., Pikunic, J., 2008. In: Bottani, E.J., Tascon, J.M.D. (Eds.), Adsorption by Carbons. Elsevier, Oxford, p. 103.

Boehm, H.P., 1966. In: Eley, D.D., Pines, H., Weisz, P.B. (Eds.), Advances in Catalysis, Vol. 16. Academic Press, New York, p. 179.

Boehm, H.P., 1994. Carbon. 32, 759.

Boehm, H.P., 2008. In: Bottani, E.J., Tascon, J.M.D. (Eds.), Adsorption by Carbons. Elsevier, Oxford, p. 301.

Bojan, M.J., Steele, W.A., 1987. Langmuir. 3 (6), 1123.

Bojan, M.J., Steele, W.A., 2008. In: Bottani, E.J., Tascon, J.M.D. (Eds.), Adsorption by Carbons. Elsevier, Oxford, p. 77.

Bonijoly, M., Oberlin, M., Oberlin, A., 1982. Int. J. Coal Geol. 1, 283.

Bottani, E.J., Tascon, J.M.D., 2004. Chem. Phys. Carbon. 29, 209.

Bottani, E.J., Tascon, J.M., 2008. Adsorption by Carbons. Elsevier, Amsterdam.

Bradley, R.H., 2011. Adsorption Sci. Technol. 29, 1.

Bradley, R.H., Rand, B., 1995. J. Colloid Interface Sci. 169, 168.

Bradley, R.H., Sutherland, I., Sheng, E., 1995. J. Chem. Soc. Faraday Trans. 91, 3201.

Bradley, R.H., Cassity, K., Andrews, R., Meier, M., Osbeck, S., Andreu, A., Johnston, C., Crossley, A., 2012. Appl. Surf. Sci. 258, 4835.

Brauer, P., Poosch, H.-R., Szombathely, M.V., Heuchel, M., Jaroniec, M., 1993. In: Suzuki, M. (Ed.), Proceedings of the 4th International Conference on Fundamentals of Adsorption. Kodansha, Tokyo, p. 67.

Breck, D.W., 1974. Zeolite Molecular Sieves. John Wiley, New York, p. 636.

Brunauer, S., 1945. The Adsorption of Gases and Vapours. Oxford University Press, London.

Cadenhead, D.A., Everett, D.H., 1958. Industrial Carbon and Graphite. S.C.I, London, 272.

Cao, L., Kruk, M., 2010. Adsorption. 16, 465.

Carrott, P.J.M., 1993. Adsorption Sci. Technol. 10, 63.

Carrott, P.J.M., Freeman, J.J., 1991. Carbon. 29, 499.

Carrott, P.J.M., Sing, K.S.W., 1987. J. Chromatography. 406, 139.

Carrott, P.J.M., Sing, K.S.W., 1988. In: Unger, K.K., Rouquerol, J., Sing, K.S.W., Kral, H. (Eds.), Characterization of Porous Solids I. Elsevier, Amsterdam, p. 77.

Carrott, P.J.M., Sing, K.S.W., 1989. Pure Appl. Chem. 61, 1835.

Carrott, P.J.M., Roberts, R.A., Sing, K.S.W., 1987. Carbon. 25, 59.

Carrott, P.J.M., Roberts, R.A., Sing, K.S.W., 1988a. Langmuir. 4, 740.

Carrott, P.J.M., Roberts, R.A., Sing, K.S.W., 1988b. In: Unger, K.K., Rouquerol, J., Sing, K.S.W., Kral, H. (Eds.), Characterization of Porous Solids I. Elsevier, Amsterdam, p. 89.

Carrott, P.J.M., Drummond, F.C., Kenny, M.B., Roberts, A., Sing, K.S.W., 1989. Colloids Surf. 37, 1.

Carrott, P.J.M., Kenny, M.B., Roberts, R.A., Sing, K.S.W., Theocharis, C.R., 1991. In: Rodriguez-Reinoso, F., Rouquerol, J., Sing, K.S.W., Unger, K.K. (Eds.), Characterization of Porous Solids II. Elsevier, Amsterdam, p. 685.

Carrott, P.J.M., Ribeiro Carrott, M.M.L., Cansado, I.P.P., Nabais, J.M.V., 2000. Carbon. 38, 465.

Carrott, P.J.M., Ribeiro Carrott, M.M.L., Cansado, I.P.P., 2001. Carbon. 39 (193), 465.

Carrott, P.J.M., Conceicao, F.L., Ribeiro Carrott, M.M.L., 2007. Carbon. 45, 1310.

Carruthers, J.D., Petruska, M.A., Sturm, E.A., Wilson, S.M., 2012. Microporous Mesoporous Mater. 154, 62.

Cazorla-Amoros, D., Alcaniz-Monge, J., Linares-Solano, A., 1996. Langmuir. 12, 2820.

Chambers, A., Park, C., Baker, R.T.K., Rodriguez, N.M., 1998. J. Phys. Chem. B102, 4253.

Choma, J., Jaroniec, M., 1998. Adsorpt. Sci. Technol. 16, 295.

Choma, J., Jaroniec, M., Kloske, M., 2002. Adsorpt. Sci. Technol. 20, 307.

Clint, J.H., 1972. J. Chem. Soc. Faraday Trans. 1. 68, 2239.

Conner, W.C., 1997. In: Fraissard, J., Conner, W.C. (Eds.), Physical Adsorption: Experiment, Theory and Applications. Kluwer, Dordrecht, p. 33.

Costa, E., Calleja, G., Marijuan, L., 1988. Adsorption Sci. Technol. 5 (3), 213.

Dacey, J.R., Clunie, J.C., Thomas, D.G., 1958. Trans. Faraday Soc. 54, 250.

Darmstadt, H., Ryoo, R., 2008. In: Bottani, E.J., Tascon, J.D.M. (Eds.), Adsorption by Porous Carbons. Elsevier, Oxford, p. 455.

Davis, B.W., Pierce, C., 1966. J. Phys. Chem. 70, 1051.

Denoyel, R., Rouquerol, F., Rouquerol, J., 2008. In: Bottani, E.J., Tascon, J.M.D. (Eds.), Adsorption by Carbons. Elsevier, Oxford, p. 273.

Derbyshire, F., Jagtoyen, M., Thwaites, M., 1995. In: Patrick, J.W. (Ed.), Porosity in Carbons. Edward Arnold, London, p. 227.

Derylo-Marczewska, A., Miroslaw, K., Marczewski, A.W., Sternik, D., 2010. Adsorption. 16, 359.

de Vooys, F., 1983. In: Capelle, A., de Vooys, F. (Eds.), Activated Carbon: A Fascinating Material. Norit N.V, Amersfoort, p. 13.

Do, D.D., Do, H.D., 2002. Adsorption. 8, 309.

Do, D.D., Do, H.D., 2003. Adsorption Sci. Technol. 21, 389.

Do, D.D., Ustinov, E.A., Do, H.D., 2008. In: Bottani, E.J., Tascon, J.D. (Eds.), Adsorption by Carbons. Elsevier, Amsterdam, p. 239.

Donnet, J.-B., 1994. Carbon. 32, 1305.

Donnet, J.-B., Bansal, R.C., 1990. Carbon Fibers. Dekker, New York.

Dubinin, M.M., 1958. Industrial Carbon and Graphite. Society of Chemical Industry, London, p. 219.

Dubinin, M.M., 1966. In: Walker, P.L. (Ed.), Chemistry and Physics of Carbon, Vol. 2. Marcel Dekker, New York, p. 51.

Dubinin, M.M., 1975. In: Cadenhead, D.A. (Ed.), Progress in Surface and Membrane Science, Vol. 9. Academic Press, New York, p. 1.

Dubinin, M.M., 1980. Carbon. 18, 355.

Dubinin, M.M., Serpinski, V.V., 1981. Carbon. 19, 402.

Ebbesen, T.W., Ajayan, P.M., 1992. Nature. 358, 220.

Economy, J., Lin, R.Y., 1971. J. Mater. Sci. 6, 1151.

Economy, J., Lin, R.Y., 1976. Appl. Polymer Symp. 29, 199.

Emmett, P.H., DeWitt, T., 1941. Ind. Eng. Chem. Anal. Ed. 13, 28.

Esteves, I.A.A.C., Cruz, F.J.A.L., Müller, E.A., Agnihotri, S., Mota, J.P.B., 2009. Carbon. 47, 948.

Everett, D.H., Powl, J.C., 1976. J. Chem. Soc. Faraday Trans. I. 72, 619.

Everett, D.H., Ward, R.J., 1986. J. Chem. Soc. Faraday Trans. 1. 82, 2915.

Fernandez-Colinas, J., Denoyel, R., Rouquerol, J., 1989a. Adsorption Sci.Technol. 6 (1), 18.

Fernandez-Colinas, J., Denoyel, R., Grillet, Y., Rouquerol, F., Rouquerol, J., 1989b. Langmuir. 5, 1205.

Figueiredo, J.L., Sousa, J.P.S., Orge, C.A., Pereira, M.F.R., Orfao, J.J.M., 2011. Adsorption. 17, 431.

Fitzer, E., Kochling, K.H., Boehm, H.P., Marsh, H., 1995. Pure Appl. Chem. 67, 473.

Forrester, S.D., Giles, C.H., 1971. Chem. Ind., 831.

Franklin, R.E., 1951. Proc. R. Soc. A209, 196.

Freeman, J.J., Sing, K.S.W., 1991. In: Suzuki, M. (Ed.), Adsorptive Separation. Institute of Industrial Science, Tokyo, p. 261.

Freeman, J.J., Gimblett, F.G.R., Roberts, R.A., Sing, K.S.W., 1987. Carbon. 25, 559.

Freeman, J.J., Gimblett, F.G.R., Roberts, R.A., Sing, K.S.W., 1988. Carbon. 26, 7.

Freeman, J.J., Sing, K.S.W., Tomlinson, J.B., 1990. In Carbone-90, Extended Abstracts, GFEC, Paris, p. 164.

Freeman, J.J., Tomlinson, J.B., Sing, K.S.W., Theocharis, C.R., 1993. Carbon. 31, 865.

Fulconis, J.M., 1996., Thèse, Université d'Aix-Marseille III, Marseille.

Furmaniak, S., Terzyk, A., Gauden, P.A., Harris, P.J.F., Wisniewski, M., Kowalczyk, P., 2010. Adsorption. 16, 197.

Futaba, D.N., Hata, K., Namai, T., Yamada, T., Mizuno, K., Hayamizu, Y., Yumura, M., Iijima, S., 2006. J. Phys. Chem. B. 110, 8035.

Gardner, L., Kruk, M., Jaroniec, M., 2001. J. Phys. Chem. 105, 12516.

Garrido, J., Linares-Solano, A., Martin-Martinez, J.M., Molina-Sabio, M., Rodriguez-Reinoso, F., Torregrosa, R., 1987. Langmuir. 3, 76.

Gay, J.-M., Suzanne, J., Wang, R., 1986. J. Chem. Soc. Faraday Trans. 2. 82, 1669.

Giles, C.H., D'Silva, A.P., Trivedi, A.S., 1970. In: Everett, D.H., Ottewill, R.H. (Eds.), Surface Area Determination. Butterworths, London, p. 317.

Gimblett, F.G.R., Freeman, J.J., Sing, K.S.W., 1989. J. Mater. Sci. 24, 3799.

Gregg, S.J., 1961. The Surface Chemistry of Solids. Chapman and Hall, London.

Gregg, S.J., Sing, K.S.W., 1976. In: Matijevic, E. (Ed.), Surface and Colloid Science, Vol. 9. Wiley, p. 231.

Gregg, S.J., Sing, K.S.W., 1982. Adsorption, Surface Area and Porosity. Academic Press, London.

Grillet, Y., Rouquerol, F., Rouquerol, J., 1979. J. Colloid Interface Sci. 70, 239.

Guillot, A., Stoeckli, F., Bauguil, Y., 2000. Adsorpt. Sci. Technol. 18, 1.

Guillot, A., Stoeckli, F., 2001. Carbon. 39, 2059.

Haines, R.S., McIntosh, R., 1947. J. Chem. Phys. 18, 28.

Hall, C.R., Holmes, R.J., 1991. Colloids Surf. 58, 339.

Hall, C.R., Holmes, R.J., 1992. Carbon. 30, 173.

Hall, C.R., Holmes, R.J., 1993. Carbon. 31, 881.

Hall, C.R., Sing, K.S.W., 1988. Chem. Br. 24, 670.

Hall, P.G., Williams, R.T., 1986. J. Colloid Interface Sci. 113, 301.

Harkins, W.D., Jura, G., 1944a. J. Am. Chem. Soc. 66, 1362.

Harkins, W.D., Jura, G., 1944b. J. Am. Chem. Soc. 66, 919.

Holmes, J.M., 1966. In: Flood, E.A. (Ed.), The Solid-Gas Interface, Vol. 1. Marcel Dekker, New York, p. 127.

Hutson, N.D., Yang, R.T., 1997. Adsorption. 3, 189.

Iijima, S., 1991. Nature. 354, 56.

Iiyama, T., Nishikawa, K., Otowa, T., Kaneko, K., 1995. J. Phys. Chem. 99, 10075.

Isirikyan, A.A., Kiselev, A.V., 1961. J. Phys. Chem. 65, 601.

Ismail, I.M.K., 1990. Carbon. 28, 423.

Ismail, I.M.K., 1992. Langmuir. 8, 360.

Ismail, I.M.K., Rodgers, S.L., 1992. Carbon. 30, 229.

Iiyama, T., Ruike, M., Kaneko, K., 2000. Chem. Phys. Lett. 331, 359.

Jagtoyen, M., Derbyshire, F., 1993. Carbon. 31, 1185.

Jaroniec, M., Madey, R., 1988. Physical Adsorption on Heterogeneous Solids. Elsevier, Amsterdam.

Johnson, J.K., Cole, M.W., 2008. In: Bottani, E.J., Tascon, J.M.D. (Eds.), Adsorption by Carbons. Elsevier, Oxford, p. 369.

Jorge, M., Seaton, N.A., 2002. Mol. Phys. 100, 3803.

Journet, C., Maser, W.K., Bernier, P., Loiseau, A., Lamy de la Chapelle, M., Lefrant, S., Deniard, P., Lee, R., Fischer, J.E., 1997. Nature. 388, 756.

Joyner, L.G., Emmett, P.H., 1948. J. Amer. Chem. Soc. 70, 2357.

Juang, R.-S., Tseng, R.-L., Wu, F.-C., 2001. Adsorption. 7, 65.

Kakei, K., Ozeki, S., Suzuki, T., Kaneko, K., 1991. In: Rodriguez-Reinoso, F., Rouquerol, J., Sing, K.S.W., Unger, K.K. (Eds.), Characterization of Porous Solids II. Elsevier, Amsterdam, p. 429.

Kaneko, K., 1996. In: Dabrowski, A., Tertykh, V.A. (Eds.), Adsorption on New and Modified Inorganic Sorbents. Elsevier, Amsterdam, p. 573.

Kaneko, K., 2000. Carbon. 38, 287.

Kaneko, K., Setoyama, N., Suzuki, T., Kuwabara, H., 1993. In: Suzuki, M. (Ed.), Proceedings of the 4th International Conference on Fundamentals of Adsorption. Kodansha, Tokyo, p. 315.

Kaneko, Y., Ohbu, K., Uekawa, N., Fujie, K., Kaneko, K., 1995. Langmuir. 11, 708.

Kaneko, K., Hanzawa, Y., Iiyama, T., Kanda, T., Suzuki, T., 1999. Adsorption. 5, 7.

Kenny, M., Sing, K., Theocharis, C., 1993. In: Suzuki, M. (Ed.), Proceedings of the 4th International Conference on Fundamentals of Adsorption. Kodansha, Tokyo, p. 323.

Kington, G.L., Beebe, R.A., Polley, M.H., Smith, N.R., 1950. J. Am. Chem. Soc. 72, 1775.

Kiselev, A.V., 1968. J. Colloid Interface Sci. 28, 430.

Krim, J., Suzanne, J., Shechter, H., Wang, R., Taub, H., 1985. Surf. Sci. 162, 446.

Kroto, H.W., Heath, J.R., O'Brien, S.C., Curi, R.F., Smalley, R.E., 1985. Nature. 318, 162.

Kruk, M., Jaroniec, M., Bereznitski, Y., 1996. J. Colloid Interface Sci. 182, 282.

Kuwabara, H., Suzuki, T., Kaneko, K., 1991. J. Chem. Soc. Faraday Trans. 87, 1915.

Lambin, P., Meunier, V., Henrard, L., Lucas, A.A., 2000. Carbon. 38, 1713.

Lamond, T.G., Metcalfe, J.E., Walker, P.L., 1965. Carbon. 3, 59.

Larher, Y., 1974. J. Chem. Soc. Faraday Trans. I. 70, 320.

Larher, Y., 1983. Surf. Sci. 134, 469.

Lee, S.M., Ann, K.H., Lee, Y.H., 2001. J. Am. Chem. Soc. 123, 5059.

Li, Z., Jaroniec, M., 2004. Anal. Chem. 76, 5479.

Linares-Solano, A., Cazorla-Amoros, D., 2008. In: Bottani, E.J., Tascon, J.M.D. (Eds.), Adsorption by Carbons. Elsevier, Oxford, p. 431.

Llewellyn, P., Rouquerol, F., Rouquerol, J., Sing, K.S.W., 2000. In: Unger, K.K. et al., (Ed.), Studies in Surface Science and Catalysis, Vol. 128. Elsevier Science BV, Amsterdam, p. 421.

Lopez-Ramon, M.V., Stoeckli, H.F., Moreno-Castella, C., Carrasco-Marin, F., 1999. Carbon. 37, 1215.

Lowell, S., Shields, J.E., Thomas, M.A., Thommes, M., 2004. Characterization of Porous Solids and Powders: Surface Area, Pore Size and Density, Kluwer, Dordrecht.

Malden, P.J., Marsh, J.D.F., 1959. J. Phys. Chem. 63, 1309.

Marsh, H., Rodriguez-Reinoso, F., 2006. Activated Carbon. Elsevier, Amsterdam.

Marsh, H., Wynne-Jones, W.F.K., 1964. Carbon. 1, 281.

Martinez-Alonso, A., Tascon, J.M.D., Bottani, E.J., 2000. Langmuir. 16, 1343.

Martinez-Alonso, A., Tascon, J.M.D., Bottani, E.J., 2001. J. Phys. Chem. B. 105, 135.

Mays, T.J., 1999. In: Burchell, T.D. (Ed.), Carbon Materials for Advanced Technologies. Pergamon, Amsterdam, p. 95.

McCallum, C.L., Bandosz, T.J., McGrother, S.C., Müller, E.A., Gubbins, K.E., 1999. Langmuir. 15, 533.

McDermott, H.L., Arnell, J.C., 1955. Can. J. Chem. 33, 913.

McEnaney, B., 2002. In: Schüth, F., Sing, K.S.W., Weitkamp, J. (Eds.), Handbook of Porous Solids, Vol. 3. Wiley-VCH, Weinheim, p. 1828.

McEnaney, B., Mays, T.J., 1991. In: Rodriguez-Reinoso, F., Rouquerol, J., Sing, K.S.W., Unger, K.K. (Eds.), Characterization of Porous Solids II. Elsevier, Amsterdam, p. 477.

Mc Kay, G., 1982. J. Chem. Technol. Biotechnol. 32, 759.

Mc Kay, G., 1984. J. Chem. Technol. Biotechnol. 34A, 294.

Medalia, A.I., Rivin, D., 1976. In: Parfitt, G.D., Sing, K.S.W. (Eds.), Characterization of Powder Surfaces. Academic, London, p. 279.

Minett, A., Atkinson, K., Roth, S., 2002. In: Schüth, F., Sing, K.S.W., Weitkamp, J. (Eds.), Handbook of Porous Solids, Vol. 3. Wiley-VCH, Weinheim, p. 1923.

Miyawaki, J., Kanda, T., Kaneko, K., 2001. Langmuir. 17, 664.

Molina-Sabio, M., Munecas, M.A., Rodriguez-Reinoso, F., McEnaney, B., 1995. Carbon. 33, 1777.

Moreno-Castilla, C., 2008. In: Bottani, E.J., Tascon, J.D.M. (Eds.), Adsorption by Carbons. Elsevier, Oxford, p. 653.

Moreno-Castillo, C., Fontecha-Camara, M.A., Alvarez-Merino, M.A., Lopez-Ramon, M.V., Carrasco-Marin, F., 2011. Adsorption. 17, 413.

Müller, E.A., 2008. J. Phys. Chem. 112, 8999.

Müller, E.A., Rull, L.F., Vega, L.F., Gubbins, K.E., 1996. J. Phys. Chem. 100, 1189.

Nandy, S.P., Walker, P.L., 1975. Fuel. 54, 169.

Neimark, A.V., Ruetsch, S., Kornev, K.G., Ravikovitch, P.I., Poulin, P., Badaire, S., Maugey, M., 2003. Nano Lett. 3, 419.

Neimark, A.V., Lin, Y., Ravikovitch, P.I., Thommes, M., 2009. Carbon. 47, 1617.

Newcombe, G., 2008. In: Bottani, E.J., Tascon, J.M.D. (Eds.), Adsorption by Carbons. Elsevier, Amsterdam, p. 679.

Oberlin, A., Bonnamy, S., Rouxhet, P.G., 1999. In: Thrower, P.A., Radovich, L.R. (Eds.), Chemistry and Physics of Carbon, Vol. 26. Marcel Dekker, New York, p. 1.

Ohba, T., Kaneko, K., 2002. J. Phys. Chem. B. 106, 7171.

Ohba, T., Kanoh, H., Kaneko, K., 2004. J. Phys. Chem. B. 108, 14964.

Olivier, J.P., 2008. In: Bottani, E.J., Tascon, J.D.M. (Eds.), Adsorption by Carbons. Elsevier, Oxford, p. 147.

Orr, C., Dalla Valle, J.M., 1959. Fine Particle Measurement. Macmillan, New York.

Otowa, T., 1991. In: Suzuki, M. (Ed.), Adsorpive Separation. Institute of Industrial Science, Tokyo, p. 273.

Otowa, T., Tanibara, R., Itoh, M., 1993. Gas Sep. Purif. 7, 241.

Otowa, T., Nojima, Y., Itoh, M., 1996. In: LeVan, M.D. (Ed.), Fundamentals of Adsorption. Kluwer, Boston, p. 709.

Palmer, J.C., Gubbins, K.E., 2012. Microporous Mesoporous Mater. 154, 24.

Papirer, E., Brendle, E., Ozil, F., Balard, H., 1999. Carbon. 37, 1265.

Parkyns, N.D., Sing, K.S.W., 1975. In: Everett, D.H. (Ed.), Specialist Periodical Report: Colloid Science, Vol. 2. The Chemical Society, London, p. 26.

Patrick, J.W., 1995. Porosity in Carbons. Edward Arnold, London.

Patrick, J.W., Hanson, S., 2002. In: Schüth, F., Sing, K.S.W., Weitkamp, J. (Eds.), Handbook of Porous Solids, Vol. 3. p. 1900.

Pekala, R.W., 1989. J. Mater. Sci. 24, 3221.

Pierce, C., 1969. J. Phys. Chem. 73, 813.

Pierce, C., Ewing, B., 1962. J. Am. Chem. Soc. 84, 4072.

Pierce, C., Ewing, B., 1967. J. Phys. Chem. 71, 3408.

Pierce, C., Smith, N., 1950. J. Phys. Colloid Chem. 54, 784.

Pierce, C., Smith, R.N., Wiley, J.W., Cordes, H., 1951. J. Amer. Chem. Soc. 73, 4551.

Pierce, C., Smith, R.N., 1953. J. Phys. Colloid Chem. 57, 64.

Polley, M.H., Schaeffer, W.D., Smith, W.R., 1953. J. Phys. Chem. 57, 469.

Poulin, P., Vigolo, B., Launois, P., 2002. Carbon. 40, 1741.

Prenzlow, C.F., Halsey, G.D., 1957. J. Phys. Chem. 61, 1158.

Puri, B.R., 1970. In: Walker, P.L. (Ed.), Chemistry and Physics of Carbon, Vol. 6. Marcel Dekker, New York, p. 191.

Qadeer, R., Saleem, M., 1997. Adsorption Sci. Technol. 15 (5), 373.

Ramos-Fernandez, J.M., Martinez-Escandell, M., Rodriguez-Reinoso, F., 2008. Carbon. 46, 384.

Ravikovitch, P.I., Vishnyakov, A., Russo, R., Neimark, A.V., 2000. Langmuir. 16, 2311.

Rodriguez-Reinoso, F., 1986. In: Figueredo, J.L., Moulijn, J.A. (Eds.), Carbon and Coal Gasification. Martinus Nijhoff, Dordrecht, p. 601.

Rodriguez-Reinoso, F., 1989. Pure Appl. Chem. 61, 1859.

Rodriguez-Reinoso, F., Martin-Martinez, J.M., Prado Burguete, C., McEnaney, B., 1987. J. Phys. Chem. 91, 515.

Rodriguez-Reinoso, F., 2002. In: Schüth, F., Sing, K.S.W., Weitkamp, J. (Eds.), Handbook of Porous Solids, Vol. 3. Wiley-VCH, Weinheim, p. 1766.

Rodriguez-Reinoso, F., Garrido, J., Martin-Martinez, J.M., Molina-Sabio, M., Torregrosa, R., 1989. Carbon. 27, 23.

Rodriguez-Reinoso, F., Molina-Sabio, M., Gonzalez, M.T., 1995. Carbon. 33, 15.

Rodriguez-Reinoso, F., Molina-Sabio, M., Gonzalez, M.T., 1997. Langmuir. 13, 2354.

Rouquerol, J., Partyka, S., Rouquerol, F., 1977. J. Chem. Soc. Faraday Trans. 1. 73, 306.

Rouquerol, J., Avnir, D., Fairbridge, C.W., Everett, D.H., Haynes, J.M., Pernicone, N., Ramsay, J.D.F., Sing, K.S.W., Unger, K.K., 1994. Pure Appl. Chem. 66, 1739.

Rouquerol, J., Llewellyn, P., Rouquerol, F. (2007) In: Characterization of Porous Solids VII (Eds. P.L. Llewellyn, F. Rodriguez-Reinoso, J. Rouquerol, N. Seaton) Elsevier, Amsterdam, p. 49.

Rouzard, J.N., Oberlin, A., 1989. Carbon. 27, 517.

Rudzinski, W., Everett, D.H., 1992. Adsorption of Gases on Heterogeneous Surfaces. Academic Press, London.

Ryoo, R., Joo, S.H., Kruk, M., 2001. Adv. Mater. 13, 677.

Schaeffer, W.D., Smith, W.R., Polley, M.H., 1953. Ind. Eng. Chem. 45, 1721.

Schlögl, R., 2002. In: Schüth, F., Sing, K.S.W., Weitkamp, J. (Eds.), Handbook of Porous Solids 3. Wiley-VCH, Weinheim, p. 1863.

Schröter, H.J., Jüntgen, H., 1988. In: Rodriguez, A.E., LeVan, M.D., Tondeur, D. (Eds.), Adsorption: Science and Technology. Kluwer, Dordrecht, p. 269.

Schüth, F., 2003. Angew. Chem. Int. Edit. 42, 3604.

Setoyama, N., Kaneko, K., 1995. Adsorption. 1, 1.

Setoyama, N., Ruike, M., Kasu, T., Suzuki, T., Kaneko, K., 1993. Langmuir. 9, 2612.

Setoyama, N., Kaneko, K., Rodriguez-Reinoso, F., 1996. J. Phys. Chem. 100 (24), 10331.

Silvestre-Albero, J., Rodriguez-Reinoso, F., 2012. In: Tascon, J.M.D. (Ed.), Novel Carbon Adsorbents. Elsevier, Amsterdam, p. 583.

Silvestre-Albero, J., Silvestre-Albero, A., Rodriguez-Reinoso, F., 2012. Carbon, 50, 3128. Elsevier, Amsterdam, p. 3.

Sing, K.S.W., 1989. Carbon. 27, 5.

Sing, K.S.W., 1991. In: Mersmann, A.B., Scholl, S.E. (Eds.), Third International Conference on Fundamentals of Adsorption. Engineering Foundation, New York, p. 69.

Sing, K.S.W., 1994. Carbon. 32, 1311.

Sing, K.S.W., 1995. In: Patrick, J.W. (Ed.), Porosity in Carbons. Edward Arnold, London and Oxford, p. 49.

Sircar, S., 2008. In: Bottani, E.J., Tascon, J.D.M. (Eds.), Adsorption by Carbons. Elsevier, Oxford, p. 565.

Smisek, M., Cerny, S., 1970. Active Carbon. Elsevier, Amsterdam, p. 10.

Starsinic, M., Taylor, R.L., Walker, P.L., 1983. Carbon. 21, 69.

Steele, W.A., 1956. J. Chem. Phys. 25, 819.

Steele, W.A., 1974. The Interaction Gases with Solid Surfaces. Pergamon, Oxford.

Steele, W.A., Bojan, M.J., 1989. Pure Appl. Chem. 61, 1927.

Stoeckli, H.F., 1981. Carbon. 19, 325.

Stoeckli, F., 1993. Adsorption Sci. Technol. 10, 3.

Stoeckli, H.F., Huguenin, D., Laederach, A., 1994a. Carbon. 32, 1359.

Stoeckli, F., Jakubov, T., Lavanchy, A., 1994b. J. Chem. Soc. Faraday Trans. 90, 783.

Stoeckli, F., Guillot, A., Slasli, A.M., Hugi-Cleary, D., 2002. Carbon. 40, 211.

Striolo, A., Chialvo, A.A., Cummings, P.T., Gubbins, K., 2003. Langmuir. 19, 8583.

Suarez-Garcia, F., Martinez-Alonso, A., Tascon, J.M.D., 2008. In: Bottani, E.J., Tascon, J.M.D. (Eds.), Adsorption by Carbons, p. 329.

Suzanne, J., Gay, J.M., 1996. In: Unertl, W.N. (Ed.), Handbook of Surface Science. Richardson, N.V., Holloway, S. (Series Eds.), Physical Structure, Vol. 1. North-Holland Elsevier, Amsterdam, p. 503.

Suzuki, M., 1994. Carbon. 32, 577.

Talu, O., Meunier, F., 1996. AIChE J. 42, 809.

Tascon, J.M.D., 2012. Novel Carbon Adsorbents. Elsevier, Amsterdam.

Thess, A., Lee, R., Nikolaev, P., Dai, H., 1996. Science. 273, 483.

Thommes, M., Morlay, C., Ahmad, R., Joly, J.P., 2011. Adsorption. 17, 653.

Thommes, M., Cychosz, K.A., Neimark, A.V., 2012. In: Tascon, J.M.D. (Ed.), Novel Carbon Adsorbents. Elsevier, London and Oxford, p. 107.

Thomy, A., Duval, X., 1969. J. Chim. Phys. 66, 1966.

Thomy, A., Regnier, J., Duval, X., 1972. In: Colloques Int CNRS. Thermochimie, Vol. 201. CNRS, Paris, p. 511.

Tolmachev, A.M., 1993. Adsorption Sci. Technol. 10, 155.

Ugarte, D., 1995. Carbon. 33, 989.

van der Plas, Th., 1970. In: Linsen, B.G. (Ed.), Physical and Chemical Aspects of Adsorbents and Catalysts. Academic Press, London, p. 425.

van Driel, J., 1983. In: Capelle, A., de Vooys, F. (Eds.), Activated Carbon: A Fascinating Material. Norit N. V, Netherlands, p. 40.

Vartapetyan, R.Sh., Voloshchuk, A.M., Buryak, A.K., Artamonova, C.D., Belford, R.L., Ceroke, P.J., Kholine, D.V., Clarkson, R.B., Odintsov, B.M., 2005. Carbon. 43, 2152.

Velasco, L.F., Ania, C.O., 2011. Adsorption. 17, 247.

Wahby, A., Ramos-Fernandez, J.M., Martinez-Escandell, M., Sepulveda-Escribano, A., Silvestre-Albero, J., Rodriguez-Reinoso, F., 2010. ChemSusChem. 3, 974.

Walker, P.L., Janov, J., 1968. J. Colloid Interface Sci. 28, 499.

Walker, P.L., Patel, R.L., 1970. Fuel. 49, 91.

Wang, Z.M., Kaneko, K., 1995. J. Phys. Chem. 99, 16714.

Wesolowski, R.P., Furmaniak, S., Terzyk, A.P., Gauden, P.A., 2011. Adsorption. 17, 1.

Wigmans, T., 1989. Carbon. 27, 13.

Wright, J.E., Freeman, J.J., Sing, K.S.W., Jackson, S.W., Smith, R.J.M., 1988. British Patent 8 723 447.

Yamamoto, M., Itoh, T., Sakamoto, H., Fujimori, T., Urita, K., Hattori, Y., Ohba, T., Kagita, H., Kanoh, H., Niimura, S., Hata, K., Takeuchi, K., Endo, M., Rodriguez-Reinoso, F., Kaneko, K., 2011. Adsorption. 17, 643.

Youssef, A.M., El-Wakil, A.M., El-Sharkawy, E.A., Farag, A.B., Tollan, K., 1996. Adsorption Sci. Technol. 13 (2), 115.

Zawadzky, J., 1989. In: Thrower, P.A. (Ed.), Chemistry and Physics of Carbon, Vol. 21. Marcel Dekker, New York, p. 147.

Zettlemoyer, A.C., Chessick, J.J., 1974. Advances in Chemistry, Vol. 43. American Chemical Society, Washington, DC, p. 58.

Zettlemoyer, A.C., Young, G.Y., Chessick, J.J., Healey, F.H., 1953. J. Phys. Chem. 57, 649.

Ziolkowska, D., Garbacz, J.K., 1997. Adsorption Sci. Technol. 15, 155.

11 Adsorption by Metal Oxides

Jean Rouquerol, Kenneth S.W. Sing, Philip Llewellyn

Aix Marseille University-CNRS, MADIREL Laboratory, Marseille, France

Chapter Contents

Adsorption by Powders and Porous Solids. http://dx.doi.org/10.1016/B978-0-08-097035-6.00011-5

11.1   Introduction

Some metal oxides (notably alumina, magnesia and silica) can be readily prepared in a stable state of high specific surface area: because of their technical importance as adsorbents, they have been featured in many fundamental and applied investigations of adsorption. Other oxides (e.g. those of chromium, iron, nickel, titanium, zinc) tend to give surfaces of lower area, but exhibit specific adsorbent and catalytic activity and therefore these oxides have also attracted considerable interest.

In their bulk properties, oxide adsorbents range from amorphous solids (e.g. silica) to various crystalline forms (e.g. anatase and rutile), but they all tend to undergo surface hydration and/or hydroxylation as a means of achieving coordinative stabilisation. The presence of other surface ligands can also be detected by infrared spectroscopy. An important consequence of the presence of surface hydroxyls is that many polar adsorptive molecules can interact in a specific manner with the surface. In some cases, surface dehydration leads to the exposure of strong cationic sites, while other parts of the surface may become less reactive. The possible types of specific interaction include hydrogen bonding and Lewis electron acceptor–donor exchange.

The most active oxide adsorbents are generally highly porous. In the past, the porosity was easy to produce but difficult to control and therefore industrial adsorbents such as many silica gels and activated aluminas had complex micropore or mesopore structures. However, considerable progress has been made in recent years in the elucidation of the mechanisms of pore formation and development and in their experimental control.

The aims of this chapter are: (a) to indicate the progress made in the development, characterisation and application of some of the more important oxide adsorbents and (b) to illustrate the procedures described in earlier chapters for the interpretation of adsorption data.

11.2   Silica

11.2.1   Pyrogenic and Crystalline Silicas

The *pyrogenic silicas* are formed at high temperatures. The most common type of 'fume silica', which is produced by the flame hydrolysis of silicon tetrachloride, is widely known as Aerosil. This term has become generic for the fume silicas, although strictly Aerosil is the trade mark of Degussa (Ferch, 1994). Other pyrogenic silicas are made by the high-temperature fusion or vaporisation of sand in an arc or plasma.

From the standpoint of adsorption, the pyrogenic silicas can be regarded as essentially non-porous. Electron microscopic examination has demonstrated that the arc silicas (e.g. Degussa TK 800 or TK 900) consist of discrete *spheroidal particles*.

However, the application of high-resolution electron microscopy has also revealed that the globules (typically of diameter 10–100 nm) are in fact *agglomerates of much smaller spheroidal units* of diameter around 1 nm (Barby, 1976). The coordination number of these primary particles is generally so high that there is *no detectable micropore structure* within the secondary globules.

In their original unaged and uncompacted state, the pyrogenic silicas give reversible Type II nitrogen isotherms at 77 K (as in Figure 11.1). When plotted in a reduced, dimensionless form (e.g. α_S vs. p/p°), the isotherms lie on a common curve over a wide range of p/p° (Carruthers et al., 1968; Bhambhani et al., 1972).

The corresponding argon isotherms at 77 K also give a common reduced isotherm (Payne et al., 1973), but in this case the 'knee' of the isotherm is more rounded and the uncertainty on the visual location of point B is larger, though the accuracy of the measurements is higher, due to a threefold decrease of the p° and of the amount of non-adsorbed gas. Another difference is that at 77 K the argon multilayer development is restricted at high p/p° (see Figure 11.1).

Standard isotherm data for nitrogen and argon adsorption at 77 K on non-porous hydroxylated silica are recorded in Tables 11.1 and 11.2, respectively. The original isotherms were determined on five different arc silicas having specific surface areas over the range 36–166 $m^2 \, g^{-1}$ (Carruthers et al., 1968; Bhambhani et al., 1972; Payne et al., 1973). Close agreement is obtained in the monolayer and lower multilayer regions with reduced isotherms on Aerosil 200 and on mesoporous silica gels (Bhambhani et al., 1972). However, in view of the wide particle size distribution and the heterogeneous nature of these materials, it would now be of interest to determine the standard isotherms for a range of carefully prepared non-porous silicas.

The early work of Kiselev (1957) revealed that the adsorption isotherms of *n*-pentane and *n*-hexane on non-porous quartz were intermediate in character between

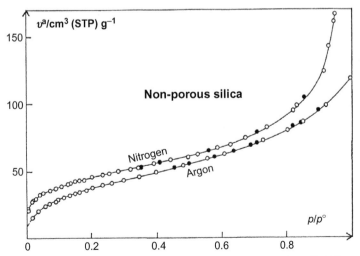

Figure 11.1 Argon and nitrogen isotherms at 77 K on non-porous silica TK 800. Courtesy Payne et al. (1973).

Table 11.1 Standard Data for the Adsorption of Nitrogen at 77 K on Non-porous Hydroxylated Silica

$p/p°$	$n/\mu\text{mol g}^{-1}$	$\alpha_S(=n/n_{0.4})$	$p/p°$	$n/\mu\text{mol g}^{-1}$	$\alpha_S(=n/n_{0.4})$
0.001	4.0	0.26			
0.005	5.4	0.35	0.28	13.6	0.88
0.01	6.2	0.40	0.30	13.9	0.90
0.02	7.7	0.50	0.32	14.2	0.92
0.03	8.5	0.55	0.34	14.5	0.94
0.04	9.0	0.58	0.36	14.8	0.96
0.05	9.3	0.60	0.38	15.1	0.98
0.06	9.4	0.61	0.40	15.5	1.00
0.07	9.7	0.63	0.42	15.6	1.01
0.08	10.0	0.65	0.44	16.1	1.04
0.09	10.2	0.66	0.46	16.4	1.06
0.10	10.5	0.68	0.50	17.0	1.10
0.12	10.8	0.70	0.55	17.8	1.14
0.14	11.3	0.73	0.60	18.9	1.22
0.16	11.6	0.75	0.65	19.9	1.29
0.18	11.9	0.77	0.70	21.3	1.38
0.20	12.4	0.80	0.75	22.7	1.47
0.22	12.7	0.82	0.80	25.0	1.62
0.24	13.0	0.84	0.85	28.0	1.81
0.26	13.3	0.86	0.90	37.0	2.40

Types II and III. Values of $C(\text{BET}) < 10$ were obtained and the differential enthalpies of adsorption decreased steeply at low surface coverage. More recently, the isotherms of isobutane (at 261 K) and neopentane (at 273 K) on TK 800 have been found to be of a similar shape (Carrott et al., 1988; Carrott and Sing, 1989). Unlike the behaviour of benzene, these alkane isotherms do not undergo a pronounced change of shape as a result of surface dehydroxylation. This is consistent with the non-specific nature of their molecular interactions (see Chapter 1).

The usefulness of the Frenkel–Halsey–Hill (FHH) equation for multilayer analysis is discussed in Chapters 5 and 10. FHH plots for nitrogen on various pyrogenic silicas are given in Figure 11.2. As expected, each FHH plot is linear over a wide range of $p/p°$, but this is rather more extensive (i.e. $p/p° \approx 0.3-0.9$) with the arc silicas, TK 800 and TK 900, when outgassed at 140 °C. Dehydroxylation of the TK 800, by outgassing the adsorbent at 1000 °C, has resulted in a deviation of the FHH plot from linearity at

Table 11.2 Standard Data for the Adsorption of Argon at 77 K on Non-porous Hydroxylated Silica

$p/p°$	$\alpha_S(=n/n_{0.4})$	$p/p°$	$\alpha_S(=n/n_{0.4})$
0.01	0.243	0.30	0.876
0.02	0.324	0.32	0.900
0.03	0.373	0.34	0.923
0.04	0.413	0.36	0.948
0.05	0.450	0.38	0.973
0.06	0.483	0.40	1.000
0.07	0.514	0.42	1.022
0.08	0.541	0.44	1.048
0.09	0.563	0.46	1.064
0.10	0.583	0.48	1.098
0.11	0.602	0.50	1.123
0.12	0.620	0.52	1.148
0.13	0.638	0.54	1.172
0.14	0.657	0.56	1.198
0.15	0.674	0.58	1.225
0.16	0.689	0.60	1.250
0.17	0.705	0.62	1.275
0.18	0.719	0.64	1.300
0.19	0.733	0.66	1.327
0.20	0.748	0.68	1.354
0.22	0.773	0.70	1.387
0.24	0.801	0.72	1.418
0.25	0.813	0.74	1.451
0.26	0.826	0.76	1.486
0.28	0.851	0.78	1.527

$p/p° < 0.5$. This deviation is likely to be associated with the reduction in the specific field gradient–quadrupole interactions as a result of the removal of surface hydroxyls. In the case of the FHH plot on the Aerosil 200, the small upward deviation from linearity, which is detectable at $p/p° > 0.8$, is due to interparticle capillary condensation.

FHH plots have also been constructed from the isotherms of isobutane and neopentane on the pyrogenic silicas (Carrott et al., 1988). Derived values of the FHH

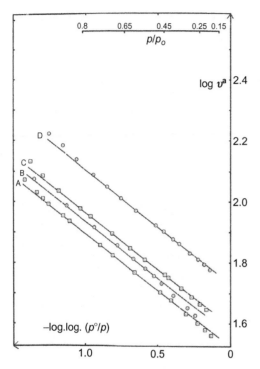

Figure 11.2 FHH plots for nitrogen adsorption on pyrogenic silicas: TK 900, 140 °C (A); TK 800, 1000 °C (B); TK 800, 140 °C (C); Aerosil 200, 140 °C (D) (Carrott et al., 1982).

Table 11.3 FHH s Values for Nitrogen (at 77 K) and Neopentane (at 273 K) on Pyrogenic Silicas, BET (N_2) Areas, BET C Values for both Gases and Apparent Cross-Sectional Area $\sigma(np)$ for Neopentane Molecule (Taking $\sigma(N_2) = 0.162$ nm^2)

Silica	$a(BET - N_2)/m^2 g^{-1}$	C (N_2)	s (N_2)	C (np)	s (np)	σ(np)/nm^2
TK 800	158	94	2.72	6	1.97	0.62
TK 900	136	101	2.71	7	1.97	0.60
TK 900[a]	120	47	2.69	8	1.96	0.58
Aeros 200	205	95	2.67	7	2.01	0.60
Aeros 200[a]	192	43	2.68	8	2.00	0.59

[a]Heated in pure N_2 for 8 h at 950 °C.
Source: Results of Carrott et al. (1988).

exponent, s, for neopentane s(np) and nitrogen $s(N_2)$ are recorded in Table 11.3. Also included in this table are the BET-nitrogen surface areas, a(BET-N_2) and the BET C values for nitrogen and neopentane, $C(N_2)$ and C(np). As expected, the high-temperature treatment of TK 800 and Aerosil 200 resulted in the removal of a high proportion of the surface OH groups: this in turn led to a significant reduction in the nitrogen–adsorbent interaction energy (see Chapter 1) and consequently a decrease in the values of $C(N_2)$.

The constancy of the values of $s(N_2)$ and $s(np)$ in Table 11.3 is striking and of course follows from the fact that the linear FHH plots are parallel in the multilayer range. These results confirm that *the multilayer* character of each adsorptive is rather *insensitive to changes in surface chemistry*.

The α_s-*method* has been used for the analysis of the isotherms of the following gases on porous and non-porous silicas: nitrogen (Bhambhani et al., 1972; Carrott and Sing, 1984), argon (Carruthers et al., 1971; Payne et al., 1973), carbon tetrachloride (Cutting and Sing, 1969) and neopentane (Carrott et al., 1988).

A consistent pattern of behaviour has emerged from the study of different samples of the pyrogenic silicas. The derived α_s-plots are all linear over the monolayer and lower multilayer range. In each case, back extrapolation gives a zero intercept, which confirms the *absence of microporosity*. The α_s-plots on samples of Aerosil 200 exhibit an upward deviation in the higher multilayer region (see Figure 11.3), which is consistent with the FHH evidence that some interparticle capillary condensation has occurred.

As to be expected from the shape of the nitrogen isotherms and the linearity of the α_s-plots in Figure 11.3, satisfactory agreement is obtained between the corresponding

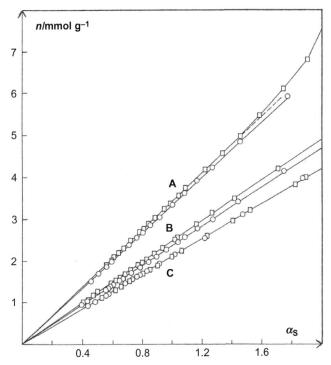

Figure 11.3 Representative α_s-plots for nitrogen adsorption on pyrogenic silicas: Aerosil 200 (A); TK 800 (B); TK 900 (C), outgassed at 298 K (circles) or at 413 K (squares) (Carrott and Sing, 1984).

Table 11.4 Comparison of Surface Areas of Pyrogenic Silicas Evaluated by the
BET and α_S Method

Silica	Outgassing T/K	$a(BET - N_2)/m^2 g^{-1}$	$a(S)/m^2 g^{-1}$
TK 800	298	153	151
	413	158	158
TK 900	298	136	136
	413	136	135
Aerosil 200	298	215	213
	413	221	218

values of the BET-area, $a(BET)$, and the area, $a(S)$, calculated from the slope of the α_S-plot – as indicated in Table 11.4. Here, the value of $a(S)$ is given by

$$a(S) = 0.0641n/\alpha_S \tag{11.1}$$

where the amount adsorbed, n, is expressed in $\mu mol\ g^{-1}$ and the factor 0.0641 has been obtained by calibration against Fransil – a well-characterised arc silica having a $BET(N_2)$ surface area of 38.7 $m^2\ g^{-1}$ (Bhambhani et al., 1972).

The situation is more complicated when Point B is either non-existent (as with the alkanes) or less well defined than for nitrogen (as with argon and krypton) since the physical significance of the BET specific monolayer capacity, n_m, is unclear. Nevertheless, it may be worthwhile analysing the n_m obtained for the above gases, in light of the $BET(N_2)$ area of the adsorbent. It is then indeed possible to derive, for the gas under study, an apparent value of its molecular area in the monolayer, σ_m, taking as a reference the $BET(N_2)$ surface area calculated with the conventional molecular area of 0.162 nm^2. The values of apparent molecular area, $\sigma(np)$, of adsorbed neopentane in Table 11.3 are calculated from its BET monolayer capacity, $n_m(np)$, and the $BET(N_2)$surface area. Thus,

$$\sigma(np)/nm^2 = \frac{1.66\,a(N_2)/m^2\,g^{-1}}{n_m(np)/\mu mol\,g^{-1}} \tag{11.2}$$

The five $\sigma(np)$ values in Table 11.3 are in close agreement (i.e. $(0.595 \pm 0.015)\ nm^2$), but they are much larger than the value, 0.40 nm^2, calculated, from the liquid density, for a close-packed monolayer of freely rotating neopentane molecules. However, it must be kept in mind that the $C(BET)$ values are very low (i.e. 6–8) and therefore one cannot be sure that $n_m(np)$ represents the true statistical monolayer capacity.

Prolonged storage of a pyrogenic silica can result in *slow ageing* as indicated by a loss of BET-area. For example, the original specific surface area of a master batch of TK 800, as evaluated from the measurements of Payne et al. (1973) and Baker and Sing (1976), was 163–166 $m^2\ g^{-1}$. A loss of area of ca. 6 $m^2\ g^{-1}$, which occurred over a period of 8 years, was evidently due to the development of some interparticle mesoporosity since this led to the development of a narrow hysteresis loop in the nitrogen isotherm at high nitrogen $p/p°$ (Carrott and Sing, 1984). A partial restoration in area was achieved by raising the outgassing temperature from 25 to 140 °C (see Table 11.4). This was probably due to the removal of water from the particle interstices: it is likely that the presence of residual water and compaction were together responsible for the loss of area.

A number of detailed studies have been made of the physisorption of gases on *compacted silica powders* (Avery and Ramsay, 1973; Gregg and Langford, 1977). The changes in the character of nitrogen adsorption as a result of increasing the compacting pressure are shown in Figure 11.4. It is striking that the isotherms in Figure 11.4 change first from Type II to IV and finally to Type I. Thus, the results of Avery and Ramsay (1973) have clearly demonstrated that mesopores and micropores can be produced by the progressive compaction of a non-porous powder. However, a drastic loss of surface area has accompanied this change (from 630 to 219 $m^2 g^{-1}$) and Avery and Ramsay point out that this was associated with a marked increase in the particle packing density.

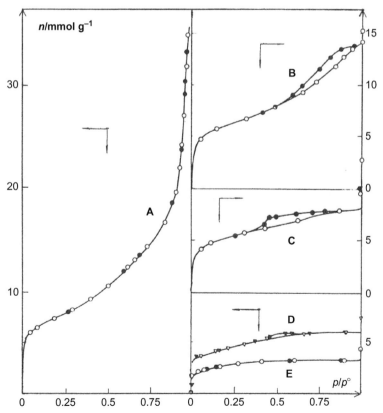

Figure 11.4 Effect of compaction of Aerosil on nitrogen adsorption isotherms at 77 K: uncompressed (a) or compressed under 1.5 (b), 6 (c), 7.5 (d) or 15 kbar (e). Courtesy Avery and Ramsay (1973).

Along the same lines, Phalippou et al. (2004), who studied aerogels (therefore obtained by super-critical drying treatment), compared the effect of thermal sintering with that of mechanical compression on the BET-nitrogen area. It was amazing to see that when the bulk density increased from 0.2 to 0.5 g cm^{-3} there was nearly no change in the BET-nitrogen area (steadily equal to ca. 425 m^2 g^{-1}) when this increase resulted from an isostatic compression (in a pressure range lower than that used by Avery and Ramsay), whereas it dropped down to ca. 275 m^2 g^{-1} when the increase in density was due to thermal sintering at 1000 °C. The authors considered that in this case mechanical pressure only induced the collapse of the largest pores, whereas thermal sintering acted on all pores, including the smallest which plaid the major part in the resulting BET-nitrogen area.

The importance of surface silanol (hydroxyl) groups in controlling the specific physisorption interactions was studied by Kiselev (1958, 1971). Kiselev and his co-workers found that *fully hydroxylated silica* has a surface OH concentration of 7–9 μmol m^{-2}, which corresponds to a surface population, N (OH), of ca. 5 OH/ nm^2 (Zhuravlev and Kiselev, 1970). More recently, Zhuravlev (1987) reported a mean N (OH) value of 4.9 OH nm^{-2} (after outgassing at 180–200 °C) for 100 fully hydroxylated amorphous silicas. This value is not far removed from N (OH) = 4.6 OH nm^{-2}, originally proposed by Vleeskens (see Okkerse, 1970).

A fully hydroxylated state in which all the OH groups are bound to Si atoms (with N (OH) = 4.6 OH/nm^2) would correspond to the octahedral, (111), face of β-cristobalite or the basal face of β-tridymite. However, as indicated in Table 11.5, the surface hydroxyl concentrations found on some arc and fume silicas are significantly lower, typical values being in the range 3.4–4 OH/nm^2 (Baker and Sing, 1976; Unger, 1979, p. 62; Gallas et al., 1991).

As was noted earlier, the level of surface OH concentration has very little effect on either the isotherms or the energetics of adsorption of non-polar molecules such as argon or the alkanes. Specific interactions become significant when the adsorptive molecules are quadrupolar (e.g. nitrogen and carbon dioxide) and even more so when hydrogen bonding is involved (e.g. with water or the lower alcohols).

Table 11.5 Levels of Hydroxylation and Sorption of Water Vapour by Pyrogenic and Precipitated Silicas and Mesoporous Silica Gels

Silica	Outgassing θ/°C	N_{OH}/nm^{-2}	Water Adsorption at $p/p° = 0.2$/μmol m^{-2}
Fransil	25	4	5.3
TK 800	25	3.6	5.0
	1000	∼0	1.3
VN3	25	14[a]	22
SiO$_2$ gels	25	3.4, 5[a]	6.4, 6.5[b]
	1000	∼0	1.3

[a]Results of Gallas et al. (1991).
[b]Results of Kiselev (1958, 1971).

Many investigations have been made of the adsorption of *water vapour on amorphous silicas* (e.g. Zettlemoyer, 1968; Baker and Sing, 1976; Iler, 1979, p. 651; Unger, 1979, p. 196; Burneau et al., 1990). There is general agreement that the removal of surface silanols by heat treatment causes a drastic reduction in the level of water adsorption (see Table 11.5). It might be expected that there would be a simple correlation between the surface OH concentration and the affinity for water vapour. However, systematic adsorption calorimetric and infrared spectroscopic investigations by Fubini et al. (1992) and Gallas et al. (1991) have revealed a more complex behaviour.

It is now apparent that isolated silanols have relatively low affinity for water. Thus, the hydrophobic nature of silica is manifested after dehydroxylation when only the siloxane bridges and some isolated silanols (giving an IR band at ca. 3750 cm^{-1}) remain. On the *dehydroxylated surface* the net adsorption enthalpy for water is negative. In this case, the enthalpy of adsorption is lower than the normal enthalpy of condensation. Application of adsorption microcalorimetry has allowed an assessment to be made of the relative extents of the hydrophilic and hydrophobic areas of the surface (Bolis et al., 1991). On the hydrophilic surface, it appears that water is adsorbed *via* two hydrogen bonds to two silanols – one acting as the hydrogen donor and the other as the acceptor. In the case of the weaker attachment to the isolated OH, the attachment involves one hydrogen bond.

Comparisons made between the properties of *crystalline and amorphous silicas* (Bolis et al., 1991; Fubini et al., 1992) have brought to light some interesting differences. For example, quartz is much less easily dehydroxylated than the pyrogenic silicas and this is consistent with its more persistent hydrophilic character (Pashley and Kitchener, 1979). In their study of the adsorption of water on α-cristobalite, Fubini and her co-workers (1992) obtained a number of irreversible water isotherms, the nature of which depended on the thermal pre-treatment. The complete rehydroxylation of a hydrophobic silica by exposure to water vapour is usually slow; but with some dehydroxylated samples a fairly rapid stage of dissociative chemisorption leads to pronounced hysteresis, which extends across the complete range of p/p° as in Figure 11.13 (Baker and Sing, 1976).

Adsorption microcalorimetry has shown that the surfaces of both amorphous and crystalline silicas are energetically heterogeneous. Furthermore, the IR spectroscopic evidence reveals that their surface structures are dependent on the conditions of preparation and treatment.

11.2.2 Precipitated Silicas

Although commercially important, the precipitated silicas have received much less attention in the scientific literature than either the Aerosils or silica gels. In certain respects, they are similar to pyrogenic silicas: indeed, at one time they were treated as alternative non-porous silicas. Thus, the reversible Type II isotherms of nitrogen and argon obtained by Bassett et al. (1968) were assumed to represent uncomplicated monolayer – multilayer adsorption. More recent work (Carrott and Sing, 1984) has shown that the apparent Type II character is here the result of adsorption both on the external surface and within some micropores, that is this isotherm should be considered in reality as a composite Type I + Type II isotherm.

In contrast to the general behaviour of the pyrogenic silicas, the level of physisorption by precipitated silicas has been found to be sensitive to changes in the conditions of outgassing. Physisorption equilibration is more difficult to attain and consequently the curves expected to be true adsorption isotherms often have a more complicated appearance. For example, the behaviour of a batch of VN3 (a Degussa product) is illustrated by the results in Figure 11.5 (Carrott and Sing, 1984). It can be seen that the increase in outgassing temperature from 25 to 110 °C has produced a significant upward movement in the nitrogen isotherm and in addition has led to the appearance of a small step in the amount adsorbed at $p/p° \approx 0.6$ and to the development of a small amount of low-pressure hysteresis. Similar features are exhibited by the isotherms determined after outgassing at the temperatures of 200 and 300 °C, but in the latter case the step is located at a much lower relative pressure ($p/p° \approx 0.2$) and now the open hysteresis extends over the complete isotherm. It is evident that these steps cannot be explained in terms of any well-defined phase transitions. They are more likely to be due to the slow diffusion of the adsorptive molecules into inner regions of the particles, which are not easily accessible, therefore indicating that certain portions of these curves are not representative of the true equilibrium and should therefore not be considered as part of a real adsorption isotherm.

To circumvent this issue, the α_s-plots for nitrogen on VN3 in Figure 11.5 have been constructed from the *desorption* isotherms. In each case, the back-extrapolated linear portion gives a positive intercept on the n-axis and an upward deviation can be seen at $p/p° \approx 0.7$. This behaviour is typical of adsorption at low $p/p°$ occurring both on the external surface and within narrow micropores. At higher $p/p°$, the upward deviation

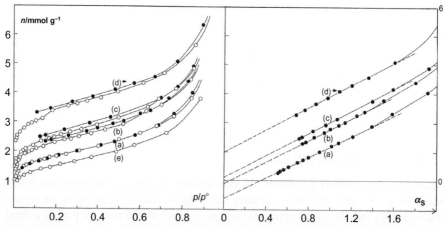

Figure 11.5 Nitrogen isotherms (left, with open circles for adsorption and closed circles for desorption) and α_S plots (right) for precipitated silica VN3 outgassed at 25 (a), 110 (b), 200 (c) and 300 °C (d), with ordinate shifted upwards of 1 mmol g^{-1}, for clarity. Run (e) after outgassing at 200 °C and n-nonane pre-adsorption (Carrott and Sing, 1984).

Table 11.6 Values of BET-Area, External Area and Micropore Volume Given by a Precipitated Silica, VN3, Outgassed at Various Temperatures and Studied by N_2 Adsorption at 77 K

Outgassing Temperature/K	$a(\text{BET}, N_2)/\text{m}^2\,\text{g}^{-1}$	$a(\text{ext, S})/\text{m}^2\,\text{g}^{-1}$	$v_p(\text{mic, S})/\text{cm}^3\,\text{g}^{-1}$
298	146	114	0.014
383	178	122	0.027
473	195	124	0.034
573	174	125	0.030

indicates that the multilayer adsorption on the external surface is accompanied by interparticle capillary condensation, which is partly responsible for the narrow hysteresis loop.

The α_s-plots in Figure 11.5 have been analysed by the method described earlier in Chapter 8. The values of the *external surface area*, $a(\text{ext, S})$ and *micropore volume*, $v_p(\text{mic, S})$, in Table 11.6 have been calculated from the slope and intercept, respectively of each α_s-plot. It can be seen that the external area does not change to any significant extent over the range 110–300 °C and that the changes in the BET-area, a (BET), are due to the development of a small micropore volume.

Further confirmation of the development of microporosity has been obtained by the application of the nonane pre-adsorption method of Gregg and Langford (1969). After it had been outgassed at 200 °C, the sample of VN3 was exposed to *n*-nonane vapour, then outgassed at 25 °C and the nitrogen isotherm redetermined. The result is isotherm (e) in Figure 11.5 and the derived value of $a(\text{BET})$ is now 122 m^2 g^{-1}, which is in good agreement with the external area, $a(\text{ext, S}) = 124$ m^2 g^{-1}.

Gallas et al. (1991) have noted the exceedingly high value of the silanol population (14 OH nm^{-2}) in their sample of precipitated silica, which they attribute to the presence of many inner hydroxyls. This is consistent with the abnormally large water uptake by VN3 – as indicated in Table 11.5. This is much greater than the amount (ca. 15.7 μmol m^{-2}) required to give a close-packed water monolayer and is further evidence of the microporous nature of precipitated silicas.

We conclude that in the precipitation process, the trapping of water has resulted in the formation of a more open intra-particle microstructure than in the pyrogenic silicas. The internal surface remains hydroxylated, but is not easily accessible to most adsorptive molecules. Water molecules, because of their small size, are able to undergo specific interactions with the internal OH groups which accounts for the abnormally high uptake of water vapour.

11.2.3 Silica Gels

Over the past 70 years a large number of physisorption studies have been reported on silica gels (Deitz, 1944; Brunauer, 1945; Okkerse, 1970; Barby, 1976; Iler, 1979, p. 488). It is not difficult to prepare stable adsorbent silica xerogels in a highly porous and fairly dense granular form. Although other more refined routes are now available, commercial silica gels are still prepared in large quantity by the dehydration of the hydrogels produced by reacting sodium silicate with acid (Patterson, 1994). These materials are widely used as relatively inexpensive adsorbents, desiccants and catalyst supports.

Many of the early physisorption measurements were undertaken on ill-defined silica gels of unknown origin. However, as a result of systematic work on the formation of the hydrogel and its ageing and conversion to the xerogel, it is now possible to control the adsorptive properties within quite narrow limits (Barby, 1976; Iler, 1979; Unger, 1979; Kenny and Sing, 1994).

Monosilicic acid is stable in aqueous solution only at low pH and very low concentration (Iler, 1979). The rate of condensation–polymerisation of silicic acid is dependent on pH, concentration and temperature. At a certain stage, the polysilicic acid sol is converted into either a precipitate (i.e. a flocculated system) or a hydrogel. Further changes take place when the hydrogel is brought into contact with an ageing medium. The mechanisms of ageing include aggregation–cementation and Ostwald ripening (i.e. growth of the larger particles at the expense of the smaller) and result in changes in the size and packing density of the colloidal particles. Particle growth and particle–particle siloxane (Si—O—Si) bonding are largely irreversible, but it is possible to achieve a certain degree of depolymerisation, for example by acid treatment (Mitchell, 1966). Simulation of silicic acid polymerisation would indicate the following succession of events: (a) dimer and small oligomer formation (b) growth of non-cyclic chains (c) ring formation and growth of spherical particles (d) Ostwald ripening and (e) cross-linking between the particles provided the silica concentration is high enough (Jin et al., 2011).

The hydrogel has an open structure (i.e. a low particle coordination number) and is mechanically weak. The removal of the aqueous liquid phase normally leads to a drastic shrinkage of the silica framework and the formation of additional siloxane bonds (Fenelonov et al., 1983). The resulting xerogel is therefore much stronger and more compact, but inevitably has a lower pore volume.

Until the availability of high-resolution electron microscopy, the structure of silica xerogels remained a matter of conjecture. Now, there is no longer any doubt that the amorphous framework is made up of very small globular units (see Figure 11.6). These primary particles are isotropic and have fairly uniform size of 1–2 nm (with a molar mass ≈ 2000 g mol^{-1}). In some xerogels, the primary particles are densely packed within secondary particles, whereas in other systems there is a more open arrangement (Barby, 1976).

The texture of the dry xerogel depends partly on the properties of its parent hydrogel and partly on the conditions of the conversion. The dependence of the final xerogel porosity on the conditions of gelation and after-treatment have been investigated in some detail (see Barby, 1976; Iler, 1979, p. 209; Kenny and Sing, 1994). In the early work of Madeley and Sing (1953, 1954, 1962), *microporous* products were obtained from hydrogels prepared by the addition of sodium silicate to sulphuric acid at relatively *low pH* (e.g. $\approx$pH 3.5). *Mesoporous* products were produced when the reaction was carried out at the *higher pH* ≈ 6. On the other hand, changes in the silicic acid

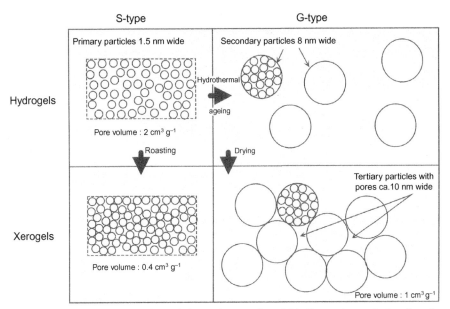

Figure 11.6 Morphology of silica gel. (For colour version of this figure, the reader is referred to the online version of this chapter.)
After Barby (1976).

concentration over a wide range had little effect, provided that the gelation was conducted in a buffered aqueous medium (Madeley and Sing, 1962).

Neimark et al. (1964) carried out extensive investigations of the effects of ageing silica hydrogels in various media. It was found that the dispersed state was stabilised when the hydrogel was left in contact with acids (at pH 2–5), which appeared to be due to the protection afforded by hydrogen-bonded water (i.e. a 'hydrate shell') around the particles. The effect that acid washing has in enhancing the adsorption activity is illustrated by the results in Figure 11.7. In this case, the original hydrogel was prepared at pH 5.4 and portions were then subjected to different forms of after-treatment (Wong, 1982). Soaking in HCl (at pH 2.0 for 24 h) resulted in a significant upward movement of the complete nitrogen isotherm: analysis of the adsorption data revealed an increase in BET-nitrogen area, a(BET), from 284 to 380 m^2 g^{-1} and an increase in pore volume, v_p, from 0.44 to 0.55 cm^3 g^{-1}.

The fact that the shape of the isotherms in Figure 11.7 has remained almost unchanged after the acid treatment, is an indication that the mesopore structure was not altered to any significant extent. However, as pointed out in Section 8.6, this form of H_2 hysteresis loop is not easy to interpret since it is associated with pronounced percolation effects in an irregular pore network.

The most striking result in Figure 11.7 was obtained when the hydrogel was washed with ethyl alcohol. The vacuum-dried material, which we shall refer to as an *alcogel*, gave a much larger uptake of nitrogen over the complete range of $p/p°$: thus, a(BET) $= 641$ m^2 g^{-1} and $v_p = 0.93$ cm^3 g^{-1}. It is evident that by replacing water

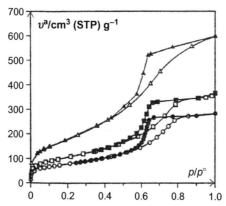

Figure 11.7 Nitrogen isotherms at 77 K for xerogel (circles), acid-washed werogel (squares; HCl, pH 2, 24 h) and alcogel (triangles) (Kenny and Sing, 1994). (For colour version of this figure, the reader is referred to the online version of this chapter.)

by ethyl alcohol as the continuous liquid phase, it was possible to *reduce the large capillary forces* which are responsible for the considerable shrinkage normally found when water is removed from the hydrogel.

An even larger pore volume can be obtained if the liquid phase is removed under *supercritical conditions* to give an *aerogel*. This type of gel has an extremely high surface area and pore volume (see Table 11.7), but it tends to be mechanically weak and unstable when exposed to water vapour – since the particle coordination number is low. The upper limiting area of a silica composed of discrete primary particles would be $\approx 2000 \; m^2 \; g^{-1}$, but specific surface areas of this magnitude are unlikely to be attained.

Barby (1976) has defined two types of conventional silica xerogels (Table 11.7). The *S-type gels* can be either microporous or mesoporous and are produced in the normal way, which allowed considerable loss of surface area and pore volume to occur during the removal of water from the hydrogel. If the hydrogel is subjected to *hydrothermal treatment*, the primary particles undergo more drastic aggregation–cementation with the result that after drying the porosity is largely confined to the interstitial space between the secondary particles as in Figure 11.6. The resulting *G-type xerogel* has a somewhat lower surface area, but a larger and more uniform pore volume (Barby, 1976).

Table 11.7 The Properties of Typical Silica Gels

Gel Type	Porosity	$a(\text{BET} - N_2)/m^2 \, g^{-1}$	$v_p/cm^3 \, g^{-1}$
Aerogel	Macro	800	2.0
G-Xerogel	Meso	350	1.2
S-Xerogel	Meso	500	0.6
	Micro	700	0.4

In view of the complexity of the structure of most silica xerogels, it is to be expected that their adsorptive behaviour would be equally complex. The following questions are pertinent:

1. How reliable are the derived values of a(BET) and pore size distribution?
2. Can we identify the various stages of micropore and mesopore filling?

The following discussion of some typical data for the adsorption of nitrogen and argon will throw light on these questions and also illustrate the value of using α_s-method of isotherm analysis.

The isotherms and corresponding α_s-plots in Figures 11.8 and 11.9 are for the adsorption of nitrogen on representative mesoporous and microporous silica gels (Bhambhani et al., 1972). The derived values of the specific surface area are given in Table 11.8. The values of the BET nitrogen area, a(BET), in Table 11.8 are based on the usual assumption that the adsorbed molecules were close-packed in the completed monolayer, with the conventional molecular area $\sigma(N_2)$ of 0.162 nm^2. The

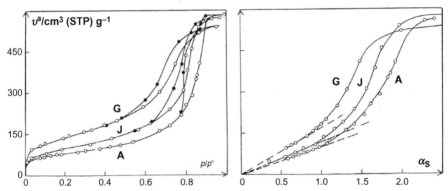

Figure 11.8 Nitrogen isotherms (left) and α_s-plots (right) for mesoporous silica gels A, J and G. From Bhambhani et al. (1972).

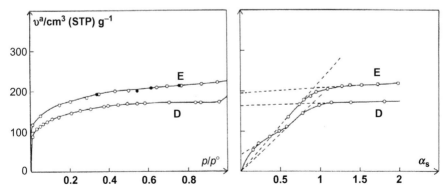

Figure 11.9 Nitrogen isotherms (left) and α_S-plots (right) for microporous silica gels D and E. From Bhambhani et al. (1972).

Table 11.8 Comparison of Values of Surface Area Derived from Nitrogen BET-Plots and α_s-Plots

Silica	Porosity	$a(BET,N_2)/m^2 g^{-1}$	$a(S,N_2)/m^2 g^{-1}$
Gel A	Meso	300	303
Gel G	Meso	504	503
Gel J	Meso	349	350
Gel D	Micro	767	810–960, 35[a]
Gel E	Micro	631	730, 20[a]

[a]Values of external area, a(ext, S).

corresponding values of $a(S, N_2)$ were calculated from the initial slope of the α_s-plots by the relation

$$a(S, N_2) = 2.89\, v^\sigma (STP/\alpha_S) \tag{11.3}$$

which is an adaptation of Equation (11.1) to the case when adsorption data are provided in volume adsorbed v^a (STP), or surface excess volume v^σ (STP), and expressed in $cm^3 (STP) g^1$.

The good agreement between the corresponding values of $a(BET, N_2)$ and $a(S, N_2)$ in Table 11.8 is an indication that gels A, G and J were only mesoporous, with no micropores, and this is confirmed by the characteristic shapes of the isotherms and α_s-plots in Figure 11.8. However, the range of linearity of each α_s-plot does not extend above $p/p° = 0.4$, which in the case of gels A and J is just below the lower limit of hysteresis. This result is consistent with the likelihood of some reversible capillary condensation occurring around the contact regions of the globular particles. The nitrogen isotherms on the microporous gels D and E in Figure 11.9 are of Type Ib in the classification given in Figure 1.2. The gradual approach to the plateau at $p/p° \approx 0.4$ is an indication that both gels contained a wide distribution of micropores (see Chapter 9). The α_s-plot on gel D confirms that there were two stages of micropore filling: the primary micropore filling at $p/p° < 0.01$ has resulted in some initial distortion of the isotherm, while the secondary (cooperative) micropore filling of supermicropores has taken place over the range $p/p° \approx 0.02$–0.2. In Table 11.8, three areas are derived from the α_S plot for gel D. The first one (810 $m^2\ g^{-1}$) derives from the slope corresponding to the primary micropore filling, by application of Equation (11.3). Nevertheless, as explained in Section 7.3.3, we consider that it is only the slope of the next portion of the α_S plot (which can be back extrapolated to the origin and corresponds the secondary micropore filling) which provides a reliable value of the *internal surface area* (960 $m^2\ g^{-1}$). Finally, as explained in Chapter 9, the *external surface area*, a(ext, S), available for multilayer adsorption can be calculated from the high $p/p°$ section of an α_s-plots, provided that there is an adequate range of linearity to confirm the absence of capillary condensation. This condition is clearly fulfilled with gel D, so that its external surface area (35 $m^2\ g^{-1}$) can also be considered as correct. For gel E, where the α_S plot shows that only secondary micropore filling occurs, we can consider as meaningful the two surface areas given, that is internal (730 m^2 g^{-1}) and external (20 m^2 g^{-1}). We should not be surprised to find, for gels E and D, BET areas different from the sum of the internal and external surface areas calculated by the α_S method. The BET method, which was not devised indeed for micropores, should better be considered in this case as a method providing a characteristic and reproducible figure but not an actual surface area (Rouquerol et al., 2007).

We turn now to the argon isotherms and α_s-plots, respectively in Figures 11.10 and 11.11 (from the data of Payne et al., 1973). Here, the experimentally determined value of p° [i.e. p°(solid)] at 77 K has been used to calculate p/p°. It is of interest to compare the shapes of the argon isotherms with those of nitrogen on the same adsorbents (mesoporous gel B and microporous gel C). The nitrogen isotherm on gel B has a well-defined Point B (C(BET) ≈ 100), while the argon isotherm has a more gradual curvature and less distinctive Point B (C(BET) < 50). In the middle range, these two isotherms run roughly parallel. It will be recalled that this behaviour is very similar to that on TK 800 (see Figure 11.1). The Type I isotherms on gel C in Figure 11.10 have the typical features of Type Ib isotherms, but the most striking aspect is the crossover of nitrogen by argon, which occurs at $p/p^\circ \approx 0.1$.

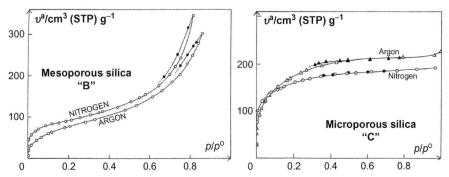

Figure 11.10 Argon and nitrogen isotherms at 77 K on mesoporous silica gel B (left) and microporous silica gel C (right) (Payne et al., 1973).

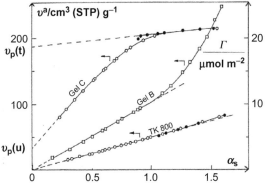

Figure 11.11 Argon α_s-plots for non-porous (TK 800), mesoporous (gel B) and microporous (gel C) silicas (Payne et al., 1973).

The deviations from linearity shown by the α_s-plots on gels B and C in Figure 11.11 confirm the respective mesoporous and microporous nature of these silica gels. Thus, the α_s-plot on gel B is linear up to $p/p° = 0.4$ (i.e. $\alpha_S = 1$). This indicates that the argon monolayer has formed on the mesopore surface in the same manner as on the non-porous surface. The upward deviation is clearly due to the onset of capillary condensation.

The behaviour of gel C is quite different. In this case, there is a strong distortion of the isotherm at $p/p° < 0.02$, which results in the positive intercept of the α_s-plot. This is followed by two linear regions, the first being associated with surface coverage and filling of supermicropores and the second with multilayer coverage of the external surface. Back extrapolation of these two linear parts provides a means of assessing the volume of the ultramicropores, $v_p(u)$ (of width ≤ 1 nm) and the total micropore volume, $v_p(t)$.

The values of specific surface area, $a(S, Ar)$, in Table 11.9 are calculated from the linear sections of the argon α_s-plots by application of the equation

$$a(S, Ar) = 3.30 v^\sigma(STP)/\alpha_S \qquad (11.4)$$

where $v^\sigma(STP)$, or v^a, is the volume $(cm^3 (STP) g^{-1})$ of argon adsorbed and the factor 3.30 has been obtained by calibration against the BET-nitrogen areas of Fransil and TK 800, which are taken as reference with the conventional nitrogen molecular area of 0.162 nm^2.

Comparison is made with the corresponding areas determined by nitrogen adsorption.

The results in Table 11.9 reveal that for non-porous and mesoporous silicas good agreement is obtained between the corresponding values of $a(S-Ar)$, $a(BET, Ar)$ and $a(BET, N_2)$, provided that the apparent molecular area of argon, $\sigma(Ar)$, is taken as either ca. 0.182 nm^2, if the argon BET-plots are based on $p°$(solid) or ca. 0.17 nm^2 if they are based on the extrapolated $p°$(liquid) at 77 K (Gregg and Sing, 1982). Alternatively, one can also obtain a good agreement by giving the argon molecule the molecular area of 0.138 nm^2 derived from its liquid density and therefore assuming for nitrogen a smaller molecular area than the conventional 0.162 nm^2. Several reasons are indeed in favour of the latter interpretation:

1. the nitrogen molecule is not spherical (at the difference from the argon molecule) and the area it covers can be as small as 0.112 nm^2 when 'standing' on the surface (Rouquerol et al., 1984)

Table 11.9 Analysis of Argon Isotherms on Silica Gels

Silica	$a(BET-N_2)/m^2 g^{-1}$	$a(S,N_2)/m^2 g^{-1}$	$a(BET-Ar)/m^2 g^{-1}$	$a(S,Ar)/m^2 g^{-1}$
Gel B	334	335	337	337
Gel K	216	217	214	222
Gel C	586	425–625, 29[a]	720	540–1150, 24[a]
Gasil	657	500–700	728	640–950

[a]Values of external area, $a(ext, S)$.

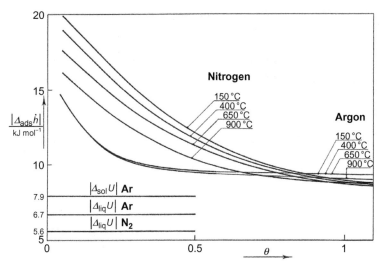

Figure 11.12 Differential enthalpies of adsorption of argon and nitrogen on a mesoporous silica gel: effect of outgassing at different temperatures (Rouquerol et al., 1979a,b).

2. specific interactions between the nitrogen quadrupole moment and the polar hydroxyls of the silica gel surface were early shown to exist (Kaganer, 1961; Aristov and Kiselev, 1963) and were directly measured by calorimetry (Rouquerol et al., 1979a,b), as illustrated hereafter in Figure 11.12.

3. if such interactions are able to provide some orientation to the nitrogen molecule their effect is necessarily to lower its molecular area, since the value of 0.162 nm^2 is a maximum corresponding to a free orientation of the molecule in all directions.

The choice between the two interpretations above (either loose argon monolayer or dense nitrogen monolayer with oriented molecules) is still to-day a matter of debate. It should be noticed that these interpretations are not exclusive from each other and that they may well hold simultaneously, each of them explaining part of the discrepancy between Ar and N$_2$ BET areas measured at 77 K.

11.2.3.1 Dehydroxylated Gels

When a silica gel is heated to progressively higher temperatures water is lost first from any mesopores and wide micropores, then from any narrow micropores and finally by the recombination of surface hydroxyls. The dehydroxylation of the surface generally takes place over the range 200–1000 °C and is accompanied by some loss of surface area (see Iler, 1979). However, the precise changes in area and pore structure depend on a number of factors including the nature of the original hydroxylated gel and the conditions of heat treatment (e.g. in air, vacuo or water vapour).

The first systematic investigations of the adsorption of gases on dehydroxylated silicas were made by Kiselev and his co-workers (Kiselev and Kiselev, 1957; Kiselev, 1958). In a study of the adsorption of argon and nitrogen, Aristov and Kiselev (1965) found that, in contrast to nitrogen, the reduced argon isotherm did not appear to depend on the degree of surface hydroxylation.

The effect of surface dehydroxylation of a mesoporous silica on the Ar and N_2 energetics of adsorption is illustrated in Figure 11.12. In the work of Rouquerol et al. (1979a,b), Tian-Calvet microcalorimetry was used to determine the variation of the differential enthalpy of adsorption as a function of surface coverage. Although strong energetic heterogeneity is a feature of the adsorption of both gases, with Ar the variation of adsorption enthalpy is nearly unchanged over the outgassing temperature range 150–900 °C. The striking difference in the behaviour of N_2 can only be due to the weakening of the specific field gradient–quadrupole interactions, which is the result of the reduction in the number of surface hydroxyls. At low N_2 coverage, the drop in differential enthalpy of adsorption between the fully hydroxylated sample (150 °C) and the nearly completely dehydroxylated sample (900 °C), which is ca. 3 kJ mol^{-1}, can be considered as a direct measurement of the nitrogen–hydroxyl specific interaction (Rouquerol et al., 1984).

As noted in Chapter 1, the specific interactions between polar molecules and silica are indeed virtually eliminated by the removal of all the surface hydroxyls and therefore the effect of partial dehydroxylation is to drastically reduce the adsorption energies of certain molecules. The polar adsorptives studied by Kiselev and his co-workers included alcohols, ketones, ethers and amines (Kiselev, 1965, 1971): with each adsorptive, the reduction in the adsorbent–adsorbate interaction energy was accompanied by a substantial change in the isotherm character.

It might be expected that a dehydroxylated silica surface would be more energetically homogeneous than the parent hydroxylated surface, and this is only partly found in practice – as exemplified in Figure 11.12. However, the effect of outgassing a silica gel at high temperature may lead to the development of ultramicroporosity. To overcome this problem, much of the later work by Kiselev's group was undertaken on hydrothermally treated silicas (see Zhuravlev, 1994). In this manner, it was possible to convert the original skeletal globular structure of silica gel into a more homogeneous spongy structure (Kiselev, 1971) and thus avoid the development of ultramicroporosity.

As noted above, many investigations have been made of the adsorption of water on non-porous silicas. Less attention has been given to the dehydroxylation of porous silicas. An early study by Dzhigit et al. (1962) of the adsorption water vapour on a mesoporous silica gel involved both isotherm and calorimetric measurements. It was found that at very low surface coverage the adsorption enthalpy was not significantly affected by dehydroxylation, but a large difference became apparent as the surface coverage increased. A slow uptake of water vapour, which occurred after dehydroxylation, was attributed to chemisorption.

The dehydroxylation of microporous and mesoporous gels was investigated by Baker and Sing (1976). The reduced water isotherms in Figure 11.13 were determined on the non-porous TK 800 and the mesoporous gel J and microporous gel E outgassed at 1000 °C. All three isotherms exhibit pronounced hysteresis extending to very low $p/p°$, but the initial sections of the three *adsorption* isotherms (at $p/p° < 0.3$) follow a similar path. This low affinity for water is clearly characteristic of the degree of

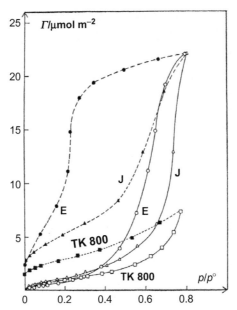

Figure 11.13 Adsorption isotherms of water vapour at 25 °C on silicas TK 800, gel E and gel J, evacuated at 1000 °C (Baker and Sing, 1976).

hydrophobicity shown by all dehydroxylated silicas. The surface rehydroxylation was very slow at $p/p° < 0.3$, but became rapid in the multilayer range and this phenomenon was evidently responsible for the low-pressure hysteresis. Indeed, it was found that the *desorption* branches of the isotherms on gels E and J were remarkably similar to the corresponding parts of the water isotherms on the two hydroxylated gels (Baker and Sing, 1976). We may conclude that the micropores in the rehydroxylated gel E regained their high affinity for water.

11.3 Aluminas: Structure, Texture and Physisorption

11.3.1 Introduction to Activated Aluminas

The name 'activated alumina' is generally applied to an adsorbent alumina (usually an industrial product) prepared by the heat treatment of some form of hydrated alumina (i.e. a crystalline hydroxide, oxide–hydroxide or hydrous alumina gel). It has been known for many years that certain forms of activated alumina can be used as powerful desiccants or for the recovery of various vapours. It was apparent at an early stage that the adsorbent activity was dependent on the conditions of heat treatment. For example, Bayley (1934) reported that the adsorption of H_2S by a commercial sample of activated alumina was affected by prior heating of the adsorbent at different temperatures, the maximum uptake being obtained after heat treatment at 550 °C. During an investigation of the catalytic dehydration of alcohols, Alekseevskii (1930) found that a calcination temperature of ca. 400 °C was required to optimise the adsorption of the alcohol reactants, whereas calcination at 600 °C was preferable for the adsorption of the olefine products.

Somewhat later it began to appear that there was a lack of agreement between the recorded dependence of surface area on the temperature of calcination (e.g. in the work of Krieger, 1941; Feachem and Swallow, 1948; Taylor, 1949; Gregg and Sing, 1951; de Boer, 1957). In fact, such differences are not really surprising. To obtain reproducible adsorbent properties, it is necessary to control: (a) the chemical and physical nature of the starting material (i.e. its structure, crystal/particle size, amount of sample, purity); (b) the conditions of heat treatment (type of furnace, atmosphere, time–temperature profile – preferably by a CRTA heating procedure, see Section 3.4); (c) the methods used to interpret the adsorption data (BET, BJH, etc.).

11.3.2   Starting Materials

11.3.2.1   Aluminium Trihydroxides

Although various modifications of aluminium trihydroxide, $Al(OH)_3$, have been described in the literature, there are only three common forms: gibbsite (originally also called hydrargillite) bayerite and nordstrandite. *Gibbsite* is the best known and most abundant. It is the main constituent of N. and S. American bauxite and is obtained as an intermediate product (i.e. 'Bayer Hydrate') in the Bayer process for the production of aluminium from bauxite.

The crystal structures of the three hydroxides are all based on a double layer of close-packed hydroxyl ions, two-thirds of the octahedral interstices being occupied by aluminium ions (see Figure 11.14). The layers are held together by hydrogen bonding between the nearest neighbour hydroxyls. Differences in structure thus reside in the inter-layer spacing, that is in the c-axis. In the case of gibbsite, the stacking of the double layers is in the sequence ABBAABBA. The distance between two adjacent A, or B, layers is 0.28 nm, whereas the A–B distance is 0.20 nm.

Bayerite does not occur in nature, but it can be made in a number of different ways (e.g. by the hydrolysis of an aluminium alkoxide). The OH layers in bayerite appear to be stacked in the order ABABAB. Within the double layer the A–B distance is 0.21 nm and between the double layers the A–B distance is 0.26 nm. The density of bayerite is correspondingly a little higher than that of gibbsite.

Although deposits of *nordstrandite* have been found, this modification is not easy to prepare in a relatively pure form. For this reason, the exact structure is still under discussion. However, the layer stacking is likely to be made up of a combination of both bayerite and gibbsite.

In its most common industrial form, *gibbsite* is a sandy material, of ca. 50–100 μm grain size. Each grain is itself a dense agglomerate of smaller hexagonal crystals, typically 5–15 μm size and the BET nitrogen surface area is usually not more than $0.2 \text{ m}^2 \text{ g}^{-1}$. Other forms of gibbsite have also been subjected to physisorption studies. These include loose, thin hexagonal crystals with BET-nitrogen areas of 5 and $15 \text{ m}^2 \text{ g}^{-1}$ corresponding to mean crystal sizes of 1 and 0.2 μm, respectively

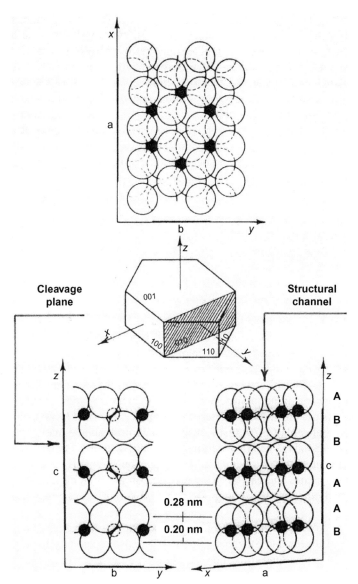

Figure 11.14 Crystalline structure of gibbsite. Small solid circles, aluminium ions; large open circles, hydroxyl ions. (For colour version of this figure, the reader is referred to the online version of this chapter.)
After Saafeld (1960).

(Rouquerol et al., 1975), and also porous aggregates. For example, Ramsay and Avery (1979) found that a batch of very pure gibbsite powder gave a Type IV nitrogen isotherm with an H1 hysteresis loop at high $p/p°$. It appeared that the gibbsite was mesoporous and possibly also macroporous, the effective pore width being mainly >20 nm.

The BET-nitrogen area was 41 $m^2 g^{-1}$, which was consistent with the mean thickness of ca. 25 nm of the thin hexagonal platelets, as determined by electron microscopy and X-ray line broadening. This rather ill-defined porosity was ascribed to the space between the gibbsite crystallites and was found to persist after heat treatment at 400 °C. A batch of gibbsite of lower BET-nitrogen area ($5.6 m^2 g^{-1}$) was used by Stacey (1987). In this case, the polycrystalline grains of mean diameter 75 μm were composed of 0.3 μm platy crystallites.

A sample of micro-crystalline nordstrandite was found to be somewhat mesoporous by Aldcroft and Bye (1967). The nitrogen adsorption isotherm was Type II(b), which indicated the existence of a non-rigid pore structure between the crystallites. The BET-nitrogen area of 34 $m^2 g^{-1}$ appeared to represent the external area of the crystallites.

A relatively low-area ($8 m^2 g^{-1}$) aged-sample of bayerite (Bye and Robinson, 1964) which gave a reversible Type II nitrogen isotherm, was shown to be essentially non-porous (Payne and Sing, 1969). Electron microscopy revealed that this sample was composed of discrete conical crystals.

11.3.2.2   Aluminium Oxide–Hydroxides

There are two well-known oxide–hydroxides (AlOOH) with closely related structures: diaspore and boehmite. Diaspore occurs in some types of clay and bauxite. It has been produced by the hydrothermal treatment of corundum α-Al_2O_3. Whereas boehmite is characterised by cubic close-packing of the anions, diaspore has a hexagonal close-packed structure. This difference probably accounts for the direct thermal transformation of diaspore to corundum at relatively low temperatures (450–600 °C).

Boehmite is of considerable interest to the surface scientist. It was pointed out by Lippens and Steggerda (1970) that a clear distinction should be made between crystalline boehmite and the gelatinous forms of pseudoboehmite, which always contains some non-stochiometric, interlamellar water. Pseudoboehmite is the main constituent of European bauxites and can be easily prepared by the neutralisation of aluminium salts, but hydrothermal conditions are required for the formation of crystalline boehmite.

In general, the BET-area of the pseudoboehmites increases with decreasing crystallinity (Lippens and Steggerda, 1970). The gels prepared by the hydrolysis of aluminium alkoxides tend to be porous, but their complex nature and the porosity are not always easy to characterise. However, by allowing the product to age in aqueous ethanol or ethanediol it is possible to obtain a well-defined mesopore structure (Bye et al., 1967; Aldcroft et al., 1971). This is believed to be the result of an aggregation–cementation type of ageing process, which involves a reduction in the solvent barrier and hence the promotion of particle–particle interaction with little particle growth (Bye and Sing, 1973).

A fibrillar type of boehmite (i.e. a du Pont product, named Baymal colloidal alumina) was first described by Bugosh et al. (1962). The BET-area was reported to be 275 $m^2 g^{-1}$. The discrete boehmite fibres of about 5 nm diameter and 100 nm length were protected by adsorbed acetate groups and were thus easily dispersible.

High-resolution adsorption measurements were undertaken by Fukasawa et al. (1994) on a porous boehmite glassy film prepared by slowly drying a boehmite sol. Small angle X-ray measurements indicated that the 2.5 nm platelets of boehmite were densely stacked in the [010] direction with the inclusion of uniform slit-shaped pores of about 0.3 nm width. The nitrogen isotherm on the boehmite film was of Type Ib (see Chapter 1) and was somewhat unusual in that the gradual increase in the amount adsorbed (in the multilayer range) extended up to $p/p° = 0.65$ (i.e. the beginning of the plateau). These results indicate that nitrogen adsorption had involved both primary and secondary micropore filling and also some multilayer adsorption on wider pores. With the aid of multiprobe adsorption measurements, it was concluded that there were indeed three groups of pores. However, the estimates of pore widths of 0.3, 0.7 and 1.3 nm must be in doubt because of the reliance placed on the Dubinin–Radushevich method of isotherm analysis.

11.3.3   Thermal Decomposition of Hydrated Aluminas

11.3.3.1   Thermal Decomposition of Trihydroxides

The structural and textural changes involved in the thermal decomposition of the three trihydroxides have been studied in considerable detail (Rouquerol, 1965; Aldcroft et al., 1968; Lippens and Steggerda, 1970; de Boer, 1972; Rouquerol et al., 1975, 1979a,b; Ramsay and Avery, 1979; Stacey, 1987). It is now clear that the dehydration sequence is dependent not only on the crystalline structure of the trihydroxide, but also on its texture and the conditions of heat treatment.

It appears to have been first noted by Achenbach (1931) that the dehydration of a *gibbsite* crystal is pseudomorphic: the crystal shape and original lattice are retained and therefore a highly porous product is formed. The fact that the loss of structural water precedes the formation of a new stable structure is obviously of great importance.

In Figure 11.15, the BET-nitrogen areas of the activated aluminas obtained by heating small crystals of bayerite and nordstrandite *in air* are plotted against the temperatures of calcination. In this work, each sample was put into the furnace at the recorded temperature and held there for 5 h (Aldcroft et al., 1968). It is evident that the trihydroxides underwent thermal decomposition at temperatures >200 °C, with the maximum BET areas being generated at 250–300 °C. Similar results had been previously reported for gibbsite, but the maximum BET areas were ≈300 m² g⁻¹ at 300–400 °C (Gregg and Sing, 1951; de Boer, 1957). The changing isotherm Type: II → I → IV provides the first indication that microporous products were obtained by the heat treatment of the trihydroxides at 250–450 °C and mesoporous products at higher temperatures (Aldcroft et al., 1968; Lippens and Steggerda, 1970).

As already explained, the α_S-method of isotherm analysis can be used to derive the external area, $a(\text{ext, S})$, and the pore volume, $v_p(\text{mic, S})$. Of course, the first requirement is to obtain an appropriate standard isotherm on a non-porous alumina. Strictly, the surface chemistry of the reference material should be exactly the same as that of the porous adsorbent, but in practice this is not easy to achieve because of the complex

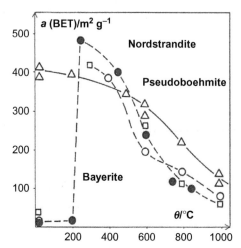

Figure 11.15 BET-area of calcined hydrated aluminas versus the temperature of calcination for 5 h (Lippens, 1961; Sing, 1972).

surface structure of active aluminas. As before, a standard isotherm data determined on the non-porous Degussa Aluminiumoxid C has been found suitable for the analysis of various isotherms on the porous aluminas (Sing, 1970).

The α_S-method was used to analyse the nitrogen isotherms determined after the calcination of very small crystals of bayerite and nordstrandite (Aldcroft et al., 1968). Thus, each value of the external area, a(ext), in Table 11.10 was obtained from the slope of the linear *multilayer* section of the α_S-plot by the application of Equation (11.6) which we give further. It is evident that the very large change in a (BET), which accompanied the thermal decomposition of the trihydroxide, was not associated with any significant change in a(ext). Indeed, the external surface area of the finely divided nordstrandite remained remarkably constant up to at least 600 °C. These results confirmed that the crystalline framework of the trihydroxide

Table 11.10 Thermal Decomposition of Bayerite and Nordstrandite

Adsorbent[a]	a(BET)/m^{2}g^{-1}	a(ext)/m^{2}g^{-1}	v_p(mic)/cm^{3}g^{-1}	Porosity
Bayerite	15	15	–	Nonpor
Bayerite (400)	382	30	0.20	Micropor
Bayerite (600)	199	23	0.21	Mesopor
Nordstrandite	34	19	0.02	(Micro)
Nordstrandite (300)	415	20	0.19	Micro
Nordstrandite (600)	265	20	0.23	Meso + micro
Nordstrandite (1000)	62	65	0.21	Meso

[a]Temperature of calcination/°C given in brackets. Each sample heated in air for 5 h at recorded temperature (Sing, 1972).

was preserved even after the thermal decomposition was virtually complete. Back extrapolation of the multilayer plot to $\alpha_S = 0$ gave each $v_p(mic)$ value recorded in Table 11.10. It can be seen that the pore volumes generated at 300 or 400 °C remained remarkably constant after further heat treatment although changes in the isotherm character indicated that pore widening had occurred at the higher temperatures.

By using fine crystals of bayerite and nordstrandite, Aldcroft et al. (1968) were able to ensure a fairly rapid release of the water produced during the removal of the structural hydroxyls. The importance of residual water in affecting the course of the thermal decomposition of gibbsite had been earlier demonstrated by Papée and Tertian (Tertian and Papée, 1953; Papée et al., 1954; Papée and Tertian, 1955). It was shown that at temperatures <400 °C, gibbsite was transformed partly into boehmite and partly into a porous form of ρ-alumina, a 'transition alumina'. The relative amounts of the products of these two parallel routes were found to depend on the controlled pressure of water vapour: over the 0–100 mbar range, low water vapour pressure favoured a direct conversion to microporous, poorly ordered ρ-alumina; while at high water pressures, up to 42% of the gibbsite was converted into boehmite.

In the 1950s, de Boer and his co-workers (de Boer et al., 1954, 1956; de Boer, 1957) used various techniques in their studies of the thermal dehydration of gibbsite and bayerite and a more detailed picture was obtained of the conditions under which the two decomposition routes were manifested. For example, it was shown that relatively well-crystallised boehmite could be produced by the treatment of gibbsite or bayerite in saturated steam at temperatures of ca. 165 °C. These and other findings provided qualitative confirmation that the formation of boehmite involved an intragranular hydrothermal transformation.

Although reproducible results can be obtained by the simple calcination of samples taken from a particular batch of trihydroxide, the chemical composition and porosity of the products are usually complex. As in other thermal decomposition studies, control of the dehydration reactions and sintering/ageing processes is considerably improved by the application of CRTA (controlled rate thermal analysis, see Section 3.4), which allows controlling accurately the residual water pressure above the sample together with the pressure and temperature gradients within the sample bed itself. As illustrated elsewhere for a number of adsorbents (Llewellyn et al., 2003), this substantially improves the homogeneity of the sample, at any stage of the thermal decomposition and makes much easier the interpretation.

Rouquerol (1965) obtained in this way, with an industrial gibbsite sample (crystal agglomerates of 80–160 μm, BET nitrogen surface area of ca. $0.1 \, m^2 \, g^{-1}$), under a controlled water vapour pressure of 0.3 mbar, the set of N_2 adsorption–desorption isotherms shown in Figure 11.16, which was interpreted as follows:

○ 183 °C: onset of the thermal decomposition, BET (N_2) area of $16 \, m^2 \, g^{-1}$, hysteresis loop with no saturation plateau, Type H4, probably due to capillary condensation in *non-rigid system* formed by partial desagglomeration of platy gibbsite crystals. A similar hysteresis loop remains up to 898 °C, probably corresponding to a nearly steady external shape and size of agglomerates

○ 207 °C: end of the thermal decomposition of gibbsite into microporous 'transition alumina' and into some boehmite. The arrow indicates the end of micropore filling. (To-day, we

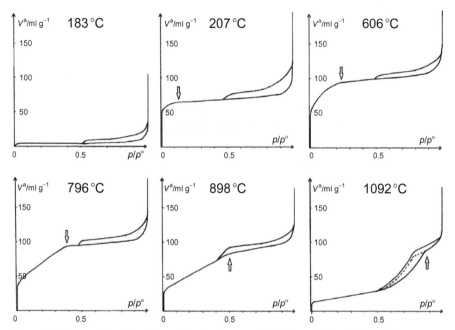

Figure 11.16 N_2 adsorption–desorption isotherms at 77 K for an industrial gibbsite sample treated by CRTA, under 0.3 mbar, up to each temperature indicated.
From Rouquerol (1965).

interpret the somewhat broader hysteresis loop by the fact that this experiment was simply carried out with a more important final condensation of N_2 than in the other experiments)

○ 606 °C: after thermal decomposition of boehmite phase, the BET (N_2) area is still larger and the arrow indicating the completion of the micropore filling is shifted towards higher pressures, as it will do up to the last treatment at 1092 °C, indicating a progressive *broadening of the pore size*

○ 1092 °C: the completion of the pore filling, as indicated by the arrow, now occurs within the hysteresis loop, which means that *the micropores have become mesopores;* the resulting quite composite isotherm looks like the addition of a Type I (due to a small amount of high-energy sites on the crystallised α-alumina), a Type IV (due to the mesopores and responsible for the lower part of the loop) and what we call to-day a Type IIb, characteristic of a non-rigid system.

This work was further developed (Rouquerol et al., 1975, 1979a,b; Rouquerol and Ganteaume, 1977) by a study of the influence of water pressure and grain size on the porous structure and on the amount of intermediate boehmite obtained. This study was started in the spirit of that of Papée and Tertian (1955), now taking advantage of the close control offered by CRTA. Figure 11.17 shows the striking effect of water vapour pressure, in the narrow range 0.04–5 mbar, on the BET (N_2) area as a function of the final temperature. These results were explained by the fact that the micropore width was directly pressure dependent, even in the pressure range of 'vacuum'. Under any of these low pressures, during the thermal decomposition, parallel micropores are 'drilled' through the gibbsite crystals, following the path of pre-existing *structural*

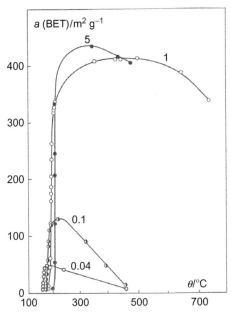

Figure 11.17 Development of the BET-nitrogen surface area during the dehydration, by CRTA, of a sample of fine gibbsite, 1 μm grain size. CRTA conditions: controlled pressure indicated in mbar on the curve; rate of dehydration, 11.4 mg h^{-1} g^{-1} (Rouquerol and Ganteaume, 1977).

channels in the z direction—that is normal to the basal (001) plane (see Figure 11.14). Their number is therefore pre-determined, so that their total N_2 adsorption capacity (as measured by the apparent BET (N_2) area) varies in the same direction as their width.

Figure 11.18a reports the same results in a different way, that is plotting the BET (N_2) area versus the mass of water lost (expressed in % of the starting sample mass). After an initial period during which some water is lost without leaving any accessibility to the nitrogen molecule, the linear increase of the BET (N_2) area is consistent with the progressive drilling of micropores of constant width. In the experiments reported in Figure 11.18b, at a given stage of the thermal decomposition of gibbsite, the water vapour pressure was suddenly dropped from 1 down to 0.04 mbar; the linear increase in BET (N_2) area then still continued, but with a smaller slope indicating that the pores were now narrower: the micropores drilled in the crystal were therefore becoming 'funnel-shaped'.

In the work of Ramsay and Avery (1979), small quantities of micro-crystalline gibbsite were heated at temperatures up to 425 °C. The product obtained by decomposition in vacuo adsorbed very little nitrogen, although it did adsorb water. In the presence of water vapour, the alumina was found to become more crystalline and to undergo pore widening: these results are consistent with the conclusions of the French investigators.

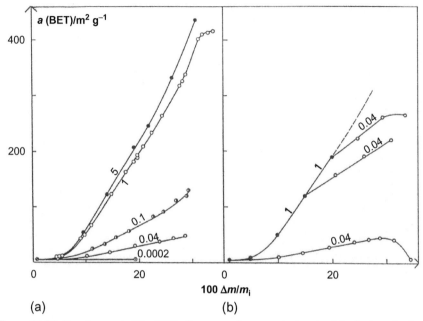

(a)   (b)

Figure 11.18 Development of the BET-nitrogen surface area during the dehydration of a fine gibbsite versus % of mass loss. (a) Same conditions as for Figure 11.17 (b) with change of controlled pressure between 1 and 0.04 mbar (Rouquerol et al., 1979a,b).

It is now clear that if the hydrothermal formation of boehmite is avoided (e.g. by using low-pressure CRTA and small crystals), the three hydroxides lose their structural water at quite low temperature ($\approx 200 °C$) to give the almost amorphous form, ρ-Al_2O_3. Complex changes occur as the temperature is increased to ≈ 250–$800 °C$ with the formation of certain members of the γ-series aluminas, γ or η and θ. At temperatures $\approx 1200 °C$, the conversion to the dense, low-area α-Al_2O_3, normally takes place.

11.3.3.2   Thermal Decomposition of Boehmite and Hydrous Alumina

Boehmite is itself decomposed at ca. 400–450 °C. As expected, the calcined products have much lower specific surface areas than the activated aluminas produced from the trihydroxides. However, the results of de Boer and his co-workers (de Boer, 1972) indicate that a sample prepared at 580 °C was highly microporous and that up to this temperature there was only a small change in the external area.

Some work has been undertaken on the decomposition of the poorly ordered forms of hydrous alumina. These materials can be prepared in the form of gels or flocs, but their surface properties are reproducible only if the conditions of gelation/flocculation, ageing, drying and storage are very carefully controlled (Bye and Sing, 1973). Exposure of microporous hydrous alumina gels to water vapour leads to the rapid and irreversible loss of BET-area, but mesoporous pseudoboehmite gels tend to be more stable (Sing, 1972). However, soaking either type of gel in liquid water generally results in its conversion to non-porous bayerite and consequently to a great reduction of specific surface area.

In contrast to the thermal activation of the well-defined hydroxides, the calcination of an amorphous or pseudoboehmite gel does not lead to any significant increase in specific surface area (see Figure 11.15). In this respect, a poorly ordered hydrous alumina is similar to silica gel. The results in Figure 11.16 also indicate that the surface areas of the calcined pseudoboehmite gels are rather more stable in the high-temperature range (700–1000 °C). For example, after being heated for 5 h at 1000 °C, a calcined hydrous alumina still possessed a specific surface area of 132 $m^2 \, g^{-1}$. Some alumina gels tend to remain poorly ordered at high temperature so that the final change to α-Al$_2$O$_3$ is delayed. Teichner and his co-workers have reported that amorphous alumina aerogels were slow to develop crystallinity when they were heated in vacuo (Teichner et al., 1976; Teichner, 1986).

11.3.4 Resulting Activated Aluminas

11.3.4.1 Structure of Activated Aluminas

Alumina occurs in nature as *corundum*, α-Al$_2$O$_3$, which is noted for its great hardness, high electrical resistance and low chemical reactivity. It can be made by the high-temperature treatment (at >1200 °C) of boehmite or gibbsite and normally has a low specific surface area ($<5 \, m^2 \, g^{-1}$).

The less compact 'transition' aluminas (γ-type) are highly porous, more reactive and do not occur in nature. They are prepared by the heat treatment of Al(OH)$_3$ or AlOOH at intermediate temperatures and undergo an irreversible change to α-Al$_2$O$_3$ at high temperature. Their BET (N$_2$)-areas are typically 300–400 $m^2 \, g^{-1}$ and they are widely used as catalysts and catalyst supports.

Various schemes have been proposed for the classification of the different alumina structures (Lippens and Steggerda, 1970). One approach was to focus attention on the temperatures at which they are formed, but it is perhaps more logical to look for differences in the oxide lattice. On this basis, one can distinguish broadly between the α-series with hexagonal close-packed lattices (i.e. ABAB$\cdots$) and the γ-series with cubic close-packed lattices (i.e. ABCABC$\cdots$). Furthermore, there is little doubt that both γ- and η-Al$_2$O$_3$ have a spinel (MgAl$_2$O$_4$) type of lattice. The unit cell of spinel is made up of 32 cubic-close-packed O^{2-} ions and therefore 21.33 Al^{3+} ions have to be distributed between a total of 24 possible cationic sites. Differences between the individual members of the γ-series are likely to be due to disorder of the lattice and in the distribution of the cations between octahedral and tetrahedral interstices.

The detailed behaviour of a model γ-Al$_2$O$_3$ surface was first discussed by Peri (1965). It was supposed that a fully hydrated (100) plane of the spinel structure would have a monolayer of OH$^-$ ions in a square lattice plane. On dehydration two adjacent OH$^-$ ions would begin to combine at random. This simple pairwise condensation would be limited, however, since only two-thirds of the surface OH could be removed without changing the surface structure. According to this picture, the remaining OH$^-$ ions would occupy different sites, which could be characterised by infrared spectroscopy and chemical activity. At high temperature, ionic migration would occur and various surface defects produced. Although this is an over-simplified model, it has been of great value for the discussion of the catalytic activity and chemical reactivity of the aluminas.

In fact, the surface structures of activated aluminas are exceedingly complex. After exposure to the air, an alumina surface is always fully hydrated, but unlike silica it is not fully hydroxylated. Thus, 'bound' water is in the form of both hydroxyls and coordinated water molecules. Removal of the latter may take place at temperatures as low as 200 °C and leave high-energy Al^{3+} sites exposed. As in the Peri model, with increase in temperature further cationic sites are formed with the combination of adjacent OH^- ions, ionic mobility and finally reorganisation of the surface structure.

11.3.4.2   Gas Physisorption by High-Temperature Aluminas

In view of the complexity of the γ-series, it is of interest to investigate the adsorption of gases by high-temperature aluminas, particularly α-Al_2O_3. Non-porous α-Al_2O_3 can be obtained by heating the spheroidal particles of a flame-hydrolysed alumina for a prolonged period at temperatures of at least 1200 °C: this was the procedure adopted by Carruthers et al. (1971). A master batch of Degussa 'Aluminiumoxid C' [designated alumina DC; $a(BET) = 111$ m^2 g^{-1}] was used as the starting material and also as the non-porous reference adsorbent. This was composed of discrete spheroidal particles and gave reversible Type II argon and nitrogen isotherms at 77 K. In the reduced α_s form, the nitrogen isotherm was virtually identical with the corresponding isotherms on samples of low-area bayerite and aged hydrous alumina (Bye et al., 1967; Payne and Sing, 1969).

Samples of alumina DC were calcined in air at the recorded temperatures and for specified times. Thus, sample DC (1200) 6 was prepared by calcination at 1200 °C for 6 h and sample DC (1200) 114 by calcination at 1200 °C for 114 h. The BET-nitrogen areas of these and other high-temperature samples are given in Table 11.11. It is noteworthy that after 6 h at 1200 °C, the specific surface area of sample DC (1200) 6 was remarkably high and there had been no detectable conversion to α-Al_2O_3, whereas after 114 h at 1200 °C, sample DC (1200) 114 had been completely converted to α-Al_2O_3.

Table 11.11 Argon and Nitrogen Adsorption on High-Temperature Aluminas (Carruthers et al., 1971).

Adsorbent	Crystal Structure	$a(BET - N_2)/$ m^2g^{-1}	$a(S, N_2)/m^2g^{-1}$	$a(S, Ar)/m^2g^{-1}$
Alumina DC	δ-Al_2O_3	111	(111)	(111)
DC (1200) 6	θ-Al_2O_3	74.8	74.6	77.0
DC (1200) 114	α-Al_2O_3	5.9	5.9	6.6
DC (1300) 6	α-Al_2O_3	4.5	4.3	4.9
DC (1400) 6	α-Al_2O_3	2.5	2.3	2.7

The α_S-plots for sample DC (1200) 6 were found to be linear over the recorded ranges of both isotherms. This correspondence of isotherm shape is to be expected since the adsorbent structure has not been appreciably changed as a result of the 6-h calcination at 1200 °C. The multilayer sections of the other α_S-plots were for the most part linear, but the monolayer sections all exhibited significant deviation. The fact that the linear multilayer plots can be back extrapolated to the origin was an indication that the multilayer development had not been affected to any great extent by the change in structure of the adsorbent.

The values of surface area, $a(S, N_2)$ and $a(S, Ar)$, in columns 4 and 5 of Table 11.11 are derived from the linear *multilayer* α_S-plots. As was pointed out in Chapter 7, if unrestricted multilayer adsorption has occurred, the specific surface area can be evaluated from the slope of the α_S-plot. The following equations have been developed for the adsorption of argon and nitrogen on alumina:

$$a(S, Ar) = 3.22 v^\sigma(STP)/\alpha_S \qquad (11.5)$$

$$a(S, N_2) = 2.87 v^\sigma(STP)/\alpha_S \qquad (11.6)$$

where $v^\sigma(STP)$, or v^a, is the volume of gas adsorbed, expressed in $cm^3(STP) g^{-1}$ and the proportionality factors have been obtained by calibration against the BET-nitrogen area of alumina DC.

The corresponding values of $a(BET-N_2)$ and $a(S, N_2)$ in Table 11.11 are evidently in good agreement. This looks logical, since both rely on the $(BET-N_2)$ area of an alumina: the alumina under study for $a(BET-N_2)$ and alumina DC for $a(S, N_2)$. The values of $a(S, Ar)$ are between 3% and 12% higher. Here again, like for silica gels, this (limited) discrepancy can be explained in two different ways, which do not exclude each other (see Section 11.2.3): changing the type of alumina surface (especially its polar sites) may influence the monolayer packing of either argon or, more probably, nitrogen (because of its permanent quadrupole moment and possible orientation).

A sample of non-porous α-Al_2O_3 [of $a(BET) = 2.7 m^2 g^{-1}$] was used in an important investigation by Barto et al. (1966) of the adsorption of a series of n-aliphatic alcohols. The isotherms, which are displayed in Figure 11.19, show very clearly the effect of 'autophobicity'. Thus, the increase in hydrocarbon chain length from C1 to C4 has resulted in a drastic reduction in the extent of multilayer adsorption. The reason for this unusual physisorption behaviour is that the molecules in the localised monolayer are oriented to allow directional hydrogen bonding between their hydroxyls and the hydrated alumina surface. The alkyl groups are therefore directed away from the surface and thereby provide an effective low energy screen against further adsorption.

Langmuir plots were constructed by Barto et al. (1966) from their alcohol isotherms: it is not surprising that the most extensive linearity was given by butanol. Values of the monolayer capacity were obtained from the linear regions of the Langmuir plots and by assuming the validity of the BET-nitrogen area it was possible to derive the apparent molecular areas for the four alcohols. For the C1–C4 series, the following values were obtained: 0.205, 0.220, 0.234, 0.248 nm^2. It is noteworthy that 0.20 nm^2 is the area occupied by a long chain alcohol in a close-packed insoluble

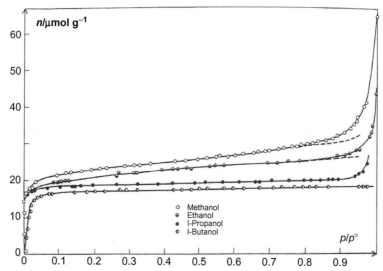

Figure 11.19 Adsorption isotherms of methanol, ethanol, propanol and butanol on α-Al$_2$O$_3$ (2.7 m^2 g^{-1}).
From Barto et al. (1966).

monolayer. As expected, it is methanol which gives the 'best' BET fit, but even in this case the range of linearity was limited to $p/p° < 0.15$.

In a further study of autophobicity by Blake and Wade (1971), adsorption isotherms were determined for water vapour and the first five n-aliphatic alcohols on the surface of oxidised aluminium foil. The results were very similar to those obtained for α-Al$_2$O$_3$. A Type II water isotherm was obtained, the initial steep slope being indicative of a high affinity of adsorption. The residual uptake of water was significantly larger than expected (equivalent to 1.5 monolayers) and it was considered likely that the oxidised surface was microporous with respect to water.

The adsorption of water vapour by various forms of alumina was studied by Carruthers et al. (1971). In order to investigate the interaction of water with α-Al$_2$O$_3$, sample DC (1200) 114 (see Table 11.11) was outgassed at 400 °C. The subsequent water uptake was much larger than that given by the same sample outgassed at 25 °C and the first adsorption–desorption cycle exhibited pronounced hysteresis over the complete range of $p/p°$. Indeed, the initial mass could not be regained by prolonged evacuation at 25 °C. The amount irreversibly held (13.1 μmol m^{-2}) corresponded to almost a close-packed monolayer of water (i.e. 15.8 μmol m^{-2}). After 25 °C evacuation, the water isotherm was reversible, but the level of uptake was lower than expected for the adsorption of a close-packed monolayer.

The interaction of water with the transitional aluminas is more complex. With the original alumina DC and sample DC (1200) 6, the water isotherms were irreversible – even after evacuation at 25 °C and it was not possible to establish thermodynamic equilibrium. It appears that there was a slow penetration of water molecules which involved the hydration of poorly ordered cations. Nevertheless, Castro and Quach (2012) made use of water adsorption microcalorimetry to assess the surface energy of transitional alumina as a function of water pre-adsorption.

As we have seen, the formation of α-Al$_2$O$_3$ is normally associated with a considerable loss of surface area. It must not be assumed, however, that this is accompanied by the removal of all porosity. The pore structure is always changed by high-temperature treatment, but a distinction must be made between open and closed pores and it is only the former that can be characterised by physisorption measurements.

The effect of high-temperature treatment of alumina fibres is an interesting example of the evolution of porosity. In the work of Stacey (1991), the alumina fibres made by sol–gel methods and calcined at 900 °C consisted of η-Al$_2$O$_3$. Further calcination at 1300 °C resulted in their conversion to α-Al$_2$O$_3$ and a considerable change in the mesopore structure. Although the BET-nitrogen area was reduced from 84 to 11 m^2 g^{-1}, the fibres continued to retain a well-defined mesopore structures. Pore widening appeared to involve a change from a bi-modal to a more normal single modal distribution. An improved picture of the changes in texture was obtained with the aid of optical and electron microscopy and small-angle neutron scattering. It was concluded that most of the residual mesoporosity in the α-Al$_2$O$_3$ fibres was oriented along the fibre axis and that the randomly distributed smaller mesopores had been preferentially eliminated during the transformation in crystal structure.

11.4   Titanium Dioxide Powders and Gels

11.4.1   Titanium Dioxide Pigments

Titanium dioxide (titania) is the major constituent of most commercial white pigments (Day, 1973; Wiseman, 1976; Solomon and Hawthorne, 1983). The physical properties of typical grades of the common white pigments are given in Table 11.12. It can be seen that the main advantages of titania pigments over the other white pigments are their *high refractive index* in the visible region of the spectrum and their relatively low densities. Two other advantages of TiO$_2$ are its chemical stability and the fact that it can be manufactured in an optimum crystal size of ca. 0.2 μm. As a consequence of their high degree of light scattering and low absorption of visible light, titanium dioxide pigments are the whitest and brightest of all the commercial white pigments.

Table 11.12 Physical Properties of Typical White Pigments (Wiseman, 1976)

Pigments	Density/g cm^{-3}	Refractive Index	a(BET)/m^2 g^{-1}	Crystal size/μm
Anatase	3.8	2.55	11	0.15
Rutile	4.2	2.76	6	0.25
ZnO	5.6	2.01	10	0.2
ZnS	4.0	2.37	6	0.25
White lead	6.9	2.0	2	1

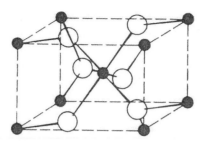

Figure 11.20 Unit cell of rutile. Black circles: titanium atoms; open circles: oxygen atoms. (For colour version of this figure, the reader is referred to the online version of this chapter.) Courtesy Adamson (1986).

There are three naturally occurring crystallographic forms of titanium dioxide: anatase, brookite and rutile. *Rutile* is the most common and stable form. Its structure, shown in Figure 11.20, is based on a slightly distorted hexagonal close-packing of oxygen atoms with the titanium atoms occupying half of the octahedral interstices. Anatase and brookite are both based on cubic packing of the oxygen atoms, but the coordination of the titanium is again octahedral.

Only anatase and rutile are manufactured on a large scale. Anatase was the first to become commercially available, but rutile is now the more important. The small pigmentary crystals of both forms are strong absorbers of UV light: this leads to photocatalysed degradation of organic molecules unless the TiO_2 surface is protected. The particularly high photoactivity of anatase renders it unsuitable for exterior finishes because of the rapid degradation of the protective film. The pigmentary rutile crystals are generally coated with alumina and/or silica and treated with organic compounds.

The early technological interest in *anatase* pigments was probably why anatase powder was for a time favoured as a non-porous adsorbent. Thus, anatase was one of the few finely divided crystalline solids used by Harkins and Jura (1944) in the development of new procedures for surface area determination. Anatase was also featured in some early adsorption calorimetric studies, for example by Kington et al. (1950). However, it was soon evident that rutile was becoming increasingly important. Finely divided rutile was adopted by Drain and Morrison (1952) as the non-porous adsorbent for an important series of calorimetric adsorption energy measurements. In spite of the energetic heterogeneity shown by the low-temperature adsorption of argon, nitrogen and oxygen on rutile (Drain, 1954), the derived values of differential and molar entropies of adsorption provided valuable supporting evidence for the validity of the BET monolayer capacities.

Somewhat later, the surface and colloidal properties of rutile were studied in considerable detail (Day, 1973; Wiseman, 1976; Parfitt, 1981; Rochester, 1986). In the early 1970s, extensive use was made of infrared spectroscopy for characterising the rutile surface and its interaction with water and other molecules. An improved understanding of the mechanisms involved in coating the rutile surface was also provided by studies of the energetics of immersion and electrokinetic behaviour together with the application of electron microscopy.

11.4.2 Rutile: Surface Chemistry and Gas Adsorption

The availability of pure rutile in a finely divided state has allowed progress to be made in the interpretation of adsorption isotherm and energy data. In particular, it has been possible to explain certain unusual features of the adsorptive properties of rutile in terms of its surface chemistry as characterised by infrared spectroscopy.

A puzzling dependence of the BET-nitrogen area on the outgassing temperature was first noted by Day and Parfitt (1967). It was found that the surface area appeared to undergo an increase of about 20% when the outgassing temperature was increased from ambient to 200 °C and thereafter remained constant over a wide temperature range. A little later, Day et al. (1971) undertook a systematic investigation of the effect of pretreating rutile with water and various alcohols. A sample was outgassed at 400 °C, equilibrated with water vapour and then outgassed at 25 °C: as a result, the BET-nitrogen area underwent a decrease from 10.2 to 7.7 $m^2 g^{-1}$, the nitrogen C value being correspondingly reduced from 450 to 180. Pretreatment with ethanol resulted in $a(BET, N_2) = 7.5$ $m^2 g^{-1}$ and $C = 39$.

It was first thought that the changes in BET-area were due to the presence of micropores in which water and other molecules could be trapped and removed only by an increase in the outgassing temperature. In the light of further adsorption and spectroscopic measurements, it now seems much more likely that these and other effects are associated with the *surface chemistry of rutile* rather than its porosity.

The external surface of a rutile crystal is almost entirely composed of the three crystal planes (110), (100) and (101). The relative area of each crystal face in a sample of finely divided rutile probably varies from one polycrystalline sample to another, but it is generally assumed that 60–80% of the overall surface area of the powder is provided by the (110) plane with the remainder divided equally between the two other planes (Jaycock and Waldsax, 1974; Boddenberg and Eltzner, 1991).

The arrangement of the Ti^{4+} and O^{2-} ions in parallel rows on an exposed (110) surface is pictured in Figure 11.21. The surface structure is clearly consistent with the composition of the unit cell (Bakaev and Steele, 1992). In this ideal model, the surface Ti^{4+} ions are coordinatively unsaturated ('cus') and on exposure to the atmosphere at ambient temperature they will be covered – either by ligand attachment or by some form of chemical bonding.

The most obvious way of removing the surface unsaturation is by reaction with water: this could involve dissociative chemisorption and/or molecular adsorption. Detailed infrared spectroscopic studies have shown that both processes occur (Griffiths and Rochester, 1977; Morishige et al., 1985; Rochester, 1986). Although the interpretation of the infrared spectra is not entirely straightforward (see Parkyns and Sing, 1975), it appears to be generally agreed that the dominant infrared bands at 3655 and 3410 cm^{-1} represent the stretching vibrations of terminal and bridged hydroxyl groups, respectively. It is likely that these hydroxyls are in the main located on the (110) face (Jaycock and Waldsax, 1974). However, other broad bands around 3400 cm^{-1} are probably due to hydrogen-bonded species. Dissociative chemisorption of water is believed to occur on the (110) face and co-ordinate bonding of water molecules on the Ti^{4+} sites exposed on the (100) and (101) faces.

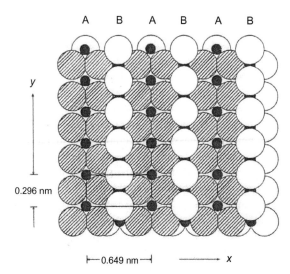

A   B   A   B   A   B

y

0.296 nm

|—0.649 nm—| ————→ x

Figure 11.21 Top view of the (110) surface of rutile. Black circles: Ti atoms; open circles, 0 atoms above plane of Ti; hatched circles, 0 atoms under plane of Ti. (For colour version of this figure, the reader is referred to the online version of this chapter.) Courtesy Rittner et al. (1995).

Dehydration of rutile crystals involves the removal of *hydrogen-bonded water*, *coordinately bonded water* and *surface hydroxyls* (dehydroxylation). The temperature ranges corresponding to these three stages overlap and depend on the rutile sample and the conditions of heat treatment; but generally the removal of molecular water occurs at temperatures of up to about 300 °C and progressive dehydroxylation over the range 200–500 °C. In a study of the interaction of water vapour with the surface of rutile, Munuera and Stone (1971) concluded that molecular water could be completely removed by evacuation at 325 °C, leaving the surface partially hydroxylated. They attributed a thermal analysis (TPD) peak at 250 °C to the removal of coordinated water and a peak at 370 °C to dehydroxylation from about 50% of the surface.

A striking difference has been found between the energetics of adsorption of argon and nitrogen on rutile (Furlong et al., 1980a,b). A Tian-Calvet microcalorimeter was used to determine differential energies of adsorption of argon and nitrogen on nonporous rutile outgassed in stages at temperatures of 150, 250 and 400 °C. The differential energies of adsorption, $\Delta_{ads}\dot{u}$, obtained after each outgassing stage, are plotted in Figure 11.22 against the surface coverage, θ.

The most striking feature of the results in Figure 11.22 is the very high initial energy of adsorption of nitrogen on rutile outgassed at 250 or 400 °C, $\Delta_{ads}\dot{u}$ being in excess of 20 kJ mol^{-1} at $\theta < 0.4$. With argon, the initial energies are closer together and the differences become more pronounced at higher coverage.

Values of $\Delta_{ads}\dot{u}$ for nitrogen adsorption at $\theta = 0.1$ are recorded in Table 11.13, which also contains the corresponding adsorption energy data for silica-coated rutile. It was independently confirmed that the surface properties of the latter sample, which had a coating of 2.6% dense silica, were very similar to those of pure silica (Furlong et al., 1980a,b).

It is noteworthy that the corresponding values of $\Delta_{ads}\dot{u}$ for nitrogen on rutile and silica-coated rutile in Table 11.13 are not very different, provided that the rutile is

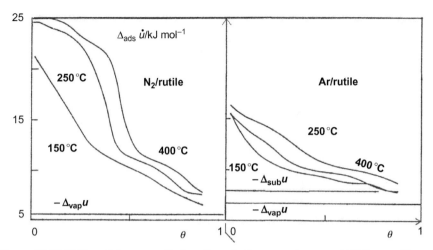

Figure 11.22 Differential energy of adsorption of nitrogen and argon, at 77 K, on rutile outgassed at 150, 250 and 400 °C (Furlong et al., 1980a,b).

Table 11.13 Differential Energies of Adsorption of Nitrogen at 77 K on Rutile and Silica-Coated Rutile

Outgassing Temperature/°C	$\Delta_{ads}\dot{u}$ (N_2 at $\theta = 0.1$)/kJ mol^{-1}	
	Rutile	**Silica-Coated Rutile**
150	18.5	19.3
250	24.0	20.0
400	24.9	20.6

outgassed at 150 °C. There is little doubt that the specificity of adsorption of nitrogen on hydroxylated silica is due to interaction between the nitrogen quadrupoles and the surface OH groups and it appears that this is also true for rutile outgassed at 150 °C. Clearly, the situation is changed when rutile is outgassed at higher temperatures. From the location of the adsorption energy curves in Figure 11.21 and the data in Table 11.13, we conclude that: (a) outgassing at 250 °C has removed most, if not all, the coordinated H_2O ligands from the cationic sites; (b) dehydroxylation at 400 °C has resulted in the exposure of more high-energy sites; (c) two types of cationic sites appear to occupy about 40% of the overall surface.

We are now in a position to explain in more detail why the BET-area is found to increase as the outgassing temperature is raised. The picture of the exposed (110) surface in Figure 11.21 reveals that even an ideal plane of the dehydrated rutile surface cannot be regarded as smooth. Although the real surface is more complex and energetically heterogeneous, infrared spectroscopy and adsorption microcalorimetry give a remarkably consistent picture of the role of the cationic sites. We conclude that the interaction of adsorbate molecules with exposed 'cus' cations results in the incorporation of over 20% of the 'monolayer' within the surface structure. These adsorbate molecules take the place of water ligands (and also some hydroxyls) and are thus located in the gaps between the oxygen ions. It is also likely that these strong cationic sites are able to orient the nitrogen molecule, (which results into a denser monolayer than conventionally assumed) and can be surrounded by clusters of nitrogen molecules more than one layer thick.

The adsorption of water vapour on rutile has been studied by a number of investigators (e.g. Dawson, 1967; Day et al., 1971; Munuera and Stone, 1971). In the work of Furlong et al. (1986), water isotherms were determined after successively outgassing the rutile adsorbent at: (1) 25 °C, (2) 100 °C, (3) 150 °C, (4) 300 °C and (5) back to 25 °C. The resulting isotherms are shown in Figure 11.23. It can be seen that isotherms (4) and (5) exhibit a second knee (Points X and X') and are nearly parallel at $p/p° > 0.1$. Isotherms (2) and (3) are almost parallel, but have no identifiable second knee.

Following the above reasoning, we arrive at the following interpretation of the isotherms in Figure 11.23. The first outgassing at 25 °C has removed the free hydrogen-bonded water so that curve (1) is the physisorption isotherm on the remaining fully hydrated surface. Ligand water has started to leave the surface at 150 and by 300 °C all the ligand water has been removed and also a high proportion of the hydroxyl groups. Isotherm (4) is therefore a composite isotherm of the specific adsorption on cationic sites and the additional physisorption. The latter has been removed by the final outgassing at 25 °C, but is taken up again in the form of isotherm (5). Thus, the vertical separation of isotherms (5) and (4) – that is between X and X' – provides a measure of the ligand water retained after outgassing at 25 °C (i.e. ~150 μmol g^{-1}).

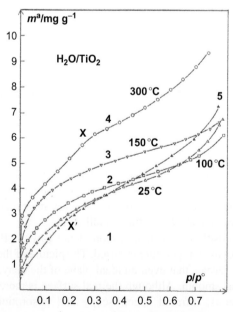

Figure 11.23 Isotherms of water vapour on rutile, outgassed between 25 and 300 °C. Run 5 follows run 4 with simple intermediate outgassing at 25 °C (Furlong et al., 1986).

The fact that isotherms (1) and (5) follow different paths is a consequence of the change in the nature of the surface, which has involved the irreversible removal of hydroxyl groups. Physisorption accounts for ca. 177 μmol g^{-1}, which corresponds to 0.21 nm^2 per water molecule, a value close to that for water on hydroxylated silica.

We have seen that cations are exposed by the surface dehydration of rutile. These 'cus' ions act as Lewis acid (acceptor) sites for the adsorption of pyridine, ammonia and other bases. On the other hand, there appears to be no indication of Bronsted acidity, although this can be produced by surface treatment (e.g. with HCl). It is not surprising that the dehydrated surface of rutile can interact specifically with a wide range of polar molecules (Parkyns and Sing, 1975).

Day et al. (1971) pointed out that there is an important difference between ethanol and isopropanol in their interactions with rutile. Whereas ethanol can displace water and undergo dissociative chemisorption to form the surface ethoxide, isopropanol is more readily adsorbed in the molecular form. This is consistent with the hydrophobic nature of ethanol-treated TiO$_2$ and the 'autophobic' nature of the ethanol monolayer. The latter effect is manifested in the form of a Type I isotherm, which is remarkably similar to that given by ethanol on alumina (see Figure 11.19).

A rutile sample of low BET(N$_2$) area (8.1 m^2 g^{-1}) was used by Grillet et al. (1985) in a combined manometric–calorimetric study of krypton adsorption at 77 K, in comparison with argon and nitrogen. Unlike nitrogen, the path of the krypton and argon isotherms was not changed to any significant extent by the increase of outgassing temperature from 140 to 400 °C. As can be seen in Figure 11.24, although the krypton

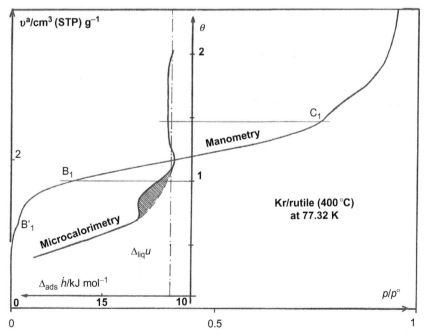

Figure 11.24 Isotherm and differential enthalpy of Kr adsorption on rutile (Grillet et al., 1985).

isotherm is essentially Type II in the IUPAC classification, it does exhibit some stepwise features. Thus, there are sub-steps at $p/p°$ of 0.016 and 0.72 (i.e. at BET coverages of 0.66 and 1.45), which indicate changes in the mode of adsorption. The first sub-step is located in the sub-monolayer range of the isotherm. The microcalorimetric measurements revealed that it was accompanied by an energy change of 610 J mol^{-1}, (hatched area) which is close to the thermal energy, RT, at 77 K (i.e. 644 J mol^{-1}). It therefore seems likely that this represents a change of state from a 2D fluid to a 2D solid.

There was no indication of any similar phase changes in the adsorbed monolayers of either nitrogen or argon on the same rutile surface. In the case of nitrogen, this is not surprising in view of its quadrupolar nature and specificity of adsorption. The difference between argon and krypton has been explained (Grillet et al., 1985) in terms of the 'dimensional incompatibility factor'. According to this view, of the two adsorbate structures, krypton is more likely to be in registry with the rutile surface.

This work has also drawn attention to the difficulties involved in deriving the surface area from the nitrogen BET monolayer capacity. Grillet et al. (1985) have shown indeed that, when the outgassing temperature of rutile increases from 140 to 400 °C the BET n_m increases of 5% for N_2 but only of 0.5% for Kr and Ar. The low dependency of the Ar BET n_m on the state of the surface plays in its favour for the surface area determination. Using for argon the usual molecular area of 0.138 nm^2 derived from its liquid density, one then finds for nitrogen a molecular area of only 0.127 nm^2 (and of 0.15 nm^2 for krypton).

A computer simulation study has been made by Rittner et al. (1995) of the adsorption of xenon on the (110) face of rutile (as depicted in Figure 11.21). The simulated isotherms, obtained by the canonical ensemble Monte Carlo technique, were in fairly good agreement with experimental data determined over the temperature range 196–273 K. The Monte Carlo calculations indicate that the surface geometry of the (110) face has a predominant influence on the adsorption of Xe. At low surface coverage, the adsorption is almost entirely confined to cationic rows, although the adsorbate structure is determined by the adsorbate–adsorbate interaction and therefore is not in registry with the 'cus' Ti-sites. Increased translational mobility of the adsorbate is associated with the occupation of the surface oxygen sites at higher coverage. It is evident that the Xe atoms are not all in the same plane when the surface is fully covered. These results again illustrate the dependence of physisorption on surface structure.

11.4.3 The Porosity of Titania Gels

Titanium dioxide gels of high surface area can be prepared in a number of different ways. In the early work of Harris and Whitaker (1962, 1963), porous gels were prepared by the steam-hydrolysis of titanium alkoxides in benzene solution. Bonsack (1973) obtained a range of microporous gels from Ti^{IV} sulphate in aqueous solution, the maximum BET-nitrogen area being ≈ 420 m^2 g^{-1}. Carefully controlled stoichiometric amounts of water were used by Teichner and his co-workers (see Teichner et al., 1976) to hydrolyze in an autoclave solutions of alkoxide in the corresponding alcohol. A variety of macroporous aerogels of anatase structure could be produced in this manner.

In a systematic investigation of the porosity of titania gels, Ragai et al. (1980) and Ragai and Sing (1982, 1984) employed aqueous solutions of titanous ions, $[Ti(H_2O)_6]^{3+}$. The addition of aqueous ammonia gave black precipitates of the hydrous Ti^{III} oxide, which were then oxidised to produce the white hydrogels of Ti^{IV} oxide. The pH of the ammonia addition was recorded and the hydrogels were thoroughly washed and the compact xerogels obtained by oven-drying at 110 °C.

A representative selection of the nitrogen isotherms and corresponding α_s-plots on the TiO_2 xerogels prepared by Ragai and Sing (1984) are displayed in Figure 11.25. Each α_s-plot is presented as the volume of nitrogen adsorbed plotted against the reduced adsorption, α_S, which had been determined on a non-porous reference TiO_2 (Ragai et al., 1980). It is evident that in each case back extrapolation of the initial linear section gives a zero intercept. This is a useful indication that pore filling was preceded by monolayer adsorption on the wide micropore walls and that there was no detectable primary micropore filling of narrow micropores.

The isotherms and α_S-plots in Figure 11.25 exemplify the adsorptive behaviour of three types of pore structure: (a) a wide range of open mesopores in gel E; (b) a well-defined mesoporous network in gel A1; (c) a distribution of wide micropores and some mesopores in gels A3 and C.

The development of porosity in many freshly prepared hydrous oxide gels (e.g. Al_2O_3, TiO_2 and ZrO_2) is associated with the removal of the ligand water. Increase

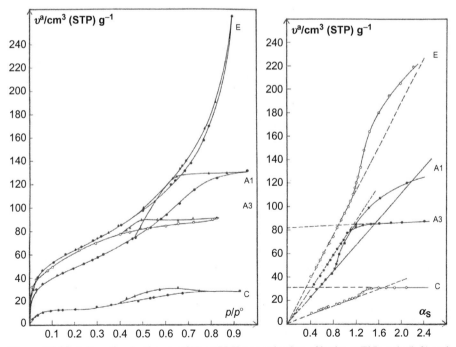

Figure 11.25 Adsorption isotherms of N_2 at 77 K on a selection of hydrous TiO_2 gels (left) and corresponding α_s-plots (right) (Ragai and Sing, 1984).

in pH generally leads to enhanced ligand displacement and consequently to the development of hydroxo and oxo bridges between neighbouring cations and ultimately to the cementation of particles. The retention of the ligand water tends to retard the development of the gel network and to leave the system in a poorly ordered state (Bye and Sing, 1973).

The observed differences between the pore structures of the TiO_2 gels featured in Figure 11.25 have been explained in this manner (Ragai and Sing, 1984). The low uptake of nitrogen by gel C was probably due to the inaccessibility of many of the cavities which remained in the vicinity of the cations when the residual water ligands were removed at low temperature. Appreciable mesoporosity appeared only after the precipitation occurred above pH $\approx$5. For example, gel E, prepared at pH 7.1, had a very open mesopore structure. It was found also that the transformation to rutile was facilitated if the gel was prepared at relatively low pH. Thus, gels C and A1 were converted directly to rutile at 500 °C, whereas anatase was formed by the same heat treatment of gel E. It seems likely that the removal of the residual water ligands has led to the formation of a defect structure and enhanced reactivity.

11.5 Magnesium Oxide

11.5.1 Physisorption of Non-polar Gases on Non-porous MgO

It is well known that magnesium oxide smoke (an aerosol) is produced when Mg ribbon is burnt in air. The dispersed material is in the form of small particles: when produced under controlled conditions these are single-crystal cubes in the size range of 20–200 nm. Non-porous MgO powders prepared in this manner appear to have unique properties as adsorbents for physisorption measurements. In particular, it is possible to prepare a highly uniform MgO surface since the (100) crystal plane of the fcc lattice is the most stable surface state (Henrich, 1976). Furthermore, although the MgO (100) surface is ionic, a relatively weak non-specific interaction is to be expected with non-polar molecules – in contrast to the strong non-specific interaction shown by the basal plane of graphite.

Remarkably uniform MgO smoke was prepared by Coulomb and Vilches (1984) by burning Mg ribbons in dry O_2/Ar mixtures. The MgO particles were collected in the form of a coating on a clean aluminium surface and were subjected to heat treatment (at ca. 950 °C and pressures $<10^{-6}$ mbar). The specific surface area of the final MgO (100) powder was ca. 8 $m^2 g^{-1}$ so that it was not difficult to undertake accurate physisorption measurements and also neutron scattering experiments.

The uniformity of the MgO (100) surface was demonstrated by the stepwise (Type VI) character of the isotherms of Kr, Xe and Ar (Coulomb et al., 1984), CH_4 (Madih et al., 1989) and C_2H_6 (Trabelsi and Coulomb, 1992). To obtain the required high degree of surface uniformity, it was necessary to exclude water vapour and control both the oxygen concentration and the conditions of thermal treatment (Coulomb and Vilches, 1984). The deleterious effect of water vapour was attributed to the

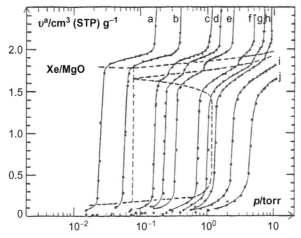

Figure 11.26 Isotherms of Xe on MgO at (a) 97, (b) 100, (c) 106, (d) 108, (e) 111, (f) 116, (g) 118, (h) 121, (i) 126 and (j) 131 K. Phase boundaries indicated by dashed lines (Coulomb et al., 1984).

formation of surface layers of $Mg(OH)_2$. This conclusion was consistent with the inferior quality (i.e. non-stepwise character) of isotherms determined on MgO prepared by the thermal decomposition of $Mg(OH)_2$.

A family of isotherms for the Xe/MgO (100) system is shown in Figure 11.26. Similar results were obtained for the Kr/MgO (100) system (Coulomb et al., 1984). The vertical risers corresponding to first and second layer formation are clearly evident as are the first-layer sub-steps. As with other systems, the sub-steps were attributed to 2D 'fluid–solid' transitions. By following the approach adopted by Larher, Coulomb et al. (1984) were able to estimate the 2D triple and critical points for Xe/MgO and Kr/MgO.

The Ar/MgO (100) isotherms determined by Coulomb et al. (1984) over the temperature range 48–69 K also exhibited well-defined first and second layer risers, but at these temperatures there appeared to be no first-layer sub-steps. This absence of monolayer phase transitions in the p–T range studied seemed to be associated with a large 2D liquid–vapour coexistence region. However, subsequent work (Coulomb, 1991) revealed the existence of a 2D solid-type structure at lower temperatures, with all the Ar atoms appearing to lie along the channels formed by the small Mg^{2+} ions. At a temperature of ca. 38 K, the long-range order along the channels is lost. According to Coulomb, there occurs a kind of '1D melting' with the formation of a 2D 'liquid crystal' state.

A neutron diffraction study was undertaken by Madih et al. (1989) alongside methane adsorption measurements on the MgO (100) powder. The well-defined stepwise character of the CH_4 isotherm at 87.4 K in Figure 11.27 is again indicative of the layer-by-layer mode of adsorption. A somewhat similar isotherm was given by C_2H_6 on the MgO (100) surface at 119.68 K, although the higher steps were not as distinctive as those for CH_4.

Coulomb and his co-workers conclude that at 87.4 K up to four adsorbed layers of methane are ordered and this is followed by a coating of disordered liquid-like layer. In a normal 2D 'solid' monolayer the CH_4 molecular area would be 0.178 nm^2. The

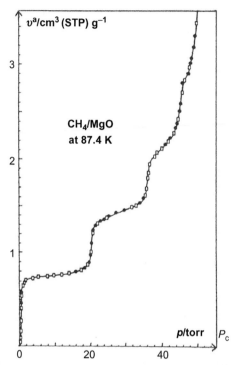

Figure 11.27 Adsorption isotherm of CH_4 on MgO (100) at 87.4 K. Open symbols, adsorption; solid symbols, desorption.
Courtesy Gay et al. (1990).

neutron diffraction patterns for CD_4 methane indicate that the commensurate 2D solid-like film is not completely melted above the triple point ($T_t = 89.7$ K), with over two statistical layers remaining ordered at 95 K. However, bulk crystallites, which are also formed at temperatures below the 2D triple point, disappear as the melting point is approached.

As might be expected, ethane adsorption on MgO (100) is more complicated and is apparently characterised by short range order. It seems that the size and shape of the C_2H_6 molecule play an important role in determining the structure of the 2D film.

11.5.2 Physisorption by Porous Forms of MgO

Magnesium oxide of high surface area can be produced by the thermal decomposition of various magnesium compounds. In the early work of Gregg and Packer (1955), a maximum specific area of about 200 m² g⁻¹ was obtained by the calcination of $Mg(OH)_2$ at 380 °C. This temperature was a little below the temperature required for the complete decomposition of the $Mg(OH)_2$ under these experimental conditions.

Table 11.14 Comparison of BET Areas Derived from Isotherms of Nitrogen and Cyclohexane on Samples of MgO Prepared by Thermal Decomposition of Oxalate (Mikhail et al., 1971)

$\theta_{decomp}/°C$	$a(BET, N_2)/m^2 g^{-1}$	$a(BET, C_6H_{12})/m^2 g^{-1}$
400	482	204
430	471	259
460	481	370
510	263	246
560	145	141
600	42	44

Vleesschauwer (1970) prepared two series of mesoporous batches of MgO by the heat treatment of crystalline $MgCO_3$ (magnesite) and crystalline $MgCO_3 \cdot 3H_2O$ (nesquehonite). A maximum surface area of about 350 $m^2 g^{-1}$ was obtained by calcination of the latter precursor at 400 °C. Although the thermal decomposition of magnesite resulted in the development of lower surface areas ($<140 m^2 g^{-1}$), the products appeared to have a more uniform mesopore structure. Thus, the nitrogen isotherm determined on a sample calcined for 24 h at 800 °C exhibited a narrow, almost vertical, hysteresis loop at $p/p° \approx 0.9$.

Mikhail et al. (1971) obtained a series of porous products by the thermal decomposition of magnesium oxalate dihydrate *in vacuo* at temperatures in the range 400–600 °C. The BET areas derived from the adsorption isotherms of nitrogen and cyclohexane, are given in Table 11.14. The values of $\sigma(N_2)$ and $\sigma(C_6H_6)$ have been taken as 0.162 and 0.39 nm^2, respectively.

It is evident that the corresponding BET areas in Table 11.14 are in fairly good agreement only after the decomposition temperature was taken above 500 °C. These results provide strong evidence for the initial formation of narrow micropores, which were inaccessible to cyclohexane molecules.

In other more recent studies of the formation of microporous MgO (Ribeiro Carrott et al., 1991a,b; Ribeiro et al., 1993), micro-crystalline $Mg(OH)_2$ was used as the precursor. The $Mg(OH)_2$ was thoroughly evacuated at room temperature before being heated *in vacuo* (10^{-3} mbar) at progressively higher temperatures. TEM revealed that the small hydroxide particles (150–7000 μm) consisted of hexagonal platelets and that this morphology was retained throughout the decomposition. The outgassed mass at 150 °C corresponded to the exact stoichiometry $Mg(OH)_2$. About 85% decomposition occurred below 300 °C, but temperatures of 700–800 °C were required to achieve 100% decomposition. X-ray diffraction showed that the thermal decomposition was accompanied by a progressive development of the cubic MgO structure and a gradual decrease in the intensity of the $Mg(OH)_2$ lines.

Representative nitrogen isotherms and the corresponding comparison plots are shown in Figure 11.28. The latter have been constructed by taking the undecomposed

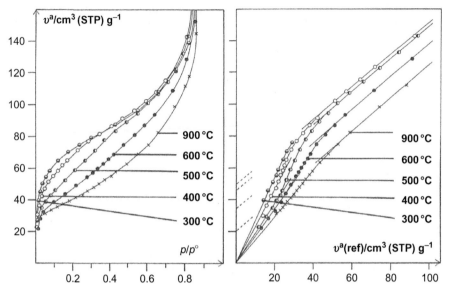

Figure 11.28 Nitrogen isotherms at 77 K on partially decomposed $Mg(OH)_2$ (left) and corresponding comparison plots with undecomposed $(Mg(OH)_2)$ as reference (right) (Ribeiro Carrott et al., 1991a,b).

$Mg(OH)_2$ as the reference material. Any change in isotherm shape is therefore manifested as a non-zero intercept and/or a deviation from linearity.

Each comparison plot in Figure 11.28 has two linear sections. Back extrapolation of the first linear section gives a zero intercept, whereas back extrapolation of the second (multilayer) section gives a positive intercept. The interpretation is based on the principles introduced in Chapter 9. An analysis of the isotherm data is given in Table 11.15.

Table 11.15 Analysis of Nitrogen Isotherms on Thermally Decomposed $Mg(OH)_2$

$\theta/°C$	% decomp.	$a(BET)/m^2 g^{-1}$	$a(com)/m^2 g^{-1}$	$v(mic)/cm^3 g^{-1}$	w_p/nm
150	0	99	99	0	0
240	12.6	110	110	0.007	1.27
250	28.0	128	128	0.017	1.13
270	73.4	230	231	0.063	0.93
400	92.8	237	230	0.078	1.15
500	95.5	191	188	0.072	1.64
600	98.7	164	163	0.053	1.77
750	100	141	142	0.033	1.57

The fact that the comparison plots in Figure 11.28 are parallel in the multilayer range indicates that the external area [$a(\text{ext}) = 99$ m^2 g^{-1}] has remained constant: this is in accordance with the absence of any detectable change in the particle morphology. The values of micropore volume, $v_p(\text{mic})$, in Table 11.15 have been calculated from the intercepts on the v^a-axis, assuming liquid-like molecular packing.

Two sets of derived surface areas are recorded in Table 11.15: the values of $a(\text{BET})$ have been obtained by the usual BET method and $a(\text{com})$ from the slope of the initial linear part of the comparison plot. The close agreement between the corresponding values of surface area is consistent with the linearity and zero intercept of the comparison plot. We conclude that the BET method appears to provide a reliable estimate of the total surface area of each adsorbent.

To obtain the values of mean pore width, w_P, in Table 11.15, we have assumed the validity of the BET-area and all pores to be slit-shaped. Then by adopting the hydraulic pore width principle, we have

$$w_p = \frac{2v_p(\text{mic})}{a(\text{BET}) - a(\text{ext})} \tag{11.7}$$

These and other results (Ribeiro Carrott et al., 1991a) indicate that there is little variation in w_p between 30% and 90% decomposition. Wide micropores of width of ca. 1 nm would be wide enough to allow monolayer adsorption of nitrogen to precede cooperative micropore filling, which is the mechanism best able to explain the features of the comparison plots in Figure 11.29.

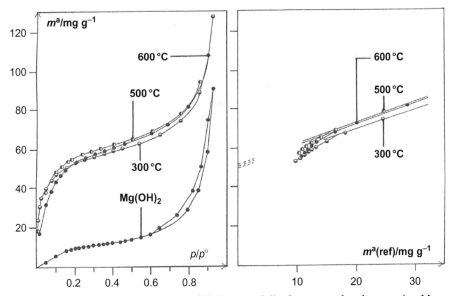

Figure 11.29 Neopentane isotherms at 273 K on partially decomposed and on starting Mg (OH)$_2$ (left) and corresponding comparison plots with undecomposed Mg(OH)$_2$ as reference (right) (Ribeiro Carrott et al., 1991a,b).

However, more bulky adsorptive molecules would be expected to behave differently, provided their diameters are large enough to give primary micropore filling. This is exactly the position with neopentane: in terms of its molecular size (0.62 nm), the MgO pore sizes extend from ca. 1.5–3 molecular diameters. This explains why the low-pressure regions of the neopentane comparison plots in Figure 10.29 are not linear and do not back-extraplolate to the origin. In this case, the physisorption forces are enhanced because of the proximity of the pore walls and the shape of the neopentane isotherm is consequently distorted at low $p/p°$.

It has been proposed that the slit-shaped micropores may be regarded as spaces between (111) planes in the freshly formed MgO structure. Thus, if the width of each micropore was equivalent to four (111) planes, its width would be 0.96 nm, which is quite close to the hydraulic values. It is not surprising to find pore widening to have occurred as the decomposition neared completion.

11.6   Miscellaneous Oxides

11.6.1   Chromium Oxide Gels

It has been known for many years that the adsorptive and catalytic properties of chromium oxide gels are very sensitive to the conditions of preparation, storage and heat treatment (Burwell et al., 1960; Deren et al., 1963; Carruthers and Sing, 1967; Baker et al., 1970, 1971).

The amorphous hydrogels produced by the neutralisation of chromium (III) salts generally retain large amounts of water. These products tend to undergo ageing (i.e. loss of surface area), but by careful drying they can be converted into highly porous xerogels. When such chromia gels are subjected to heat treatment in air or oxygen at temperatures above ca. 200 °C, there occurs an oxidation–reduction cycle, $Cr^{3+} \rightarrow Cr^{6+} \rightarrow Cr^{3+}$, which finally results in the formation of low-area, crystalline α-Cr_2O_3 (Baker et al., 1971).

The crystallisation process is a highly exothermic transformation (or 'glow phenomenon'), which normally occurs at remarkably low temperatures (350–400 °C); this can be delayed and minimised if the gel is heated in an inert atmosphere (Carruthers and Sing, 1967). The changes in BET-nitrogen area brought about by the heat treatment of a chromia gel in, vacuo and dry nitrogen are shown in Figure 11.30.

The results in Figure 11.30 illustrate the important influence of the surrounding atmosphere during heating. Particularly striking is the protection afforded by nitrogen until the temperature approached 500 °C, which is consistent with the upward displacement of the glow temperature. It is also of interest that the drastic ageing at 100 °C in air and nitrogen was eliminated by heating in vacuo.

The nitrogen isotherms in Figure 11.31 were determined on samples of another chromia preparation, gel B, which had been heated for different periods in air (A) or in vacuo (V). Temperature and duration of heat treatment are indicated for each sample.

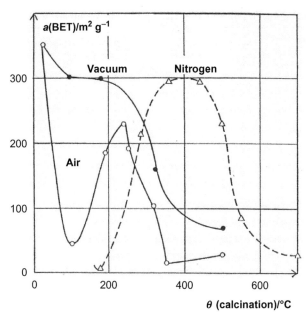

Figure 11.30 BET nitrogen area of chromia gel versus temperature of calcination (in air, in vacuo or in stream of dry nitrogen) (Carruthers and Sing, 1967).

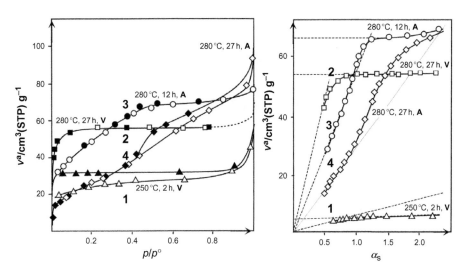

Figure 11.31 Nitrogen isotherms at 77 K (left) and α_S-plots (right) for chromia gel B heated in air (A) or in vacuo (V) (temperature and duration indicated on each curve) (Baker et al., 1971).

For sample B (250 °C, 2 h, V), the nitrogen isotherm is irreversible, a long part of the desorption branch lying parallel to the $p/p°$-axis. Prolonged outgassing (280 °C, 27 h, A) has resulted in a reversible Type I isotherm. Heat treatment in air (280 °C, 12 h, A and 280 °C, 27 h, A) has led to further pore widening and a decrease in surface area (see Table 11.16). The curves 1, 2, 3 and 4 reveal a progressive change

Table 11.16 Surface Areas and Porosities of Chromia Gel B Heated in Vacuo and in Air

Sample	Porosity	$a(BET)/m^2g^{-1}$	$a(S)/m^2g^{-1}$	$v(mic)/cm^3g^{-1}$
B (250 °C, 2 h, V)	narrow micro.	18		0.008
B (280 °C, 12 h, V)	narrow micro.	235	(280)	0.081
B (280 °C, 27 h, V)	narrow micro.	240	(280)	0.084
B (280 °C, 12 h, A)	wide micro.	167	173	0.112
B (280 °C, 27 h, A)	mesopores	91	94	0.113 (mes)

in the isotherm character from Type I to Type IV, which is indicative of the change from a microporous to a mesoporous structure.

The α_S-plots in Figure 11.31 provide a basis for the analysis of the corresponding nitrogen isotherms. The standard isotherm was determined on a non-porous sample of α-Cr_2O_3 (Baker et al., 1971). As previously explained, curves 1 and 2 are characteristic of micropore filling followed by multilayer adsorption on a small external surface (<5 m^2 g^{-1}). The values of micropore volume, $v_p(mic)$ in Table 11.16 have been obtained by back extrapolation of the linear multilayer branch to the adsorption axis and by assuming liquid-like molecular packing.

The long range of near-linearity of the α_S-curve 3 (for sample B (280 °C, 12 h, A)) can be attributed to the reversible filling of wide micropores. In the case of curve 4, the upward deviation from linearity is due to capillary condensation in mesopores. In both cases, the initial linear region can be back extrapolated to the origin and we may conclude that pore filling was preceded by surface coverage of the pore walls. The values of total surface area, $a(S)$ in Table 11.16 are calculated from the slope of this linear region by using Equation (11.3). It can be seen that fairly good agreement is obtained between the corresponding values of $a(BET)$ and $a(S)$ only in the case of sample with mesopores and wide micropores.

It was established that the oxidation process $Cr^{3+} \rightarrow Cr^{6+}$ is generally accompanied by pore widening; but that the reduction stage $Cr^{6+} \rightarrow Cr^{3+}$, which involves the crystallisation of α-Cr_2O_3, is associated with the removal of pores and a loss of surface area. It was also evident that the formation of the higher oxidation state is facilitated by the removal of the ligand water, thus reducing the stability of the Cr^{3+} ions (Baker et al., 1971).

It was noted that the orthorhombic CrOOH and ferromagnetic CrO_2 structures were present in some chromia gels, which had been calcined under hydrothermal and oxidising conditions (Carruthers et al., 1967, 1969). This discovery prompted an investigation of the topotactic interconversion of the oxy-hydroxide and the dioxide (Alario Franco and Sing, 1972, 1974), which involved the thermal decomposition in vacuo of small crystals of CrOOH. Gas adsorption and electron microscopy revealed that slit-shaped pores are first formed with little change in the external dimensions of the crystals (Alario Franco et al., 1973). The CrO_2, which is formed as an intermediate product, is finally decomposed with the formation of larger pores and the growth of

α-Cr$_2$O$_3$ crystals. As might be expected, the reduction of CrO$_2$ in H$_2$ similarly results in the formation of CrOOH with very little change in external area and with the generation of a relatively small micropore volume.

11.6.2 Ferric Oxide: Thermal Decomposition of FeOOH

In the early work of Goodman and Gregg (1959), it was found that a highly active hydrous ferric oxide [a(BET) $\approx$ 300 m^2 g^{-1}] underwent a progressive loss of surface area when calcined in air. Thus, after heat treatment at 400 °C for 5 h the BET-area was reduced to ca. 50 m^2 g^{-1}. One of the first systematic investigations of the thermal decomposition of a well-defined iron oxide precursor was undertaken by Bye and Howard (1971), who reported the formation of microporosity from goethite. More recently, the thermal transformation of goethite, α-FeOOH, into hematite, α-Fe$_2$O$_3$, has been investigated in a number of laboratories.

In the work of Naono et al. (1987), samples of the mineral form of goethite were heated in vacuo at different temperatures between 200 and 700 °C (i.e. corresponding to the stoichiometric range of decomposition). Nitrogen adsorption measurements revealed that the BET-area attained a maximum value of 151 m^2 g^{-1} after the goethite had been outgassed for 10 h at 300 °C. The nitrogen isotherms can be broadly separated into two groups, as arranged in Figure 11.32. Most of those in (a) clearly have fairly pronounced Type I character, whereas those in group (b) are either Type IV or

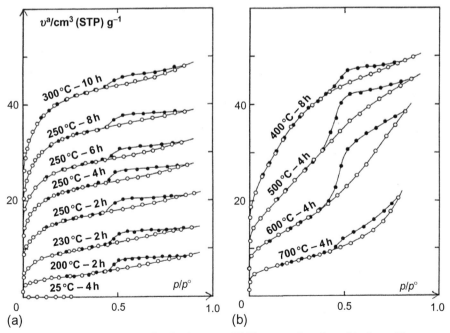

Figure 11.32 Nitrogen adsorption isotherms at 77 K on samples of goethite heated between 200 and 700 °C.
Courtesy Naono et al. (1987).

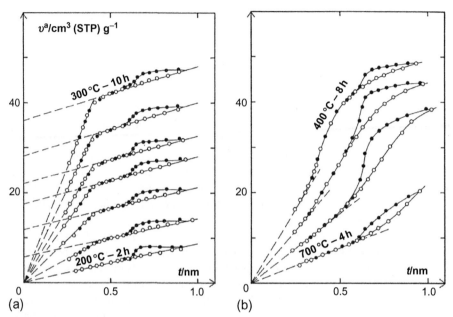

Figure 11.33 Nitrogen *t*-plots constructed from the isotherms in Figure 11.32, for samples of goethite heated between 200 and 700 °C.
Courtesy Naono et al. (1987).

non-reversible Type II isotherms (now classified as Type IIb). All the isotherms exhibit hysteresis loops, the shape changing from H4 to H3.

The *t*-plots in Figure 11.33 were constructed from the isotherms in Figure 11.32. For this purpose, the standard *t* curve was derived from nitrogen isotherms on two samples of non-porous goethite. As with the isotherms, the *t*-plots are broadly divided into two groups: with the exception of the plot for the sample heated 2 h at only 200 °C, those in group (a) are associated with wide micropore filling; while the group (b) plots indicate mainly capillary condensation in mesopores.

The *t*-plots in group (a) are of particular interest because of the linearity of the reversible region at low $p/p°$ and also of the *adsorption* branch of the hysteresis loop. Moreover, the first linear sections can be back extrapolated to the origin. We may therefore conclude that monolayer adsorption has occurred on both an external surface and the walls of wide micropores, but multilayer adsorption has occurred only on an external surface. This evidence together with the H4 hysteresis points to a range of slit-shaped pores, which was confirmed by high-resolution electron microscopy (Naono et al., 1987). Indeed, the electron micrographs of the porous samples have a remarkably similar appearance to the micrographs obtained with partially decomposed CrOOH (Alario Franco and Sing, 1972).

The change in the character of the isotherms and *t*-plots reveals that the loss of surface area, which occurred when the outgassing temperature of the goethite was taken above 400 °C, was associated with pore widening. The sample obtained at 400 °C appears to have a broad range of pores extending from the wide micropore into the mesopore range. The oxide has become predominantly mesoporous at higher temperature, but it is evident that the limits of pore size are not discernible.

Naono and his co-workers have also studied the changes in texture produced by the thermal decomposition of γ-FeOOH (Naono and Nakai, 1989) and β-FeOOH (Naono et al., 1993). In the latter investigation, a partially decomposed sample gave reversible Type II nitrogen isotherms. However, the t-plots provided unambiguous evidence that a micropore structure was developed and that the external area remained virtually unchanged. Since the initial linear region could be back extrapolated to the origin, it appears that there was no detectable distortion of isotherm shape and therefore no significant filling by narrow micropores.

11.6.3   Micro-crystalline Zinc Oxide

The surface chemistry of zinc oxide is of particular interest in relation to its catalytic and photocatalytic properties. For example, the (0001) hexagonal crystal plane appears to have a special role in the catalytic methanol-synthesis reaction (Bowker et al., 1983). The chemisorption of CO and dissociative chemisorption of H_2 occur on the exposed Zn^{2+} cations: Bolis et al. (1986) have found that the relative magnitude of this 'active' area of ZnO was highly dependent on the nature of the precursor (oxalate, carbonate of Zn). Similar conclusions can be drawn from the infrared spectroscopic measurements of Chauvin et al. (1986).

The adsorption isotherm of krypton determined at 77 K on a sample of zinc oxide powder (Grillet et al., 1989) is shown in Figure 11.34. The adsorbent had been outgassed at temperatures up to 450 °C by the CRTA technique. As first pointed out by Bonnetain and his co-workers (Audier et al., 1981), it is possible to estimate the extent of surface homogeneity by taking the ratio of the step heights, Y_2/Y_1. In

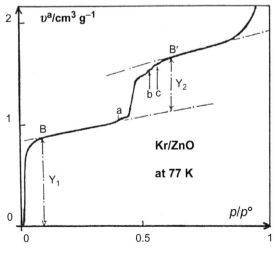

Figure 11.34 Adsorption isotherm of krypton on ZnO at 77 K (Grillet et al., 1989).

the case of an ideally homogeneous surface, one would expect to obtain $Y_2/Y_1 = 1$: the ZnO in Figure 11.34 gives 0.59 (cf. 0.9 for exfoliated graphite).

An unusual feature of the Kr isotherm in Figure 11.34 is the presence of the three small sub-steps (a, b and c), which could be detected by the use of the continuous quasi-equilibrium procedure for the determination of the adsorption isotherm. At present, the significance of these sub-steps is not clear, but it seems more likely to be related to the adsorption on the different crystal faces rather than to any phase changes of the adsorbate. BET-analysis of the adsorption isotherms of Kr, Ar and N_2 on the sample of ZnO gave the following values of surface area: 3.53, 3.56 and 3.76 $m^2 \, g^{-1}$ (with σ taken as 0.143, 0.138 and 0.162 nm^2, respectively).

The differential enthalpies for the adsorption of N_2 and Ar are plotted against the surface coverage θ, in Figure 11.35. It is significant that both N_2 and Ar give an almost constant differential enthalpy of adsorption up to ca. half-coverage. This degree of energetic homogeneity is consistent with the stepwise character of the Kr and Ar isotherms. As with TiO_2, the initial N_2 adsorption energy is much greater than the corresponding Ar energy. The very high N_2 value is mainly due to its strong specific interaction with cationic sites. Thus, the outgassing at 450 °C has removed protecting ligands such as H_2O or CO_2 leaving the unscreened Zn^{2+} ions freely exposed.

It can be seen that the N_2 plateau in Figure 11.35 is somewhat longer than the corresponding Ar plateau. It seems likely that the strong N_2 specific interaction is not confined to the (0001) Zn^{2+} sites; it may also involve surface defects and edge sites. If we assume that the length of the Ar plateau provides a more reliable indication of the extent of the cationic surface, we arrive at estimated fractions of about 40% and 10%, respectively, for the effective areas of Zn^{2+} and other high-energy sites. There is a peak in the Ar differential energy curve at high surface coverage. The associated adsorbate–adsorbate interaction is probably the result of a 2D phase change (possibly a 2D fluid–solid transformation), but the smooth Ar isotherm shows no indication of a sub-step. These results illustrate the value of adsorption microcalorimetry for investigating the surface properties of micro-crystalline powders.

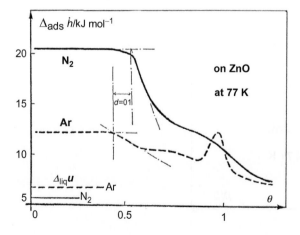

Figure 11.35 Differential enthalpy of N_2 and Ar on ZnO at 77 K (Grillet et al., 1989).

11.6.4   Hydrous Zirconia Gels

Zirconium dioxide is notable for its chemical inertness and refractory nature. The first physisorption measurements on ZrO_2 appear to have been made around 1970 (Rijnten, 1970; Holmes et al., 1972). The fact that relatively few adsorption studies have been undertaken since then may be due to the complexity of zirconium chemistry. For instance, a theoretical investigation on the adsorption of CO_2 on ZrO_2 concludes that the CO_2 molecule can be linked to the surface in four different ways, that is (i) by exothermic dissociation on two adjacent unsaturated Zr atoms, (ii) by strong physisorption on a similar pair of Zr atoms, (iii) by 'apical' adsorption on a single unsaturated Zr atom and (iv) by weak adsorption on surface OHs (Boulet et al., 2012). Thus, in the preparation of a well-defined oxide–hydroxide structure it is necessary to take account of the exceptionally high coordination number generally exhibited by Zr^{IV} and its tendency to form polymeric ionic networks.

Non-porous ZrO_2 powders can be produced by high-temperature vapour phase condensation methods: in this manner discrete spherical particles of ca. 4 nm diameter have been obtained (Avery and Ramsay, 1973). It is also possible to prepare colloidal dispersions of sub-micron sized, spheroidal particles of basic salts such as $Zr_2(OH)_6CO_3$ and $Zr_2(OH)_6SO_4$ with the aid of the carefully controlled sol–gel techniques developed by Matijevic (1988).

In the work of Rijnten (1970), a number of procedures were used for the preparation of hydrous zirconia gels. For example, a dilute solution of NH_4OH was added dropwise to a vigorously stirred solution of $ZrCl_4$ until a particular pH was attained. The nitrogen isotherms determined on the aged, washed and dried gels prepared at pH 4, 6 and 8 were of Types Ib, I + IV and IVb, respectively – indicating a change in the pore structure from wide micropores to mesopores. The corresponding BET areas were 244, 308 and 320 $m^2 \, g^{-1}$.

In the more recent adsorption studies by Gimblett et al. (1981, 1984) and Ragai (1989, 1994), zirconyl chloride, $ZrOCl_2 \bullet 8H_2O$, was adopted as the starting material. In one series of preparations, NH_3 gas was bubbled at a controlled rate through the $ZrOCl_2$ solution, the pH being continuously monitored. Nitrogen isotherms determined on gel ARZr3, prepared at pH 5, are shown in Figure 11.36a. The isotherm obtained after outgassing at 25 °C is of Type Ib, which as we have seen is indicative of wide microporosity. However, outgassing at 200 °C resulted in the removal of a substantial fraction of the micropore capacity.

The corresponding water isotherms in Figure 11.36b were determined after a sample of ARZr3 has been outgassed at 25 °C. Three successive sorption/desorption runs are depicted. The general Type I character again indicates microporosity, but now low-pressure hysteresis is observed. Outgassing after the first desorption run resulted in a substantial loss of weight (i.e. change from A to B). This behaviour extended into the second and third adsorption/desorption cycles, but smaller losses were involved.

Increasing the pH at which the gel was prepared led to the development of low-pressure hysteresis with both nitrogen and water. The gels prepared over the pH range 5–11 were all essentially microporous, although the proportion of wider micropores increased at higher pH. The BET-nitrogen areas and apparent pore volumes are

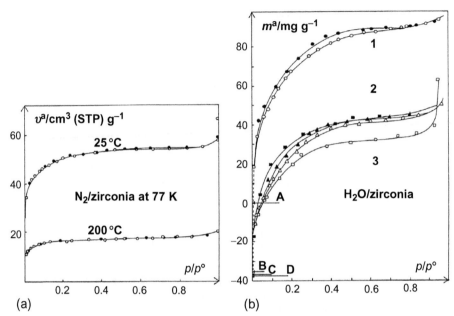

Figure 11.36 (a) Nitrogen adsorption isotherms at 77 K on zirconia gel ArZr3 outgassed at 25 and 200 °C (closed circles for desorption). (b) Water sorption isotherms on same gel outgassed at 25 °C. Three successive runs. Runs 1, 2 and 3 start at point are A, B and C, respectively. D is the final desorption point of run 3 (solid symbols for desorption) (Gimblett et al., 1981).

summarised in Table 11.17. It can be seen that outgassing at 200 °C has resulted in a drastic reduction in the BET-area of each gel.

These and other results (Gimblett et al., 1981; Ragai et al., 1994) indicate that the microporosity of the hydrous ZrO_2 gels was largely dependent on the microstructure

Table 11.17 BET-Nitrogen Areas and Apparent Pore Volumes of Hydrous Zirconia Gels

Sample	pH (Prep)	Outgassing $\theta/°C$	a(BET)/m^{2}g^{-1}	v(mic,N$_2$)/ cm^{3}g^{-1}	v(mic,H$_2$O)/ cm^{3}g^{-1}
ARZr3	5	25	162	0.093	0.091
		200	63	0.030	–
ARZr4	7	25	210	0.119	0.110
		200	88	0.042	0.050
		400	18	0.013	0.003
ARZr5	9	25	220	0.124	0.075
		200	100	0.056	0.056
ARZr6	11	25	226	0.136	–

generated through the displacement of coordinated water. Thus, a proportion of the water removed by outgassing at 25 °C was present in the form of weakly bound ligands in the coordination spheres of the Zr. Their removal at low temperature has led to the creation of vacancies and hence to the generation of microporosity. This mechanism is very similar to that proposed for the low-temperature formation of microporosity in chromia gels.

The similarity with chromia and other hydrous oxides is also evident in the loss of surface area which can occur through the removal (i.e. slow outgassing) and addition of water at quite low temperature. It is evident that the presence of the water ligands helps to stabilise the poorly ordered structure, whereas their removal renders the system more liable to undergo further change to a more thermodynamically favourable state. The sorption of water vapour makes this possible.

The use of bicarbonates as the precipitating agents has been found (Gimblett et al., 1984) to lead to mixed microporous–mesoporous zirconia gels. The nitrogen adsorption and water vapour sorption measurements revealed that the microporosity obtained after outgassing the hydrous gel at low temperature (<200 °C) was due to removal of H_2O and NH_3 ligands, but that the removal of the bidentate carbonato groups required higher outgassing temperatures and resulted in the development of mesoporosity.

The hydrous ZrO_2 gel particle may be pictured as a 3D agglomerate of indefinite size, shape and structure. It contains a large number of water molecules, many of which are coordinatively bound to the Zr^{4+} ions. The latter are linked in tetrameric units through hydroxo and oxo bridging. As the pH is increased the extent of these linkages also increases so that the gel becomes more rigid and compact. Spectroscopic and thermogravimetric measurements show the absence of any well-defined hydrates and instead indicate the presence of hydrogen-bonded and coordinated water (Gimblett et al., 1980).

An interesting approach has been used by Hudson and Knowles (1996) to obtain mesoporous high-area ZrO_2. This involves the incorporation of cationic quaternary ammonium cations in the hydrous oxide and the subsequent calcination of the inorganic/organic intermediate. The incorporation is achieved by cation exchange at a pH above the isoelectric point of the hydrous oxide. After calcination at 723 K, ordering of the structure is detectable and appears to be a linear function of the chain length. Well-defined Type IVb nitrogen isotherms have been obtained on the calcined products, many of which have BET(N_2)-areas above 300 $m^2 \, g^{-1}$. A similar approach was used by Knöfel et al. (2008) who, subsequently, carried out calorimetric determinations on CO_2 adsorption which could be compared with the results of a theoretical study (Hornebecq et al., 2011). There was a satisfactory consistency, except for the highest initial experimental value of 124 kJ mol^{-1} which could not be accounted for by the model use in simulation: a further indication of the complexity of the zirconia surface.

11.6.5 Beryllium Oxide

Beryllium oxide, when sintered, is a light refractory ceramic with a high thermal conductivity which makes it an interesting material for special electronic components, rockets and the nuclear industry, since it is a good neutron moderator. Unfortunately, the fine powder, when breathed, is highly carcinogenic, which limits its application.

Nevertheless, examining the thermal modification of its adsorbent behaviour lent to a most meaningful case study. Rouquerol (1965) and Rouquerol et al. (1970, 1985) started from a precipitated α-Be (OH)$_2$ sample made up of sheet-like crystals, micronsized and thin (a few tens of a nm), with a BET-nitrogen surface area of 46 m^2 g^{-1}. This hydroxide was decomposed by controlled rate thermal analysis (CRTA, see Section 3.4) under a constant water vapour pressure of 0.1 mbar, at a constant rate of water evolution of 11.2 mg h^{-1} g^{-1}. This CRTA heat treatment was carried out up to 1075 °C, but in successive steps. The succession of N$_2$ adsorption–desorption isotherms determined at 77 K is shown in Figure 11.37.

We see clearly the following features:

○ The untreated Be(OH)$_2$ (simply outgassed at 20 °C) provides a typical H3 hysteresis, with no saturation plateau; the large hysteresis loop is understood as totally due to *condensation between the loose, sheet-like crystals*
○ For the oxide obtained at 200 °C (i.e. after the total decomposition of Be(OH)$_2$), a typical *Ib + IIb composite isotherm* is obtained, indicating a filling of micropores (Ib) followed by condensation between the crystals (IIb, with H3 hysteresis)
○ The *homogeneity* of the CRTA treatment produces in the crystals *pores of a uniform size*, whose complete filling gives rise to a typical break (marked with an arrow) on the adsorption isotherm

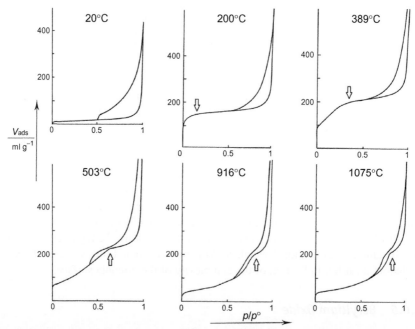

Figure 11.37 Set of N$_2$ adsorption–desorption isotherms on the products of controlled rate thermal analysis (CRTA) of crystalline α-Be(OH)$_2$ under 0.1 mbar, with a loss rate of 11.2 mg h^{-1} g^{-1}.
From Rouquerol (1965).

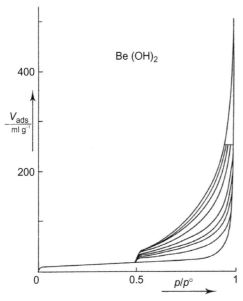

Figure 11.38 'Scanning' experiments within the hysteresis loop region of the N_2 adsorption–desorption isotherm determined at 77 K on crystalline α-Be $(OH)_2$ (Rouquerol et al., 1985).

- The shift of this break towards the higher pressures, as the thermal treatment proceeds, corresponds to an increase in width evaluated by the authors to be from 0.8 (200 °C sample) to 7.6 nm (1075 °C sample)
- From 503 °C upwards, the adsorption isotherm is still composite, but now of IVa + IIb type, with a *double hysteresis loop*, indicating that capillary condensation occurs in the mesopores (lower part of the loop) and then between the sheet-like crystals (upper part).

Most of these features could also be seen in the case of gibbsite thermal decomposition (see Section 11.3.3 and Figure 11.17), though the grains of the starting material were not as regular as here and the resulting isotherms not as nicely shaped.

Figure 11.38 shows a number of scanning experiments carried out within the hysteresis loop, for the starting Be $(OH)_2$ material. The adsorption branch is common to all experiments, which are simply carried out up to different amounts condensed. The dependence of each desorption branch on the amount condensed, together with the homothetic shape of all branches is consistent with the assumption of an inter-sheet condensation with no saturation limit and therefore variable final intersheet distance (Rouquerol et al., 1985).

11.6.6 Uranium Oxide

In the processes of separation and preparation of radioactive uranium oxide, there are intermediate steps where the development of a surface area is required. One of the various ways to obtain divided uranium oxide UO_3 is from the thermolysis of hexahydrated uranyle nitrate $UO_2 (NO_3)_2$ $6H_2 O$. This 'dry route' is interesting for the compactness of the equipment and for not producing liquid wastes. The mechanism of thermal decomposition and of development of a porous structure under controlled conditions of CRTA (see Section 3.4) was studied by Bordère et al. (1990, 1993, 1998).

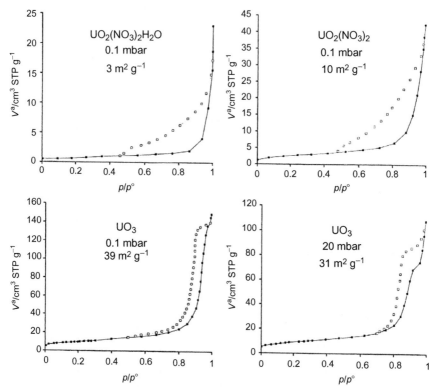

Figure 11.39 N_2 adsorption–desorption isotherms at 77 K on products of controlled rate thermal decomposition (CRTA) of $UO_2 (NO_3)_2$ $6H_2 O$.
From Bordère et al. (1998).

Figure 11.39 shows a few typical N_2 adsorption–desorption isotherms obtained on a few intermediates or end-products obtained in this study and about which we can make the following comments:

- Three of these products were obtained under a *constant residual pressure of 0.1 mbar* and at a low rate of mass loss which were found to best separate the six successive steps of the decomposition and to produce the best final product, that is an UO_3 sample with a BET-nitrogen-specific surface area of 39 m^2 g^{-1}: given the high molar mass of uranium, this is considered to be a very high specific area for this type of material
- The monohydrate (which, incidentally, could only be prepared and isolated by this CRTA technique) has a typical H3 hysteresis (without any saturation plateau) due to capillary condensation in its *sheet-like and non-rigid structure*
- The uranyl nitrate anhydrate has the same sheet-like structure and therefore the same Type IIb isotherm: the increase in surface area from 3 to 10 m^2 g^{-1} is essentially due to a cleavage of the sheets during the departure of the last water molecule.
- The uranium oxide obtained under 0.1 mbar now has a Type IV isotherm (although with a very short a saturation plateau) with a type H1 hysteresis; assuming the formation of *slit-shaped mesopores*, one finds, by applying the BJH method, a narrow pore width distribution around 8 nm

○ Finally, the other uranium oxide, obtained under a higher controlled pressure of 20 mbar, has
   a double hysteresis loop but, at the difference from those observed with alumina (see
   Section 11.3.3) or beryllium oxide (see Section 11.6.5) it is *not partly due to a non-rigid
   structure*: if the structure was non-rigid the lower point of the hysteresis loop would not
   be at a relative pressure of 0.7 but somewhere between 0.42 and 0.5, as it is for the two isotherms on top of the figure. We simply have a *bi-modal porosity*, because the route under
   20 mbar has produced two distinct types of uranyl nitrate anhydrate which are responsible
   for the more complex structure of the final uranium oxide; the absence of an upper saturation
   plateau is here due to a continuous distribution up to the macropore range.

11.7   Applications of Adsorbent Properties of Metal Oxides

11.7.1   Applications as Gas Adsorbents and Desiccants

SO_2 and SO_3 adsorption: the use of metal oxides is being developed for elimination of
SO_x produced by thermal power plants. These oxides are used as single oxides (CaO or
MgO) but most often as supported oxides (e.g. Cu, Mn, etc.) The supports should have
a large surface area and a good textural stability. The supports are usually carbon
(Tseng and Wey, 2004), alumina, or silica-based supports which are now being
favoured, with BET-nitrogen areas in the range of 700–1200 $m^2 g^{-1}$ and a thermal
stability in the range of 700–1200 °C (Mathieu et al., 2013).

   H_2S adsorption seems to be effective, even at near-ambient temperature
(25–50 °C), by Zn—Ti based mixed metal oxides (Polychronopoulou et al., 2005).

   CO_2 adsorption: CaO is the cheapest adsorbent available for CO_2 but it suffers
from a significant degradation of its performance when recycled after regeneration
(i.e. decarbonation). Recent theoretical work demonstrates that, in a broad range of
working conditions, adsorption of CO_2 only occurs at low coverage and calcite nucleation takes place by a localised mechanism, not over the entire surface (Besson et al.,
2012).

   MgO is already considered as an efficient adsorbent for CO_2, with high adsorption
capacity and low energy requirement for regeneration (Bhagiyalakshmi et al., 2010;
Bian et al., 2010) and was prepared for that purpose by many routes, with CO_2 adsorption capacities which seemed to be in good agreement with BET-nitrogen areas (Zhao
et al., 2011), it seems that template mixed MgO—TiO_2 can have an appreciably higher
capacity, due its increased pore volume and BET-nitrogen area (Jeon et al., 2012).

   Influence of water pre-adsorption on the capture of CO_2 by metal oxides received
appreciable attention (e.g. in the case of TiO_2 or of mixed hydroxylated Fe_2O_3,
γ-Al_2O_3; Baltrusaitis et al., 2011).

   Fluorine adsorption by alumina allowed solving the environmental issue due to
fluorine fumes escaping from the cryolith baths used to melt alumina in the electrol-
ysis cells used to produce aluminium. An appropriate pore-sized alumina (selected by
water adsorption, since water and fluorine have molecules of comparable size) allows
capturing fluorine and re-incorporating it, together with the alumina adsorbent, into
the cryolith bath.

Activated alumina is also considered for *ammonia adsorption*, with similar environmental objectives (Saha and Deng, 2010).

Dessication silica xerogels and activated aluminas are largely used as desiccants, with the double interest of being cheap and easy to regenerate by simple heating.

Silica, sometimes mixed with a small amount of coloured indicator of humidity, is extensively used as a cheap desiccant, in the packing of electronic equipment and of drugs (e.g. in the tap of tubes of tablets). Nevertheless, because of a certain width of its pore size range, it cannot guarantee a constant value of the residual vapour pressure of water, at the difference, for instance, from 4A zeolites.

Alumina is used as a desiccant in many industrial processes, for example at the entrance of Pressure Swing Adsorption (PSA) units for air separation. Alumina gel is also used in pharmacy as a gastric adsorbent.

11.7.2 Applications as Gas Sensors

When used as gas sensors (for CO, O_2, NO_2, H_2S, H_2, CH_4, CH_4OH, etc.) metal oxides are evidently used for their adsorbent properties but, if they are still highly divided, it is less in the form of porous materials than in the form of very thin deposits.

Semiconducting *microsensors* are indeed most often in the form of thin films deposited on a micro-hotplate heated at a few hundred °C. The variations in the electrical properties of the metal oxide upon adsorption (resistance, capacitance and also electrical noise; Contaret et al., 2011) are analysed to provide specific information on the gas adsorbed and its partial pressure. Other sensors may use a thick film (Srivastava et al., 1994) and detect either the electrical conductance or, more rarely, simply the thickness with a an appropriate transducer.

The *selection of the metal oxide* is to be made among a large choice (e.g. WO_3, ZnO, SnO_2, TiO_2, Ti_2O_3, $Cr_2 O_3$, NiO, V_2O_3, La_2O_3, Sb_2O_3, Y_2O_3, etc.), following a number of criteria which were well reviewed by Korotcenko (2007). Mixed oxides are also found interesting, for example TiO_2- Ga_2O_3 or TiO_2- Er_2O_3, where the effect of the second oxide is to stabilise the more active anatase form and to limit the grain growth, that is the loss of surface area and of active sites (Mohammadi et al., 2008). A comprehensive review on the experimental techniques allowing to study metal oxides as conductometric gas sensors can be found in Barsan et al. (2007).

11.7.3 Applications as Catalysts and Catalyst Supports

Alumina gels and, even more, *silica–alumina* gels were much used in the petrochemical industry, as catalysts and catalyst supports, essentially before the use of zeolites and similar materials (zeotypes) for that purpose.

Zirconia interest in catalysis comes altogether from its combination of acidic, basic, oxidising and reducing properties and from its resistance to corrosion and to high temperatures (Tanabe, 1985). It is especially suited for conversion of alcohols, isomerisation and methanol synthesis (Bowker et al., 1983; He and Ekerdt, 1984).

Rutile (TiO_2) has applications in catalysis, photocatalysis and microelectronics (Linsevigler et al., 1995; Haruta, 1997; Diebold, 2003). Adsorption of volatile organic compounds (VOC) on TiO_2 in view of their degradation by photocatalysis is of particular interest, together with the role – unfortunately negative – of preadsorbed water (Demeestere et al., 2003). Films of mesoporous *anatase* look promising for use as a coating of buildings in cities, to produce the photocatalytic degradation, by ambient light, of NO and also of oleic acids, which model typical urban pollutants (Kalousek et al., 2008).

The surface chemistry of *zinc oxide* is also of particular interest in relation to its catalytic and photocatalytic properties. For example, the (0001) hexagonal crystal plane appears to have a special role in the catalytic methanol-synthesis reaction (Bowker et al., 1983). The chemisorption of CO and dissociative chemisorption of H_2 occur on the exposed Zn^{2+} cations.

Iron oxide Fe_3O_4 is the basic catalyst in the Haber process for ammonia synthesis; one of its other catalytic uses is ethylbenzene dehydrogenation to produce styrene, its efficiency being enhanced by use of an appropriate support like activated carbon (Pereira Barbosa et al., 2008).

11.7.4   Applications as Pigments and Fillers

Applications as pigments or fillers require, at a stage of the process, a good interaction with the surrounding liquid medium, that is good adsorbent properties and also an optimum grain size.

Titanium oxide is the major and brightest white pigment (see Section 11.4.1). The strong photocatalytic activity of rutile (and even more of anatase) makes it necessary to coat their crystals with either silica or alumina to avoid degrading the organic components of the painting or the polymeric paste in which they are incorporated.

Although in a lower extent, *zinc oxide* is also broadly used as a white pigment.

Precipitated silica is used as a filler in polymers: for example in tyres, it can replace carbon black when white or even coloured rubber is desired.

Aerosil, is able, added in small percentage to an organic liquid, to produce a spontaneous network through the liquid (the silica particles being then inter-linked by hydrogen bonds), which drastically increases the viscosity, so that it is much used to formulate non-drip paintings and glues (after having been also used with gasoline to make napalm..).

Nanoparticles of Fe_3O_4 (*magnetite*) are synthesised in uniform size to be used as black pigment (Cornell and Schwertmann, 2007).

11.7.5   Applications in Electronics

Zinc oxide is considered to have a large potential for applications in optics, electronics and piezoelectricity (Wöll, 2007). Its surface has a bi-modal photoluminescence spectrum (one visible emission peak and another one in the UV) which is influenced by adsorption (e.g. of O_2, CO, H_2, HCOOH, CH_3OH, etc.; Idriss and Barteau, 1992).

References

Achenbach, H., 1931. Chem. Erde 6, 307.

Adamson, A.W., 1986. Textbook of Physical Chemistry. Academic Press, Orlando, p. 875.

Alario Franco, M.A., Sing, K.S.W., 1972. J. Therm. Anal. 4, 47.

Alario Franco, M.A., Sing, K.S.W., 1974. An. Quim. 70, 41.

Alario Franco, M.A., Baker, F.S., Sing, K.S.W., 1973. In: Bevan, S.C., Gregg, S.J., Parkyns, N.D. (Eds.), Progress in Vacuum Microbalance Techniques, Vol. 2. Heyden, London, p. 51.

Aldcroft, D., Bye, G.C., 1967. Science of Ceramics, Vol. 3. Academic Press, London, p. 75.

Aldcroft, D., Bye, G.C., Robinson, J.G., Sing, K.S.W., 1968. J. Appl. Chem. 18, 301.

Aldcroft, D., Bye, G.C., Chigbo, G.O., 1971. Trans. Br. Ceram. Soc. 70, 19.

Alekseevskii, E.V., 1930. J. Russ. Phys. Chem. Soc. 62, 221.

Aristov, B.G., Kiselev, A.V., 1963. Russ. J. Phys. Chem. 37, 1359.

Aristov, B.G., Kiselev, A.V., 1965. Kolloid Z 27, 299.

Audier, M., Guinot, J., Coulon, M., Bonnetain, L., 1981. Carbon 19, 99.

Avery, R.G., Ramsay, J.D.F., 1973. J. Colloid Interface Sci. 42 (3), 597.

Bakaev, V.A., Steele, W.A., 1992. Langmuir 8, 1372.

Baker, F.S., Sing, K.S.W., 1976. J. Colloid Interface Sci. 55 (3), 605.

Baker, F.S., Sing, K.S.W., Stryker, L.J., 1970. Chem. Ind., 718.

Baker, F.S., Carruthers, J.D., Day, R.E., Sing, K.S.W., Stryker, L.J., 1971. Disc. Faraday Soc. 52, 173.

Baltrusaitis, J., Schuttlefield, J., Zeitler, E., Grassian, V., 2011. Chem. Eng. J. 170, 471.

Barby, D., 1976. In: Parfitt, G.D., Sing, K.S.W. (Eds.), Characterization of Powder Surfaces. Academic Press, London, p. 353.

Barsan, N., Koziej, D., Weimar, U., 2007. Sens. Actuators B121, 18.

Barto, J., Durham, J.L., Baston, V.F., Wade, W.H., 1966. J. Colloid Interface Sci. 22, 491.

Bassett, D.R., Boucher, E.A., Zettlemoyer, A.C., 1968. J. Colloid Interface Sci. 27, 649.

Bayley, C.H., 1934. Can. J. Res. 10, 19.

Besson, R., Rocha Vargas, M., Favergeon, L., 2012. Surf. Sci. 606 (3-4), 490.

Bhagiyalakshmi, M., Lee, J.Y., Jang, H.T., 2010. Int. J. Greenhouse Gas Control 4, 51.

Bhambhani, M.R., Cutting, P.A., Sing, K.S.W., Turk, D.H., 1972. J. Colloid Interface Sci. 38, 109.

Bian, S.W., Baltrusaitis, J., Galhotra, P., Grassian, V.H., 2010. J. Mater. Chem. 20, 8705.

Blake, T.D., Wade, W.H., 1971. J. Phys. Chem. 75, 1887.

Boddenberg, B., Eltzner, K., 1991. Langmuir 7, 1498.

Bolis, V., Fubini, B., Giamello, E., 1986. Actes des XVIIèmes Journées de Calorimétrie et d'Analyse Thermique, Ferrara, p. 33.

Bolis, V., Fubini, B., Marchese, L., Martra, G., Costa, D., 1991. J. Chem. Soc. Faraday Trans. 87, 497.

Bonsack, J.P., 1973. J. Colloid Interface Sci. 44, 430.

Bordère, S., Floreancig, A., Rouquerol, F., Rouquerol, J., 1993. Solid State Ionics. 63-65, 229.

Bordère, S., Llewellyn, P., Rouquerol, F., Rouquerol, J., 1998. Langmuir, 4217.

Bordère, S., Rouquerol, F., Rouquerol, J., Estienne, J., Floreancig, A., 1990. J. Therm. Anal. 36, 1651.

Boulet, P., Knöfel, C., Kuchta, B., Hornebecq, V., Llewellyn, P., 2012. J. Mol. Model. 18, 4819.

Bowker, M., Houghton, H., Waugh, K.C., Giddings, T., Green, M., 1983. J. Catal. 84, 252.

Brunauer, S., 1945. The Adsorption of Gases and Vapours. University Press, Princeton.

Bugosh, J., Brown, R.L., McWhorter, J.R., Sears, G.W., Sippel, R.J., 1962. Ind. Eng. Chem. Prod. Res. Dev. 1, 157.

Burneau, A., Barres, O., Gallas, J.P., Lavalley, J.C., 1990. Langmuir 6, 1364.

Burwell, R.L., Littlewood, A.B., Cardew, M., Pass, G., Stoddart, C.T.H., 1960. J. Am. Chem. Soc. 82, 6272.

Bye, G.C., Howard, C.R., 1971. J. Appl. Chem. Biotechnol. 21, 324.

Bye, G.C., Robinson, J.G., 1964. Kolloid Z 198, 53.

Bye, G.C., Sing, K.S.W., 1973. In: Smith, A.L. (Ed.), Particle Growth in Suspensions. Academic Press, London, p. 29.

Bye, G.C., Robinson, J.G., Sing, K.S.W., 1967. J. Appl. Chem. 17, 138.

Carrott, P.J.M., Sing, K.S.W., 1984. Adsorption Sci. Technol. 1, 31.

Carrott, P.J.M., Sing, K.S.W., 1989. Pure Appl. Chem. 61, 1835.

Carrott, P.J.M., McLeod, A.I., Sing, K.S.W., 1982. In: Rouquerol, J., Sing, K.S.W. (Eds.), Adsorption at the Gas–Solid and Liquid–Solid Interface. Elsevier, Amsterdam, p. p. 403.

Carrott, P.J.M., Roberts, R.A., Sing, K.S.W., 1988. Langmuir 4, 740.

Carruthers, J.D., Sing, K.S.W., 1967. Chem. Ind., 1919.

Carruthers, J.D., Fenerty, J., Sing, K.S.W., 1967. Nature 213, 66.

Carruthers, J.D., Cutting, P.A., Day, R.E., Harris, M.R., Mitchell, S.A., Sing, K.S.W., 1968. Chem. Ind., 1772.

Carruthers, J.D., Fenerty, J., Sing, K.S.W., 1969. In: Mitchell, J.W. (Ed.), 6th International Symposium on Reactivity of Solids. John Wiley, New York, p. 127.

Carruthers, J.D., Payne, D.A., Sing, K.S.W., Stryker, L.J., 1971. J. Colloid Interface Sci. 36 (2), 205.

Castro, R.H.R., Quach, D.V., 2012. J. Phys. Chem. C 116, 24726.

Chauvin, C., Saussey, J., Rais, T., 1986. Appl. Catal. 25, 59.

Contaret, T., Florido, T., Seguin, J.L., Aguir, K., 2011. Procedia Eng. 25, 375.

Cornell, R.M., Schwertmann, U., 2007. The Iron Oxides: Structure, Properties, Reactions, Occurrences and Uses. Wiley-VCH, Weinheim.

Coulomb, J.P., 1991. In: Phase Transitions in Surface Films II. NATO-ASI Series B, Vol. 267. Plenum Press, Inc., New York, p. 113.

Coulomb, J.P., Vilches, O.E., 1984. J. Phys. 45, 1381.

Coulomb, J.P., Sullivan, T.S., Vilches, O.E., 1984. Phys. Rev. B. 30, 4753.

Cutting, P.A., Sing, K.S.W., 1969. Chem. Ind., 268.

Dawson, P.T., 1967. J. Phys. Chem. 71, 838.

Day, R.E., 1973. Progress in Organic Coatings 2. Elsevier Sequoia, Lausanne, p. 269.

Day, R.E., Parfitt, G.D., 1967. Trans. Faraday Soc. 63, 708.

Day, R.E., Parfitt, G.D., Peacock, J., 1971. Disc. Faraday Soc. 52, 215.

de Boer, J.H., 1957. In: Schulman, J.H. (Ed.), 2nd International Congress on Surface Activity II. Butterworths, London, p. 93.

de Boer, J.H. (Ed.), 1972. Thermochimie. Colloques Internationaux du CNRS, No. 201. CNRS, Paris, p. 407.

de Boer, J.H., Fortuin, J.M.H., Steggerda, J.J., 1954. Proc. Kon. Ned. Akad. Wetensch. 57B, 170.

de Boer, J.H., Steggerda, J.J., Zwietering, P., 1956. Proc. Kon. Ned. Akad. Wetensch. 59B, 435.

Deitz, V.R., 1944. Bibliography of Solid Adsorbents. National Bureau of Standards, Washington, p. 156.

Demeestere, K., Dewulf, J., van Langenhove, H., Sercu, B., 2003. Chem. Eng. Sci. 58, 2255.

Deren, J., Haber, J., Podgorecka, A., Burzyk, J., 1963. J. Catal. 2, 161.

Diebold, U., 2003. Surf. Sci. Rep. 48, 53.

Drain, L.E., 1954. Sci. Progr. 42, 608.

Drain, L.E., Morrison, J.A., 1952. Trans. Faraday Soc. 48, 840.

Dzhigit, O.M., Kiselev, A.V., Muttik, G.G., 1962. Kolloid Z 24, 15.

Feachem, G., Swallow, H.T.S., 1948. J. Chem. Soc. 267.

Fenelonov, V.A., Gavrilov, V.Y., Simonova, L.G., 1983. In: Poncelet, G., Grange, P., Jacobs, P.A. (Eds.), Preparation of Catalysts III. Elsevier, Amsterdam, p. 665.

Ferch, H.K., 1994. In: Bergna, H.E. (Ed.), The Colloid Chemistry of Silica. American Chemical Society, Washington, p. 481.

Fubini, B., Bolis, V., Cavenago, A., Ugliengo, P., 1992. J. Chem. Soc. Faraday Trans. 88, 277.

Fukasawa, J., Tsutsumi, H., Sato, M., Kaneko, K., 1994. Langmuir 10, 2718.

Furlong, D.N., Rouquerol, F., Rouquerol, J., Sing, K.S.W., 1980a. J. Chem. Soc. Faraday I 76, 774.

Furlong, D.N., Rouquerol, F., Rouquerol, J., Sing, K.S.W., 1980b. J. Colloid Interface Sci. 75, 68.

Furlong, D.N., Sing, K.S.W., Parfitt, G.D., 1986. Adsorption Sci. Technol. 3, 25.

Gallas, J.P., Lavalley, J.C., Burneau, A., Barres, O., 1991. Langmuir 7, 1235.

Gay, J.M., Suzanne, J., Coulomb, J.P., 1990. Phys. Rev. B 41, 11346.

Gimblett, F.G.R., Rahman, A.A., Sing, K.S.W., 1980. J. Chem. Technol. Biotechnol. 30, 51.

Gimblett, F.G.R., Rahman, A.A., Sing, K.S.W., 1981. J. Colloid Interface Sci. 84, 337.

Gimblett, F.G.R., Rahman, A.A., Sing, K.S.W., 1984. J. Colloid Interface Sci. 102, 483.

Goodman, J.F., Gregg, S.J., 1959. J. Chem. Soc. 3612.

Gregg, S.J., Langford, J.F., 1969. Trans. Faraday Soc. 65, 1394.

Gregg, S.J., Langford, J.F., 1977. J. Chem. Soc. Faraday Trans. I 73, 747.

Gregg, S.J., Packer, R.K., 1955. J. Chem. Soc., 51.

Gregg, S.J., Sing, K.S.W., 1951. J. Phys. Colloid Chem. 55, 592.

Gregg, S.J., Sing, K.S.W., 1982. Adsorption, Surface Area and Porosity. Academic Press, London, p. 74.

Griffiths, D.M., Rochester, C.H., 1977. J. Chem. Soc. Faraday Trans. I 73, 1510.

Grillet, Y., Rouquerol, F., Rouquerol, J., 1985. Surf. Sci. 162, 478.

Grillet, Y., Rouquerol, F., Rouquerol, J., 1989. Thermochim. Acta. 148, 191.

Harkins, W.D., Jura, G., 1944. J. Am. Chem. Soc. 66, 1362.

Harris, M.R., Whitaker, G., 1962. J. Appl. Chem. 12, 490.

Harris, M.R., Whitaker, G., 1963. J. Appl. Chem. 13, 198.

Haruta, M., 1997. Catal. Today. 36, 153.

He, M.Y., Ekerdt, J.C., 1984. J. Catal. 90, 17.

Henrich, V.E., 1976. Surf. Sci. 57, 355.

Holmes, H.F., Fuller, E.L., Gammage, R.B., 1972. J. Phys. Chem. 76, 1497.

Hornebecq, V., Knöfel, C., Boulet, P., Kuchta, B., Llewellyn, P., 2011. J. Phys. Chem. C 115, 10097.

Hudson, M.J., Knowles, J.A., 1996. J. Mater. Chem. 6, 89.

Idriss, H., Barteau, M.A., 1992. J. Phys. Chem. 96, 3382.

Iler, R.K., 1979. The Chemistry of Silica. John Wiley, New York.

Jaycock, M.J., Waldsax, J.C.R., 1974. J. Chem. Soc. Faraday Trans. I 70, 1501.

Jeon, H., Min, Y.J., Ahn, S.H., Hong, S.M., Shin, J.S., Kim, J.H., Lee, K.B., 2012. Colloids Surf. A Physicochem. Eng. Aspects 414, 75.

Jin, L., Auerbach, S.M., Monson, P.A., 2011. J. Chem. Phys. 134, 134703.

Kaganer, M.G., 1961. Dokl. Akad. Nauk SSSR 138, 405.

Kalousek, V., Rathousky, J., Tschirch, J., Bahnemann, D., 2008. In: Sayari, A., Jaroniec, M. (Eds.), Nanoporous Materials. Proceedings of the 5th International Symposium. World Scientific, New Jersey, p. 553.

Kenny, M.B., Sing, K.S.W., 1994. In: Bergna, H.E. (Ed.), The Colloid Chemistry of Silica. American Chemical Society, Washington, p. 505.

Kington, G.L., Beebe, R.A., Polley, M.H., Smith, W.R., 1950. J. Am. Chem. Soc. 72, 1775.

Kiselev, A.V., 1957. In: Schulman, J. (Ed.), Second International Congress Surface Activity. Butterworths, London, p. 229.

Kiselev, A.V., 1958. In: Everett, D.H., Stone, F.S. (Eds.), Structure and Properties of Porous Materials. Butterworths, London, p. 195.

Kiselev, A.V., 1965. Disc. Faraday Soc. 40, 205.

Kiselev, A.V., 1971. Disc. Faraday Soc. 52, 14.

Knöfel, C., Hornebecq, V., Llewellyn, P., 2008. Langmuir 24, 7963.

Korotcenko, G., 2007. Mater. Sci. Eng. B 139, 1.

Krieger, K.A., 1941. J. Am. Chem. Soc. 63, 2712.

Linsevigler, A.L., Lu, G., Yates Jr., J.T., 1995. Chem. Rev. 95, 735.

Lippens, B.C., 1961. Thesis. University of Delft.

Lippens, B.C., Steggerda, J.J., 1970. In: Linsen, B.G. (Ed.), Physical and Chemical Aspects of Adsorbents and Catalysts. Academic Press, London, p. 171.

Llewellyn, P., Rouquerol, F., Rouquerol, J., 2003. In: Toft Sörensen, O., Rouquerol, J. (Eds.), Sample Controlled Thermal Analysis. Kluwer Academic Publishers, Dordrecht, Boston, London, p. 135 (Chapter 6).

Madeley, J.D., Sing, K.S.W., 1953. J. Appl. Chem. 3, 549.

Madeley, J.D., Sing, K.S.W., 1954. J. Appl. Chem. 4, 365.

Madeley, J.D., Sing, K.S.W., 1962. J. Appl. Chem. 12, 494.

Madih, K., Croset, B., Coulomb, J.P., Lauter, H.J., 1989. Europhys. Lett. 8, 459.

Mathieu, Y., Tzanis, L., Soulard, M., Patarin, J., Vierling, M., Molière, M., 2013. Fuel Proc. Technol. 114, 81.

Matijevic, E., 1988. Pure Appl. Chem. 60, 1479.

Mikhail, R.S., Nashed, S., Kahlil, A.M., 1971. Disc. Faraday Soc. 52, 187.

Mitchell, S.A., 1966. Chem. Ind. 924.

Mohammadi, M.R., Fray, D.J., Ghorbani, M., 2008. Solid State Sci. 10, 884.

Morishige, K., Kanno, F., Ogawara, S., Sasaki, S., 1985. J. Phys. Chem. 89, 4404.

Munuera, G., Stone, F.S., 1971. Disc. Faraday Soc. 52, 205.

Naono, H., Nakai, K., 1989. J. Colloid Interface Sci. 128, 146.

Naono, H., Nakai, K., Sueyoshi, T., Yagi, H., 1987. J. Colloid Interface Sci. 120 (2), 439.

Naono, H., Sonoda, J., Oka, K., Hakuman, M., 1993. In: Suzuki, M. (Ed.), Fundamentals of Adsorption IV. Kodansha, Tokyo, p. 467.

Neimark, I.E., Sheinfain, R.Y., Lipkind, B.A., Stas, O.P., 1964. Kolloid Z 26, 734.

Okkerse, C., 1970. In: Linsen, B.G. (Ed.), Physical and Chemical Aspects of Adsorbents and Catalysts. Academic Press, London, p. 213.

Papée, D., Tertian, R., 1955. Bull. Soc. Chim. France. 983.

Papée, D., Charrier, J., Tertian, R., Houssemaine, R., 1954. In: Proceedings of the Congrès de l'Aluminium, p 31.

Parfitt, G.D., 1981. Dispersion of Powders in Liquids. Applied Science, London, p. 1.

Parkyns, N.D., Sing, K.S.W., 1975. Specialist Periodical Report Colloid Science 2. Chemical Society, London, p. 1.

Pashley, R.M., Kitchener, J.A., 1979. J. Colloid Interface Sci. 71, 491.

Patterson, R.E., 1994. In: Bergna, H.E. (Ed.), The Colloidal Chemistry of Silica. American Chemical Society, p. 617.

Payne, D.A., Sing, K.S.W., 1969. Chem. Ind., 918.

Payne, D.A., Sing, K.S.W., Turk, D.H., 1973. J. Colloid Interface Sci. 43 (2), 287.

Pereira Barbosa, D., Do Carmo Rangel, M., Rabelo, D., 2008. In: Sayari, A., Jaroniec, M. (Eds.), Nanoporous Materials. Proceedings of the 5th International Symposium. World Scientific, New Jersey, p. 607.

Peri, J.B., 1965. J. Phys. Chem. 69, 211.

Phalippou, J., Despetis, F., Calas, S., Faivre, A.L., Dieudonné, P., Sempéré, R., Woignier, T., 2004. Opt. Mater. 26 (2), 167.

Polychronopoulou, K., Fierro, J.L.G., Efstathiou, A.M., 2005. Appl. Cat. B Environ. 57, 125.

Ragai, J., 1989. Adsorption Sci. Technol. 6, 9.

Ragai, J., Sing, K.S.W., 1982. J. Chem. Technol. Biotechnol. 32, 988.

Ragai, J., Sing, K.S.W., 1984. J. Colloid Interface Sci. 101 (2), 369.

Ragai, J., Sing, K.S.W., Mikhail, R., 1980. J. Chem. Technol. Biotechnol. 30, 1.

Ragai, J., Selim, S., Sing, K.S.W., Theocharis, C., 1994. In: Rouquerol, J., Rodriguez-Reinoso, F., Sing, K.S.W., Unger, K.K. (Eds.), Characterization of Porous Solids III. Elsevier, Amsterdam, p. 487.

Ramsay, J.D.F., Avery, R.G., 1979. In: Gregg, S.J., Sing, K.S.W., Stoeckli, H.F. (Eds.), Characterization of Porous Solids. Society of Chemical Industry, London, p. 117.

Ribeiro Carrott, M., Carrott, P., Brotas de Carvalho, M.M., Sing, K.S.W., 1991a. J. Chem. Soc. Faraday Trans. 87 (1), 185.

Ribeiro Carrott, M., Carrott, P.J.M., Brotas de Carvalho, M.M., Sing, K.S.W., 1991b. In: Rodriguez-Reinoso, F., Rouquerol, J., Sing, K.S.W., Unger, K.K. (Eds.), Characterization of Porous Solids II (1961). Elsevier, Amsterdam, p. 635.

Ribeiro, Carrott M., Carrott, P., Brotas de Carvalho, M.M., Sing, K.S.W., 1993. J. Chem. Soc. Faraday Trans. 89, 579.

Rijnten, H.T., 1970. In: Linsen, B.G. (Ed.), Physical and Chemical Aspects of Adsorbents and Catalysts. Academic Press, London, p. 315.

Rittner, F., Paschek, D., Boddenberg, B., 1995. Langmuir 11, 3097.

Rochester, C.H., 1986. Colloids Surf. 21, 205.

Rouquerol, F., 1965. Thesis. Paris-Sorbonne University.

Rouquerol, J., Ganteaume, M., 1977. J. Therm. Anal. 11, 201.

Rouquerol, F., Rouquerol, J., Imelik, B., 1970. Bull. Soc. Chim. France 10, 3816.

Rouquerol, J., Rouquerol, F., Ganteaume, M., 1975. J. Catal. 36, 99.

Rouquerol, J., Rouquerol, F., Ganteaume, M., 1979a. J. Catal. 57, 222.

Rouquerol, J., Rouquerol, F., Peres, C., Grillet, Y., Boudellal, M., 1979b. Gregg, S.J., Sing, K.S.W., Stoeckli, H.F. (Eds.), Characterization of Porous Solids. Society of Chemical Industry, London, p. 107.

Rouquerol, J., Rouquerol, F., Grillet, Y., Torralvo, M.J., 1984. Fundamentals of Adsorption. In: Myers, A., Belfort, G. (Eds.), Proceedings of 1st Conference on Fundamentals of Adsorption, Schloss Elmau, Bavaria, Germany. Engineering Foundation, New York, p. 501.

Rouquerol, F., Rouquerol, J., Imelik, B., 1985. In: Haynes, J.M., Rossi-Doria, P. (Eds.), Principles and Applications of Pore Structural Characterization. Bristol, Arrowsmith, p. 213.

Rouquerol, J., Llewellyn, P., Rouquerol, F., 2007. In: Llewellyn, P., Rodriguez-Reinoso, F., Rouquerol, J., Seaton, N. (Eds.), Characterization of Porous Solids VII, Studies in Surface Science and Catalysis, Vol. 160. Elsevier, Amsterdam/Oxford, p. 49.

Saafeld, H., 1960. Neues Jahrb. Miner. Abh. 95, 1.

Saha, D., Deng, S., 2010. J. Chem. Eng. Data 55, 5587.

Sing, K.S.W., 1970. In: Everett, D.H., Ottewill, R.H. (Eds.), Surface Area Determination. Butterworths, London, p. 25.

Sing, K.S.W., 1972. In: Thermochimie. Colloques Internationaux No. 201. CNRS, Paris, p. 601.

Solomon, D.H., Hawthorne, D.G., 1983. Chemistry of Pigments and Fillers. John Wiley, New York, p. 51.

Srivastava, R.K., Lal, P., Dwivedi, R., Srivastava, S.K., 1994. Sens. Actuators B 21, 213.

Stacey, M.H., 1987. Langmuir 3, 681.

Stacey, M.H., 1991. In: Rodriguez-Reinoso, F., Rouquerol, J., Sing, K.S.W., Unger, K.K. (Eds.), Characterization of Porous Solids II. Elsevier, Amsterdam, p. 615.

Tanabe, K., 1985. Mater. Chem. Phys. 13, 347.

Taylor, R.J., 1949. J. Soc. Chem. Ind. 68, 23.

Teichner, S.J., 1986. In: Fricke, J. (Ed.), Aerogels. Springer-Verlag, Berlin, p. 22.

Teichner, S.J., Nicolaon, G.A., Vicarini, M.A., Gardes, G.E.E., 1976. Adv. Colloid Interf. Sci. 5, 245.

Tertian, R., Papée, D., 1953. Comp. Rend. Acad. Sci. 236, 1565.

Trabelsi, M., Coulomb, J.P., 1992. Surf. Sci. 272, 352.

Tseng, H.H., Wey, M.Y., 2004. Carbon 42, 2269.

Unger, K.K., 1979. Porous Silica. Elsevier, Amsterdam.

Vleesschauwer, W.F.M., 1970. In: Linsen, B.G. (Ed.), Physical and Chemical Aspects of Adsorbents and Catalysts. Academic Press, London, p. 265.

Wiseman, T.J., 1976. In: Parfitt, G.D., Sing, K.S.W. (Eds.), Characterization of Powder Surfaces. Academic Press, London, p. 159.

Wöll, C., 2007. Prog. Surf. Sci. 82, 55.

Wong, W.K., 1982. PhD Thesis. Brunel University, UK.

Zettlemoyer, A.C., 1968. J. Colloid Interface Sci. 28, 343.

Zhao, Z., Dai, H., Du, Y., Deng, J., Zhang, L., Shi, F., 2011. Mater. Chem. Phys. 128, 348.

Zhuravlev, L.T., 1987. Langmuir 3, 316.

Zhuravlev, L.T., 1994. In: Bergna, H.E. (Ed.), The Colloid Chemistry of Silica. American Chemical Society, Washington, p. 629.

Zhuravlev, L.T., Kiselev, A.V., 1970. In: Everett, D.H., Ottewill, R.H. (Eds.), Surface Area Determination. Butterworths, London, p. 155.

12 Adsorption by Clays, Pillared Clays, Zeolites and Aluminophosphates

Jean Rouquerol, Philip Llewellyn, Kenneth Sing

Aix Marseille University-CNRS, MADIREL Laboratory, Marseille, France

---

Chapter Contents

Adsorption by Powders and Porous Solids. http://dx.doi.org/10.1016/B978-0-08-097035-6.00012-7

12.1   Introduction

It is well known that natural clays are the products of the weathering of rocks and are widely distributed. Their overall chemical compositions and textures vary from one location to another, being dependent on their geological origin and the presence of organic and inorganic impurities. Experts in the field like to reserve the term 'clay' to a naturally occurring material (e.g. Kaolin, Ball clay, Fuller's earth, China Clay, etc.) composed of minerals finer than ca. 4 μm, plastic when hydrated and able to harden on drying (Bergaya and Lagaly, 2006). The 'clay minerals' are their well-defined mineralogical constituents (e.g. Kaolinite, Montmorillonite, Sepiolite, etc.). Nevertheless, for the sake of simplicity, the latter are often called 'clays' and this is what will be done in the rest of this chapter which essentially deals, strictly speaking, with clay minerals.

Clay minerals are hydrous layer silicates of colloidal dimensions, with most if not all of the individual platy particles in the colloidal range of ca. 1 nm–1 μm (van Olphen, 1976; Van Damme et al., 1985). The term 'phyllosilicate' (phyllo = leaf like) is applied to the broad group of hydrous silicates with layer structures. The essential components of the phyllosilicate structure are 2D tetrahedra and octahedra of oxygen atoms (or ions). The coordinating atoms (or cations) in the centre of the tetrahedra are for the most part Si, but Al^{3+} or Fe^{3+} may also be present. The coordinating cations in the octahedra are usually Al^{3+}, Mg^{2+}, Fe^{3+} or Fe^{2+}. Some clay structures (e.g. hectorite) can be synthesised in a reproducible and relatively homogeneous form.

Certain properties of clays were known and exploited in ancient times: in particular, clays were used for the fabrication of pottery, bricks and tiles. The chief constituent of china clay (or kaolin) is kaolinite, which is still used on a very large scale in the manufacture of paper and refractories. Ball clay, a fine-grained and highly plastic form of kaolinite, contains some mica and quartz and is now favoured for crockery, porcelain and floor tiles.

The swelling and thixotropic properties of the smectic clays have long been known and are of great importance in agriculture and civil engineering. Fuller's earth (mainly calcium montmorillonite) has high adsorbent and cation exchange properties, while bentonite (sodium montmorillonite) is extensively used as a constituent of drilling mud, mortar and putty—to provide the required degree of plasticity. The acidic nature of activated bentonite was exploited in an early catalytic cracking process (the Houdry process) for the production of gasoline from high molar mass oils.

More recently, various attempts have been made to develop cracking catalysts from pillared smectite clays, in which the layers are separated and held apart by the intercalation of large cations. *Pillared clays* (PILCs) have large surface areas within fairly well-ordered micropore structures (pore widths in the approximate range 0.6–1.2 nm). It is not surprising that these materials have attracted considerable interest with the prospect of an alternative type of catalytic shape selectivity (Thomas, 1994; Fripiat, 1997; Thomas et al., 1997).

Zeolites have a connection with clays, being also silicates and generally sharing with them a large water adsorption capacity, an open structure and the presence of exchangeable cations; moreover, the natural aluminosilicate zeolites often occur together with the clays (Bish, 2006). Zeolites may be regarded as the most important and well-established members of a special class of microporous adsorbents in which the porosity is intracrystalline. Although zeolites have been known for over 200 years, their potential value as highly selective adsorbents was first realised about 50 years ago (Barrer, 1945, 1966, 1978). Interest was further stimulated by the announcement by Breck et al. (1956) of the synthesis of the hitherto unknown zeolite A (i.e. Linde sieve A). Since then, more than 200 new porous zeolites have been synthesised.

The molecular sieve zeolites have attained great technological importance for catalysis, gas separation and drying and many other applications. They are now used as industrial catalysts for such reactions as the cracking of paraffins and the isomer-isation and disproportionation of aromatic compounds (Thomas and Theocharis, 1989; Thomas, 1995; Martens et al., 1997).

Phosphate-based molecular sieves started to be synthetized in 1982 and more than 60 structures of them are now listed by the International Zeolite (IZA) Association Structure Commission (see http://www.iza-structure.org/databases/). Their chemical composition is different from that of a zeolite, since the SiO_4 tetrahedra are replaced by PO_4 tetrahedra, but their structure, properties and applications are comparable to those of zeolites. Substituting further can lead to crystalline zeolite-like structures with extremely large pores with rings of 24 and 28 to 40, 48, 56, 64, and 72 T atoms able to be prepared in the case of gallium zincophosphites (Lin et al., 2013). Substitut-ing germanium into the structure of zeolite-like materials can also lead to large-pore structures as well as hierarchical micro-mesoporous solids (Jiang et al., 2011).

In a single chapter it would be impossible to consider all aspects of the physisorp-tion properties of clays, zeolites and phosphate-based molecular sieves. As in other Chapters from 10 to 14, our aim here is to apply and discuss the significance of the general principles propounded in Chapters 4–9. For this purpose, it is expedient to focus attention on particular systems (e.g. industrially important synthetic zeolites). However, since the adsorbent behaviour of clays, zeolites and phosphate-based molecular sieves is to a large extent dependent on their solid structures, some attention is given in this chapter to structural chemistry.

12.2 Structure, Morphology and Adsorbent Properties of Layer Silicates

12.2.1 *Structure and Morphology of Layer Silicates*

As already indicated, a feature common to all clay minerals is the 2D polymeric sheet which is composed of interlinked SiO_4 tetrahedra. These siloxane, O–Si–O, hexagonal (or 'tetrahedral') sheets are constructed from three of the four available oxygen atoms at the corners of the tetrahedra. There remain the spare apical oxygens, which are directed away (upwards or downwards) from each sheet.

The other main component is an 'octahedral sheet': this is made up of oxygen and metal atoms, which are typically aluminium or magnesium (since Al is present in the majority of clay minerals, they are sometimes referred to as aluminosilicates). The junction plane between the hexagonal and octahedral sheets consists of the shared apical oxygens of the tetrahedra and also some hydroxyl groups. In one group of clay minerals, one octahedral sheet is directly attached to one silica sheet, thus giving a 1:1 type of two-sheet elemental structure. In another main group of clays, one octahedral sheet is sandwiched between two silica sheets, giving a 2:1 type of three-sheet layer structure.

Individual particles of clay minerals consist of stacks of layers, which in some systems are separated by interlayer materials. The individual layers are held together by secondary forces (i.e. van der Waals attractive forces, hydrogen bonding or weak electrostatic attraction).

12.2.1.1   Kaolinite

The best known example of a 1:1 (two-sheet) type clay is kaolinite. A pictorial impression of the ideal kaolinite structure is given in Figure 12.1. The upper and lower basal surfaces of the two-sheet layer are clearly quite different. The layer repeat distance, or c-spacing, is ca. 0.72 nm, which is the distance between atom centres in two

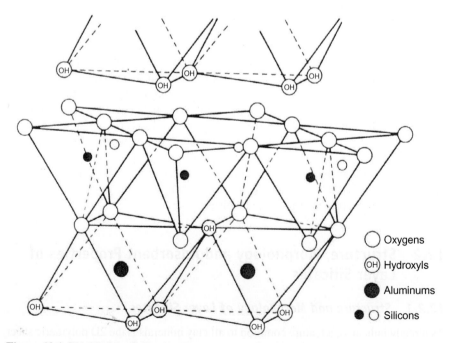

Figure 12.1 The ideal kaolinite structure.

contiguous layers. In fact, this is approximately the same as the sum of the atomic radii and therefore in an ideal structure there is insufficient space to accommodate any interlayer material such as intercalated water.

In the perfect two-layer crystal of kaolinite the composition of the unit cell is $[Al_2(OH)_4(Si_2O_5)]_2$, but in most kaolitic clays there are defects due in part to isomorphous substitution of Si by Al and other small atoms. The excess negative charge is compensated by cations located on the outer surface of the crystallites.

12.2.1.2   Smectites and Vermiculites

The 2:1 layer clays (i.e. with three-sheet layers) include a group of expanding or swelling clays, which comprise the smectites (e.g. montmorillonite, saponite and hectorite) and the vermiculites. The basic structure of a smectite is shown in Figure 12.2.

Pyrophyllite is the simplest layer aluminosilicate in which two tetrahedral SiO_4 layers are condensed onto the octahedral AlO_6 layer to produce a three-sheet layer, the composition of the unit cell being $[Al_2(OH)_2(Si_2O_5)_2]_2$. Another 'ideal' structure in which Al is replaced by Mg is that of talc. In both cases the three-sheet layer is electrically neutral and the layers are stacked in the ABAB $\cdots$ sequence. Because of the cohesive strength of this ideal structure, neither pyrophyllite nor talc occurs in the form of the very fine particles which generally characterise clay minerals.

An important feature of the smectites, vermiculites and other 2:1 layer silicates is that isomorphous substitutions can occur in both the tetrahedral and octahedral sheets. Thus, substitution of Si by Al occurs in the tetrahedral sheet together with replacement of Al by Mg, Fe, Li or other small atoms in the octahedral sheet. The substitutions lead to a deficit of positive charge, which is compensated by the presence of exchangeable, interlayer cations.

Because of the presence of the cations, the c-spacing in the smectites and vermiculites is somewhat larger than in the uncharged pyrophyllite (ca. 0.92 nm). Water molecules are able to penetrate between the layers, causing an expansion of the layer lattice—as indicated by an increase in the c-spacing. With some smectites, the expansion appears to occur in discrete steps corresponding to the formation of one to four layers of water between the lattice layers (van Olphen, 1976).

12.2.1.3   Palygorskites

Attapulgite and sepiolite are members of the palygorskite group of fibrous clay minerals. As before, the SiO_4 tetrahedra are linked together to form the polymeric silica layer, but now the vertices do not all point in the same direction. Instead, they are arranged in strips: in one strip all the vertices point up and in the next strip, they all point down. In attapulgite the strip width is of four tetrahedra and in sepiolite it is of six tetrahedra. The MgO_6 octahedra are arranged to give threefold strips leaving channels parallel to the fibre axis containing water molecules (Barrer, 1978).

The idealised attapulgite composition for a half unit cell is $Mg_5Si_8O_{20}(OH)_2$ $(H_2O)_4, 4H_2O$. Thus, four H_2O molecules are present in the channels (i.e. the 'zeolitic

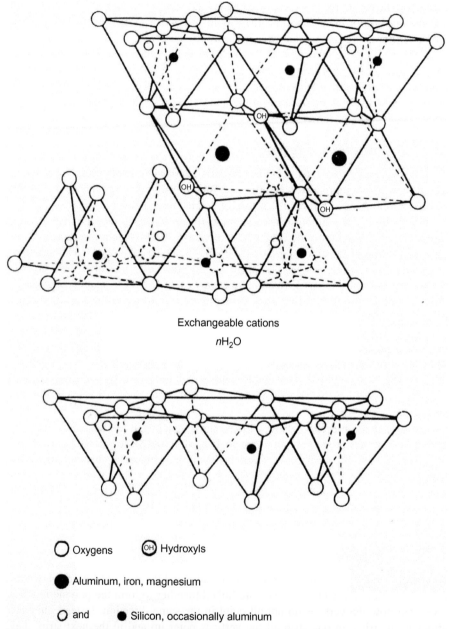

Exchangeable cations

$n\mathrm{H_2O}$

○ Oxygens   ⊙ Hydroxyls

● Aluminum, iron, magnesium

○ and   ● Silicon, occasionally aluminum

Figure 12.2 The ideal smectite structure.

water') and four others are bound to the octahedral cations. On heating, atmosphere water is lost in three stages: (a) zeolitic water and water adsorbed on the external surface are removed at temperatures <75 °C; (b) the removal of the coordinated water over the range 75–370 °C; (c) the final removal of structural water. An irreversible collapse of the structure begins to occur at temperatures of ca. 130 °C, corresponding to the loss of about half of the coordinated water (Grillet et al., 1988; Cases et al., 1991). The detailed study of these changes was made possible by the application of Controlled Rate Thermal Analysis (see Section 3.4.3) whose specific interest for clays is detailed by Rouquerol et al. (2006).

12.2.1.4 Morphology of Clay Particles and Aggregates

Kaolinite particles (platelets) are relatively thick and rigid, usually containing 100 or more stacked layers. The platy particles tend to be of hexagonal shape with diameters of up to 1 μm The crystal shape is dependent on the basal (001) face and the prismatic edges: (110), etc. The interlayer attraction (mainly hydrogen bonding) is sufficiently strong to prevent cleavage under ordinary conditions. However, the crystallites do exhibit stacking faults. There appears to be an inverse relation between the number of structural defects and the particle size (Cases et al., 1982). The BET-nitrogen areas of refined natural kaolinites are generally in the range $10–20 \text{ m}^2 \text{ g}^{-1}$ (Gregg et al., 1954; Cases et al., 1986). Comminution occurs under continuous grinding, although the extent of breakdown has been found to reach a maximum after a certain period of time (Gregg, 1968). For example, by the prolonged or repeated grinding of certain china clays, it is possible to reach maximum areas of ca. $50 \text{ m}^2\text{g}^{-1}$.

Smectite platelets (tactoids) are thin and flexible: the diameter is relatively large (up to 2 μm), but the individual platelet width is much smaller (e.g. 1 nm). It is evidently quite difficult to specify an unambiguous value of particle size. The BET-nitrogen areas of the *montmorillonites* are often in the region of $30 \text{ m}^2\text{g}^{-1}$, while their particle sizes may appear to be 1 μm. In the case of *laponite*, the BET-nitrogen area may be as high as $300 \text{ m}^2\text{g}^{-1}$ (Bergaya, 1995). *Sepiolites* have a fibrous morphology. Typical fibres have a length of 2–3 μm and a diameter of 0.1 μm. The more crystalline forms of sepiolite have a more rigid needle-like appearance.

In dilute suspensions clays tend to form gels. The classical model is the 'house of cards' structure of kaolinite in which the face-to-edge association leads to an open 3D structure (van Olphen, 1965). In the case of smectite-water systems it now seems more likely that the microstructure is mainly controlled by the face-to-face interactions (Van Damme et al., 1985).

12.2.2 Physisorption of Gases by Layer Silicates

12.2.2.1 Physisorption by Kaolinite

12.2.2.1.1 Nitrogen Isotherms

In the uncompacted state, a sample of natural kaolinite was found to give the reversible Type II nitrogen isotherm shown in Figure 12.3 (Gregg, 1968). It is apparent that the initial part of the isotherm (up to $p/p° = 0.4$) was not changed to any detectable extent by the compaction of the kaolinite at a pressure of 1460 MPa, but a narrow hysteresis loop appeared at higher relative pressure. The adsorption isotherms in Figure 12.3

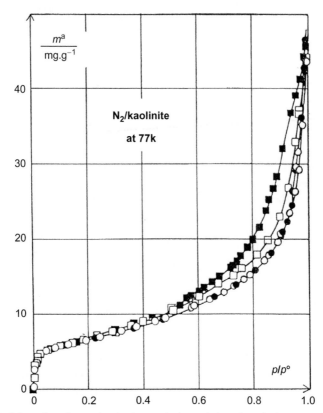

Figure 12.3 Adsorption–desorption isotherms (open and closed symbols, respectively) of N_2 at 77 K on kaolinite, before (circles) and after (squares) compaction at 1460 MPa. After Gregg (1968).

are of interest for several reasons. First, it may seem surprising that an assemblage of kaolinite platelets should give a reversible isotherm. The adsorbent had a BET-nitrogen-specific surface area of 17 m^2 g^{-1}, which would appear to correspond to a platelet thickness of ca. 50 nm. The particle rigidity and the house-of-cards packing have probably resulted in the formation of a macroporous aggregate which accounts for the appearance of the reversible Type II(a) isotherm (open and closed circles).

It is striking that the high compaction pressure, which was sufficient to convert assemblages of spheroidal particles into well-defined mesopore structures (Gregg, 1968), had a moderate effect on the course of the *adsorption* isotherm (open squares). In contrast, the *desorption* curve (closed squares) was displaced in the multilayer range, then producing an hysteresis characteristic of a type II(b) isotherm, with a weak aspect of type IV(a) (upper slightly convex part of the desorption branch). We conclude that this hysteresis loop is associated with both the capillary condensation in a non-rigid system of platelets (since compaction has rendered the kaolinite platelets closer and more parallel to each other) and some development of a pore network.

In an early study of the effect of calcination on the surface area of kaolinite, Gregg and Stephens (1953) found a small but progressive decline in the BET-area over the temperature range 100–800 °C. These results were in contrast to a 12% loss of structural water at 450 °C, which could have been expected to break the structure and activate the solid: apparently, this was not the case.

12.2.2.1.2   Energetics of Argon and Nitrogen Adsorption

Cases et al. (1986) have used adsorption microcalorimetry of N_2 and Ar at 77 K along with other techniques in a study of the crystallographic and morphological properties of kaolinite. The differential adsorption energy curves in Figure 12.4 were determined on two different forms of kaolinite: sample GB3 was a well-ordered English china

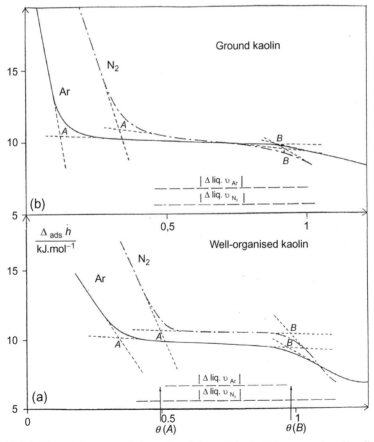

Figure 12.4 Differential energy of adsorption of Ar and N_2 at 77 K on samples of kaolinite as a function of surface coverage. (a) On sample GB3 and (b) on sample FU7.
Courtesy Cases et al. (1986).

Table 12.1 Adsorption of Ar and N_2 on Samples of Kaolinite at 77 K

Adsorbent	Adsorptive	$\dfrac{a(\text{BET})}{\text{m}^2\ \text{g}^{-1}}$	θ_A	θ_B	$100\ (\theta_B - \theta_A)$
GB3	Ar	11.6	0.34	0.97	63
	N_2	11.4	0.49	0.98	
FU7	Ar	47.3	0.12	0.91	79
	N_2	46.8	0.33	0.92	

clay; whereas sample FU7 was a French kaolinite subjected to repeated dry grinding and fractionation. The BET-areas, as also determined by argon and nitrogen adsorption at 77 K, are given in Table 12.1.

For our present purpose, the results in Figure 12.4 and Table 12.1 can be summarised as follows:

1. The corresponding BET-areas derived from the argon and nitrogen isotherms are in fairly good agreement (as also found with other samples of kaolinite).
2. Each adsorption energy curve can be divided into three distinctive parts: (a) an initial steep decrease; (b) a long middle decline, AB; (c) a multilayer declination.
3. Values of BET-coverage, θ_A and θ_B, corresponding to the locations A and B are given respectively in columns 4 and 5 of the table.

We are now in a position to discuss the significance of these findings. The strong energetic heterogeneity at low coverage is probably due to adsorption on the *edge sites* (lateral faces) and *defect sites* (e.g. crevices) in the basal plane. In view of its specificity of interaction, it is not surprising to find that the nitrogen energies are appreciably greater than the corresponding argon values. The much smaller variation of adsorption energy in the middle region (along AB) is consistent with the energetic homogeneity of the (001) *basal faces*. It is of interest that the constant adsorption energy for argon on ground kaolin (FU7) extends over 79% of the surface coverage, whereas the comparable surface coverage is about 63% in the case of well-organised kaolin (GB3). These results appear to confirm that grinding has led to an appreciable increase in the fractional contribution of the basal faces to the total surface area.

Cases et al. (1986) found that the percentage areas corresponding to the lateral domains (i.e. the high-energy edge sites) evaluated from the differential energies of argon adsorption agreed quite well with the corresponding values obtained from the adsorption isotherms of alkyldodecylammonium ions. These results illustrate the value of adsorption microcalorimetry for the characterisation of clay minerals.

12.2.2.2 *Physisorption by Smectites and Vermiculites*

Fuller's earth (whose main constituent is usually montmorillonite) was traditionally used for the removal of grease from cloth and as a 'bleaching' or decolorizing agent. It has long been known that these sorption properties, although essentially physical in nature, are mainly dependent on the smectite structure. A number of attempts were made in the 1930s to explore the surface properties of natural and acid-activated bleaching clays, but the first important advances were made by Barrer and his co-workers in the 1950s.

12.2.2.2.1 Adsorption of Non-polar Molecules

The physisorption measurements by Barrer and McLeod (1954) were undertaken on natural *montmorillonite*. The isotherms for the relatively non-polar molecules oxygen, nitrogen and benzene in Figure 12.5 are very similar in form to those reported subsequently (e.g. by Cases et al., 1992). It is evident that the hysteresis loops are of Type H3 in the IUPAC classification (see Section 8.6), with no indication of a plateau at high $p/p°$ and therefore they should *not* be regarded as Type IV isotherms. Moreover, each adsorption branch appears to have the same typical Type II(a) character as the nitrogen isotherms on kaolinite in Figure 12.3. The desorption branch follows a different path until a critical $p/p°$ is reached.

The isotherms in Figure 12.5 are indeed clearly Type II(b) after the classification of Figure 1.2. Such isotherms are given by either non-rigid slit-shaped pores or, as in the present case, non-rigid assemblages of platy particles (see Rouquerol et al., 1970, 1985). The fact that the montmorillonite particles are thin and flexible may be responsible for the closer proximity of the basal faces than in uncompacted kaolinite.

The properties of a well-characterised sample of sodium montmorillonite were investigated in some detail by Cases et al. (1992). As expected, the nitrogen isotherm

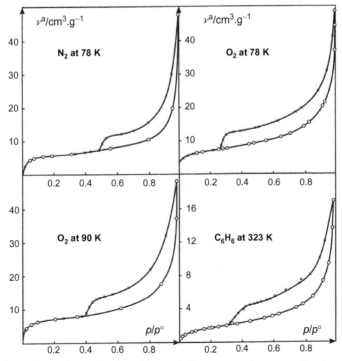

Figure 12.5 Isotherms for molecules of low polarity on natural montmorillonite. From Barrer (1989).

at 77 K had again a well-defined H3 hysteresis loop and was therefore a good example of a Type II(b) isotherm. The measurements were taken to a high $p/p°$, which confirmed that there was no plateau and therefore no indication of the completion of mesopore filling. Thus, in our view the isotherm was mistakenly referred to a Type IV isotherm, at a time when the meaning of H3 hysteresis loop and Type II(b) isotherm where not yet generally recognised.

The nitrogen isotherm was replotted by Cases et al. (1992) in the usual BET coordinates and as a t-plot. The derived BET-area of 43.3 $m^2 g^{-1}$ appeared to be not far removed from the value of 45.9 $m^2 g^{-1}$ obtained from the amount adsorbed at Point B. The t-plot was constructed in the manner originally proposed by de Boer et al. (1966), which involved adopting a standard isotherm with the same value of C, which in this case was 485. It was not easy to interpret the t-plot, although three short linear sections were identified. From the initial slope, the total surface area appeared to be ca. 50 $m^2 g^{-1}$. Back-extrapolation of a linear region at higher $p/p°$ gave an apparent micropore volume of ca. 0.01 $cm^3 g^{-1}$. The existence of this small micropore filling contribution is consistent with the high BET C value.

One could think of a more realistic quantitative assessment of the microporosity by comparison with nitrogen isotherm data on a truly nonporous form of Na-montmorillonite. In practice, however, this may be difficult to accomplish and a more pragmatic approach would be to construct a series of comparison plots for the adsorption of N_2 (and preferably also Ar) on pairs of samples of differing particle sizes and defect structures. In this way it should be possible to establish quantitative differences in the micropore capacities.

In another investigation, nitrogen isotherms were determined on samples of acid-activated bentonite (Srasra et al., 1989). The acid activation produced an increase in the BET-area from 80 to 250 $m^2 g^{-1}$. From the change in shape of the nitrogen isotherm, it would appear that pore widening was associated with the removal of some microporosity in the original bentonite. The adsorption of β-carotene was also investigated, but there appeared to be no correlation with the adsorption of nitrogen at 77 K. This is hardly surprising in view of the differences in modes of adsorption and the complexity of the clay.

As illustrated above, it is important to define the preparation of the montmorillonite sample in such a way as to obtain a reproducible extent of delamination. Differences in preparation can explain large differences between surface areas assessed by different methods. This is what happened in a comparative measurement of a montmorillonite surface area by the BET-nitrogen method (61 $m^2 g^{-1}$) and by atomic force microscopy (AFM), where 62 particles were examined, after dispersion in NaOH solution and sonication: the AFM measurement led to a surface area of 346 $m^2 g^{-1}$ (Macht et al., 2011).

Michot et al. (1994) and Michot and Villiéras (2002) showed, most interestingly, by a simple derivation of the argon adsorption isotherm at 77 K, how adsorption occurred at different pressures on lateral and basal surfaces of the talc particles.

12.2.2.2.2 Sorption of Polar Molecules

The character of the isotherms for the polar molecules on natural montmorillonite in Figure 12.6 is quite different to that of the isotherms in Figure 12.5. The hysteresis now extends across the entire $p/p°$ range and is associated with the expansion and

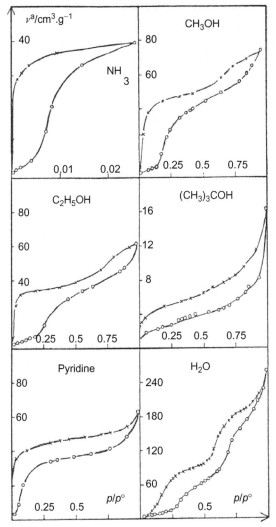

Figure 12.6 Isotherms for polar molecules on natural montmorillonite. From Barrer and Reay (1957).

contraction of the layer structure (Barrer, 1978). The interlayer sorption, which is a form of intercalation, can be treated from the standpoint of a delayed phase change (Barrer, 1989).

An interesting approach was adopted by Annabi-Bergaya et al. (1979), who determined methanol and isopropanol desorption isotherms on a series of charge-deficient Ca-montmorillonites (prepared from Na- and Li-saturated montmorillonite). Each desorption isotherm was determined after a 'surface cleaning' process with the particular alcohol in order to prevent the irreversible collapse of the interlamellar space. The adsorbent was exposed to the alcohol vapour at $p/p° = 0.9$ and the stepwise mass loss recorded as the $p/p°$ was reduced. At each stage, X-ray diffraction provided an independent measure of the change in the d_{001} spacings. It was reasonably assumed that the experimental isotherm was composed of two parts: the 'internal' and 'external' isotherms. The external isotherm was defined as the isotherm on the external surface of the collapsed material, which remained independent of the extent of charge-deficiency. The computed external isotherms for methanol and isopropanol turned out to have different shapes: the former having Type II(a) character, whereas the latter was predominantly Type I. This difference is not surprising, but it is more difficult to explain the magnitude of the derived external area. Thus, a value of ca. 300 m^2 g^{-1} was obtained from the methanol isotherms, whereas the BET-nitrogen value was only 140 m^2 g^{-1}. It seems therefore that there may be an overestimate of the extent of the methanol adsorption on the external surface.

The mechanisms of methanol adsorption were discussed in a further paper by Annabi-Bergaya et al. (1981). The authors drew particular attention to the importance of the hydrogen bonding between the adsorbed molecules, which is in competition with the dipole-cation interaction. In some cases, the latter is relatively strong and the cation may then adopt the structure-forming role; but with other smectites (e.g. Li-montmorillonite), the specific adsorbate–adsorbate interactions allow the formation of a continuous network of adsorbed species (analogous to the structure of crystalline CH_3OH).

Many investigations have been made of the sorption of water vapour by various forms of montmorillonite and vermiculite. In a study of the effect of increasing the outgassing temperature of natural vermiculite on the sorption of water vapour, Gregg and Packer (1954) obtained a set of unusual stepwise Type I isotherms, all of which displayed low-pressure hysteresis. It is of interest that the location of the step riser, at about $p/p° = 0.02$, appeared to be almost independent of the outgassing temperature. The water sorption capacities were reported to be far greater than would be expected for adsorption on the external areas, as determined by the BET-nitrogen method (i.e. 1–2 m^2 g^{-1}).

The water isotherm on natural montmorillonite in Figure 12.6 (Barrer and Reay, 1957) has an ill-defined double step. Similar results were reported by van Olphen (1965) for the water/vermiculite system. After further work, van Olphen (1976) came to the conclusion that the sorption of water molecules produces a *stepwise expansion of the layer lattice* of smectites and vermiculites with the interlayer formation of *one to four monolayers* of water.

A clearer picture of the sorption of water vapour by montmorillonite was obtained by Cases et al. (1992). Their adsorption–desorption isotherms of water on sodium montmorillonite are shown in Figure 12.7. The wavy nature of the adsorption and desorption branches (A_1 and D_1, respectively) of the full hysteresis loop in Figure 12.7 is evidently similar to that of the water isotherm in Figure 12.6 and is indicative of a complex mechanism. However, it was established that the scale of the hysteresis loop depended on the maximum relative pressure reached before the pressure was reduced. This dependency is illustrated by the appearance of the partial sorption isotherms also plotted in Figure 12.7. Here, a small hysteresis loop (desorption branch D3) was the result of $(p/p°)_{max} < 0.25$, in contrast to much larger loop (desorption branch D2) when the adsorption was taken to $(p/p°)_{max} = 0.35$. The results in Figure 12.7 were consistent with the movement of the c-axis spacing, d_{001}: at

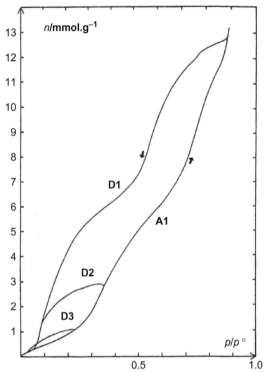

Figure 12.7 Adsorption–desorption isotherm for water vapour on sodium montmorillonite at 25 °C— desorption branches D1, D2 and D3 after adsorption up to relative pressures of 0.88, 0.35 and 0.25, respectively.
After Cases et al. (1992).

$p/p° < 0.25$, this remained close to the dry-state value of 0.96 nm; but as the water vapour pressure was increased, d_{001} underwent a stepwise change to ca. 1.8 nm. The initial sharp increase in the region of $p/p° = 0.25$ confirmed that at 25 °C this represented the threshold relative pressure (and corresponding chemical potential) for the sorption of water within the interlamellar space.

Immersion microcalorimetry was one of the techniques used by Cases et al. (1992) to provide additional information on the nature of the microstructural changes produced by the sorption of water vapour. The approach was based on the determination of the energy of immersion after the progressive pre-adsorption of water vapour. The change in the energy of immersion as a function of the pre-coverage $p/p°$ is shown in Figure 12.8. In contrast to the behaviour of kaolinite, a relative pressure of ca. 0.75 was required in order to reduce the immersion energy to its final constant level of 12.6 J g^{-1}. By assuming that this corresponds to the immersion of particles coated with liquid water, we obtain, by applying the Harkins and Jura method (see Section 4.2.3) a value of 10^5 m^2 g^{-1} for the external area (since the surface internal energy of pure liquid water is 0.119 J m^{-2}).

As already indicated, the BET-nitrogen surface area of the dry clay appeared to be about 50 m^2 g^{-1}. The platy particles (the tactoids) were therefore about 20 layers thick: during the first stage of water sorption, the particles were split into smaller tactoids of around six clay platelets (surface area of 10^5 m^2 g^{-1}). The external dimensions at $p/p° > 0.25$ remained fairly constant, but the interlamellar sorption was accompanied

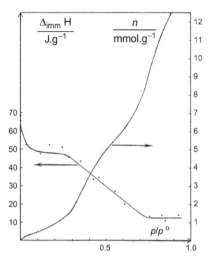

Figure 12.8 Energy of immersion in water versus pre-coverage relative pressure of water vapour.
Courtesy Cases et al. (1992).

by swelling and the development of an accessible internal area (possibly as high as $800 \, m^2 \, g^{-1}$). Cases et al. (1992) concluded that on the adsorption branch two-layer and three-layer hydrates were formed over the range $p/p° = 0.5{-}0.9$. On the desorption branch, the initial stage of the loss of some water from the external surface and meso-pores was followed by a reduction in the amount of interlayer water.

A Monte Carlo simulation study of the adsorption of water was undertaken by Delville and Sokolowski (1993). The results appear to confirm that water molecules confined in 2 nm slits between montmorillonite sheets do not have the same prop-erties as liquid water. The calculated water isotherms on open and confined clay surfaces appeared to be quite different and the results indicated a strong correlation between the wetting properties of a clay surface and its porosity and ionic nature.

An interesting study was carried out by Lantenois et al. (2007) on the swelling of a synthetic beidelite (whose structure is similar to that of montmorillonite, the differ-ence being essentially in a different location of the deficit charge). They analysed the water adsorption data in such a way as to derive, all along the swelling process, as the partial pressure of water increased up to 0.8, what can be called an apparent BET-water surface area. Their results showed that interlayer adsorption occurred after external adsorption and that the first internal monolayer existed in the form of clusters or pillars and did not fill more than 30% of the available volume at the end of the first step shown by the experimental isotherm.

12.2.2.2.3 Physisorption by Expanded Smectites
It was already known in 1955 that the exchangeable cations in montmorillonite lie between the negatively charged layers, the degree of separation being dependent on the size of the cation and its state of hydration. Barrer and MacLeod (1955) conceived the idea that by replacing small cations by larger ones, it should be possible

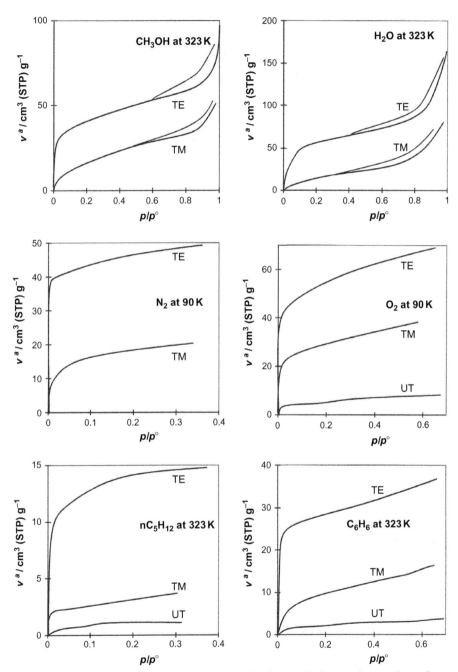

Figure 12.9 Isotherms for polar and non-polar molecules on alkylammonium exchanged montmorillonite.
From Barrer and Reay (1957).

to hold the layers permanently apart so that the physisorption capacity would be considerably enhanced and possibly become selective. It was therefore decided to exchange the Na^+ ions for various alkylammonium ions [e.g. $(CH_3)_4N^+$ and $(C_2H_5)_4N^+$]. The capacity of the clay for the non-polar molecules was greatly increased, as can be seen from the results in Figure 12.9. It is of particular interest that polar molecules were freely adsorbed, the low-pressure hysteresis in Figure 12.6 now being removed.

In an investigation of the adsorbent activities of other alkylammonium montmorillonites, Barrer and Reay (1957) found that the $CH_3NH_3^+$ form exhibited molecular sieve properties. However, the uptake of benzene was higher than expected which confirmed that the expanded clays did not behave as completely rigid molecular sieves.

It was shown by Barrer and his co-workers that a great variety of organic cations can be introduced into the interlayer regions of smectites and vermiculites. The

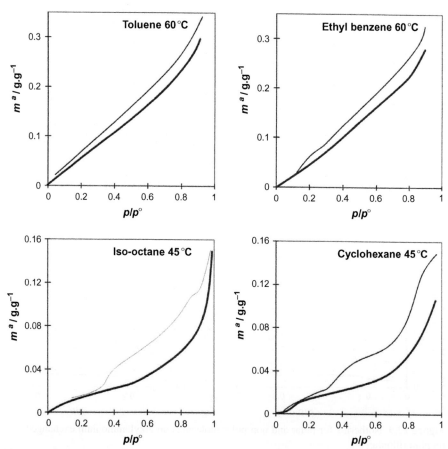

Figure 12.10 Hydrocarbon isotherms obtained with dimethyldioctadecylammonium bentonite. From Barrer (1978).

products have been referred to as *organo-clays* (Barrer, 1978) and some have been found to be useful thickeners in paints, inks, etc. A selection of hydrocarbon vapour isotherms obtained with dimethyldioctadecylammonium bentonite is displayed in Figure 12.10. The toluene and ethyl benzene isotherms are almost linear over a wide range of $p/p°$ and are more like the sorption isotherms given by organic polymers than by inorganic porous adsorbents. On the other hand, the Type II(b) character of the iso-octane and cyclohexane isotherms is apparent (Barrer, 1978).

With the two aromatic hydrocarbons, Barrer and Kelsey (1961) found that the d_{001} spacing increased steadily with the increase in $p/p°$, but with the alkanes there was very little change. In the former case, it appeared that most of the uptake was in the interlamellar region. As indicated by the shape of the Type II(b) isotherms, the sorption of the other organic vapours probably included an appreciable amount of multilayer adsorption on, and between, the clay platelets (see Figure 12.5).

12.3 Pillared Clays: Structures and Properties

12.3.1 Formation and Properties of Pillared Clays

As we have already seen, swelling is the direct result of the interlayer sorption of polar molecules by smectites. The generation of an internal area of about 800 $m^2 g^{-1}$ by the sorption of water vapour is associated with the development of an interlayer width of at least 0.6 nm, but the expanded structure lacks thermal stability. It is perhaps surprising that the work of Barrer on the sorption properties of expanded smectites, described in the previous section, did not immediately attract more attention. Twenty years were to elapse before the first successful attempts to produce stable permanently expanded smectites by the introduction of inorganic pillars were made by Vaughan et al. (1974) together with Brindley and Sempels (1977).

The work by Vaughan's group in the laboratories of W.R. Grace and Co., was prompted by the need to develop new catalysts for the processing of heavy crude oils (Vaughan, 1988). The initial aim was to produce stable wide-pore cracking catalysts, but so far progress in this direction has been disappointing. The present revival of interest in pillared clays (PILCs) is largely in the hope that they may find application in other areas such as separation technology (De Stefanis et al., 1994; Fripiat, 1997).

12.3.1.1 Pillaring

A simplified picture of the stages involved in pillaring is shown in Figure 12.11. It is apparent that the replacement of the exchangeable cations by large inorganic ions is responsible for creation of the pore structure, which is then stabilised by thermal treatment with the removal of H_2O and OH groups. In this manner, dense nanoscale oxide pillars are inserted into the interlayer 'galleries'.

It was established by Vaughan and his co-workers that certain inorganic polymeric cations could intercalate and thereby expand mineral and synthetic forms of smectic clays (Vaughan, 1988). Much of the early work involved the application of a large hydroxyaluminium cation, which can be readily prepared from aqueous aluminium

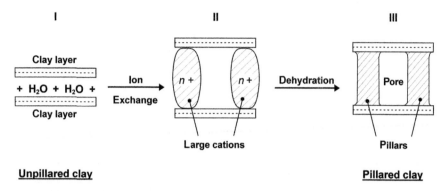

Figure 12.11 Simplified picture of pillaring.

chloride. The structure of the Al_{13} cation, $[Al_{13}O_4(OH)_{24}(H_2O)_{12}]^{7+}$, was already known and it was originally thought that this polymeric ion would retain its identity in the intercalated state. However, it became evident that complex changes accompany the exchange and ageing of the intercalated material (Vaughan, 1988; Fripiat, 1997).

In the initial stages, Vaughan and his co-workers were encouraged to find that their samples of PILC were catalytically active and also appeared to be thermally stable. However, an irreversible collapse of the pore volume was found to occur when the products were exposed to mild hydrothermal conditions at 600 °C. The first successful way of stabilising the pillar structure involved increasing the molar mass of the Al_{13} cation by further polymerisation to give, for example, $[Al_{26}O_8(OH)_{52}(H_2O)_{20}]^{10+}$. Other procedures included the formation of larger co-polymers by reacting the Al_{13} ions with compounds of Mg, Si, Ti or Zr.

Experiments in the Grace laboratories were also carried out with solutions of $ZrOCl_2 \times H_2O$, which tended to give more stable PILCs than those obtained by the Al_{13} route. On the other hand, it was more difficult to obtain reproducible products because of the complexity of the zirconia hydroxypolymers (Vaughan, 1988). Progress in this direction has been made by Burch (1987), Farfan-Torres et al. (1991) and Ohtsuka et al. (1993).

The possibility of pillaring smectites with various other large polyhydroxides (e.g. of titanium and various transition metals) has been under active investigation. For example, the preparation of thermally stable TiPILCs (e.g. with BET-areas $>300 \text{ m}^2 \text{ g}^{-1}$ at 600 °C) has been described by Sterte (1991) and characterised by Bernier et al. (1991). The use of different pillaring agents offers the prospect of the development of a broad spectrum of different PILCs with pore sizes extending into the mesopore range (Barrer, 1989; Fripiat, 1997).

12.3.1.2 Chemical and Physical Nature of Pillared Clays

It is well known that the smectic clays exhibit both Bronsted and Lewis acidity: the former is associated with the clay surface and the latter with the exchangeable cations. It has been pointed out by Fripiat (1997) that pillaring the clay will modify the clay acidity in at least two ways: (i) the polycationic pillars replace a high proportion of the original cations and (ii) the pillars convey additional acidity. However, because of the pillar-smectite interaction, the net result is unlikely to be a simple substitution of one form of acidity by another. Further changes, which involve dehydroxylation and cationic dehydration, are brought about by thermal activation and lead to various forms of cross-linking between pillars and smectite framework. These changes have been followed by the application of IR spectroscopy, MAS NMR and thermal analysis.

Detailed investigations of clay morphology have revealed that Figure 12.11 gives an oversimplified impression of the microstructure (Van Damme et al., 1985). Thus, there is normally an appreciable amount of disorder in the arrangement of aggregated clay platelets, resulting in a broad distribution of ill-defined interparticle pores. The BET-areas of the calcined pillared clays are generally in the range $200–400 \text{ m}^2 \text{ g}^{-1}$, although values of over $600 \text{ m}^2 \text{ g}^{-1}$ have been recorded (Dailey and Pinnavaia, 1992).

12.3.2 Physisorption of Gases by Pillared Clays

Nitrogen isotherm measurements (at 77 K) have been used by many investigators for the evaluation of the surface area and porosity of pillared clays. Although the exact dimensions of the pillared pore structures are uneasy to establish (Fripiat, 1997), there is general agreement that the gallery (interlayer) pores are for the most part within the micropore range. Therefore, one might expect pillared clays to give Type I(a) or I(b) nitrogen isotherms. In fact, a few Type I isotherms have been reported (e.g. by Diano et al., 1994; Cool and Vansant, 1996), but these are rarely fully reversible. The most common types of hysteresis loops are H3 and H4 in the IUPAC classification. With the majority of PILCs, the nitrogen isotherm turns out to be a combination of Types I and II(b), the exact shape depending on the relative extents of the interlamellar and external areas.

The values of BET-nitrogen area and apparent pore volume given by a particular PILC are of little use unless the external surface coverage and micropore filling contributions can be resolved. However, a useful initial step is to look for similarities and changes in the isotherm shape. For example, the nitrogen isotherms in Figure 12.12 were given by (a) the montmorillonite precursor, (b) the PILC and (c) after autoclaving (Sterte, 1991).

It is evident that the multilayer sections of the isotherms in Figure 12.12 have a remarkably similar shape and indeed appear to lie parallel. Clearly, this suggests that the secondary pore structure of the smectic clay (i.e. its interparticle porosity) and its external area underwent little change. The steep low-pressure region of isotherm (b) indicates that the PILC contained narrow micropores, which were widened by the hydrothermal treatment—as can be seen by the appearance of isotherm (c).

Various procedures have been adopted for the analysis of the nitrogen isotherms. Some investigators have used the t-method of Lippens and de Boer in its original form. Other authors have preferred to follow the IUPAC recommendations of comparing the shape of a given isotherm with that of a standard on an appropriate nonporous adsorbent.

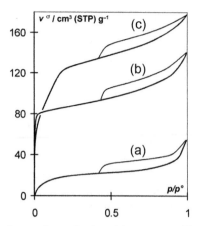

Figure 12.12 Nitrogen isotherms determined on (a) a montmorillonite, (b) the untreated La–Al-pillared montmorillonite, (c) after autoclaving treatment. After Sterte (1991).

The latter approach was adopted by Trillo et al. (1993) in their study of the effect of thermal and hydrothermal treatment on the accessible microporosity of alumina-pillared montmorillonite. This work revealed that X-ray measurements of the d_{001} spacing taken alone may give a misleading impression of the thermal stability of the PILC micropore structure. For example, after heat-treatment of the Al-PILC at 300 °C, it was found that the apparent micropore volume available for nitrogen adsorption amounted to only ca. 30% of that indicated by the d_{001} spacing.

The α_s-method was also used by Grange and his co-workers (Gil et al., 1995; Gil and Grange, 1996) for analysing nitrogen isotherms on a series of pillared clays prepared from Na-montmorillonite. Hysteresis loops of Type H4 were associated with the secondary porosity and high values of the Langmuir constant b (see Equation 5.14) indicated microporosity. In the case of a sample of Al-PILC, the micropore capacity was estimated to contribute about 60% to the total uptake at $p/p° = 0.99$.

Gil and Grange (1996) also attempted to apply the DR and DA equations (i.e. Equations 5.41 and 5.46)—the latter is given later in this chapter as Equation (12.1). The pillared clays appeared to give bimodal adsorption potential distributions, but the significance of these findings is not entirely clear. Gil and Gandia (2003) carefully studied the porosity of an alumina-pillared saponite, by N_2 and CO_2 adsorption and by various methods of data processing. It was surprising that the Horvath–Kawazoe method could be indifferently applied with cylindrical and slit-like pore geometries, whereas the DFT model used was not considered to be suited for this type of material: this shows the difficulties of such a micropore characterisation.

Some aspects of the interpretation of adsorption data were discussed by Bergaya et al. (1993), with the useful reminder that the packing of adsorbed molecules in narrow pores is strongly dependent on the pore width. It was suggested that the molecular confinement in interlamellar pores is a major source of underestimation of the gallery pore volume. These comments reinforce the IUPAC recommendation that no experimental method should be expected to provide an absolute assessment of the surface area or porosity of highly porous materials (Rouquerol et al., 1994). The following will also illustrate the importance of this recommendation.

In a study of the porosity of alumina-pillared montmorillonites (Al-PILCs), Zhu et al. (1995) have obtained values of the mean slit-width of 0.8–0.9 nm from the volume/surface ratio. In this case, the nitrogen adsorption values were in agreement with the corresponding d_{001} values of ca. 0.8 nm. However, effective micropore volumes obtained from the nitrogen isotherms and from water sorption data were significantly different and it was suggested that the density of the sorbed water was lower than that of liquid water.

A different opinion has been expressed by Bergaoui et al. (1995), who consider that the characterisation of Al-PILCs by nitrogen adsorption can lead to misleading results. These investigators have found that the Dubinin–Radushkevich (DR) analysis of carbon dioxide isotherms gave values of micropore volume in good agreement with the 'theoretical' values derived from the amounts of intercalated Al_{13} polymer in synthetic saponites. It is of interest that a quasi-equilibrium technique was used to determine the CO_2 isotherms at 273 and 293 K. The application of CO_2 adsorption for micropore characterisation, originally devised by Rodriguez-Reinoso, is undoubtedly useful when applied to activated carbons, but the likelihood of a relatively strong field gradient–quadrupole interaction of CO_2 with clays and PILCs may complicate the interpretation of the adsorption data. This may explain the curvature of the DR plots obtained with intercalated saponites (Bergaoui et al., 1995).

The importance of using a number of adsorptive molecules for the characterisation of the porosity of PILCs is demonstrated by the recent work of Cool and Vansant (1996). The Zr-pillaring of natural hectorite and synthetic laponite was investigated with the aid of various techniques including measurements of the adsorption of nitrogen, oxygen and cyclohexane. The low-temperature N_2 isotherm on an uncalcined sample of Zr-laponite was partly of Type I(b), but also had a small hysteresis loop. These features are indicative of a wide range of pores, which extended from narrow micropores (<0.7 nm) to supermicropores and narrow mesopores (ca. 1.4–2.1 nm).

According to Cool and Vansant (1996), pores between 0.7 and 1.1 nm are probably present in all pillared clays, whereas the narrow and wider pores are particular features of the Zr-laponite and Zr-hectorite. A relatively high adsorption affinity (i.e. the low-pressure capacity) of Zr-laponite for cyclohexane was attributed to the presence of a large number of narrow pores, giving rise to enhanced adsorbate-adsorbent interactions.

Reversible Type I(b) isotherms have been reported by Galarneau et al. (1995) for the adsorption of nitrogen on some novel 'porous clay heterostructures' (PCHs). A so-called 'intra-gallery templating process' was used to produce thermally stable pores of effective width of 1.4–2.2 nm (as evaluated by the Horvath–Kawazoe method). This interesting approach involved the use of intercalated quaternary ammonium cations and neutral amines as co-surfactants to direct the interlamellar hydrolysis and condensation of an organo-Si compound such as tetraethylorthosilicate. Removal of the surfactant by calcination then left a stable and highly developed supermicroporous/mesoporous structure.

12.4   Zeolites: Synthesis, Pore Structures and Molecular Sieve Properties

12.4.1   Zeolite Structure, Synthesis and Morphology

12.4.1.1   Zeolite Structures

The basic unit of a zeolitic structure is the TO_4 tetrahedron, where T is normally a silicon or aluminium atom/ion (or phosphorus in an aluminophosphate). In this section we deal with the aluminosilicate zeolites, which have the general formula $M_{x/n}[(AlO_2)_x(SiO_2)_y]\cdot mH_2O$. The zeolite framework is composed of $[(AlO_2)_x(SiO_2)_y]$ and M is a non-framework, exchangeable cation.

Although some pure silica zeolites (notably Silicalite and also many 'ITQ' structures obtained at the Instituto de Tecnologia Quimica de Valencia) are known, it is not possible to obtain a zeolitic alumina. Indeed, according to Loewenstein's rule, to avoid any direct Al–O–Al linkage, the permitted Si/Al ratio is at least 1.0 (i.e. $y > x$).

The great variety of zeolites is made possible by the different arrangements of linked TO_4 tetrahedra within secondary building units (SBUs), which are themselves linked together in numerous 3D networks. The two simplest SBUs are rings of four and six tetrahedra and others comprise larger single and double rings—up to 16 T atoms. The unit cell always contains an integral number of SBUs.

A zeolitic structure can be described in various crystallographic terms. For many systems it is now possible to specify the following structural features: the SBUs, the framework density, the coordination sequences, the unit cell dimensions and composition, the direction of the channels and the aperture (window) dimensions (Atlas of Zeolite Structure Types, 1992; Thomas et al., 1997). The framework density, FD, is defined as the number of T atoms per 1000 $\mathring{A}^3$ (i.e. per 1 nm^3) of the structure.

The sodalite unit (or β cage), which is a characteristic feature of the A, X and Y zeolites (see Figure 12.13), is made up of both four and six rings arranged in the form of a cubo-octahedron (i.e. a truncated octahedron). The cage has an internal effective diameter of about 0.6 nm.

Some of the most common structures are generated by linking the sodalite units in different ways. Thus, by bridging the sodalite units via the four rings we obtain zeolite A (Linde Type A). If the six rings are linked, on the other hand, the faujasite (FAU) and EMT structures are produced. It is evident that this mode of bridging is responsible for the generation of the 'supercages', that is the large cavities. Access to these regions is through the apertures (i.e. the 'windows'). The number of tetrahedral atoms T (i.e. Si or Al) surrounding the aperture determines its size. It is common, in catalysis, to speak of *small pore zeolite* in case of an eight-membered-ring aperture (i.e. eight T atoms and eight oxygen atoms, which give an aperture of 0.3–0.45 nm), of *medium pore zeolite* with a 10-membered-ring aperture (0.45–0.60 nm) and of *large-pore zeolite* with a 12-membered-ring aperture (0.60–0.80 nm) (Guisnet and Gilson, 2002).

The intracrystalline porosity is often taken as the fraction of volume occupied by liquid water evolved on heating and evacuating the zeolite. Typical values are: mordenite, 0.26; zeolite L, 0.28; zeolite A, 0.47; faujasite, 0.53 (Barrer, 1981). Because of

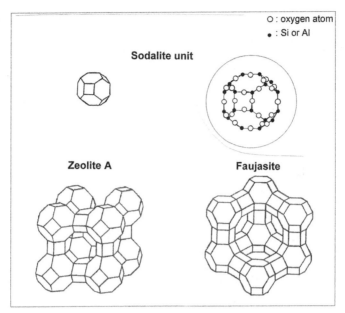

Figure 12.13 The composition of the sodalite cage and the structures of sodalite, zeolite A and faujasite (zeolites X and Y).

the rigidity of the framework, there is little swelling or shrinking of the framework during adsorption or desorption. This behaviour is in contrast to that of clays and other aluminosilicates.

12.4.1.1.1 Zeolite A

The Linde Type A zeolite (Breck et al., 1956; Reed and Breck, 1956) has a typical unit cell composition of $[Na_{12}\{Al_{12}Si_{12}O_{48}\}\cdot 27H_2O]_8$, the Si/Al ratio being always close to 1.0. The pseudo cell pictured in Figure 12.13 consists of eight sodalite units (β cages). The SBUs are double four rings and eight rings. The FD is 12.9 nm^{-3}. Zeolite A crystallises with cubic symmetry. The supercage (α cage) has a free diameter of about 1.14 nm and the eight-membered windows have free apertures of ca. 0.42 nm, providing access to a 3D isotropic channel structure. However, as discussed later, this window size is reduced by the presence of certain cations.

12.4.1.1.2 Zeolites X and Y

The faujasite-type zeolites all have the same framework structure, as indicated in Figure 12.13, and they crystallise with cubic symmetry. The general composition of the unit cell of faujasite is $(Na_2, Ca, Mg)_{29}[Al_{58}Si_{134}O_{384}]\cdot 240H_2O$. The SBUs are double six rings and the FD is 12.7 nm^{-3}. The unit cell contains eight cavities, each of diameter ≈ 1.3 nm. The 3D channels, which run parallel to [110], have 12-ring windows with free apertures of about 0.74 nm. The difference between zeolites X and Y is in their Si/Al ratios which are 1–1.5 and 1.5–3, respectively.

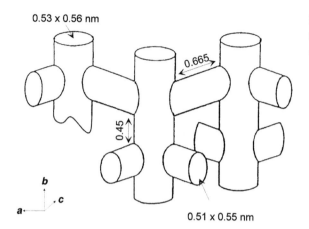

0.53 x 0.56 nm

0.665

0.45

b

, c

a

0.51 x 0.55 nm

Figure 12.14 Channel structures of ZSM-5. After Thamm (1987).

12.4.1.1.3 Pentasil Zeolites

The pentasil structure is based on double five-ring SBUs. The most important member is the MFI zeolite ZSM-5 (see Figure 12.14), called Silicalite-1 when in the pure silica form). This has the general formula $Na_n[Al_nSi_{96-n}O_{192}].16\,H_2O$, where $n < 27$ and crystallises with orthorhombic symmetry. The FD of ZSM-5 is $17.9\ nm^{-3}$. A closely related structure is the MEL zeolite ZSM-11 with the same overall formula (but now $n < 16$). As is illustrated in Figure 12.15, ZSM-5 has intersecting straight and sinusoidal channels. Instead, ZSM-11 has intersecting straight channels. Intergrowths of the MFI and MEL forms are common and affect the catalytic properties (Thomas et al., 1997).

In the orthorhombic ZSM-5 and Silicalite I crystal the sinusoidal (zig-zag) channel runs along [100] and the other intersecting straight channel system along [010] (Kokotailo et al., 1978). The pore openings are defined by 10-membered elliptical rings, the free diameters being $0.51 \times 0.55\ nm$ for the straight channels and $0.53 \times 0.56\ nm$ for the sinusoidal channels.

There are four interconnected cavities in the unit cell, each of ca. 0.8 nm width. Each of the four interconnecting straight channels are 0.46 nm long and each of the four sinusoidal channels are 0.66 nm long.

12.4.1.1.4 Role of Exchangeable Cations

So far, we have not considered the role of the non-framework cations, M. Since the aluminosilicate framework is composed of $(AlO_2)_x$ and $(SiO_2)_y$, it is anionic, the net negative charge being governed by the number of Al atoms in the T-positions. It is evident that a corresponding number of M cations are required to provide an overall balance of electric charge. A zeolite with a given Si/Al ratio normally has a certain number of exchangeable cations, which may be of different types, located at various sites within the cavities and channel structure.

In the case of zeolite A, three different cation sites have been identified: most of the cations occupy corner sites in the central cavity (Type I sites), but some of the total of 12 univalent ions (e.g. Na^+ or K^+) ions per cage must occupy sites within the eight-ring windows and therefore *partially obstruct* the channels. The effective window size of NaA (i.e. the 4A sieve) is thereby reduced from 0.42 to ca. 0.38 nm. Since the K^+ ion is somewhat larger, the window size becomes even smaller (i.e. the 3A sieve). When the Na^+ cations are exchanged for Ca^{2+} or Mg^{2+}, the number of requisite cations is reduced and the effective aperture size and pore volume are both increased (i.e. the 5A sieve).

The cation distribution in the faujasite zeolites is much more complex than in zeolite A. Five different sites have been identified and it is apparent that the distribution is dependent on the nature of the cations and the presence of water.

If the framework structure of a zeolite remains constant, the cation exchange capacity is inversely related to the Si/Al ratio. Furthermore, 'fine tuning' of the adsorptive and catalytic properties can be achieved by adjustment of the size and valency of the exchangeable cations. Dealumination of certain silica-rich zeolites can be achieved by acid treatment and the resulting 'hydrophobic' zeolites then become suitable for the removal of organic molecules from aqueous solutions or from moist gases.

12.4.1.2   Zeolite Synthesis

Obviously, the time scales involved in the formation of natural zeolites cannot be reproduced in the laboratory, but in the 1940s it was shown by Barrer and his co-workers (see Barrer, 1982) that a number of the natural zeolites could be synthesised under hydrothermal conditions. The essential synthetic ingredients are suitably reactive forms of alumina, silica and base.

The initial formation of a poorly ordered aluminosilicate hydrogel is followed by the growth of oligomer precursors and the development of crystallinity. The crystalline zeolite may not be the most stable product and its formation will depend on the nature of the reactants, the reaction conditions and 'kinetic control'. The generally recognised stages of crystallisation are supersaturation, nucleation and crystal growth. The key parameters governing zeolite formation are specified by Feijen et al. (1997) as hydrogel composition, pH, reaction conditions (temperature and time) and template.

Often organic ions (e.g. tetraalkylammonium cations such as tetramethylammonium, TMA^+, ions) or neutral molecules are used as templates (i.e. structure-directing agents). In the past certain templates were introduced empirically and although they are now widely used, their action is still not fully understood.

Crystals of synthetic zeolites often have a non-uniform distribution of their trivalent T ions (i.e. they exhibit some degree of T-III zoning). The extent of this form of zoning appears to be the result of the synthesis procedure (Jacobs and Martens, 1987). Isothermal gas adsorption microcalorimetry was used in a novel manner to investigate the T-III zoning in MFI-type zeolites (Llewellyn et al., 1994a).

12.4.1.3   Zeolite Morphology

Crystals of the synthetic zeolites are usually quite small and often exhibit various forms of twinning and intergrowth. With some zeolites, individual crystallites (e.g. cube-shaped NaA) of size <50 nm have been identified by electron microscopy, but the agglomerate sizes are generally in the approximate range 1–10 μm. For example, the particle size distribution over this range of a typical NaA powder was reported to be of a broad log-normal character (Breck, 1974, p. 388).

Whilst some current research is aimed at the preparation of nanosized particles, the particle size of many synthetic zeolites is too small for some applications and therefore they must be formed into polycrystalline aggregates (e.g. by pelletisation). Binders are often added to improve the aggregate strength and durability. It must be kept in mind that these or other changes in the particle or aggregate morphology may significantly affect the equilibria or dynamics of adsorption.

12.4.2   Adsorbent Properties of Molecular Sieve Zeolites

It is evident that the channels and cavities within many zeolitic structures are of molecular dimensions and that their size and configuration are intrinsic properties of the particular crystalline framework. In addition, the local electrostatic fields, which emanate from the exchangeable cations, are to a large extent responsible for the strong affinity for water and other polar molecules. It follows that for a given zeolitic composition and structure, the adsorptive behaviour of a well-defined zeolite crystal is remarkably uniform and stable. Furthermore, within certain limits, the adsorbent and ion exchange properties can be varied in a controlled manner by changing the framework structure, the Si/Al ratio and the nature of the exchangeable cations.

In the submonolayer range, the amount adsorbed on the external area of a 1-μm cubic zeolite crystal is very small in comparison with the adsorption within the micropore structure (the intracrystalline sorption). Also, apart from a small multilayer adsorption on the external surface, there should be no additional uptake at higher $p/p°$. However, there are three ways in which the non-zeolitic contribution may be increased: (a) the binder (most often a clay) may have a relatively large specific surface; (b) the zeolite crystallite size may be much smaller than 1 μm; (c) the zeolite may contain some amorphous aluminosilicate or silica. In practice, one or more of these effects can result in a significant distortion from the form of a Type I isotherm (see Sayari et al., 1991).

12.4.2.1   Physisorption of Gases by Zeolite A

Certain forms of zeolite A (e.g. the 3A and 4A sieves) exhibit pronounced molecular sieving together with packing limitations and consequently their physisorption capacities are not as large as one would calculate by assuming that their pores are completely filled by the adsorbate in the liquid state (see the Gurvich rule, Section 8.2.1). At 77 K the uptake of argon or nitrogen by NaA is very low to measure. The amounts adsorbed are appreciably increased as the temperature is raised, reaching maxima at ca. 120 K for Ar and at ca. 200 K for N_2. At 273 K the Ar adsorption is very small, whereas the N_2 adsorption is still significant.

These results reveal that at low temperature the rate of diffusion of Ar and N_2 into the intracrystalline pore structure is extremely slow. The increase in the adsorption with temperature is not thermodynamically controlled but is instead dependent on the molecules gaining enough kinetic energy to allow their passage through some of the 4A apertures. This process is probably assisted by enhanced vibrational amplitude of the oxygen ring structure.

Slightly smaller molecules such as O_2 are able to move more freely through the eight-ring apertures and in consequence the amount adsorbed decreases as temperature increases, in the normal manner. However, as indicated in Table 11.2, the derived values of v_p are not in close agreement (Breck, 1974, p. 428).

The exchange of sodium by calcium has a significant effect on the adsorbent properties of zeolite A. An abrupt change in the adsorptive properties occurs when between 3 and 5 Na^+ ions are replaced. Thus, Ar and N_2 are now both able to enter the channels at low temperature, although the lack of agreement between the different values of v_p in Table 12.2 is still evident.

The isotherms for the adsorption of oxygen and nitrogen by a 5A zeolite at the much higher temperatures of 273 and 293 K are shown in Figure 12.15. Of course, at these temperatures the isotherm curvature is much reduced: indeed, the oxygen isotherms are almost linear up to 10 bar (i.e. obeying Henry's law). The fact that the levels of nitrogen adsorption are significantly greater than those of oxygen is due mainly to its specific field gradient–quadrupole interaction. This enhancement of the nitrogen adsorbent–adsorbate interaction is responsible for the higher affinity

Table 12.2 Derived Values of Pore Volume for Zeolites A and X

Zeolite	Adsorptive	θ /°C	V_p / cm^3 g^{-1}
Na$_{12}$A	H_2O	25	0.29
	CO_2	−75	0.25
	O_2	−183	0.21
Ca$_6$A	H_2O	25	0.31
	O_2	−183	0.24
	Ar	−183	0.26
	N_2	−196	0.30
	n-Butane	25	0.23
NaX (Si/Al = 1.25)	H_2O	25	0.36
	CO_2	−78	0.33
	Ar	−183	0.30
	O_2	−183	0.31
	N_2	−196	0.35
	n-Pentane	25	0.30
	Neopentane	25	0.26
	Benzene	25	0.30

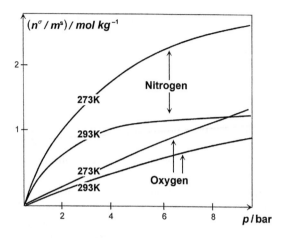

Figure 12.15 Adsorption isotherms for oxygen and nitrogen on a 5A zeolite at 273 and 293 K. After Kirkby (1986).

of adsorption, which is indicated by the difference in the slopes of the isotherms at very low loading.

The 5A zeolite gives a N_2/O_2 selectivity ratio of 2–3 in the normal pressure swing adsorption (PSA) working range; although this is evidently reduced as the partial pressures are increased. At ambient temperature, adsorption equilibration is very rapid. At lower temperatures, the rates of adsorption are decreased and the separation becomes less efficient. At 293 K, argon gives an isotherm which is very close to that of oxygen and this component therefore tends to remain in the oxygen fraction. Since a selective adsorption of oxygen cannot be achieved with a zeolite, for nitrogen generation it is necessary to use a special type of molecular sieve carbon.

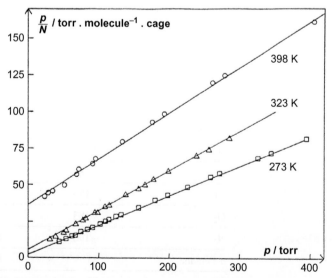

Figure 12.16 Langmuir plots for the adsorption of program by 5A zeolite. Courtesy Ruthven and Loughlin (1972).

If the operational temperatures are not too high, many zeolites give reversible Type I isotherms. It might be expected that the Langmuir equation could be applied to these systems—at least over a certain range of pressure. The long ranges of linearity of the Langmuir plots in Figure 12.16 may appear to support the applicability of the Langmuir equation. However, as Ruthven (1984) has pointed out, the apparent conformity is deceptive. The three linear plots are given over different ranges of apparent surface coverage, θ, and the derived values of Henry's law constant and monolayer capacity, n_m, turn out to be incompatible with each other and with values obtained by a more detailed analysis of the adsorption data (e.g. virial treatment).

A more searching test of conformity to the Langmuir equation is obtained by plotting the adsorption data as a function of θ rather than p (Barrer, 1978). Some pronounced deviations then appear, which are consistent with the inadequacy of the simple Langmuir model.

12.4.2.2 Physisorption of Gases by Zeolites X and Y

Values of the effective pore volume of zeolite NaX, as determined by the adsorption of a selection of molecules, are included in Table 12.2. By omitting the water, nitrogen and neopentane values, we arrive at the value $v_p = (0.31 \pm 0.02)\,\mathrm{cm^3 g^{-1}}$, which is in agreement with a calculated supercage volume of $0.30\,\mathrm{cm^3 g^{-1}}$. This supports Breck's (1974, p. 428) conclusion that, with the possible exception of water, only the large supercages are available for physisorption and is consistent with a supercage volume of about $6.7\,\mathrm{nm^3}$ per unit cell.

The anomalous behaviour of water in both zeolite A and X is not surprising in view of its abnormal specificity, strong interactions with exchangeable cations and also the possibility of some penetration into the small β-cages. The high nitrogen value is of interest since it provides further evidence that in narrow micropores adsorbed nitrogen does not adopt the normal liquid structure.

Various attempts were made by Dubinin and his co-workers to apply the fractional volume filling principle and thereby obtain a characteristic curve for the correlation of a series of physisorption isotherms on a zeolite (Dubinin, 1975). As was noted in Section 5.2.4, the original DR equation (i.e. Equation 5.41) was found to be inadequate and in its place the more general Dubinin–Astakhov (DA) equation was applied (i.e. Equation 5.46).

A convenient form of the DA equation is

$$n/n_p = \exp\left[-(A/E)^N\right] \tag{12.1}$$

where n_p is now the specific amount adsorbed when all the channels and cavities are full (i.e. the micropore capacity). The terms A and E are as defined in Section 5.2.4: A is a measure of the adsorption affinity (so-called adsorption potential) and E is a characteristic energy for the given system.

Dubinin (1975) found it necessary to distinguish between the adsorption of relatively large and small molecules by NaX and other faujasite zeolites. With *large molecules*, such as benzene and cyclohexane, it was apparently possible to apply

Equation (12.1) in a fairly straightforward way: by using successive approximations the best values of n_p and E were obtained. For each system, this procedure gave a temperature-invariant characteristic curve, by which the fractional filling, n/n_p, was expressed as a function of the potential, A. For example, Equation (12.1) was applicable with $N=4$ to the adsorption isotherms of cyclohexane on NaX over the remarkably wide fractional filling range $n/n_p = 0.10$–0.98 and within the temperature range 80–140 °C (maximum deviation $<10\%$).

The adsorption of *smaller polar molecules*, such as water and carbon dioxide, was more complex and Dubinin (1975) concluded that the overall pore-filling process could be expressed as a two-term equation, each term having the mathematical form of Equation (12.1). In the low-filling region, the interaction with the cationic sites was considered to be the most important contribution, with the normal dispersion interactions becoming more important at higher loadings.

Although many experimental isotherms appear to obey the DA equation over appreciable ranges of pressure, the theoretical basis of this conformity is highly questionable. However, as Ruthven (1984) points out, even with NaX and other zeolites the temperature-invariant characteristic curve can provide a useful empirical means of correlating engineering data.

It is generally agreed that a virial form of isotherm equation is of greater theoretical validity than the DA equation. As explained in Section 5.2.1, a virial equation has the advantage that since it is not based on any model it can be applied to isotherms on both nonporous and microporous adsorbents. Furthermore, unlike the DA equation, a virial expansion has the particular merit that as $p \rightarrow 0$, it reduces to Henry's law.

The exponential form of the virial isotherm favoured by Kiselev and his co-workers (e.g. Avgul et al., 1973) was Equation (5.6), that is

$$p = n \exp\left(C_1 + C_2 n + C_3 n^2 + C_4 n^3 + \cdots\right) \qquad (12.2)$$

By using the first three or four coefficients, Avgul et al. (1973) were able to satisfactorily apply Equation (12.2) up to 70–80% of the total filling of zeolites NaX and LiNaX by Ar and Xe.

The two-constant versions of Equation (12.2) and other virial expansions can be applied to the low fractional filling section of isotherms on the faujasite zeolites, provided that the temperature is not too low. In this manner it is then possible to obtain the Henry's law constant, k_H.

An alternative way of determining k_H is by a gas chromatographic method. This is the generally preferred approach at higher temperatures where the isotherm curvature is reduced. A perturbation chromatographic technique was adopted by Denayer and Baron (1997) in their study of the adsorption of a range of normal and branched paraffins by various forms of zeolite Y. By measuring the retention times corresponding to the perturbation of the adsorption system at different loadings, it was possible to derive the adsorption isotherm for each component. The measurements were made over the range 275–400 °C. This study of the effect of chain length and branching of alkanes (from C_6 to C_{12}) followed earlier investigations of the adsorption of lower hydrocarbons by the faujasite zeolites (e.g. the work of Atkinson and Curthoys, 1981;

Thamm et al., 1983). Our present interest is in the behaviour of NaY and HY. Selected values of Henry's law constant, k_H, and low-coverage energy of adsorption, E_0, are given in Table 12.3.

Inspection of Table 12.3 reveals that there are relatively small differences between the corresponding values of k_H and E_0 for NaY and HY. This is to be expected since the adsorbent–adsorbate interactions are essentially non-specific (see Section 1.6). Decationisation of zeolite Y thus has a minimal effect on the energetics of adsorption of the paraffins. Molecular shape of the adsorptive is also unimportant. In accordance with the results in Figures 1.3 and 1.4, the molar mass (number of carbon atoms) is much more important than the molecular shape. As before, there is a linear relation between E_0 and N_c. In these conditions, an exponential increase of k_H with N_c is consistent with the form of Equation (5.3).

The interaction of polar molecules with ionic and polar surfaces was briefly discussed in Section 1.6. In a simplified form one can write:

$$E_0 = E_{ns} + E_{sp} \qquad (12.3)$$

where E_{ns} represents the non-specific adsorbent–adsorbate interactions and E_{sp} the various specific contributions. When zeolites are used as adsorbents the E_{sp} term becomes extremely important (Kiselev, 1967; Barrer, 1978). The magnitude of the specific contributions is illustrated by the low-coverage adsorption calorimetric data in Table 12.4.

Table 12.3 Henry's Law Constants and Low-Coverage Energies of Adsorption for Various Alkanes on NaY and HY (Denayer and Baron, 1997)

Adsorptive	NaY		HY	
	$k_H \cdot 10^5$/mol g^{-1}	E_0/kJ mol^{-1}	$k_H \cdot 10^5$/mol g^{-1}	E_0/kJ mol^{-1}
n-Hexane	1.9	45.5	1.7	44.2
2-Methylpentane	2.0	45.3	1.7	44.2
3-Methylpentane	2.0	44.5	1.7	43.5
2,3-Dimethylbutane	2.0	44.1	1.8	43.5
2.2-Dimethylbutane	2.1	43.2	1.8	43.5
n-Heptane	4.4	51.9	3.6	50.1
2,3-Dimethylpentane	5.1	50.6	3.6	50.1
n-Octane	10	57.5	7.9	56.0
2-Methylheptane	10	57.2	7.9	55.7
2,5-Dimethylexane	11	57.1	8.3	56.0
n-Nonane	23	63.4	17	62.0

Table 12.4 Adsorption Energies E_0 at Zero Coverage for Different Adsorptives on NaX Zeolite, with Derived Specific Contribution E_{sp} (Kiselev, 1967)

Adsorptive	E_0/kJ mol^{-1}	E_{sp}/kJ mol^{-1}
Argon	13.0	
Nitrogen	21.7	9
Ethane	25.9	
Ethylene	38.5	13
n-Hexane	61.4	
Benzene	75.2	14
n-Pentane	51.8	
Diethyl ether	87.8	36

Comparison of the values of E_0 is made in Table 12.4 for pairs of molecules of very similar polarisabilities. By assuming that for each pair the corresponding E_{ns} values are approximately equal, we are able to obtain a rough estimate of the E_{sp} contribution given in column 3. On this basis, the most striking result is the very large E_{sp} contribution for diethyl ether.

Large E_{sp} contributions were also reported by Barrer (1978) for the adsorption of carbon dioxide, ammonia and water vapour on NaX. Indeed, in the case of water, over 90% of the low-coverage adsorption energy was attributed to E_{sp}. With these highly polar molecules it is likely that the cation-adsorbate interaction provides a major contribution to E_{sp}.

In the work of Schirmer et al. (1980), a Tian-Calvet type microcalorimeter was used to determine the energetics of adsorption for n-hexane, cylohexane and benzene on NaY zeolite.

The results in Figure 12.17 are representative of the adsorption energy plots for various alkanes and aromatic hydrocarbons on faujasite zeolites. Thus, at low fractional filling by NaY, the benzene adsorption energy is greater than the n-hexane energy, although the difference (≈ 6 kJ mol^{-1}) is much smaller than the corresponding difference (≈ 14 kJ mol^{-1}) for NaX in Table 12.4. As already indicated, this is in accordance with the larger number of exchangeable cations in NaX. The pronounced maximum in the n-hexane curve in Figure 12.17 is indicative of strong adsorbate–adsorbate interactions at high loading, in contrast to benzene on graphitized carbon (see Figure 10.10a).

In view of the difference in the benzene adsorption energies on NaY and NaX, we would expect to find a difference in the benzene isotherms—especially at low loadings. The results of Kacirek et al. (1980), confirm that the benzene adsorption affinity of NaX is indeed significantly higher than that of NaY.

Generally, those polar adsorptives which have been found to exhibit the strongest specificity at very low coverage (e.g. H_2O and CO_2 on NaX) also give pronounced

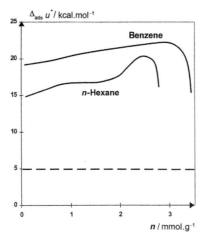

Figure 12.17 Differential energy of adsorption of n-hexane or benzene versus amount adsorbed by NaY zeolite.
After Schirmer et al. (1980).

energetic heterogeneity: their differential adsorption energies decrease sharply with increased fractional pore filling (Kiselev, 1965; Barrer, 1978). On the other hand, the adsorption energies of non-polar and weakly polar molecules tend not to undergo much initial change. As we have seen, the cation density is controlled by the Si/Al ratio and therefore a change from X to Y or dealumination generally leads to a higher degree of energetic uniformity (Barrer, 1978; Schirmer et al., 1980).

12.4.2.3   Physisorption of Gases by ZSM-5 and Silicalite I

In the early work on both ZSM-5 and Silicalite (e.g. by Flanigen et al., 1978; Ma, 1984) the adsorption isotherms of aliphatic and aromatic hydrocarbons and other vapours appeared to have an overall Type I appearance. However, the individual adsorption uptakes were widely spaced and generally desorption measurements appeared not to have been undertaken. More recent measurements by Rouquerol and Unger and their co-workers (e.g. Reichert et al., 1991) have revealed that nitrogen and argon isotherms on well-defined crystals of Silicalite I exhibit sub-steps within the micropore filling range of $p/p°$. Even more remarkable is the existence of a hysteresis loop in the pre-capillary condensation region of the nitrogen isotherm (Carrott and Sing, 1986; Muller and Unger, 1986).

The values of pore volume, v_p, of Silicalite I in Table 12.5 were obtained from the saturation adsorption capacities in the usual manner (i.e. after the Gurvich rule): in each case, the uptake at $p/p° \rightarrow 1$ was converted into the adsorbed volume by assuming the adsorbate to have the normal liquid density at the operational temperature.

There are a number of possible explanations to account for the lack of agreement between various values v_p in Table 12.5. First, it must be kept in mind that the total uptake at high $p/p°$ is controlled by three mechanisms: (i) the intracrystalline filling at

Table 12.5 Derived Values of Pore Volume for Silicalite I

Adsorptive	$\dfrac{T}{K}$	$\dfrac{V_p}{cm^3 \ g^{-1}}$	Reference
Nitrogen	77	0.190	Kenny and Sing (1990)
Oxygen	90	0.185	Flanigen et al. (1978)
n-Butane	293	0.190	Flanigen et al. (1978)
n-Hexane	293	0.199	Flanigen et al. (1978)
n-Hexane	293	0.185	Ma (1984)
Benzene	293	0.134	Flanigen et al. (1978)
Benzene	293	0.126	Ma (1984)
p-Xylene	293	0.13	Ma (1984)
m-Xylene	293	0.085	Ma (1984)
o-Xylene	293	0.062	Ma (1984)
Neopentane	293	0.029	Ma (1984)
Water	293	0.019	Kenny and Sing (1990)

low $p/p°$, with a possible molecular sieving effect and with a packing depending on the size and shape of the molecule and also on the geometry of the pores, (ii) the multilayer adsorption on the external surface and (iii) capillary condensation within a secondary pore structure. Processes (ii) and (iii) are manifested in the form of a finite multilayer slope and a hysteresis loop in the capillary condensation range (Kenny and Sing, 1990).

Various procedures have been proposed for separating the evaluation of the sole intracrystalline capacity from that of the external surface area (see Sayari et al., 1991). The α_s-method is one way of analysing composite isotherms, which has been applied to nitrogen isotherms on different samples of ZSM-5 (Sing, 1989). This approach was used by Gil et al. (1995) in their study of the microporosity of pillared clays and zeolites. By this means, mesopores were estimated to have contributed about 25% to the total pore volume of a commercial sample of HZSM-5.

By using relatively large crystals of HZSM-5 (of length ca. 350 μm), Muller and Unger (1988) were able to obtain the isotherms of Ar and N_2 shown in Figure 12.18. These results demonstrate the advantage of studying larger crystals than had been possible in previous work. With each adsorptive, the very low slope in the multilayer range provided unambiguous confirmation that the external surface area was very small and therefore that the amount adsorbed at the plateau corresponded to the micropore capacity.

In Figure 12.18 the argon isotherm is apparently a classical Type I(b) isotherm, whereas the nitrogen isotherm has a well-defined hysteresis loop in the region of $p/p° = 0.12–0.15$. The nitrogen loop has upper and lower closure points and is quite stable and reproducible. This phenomenon must not be confused with the more

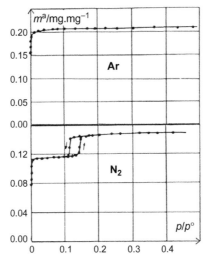

Figure 12.18 Adsorption isotherms of Ar and N_2 on HZSM-5.
Courtesy Muller and Unger (1988).

common form of low-pressure hysteresis, which is much less well-defined and persists to the lowest attainable pressures. However, a loop in this region of a nitrogen isotherm is not associated with capillary condensation, since at 77 K this can occur only at $p/p° > 0.4$ (see Chapter 8).

Muller and Unger (1988) decided to evaluate the micropore volume from the uptakes of nitrogen and argon at $p/p° = 0.1$ and from the liquid densities. The nitrogen capacity was therefore taken near the end of the first plateau (i.e. before the hysteresis loop). The resulting values of intracrystalline pore volume are 0.144 cm^3 g^{-1} (by nitrogen) and 0.147 cm^3 g^{-1} (by argon), that is much lower than the previously estimated nitrogen value. Their similarity is consistent with the apparent liquid-like character of both adsorbates. Also, by supposing that nitrogen has a solid-like packing at the second plateau, we can explain why the ratio of the uptakes (ca. 0.78) is extremely close to the ratio of the nitrogen solid and liquid densities.

Detailed investigations (Muller and Unger, 1988) of the novel hysteresis loop revealed that it became more pronounced as the crystal size of the HZSM-5 was increased and its aluminium content was reduced. In fact, the most prominent loop was given by uniform crystals (of ca. 150 μm) of pure Silicalite I. In this case, the almost vertical riser of the associated sub-step corresponded to loadings of ca. 25–30 molec uc^{-1} (molecules per unit cell).

Microcalorimetric and then high-resolution adsorption measurements (Muller et al., 1989a,b; Llewellyn et al., 1993a,b) have revealed the presence of smaller sub-steps in the isotherms of both argon and nitrogen at ca. 22 molec uc^{-1}. The results for argon and nitrogen adsorption on Silicalite I are given along with the corresponding isotherms in Figures 12.19 and 12.20. With each adsorptive on pure Silicalite, the differential adsorption enthalpy, $\Delta_{ads}\dot{h}$, remains almost constant over a wide range of loading, until $N^\sigma \approx 20$ molec uc^{-1}.

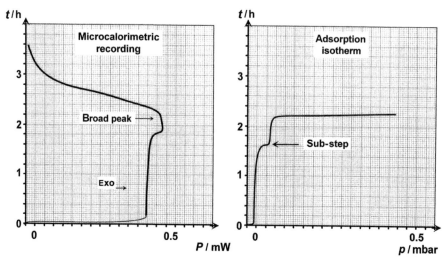

Figure 12.19 Adsorption isotherm and corresponding microcalorimetric recording for *argon* at 77 K on Silicalite I.
Courtesy Grillet Y. and Llewellyn P. (personal communication).

In the case of *argon*, this is followed by a single broad peak in $\Delta_{ads}\dot{h}$ over the range $N^{\sigma} = 22-30$ molec uc^{-1}, which corresponds to the riser of the sub-step.

As might be expected, the behaviour of *nitrogen* on Silicalite I is more complex. There are now two peaks, the first being located between two regions of minima of $\Delta_{ads}\dot{h}$. The first peak, 1, is at $N^{\sigma} = 22-25$ molec uc^{-1} and the second broad peak, 2, is over the range $N^{\sigma} = 25-30$ molec uc^{-1}—again corresponding to the locations of the isotherm sub-steps.

Neutron diffraction experiments (Reichert et al., 1991; Llewellyn et al., 1993a,b) have confirmed that the sub-steps and associated energy changes are due to *phase transitions in the adsorbate*. In the case of argon, the sharp sub-step and exothermic change appeared to be due to a transition from a disordered phase to a solid-like structure with diffraction peaks which remained stable over the temperature range 10–100 K. Nitrogen underwent a similar overall change, but this has taken place in two stages: the first, transition 1 involved a change from a disordered mobile phase to a localised state (or lattice fluid-like phase), whereas the second and larger, transition 2 led to the formation of a solid-like commensurate structure. The zeolite itself undergoes a change in structure from monoclinic to orthorhombic during adsorption as viewed by the comparison between ^{36}Ar and ^{40}Ar in Neutron Diffraction experiments (Coulomb et al., 1994a,b) which mirrors that observed during the adsorption of *p*-xylene (van Koningsveld et al., 1989). This structure transition may be initiated by the adsorption process, but this alone would seem unlikely to account for the change in state of the argon or nitrogen within the pores (Douguet et al., 1996). Molecular simulation studies of argon adsorption in silicalite suggest that the isotherm step can be attributed to an in-site/off-site phase transition of the adsorbed phase and that a

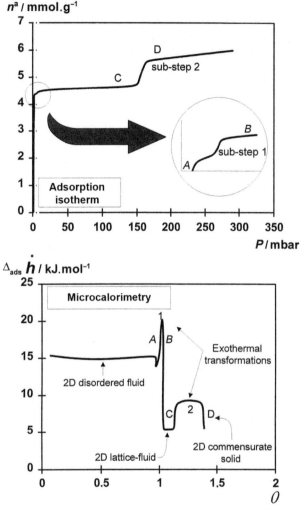

Figure 12.20 Isotherm and enthalpy of adsorption for nitrogen at 77 K on Silicalite I. After Llewellyn et al. (1993b).

distortion of the zeolite framework may occur under the stress of the adsorbed fluid at high loading (Douguet et al., 1996).

Because the 77 K isotherms are very steep (i.e. high adsorption affinity), it is very difficult to undertake any form of virial analysis. The few detailed nitrogen isotherms so far determined (Reichert et al., 1991) at higher temperatures (i.e. 293–373 K) on Silicalite and HZSM-5 have indicated that Henry's law is obeyed at low fractional loading. The derived values of isosteric enthalpies of adsorption are (15.0 ± 1.3) kJ mole^{-1} and, within experimental error, appear to be very similar for Silicalite I and HZSM-5. These values are consistent with the microcalorimetric measurements of the energies of adsorption at 77 K.

The work of Llewellyn et al. (1993a,b) also showed the effect of changing the Si/Al ratio in the MFI structure. As noted in the earlier work of Muller et al. (1989a,b), the sub-steps become less distinctive and finally almost disappear completely as the Al content is increased. Furthermore, the nitrogen adsorption calorimetric measurements reveal a significant increase in energetic heterogeneity, which is due to the development of field gradient–quadrupole interactions between N_2 molecules and the Al and cationic sites.

Adsorbent–adsorbate potential energy calculations have been made for the adsorption of argon in the channels and intersections of Silicalite I (Muller et al., 1989a,b). The most favourable sites for localised adsorption are within the straight and sinusoidal channels, which together should be able to accommodate 20 molec uc^{-1}. At a loading of 24 molec uc^{-1} all the available sites in the channels and intersections are probably occupied by localised molecules.

As predicted, at low loadings, argon and nitrogen are adsorbed in a very similar manner on pure Silicalite. Thus, in each case the adsorption energy remains almost constant until $N^{\sigma} = 20$ molec uc^{-1}. This suggests that localised adsorption is taking place with very little adsorbate–adsorbate interaction. The adsorbed molecules are mainly located in the channels and at a lower concentration in the intersections.

A small increase in adsorption energy may be due to cooperative interactions within the intersections, but this is quickly followed by the first phase transition, which involves a more drastic change in the packing density of argon than of nitrogen. The fact that the Ar transition 1 can take place at a much lower p/p° than the corresponding N_2 transition 2 is associated with the difference in the electronic properties of the two adsorbates. The non-polar nature of Ar must allow the adsorbed molecules to more readily undergo adsorbate–adsorbate interactions and hence give the opportunity for close-packing and hence densification of the adsorbate structure.

It seems likely that repulsion between the ends of the quadrupolar nitrogen molecules is responsible for the sharp fall in adsorption energy which precedes its transition 2. The sub-step 2 is probably accompanied by the reorientation of the molecules to permit a more favourable quadrupole–quadrupole interaction: that is to allow the end of one molecule to approach the centre of its neighbour. This transformation could lead to the development of the quasi-crystalline order by the formation of a chain-like structure (Sing and Unger, 1993). It is not surprising to find that hysteresis is involved in the more drastic molecular rearrangement of adsorbed nitrogen. An energy barrier must be overcome and since each molecular domain is so uniform, the macroscopic result is the appearance of a well-defined hysteresis loop with reproducible boundaries and scanning behaviour (Reichert et al., 1991).

It is of interest to compare the behaviour of argon and nitrogen with that of other adsorptives on Silicalite I. Llewellyn et al. (1993a,b) have shown that in certain respects krypton and argon behave in a similar way. Thus, up to the loading $N^{\sigma} \approx 20$ molec uc^{-1} the adsorption energies at 77 K are both constant and almost identical. The type 2 sub-steps are also similar in character; but, in contrast to Ar, the Kr sub-step is associated with an endothermic change. At present, the explanation for this difference is not clear, but it must be kept in mind that the Kr molecule is rather more bulky than Ar and therefore the endothermic phenomenon may be the result of confinement effects within the micropore network (Derycke et al., 1991). No indication of a phase transformation has so far been found with methane at 77 K. In this case, the adsorption energy also remains constant up to $N^\sigma \approx 20$ molec uc^{-1}, but thereafter it decreases sharply (Llewellyn et al., 1993a,b).

It is perhaps not surprising to find that the adsorption of carbon monoxide at 77 K on the series of MFI-type zeolites is very similar to nitrogen adsorption (Llewellyn et al., 1993a,b). On Silicalite I, CO also gives the transitions 1 and 2 and increase of Al in the MFI structure has the same effect of smoothing the isotherm and producing energetic heterogeneity. Indeed, because of the more polar character of CO, there is a somewhat larger change in the energetics of adsorption.

The Al content has an even greater influence on the level of the adsorption of water vapour by the MFI-type zeolites. In their original work, Flanigen et al. (1978) drew attention to the similarity of Silicalite to adsorbent carbons in having a low affinity for water. It has been found (Kenny and Sing, 1990; Carrott et al., 1991) that the low uptake of water vapour by Silicalite I (at say, 293 K) extends over virtually the complete range of $p/p°$. To illustrate this behaviour, the apparent fractional pore filling by water and nitrogen is compared in Figure 12.21.

A number of interesting features can be seen in Figure 12.21. First, the level of water adsorption at $p/p° = 0.90$ by Silicalite I is only about 10% of the capacity available for nitrogen and other small adsorptive molecules (see Table 12.5). This is increased to about 18% for HZSM-5, when the Si/Al ratio is reduced to 90. The presence of the hysteresis loops in the capillary condensation range indicates that

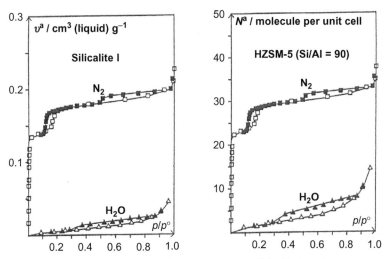

Figure 12.21 Adsorption of nitrogen and water vapour on Silicalite I and water vapour on HZSM-5 (Si/Al = 90) (Sing, 1991).

a high proportion of the water adsorption has occurred within the secondary pore structure or defect structure rather than in the zeolitic channel structure. Similar findings have been reported by Llewellyn et al. (1996).

Another interesting feature is the reversibility of the water isotherms at low $p/p°$. This is in marked contrast to the low-pressure hysteresis exhibited by water isotherms determined on most other forms of dehydroxylated silicas. The fact that there is no apparent tendency for rehydroxylation suggests that water does not easily penetrate into the intracrystalline pores of Silicalite or HZSM-5 to any great extent. However, in the work of Llewellyn et al. (1994a,b,c), water was condensed on the Silicalite I sample at $p/p° = 1.0$ and this did produce an irreversible change in the low-pressure region of the water isotherm.

In seeking an explanation for these findings, we must take into account the geometry of the pores in addition to their size. As we have seen, the intracrystalline pores of the Silicalite/ZSM-5 system are for the most part tubular and of ≈ 0.55 nm diameter. In such a confined space, a 3D array of the hydrogen-bonded water structure cannot be accommodated without some considerable distortion of the directional hydrogen bonds. The situation is quite different in the case of carbon molecular sieves, which have slit-shaped pores.

The quality of the silicalite crystals has a large effect on its hydrophobicity and consequent water uptake. In an experimental and molecular modelling study, three samples with different types of defects were compared (Trzpit et al., 2007). A shift in the condensation transition to higher pressures was observed because this decreased the amount of defects and therefore increased the hydrophobicity, as previously suggested by Ramachandran et al. (2006). Indeed, pressures as high as 80–100 MPa need to be applied for pore filling to occur (Eroshenko et al., 2001). Nevertheless, the work of Trzpit et al. equally showed that a small water uptake already takes place at low relative pressure (below 0.1), probably because of hydrophilic patches connected, at least in part, with the crystal defects. It was calculated that the water confined in the silicalite pore environment has a 10% lower dipole moment, as compared to the bulk (Puibasset and Pellenq, 2008). Because of the very high pressures (and therefore the notable work) required to completely fill with water the pores of hydrophobic silicalite, it was proposed to use this feature for energy related applications (Eroshenko et al., 2001).

The molecular sieving behaviour of Silicalite I, as illustrated in Table 12.5 by the low saturation uptakes of neopentane and o-xylene, is primarily dependent on size exclusion. It is of interest that n-nonane has been found to give an isotherm of essentially Type I character at 296 K (Grillet et al., 1993a,b). The initial part of this isotherm was completely reversible, but a small sub-step at $p/p° \approx 0.2$ was followed by a long plateau and associated narrow, Type H4, hysteresis loop. The plateau was located at $N^\sigma \approx 4$ molec uc^{-1}. This level of pre-adsorption was sufficient to block the whole of the intracrystalline pore structure. The accessibility to nitrogen was gradually restored by the progressive removal of the nonane. These results confirm the complexity of the nonane pre-adsorption and entrapment in relation to the pore network and indicate that there is no simple relation between the thermal desorption of n-nonane and the adsorbent pore structure.

12.5   Phosphate-Based Molecular Sieves: Background and Adsorbent Properties

12.5.1   Background of Phosphate-Based Molecular Sieves

In reporting the results of a spectroscopic study of aluminium phosphate, in 1971 Peri drew attention to the isostructural nature of $AlPO_4$ and SiO_2 and the likely value of $AlPO_4$ as an adsorbent and catalyst support. Stable high-area $AlPO_4$ gels could readily be prepared in 1971, but at that time there was no indication in the open or patent literature that zeolitic forms of aluminophosphate could be synthesised.

The synthesis of the first members of a new family of aluminophosphate molecular sieves (the AlPOs) was disclosed by Union Carbide scientists in the early 1980s (Wilson et al., 1982a,b). The zeotype frameworks of the AlPO structures can be pictured as alternating $[AlO_2]^-$ and $[PO_2]^+$ units and so are electrically neutral with both Al and P occupying adjoining T-sites.

Great expectations were aroused by the discovery of a large-pore AlPO, VPI-5 (VPI after Virginia Polytechnic Institute), by Davis et al. (1988). It had been tacitly assumed by most investigators that the channel diameter of a zeolite structure could not exceed about 1 nm and therefore the discovery of a 1.2-nm channel structure was regarded as a kind of 'psychological breakthrough' (Ozin, 1992).

The two relatively simple AlPOs, which were the first studied in most detail, are $AlPO_4$-5 and VPI-5. Both structures can be considered models for physisorption studies. The framework (001) plane projections of their structures are shown in Figure 12.22.

In $AlPO_4$-5 and VPI-5, the largest channels are essentially straight cylinders; but, as indicated in Table 12.6, there is an appreciable difference in the ring size, the number of T atoms being 12 in $AlPO_4$-5 and 18 in VPI-5. The pore dimensions are derived from crystallographic data of Davis et al. (1989a,b).

Chemical modification of the AlPOs is required to create a new class of catalysts. In the *metalloaluminophosphates* (MeAlPOs), the framework contains metal (Me), aluminium and phosphorus. Thus, it becomes possible to produce a wide range of active catalysts with Lewis and Bronsted acid sites and redox properties by the partial replacement of Al^{3+} by Me^{2+} ions (e.g. Co, Cu, Mg, Zn, etc.) in an AlPO

AlPO₄-5   **VPI-5**

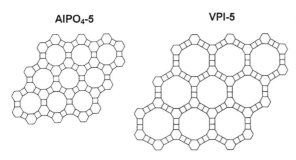

Figure 12.22 The framework (001) plane projections of $AlPO_4$-5 and VPI-5. From Davis et al. (1988).

Table 12.6 Properties of AlPO$_4$-5 and VPI-5

	Ring sizes	Pore width/nm	Pore volume/cm^3 g^{-1}	
AlPO$_4$	4, 6, 12	0.73	0.18	0.15[a]
VPI	4, 6, 18	1.21	0.31	0.26[a]

[a]Unidimensional pores.

framework (Thomas, 1995; Martens et al., 1997; Thomas et al., 2001; Thomas and Raja, 2005).

In the early 1980s, the *silicoaluminophosphates* (SAPOs) with small windows size were discovered to be effective in the conversion of methanol to ethylene and propylene. The most efficient and still most used was the SAPO-34, with cavities of ca. 0.94 nm diameter and eight-membered-ring windows of 0.38 nm diameter. Such a window size allows penetration of linear hydrocarbons (as studied for instance by pulse chromatography, for *n*-alkanes from C$_1$ to C$_{14}$, by Denayer et al., 2008), but not of branched molecules. The partial substitution of Si for P results in a negative charged framework allowing to introduce Bronsted acidity. The acid strength and the acid density, which are as important as the structural features for catalytic activity, can be modified and tuned in various ways (Mees et al., 2003).

The synthesis and properties of another group of acidic metallophosphates, having the general formula Al$_{100}$P$_x$Fe$_y$, has been discussed by Petrakis et al. (1995). Products containing at least 15% phosphorus were found to be X-ray amorphous, mesoporous solids, but at a lower P content there was evidence of phase separation of the Fe and Al.

12.5.2 Adsorbent Properties of Phosphate-Based Molecular Sieves

12.5.2.1 Physisorption of Gases by AlPO$_4$-5

A series of high-resolution adsorption studies have been made at Marseille on different lots of 150 μm hexagonal crystals of AlPO$_4$-5, which had been synthesised by Unger and his co-workers (Muller and Unger, 1988). With the aid of CRTA (see Section 3.4) it was first established that an outgassing temperature of 200 °C was high enough for the removal of physisorbed water without any significant structural change (Tosi-Pellenq et al., 1992; Grillet et al., 1993a,b).

The initial measurements of nitrogen and argon adsorption at 77 K gave smooth Type I isotherms (Muller et al., 1989a,b). The corresponding differential energies of adsorption are shown versus molecules per unit cell in Figure 12.23. As expected, the nitrogen curve is above that of argon, but both systems apparently exhibit a high degree of energetic homogeneity. These results were therefore considered to confirm the theoretical prediction that AlPO$_4$-5 behaves as an essentially homogeneous adsorbent with a micropore capacity of about four molecules of nitrogen or argon per unit cell.

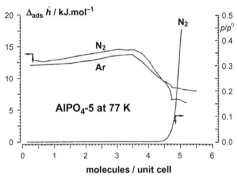

Figure 12.23 Nitrogen isotherm and differential enthalpy of adsorption of nitrogen and argon on AlPO$_4$-5 crystals (Muller et al., 1989a,b).

A nitrogen isotherm, determined at 77 K by static manometry on a second sample of AlPO$_4$-5, is shown in Figure 12.24 (Tosi-Pellenq et al., 1992). In view of the crystalline nature of the AlPO$_4$-5, it was surprising to find that this adsorbent gave a fairly large hysteresis loop in the capillary condensation range. The shape of the loop indicates that the hysteresis was more closely associated with delayed condensation than with network percolation effects. Application of the BJH method of mesopore size analysis to the desorption branch gives a broad distribution of pore widths between 2 and 20 nm. However, this form of mesoporosity cannot be an intrinsic property of the AlPO structure: it is probably the result of structural defects and voids between the crystals.

The same sample of AlPO$_4$-5 was used to obtain the results shown in Figure 12.25. Here, the isotherms of methane, argon, nitrogen and carbon monoxide and the corresponding net differential adsorption enthalpies were determined at 77 K after the adsorbent had been outgassed by CRTA to 353 K.

The adsorption isotherms of methane, nitrogen and carbon monoxide in Figure 12.25a all seem to follow an almost common path up to $p/p° = 0.2$. In fact, there are significant differences in the isotherms at very low $p/p°$ (Coulomb et al., 1997a,b)—as would be expected from the appearance of the net adsorption energy curves in Figure 12.25b. The uptake at the plateau corresponds to 3.5 molecules of argon, nitrogen

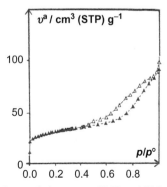

Figure 12.24 Adsorption isotherm of nitrogen at 77 K on AlPO$_4$-5 (Tosi-Pellenq et al., 1992).

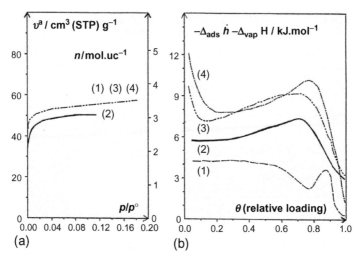

Figure 12.25 (a) Adsorption isotherms on AlPO$_4$-5 at 77 K of methane (1), argon (2), nitrogen (3) and carbon monoxide (4). (b) Corresponding net differential enthalpies of adsorption.

or carbon monoxide adsorbed per unit cell and three molecules of methane. These values are slightly lower than the micropore capacity of four molecules of argon or nitrogen per unit cell observed earlier by the quasi-equilibrium procedure.

The energetic heterogeneity shown by nitrogen and carbon monoxide at low loading (but not by argon and methane) is clearly evident and is probably due to specific interactions with hydroxyl groups and defect sites. With argon, nitrogen and carbon monoxide, the adsorbate–adsorbate interactions are responsible for the enthalpy maxima at loadings of 70–80%. The methane net differential enthalpy curve is quite different and is indicative of an adsorbate phase change, which was originally thought to involve a change from a type of 'fluid' to a 'solid' phase.

Neutron diffraction studies (Coulomb et al., 1994a,b, 1997a,b) have provided more information on the methane phase transition. Adsorption isotherm and neutron diffraction measurements were made at different temperatures for D$_2$, ^{36}Ar, CD$_4$ and CF$_4$ on two samples: one of pure AlPO$_4$-5 and another, SAPO-5 (containing 5% Si). A characteristic sub-step was observed only with the CD$_4$/AlPO$_4$-5. The neutron diffraction patterns have indicated that two different solid-type phases are involved in the phase transition. It appears that the adsorbed phases are comprised of ordered chains of CD$_4$ molecules and that a structural change occurs at the sub-step. It has been tentatively suggested that the chains are changed from a 'dimeric' form (at a loading of 4 molec uc^{-1}) to 'trimers' (at 6 molec uc^{-1}). However, since the kinetic diameter of methane is ca. 0.38 nm, it would seem unlikely that the change in ordering is quite as simple as this would suggest ! Another viewpoint has been provided by a related computer simulation study, which is discussed later.

Other neutron diffraction spectra obtained by Coulomb et al. (1997a,b) have revealed different phase 'signatures'. For example, with SAPO-5 (derived from AlPO$_4$-5 by isomorphous substitution of aluminium and phosphorus by silicon) it was observed that the diffraction pattern for the Ar/SAPO-5 system underwent a change at high Ar loading, which was interpreted as an increase in short range order. These results indicate that at low and medium loadings the adsorbed Ar has a relatively high mobility while at high loading it 'solidifies' in a vitreous form. The preservation of a high degree of disorder would therefore appear to be responsible for the absence of a phase transformation. The presence of a weak sub-step in the Kr/AlPO$_4$-5 isotherm at 77.3 K is an indication of the intermediate behaviour of this system at the operational temperature.

Molecular simulation of the adsorption of gases by the AlPOs was pioneered by Cracknell and Gubbins (1993), who pointed out that the aluminophophates should be easier to model than the aluminosilicates. There are two important advantages: first, the charge neutrality of the framework means that there are *no exchangeable cations* to be taken into account (this is, of course, also true for pure Silicalite); secondly, the modelling is simpler because the pores are unidirectional with *no interconnections*.

Adsorption isotherms for argon in AlPO$_4$-5 and other AlPOs were simulated at 77 and 87 K by the Grand Canonical Monte Carlo (GCMC) method. Refined Lennard-Jones (12-6) potentials were used to model the argon-oxygen interactions with some adjustment made for the less important contribution from the argon interactions with the T atoms. The polarisation contribution, E_p, was considered to be very small and was ignored. The procedures for the GCMC simulations were based on the generally accepted approach for bulk fluids (Allen and Tildesley, 1987). The acceptance probabilities for the movement, creation and deletion of particles were chosen to allow the probability of a given state of the system, after a sufficiently high number of trials, to correspond to that of the Grand Canonical Ensemble at constant μ, V and T. The simulations thus gave the ensemble average of the number of argon atoms per unit cell, which could be plotted against the reduced pressure thereby giving a model isotherm for each ALPO.

Cracknell and Gubbins (1993) found that the low-pressure region of their simulated isotherm for Ar/AlPO$_4$-5 agreed well with experimental data at 87 K, but that the curves began to diverge at $p/p^\circ \approx 10^{-4}$ with the simulated maximum loading lying about 25% above the experimental capacity determined by Hathaway and Davis (1990). The simulated variation of adsorption energy with uptake, although similar in general form to the argon curve in Figure 12.25, also deviated from the experimental microcalorimetric data. In seeking an explanation for these discrepancies, Cracknell and Gubbins drew attention to the assumptions made in the simulations including the use of the Lennard-Jones parameters and the supposed rigidity of the AlPO$_4$-5 structure. However, it was considered more likely that a more serious problem was an experimental one: that the unidimensional pores are particularly susceptible to pore blocking.

A detailed molecular simulation study of CH$_4$/AlPO$_4$-5 was undertaken by Lachet et al. (1996). The standard GCMC procedure was again used to equilibrate the system at a series of pressures and thereby allow an assessment to be made of the number of molecules per unit cell and the differential energies of adsorption. It was also assumed that the AlPO$_4$-5 structure remained rigid but in this study various potentials were used to model the adsorbent–adsorbate and adsorbate–adsorbate interactions. Sub-steps could be generated with certain sets of parameters, but the position of the sub-step and the agreement with microcalorimetric data depended critically on the repulsive oxygen-methane parameter and therefore indirectly on the pore size. Lachet et al. (1996) have concluded that the experimental sub-step can be interpreted as a structural rearrangement of the adsorbed methane. On the other hand, they are undecided whether the phenomenon should be regarded as a true 'phase transition' and they point out that the exact structure at maximum loading is still unknown.

We cannot avoid remarking from the foregoing discussion that significant differences have been noted in the properties of the various samples of AlPO$_4$-5.

It is interesting that, unlike the situation with Silicalite I, argon and nitrogen do not appear to undergo well-defined phase transitions in the unidirectional channels of AlPO$_4$-5. This is consistent with the results of a careful study carried out by Boddenberg et al. (2004) on N$_2$ adsorption, over seven decades of pressure, at 77 K, by a relatively similar SAPO-5 sample, with unidimensional channels of 0.73 nm diameter, They conclude that adsorption takes place in three successive steps, namely (i) unidimensional micropore filling, with a site-adsorption explaining—better than any pore blocking—the incomplete pore filling of only 2 N$_2$ molecules per unit cell, (ii) multilayer adsorption on the outer surface of ca. 200 nm sized agglomerates (corresponding to a BET-area of 63 m^2 g^{-1}) and (iii) capillary condensation between the agglomerates, similar to that observed by Tosi-Pellenq et al. (1992) on AlPO$_4$-5 (see Figure 12.24).

The adsorption–desorption isotherm of water vapour in Figure 12.26 exhibits hysteresis over the entire range of $p/p°$. This form of hysteresis, which extends down to very low $p/p°$, cannot be attributed solely to the filling and emptying of mesopores. Low-pressure hysteresis is a common feature of water isotherms: the most likely explanation for the strong retention of water by the AlPO$_4$-5 is dissociative chemisorption of water on the defect sites to give Al-OH and P-OH. Indeed, the amount of water retained at an outgassed pressure of 10^{-3} hPa is consistent with the high-energy uptakes of nitrogen and carbon monoxide.

By using quasi-equilibrium adsorption–desorption gravimetry, Grillet et al. (1993a,b) were able to detect the sub-step, AB, in the water isotherm at $p/p° \approx 0.15$.

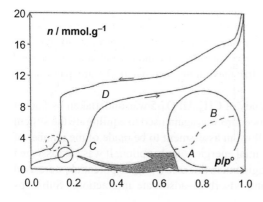

Figure 12.26 The adsorption–desorption isotherm of water vapour at 290 K on AlPO$_4$-5 (Grillet et al., 1993a,b).

The fact that the sub-step can be seen in both the adsorption and desorption isotherms (although at slightly different positions) suggests that it is associated with an adsorbate phase change. This explanation is consistent with a small change in the isosteric enthalpy of adsorption and with the changes in NMR spectra, but at this stage it is not possible to explain the nature of the transition.

The much larger step, CD, in the adsorption branch at $p/p° ≈ 0.3$ is unlikely to be due to a simple form of capillary condensation. It is significant that the corresponding—very sharp—desorption step is located at a much lower $p/p°$. Stach et al. (1986) have suggested that it corresponds to both capillary condensation and a slow coordinative hydration giving *inter alia* $Al(OP_4)(OH_2)_2$. This explanation is supported by the results of NMR spectroscopy.

12.5.2.2 Physisorption of Gases by VPI-5

The physisorption isotherms obtained by Davis et al. (1988, 1989a,b) in their original investigations of VPI-5 were reported to be of Type I. However, inspection of the isotherms of argon, oxygen and water vapour reveals that a significant amount of additional adsorption of each adsorptive occurred $p/p° > 0.4$. No desorption measurements were reported and therefore it is not possible to deduce whether the further uptake at high $p/p°$ was due to multilayer adsorption, capillary condensation or an irreversible change in the adsorption system.

The more recent adsorption–desorption isotherms of some organic vapours of different molecular diameter are shown in Figure 12.27. These measurements were made on different samples of VPI-5, each having been outgassed for 16 h at 673 K. Although essentially of Type I, the isotherms reveal some degree of thermodynamic irreversibility with hysteresis extending back to very low $p/p°$. Evidently, this form of hysteresis cannot be attributed to capillary condensation and is instead indicative of a more complex change in the adsorption system (e.g. activated entry of adsorptive molecules or swelling of the adsorbent).

Values of apparent specific micropore volume, v_p, of VPI-5 evaluated from the uptakes determined by Kenny et al. (1992) of various adsorptives at $p/p° = 0.4$ are recorded in Table 12.7. As before, the adsorbate densities are assumed equal to

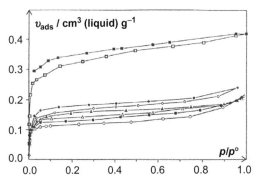

Figure 12.27 Adsorption–desorption isotherms on VPI-5 of methanol (squares), isobutane (diamonds), neopentane (circles) and propane (triangles) determined at 298, 261, 273 and 196 K, respectively (Kenny et al., 1992).

Table 12.7 Values of Apparent Micropore volume, v_p, of VPI-5 Evaluated from the Adsorption Capacities of Different Vapours at $p/p° = 0.4$ (Kenny et al., 1992)

Adsorptive	Kinetic diameter nm	Temperature K	Apparent v_p $cm^3 \, g^{-1}$
Water	0.27	298	0.31–0.35[a]
Methanol	0.38	298	0.32
Oxygen	0.35	77	0.23[a]
Argon	0.34	77	0.17[b]
Nitrogen	0.36	77	0.187, 0184[c], 0.146[b]
Methane	0.38	77	0.17[b]
Propane	0.43	196	0.17
Isobutane	0.50	261	0.18
Neopentane	0.62	273	0.12, 0.15[a]

[a]Davis et al. (1989a,b).
[b]Reichert et al. (1994).
[c]Schmidt et al. (1992).

the respective liquid densities at the operational temperatures. Also included in Table 12.7 are values of v_p derived from the measurements of Davis et al. (1989b), Schmidt et al. (1992) and Reichert et al. (1994). The complexity of the behaviour of VPI-5 is illustrated by the poor agreement between some individual values of v_p and also the apparent molecular sieving effects.

In attempting to explain the anomalous results in Figure 12.27 and Table 12.7, Kenny et al. (1991a, 1992) undertook a systematic study of the stability of VPI-5. First, successive BET-nitrogen areas were determined (by the single-point method) on two samples taken from the same batch of VPI-5. Sample A was kept in a dry condition between successive nitrogen adsorption measurements, while sample B was exposed to water vapour. The same conditions of outgassing were adopted for both samples, namely 16 h at 673 K. The BET-area of sample A remained almost constant at $394 \pm 7 \, m^2 \, g^{-1}$, whereas that of sample B underwent the following progressive decrease: 404, 309, 179, 147 $m^2 \, g^{-1}$.

In another set of experiments, the effect of changing the outgassing conditions was studied by outgassing one sample at increasingly higher temperatures over the range 273–1173 K and outgassing separate samples at the same temperatures. The separate samples had consistently higher BET-areas, differences of ca. 100 $m^2 \, g^{-1}$ being recorded.

With the aid of thermal analysis, X-ray diffraction and diffuse reflectance FTIR, it was possible to deduce that the removal of the zeolitic water brought about a partial structural transformation of the VPI-5 to $AlPO_4$-8. Schmidt et al. (1992) have studied this phase transition in some detail and have shown that it can be avoided if the sample is heated slowly under reduced pressure (10^{-5} hPa). In this manner, VPI-5 can be heated to 450–500 °C without any detectable change of structure.

Since the $AlPO_4$-8 structure has a ring opening of about 0.8 nm, one would expect it to accommodate all the molecules in Table 12.7. The low uptake of neopentane and the low-pressure hysteresis indicate that the aged material has an imperfect $AlPO_4$-8 structure. It seems likely that the phase transition has resulted in the development of constrictions and some blockage of the unidirectional channels.

The results in Table 12.7 appear to show that a considerably larger micropore volume is available for the adsorption of water and methanol than for the other adsorptives. A somewhat larger uptake of water is to be expected in view of its molecular size and apparent ability to enter the narrow six-membered rings (Davis et al., 1989a,b), but the high uptake of methanol is unexpected and requires further investigation.

The unusual character of the water isotherm on VPI-5 can be seen in Figure 12.28. There are three distinct steps at low $p/p°$. Step 1 occurs at $p/p° < 0.001$, step 2 at $p/p° = 0.013$ and step 3 at $p/p° = 0.060$. Between steps 2 and 3 there is a narrow hysteresis loop, which in this region of the isotherm cannot be associated with capillary condensation.

Step 1 in Figure 12.28 must be the result of interactions between water molecules and highly active regions of the VPI-5 surface, which may be exposed P–OH groups and possibly other defect sites (Kenny et al., 1991b). At present, the explanation for the two pore-filling steps, 2 and 3, must also be speculative. It is of interest that these risers are located at much lower relative pressures than the corresponding single pore-filling step for $AlPO_4$-5 (Carrott and Freeman, 1991). This suggests that the 3D, hydrogen-bonded water structure can be more readily formed in the wider channels of VPI-5.

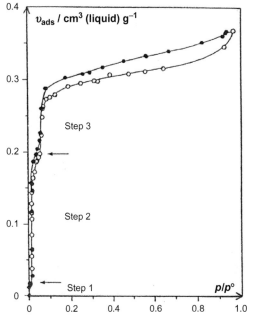

Figure 12.28 The adsorption–desorption isotherm of water vapour determined at 298 K on VPI-5 outgassed at 673 K for 16 h (Kenny et al., 1991b).

It seems likely that a rearrangement of water molecules at the pore walls is responsible for the separation of steps 2 and 3 and the presence of the associated hysteresis loop.

The difficulty of arriving at an unambiguous assessment of the micropore capacity is immediately apparent when we see the shape of the water isotherm in Figure 12.28. Since the micropore filling process is confined to the range $p/p° <\approx 0.1$, there is an overestimate of $\approx 20\%$ in v_p if the molecular sieve capacity is taken as the uptake at $p/p° = 0.4$.

Finally, we note that the maximum $p/p°$ attained on the adsorption branch in Figure 12.28 is ca. 0.95. As indicated earlier, exposure of the adsorbent to water vapour at $p/p° \approx 1$ caused an irreversible partial conversion of the VPI-5 structure to $AlPO_4$-8. Furthermore, a significant decrease in the water adsorption capacity was found when the outgassing temperature was taken to 673 K in a stepwise manner (Kenny et al., 1992). We conclude that CRTA would be admirably suitable for a more rigorous study of the properties and thermal stability of VPI-5.

12.6 Applications of Clays, Zeolites and Phosphate-Based Molecular Sieves

12.6.1 Applications of Clays

The adsorbent properties of clays, their availability, their inertness and stability and their low cost have led to a large number of industrial applications (Harvey and Lagaly, 2006). The applications hereafter are those directly related to adsorbent properties.

Adsorption from the liquid phase: clays are used for *refining* of mineral oils, regeneration of solvents for dry cleaning, *decolouration*, dye adsorption (Errais et al., 2012; Günay et al., 2013) stain-removing pastes, carbonless copying paper (where bentonite adsorbs the dye expelled by the microcapsules crushed by the ball-pen), adsorption of radioactive materials (e.g. cesium), adsorption of heavy metals: lead (Oubagaranadin and Murthy, 2009), chromium and cadmium (Ghorbel-Abid and Trabelsi-Ayadi, 2011), cadmium, nickel, zinc, lead and copper (Padilla-Ortega et al., 2013), de-inking in waste paper recycling, and also as a drug carrier.

Adsorption from the gas phase: clays are also used for odour control, as a gastric gas adsorbent, as a cigarette filter (sepiolite) and have other potential uses: SO_2 adsorption by montmorillonite or by thermally and acid-activated kaolin (Volzone and Ortiga, 2009, 2011), H_2 adsorption by pillared montmorillonite (Gil et al., 2009), natural gas purification, with help of a silica-smectite heterostructure (Pires et al., 2008).

Catalysis: the catalytic activity of smectites, usually acid-treated and often cation-exchanged, results all at once from their surface area, from their chemical nature (acidity, cations) and, also from the 2D interlayer space which lends itself to an easier diffusion or reactants and products than a 1D pore and which gives the molecules a greater chance to meet, on a random diffusion, than a 3D space (Laszlo, 1987).

Pillaring allows tuning the interlayer distance, introducing catalytic active pillars and increasing the thermal stability of the clay. The many applications are those of Bronsted acid catalysis, Lewis acid catalysis and redox reactions, plus of course those of simple catalyst support. Nevertheless, it seems that the further development of their industrial use still depends on a predictive capability based on a better understanding of the mechanisms and on the preparation of very well characterised and highly reproducible materials (Adams and McCabe, 2006). This is in part made difficult by the natural origin of the clays.

12.6.2   Applications of Zeolites

The exceptional properties of zeolites, not only as molecular sieves but also as thermally and mechanically stable materials, lending themselves to ion-exchange and to the tuning of their windows aperture, makes them a unique material in chemical engineering and for practical applications (Guisnet and Gilson, 2002; Maesen, 2007).

12.6.2.1   Gas Physisorption

In *gas physisorption*, zeolites are extremely efficient *dessicants* and they provide a very low residual vapour pressure (especially, of course, the 3A and 4A types) as long as they are not saturated: as early as 1955 they were used in the *drying of refrigerant gas and natural gas*. One of their important uses (in tonnage) is for *double glazed windows*, where they are hidden in the frame, to avoid any condensation on the cold side.

Nevertheless, it is in the field of *gas separation* that the unique properties of zeolites for gas physisorption are better exploited. Over the past 30 years, *pressure swing adsorption* (PSA) has become indeed one of the most important procedures for the separation and enrichment of industrial gases (Sircar, 1993). PSA technology is dependent on the selective adsorption of one, or more than one, of the components of a gas mixture and involves both the chemical and diffusion properties of the zeolite (Ruthven and Brandani, 2000). A change in composition occurs when the gas mixture is brought into contact with an adsorbent. Desorption of a selectively adsorbed component occurs when its partial pressure is sufficiently reduced and the adsorbent is thereby cleaned and ready for the next stage of adsorption. In this manner, the PSA process consists of a cycle of stages of adsorption at high partial pressures and desorption at low partial pressures. PSA technology is used for the separation of gases produced by oil refineries and for the recovery and purification of hydrogen. It is extensively used for the fractionation of air, with a number of different processes in industrial operation; the decision to use a certain design of plant and a particular adsorbent is governed by whether oxygen or nitrogen is required and by the level of purity. A zeolite (e.g. a 5A, NaX (13X) or, still, LiX) is the preferred adsorbent for the enrichment (up to about 95%) of oxygen (Yang, 1987; Hutson et al., 2000). It is not within the scope of this book to discuss the chemical engineering principles of PSA, but it is important to note that the application of a zeolite as the working adsorbent does not depend on molecular sieving. In this case, the achievable level of separation is closely related to the separation factor (or selectivity ratio)—defined in terms of the equilibrium distribution of each component between the gas and the adsorbed phase (Yang, 1987).

12.6.2.2   Liquid Adsorption

In *liquid adsorption* a simple but most important application in tonnage is in *detergent powders*, where zeolites replace the former phosphates as ion exchangers to adsorb Ca^{++} and Mg^{++} in place of Na^+, in order to soften the water and prevent the precipitation of calcium or magnesium surfactant salts which give an unclean aspect. Zeolites are also used as a nutrient release agent in agriculture and horticulture. The use of MFI zeolites as an alternative to blood dialysis by microporous polymeric membranes, in order to eliminate uremic toxins, looks promising (Wernert et al., 2005, 2006; Bergé-Lefranc et al., 2008, 2009, 2012). Hydrophobic zeolites were also shown to store energy, by water adsorption under high pressure, so that it can be envisaged to use the zeolite-water system as a low volume damper (Eroshenko et al., 2001).

12.6.2.3   Catalysis

Catalytic applications of zeolites attracted the major research effort. This started in 1967, with one of the most significant stages in the development of zeolite catalysts, which was the synthesis by Mobil scientists (U.S. Patent 3,702,866) of the zeolite now universally known as *ZSM-5* (i.e. Zeolite Socony Mobil-5). This was the first—and most important—member of a new class of shape-selective catalysts, which have made viable the production of 'synthetic gasoline'. In this process, high-octane gasoline is produced by the catalytic conversion of methanol to a mixture of aromatic and aliphatic hydrocarbons (Derouane, 1980). Because of its unique combination of chemical nature and pore structure, ZSM-5 is a highly effective dehydration, isomerisation and polymerisation catalyst. It opened the way to a family of catalysts whose highly similar cavities can be considered as a large set of nano-reactors operating simultaneously.

In 1978, Flanigen and her co-workers reported the synthesis, structure and properties of 'a new hydrophobic crystalline silica molecular sieve' (Flanigen et al., 1978). The new material, named *Silicalite* (now generally called Silicalite I), has a remarkably similar channel structure to that of ZSM-5 but contains no aluminium. It was pointed out by the Union Carbide scientists that, unlike the aluminium-containing zeolites, Silicalite has no cation exchange properties and consequently exhibits a low affinity for water. In addition, it was reported to be unreactive to most acids (but not HF) and stable in air to over 1100 °C.

To-day, the properties which make the success of zeolites in catalysis are (i) the possibility of a *high concentration* of *active sites* (ii) their high *thermal and hydrothermal stability* and (iii) their selectivity, based in great part on *size exclusion* (Corma and Martinez, 2002). The catalytic applications favour zeolites with well-separated locations of the cationic protons and with a high silica content, which make them able to withstand the high temperatures required during catalytic reactions and during the regeneration. The great choice of zeolite structures and their fine tuning (in size and in cationic content) provides many possibilities for *shape-selective conversions*. They are largely used in *oil cracking*, in the *petrochemical industry* and in *fine chemistry* (drugs, fragrances, of which most are synthetic to-day).

12.6.2.4   Zeolitic Membranes

One special application of zeolites is in the form of *membranes*, where zeolites are usually supported by a porous ceramic, most often alumina, on which the zeolite is grown by various ways (Julbe, 2005).

Their main industrial application is the *dehydration* of organic solvents (alcohols, ketones, ethers, cyclic hydrocarbons, etc.) by means of a continuous pervaporation process making use of hydrophilic zeolites (Morigami et al., 2001; Kita, 2006).

They can also be used for *gas separation* (Gavalas, 2006), like air separation (Wang et al., 2002) but have to compete, in most applications, with the cheaper and less mechanically fragile polymeric membranes.

The interest of zeolite membranes is then still, their chemical and thermal properties, especially to process fluids at temperatures which are not withstood by the polymers. Their performances are highly dependent on the *absence of defects* (anomalous pore sizes), whose necessary detection is made with help of either dynamic or static methods (Julbe and Ramsay, 1996).

12.6.3   Applications of Phosphate-Based Molecular Sieves

The applications of phosphate-based molecular sieves are of same nature as those of zeolites, though mainly centred on catalysis.

Two of them deserve a special mention for use as *catalysts* at an industrial scale: SAPO-11 with a Si mole fraction of ca.15% and a micropore size around 0.6 nm (Jacobs and Martens, 1986) is commonly used for *dewaxing of lubricating oil*, whereas SAPO-34, with a Si mole fraction of ca. 3% and a micropore size around 0.4 nm is much used to *convert methanol into olefins* (Wilson, 2007).

Also, the use of MeAPO molecular sieves containing a few percent of transition metal ions (of Co, Mn, Fe) in place of Al^{+++} sites for *shape-selective oxidation* of organic molecules was extensively studied by Thomas et al. (2001, 2005): the size of the pores allows the oxygen molecule to freely diffuse, whereas the selection of appropriate cage dimensions, mean distance between metal ions and windows aperture allows a precise orientation of the reaction towards the desired part of the organic molecule to be oxidised.

These phosphate-based molecular sieves also lend themselves to the preparation of *ceramic membranes* for the *separation of small molecules*: ceramics based on SAPO-34 were found to separate CO_2/CH_4 with a selectivity as high as 270 (Li et al., 2005). Mixed membranes of SAPO-34 with a polymer matrix were also made and characterised, and led to a CO_2/CH_4 selectivity of 67 (Peydayesh et al., 2013).

References

Adams, J.M., McCabe, R.W., 2006. In: Bergaya, F., Theng, B.K.G., Lagaly, G. (Eds.), Handbook of Clay Science. Elsevier, Amsterdam, p. 541 (Chapter 10.2).

Allen, M.P., Tildesley, D.J., 1987. Computer Simulation of Liquids. Oxford University Press, Oxford.

Annabi-Bergaya, F., Cruz, M.I., Gatineau, L., Fripiat, J.J., 1979. Clay Miner. 14, 249.

Annabi-Bergaya, F., Cruz, M.I., Gatineau, L., Fripiat, J.J., 1981. Clay Miner. 16, 115.

Atkinson, D., Curthoys, G., 1981. J. Chem. Soc. Faraday Trans. I 77, 897.

Atlas of Zeolite Structure Types, 1992. In: Meier, W.M., Olson, D.H. (Eds.), International Zeolite Association. Butterworth–Heinemann, London.

Avgul, N.N., Bezus, A.G., Dobrova, E.S., Kiselev, A.V., 1973. J. Colloid Interface Sci. 42, 486.

Barrer, R.M., 1945. J. Soc. Chem. Ind. 64, 130.

Barrer, R.M., 1966. J. Colloid Interface Sci. 21, 415.

Barrer, R.M., 1978. Zeolites and Clay Minerals. Academic Press, London.

Barrer, R.M., 1981. J. Chem. Technol. Biotechnol. 31, 71.

Barrer, R.M., 1982. Hydrothermal Chemistry of Zeolites. Academic Press, London.

Barrer, R.M., 1989. Pure Appl. Chem. 61, 1903.

Barrer, R.M., Kelsey, K.E., 1961. Trans. Faraday Soc. 57, 625.

Barrer, R.M., MacLeod, D.M., 1955. Trans. Faraday Soc. 51, 1290.

Barrer, R.M., McLeod, D.M., 1954. Trans. Faraday Soc. 50, 980.

Barrer, R.M., Reay, J.S.S., 1957. In: Schulman, J.H. (Ed.), Proc. Second Int. Cong. Surface Activity, II. Butterworths, London, p. 79.

Bergaoui, L., Lambert, J.F., Vicente-Rodriguez, M.A., Michot, L.J., Villieras, F., 1995. Langmuir. 11, 2849.

Bergaya, F., 1995. J. Porous Mater. 2, 91.

Bergaya, F., Lagaly, G., 2006. In: Bergaya, F., Theng, B.K.G., Lagaly, G. (Eds.), Handbook of Clay Science. Elsevier, Amsterdam, p. 1 (Chapter 1).

Bergaya, F., Gatineau, L., Van Damme, H., 1993. In: Sequeira, C.A.C., Hudson, M.J. (Eds.), Multifunctional Mesoporous Inorganic Solids. Kluwer, Dordrecht, p. 19.

Bergé-Lefranc, D., Pizzala, H., Paillaud, J.L., Schaef, O., Vagner, C., Boulet, P., Kuchta, B., Denoyel, R., 2008. Adsorption. 14 (2/3), 377.

Bergé-Lefranc, D., Pizzala, H., Denoyel, R., Hornebecq, V., Berge-Lefranc, J.L., Guieu, R., Brunet, P., Ghobarkar, H., Schaef, O., 2009. Microp. Mesop. Mater. 119 (1–3), 186.

Bergé-Lefranc, D., Vagner, C., Calaf, R., Pizzala, H., Denoyel, R., Brunet, P., Ghobarkar, H., Schaef, O., 2012. Microp. Mesop. Mater. 153, 288.

Bernier, A., Admaiai, L.F., Grange, P., 1991. Appl. Catal. 77, 269.

Bish, D.L., 2006. In: Bergaya, F., Theng, B.K.G., Lagaly, G. (Eds.), Handbook of Clay Science. Elsevier, Amsterdam, p. 1097 (Chapter 13.2).

Boddenberg, B., Rani, V.R., Grosse, R., 2004. Langmuir. 20, 10962.

Breck, D.W., 1974. Zeolite Molecular Sieves. Wiley, New York.

Breck, D.W., Eversole, W.G., Milton, R.M., Reed, T.B., Thomas, T.L., 1956. J. Am. Chem. Soc. 78, 5963.

Brindley, G.W., Sempels, R.E., 1977. Clay Mineral. 12, 229.

Burch, R., 1987. Catal. Today. 2, 185.

Carrott, P.J.M., Freeman, J.J., 1991. Carbon. 29, 499.

Carrott, P.J.M., Sing, K.S.W., 1986. Chem. Ind. 786.

Carrott, P.J.M., Kenny, M.B., Roberts, R.A., Sing, K.S.W., Theocharis, C.R., 1991. In: Rodriguez-Reinoso, F., Rouquerol, J., Sing, K.S.W., Unger, K.K. (Eds.), Characterization of Porous Solids II. Elsevier, Amsterdam, p. 685.

Cases, J.M., Lietard, O., Yvon, J., Delon, J.F., 1982. Bull. Mineral. 105, 439.

Cases, J.M., Cunin, P., Grillet, Y., Poinsignon, C., Yvon, J., 1986. Clay Miner. 21, 55.

Cases, J.M., Grillet, Y., François, M., Michot, L., Villieras, F., Yvon, J., 1991. In: Rodriguez-Reinoso, F., Rouquerol, J., Sing, K.S.W., Unger, K.K. (Eds.), Characterization of Porous Solids II. Elsevier, Amsterdam, p. 591.

Cases, J.M., Berend, I., Besson, G., Francois, M., Uriot, J.P., Thomas, F., Poirier, J.E., 1992. Langmuir. 8, 2730.

Cool, P., Vansant, E.F., 1996. Microp. Mater. 6, 27.

Corma, A., Martinez, A., 2002. In: Guisnet, M., Gilson, J.P. (Eds.), Zeolites for Cleaner Technologies. Imperial College Press, London, p. 29 (Chapter 2).

Coulomb, J.P., Llewellyn, P., Grillet, Y., Rouquerol, J., 1994a. In: Rouquerol, J., Rodriguez-Reinoso, F., Sing, K.S.W., Unger, K.K. (Eds.), Characterization of Porous Solids III. Elsevier, Amsterdam, p. 535.

Coulomb, J.-P., Martin, C., Grillet, Y., Tosi-Pellenq, N., 1994b. In: Weitkamp, J., Karge, H.G., Pfeifer, H., Holderich, W. (Eds.), Zeolites and Related Microporous Materials: State of the Art. Elsevier, Amsterdam, p. 445.

Coulomb, J.-P., Martin, C., Grillet, Y., Llewellyn, P.L., André, J., 1997a. In: Chon, H., Ihm, S.K., Uh, Y.S. (Eds.), Progress in Zeolites and Microporous Materials, Studies in Surface Science and Catalysis, Vol. 105. p. 1827.

Coulomb, J.-P., Martin, C., Llewellyn, P.L., Grillet, Y., 1997b. Progress in Zeolites and Microporous Materials, Studies in Surface Science and Catalysis, Vol. 105. p. 2355.

Cracknell, R.F., Gubbins, K.E., 1993. In: Suzuki, M. (Ed.), Proc IVth Int. Conf. on Fundamentals of Adsorption. Kodansha, Tokyo, p. 105.

Dailey, J.S., Pinnavaia, T.J., 1992. Chem. Mater. 4, 855.

Davis, M.E., Saldarriaga, C., Montes, C., Garces, J., Crowder, C., 1988. Zeolites. 8, 362.

Davis, M.E., Hathaway, P.E., Montes, C., 1989a. Zeolites. 9, 436.

Davis, M.E., Montes, C., Hathaway, P.E., Arhancet, J.P., Hasha, D.L., Garces, J.M., 1989b. J. Am. Chem. Soc. 111, 3919.

de Boer, J.H., Lippens, B.C., Linsen, B.G., Broekhoff, J.C.P., van den Heuvel, A., Osinga, Th.J., 1966. J. Colloid Interface Sci. 21, 405.

Delville, A., Sokolowski, S., 1993. J. Phys. Chem. 97, 6261.

Denayer, J.F.M., Baron, G.V., 1997. Adsorption. 3, 251.

Denayer, J.F.M., Devriese, L.I., Couck, S., Martens, J., Singh, R., Webley, P.A., Baron, G.V., 2008. J. Phys. Chem. C. 112, 16593.

Derouane, E.G., 1980. Catalysis by Zeolites. In: Imelik, B., Naccache, C., Ben Taarit, Y., Vedrine, J.C., Coudurier, G., Praliaud, H. (Eds.), Elsevier, Amsterdam, p. 5.

Derycke, L., Vigneron, J.P., Lambin, P., Lucas, A.A., Derouane, E.G., 1991. J. Chem. Phys. 94, 4620.

de Stefanis, A., Perez, G., Tomlinson, A.A.G., 1994. J. Mater. Chem. 4, 959.

Diano, W., Rubino, R., Sergio, M., 1994. Microp. Mater. 2, 179.

Douguet, D., Pellenq, R.J.M., Boutin, A., Fuchs, A.H., Nicholson, D., 1996. Mol. Sim. 17 (4–6), 255.

Dubinin, M.M., 1975. Progress in Surface and Membrane Science, Vol. 9. Academic Press, New York, p. 1.

Eroshenko, V., Regis, R.C., Soulard, M., Patarin, J., 2001. J. Am. Chem. Soc. 123 (33), 8129.

Errais, E., Duplay, J., Elhabiri, M., Khodja, M., Ocampo, R., Baltenweck-Guyot, R., Darragi, F., 2012. Colloids Surf. A Physicochem. Eng. Asp. 403, 69.

Farfan-Torres, E.M., Dedeycker, O., Grange, P., 1991. In: Poncelet, G., Grange, P., Delmon, B. (Eds.), Preparation of Catalysts V. Elsevier, Amsterdam, p. 337.

Feijen, E.J.P., Martens, J.A., Jacobs, P.A., 1997. In: Ertl, G., Knozinger, H., Weitkamp, J. (Eds.), Handbook of Heterogeneous Catalysis, Vol. 1. Wiley-VCH, Weinheim, p. 311.

Flanigen, E.M., Bennett, J.M., Grose, R.W., Cohen, J.P., Patton, R.L., Kirchner, R.M., Smith, J.V., 1978. Nature. 271, 512.

Fripiat, J.J., 1997. In: Ertl, G., Knozinger, H., Weitkamp, J. (Eds.), Handbook of Heterogeneous Catalysis, Vol. 1. Wiley-VCH, Weinheim, p. 387.

Galarneau, A., Barodawalla, A., Pinnavaia, T.J., 1995. Nature. 374, 529.

Gavalas, G.R., 2006. In: Pinnau, I., Yampolskii, Y., Freeman, B.D. (Eds.), Materials Science of Membranes for Gas and Vapor Separation. Wiley, Chichester, UK, p. 307.

Ghorbel-Abid, I., Trabelsi-Ayadi, M., 2011. Arabian J. Chem.

Gil, A., Gandia, L.M., 2003. Chem. Eng. Sci. 58, 3059.

Gil, A., Grange, P., 1996. Colloids Surf. 113, 39.

Gil, A., Massinon, A., Grange, P., 1995. Microp. Mater. 4, 369.

Gil, A., Trujillano, R., Vicente, M.A., Korili, S.A., 2009. Int. J. Hydrogen Energy. 34, 8611.

Gregg, S.J., 1968. Chem. Ind., 611.

Gregg, S.J., Packer, R.K., 1954. J. Chem. Soc., 3887.

Gregg, S.J., Stephens, M.J., 1953. J. Chem. Soc., 3951.

Gregg, S.J., Parker, T.W., Stephens, M.J., 1954. J. Appl. Chem. 4, 666.

Grillet, Y., Cases, J.M., François, M., Rouquerol, J., Poirier, J.E., 1988. Clays Clay Miner. 36, 233.

Grillet, Y., Llewellyn, P.L., Kenny, M.B., Rouquerol, F., Rouquerol, J., 1993a. Pure Appl. Chem. 65, 2157.

Grillet, Y., Llewellyn, P.L., Tosi-Pellenq, N., Rouquerol, J., 1993b. In: Suzuki, M. (Ed.), Proc IVth Int. Conf. on Fundamentals of Adsorption. Kodansha, Tokyo, p. 235.

Guisnet, M., Gilson, J.P., 2002. In: Guisnet, M., Gilson, J.P. (Eds.), Zeolites for Cleaner Technologies. Imperial College Press, London, p. 1 (Chapter 1).

Günay, A., Ersoy, B., Dikmen, S., Evcin, A., 2013. Adsorption. 19, 757.

Harvey, C.C., Lagaly, G., 2006. In: Bergaya, F., Theng, B.K.G., Lagaly, G. (Eds.), Handbook of Clay Science. Elsevier, Amsterdam, p. 501 (Chapter 10.1).

Hathaway, P.E., Davis, M.E., 1990. Catal. Lett. 5, 333.

Hutson, N.D., Zajic, S.C., Yang, R.T., 2000. Ind. Eng. Chem. Res. 39, 1775.

Jacobs, P.A., Martens, J.A., 1986. Pure Appl. Chem. 10, 1329.

Jacobs, P.A., Martens, J.A., 1987. Synthesis of High Silica Aluminophosphate Zeolites. Elsevier, Amsterdam.

Jiang, J., Jorda, J.L., Yu, J., Baumes, L.A., Mugnaioli, E., Diaz-Cabanas, M.J., Kolb, U., Corma, A., 2011. Science. 333 (6046), 1131.

Julbe, A., 2005. In: Cejka, J., van Bekkum, H. (Eds.), Zeolites and Ordered Mesoporous Materials Progress and Prospects. Stud. Surf. Sci. Catal., Vol. 157. Elsevier, Amsterdam, p. 135.

Julbe, A., Ramsay, J.D.F., 1996. In: Cot, L., Burggraaf, A.J. (Eds.), Fundamentals of Inorganic Membrane Science and Technology. Membrane Science and Technology Series, Vol. 4. Elsevier, Amsterdam, p. 67.

Kacirek, H., Lechert, H., Schweitzer, W., Wittern, K.-P., 1980. In: Townsend, R.P. (Ed.), Properties and Applications of Zeolites. The Chemical Soc, London, p. 164.

Kenny, M.B., Sing, K.S.W., 1990. Chem. Ind., 39.

Kenny, M.B., Sing, K.S.W., Theocharis, C.R., 1991a. Chem. Ind., 216.

Kenny, M.B., Sing, K.S.W., Theocharis, C.R., 1991b. J. Chem. Soc. Chem. Commun. 974.

Kenny, M.B., Sing, K.S.W., Theocharis, C.R., 1992. J. Chem. Soc. Faraday Trans. 88, 3349.

Kirkby, N.F., 1986. Membranes in Gas Separation and Enrichment. Royal Soc. Chem, London, p. 221 (Special Publication 62).

Kiselev, A.V., 1965. Discuss. Faraday Soc. 40, 205.

Kiselev, A.V., 1967. Adv. Chromatogr. 4, 113.

Kita, H., 2006. In: Pinnau, I., Yampolskii, Y., Freeman, B.D. (Eds.), Materials Science of Membranes for Gas and Vapor Separation. Wiley, Chichester, UK, p. 373.

Kokotailo, G.T., Lawton, S.L., Olson, D.H., Meier, W.M., 1978. Nature. 272, 438.

Lachet, V., Boutin, A., Pellenq, R.J.M., Nicholson, D., Fuchs, A.H., 1996. J. Phys. Chem. 100, 9006.

Lantenois, S., Nedellec, Y., Prélot, B., Zajac, J., Muller, F., Douillard, J.M., 2007. J. Colloid Interface Sci. 316, 1003.

Laszlo, P., 1987. Preparative using Supported Reagents. Academic Press, New York.

Li, S., Martinek, J.G., Falconer, R.D., Noble, R.D., Gardner, T.Q., 2005. Indus. Eng. Chem. Res. 44, 3220.

Lin, H.Y., Chin, C.Y., Huang, H.L., Huang, W.Y., Sie, M.J., Huang, L.H., Lee, Y.H., Lin, C.H., Lii, K.H., Bu, X., Wang, S.L., 2013. Science. 339 (6121), 811.

Llewellyn, P.L., Coulomb, J.-P., Grillet, Y., Patarin, J., Lauter, H., Reichert, H., Rouquerol, J., 1993a. Langmuir. 9, 1846.

Llewellyn, P.L., Coulomb, J.-P., Grillet, Y., Patarin, J., Andre, G., Rouquerol, J., 1993b. Langmuir. 9, 1852.

Llewellyn, P.L., Grillet, Y., Rouquerol, J., 1994a. Langmuir. 10, 570.

Llewellyn, P.L., Pellenq, N., Grillet, Y., Rouquerol, F., Rouquerol, J., 1994b. J. Thermal Anal. 42, 855.

Llewellyn, P.L., Grillet, Y., Schüth, F., Reichert, H., Unger, K.K., 1994c. Microp. Mater. 3, 345.

Llewellyn, P.L., Grillet, Y., Rouquerol, J., Martin, C., Coulomb, J.-P., 1996. Surface Sci. 352, 468.

Ma, Y.H., 1984. In: Myers, A.L., Belfort, G. (Eds.), Fundamentals of Adsorption. Engineering Foundation, New York, p. 315.

Macht, F., Eusterhues, K., Pronk, G.J., Totsche, K.U., 2011. Appl. Clay Sci. 53, 20.

Maesen, T., 2007. In: Ceijka, J., van Bekkum, H., Corma, A., Schüth, F. (Eds.), Introduction to Zeolite Science and Practice. Elsevier, Amsterdam, p. 1 (Chapter 1).

Martens, J.A., Souverijns, W., van Rhijn, W., Jacobs, P.A., 1997. In: Ertl, G., Knözinger, H., Weitkamp, J. (Eds.), Handbook of Heterogeneous Catalysis, I. Wiley-VCH, Weinheim, p. 324.

Mees, F.D.P., van des Voort, P., Cool, P., Martens, R.M., Janssen, M.J.G., Verberckmoes, A.A., Kennedy, G.J., Hall, R.B., Wang, K., Vansant, E.F., 2003. J. Phys. Chem. B. 107, 3161.

Michot, L.J., Villieras, F., François, M., Yvon, J., Le Dred, R., Cases, J.M., 1994. Langmuir. 10, 3765.

Michot, L.J., Villiéras, F., 2002. Clay Miner. 37, 39.

Morigami, Y., Kondo, M., Abe, J., Kita, H., Okamoto, K., 2001. Sep. Purif. Technol. 25, 251.

Muller, U., Unger, K.K., 1986. Fortschr. Mineral. 64, 128.

Muller, U., Unger, K.K., 1988. In: Unger, K.K., Rouquerol, J., Sing, K.S.W., Kral, H. (Eds.), Characterization of Porous Solids I. Elsevier, Amsterdam, p. 101.

Muller, U., Reichert, H., Robens, E., Unger, K.K., Grillet, Y., Rouquerol, F., Rouquerol, J., Pan, D., Mersmann, A., 1989a. Fresenius Z. Anal. Chem. 1.

Muller, U., Unger, K.K., Pan, D., Mersmann, A., Grillet, Y., Rouquerol, F., Rouquerol, J., 1989b. In: Karge, H.G., Weitkamp, J. (Eds.), Zeolites as Catalysts, Sorbents and Detergent Builders. Elsevier, Amsterdam, p. 625.

Ohtsuka, K., Hayashi, Y., Suda, M., 1993. Chem. Mater. 5, 1823.

Oubagaranadin, J.U.K., Murthy, Z.V.P., 2009. Ind. Eng. Chem. Res. 48, 10627.

Ozin, G.A., 1992. Adv. Mater. 4, 612.

Padilla-Ortega, E., Levya-Ramos, R., Flores-Cano, J.V., 2013. Chem. Eng. J. 225, 535.

Petrakis, D.E., Hudson, M.J., Pomonis, P.J., Sdoukos, A.T., Bakas, T.V., 1995. J. Mater. Chem. 5, 1975.

Peydayesh, M., Asarehpour, S., Mohammadi, T., Bakhtiari, O., 2013. Chem. Eng. Res. Design. 91,1335.

Pires, J., Bestilleiro, M., Pinto, M., Gil, A., 2008. Sep. Purif. Technol. 61, 161.

Puibasset, J., Pellenq, R.J.M., 2008. J. Phys. Chem. B. 112 (20), 6390.

Ramachandran, C.E., Chempath, S., Snurr, R.Q., 2006. Microp. Mesop. Materials. 90, 293.

Reed, T.B., Breck, D.W., 1956. J. Am. Chem. Soc. 78, 5972.

Reichert, H., Muller, U., Unger, K.K., Grillet, Y., Rouquerol, F., Rouquerol, J., Coulomb, J.P., 1991. In: Rodriguez-Reinoso, F., Rouquerol, J., Sing, K.S.W., Unger, K.K. (Eds.), Characterization of Porous Solids II. Elsevier, Amsterdam, p. 535.

Reichert, H., Schmidt, W., Grillet, Y., Llewellyn, P., Rouquerol, J., Unger, K.K., 1994. In: Rouquerol, J., Rodriguez-Reinoso, F., Sing, K.S.W., Unger, K.K. (Eds.), Characterization of Porous Solids III. Elsevier, Amsterdam, p. 517.

Rouquerol, J., Rouquerol, F., Llewellyn, P., 2006. In: Bergaya, F., Theng, B.K.G., Lagaly, G. (Eds.), Handbook of Clay Science. Elsevier, Amsterdam, p. 1003 (Chapter 12.11).

Rouquerol, J., Rouquerol, F., Imelik, B., 1970. Bull. Soc. Chim. France. 10, 3816.

Rouquerol, J., Rouquerol, F., Imelik, B., 1985. In: Haynes, J.M., Rossi-Doria, P. (Eds.), Principles and Applications of Pore Structural Characterization. Arrowsmith, Bristol, p. 213.

Rouquerol, J., Avnir, D., Fairbridge, C.W., Everett, D.H., Haynes, J.M., Pernicone, N., Ramsay, J.D.F., Sing, K.S.W., Unger, K.K., 1994. Pure Appl. Chem. 66, 1739.

Ruthven, D.M., 1984. Principles of Adsorption and Adsorption Processes. Wiley, New York (p. 51).

Ruthven, D.M., Brandani, S., 2000. In: Kanellopoulos, N.N. (Ed.), Recent Advances in Gas Separation by Microporous Ceramic Membranes. Elsevier, Amsterdam, p. 187.

Ruthven, D.M., Loughlin, K.F., 1972. J. Chem. Soc. Faraday Trans. I. 68, 696.

Sayari, A., Crusson, E., Kaliaguine, S., Brown, J.R., 1991. Langmuir. 7, 314.

Schirmer, W., Thamm, H., Stach, H., Lohse, U., 1980. In: Townsend, R.P. (Ed.), Properties and Applications of Zeolites. The Chemical Society, London, p. 204.

Schmidt, W., Schüth, F., Reichert, H., Unger, K., Zibrowius, B., 1992. Zeolites 12, 2.

Sing, K.S.W., 1989. Colloids Surf. 38, 113.

Sing, K.S.W., 1991. In: Mersmann, A.B., Scholl, S.E. (Eds.), 3rd Fundamentals of Adsorption. Engineering Foundation, New York, p. 78.

Sing, K.S.W., Unger, K.K., 1993. Chem. Ind. 165.

Sircar, S., 1993. In: Suzuki, M. (Ed.), Fundamentals of Adsorption IV. Kodansha, Tokyo, p. 3.

Srasra, E., Bergaya, F., Van Damme, H., Ariguib, N.K., 1989. Appl. Clay Sci. 4, 411.

Stach, H., Thamm, H., Fiedler, K., Grauert, B., Wieker, W., Jahn, E., Ohlmann, G., 1986. Stud. Surface Sci. Catal. 28, 539.

Sterte, J., 1991. Clays Clay Miner. 39, 167.

Thamm, H., 1987. Zeolites. 7, 341.

Thamm, H., Stach, H., Fiebig, W., 1983. Zeolites 3, 94.

Thomas, J.M., 1994. Nature 368, 289.

Thomas, J.M., 1995. Faraday Discuss. 100, C9.

Thomas, J.M., Raja, R., 2005. Ann. Rev. Mater. Res. 35, 315.

Thomas, J.M., Theocharis, C.R., 1989. In: Scheffold, R. (Ed.), Modern Synthetic Methods, Vol. 5. Springer-Verlag, Berlin, p. 249.

Thomas, J.M., Bell, R.G., Catlow, C.R.A., 1997. In: Ertl, G., Knozinger, H., Weitkamp, J. (Eds.), Handbook of Heterogeneous Catalysis, Vol. 1. Wiley-VCH, Weinheim, p. 286.

Thomas, J.M., Raja, R., Sankar, G., Bell, R.G., 2001. Acc. Chem. Res. 34, 191.

Tosi-Pellenq, N., Grillet, Y., Rouquerol, J., Llewellyn, P., 1992. Thermochim. Acta. 204, 79.

Trillo, J.M., Alba, M.D., Castro, M.A., Poyato, J., Tobias, M.M., 1993. J. Mater. Sci. 28, 373.

Trzpit, M., Soulard, M., Patarin, J., Desbiens, N., Cailliez, F., Boutin, A., Demachy, I., Fuchs, A.H., 2007. Langmuir 23 (20), 10131.

van Damme, H., Levitz, P., Fripiat, J.J., Alcover, J.F., Gatineau, L., Bergaya, F., 1985. In: Boccara, N., Daoud, M. (Eds.), Physics of Finely Divided Matter. Springer-Verlag, Berlin, p. 24.

van Koningsveld, H., Tuinstra, F., van Bekkum, H., Jansen, J.C., 1989. *Acta Cryst. B.* 45, 423.

van Olphen, H., 1965. J. Colloid Sci. 20, 822.

van Olphen, H., 1976. In: Parfitt, G.D., Sing, K.S.W. (Eds.), Characterization of Powder Surfaces. Academic Press, London, p. 428.

Vaughan, D.E.W., 1988. Catalysis Today. Elsevier, Amsterdam, p. 187.

Vaughan, D.E.W., Maher P.K., Albers E.W., 1974. U.S. Patent 3,838,037 (to WR Grace and Co).

Volzone, C., Ortiga, J., 2009. Appl. Clay Sci. 44, 251.

Volzone, C., Ortiga, J., 2011. J. Environ. Manag. 92, 2590.

Wang, H., Huang, L., Holmberg, B.A., Yan, Y., 2002. Chem. Commun. 16, 1708.

Wernert, V., Schaef, O., Ghobarkar, H., Denoyel, R., 2005. Microp. Mesop. Mater. 83 (1–3), 101.

Wernert, V., Schaef, O., Faure, V., Brunet, P., Dou, L., Berland, Y., Boulet, P., Kuchta, B., Denoyel, R., 2006. J. Biotechnol. 123 (2), 164.

Wilson, S.T., 2007. In: Ceijka, J., van Bekkum, H., Corma, A., Schüth, F. (Eds.), Introduction to Zeolit Science and Practice. Elsevier, Amsterdam, p. 105 (Chapter 4).

Wilson, S.T., Lok, B.M., Messina, C.A., Cannan, T.R., Flanigen, E.M., 1982a. J. Am. Chem. Soc. 104, 1146.

Wilson, S.T., Lok, B.M., Flanigen, E.M., 1982b. US Patent 4310440.

Yang, R.T., 1987. Gas Separation by Adsorption Processes. Butterworths, Boston (p. 263).

Zhu, H.Y., Gao, W.H., Vansant, E.F., 1995. J. Colloid Interface Sci. 171, 377.

13 Adsorption by Ordered Mesoporous Materials

Philip Llewellyn

Aix Marseille University-CNRS, MADIREL Laboratory, Marseille, France

Chapter Contents

13.1 Introduction

The different families of ordered mesoporous materials form an important complement to the microporous zeolites. While one can argue that these materials were discovered 20 years before (Chiola et al., 1971), it is the work by a number of scientists from Mobil in the early 1990s (Beck et al., 1992; Kresge et al., 1992) that opened the road on which an extensive amount of research has been pursued (Di Renzo et al., 1997). From a synthesis standpoint, this area has brought together the two communities from zeolite and sol–gel science. From an applications point of view, research is carried in many areas of physics, chemistry and biology.

Interestingly, the research that lead to these materials was motivated by the need to convert high molecular weight petroleum products (Kresge and Roth, 2013) in which work was being carried out on how to introduce large pores between layered zeolites using surfactant molecules. An approach equally used in the synthesis of pillared clays

Adsorption by Powders and Porous Solids. http://dx.doi.org/10.1016/B978-0-08-097035-6.00013-9

(Chapter 12) and in the case of zeolites, the research from the Mobil laboratories lead to the synthesis of MCM-36 with MCM-22 sheets (Roth et al., 1995).

The second aspect of this pioneering work was the formation of ordered mesoporous materials with the hexagonally arranged MCM-41 and subsequently the cubic MCM-48 and layered MCM-50, collectively known as the M41S family (Beck et al., 1992; Kresge et al., 1992). A cooperative self-assembly mechanism is used to generally describe the synthesis of these materials in which cationic surfactants were used (Monnier et al., 1993). Later work showed the possibility to prepare materials using non-ionic surfactants (Tanev et al., 1994) with notably materials such as SBA-15 (hexagonal) and SBA-16 (cubic) well-known members (Zhao et al., 1998a,b). Most materials are silica based but it is possible to introduce other oxides into the walls to increase the acidic properties (Zeleňàk et al., 2006; Zukal et al., 2008). Functionalisation of the surface is possible via the grafting of organic groups (Brunel, 1999; Huang et al., 2003; Drese et al., 2009; Knofel et al., 2009). It is equally possible to have these organic groups as the walls themselves with the use of silica 'glue' (Asefa et al., 1999).

A liquid crystal phase approach can equally be used to prepare ordered mesoporous materials which opens the possibility to prepare mesoporous metals (Attard et al., 1995). Finally, as mentioned in Chapter 10, it is possible to use ordered mesoporous silicas as hard templates for the preparation of other materials including ordered mesoporous carbons and carbon nitride for example (Ryoo et al., 1999; Jun et al., 2000). However, it is also possible to prepare ordered mesoporous carbons using a direct synthesis route (Ma et al., 2013).

It can thus be appreciated that a wide range of synthesis pathways and conditions can be used to prepare these ordered mesoporous materials with a variety of chemical properties. The aim of this chapter is to give the reader an overview of some of the adsorption properties of these materials bearing in mind that several of these can be considered as models to understand capillary condensation phenomena.

13.2   Ordered Mesoporous Silicas

13.2.1   The M41S Family

As mentioned above, the synthesis of the M41S family of silicate/aluminosilicate mesoporous adsorbents was disclosed by Mobil scientists in 1992 (Beck et al., 1992; Kresge et al., 1992). The most thoroughly investigated member of this family is MCM-41 (Mobil Catalytic Material, number 41), which consists of an assemblage of non-intersecting tubular pores. As can be seen in Figure 13.1, the pore structure is composed of a hexagonal array of uniform channels of controlled size.

In their original synthesis of MCM-41, Beck et al. (1992) achieved the hydrothermal conversion of aluminosilicate gels in the presence of quaternary ammonium surfactants (e.g. hexadecyltrimethylammonium ions). The washed and air-dried products were calcined at 550 °C to remove residual organic material.

It was considered by the Mobil scientists that the mesoporous structure was formed as a result of a 'liquid crystal templating' mechanism (see Figure 13.1) since the electron micrographs and diffraction patterns were remarkably similar to those given

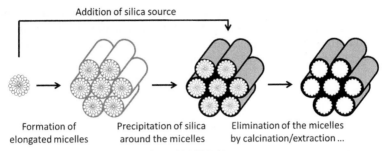

Addition of silica source

Formation of   Precipitation of silica   Elimination of the micelles
elongated micelles   around the micelles   by calcination/extraction ...

Figure 13.1 Schematic of the formation of MCM-41.

by the micellar phases of certain surfactant/water liquid crystals. The observed dependence of the pore width (within the range ca. 1.5–10 nm) on the alkyl chain length and solubilisation of organic oleophilic species was also consistent with the idea of a liquid crystal templating mechanism.

However, the exact nature of the intermediate products was unknown in 1992 and it was Schüth, Stucky and their co-workers who studied in more detail the formation and morphology of various intermediate mesostructures (Monnier et al., 1993; Stucky et al., 1994; Huo et al., 1995; Firouzi et al., 1995). It now appears that a simple liquid crystal templating mechanism cannot explain the fact that the hexagonal surfactant/silicate phase can be produced at very low level surfactant concentrations, this being inconsistent with the pure surfactant phase diagram. It is therefore evident that the mesophase morphology is dependent inter alia on the nature of the lyotropic transformations and the interactions between the surfactant and the silicate phase. A generalised model, which makes use of the cooperative organisation of inorganic and organic species into 3D structured arrays, was proposed for the synthesis of mesostructured oxides and nanocomposite materials (Ciesla et al., 1994; Firouzi et al., 1995).

Various procedures have been proposed for the preparation of M41S and other related materials. A comprehensive review of verified synthesis procedures can be found in the publication by Meynen et al. (2009). For example, Edler and White (1995) have described a method for the low-temperature synthesis of a pure silica form of MCM-41. An essential stage in their preparative route was the controlled ageing of the mixed solution of sodium silicate and cetyltrimethylammonium bromide (CTAB) before the intermediate product was filtered, washed, dried and finally calcined in air at 550 °C. Another simple method to prepare high quality MCM-41 is given by Grun et al. (1999). The pseudomorphic synthesis is equally of interest where the as-prepared MCM-41 keeps the same overall morphology of the reactant silica gel particles (Martin et al., 2002a). The optimised synthesis of MCM-41 materials prepared with different surfactant chain lengths is given in the work of Kruk et al. (2000).

The calcination stage can have a strong effect on the amount of surface hydroxyl groups (Keene et al., 1999) and pore size (Keene et al., 1998). While thermal treatment is often used, it is also possible to use other treatments such as ozonation (Keene et al., 1998; Buchel et al., 2001) and solvent extraction (Zhao et al., 1998a,b; Knofel et al., 2010).

Inagaki et al. (1993) have reported the formation of highly ordered mesoporous materials from kanemite, a layered polysilicate, which was itself prepared from hydrated sodium silicate. The kanemite was dispersed in an aqueous solution of hexadecyltrimethylammonium chloride and the mixture then heated at 70 °C for several hours. HCl was added to bring the pH to 8.5. Finally, after it had been filtered and washed, the product was calcined in air at 700 °C. Electron microscopy revealed the same type of hexagonal channel structure as that of MCM-41.

Since 1992, physisorption isotherms have been reported on many different samples of MCM-41. Although all these isotherms have the general Type IV appearance, they are not all of exactly the same shape: a few are completely reversible, while others exhibit hysteresis. A series of systematic investigations (Branton et al., 1993, 1995a,b, 1997) has shown that at certain temperatures the same adsorbent sample can give well-defined hysteresis with some adsorptives and reversible, or nearly reversible, isotherms with others. Furthermore, with these systems, the hysteresis loop can be removed by increasing the operational temperature (Ravikowitch et al., 1995; Branton et al., 1997; Morishige and Shikimi, 1998).

To give an example, the nitrogen isotherm in Figure 13.2 was determined on a sample taken from a master batch prepared by Keung (1993) in accordance with the original Mobil recipe. The most striking features are the sharp step over a narrow range of relative pressure, $p/p° = 0.41–0.46$, and the overall reversibility of the isotherm. This is an example of a truly reversible Type IV isotherm – now designated a Type IVb isotherm (see Chapter 1). If the amount adsorbed is measured discontinuously (i.e. the usual procedure), it is necessary to allow at least 1 h for the equilibration of each desorption point in the region of the step (Branton et al., 1993). The absence of hysteresis with this particular grade of MCM-41 has been confirmed in various laboratories (e.g. Branton et al., 1994; Llewellyn et al., 1994;

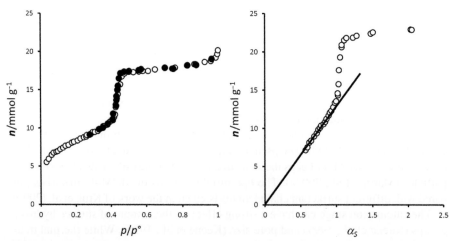

Figure 13.2 Left: Adsorption isotherm of nitrogen at 77 K on 4 nm MCM-41. Right: corresponding α_s-plot.
Adapted from Branton et al. (1994).

Ravikowitch et al., 1995). In the following discussion, this grade is referred to as '4 nm MCM-41' or simply as the 4 nm sample of MCM-41.

The nitrogen isotherm data on non-porous hydroxylated silica in Table 11.1 (Bhambhani et al., 1972) have been taken as reference to construct the α_s-plot in Figure 13.2. Since the initial linear section can be back-extrapolated to the origin, we are reasonably sure that monolayer–multilayer adsorption has occurred on the mesopore walls before the onset of pore filling at $p/p^\circ = 0.41$ and therefore that there was no detectable primary micropore filling at low p/p°.

The argon isotherm (at 77 K) on the 4 nm MCM-41, together with the corresponding α_s-plot are given in Figure 13.3. The shape of the argon α_s-plot is again indicative of an abrupt change from monolayer–multilayer adsorption to pore filling. The adsorption isotherm of oxygen (at 77 K) on the same sample of 4 nm MCM-41 is also shown in Figure 13.3. It is evident that the argon and oxygen isotherms have very similar hysteresis loops (of Type H1), in contrast to nitrogen.

In calculating the surface area of these materials, it is possible to use argon as the reference probe and then calculate values of apparent molecular area, σ, of each other adsorptive in the BET monolayer. Indeed, one accepts that the nitrogen molecule, which interacts via its quadrupole to the surface, can take an orientation corresponding to a smaller σ than that conventionally assumed (e.g. in the BET method). This is not the case with the non-polar and spherical argon molecule, even though the argument can be made that argon does not homogeneously wet some surfaces. This recalculation of the cross sectional area of nitrogen has been made in the case of hydroxylated silica (see Section 10.2.3; Rouquerol et al., 1979; Ismaïl, 1992; Jelinek and Kováts, 1994). In the present case, for various MCM-41, nitrogen cross sectional areas of 0.135–0.140 nm^2 are obtained and confirm those recalculated in previous studies (Llewellyn et al., 1997; Galarneau et al., 1999).

A further consideration to be made with such mesoporous systems is that in calculating a BET area, one considers the cross sectional area of the molecule on a flat surface via an area drawn through the centre of the molecule. In cylindrical pores, due to the pore curvature, this area may be underestimated with respect to the probe-accessible area. A difference of around $w/(w-d)$ where w is the pore width

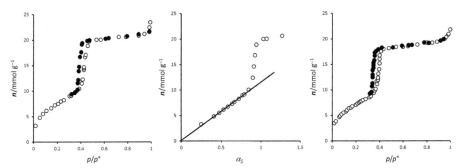

Figure 13.3 Left: Argon isotherm at 77 K on 4 nm MCM-41. Middle: Corresponding α_s-plot. Right: Oxygen isotherm at 77 K on 4 nm MCM-41.
Adapted from Branton et al. (1994).

and d the molecular diameter can be deduced which, in the case of nitrogen, can be over 10% for cylindrical pores below 4 nm (Kruk and Jeroniek, 2001).

Values of the total-specific mesopore volume v_p(mes) can be obtained from the amounts adsorbed at $p/p° = 0.95$, by making the usual assumption that the pores are filled with the condensed liquid adsorptive (i.e. assuming the validity of the Gurvich law). This seems to be the case for a number of adsorptives at this temperature including nitrogen, argon, oxygen, carbon monoxide and methane. The fairly good agreement between the values of v_p(mes) is consistent with the assumption that the capillary condensed state of argon is the supercooled liquid rather than the solid. This may not be the case for krypton at 77 K where the fluid may be closer to be in the solid state (Llewellyn et al., 1997).

The sharp steps in the isotherms and α_s-plots suggest that capillary condensation occurs in a narrow range of mesopores. When calculating the mesopore size, it is possible to apply the Kelvin equation (assuming hemispherical meniscus formation) and correcting for the adsorbed layer thickness. It must be re-emphasised that it is unlikely that the Kelvin equation provides a reliable basis for the calculation of pore widths of around 6 molecular diameters. Also, as pointed out in Chapter 8, the application of the standard statistical multilayer thickness correction may be an oversimplification. Several groups have highlighted the difference between pore sizes directly observed by TEM and those calculated by the BJH method. To overcome this, various approaches have been taken. Kruk et al. (1997) used XRD analysis of the structures to estimate the pore sizes of a series of MCM-41 materials and suggested adding a value of 0.3 nm to the BJH analysis. Other models such as those proposed by Derjaguin or the use of the Broekhoff/de Boer reference curve are discussed in the review by Coasne et al. (2013). Indeed, for MCM-41, it is to be expected that the tubular pore shape and absence of network-percolation effects should at least minimise any deviation from the 'ideal' capillary condensation model. However, the more widespread use of NLDFT computation (see Section 8.5) seems to allow the user to obtain a more precise estimation of mesopore size as long as the appropriate kernel is available and can be used.

As already noted, although the nitrogen isotherm is reversible, the argon and oxygen isotherms exhibit well-defined hysteresis loops at 77 K. Similar loops have been obtained with the following adsorptives on the same 4 nm sample of MCM-41: methanol, ethanol, propan-1-ol, butan-1-ol and water vapour (Branton et al., 1995a) and carbon dioxide and sulphur dioxide (Branton et al., 1995b). All of these loops are of Type H1 in the IUPAC classification (see Section 8.6), but there are significant differences in their widths. This is illustrated in Figure 13.4, where it can be seen that methanol at 290 K has given a narrower loop than propan-1-ol at 298 K. Depending on the hydrophobicity of the sample, water gives rise to a Type V isotherm (Llewellyn et al., 1995) with a shape much like that for carbon tetrachloride (Figure 13.4).

Many investigations have been made of the dependence of isotherm shape on the pore size of MCM-41 (e.g. Llewellyn et al., 1994; Morishige and Shikimi, 1998; Gelb et al., 1999). The normalised nitrogen isotherms in Figure 13.5 exemplify the three types of isotherms given by MCM-41 adsorbents of different pore size. In accordance with previous work, a very steep and reversible pore-filling riser is given by the 4 nm sample (over the range $p/p° \approx 0.40$–0.44). The isotherm obtained on the 2.5 nm

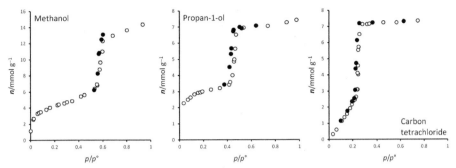

Figure 13.4 Adsorption of organic vapours on 4 nm MCM-41. Left: Methanol isotherm at 290 K. Middle: Propanon-1-ol at 290 K. Right: Carbon tetrachloride at 303 K. Adapted from Branton et al. (1995a, 1997).

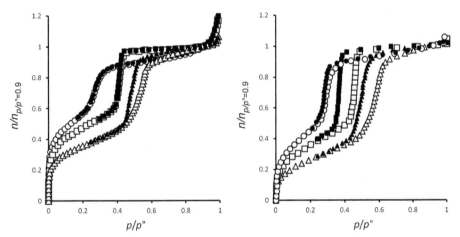

Figure 13.5 Nitrogen (left) and argon (right) isotherms (normalised to the amount adsorbed at $p/p° = 0.9$) of MCM-41 materials with average pore diameters of 2.5 (circles), 4.0 (squares) and 4.5 nm (triangles) (open symbols, adsorption; closed symbols, desorption; Llewellyn et al., 1994).

sample is also reversible, but in this case the pore filling has evidently occurred over a much wider range of $p/p°$ (ca. 0.22–0.34). In contrast, the isotherm on the 4.5 nm sample is of a more usual Type IVa shape with a well-defined hysteresis loop. Samples of modified kanemite referred to earlier gave reversible pore filling by nitrogen over an appreciable range of $p/p°$ (Branton et al., 1996), and the isotherm shape was similar to that given by the 2.5 nm MCM-41 in Figure 13.5.

Somewhat similar variations in isotherm shape and position of the capillary condensation step with pore size have been reported by other investigators for other adsorptives (Franke et al., 1993; Ravikowitch et al., 1995; Rathousky et al., 1994; Schmidt et al., 1995; Branton et al., 1997; Kruk et al., 2000). Carbon tetrachloride isotherms were determined on the 3.4 nm siliceous MCM-41 at the temperatures of

273, 288, 303 and 323 K. These isotherms were essentially of Type V, like for water (see below): that at 323 K was completely reversible, while the others had narrow, almost vertical hysteresis loops of Type H1 (see Figure 13.4).

We are now in a position to discuss the significance of the isotherm shape in relation to the appearance, or absence, of hysteresis. First, we note that all those hysteresis loops that have been given by various samples of MCM-41 are essentially of Type H1. We believe this to be a useful indication that network-percolation effects are not playing a major role in the emptying of the mesopores (i.e. on the desorption branch). Thus, the narrow and almost vertical loops in Figure 13.2 are more likely to be associated with delayed condensation rather than the more complex percolation pore-blocking phenomena (see Chapter 8). This can be confirmed by the use of scanning curves in the hysteresis loop where desorption scanning curves remain essentially horizontal (Coasne et al., 2013). Of course, this is to be expected in view of the non-intersecting tubular pore structure of the model MCM-41.

Next, we must consider the question of the lower limit of capillary condensation hysteresis. As pointed out in Chapter 8, a considerable amount of previous work has indicated that the lower closure point of this form of hysteresis loop is never located below a critical relative pressure, which is dependent on the adsorptive and temperature (but not on the adsorbent). In the case of nitrogen at 77 K, the lower limit of hysteresis appears to be at $p/p° = 0.42$. This capillary critical point has been investigated in function of pore size, temperature and nature of the adsorptive gases in detail by Morishige and co-workers (Morishige et al., 1997; Morishige and Shikimi, 1998).

The nitrogen isotherm on the 4 nm sample is of special interest because at first sight it appears to represent an almost 'ideal' case of reversible capillary condensation/ evaporation in a narrow distribution of uniform cylindrical pores (cf. Figure 8.8). However, it must be kept in mind that the steep pore-filling riser is located at $p/p° \approx 0.42$. Because of the pore geometry, there is no detectable interparticle condensation, which often results in a reversible deviation from the standard isotherm at $p/p° < 0.42$. We may therefore conclude that the condensate has become unstable at the critical chemical potential corresponding to $p/p° = 0.42$, while leaving the adsorbed multilayer (under the influence of surface forces) on the pore walls.

This system has been discussed in some detail by Ravikowitch et al. (1995). Model nitrogen isotherms at different temperatures (70–82 K) were computed for a wide range of pore sizes (1.8–8 nm) by using non-local density functional theory, NLDFT. Although it was necessary to assume energetic homogeneity of the adsorbent surface and to adopt a simplified treatment of the solid–fluid intermolecular potential, the computed isotherms were found to exhibit the same stepwise Type IV character as the experimental isotherms. However, wide hysteresis loops were generated and it was concluded that the magnitude of grand potential difference can be regarded as a measure of the degree of metastability. As observed experimentally, for a given pore size, the size of the loop becomes smaller with an increase in temperature. Furthermore, at a given temperature, there is a critical pore size, below which irreversible pore filling – giving rise to hysteresis – is transformed into reversible pore filling.

While the pore size distribution of MCM-41 samples can be very narrow, the surface chemistry of these materials is quite heterogeneous. This is shown in the

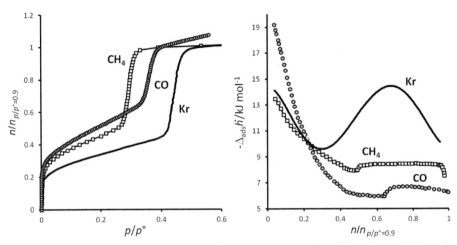

Figure 13.6 Adsorption isotherms at 77 K (left) and corresponding differential enthalpies of adsorption (right) for methane, carbon monoxide and krypton on MCM-41 of 4 nm pore width. Adapted from Llewellyn et al. (1997).

differential enthalpies of adsorption (Figure 13.6) for methane, carbon monoxide and krypton at 77 K (Llewellyn et al., 1996, 1997). In a first instance, one can compare the adsorption of the non-polar molecule (methane) with that of carbon monoxide which has a permanent moment (dipole moment $= 0.39 \times 10^{-30}$ Cm). In both cases, the initial enthalpy curves decrease which is typical for heterogeneous surfaces and often observed with silica samples (e.g. in Rouquerol et al., 1979). The initial value of the differential enthalpy of adsorption varies with the type of probe molecule used. Thus molecules with permanent molecular moments such as carbon monoxide and nitrogen give rise to larger interactions in this region. Kruk et al. (1997) have analysed their detailed nitrogen isotherms determined on a series of siliceous MCM-41 samples of different pore size. The low-pressure data were used to calculate the adsorption energy distribution, which in general appeared to show little variation from one sample to another. However, it was found that the adsorption at the lowest accessible pressures ($5 \times 10^{-7} < p/p° < 10^{-3}$) was significantly enhanced by the decrease in pore size. This was concluded to be due to either corrugations in the pore walls or an overall increased curvature of the pore wall.

The differential enthalpy of adsorption for all of the gases increases slightly during the capillary step in the corresponding isotherms (Figure 13.6). By comparison, isosteric heats of adsorption have been calculated from experimental data for the adsorption of cyclopentane on MCM-41 at 253–293 K (Franke et al., 1993; Rathousky et al., 1995) which show a similar type of behaviour.

For the adsorption of krypton on the MCM-41 (Figure 13.6), a much larger peak in the differential enthalpy curve is observed during capillary condensation. The value of around 4 kJ mol^{-1} above the enthalpy of solidification is higher than would be expected for a simple condensation type mechanism. This, and considering the fact that to calculate the filling of the mesopores using the Gurvich rule the solid density of krypton is required, would suggest that maybe a solid-like krypton phase is formed in this region.

The effect of allowing for surface energetic heterogeneity has been studied by molecular simulation of the adsorption of nitrogen by MCM-41, Maddox et al. (1997). This approach was prompted by the fact that models in which the solid–fluid interaction potential was assumed homogeneous were incapable of reproducing an experimental isotherm, which had been accurately determined in the region of low surface coverage (at $p/p°$ down to 2.7×10^{-6}). Maddox et al. (1997) point out that hysteresis loops obtained as a result of molecular simulation arise from metastable states and may depend on the length of the simulation run; therefore, they are not directly comparable with experimental hysteresis loops.

To be able to accurately model the adsorption isotherms of argon and krypton (Kuchta et al., 2004) also showed that heterogeneity effects are immense. They were able to model structural heterogeneity (roughness) and chemical disorder into the walls to be able to match experimental isotherms. This adsorption site distribution is indispensable for modelling low-pressure adsorption on these mesoporous silicas.

In order to follow the order of the fluid adsorbed inside the mesopores, neutron diffraction patterns measured for different quantities of deuterated methane (CD_4) adsorbed at 77 K on the MCM-41 of around 2.5 nm in pore width (Figure 13.7 left). The quantities adsorbed correspond to: (a) the diffraction pattern spectrum on the outgassed material before adsorption, (b) just *before* the capillary condensation step, (c) *during* the capillary and (d) *after* the condensation step. The diffraction signal of the deuterated methane adsorbed on this MCM-41 sample is characterised by a very broad peak in the scattering wave range vector $1.7 \leq Q \leq 1.9 \ \text{Å}^{-1}$. This is illustrative of the absence of long distance order. A similar result is equally

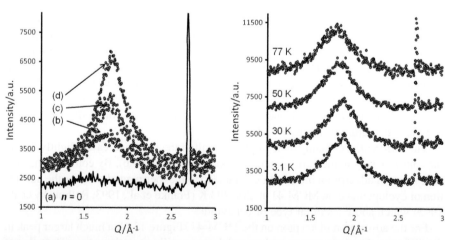

Figure 13.7 Neutron diffraction patterns obtained with CD_4 on 2.5 nm MCM-41 for different loadings at 77 K (left) and (right) at different temperatures for a sample whose mesopores are half filled with CD_4 (point (c) in the left plot).
From Llewellyn et al. (1996).

obtained for the adsorption of deuterated hydrogen carried out at 16.4 K. Thus only a short range order ($20 \leq L_{coher} \leq 30$ Å) is observed for the adsorption of these gases.

Figure 13.7 right shows the neutron diffraction spectra of the deuterated methane for the point during the condensation step (point (c) in Figure 13.7 left) in the temperature region between 3.1 and 77.5 K. It can be seen that no change in the shape of the peak occurs which suggests that the molecular organisation of the methane rests unchanged, that is of short range order. This short range order is characteristic of fluid or amorphous solid phases. These experiments are thus not able to differentiate between this static disorder (amorphous solid) or dynamic disorder (fluid phase) even at 3.1 K (Llewellyn et al., 1996).

A striking feature of MCM-41 and other related ordered mesoporous materials is the magnitude of their specific surface areas. It is remarkable that it is possible to produce non-microporous structures with stable surface areas well in excess of $1000 \ m^2 \ g^{-1}$. Of course, the implication of the combination of high-area and high-pore volume is that the pore walls must be quite thin – in some cases the walls may be no more than two oxygen atoms thick. This would seem to be the origin of the poor long term stability (Cassiers et al., 2002) of this material and therefore, to the lack of widespread applications. This is equally observed in the case of water, see below.

In view of the surface heterogeneity of MCM-41, it might be expected that the adsorption of water vapour would involve some degree of specificity. It was found (Branton et al., 1995b; Llewellyn et al., 1995), however, that water isotherms on the 4 and 2.5 nm samples were of Type V in the IUPAC classification (cf. the carbon tetrachloride isotherm in Figure 13.4). Indeed, the initial differential enthalpies of adsorption were found to be below those of the enthalpy of liquefaction, indicative of a hydrophobic surface (Llewellyn et al., 1995). The initial section of the isotherm on the 4 nm material was almost reversible – indicating that exposure to water vapour had led to very little rehydroxylation. Ribeiro Carrott et al. (1999) have found that a high-area sample of MCM-41 underwent pronounced ageing when exposed to water vapour at high $p/p°$. Inagaki and Fukushima (1998) also found that their sample of modified kanemite, FSM-16, which has similar properties to those of MCM-41, underwent a significant amount of rehydroxylation when exposed to water vapour. This behaviour is similar to that of most dehydroxylated silicas.

13.2.1.1 MCM-48

Of the other M41S materials, MCM-48 is of note (Vartuli et al., 1994). As mentioned above, MCM-48 has a 3D bicontinuous pore system which can be modelled by a minimal gyroid surface. Crystallographically, the sample has Ia3d symmetry (Monnier et al., 1993). The synthesis composition window of this material is narrower, and requires higher surfactant to silica ratios compared to MCM-41 (Meynen et al., 2009). Schmidt et al. (1996) have described the synthesis of an MCM-48 of Si/Al ratio 22, in which the Al appeared to be only in the tetrahedral form. However, thickness of the walls of MCM-48 is comparable to MCM-41 and as such the intrinsic hydrothermal stability of the two materials is similar.

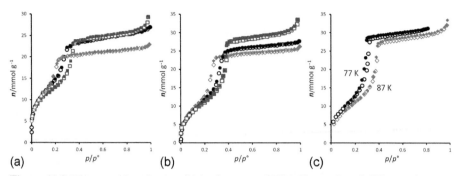

Figure 13.8 Nitrogen (a) and argon (b) isotherms on MCM-48 samples of different sizes.
(c) Argon isotherms on the same MCM-48 sample obtained at 77 and 87 K.
Adapted from Thommes et al. (2000).

Another bicontinuous cubic (Ia3d) mesoporous silica with a similar pore structure
to MCM-48 is KIT-6 (Kleitz et al., 2003). This sample has equally received some
attention with respect to its adsorption properties and characterisation of the pore system using scanning hysteresis (Cimino et al., 2013).

From an adsorption standpoint, MCM-48 can be regarded as a cage-like structure
whose cavities and windows are of similar size. As such, this sample is of interest as a
model to understand the effects of networking and interconnections among pores of
similar size.

Figure 13.8 shows nitrogen and argon isotherms obtained on three different samples of MCM-48 (Thommes et al., 2000). The isotherms in Figure 13.8a and b, are
essentially reversible which is certainly due to the small pore size of the materials prepared in this study. However, it was also shown that the temperature has an effect on
the isotherms as can equally be seen in Figure 13.8c. Indeed, for this sample of average
pore diameter estimated at around 3 nm, a small hysteresis is observed with argon at
77 K which almost disappears at 87 K.

Morishige et al. (2003) systematically followed the effect of temperature on nitrogen adsorption, in the range 55–113 K, with an MCM-48 sample of estimated diameter of 2.4 nm. Here, the hysteresis loop was observed to disappear at around 67 K,
a point described as the hysteresis critical temperature (T_{ch}) (Ravikowitch et al.,
1995). This value can be compared with the pore critical temperature (T_{cp}), defined
as the critical temperature of vapour–liquid equilibrium in a confined media (i.e.
for unconnected cylindrical pores). Thus, capillary condensation is accompanied by
hysteresis below T_{ch}, the first-order transition is reversible between T_{ch} and T_{cp}. Above
T_{cp}, adsorption is considered to take place continuously without capillary condensation. Gelb et al. (1999) had suggested that interconnections between pores may affect
the location of T_{cp} as at connection points, correlation lengths may be greater than the
pore width. Morishige et al. (2003) estimated a T_{cp} of 98 K for the MCM-48 sample
under consideration. He compared these values with those in the literature, notably
with MCM-41 and found that the values obtained with MCM-48 did not deviate
from a linear relationship between the critical point shifts and reciprocal pore radius.

He thus concluded that for MCM-48, whose pore size does not greatly vary throughout the 3D network, the effect of interconnections within the pores does not affect hysteresis and pore critical temperature.

Indeed, quite simply, the nitrogen isotherms shown in Figure 13.8 for MCM-48 resemble closely those in Figure 13.5 for MCM-41. This further suggests that the difference in network geometries (regular cylindrical pores in MCM-41 and bicontinuous pores in MCM-48) does not greatly affect the adsorption phenomena themselves. However, a difference between essentially the 2D MCM-41 pore system and a 3D MCM-48 system may arise in terms of transport. Indeed any pore blocking, leading to restrictions in gas diffusion would affect the MCM-41 pores greater than in MCM-48.

There appears to be considerable scope for the application of ordered mesoporous structures in such fields as biochemical separations and nanocomposite materials. Using these high-specific area silicas as frameworks for anchoring specific functions (organic ligands, enzymes, metal or metal oxides, etc.) is of interest as we shall see in later sections. It is thus possible to have high concentrations of structurally well-defined active sites. Indeed, the potential value of MCM-41- and MCM-48-based catalysts has been stressed by various several authors (e.g. Thomas 1994, 1995; Øye et al., 2001).

Nevertheless, as in the above comparison between MCM-41 and MCM-48, it is still not clear whether regularity in pore structure is the key to applications or whether regularity in pore size and high surface area are key material parameters.

In terms of applications, however, the major problem observed by many is the relatively poor hydrothermal stability of these phases which seems to be linked to the narrow pore wall thickness (Cassiers et al., 2002).

13.2.2 The SBA Family

The SBA (Santa Barbara acid) family of materials emerged in the late 1990s from the group of Galen Stucky (e.g. Huo et al., 1996; Zhao et al., 1998a,b; Sakamoto et al., 2000). A number of different materials with different pore sizes and architectures have been prepared (see Table 13.1) and a wide variety of surfactants can be used for their preparation. In many cases, non-ionic surfactants can be used which opens the possibility to use simple solvent extraction methods for the removal of the organic phase. One general feature of interest with respect to the M41S family is the thicker pore walls which confer greater stability.

Of these different phases, two are of particular interest as models to understand physisorption phenomena in comparison with MCM-41 and MCM-48. The first of these is SBA-15 (Zhao et al., 1998a) which has a hexagonal pore arrangement like that of MCM-41. While MCM-41 has a non-intersection network of cylinders, SBA-15 can be prepared under different conditions to include varying amounts of micropores which can intersect with the cylindrical mesopores (Galarneau et al., 2001, 2003). SBA-16 (Zhao et al., 1998b) on the other hand, can be compared with MCM-48. Both systems have bicontinuous pore structures which can be modelled by gyroid minimal surfaces into cubic structures. As we have seen above, the pores of

Table 13.1 List of Different SBA Materials

Name	Pore Network Type	Reference
SBA-1	Cubic P*m3n* structure, 3D cage-type pore system with smaller width windows	Huo et al. (1996) and Kim and Ryoo (1999)
SBA-2	P6₃/*mmc* structure, hcp array of spherical pore structure connected by cylindrical channels	Huo et al. (1996) and Pérez-Mendoza et al. (2004)
SBA-3	P6*mm* hexagonal structure with cylindrical pores	Huo et al. (1996)
SBA-6	Cubic P*m3n structure*, 3D cages and bimodal pore structure	Sakamoto et al. (2000)
SBA-11	*Pm3m*, cubic pore system	Zhao et al. (1998a,b)
SBA-12	*P63/mmc*, 3D hexagonal	Zhao et al. (1998a,b)
SBA-15	*P6mm*, 2D hexagonal pore system (like MCM-41)	Zhao et al. (1998a,b)
SBA-16	*Im3m* a, Cubic caged 3D pore structure (like MCM-48)	Zhao et al. (1998a,b) and Sakamoto et al. (2000)

MCM-48 have similar widths throughout the pore network. This is not the case with SBA-16 where the pore bodies are larger than the entrances which confer to this sample an almost ideal scenario of ink-bottle pores.

13.2.2.1   SBA-15

SBA-15 consists of a 2D hexagonal mesopore structure (P6mn space group) (Zhao et al., 1998a,b). The pore size can be varied via the use of block co-polymers of different sizes. The pore geometry is an arrangement of cylinders and as such can be considered as a model to understand capillary condensation phenomena much like MCM-41. However, depending on the synthesis conditions (especially the temperature), there exists the possibility for one of the block copolymer chains to be occluded in the silica walls, which, when removed, leads to the possibility of the materials to have a certain degree of microporosity (Zhao et al., 1998a,b; Galarneau et al., 2001, 2003). It is also possible to influence on the degree of microporosity by varying the calcination temperature (Ryoo et al., 2000).

An example of the isotherms given by various samples of SBA-15 is given in Figure 13.9a. Here, nitrogen at 77 K is used to probe samples prepared at 60, 80, 110 and 130 °C (Galarneau et al., 2003). It can be seen directly from these isotherms that as the sample synthesis temperature is increased, the pore size and mesopore volume equally increase. To gain a better insight into the effect of synthesis temperature on micropore volume, the authors carried out argon adsorption and used comparison plots (Figure 13.9b) to show deviations at low loadings. Indeed, any instance of non-linearity in the *t*-plot results from a deviation from the mechanism of monolayer–multilayer adsorption compared to the reference non-porous solid. The significant downward deviation at low thickness is indicative of microporosity as observed

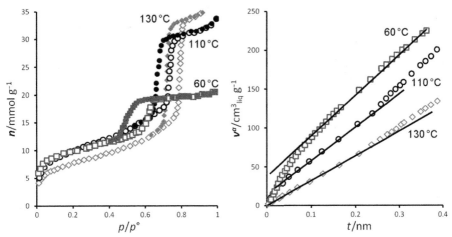

Figure 13.9 Left: Nitrogen adsorption/desorption isotherms at 77 K for SBA-15 synthesised at 60, 110 and 130 °C and calcined at 550 °C. Right: Comparison plots of Ar adsorption at 77 K on SBA-15 prepared at 60, 110, 130 °C and calcined at 550 °C.
From Galarneau et al. (2003).

for the samples prepared at 60 and 110 °C. However, it can be seen that synthesis at 130 °C leads to a sample with little or no microporosity (Galarneau et al., 2003). In other work, scanning curves of the capillary condensation step and hysteresis were carried out on SBA-15 (Cimino et al., 2013). They showed, that even if an intersecting micropore system is present, the mesopores can be considered as independent.

This presence of both microporosity and mesoporosity in a same sample can pose a problem when estimating pore widths. Indeed, one may question the validity of a method where the micropore and mesopore sizes are not able to be estimated with a single classical approach. In such cases, the use of NLDFT or QSDFT methods can be of interest. An example is given for the treatment of the SBA-15 isotherm in Figure 13.10a. Here, the NLDFT kernel was used for nitrogen at 77 K on a cylindrical silica surface using the desorption branch of the isotherm. From Figure 13.10b, it can be seen that the comparison between the experimental isotherm and reconstructed isotherm is very good. The resulting pore size distribution suggests a majority of mesopores which are centred on a diameter of 8.5 nm. In this sample, there is no visible microporosity.

13.2.2.2 SBA-16

SBA-16 can be prepared using block co-polymers with quite long ethylene oxide chains such as F-127. These chains favour the formation of globular aggregated structures which can then lead to a pore formed of intersections connected through smaller windows reminiscent of the classical view of ink-bottle pores. The Type IVa nitrogen adsorption isotherm obtained at 77 K (Figure 13.11) is characteristic of a gradual pore

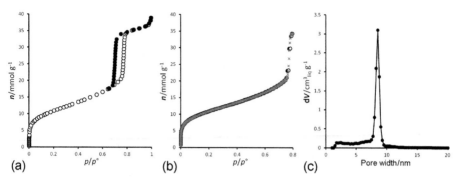

Figure 13.10 N$_2$ adsorption on SBA-15 at 77 K: (a) experimental isotherm, (b) comparison of the experimental isotherm and fit from NLDFT analysis (adsorption branch, silica cylinders) and (c) pore size distribution.

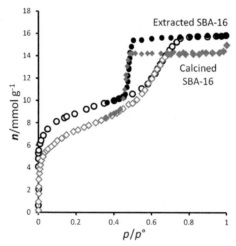

Figure 13.11 Nitrogen isotherms at 77 K for SBA-16 samples activated via solvent extraction or calcination to 550 °C.
From Knöfel et al. (2010).

filling on adsorption and a more rapid pore emptying process which is typified by a Type H2 hysteresis loop.

As can be appreciated from Figure 13.11, the pore size can be varied by changing the extraction conditions (Knofel et al., 2010). Indeed, calcination at 550 °C leads to a contraction in the structure which is not observed if solvent extraction is used. This observation is equally observed for many of the ordered mesoporous silicas.

The form of the pore structure of SBA-16 can depend on the synthesis temperature as shown by Kleitz et al. (2006). Indeed, an increase in pore size occurs with increasing synthesis temperature while typified by a shifting in the capillary condensation step to high relative pressures. However, the pore emptying step remains in the same region suggesting that the pore entrances remain 'small'. Further to this,

Kleitz et al. (2006) was able to characterise the pore size using X-ray diffraction analysis. In comparison to this, these authors showed that an estimation of the pore size using NLDFT was best carried out using specific kernels based on cage-like pores using the adsorption branch. The use of NLDFT calculations with cylindrical pores lead to an underestimation of the pore size by around 30%. Further to this, the use of BJH calculations leads to errors of up to about 45% for pores narrower than 4 nm.

The traditional point of view for such a cage-like pore model system is based on the concept of desorption from the cavity which is delayed until the vapour pressure is reduced below the equilibrium desorption pressure from the pore window (pore-blocking effects) (Mason, 1988). However, both molecular dynamics (Sarkisov and Monson, 2001) and non-local density functional theory (NLDFT) (Ravikowitch and Neimark, 2002) suggest that this classical picture of adsorption and desorption in this geometry based on the concept of pore blocking does not necessarily hold.

Morishige et al. (2003), in the same study mentioned above for MCM-48, equally looked at the temperature dependence on the hysteresis of an SBA-16 sample. The sample was prepared using the same procedure as that which was also studied by high resolution electron microscopy (Sakamoto et al., 2000) which study suggested a pore entrance diameter of 2.3 nm and a pore body diameter of 9.5 nm for this given sample. Comparing the adsorption branch of the isotherm with NLDFT, Morishige et al. (2003), calculated a pore diameter of 9.2 nm in good agreement with the previous work. However, the desorption of nitrogen at 77 K at a relative pressure of around 0.47 is well above that expected for 2.3 nm diameter pore entrances. In this study, the temperature dependence of the capillary condensation and evaporation pressures of nitrogen onto SBA-16 was also followed. The temperature at which the hysteresis was no longer observed (T_{ch}) was 98 K. Interestingly, the temperature dependence of the irreversible capillary evaporation pressure suddenly changed at around 72 K which was not linked to any phase change of liquid nitrogen within the bulk porosity. Overall, in this study, the behaviour of the system confirmed the view that the pore emptying on desorption occurs via a cavitation mechanism (Sarkisov and Monson, 2001). Cavitation is believed to occur when the size of blocking pores is so small that the condensed fluid approaches the limit of metastability and evaporates spontaneously even though the neighbouring pores are still filled. The cavitation phenomena takes places in the range of relative pressures 0.50–0.42 for nitrogen adsorption at 77.4 K and it is characterised by the abrupt step on the desorption isotherm (Rasmussen et al., 2010). It would seem from experiments and computer simulation that is possible to relate the relative pressure at which cavitation takes place and the pore size (Rasmussen et al., 2010).

13.2.3 Large Pore Ordered Mesoporous Silicas

Various groups have looked to control the pore size of the silica materials outside the range offered by the use of commercial surfactant templates (cf. the review of Deng et al., 2013). In the work of Corma et al. (1997), the pore size of MCM-41 was varied by the use of different surfactant/silica ratios as well as using different crystallisation times. Post-synthesis hydrothermal retreatment can also be used to increase the size and volume of MCM-41 mesopores (Sayari et al., 1997). In their original work, Mobil scientists suggested the use of swelling agents to increase the size of the surfactant micelles (Beck et al., 1992). In later work, a two-step synthesis procedure was developed to increase both the size of the pores and their volume (Sayari et al., 1999).

An example of the comparison between an original MCM-41 and a pore expanded material is given in Figure 13.12 (Franchi et al., 2005). Here, the pore expansion was obtained by using a post-synthesis hydrothermal treatment of non-calcined MCM-41 in the presence of N-dimethyldecylamine at a temperature of 120 °C for 72 h. It can be seen that a mesopore volume three times greater is obtained with the pore expanded material along with a significant increase in pore size.

In the synthesis of mesoporous materials with cationic surfactants, it is possible to easily purchase surfactants of varying chain length. When using block co-polymers it is not so simple to be able to systematically study the effect of each chain type on the porosity of the materials prepared. This has been done in cases when using diblock co-polymers (Bloch et al., 2009) and triblock co-polymers (Bloch et al., 2008) as can be seen in Figures 13.13 and 13.15, respectively.

The nitrogen isotherms obtained with the series of materials prepared with the same PEO (polyethylene oxide) block length and different PS (polystyrene) unit lengths (Figure 13.13) are all characterised by an initial sharp uptake at a relative pressure below 0.05 typical of micropore filling. This is followed by a pseudo-linear region and capillary condensation step at relative pressures between 0.65 and 0.85 depending on the sample. On desorption, large hysteresis loops are observed with pore emptying occurring from quite small entrances at similar relative pressures between 0.48 and 0.5, due to cavitation. Similar isotherms are obtained with KLE silicas (Thommes et al., 2006). It is interesting to note that the position of the desorption steps seem to equally be in relation to the position of the capillary condensation (i.e. to the pore

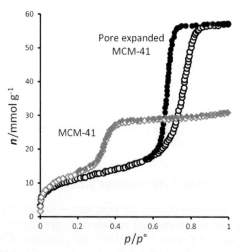

Figure 13.12 Nitrogen isotherms at 77 K for MCM-41 and pore-expanded MCM-41 (Franchi et al., 2005).

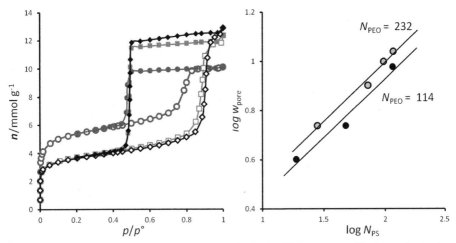

Figure 13.13 Left: Nitrogen isotherms (77 K) obtained with large pore silica's synthesised with diblock co-polymers having the same PEO length (232 units), but different PS lengths 28, 72 and 115 units. Right: logarithmic variation of the pore width as a function of N_{PS} for two series of silica matrices with $N_{PEO}=114$ and $N_{PEO}=232$ (Bloch et al., 2009).

size). This confirms the study of Rasmussen et al. (2010) which relates the relative pressure at which cavitation takes place with the pore size.

This study by Bloch et al. (2009) allows the evolution of the pore diameter as a function of the PS block to be followed. The effect of the PEO block length was equally followed as two series of samples were prepared. The PS blocks participate solely to the mesopores and in the evolution of the pore size, for each series a general trend is confirmed: when N_{PS} increases the pore diameter increases. Interestingly, the PEO block participates to the mesoporosity as well as to the microporosity. The authors were able to relate the size of the mesopores to the lengths, N, of each block with the following expression:

$$r_P(\text{nm}) = 0.36 \cdot N_{PEO}^{0.19} \cdot N_{PS}^{0.5}$$

This opens the possibility to engineer the size of the pores prepared.

In order to create more open porosity, Bloch et al. (2009) simply changed the synthesis temperature (Figure 13.14). It can be seen that the increase in synthesis temperature lead to a slight increase in mesopore size. More interesting though, is that the increase in synthesis temperature from 25 to 60 °C leads to a significant shift, to higher $p/p°$, in the position of the pore emptying step. This suggests an opening of the pore entrances. Interesting here is the sample that was prepared at 45 °C in which the two types of openings are present independently.

It is equally possible to systematically follow the influence of block length on mesopore size when using triblock co-polymers (Bloch et al., 2008). Here, polystyrene (PS) and polyethylene oxide (PEO) blocks can be used in a copolymer of formula $PS_x-PEO_y-PS_x$ where x and y are the number of block units. As in the case of the

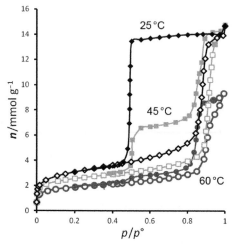

Figure 13.14 Nitrogen isotherms (77 K) obtained with mesoporous silica's synthesised at different temperatures. The formula of the diblock copolymer used for these synthesis is $N_{PEO} = 114$ and $N_{PS} = 115$ (Bloch et al., 2009).

diblocks described above, the polystyrene units uniquely participate to the formation of the mesoporosity whereas the PEO units seem to participate to the formation of both the mesopores and any microporosity. The extent of microporosity can be related to the PS–PEO interface which has to be stabilised. Indeed the ratio of PEO to PS lengths leads to micelles in which the PEO chain is stretched to varying degrees. It would seem that micropore volume increases with increasing the length of the PEO block, which forms the corona of the micelles. The isotherms in Figure 13.15 again are indicative of Type I and Type IV behaviour. The mesopore size can be directly related to the length of the PS block (Figure 13.15 right). On desorption, cavitation is again observed whose position seems to shift slightly with pore size.

13.3   Effect of Surface Functionalisation on Adsorption Properties

The ordered mesoporous silica materials can be used as high surface area frameworks on which is it possible to add specific functions, often with a given application in mind. This section aims to give an insight into the modification of the adsorption properties related to this functionalisation.

13.3.1   Incorporation of Metal Oxides into the Walls

The surfaces of ordered mesoporous silicas are acidic which is related to the amount of hydroxyl groups. It is possible to further vary the acidity of the surface by the incorporation of other metal oxides onto or into the pore walls. This has often been carried

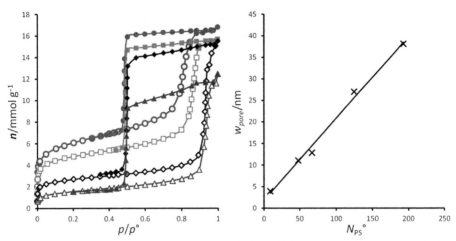

Figure 13.15 Left: Nitrogen isotherms (77 K) obtained with large pore silica's synthesised with PS–PEO–PS triblock co-polymers having the same PEO length (227 units), and different PS lengths 48 (circle), 64 (square), 125 (diamond) and 192 (triangle) units. Right: Variation of the BJH pore width as a function of number of PS units (Bloch et al., 2008).

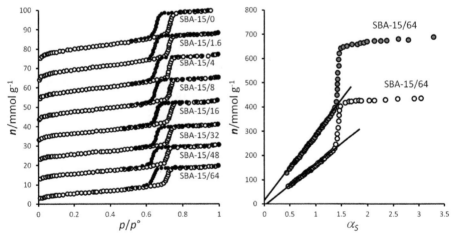

Figure 13.16 Left: Nitrogen isotherms (77 K) on SBA-15 and alumina coated samples (the coated samples are labelled as SBA-15/c, where c denotes the concentration of aluminium in the grafting solution) Right: α_s-plots of the original SBA-15 and the one with maximum alumina coating.
From Zukal et al. (2008).

out with the aim to improve the catalytic properties of the solid (Taguchi and Schüth, 2005). It is equally possible to prepare mesoporous materials made of the pure oxide without the incorporation of silica (Yang et al., 1998).

Zukal et al. (2008) studied the effect of covering the SBA-15 surface with various amounts of alumina. The nitrogen isotherms (77 K) are shown in Figure 13.16 along with the α_s-plots for the pure silica SBA-15 and sample most coated with alumina. The isotherms show that the addition of the alumina does not greatly modify the mesoporous structure although a slight decrease in mesopore size and volume is observed. Additional ^{27}Al MAS NMR measurements showed that at high Si/Al ratios most of the aluminium atoms are tetrahedral, indicating the formation of framework aluminium species. However, at lower Si/Al ratios, the formation of loosely bonded species is clearly shown by the presence of octahedral and pentacoordinated aluminium.

The main variation that is observed is a decrease in surface area with increasing alumina coating as can be observed with the different slopes in the α_s-plots. Any small amount of microporosity was equally filled with the alumina coating. This suggests that for this series of samples, that any surface roughness of the initial silica was filled with the alumina creating a smoother pore wall corona without any blocking of the mesopores.

In terms of adsorption-related applications, it is possible to introduce cations to compensate the acidic surface of the alumina coated SBA-16 (Zukal et al., 2010). Much like for the case of zeolites, the compensation cation has a strong effect on the CO_2 uptakes at low pressure which is reflected in the isosteric enthalpies of adsorption (Figure 13.17) (Zukal et al., 2010). Indeed, it can be seen from this figure that an alumina surface induces a greater interaction with respect to a pure silica SBA-15. The addition of cations to the alumina increases further the interaction with CO_2 as typified by the isosteric enthalpies of adsorption of around 34 kJ mol^{-1} for the sodium and potassium exchanged samples.

It is equally possible to co-precipitate the oxide with the mesoporous silica in the presence of a surfactant. This was carried out by Zeleňàk et al. (2006) where various amounts of TiO_2 were introduced into the SBA-16 synthesis mixture. The resulting nitrogen isotherms (77 K) are given in Figure 13.18.

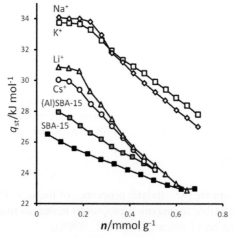

Figure 13.17 Isosteric adsorption enthalpies of carbon dioxide on SBA-15, (Al)SBA-15 and cation exchanged forms of (Al)SBA-15 (Zukal et al., 2010).

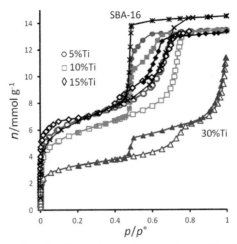

Figure 13.18 Nitrogen adsorption/desorption isotherms of pure silica SBA-16 and samples prepared with varying amounts of TiO$_2$ (Zeleňàk et al., 2006).

The five isotherms in Figure 13.18 correspond to the pure SBA-16 sample as well as the samples prepared with 5%, 10%, 15% and 30% of titania. In general, these isotherms correspond to type I and type IV behaviour indicative of the presence of both micropores and mesopores. The adsorption at low pressures, below a relative pressure of 0.05, indicates microporosity that probably results from the partial penetration of the poly(etheleneoxide) chains of the triblock copolymer in the silica walls during the synthesis (Galarneau et al., 2003; Bloch et al., 2008). An overall trend shows that an increasing amount of Ti in the sample leads to a decrease in the amount of microporosity. The mesoporosity is characterised by the capillary condensation step in the isotherm between relative pressures of 0.5 and 0.8. This capillary condensation step is much reduced for the sample with 30% Ti; however, an additional filling step at a relative pressure above 0.9 is observed. One can speculate about this macroporosity which is most probably due to interparticular voids, maybe between TiO$_2$ particles formed separately from the silica. For the other samples, the height of the capillary condensation step does not vary significantly with titania content in the sample. However, one can note that the onset of this step occurs at increasing relative pressures with increasing Ti content in the sample indicative of an increasing pore width.

Further differences can be seen in the desorption isotherms. Samples with 10% and 15% of Ti show a distinct emptying of the pores in two steps. The pore emptying initially occurs over a range of relative pressures which is indicative of a heterogeneous distribution of pore entrance widths. The sharp desorption step at $p/p°$ around 0.48 is indicative of cavitation.

These results suggest that the introduction of up to 15% titania into the synthesis leads to a decrease in the micropore volume with no influence on the mesopore volume. However, a small increase in mesopore size is observed with increasing titiania content. The addition of 30% of titania leads to an isotherm much like that of the pure silica SBA16 with much reduced porosity. This would suggest that the titania does not

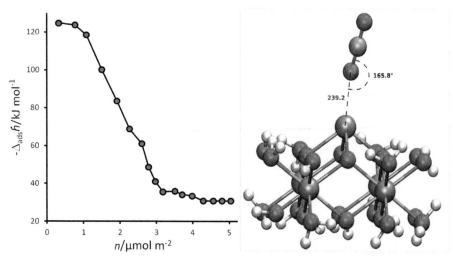

Figure 13.19 Differential enthalpies of adsorption for CO_2 on mesoporous titania (left) and (right) one configuration for the CO_2 adsorbing on a $Zr_5O_{24}H_{28}$ titania cluster. (For colour version of this figure, the reader is referred to the online version of this chapter.)
From Hornebecq et al. (2011).

participate in the mesoporosity but may be in the form of particles creating macroporous domains in the sample.

As with the alumina and cation exchanged SBA-15 samples, it is possible to follow the CO_2 adsorption on these polar titania surfaces (Hornebecq et al., 2011; Boulet et al., 2012). The differential enthalpies of adsorption measured with CO_2 at 30 °C are shown in Figure 13.19. These show an initial very strong (>100 kJ mol^{-1}) differential enthalpy of adsorption which is characteristic of chemisorption. These enthalpies then gradual decrease to $\sim-40/-35$ kJ mol^{-1} (in the range of physisorption) and remain stable at this value. This suggests a heterogeneous surface.

In order to further understand the CO_2 adsorption properties on the mesoporous titania, a DFT approach was used (Boulet et al., 2012). Various different titania clusters were explored and the interaction of a CO_2 molecule with each of these configurations was calculated. Table 13.2 gives the molar enthalpies of adsorption of CO_2 onto the different clusters. The molar enthalpy of adsorption ranges between -65 and -25 kJ mol^{-1} depending on the nature of the atom with which CO_2 interact. The weakest energy (-24.9 kJ mol^{-1}) is obtained when carbon dioxide adsorbs on an oxygen of the surface, which is observed for $Zr_2O_{14}H_{20}$. In this case, CO_2 adsorbs parallel to the surface and it is aligned with the Zr–O–Zr axis in order to maximise the interactions between the CO_2 oxygen atoms. The adsorption of CO_2 onto an oxidised surface that exhibits hydroxyl groups is slightly stronger than when no OH groups are involved. It is the case of $Zr_2O_{14}H_{20}$ and $Zr_3O_{19}H_{26}$ clusters where the molar adsorption enthalpy amounts to -32.6 and -33.9 kJ mol^{-1}, respectively. This is typical of a physisorbed form of CO_2 and, as expected, neither the sorbate nor the cluster are strongly electronically perturbed during the adsorption process.

Table 13.2 Calculated Differential Enthalpies of Adsorption of CO_2 on Various Zirconium Oxide Clusters (Hornebecq et al., 2011; Boulet et al., 2012)

Cluster	Molar Enthalpy of Adsorption/-kJ mol^{-1}
$Zr_2O_{14}H_{20}$	32.6
$Zr_3O_{16}H_{20}$	24.9
$Zr_3O_{19}H_{26}$	33.9
$Zr_5O_{24}H_{28}$	64.6

On the $Zr_5O_{24}H_{28}$ cluster, carbon dioxide adsorbs in apical conformation (see Figure 13.19). The position where the carbon atom would interact with the zirconium one, hence forming a carbonate-like structure $Zr - CO_2^{\delta-}$, is unstable and the optimisation leads to the apical form. The differential enthalpy of adsorption is the strongest one among the studied structures, and amounts to about -65 kJ mol^{-1}. The Zr–O (CO) distance is 239 pm and CO_2 is slightly tilted by 16° with respect to normal of the plane formed by the four underlying zirconium atoms. The energy calculated is quite high for a physisorption mechanism, though not as high as the highest measured experimentally. This further suggests a certain degree of chemisorption occurs during the interaction of this mesoporous titania with CO_2.

13.3.2   Occlusion of Metal Nanoparticles into the Pores

The ordered mesoporous silicas that have been described in this chapter have also been used as supports for metal nanoparticles (Shephard et al., 1998; Rioux et al., 2005; Chatterjee et al., 2006; Sacaliuc et al., 2007). This can be understood as these materials have high surface area and the mesopores allow for increased accessibility to active sites with respect to the equivalent microporous zeolites.

In terms of adsorption on these materials, it can be of interest to use metal nanoparticles for the adsorption of specific gases. This is certainly the case when considering applications with carbon monoxide. One example is the inclusion of silver nanoparticles into mesoporous silicas (Bloch et al., 2010). In this study, samples were used which are described above in Section 13.2.3.

The nitrogen isotherms obtained at 77 K for the parent silica and silver impregnated sample are given in Figure 13.20. From these isotherms it is possible to see that the presence of silver nanoparticles in the mesoporous silica leads to a decrease of the BET surface area, the total pore volume and the total adsorption capacity. However, after incorporation of silver nanoparticles in the host silica matrix, the pore diameter remains unchanged which suggests that this does not significantly affect the pore diameter in contrast with the other structural parameters.

The adsorption of carbon monoxide can be contrasted to that of another polar molecule, carbon dioxide at 30 °C. However, to compare the adsorption on the parent silica and silver silica composite is not simple as the surface areas are different. In this

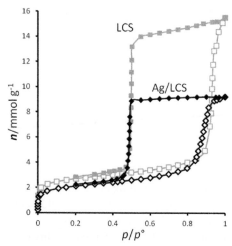

Figure 13.20 Nitrogen adsorption isotherms (77 K) of the parent large cage silica (LCS) and the sample in which silver nanoparticles have been impregnated (Bloch et al., 2010).

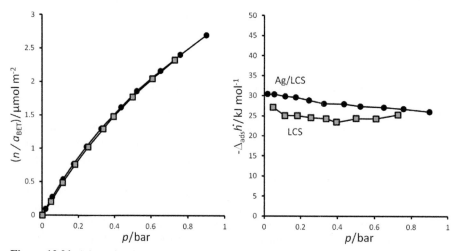

Figure 13.21 Adsorption isotherms (left) and corresponding differential enthalpies of adsorption (right) of CO_2 on the parent large cage silica (LCS) and silver/silica composite (Bloch et al., 2010).

case, one can plot the isotherms as the amount of probe gas adsorbed *per unit surface area* as a function of the pressure as shown in Figure 13.21. In this figure, the isotherms overlap which suggests that the presence of silver does not affect the CO_2 uptake in terms of unit surface, that is of available BET area.

To further investigate the adsorption phenomenon from an energetic point of view, the differential enthalpies of adsorption for both samples are plotted as a function of pressure (Figure 13.21). Small variations in these enthalpies are observed for the

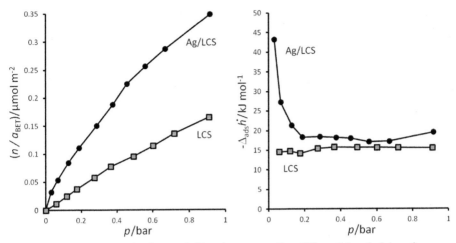

Figure 13.22 Adsorption isotherms (left) and corresponding differential enthalpies of adsorption (right) of CO on the parent large cage silica (LCS) and silver/silica composite (Bloch et al., 2010).

parent mesoporous silica matrix and on the silver/silica composite. A differential enthalpy of adsorption of around -25 to -26 kJ mol^{-1} is obtained for the parent silica whereas a value of around -28 to -29 kJ mol^{-1} is obtained for the silver/silica composite. For both samples, the differential enthalpies of adsorption do not decrease significantly indicating that the surfaces are energetically fairly homogeneous towards CO_2. Nevertheless, the silver containing sample seems to show a slightly greater interaction with the CO_2.

The adsorption of carbon monoxide on the same samples is shown in Figure 13.22. In contrast with CO_2 adsorption, there is a significant difference in behaviour during the adsorption of carbon monoxide on these two samples. Indeed, the areal amount of carbon monoxide adsorbed on the silver/silica composite sample is twice that obtained with the parent silica. This suggests that the doubling in CO uptake observed with the composite material with respect to the parent sample is only due to the presence of the silver. This is supported by the large increase in the initial differential enthalpy of adsorption measured for CO adsorbed on the silver/silica composite is of around -45 kJ mol^{-1}. This value rapidly decreases after 0.2 bar to a value similar to the parent silica matrix (-15 kJ mol^{-1}).

13.3.3 Grafting of Organic Ligands on the Surface

The silica surface is well adapted to grafting with silanes. For example, the surface of MCM-41 has been grafted with chlorodimethyloctyl silane groups to render the pore walls hydrophobic (Martin et al., 2002b). In other studies, specific organic groups have been grafted in order to increase the catalytic properties of these materials (Brunel, 1999). One current adsorption-related application is the capture of greenhouse gases such as carbon dioxide. In this respect, several groups have used amine containing ligands to graft onto different forms of ordered mesoprous silica (e.g. Huang et al., 2003; Franchi et al., 2005; Harlick and Sayari, 2006; Heydari-Gorji et al., 2011). Interesting work has shown the effectiveness to CO_2 capture of almost completely filling the mesoporosity with hyperbranched amine containing species (Hicks et al., 2008; Drese et al., 2009, 2012) although diffusion effects may occur.

The type of amine group that is grafted on the silica surface affects the basicity of the surface and thus the CO_2 adsorption behaviour. In one example, SBA-12 mesoporous silica was grafted with different amine functionalities resulted in different base strengths of the surface (Zeleňàk et al., 2008). These played an important role in the chemical fixation of carbon dioxide. The results showed that higher basicity of amino ligand increases efficiency of sorbents with respect to CO_2. On the other hand, the regeneration was faster for the sorbent containing active sites of lower basicity.

An example of the effect of grafting amine groups on SBA-16 with respect to carbon dioxide adsorption properties is given in Figure 13.23 (Knöfel et al., 2007; Knofel et al., 2009). The SBA-16 was grafted with a diamine TEDA $[(CH_3O)_3$-$Si-(CH_2)_3-NH-(CH_2)_2-NH_2]$. The carbon dioxide isotherms are given to high pressures and show that above around 8 bar, more CO_2 adsorbs in the non-functionalised material suggesting a larger pore volume available for adsorption. The grafting process can be expected to lead to a decrease in pore volume which results in a decrease of adsorption capacity for the grafted samples.

However, at the initial loading (<1 bar), the enthalpies of adsorption are much higher for the amine grafted sample. This behaviour can be explained by the initial adsorption of carbon dioxide on the more reactive amine sites at low pressures. Indeed, differential enthalpies of adsorption above 100 kJ mol^{-1} are measured which suggests that maybe chemisorption occurs. Complementary in situ infrared spectroscopy has been carried out to study this reactivity and the formation of three products (carbamate, carbamic acid and bidentate carbonate) was observed (Knofel et al., 2009).

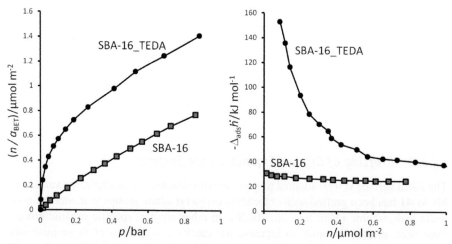

Figure 13.23 Differential enthalpies of adsorption for carbon dioxide at 30 °C on pure SBA-16 and amine grafted SBA-16 (Knöfel et al., 2009).

13.4   Ordered Organosilica Materials

As mentioned above, one direction taken with these ordered mesorporous materials has been to functionalise the surface with various groups. Organic groups have been grafted to the surface and, in view of increasing the density of the desired functions, the mesopore volume has equally been filled to various extents. It would thus seem natural for researchers to prepare ordered mesoporous materials whose pore walls are purely organic. As the silane/silica chemistry is well understood, it is in this domain that most of the periodic mesoporous organosilicas (PMOs) have been developed (Asefa et al., 1999; Mizoshita et al., 2011; van der Voort et al., 2013).

PMO materials are synthesised by the simultaneous use of a soft surfactant template and a hydrolysable bis-silane, condensing around the template. The bis-silane holds an organic functional linker between the silicon atoms giving a general formula of $Z_3Si–R–SiZ_3$, where Z represents a hydrolysable group (usually ethoxy or methoxy groups) and R is the functional linker (Mizoshita et al., 2011; van der Voort et al., 2013). The synthesis conditions (temperature, acidity, etc.) that can be used to prepare these materials can be quite varied and mirror much of those that can be used for the preparation of other ordered mesoporous silicas (M41S, SBA, etc.). Liberation of the porosity is obtained by solvent extraction of the surfactant template.

A comprehensive review of the synthesis of the various members of the PMO family of materials is given by van der Voort et al. (2013). The chemistry of frameworks of this family of materials can be varied to a very large extent. As such, organic groups containing hetero-elements (e.g. N, S, P), metal complexes/nanoparticles, chiral bridges, etc. have been used which has opened up the possibility to develop materials for many application areas. These include catalysis, chromatography, chiral separations, adsorption of metal ions, organic vapours, acid gases, low-k films, light harvesting devices and biological supports (Mizoshita et al., 2011; van der Voort et al., 2013). Hydrogen adsorption has been studied in these families (Jung et al., 2006).

The pore structure of these periodic mesoporous organosilicas is quite similar to the pure silica counterparts. This is due to the fact that the synthesis procedures can be quite similar. Some examples of nitrogen isotherms are given in Figure 13.24 which shows the resemblance between these samples and M41S solids or SBA materials.

13.5   Replica Materials

With so much effort devoted to the synthesis of ordered mesoporous silicas, and the extensions to this line of study, it was only logical that great effort would be devoted to the preparation of ordered mesoporous carbon materials.

The first route taken to prepare these materials has been to use ordered silica materials as 'hard' templates in which a carbon source (e.g. resorcinol, sucrose, etc.) is

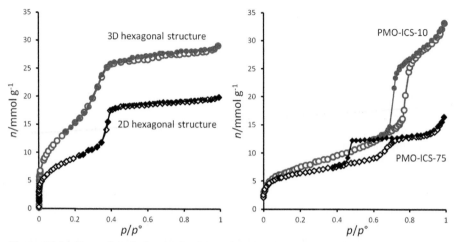

Figure 13.24 Examples of nitrogen isotherms (77 K) obtained with periodic mesoporous organosilicas with containing: (a) 1,2-bis(trimethoxysilyl)ethane with two different crystal symmetries leading to pore topologies and isotherms similar to the M41S family (Inagaki et al., 1999). (b) Tris[3-(trimethoxysilyl)propyl]isocyanurate (ICS) having pore topologies and isotherms similar to the large caged silicas (Olkhovyk and Mietek, 2005).

introduced into the pores (Lee et al., 1999; Ryoo et al., 1999). This is then carbonised and the silica removed by dissolution as mentioned in Chapter 10. Several materials, from the 'CMK' family are of note:

- CMK-1. This is made using MCM-48 as the template (Ryoo et al., 1999). Two bicontinuous rods of carbon material are prepared which fuse to a certain extent during the template removal. In the case where the carbon rods do not fuse, the material is known as CMK-4 (Kaneda et al., 2002).
- CMK-2 is made using SBA-1 as the silica source (Ryoo et al., 1999).
- CMK-3 is made from SBA-15 where the pores are completely filled (Jun et al., 2000).
- CMK-5 is prepared from SBA-15 but here the pores are not completely filled which leads to two distinct pore systems (Joo et al., 2001).

Note that MCM-41 does not lead to replica materials of interest. Furthermore, this 'nanocasting' route takes several synthesis steps and can be considered quite complicated and time consuming. In terms of synthesis upscale for any applications of these materials, the use of an ordered mesoporous silica (M41S, SBA, etc.) casts along with the previous considerations, make these materials rather unsuitable for anything else than for lab scale experiments.

These materials have been shown to be of interest, as 'negatives' in order to understand the silica framework structure. This was especially the case to show the presence of micropores which connect between the mesopores of SBA-15 (Jun et al., 2000).

It has been of interest to note that, even though the parent mesoporous silica casts can show sharp capillary condensation steps indicative of very narrow pore size distributions (PSDs), the resulting carbon materials often show more gradual pore filling

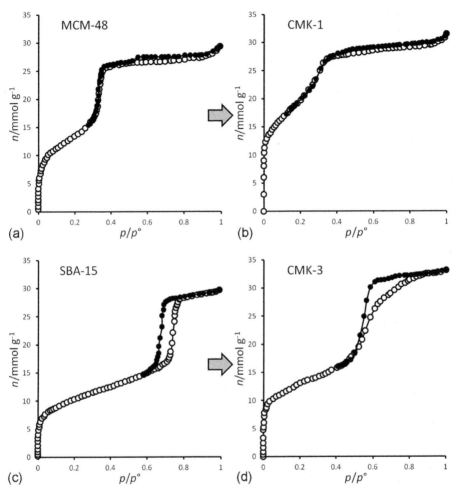

Figure 13.25 Nitrogen isotherms (77 K) of parent MCM-48 (a) and resultant CMK-1 (b) as well as the parent SBA-15 (c) and resultant CMK-3 (d).
From Kruk et al. (2000) and Kruk et al. (2003).

suggesting a much larger PSD (Figure 13.25). However, electron microscopy studies suggest materials with some degree of perfect order as suggested in the study of CMK-5 (Solovyov et al., 2004).

The use of other templates such as latex spheres or silica nanospheres can lead to well-defined spherical caged pore structures. Scanning of the hysteresis loops has been used to explore what can be a quite interesting pore structure (Cychosz et al., 2012).

The hard templating route has been used to make other materials such as oxides for catalysis (Lu and Schüth, 2006; Valdés-Solís and Fuertes, 2006; Tiemann, 2008). Boron nitride (Dibandjo et al., 2005) and silicon carbide (Krawiec et al., 2006) have equally been prepared.

However, due to the limitations of the hard templating route, the direct synthesis of mesoporous carbons has been developed. A review of the synthesis of these materials has been made by Ma et al. (2013). As for the periodic mesoporous organosilicas, synthesis procedures mimicking those used for the synthesis of ordered silica materials can be used for the preparation of these ordered mesoporous carbons. As such, the physisorption isotherms obtained with these materials are similar to those of the equivalent silica counterparts.

13.6   Concluding Remarks

A fair amount of interest has been aroused since the initial development of the first ordered mesoporous silicas. From a physisorption point of view, many of these materials can be considered as possessing model pore geometries:

MCM-41 with non-intersecting cylindrical pores.
SBA-15 with either non-intersecting cyclindrical mesopores or with micropores connecting between the mesopores, depending on the synthesis conditions.
MCM-48 with 3D cylinder-like network having intersections with essentially the same width as the pores. KIT-6 is another example of this type of pore system.
SBA-16 with cage-like structures resembling the traditional ink-bottle pore scenario.

The pore size can be modulated with the use of surfactants of different lengths or by varying specific synthesis conditions (temperature, ageing, post-synthesis restructuration). Several research groups have prepared their own surfactants in order to explore the possibility to make other pore geometries or sizes.

Similar synthesis strategies have been used to prepare materials of different chemical compositions with notably the preparation of ordered mesoporous organosilicas or ordered mesoporous carbons. The low-temperature physisorption properties of these materials are quite similar to those of equivalent mesoporous silicas.

To extend the chemistry of these adsorbents in view of various applications, functionalisation of these materials has been made with the grafting of organic groups, the introduction of metal nanoparticles of metal oxides. Samples have been made of mesoporous oxides other than silica either by coating the silica or using the mesoporous silica (or carbon) as template. Non-oxide materials have equally been prepared using this nanocasting methodology.

In terms of applications, the cost of the preparation of these materials can be seen as an obstacle to overcome for basic storage or separations. Thus high added value applications could be of interest but the question arises as to whether the high order of the pore structure is of interest with respect to the relatively high surface area of many of these materials.

Nevertheless, the leaps and bounds in the understanding of capillary condensation-related phenomena that have been made in recent years, which has been accompanied by the development of NLDFT approaches to pore size characterisation, would not have been possible without the development of this family of materials.

References

Asefa, T., MacLachlan, M.J., Coombs, N., Ozin, G.A., 1999. Nature. 402, 867.

Attard, G.S., Glyde, J.C., Goltner, C.G., 1995. Nature. 378, 366.

Beck, J.S., Vartuli, J.C., Roth, W.J., Leonowicz, M.E., Kresge, C.T., Schmitt, K.D., Chu, C.T.W., Olson, D.H., Sheppard, E.W., McCullen, S.B., Higgins, J.B., Schlenker, J.L., 1992. J. Am. Chem. Soc. 114, 10834.

Bhambhani, M.R., Cutting, P.A., Sing, K.S.W., Turk, D., 1972. J. Coll. Interf. Sci. 38, 109.

Bloch, E., Phan, T., Bertin, D., Llewellyn, P.L., Hornebecq, V., 2008. Micro. Meso. Mat. 112 (1–3), 612.

Bloch, E., Llewellyn, P.L., Phan, T., Bertin, D., Hornebecq, V., 2009. Chem. Mater. 21, 48.

Bloch, E., Llewellyn, P.L., Vincent, D., Chaspoul, F., Hornebecq, V., 2010. J. Phys. Chem. C. 114 (51), 22652.

Boulet, P., Knöfel, C., Kuchta, B., Hornebecq, V., Llewellyn, P.L., 2012. J. Mol. Model. 18 (11), 4819.

Branton, P.J., Hall, P.G., Sing, K.S.W., 1993. Chem. Commun. 1257.

Branton, P.J., Hall, P.G., Sing, K.S.W., Reichert, H., Schüth, F., Unger, K.K., 1994. Faraday Trans. 90, 2965.

Branton, P.J., Hall, P.G., Sing, K.S.W., 1995a. Adsorption. 1, 77.

Branton, P.J., Hall, P.G., Treguer, M., Sing, K.S.W., 1995b. Faraday Trans. 91, 2041.

Branton, P.J., Kaneko, K., Setoyama, N., Sing, K.S.W., Inagaki, S., Fukusima, Y., 1996. Langmuir. 12, 599.

Branton, P.J., Sing, K.S.W., White, J.W., 1997. Faraday Trans. 93, 2337.

Brunel, D., 1999. Micro. Meso. Mat. 27 (2–3), 329.

Buchel, G., Denoyel, R., Llewellyn, P.L., Rouquerol, J., 2001. J. Mat. Chem. 11 (2), 589.

Cassiers, K., Linssen, T., Mathieu, M., Benjelloun, M., Schrijnemakers, K., Van Der Voort, P., Cool, P., Vansant, E.F., 2002. Chem. Mater. 14, 2317.

Chatterjee, M., Ikushima, Y., Hakuta, Y., Kawanami, H., 2006. Adv. Synth. Catal. 348 (12–13), 1580.

Chiola, V., Ritsko, J.E., Vanderpool, C.D., 1971. US Patent 3 556 725.

Ciesla, U., Demuth, D., Leon, R., Petroff, P., Stucky, G., Unger, K.K., Schüth, F., 1994. Chem. Commun. 1387.

Cimino, R., Cychosz, K.A., Thommes, M., Neimark, A.V., 2013. Coll. Surf. A. http://dx.doi. org/10.1016/j.colsurfa.2013.03.025 (in press).

Coasne, B., Galarneau, A., Pellenq, R.J.M., Di Renzo, F., 2013. Chem. Soc. Rev. 42, 4141.

Corma, A., Kan, Q., Navarro, M.T., Pérez-Pariente, J., Rey, F., 1997. Chem. Mater. 9, 2123.

Cychosz, K.A., Guo, X., Fan, W., Cimino, R., Gor, G.Y., Tsapatsis, M., Neimark, A.V., Thommes, M., 2012. Langmuir. 28, 12647.

Deng, Y., Wei, J., Sun, Z., Zhao, D., 2013. Chem. Soc. Rev. 42, 4054.

Di Renzo, F., Cambon, H., Dutartre, R., 1997. Micro. Meso. Mat. 10 (4–6), 283.

Dibandjo, P., Chassagneux, F., Bois, L., Sigala, C., Miele, P.J., 2005. Mater. Chem. 15, 1917.

Drese, J.H., Choi, S., Lively, R.P., Koros, W.J., Fauth, D.J., Gray, M.L., Jones, C.W., 2009. Adv. Funct. Mat. 19 (23), 3821.

Drese, J.H., Choi, S., Didas, S.A., Bollini, P., Gray, M.L., Jones, C.W., 2012. Micro. Meso. Mat. 151, 231.

Edler, K.J., White, J.W., 1995. Chem. Commun., 155.

Firouzi, A., Kumar, D., Bull, L.M., Besier, T., Sieger, P., Huo, Q., Walker, S.A., Zasadzinski, J.A., Glinka, C., Nicol, J., Margolese, D., Stucky, G.D., Chmelka, B.F., 1995. Science 267, 1138.

Franchi, R.S., Harlick, P.J.E., Sayari, A., 2005. Ind. Eng. Chem. Res. 44, 8007.

Franke, O., Schulz-Ekloff, G., Rathousky, J., Starek, J., Zukal, A., 1993. Chem. Commun. 9, 724.

Galarneau, A., Desplantier, D., Dutartre, R., Di Renzo, F., 1999. Micro. Meso. Mat. 27 (2–3), 297.

Galarneau, A., Cambon, H., Di Renzo, F., Fajula, F., 2001. Langmuir 17, 8328.

Galarneau, A., Cambon, N., Di Renzo, F., Ryoo, R., Choi, M., Fajula, F., 2003. New J. Chem. 27 (1), 73.

Gelb, L.D., Gubbins, K.E., Radhakrishnan, R., Sliwinska-Bartkowiak, M., 1999. Rep. Prog. Phys. 62, 1573.

Grun, M., Unger, K.K., Matsumoto, A., Tsutsumi, K., 1999. Micro. Meso. Mat. 27 (2–3), 207.

Harlick, P.J.E., Sayari, A., 2006. Ind. Eng. Chem. Res. 45 (9), 3248.

Heydari-Gorji, A., Belmabkhout, Y., Sayari, A., 2011. Langmuir 27 (20), 12411.

Hicks, J.C., Drese, J.H., Fauth, D.J., Gray, M.L., Qi, G.G., Jones, C.W., 2008. J. Am. Chem. Soc. 130 (10), 2902.

Hornebecq, V., Knöfel, C., Boulet, P., Kuchta, B., Llewellyn, P.L., 2011. J. Phys. Chem. C. 115 (20), 10097.

Huang, H.Y., Yang, R.T., Chinn, D., Munson, C.L., 2003. Ind. Eng. Chem. Res. 42 (12), 2427.

Huo, Q., Leon, R., Petroff, P.M., Stucky, G.D., 1995. Science 268, 1324.

Huo, Q., Margolese, D.I., Stucky, G.D., 1996. Chem. Mater. 8, 1147.

Inagaki, S., Fukushima, Y., Kuroda, K., 1993. Chem. Commun., 680.

Inagaki, S., Fukushima, Y., 1998. Micro. Meso. Mat. 21 (4–6), 667.

Inagaki, S., Guan, S., Fukushima, Y., Ohsuna, T., Terasaki, O., 1999. J. Am. Chem. Soc. 121, 9611.

Ismaïl, I.M.K., 1992. Langmuir 8, 360.

Jelinek, J.L., Kováts, E., 1994. Langmuir 10, 4225.

Joo, S.H., Choi, S.J., Oh, I., Kwak, J., Liu, Z., Terasaki, O., Ryoo, R., 2001. Nature 412, 169.

Jun, S., Joo, S.H., Ryoo, R., Kruk, M., Jaroniec, M., Liu, Z., Ohsuna, T., Terasaki, O., 2000. J. Am. Chem. Soc. 122 (43), 10712.

Jung, J.H., Han, W.S., Rim, J.A., Lee, S.J., Cho, S.J., Kim, S.Y., Kang, J.K., Shinkai, S., 2006. Chem. Lett. 35 (1), 32.

Kaneda, M., Tsubakiyama, T., Carlsson, A., Sakamoto, Y., Ohsuna, T., Terasaki, O., Joo, S.H., Ryoo, R., 2002. J. Phys. Chem. B. 106 (6), 1256.

Keene, M.T.J., Denoyel, R., Llewellyn, P.L., 1998. Chem. Commun., 20, 2203.

Keene, M.T.J., Gougeon, R.D.M., Denoyel, R., Harris, R.K., Rouquerol, J., Llewellyn, P.L., 1999. J. Mat. Chem. 9 (11), 2843.

Keung, M., 1993. PhD Thesis, Brunel University.

Kim, M.J., Ryoo, R., 1999. Chem. Mater. 11 (2), 487.

Kleitz, F., Choi, S.H., Ryoo, R., 2003. Chem. Commun. 2136.

Kleitz, F., Czuryszkiewicz, T., Solovyov, L.A., Lindén, M., 2006. Chem. Mater. 18, 5070.

Knöfel, C., Descarpentries, J., Benzaouia, A., Zeleňák, V., Mornet, S., Llewellyn, P.L., Hornebecq, V., 2007. Micro. Meso. Mat. 99 (1–2), 79.

Knofel, C., Martin, C., Hornebecq, V., Llewellyn, P.L., 2009. J. Phys. Chem. C 113 (52), 21726.

Knofel, C., Lutecki, M., Martin, C., Mertens, M., Hornebecq, V., Llewellyn, P.L., 2010. Micro. Meso. Mat. 128 (1–3), 26.

Krawiec, P., Geiger, D., Kaskel, S., 2006. Chem. Commun. 23, 2469.

Kresge, C.T., Roth, W.J., 2013. Chem. Soc. Rev. 42 (9), 3663.

Kresge, C.T., Leonowiz, M.E., Roth, W.J., Vartuli, J.C., Beck, J.S., 1992. Nature. 359, 710.

Kruk, M., Jaroniec, M., Sayari, A., 1997. Langmuir 13 (23), 6267.

Kruk, M., Jaroniec, M., Sakamoto, Y., Terasaki, O., Ryoo, R., Ko, C.H., 2000. J. Phys. Chem. B. 104, 292.

Kruk, M., Jaroniek, J., 2001. Chem Mater. 13, 3169.

Kruk, M., Jaroniec, M., Joo, S.H., Ryoo, R., 2003. J. Phys. Chem. B. 107, 2205.

Kuchta, B., Llewellyn, P., Denoyel, R., Firlej, L., 2004. Coll. Surf. A 241 (1–3), 137.

Lee, J., Yoon, S., Hyeon, T., Oh, S.M., Kim, K.B., 1999. Chem. Commun., 2177.

Llewellyn, P.L., Grillet, Y., Schüth, F., Reichert, H., Unger, K.K., 1994. Micro. Mater. 3, 345.

Llewellyn, P.L., Schüth, F., Grillet, Y., Rouquerol, F., Rouquerol, J., Unger, K.K., 1995. Langmuir 11 (2), 574.

Llewellyn, P.L., Grillet, Y., Rouquerol, J., Martin, C., Coulomb, J.P., 1996. Surf. Sci. 352, 468.

Llewellyn, P.L., Sauerland, C., Martin, C., Grillet, Y., Coulomb, J.P., Rouquerol, F., Rouquerol, J., 1997. In: McEnaney, B., Mays, T.J., Rouquerol, J., Rodríguez-Reinoso, F., Sing, K.S.W., Unger, K.K. (Eds.), Characterisation of Porous Solids IV. The Royal Society of Chemistry, Cambridge, p. 111.

Lu, A.H., Schüth, F., 2006. Adv. Mater. 18, 1793.

Ma, T.Y., Liu, L., Yuan, Z.Y., 2013. Chem. Soc. Rev. 42, 3977.

Maddox, M.W., Olivier, J.P., Gubbins, K.E., 1997. Langmuir. 13 (6), 1737.

Martin, T., Galarneau, A., Di Renzo, F., Fajula, F., Plee, D., 2002a. Angew. Chemie. Int. Ed. 41 (14), 2590.

Martin, T., Lefevre, B., Brunel, D., Galarneau, A., Di Renzo, F., Fajula, F., Gobin, P.F., Quinson, J.F., Vigier, G., 2002b. Chem. Commun., 24

Mason, G., 1988. Proc. R. Soc. Lond. A 415, 453.

Meynen, V., Cool, P., Vansant, E.F., 2009. Micro. Meso. Mat. 125 (3), 170.

Mizoshita, N., Tania, T., Inagaki, S., 2011. Chem. Soc. Rev. 40, 789.

Monnier, A., Schüth, F., Huo, Q., Kumar, D., Margolese, D., Maxwell, R.S., Stucky, G.D., Krishnamurty, M., Petroff, P., Firouzi, A., Janicke, M., Chmelka, B.F., 1993. Science. 261, 1299.

Morishige, K., Shikimi, M., 1998. J. Chem. Phys. 108, 7821.

Morishige, K., Fujii, H., Uga, M., Kinukawa, D., 1997. Langmuir 13, 3494.

Morishige, K., Tateishi, N., Fukuma, S., 2003. J. Phys. Chem. B. 107, 5177.

Olkhovyk, O., Mietek, J.M., 2005. J. Am. Chem. Soc. 127, 60.

Øye, G., Sjöblom, J., Stöcker, M., 2001. Adv. Coll. Interf. Sci. 89–90, 439.

Pérez-Mendoza, M., Gonzalez, J., Wright, P.A., Seaton, N.A., 2004. Langmuir 20, 7653.

Rasmussen, C.J., Vishnyakov, A., Thommes, M., Smarsly, B.M., Kleitz, F., Neimark, A.V., 2010. Langmuir 26 (12), 10147.

Rathousky, J., Zukal, A., Franke, O., Schulz-Ekloff, G., 1994. Faraday Trans. 90, 2821.

Rathousky, J., Zukal, A., Franke, O., Schulz-Ekloff, G., 1995. Faraday Trans. 91, 937.

Ravikowitch, P., Neimark, A.V., 2002. Langmuir 18, 1550.

Ravikowitch, P.I., Domhnail, S.C.O., Neimark, A.V., Schuth, F., Unger, K.K., 1995. Langmuir. 11, 4765.

Ribeiro Carrott, M.M.L., Estêvão Candeias, A.J., Carrott, P.J.M, Unger, K.K., 1999. Langmuir. 15 (26), 8895.

Rioux, R.M., Song, H., Hoefelmeyer, J.D., Yang, P., Somorjai, G.A., 2005. J. Phys. Chem. B 109 (6), 2192.

Roth, W.J., Kresge, C.T., Vartuli, J.C., Leonowicz, M.E., Fung, A.S., McCullen, S.B., 1995. In: Beyer, H.K., Karge, H.G., Kiricsi, I., Nagy, J.B. (Eds.), Catalysis by Microporous Materials, Stud. Surf. Sci. Catal., vol. 94, p. 301.

Rouquerol, J., Rouquerol, F., Peres, C., Grillet, Y., Boudellal, M., 1979. In: Gregg, S.J., Sing, K.S.W., Stoeckli, H.F. (Eds.), Characterisation of Porous Solids. Soc. Chem. Ind. London, p. 107.

Ryoo, R., Joo, S.H., Jun, S., 1999. J. Phys. Chem. B 103 (37), 7743.

Ryoo, R., Ko, C.H., Kruk, M., Antochshuk, V., Jaroniec, M., 2000. J. Phys. Chem. B 104, 11465.

Sacaliuc, E., Beale, A.M., Weckhuysen, B.M., Nijhuis, T.A., 2007. J. Catal. 248 (2), 235.

Sakamoto, Y., Kaneda, M., Terasaki, O., Zhao, D.Y., Kim, J.M., Stucky, G.D., Shin, H.J., Ryoo, R., 2000. Nature 408, 449.

Sarkisov, L., Monson, P.A., 2001. Langmuir 17, 7600.

Sayari, A., Liu, P., Kruk, M., Jaroniec, M., 1997. Chem. Mater. 9, 2499.

Sayari, A., Yang, Y., Kruk, M., Jaroniec, M., 1999. J. Phys. Chem. B. 103, 3651.

Schmidt, R., Stöcker, M., Hansen, E., Akporiaye, D., Ellestad, O.H., 1995. Micro. Mater. 3, 443.

Schmidt, R., Junggreen, H., Stocker, M., 1996. Chem. Commun 875.

Shephard, D.S., Maschmeyer, T., Sankar, G., Thomas, J.M., Ozkaya, D., Johnson, B.F.G., Raja, R., Oldroyd, R.D., Bell, R.G., 1998. Chem. Eur. J. 4 (7), 1214.

Solovyov, L.A., Kim, T.W., Kleitz, F., Terasaki, O., Ryoo, R., 2004. Chem. Mater. 16, 2274.

Stucky, G.D., Monnier, A., Schüth, F., Huo, Q., Margolese, D., Kumar, D., Krishnamurty, M., Petroff, P., Firouzi, A., Janicke, M., Chmelka, B.F., 1994. Mol. Cryst. Liq. Cryst. 240, 187.

Taguchi, A., Schüth, F., 2005. Micro. Meso. Mat. 77 (1), 1.

Tanev, P.T., Chibwe, M., Pinnavaia, T.J., 1994. Nature 368, 321.

Thomas, J.M., 1994. Nature. 368, 289.

Thomas, J.M., 1995. Faraday Discuss. 100, C9.

Thommes, M., Köhn, R., Fröba, M., 2000. J. Phys. Chem. B. 104, 7932.

Thommes, M., Smarsly, B., Groenewolt, M., Ravikovitch, P.I., Neimark, A.V., 2006. Langmuir 22, 756.

Tiemann, M., 2008. Chem. Mater. 20, 961.

Valdés-Solís, T., Fuertes, A.B., 2006. Mat. Res. Bull. 41, 2187.

Vartuli, J.C., Schmitt, K.D., Kresge, C.T., Roth, W.J., Leonowicz, M.E., McCullen, S.B., Hellring, S.D., Beck, J.S., Schlenker, J.L., Olson, D.H., Sheppard, E.W., 1994. Chem. Mat. 6 (12), 2317.

Van der Voort, P., Esquivel, D., De Canck, E., Goethals, F., Van Driessche, I., Romero-Salguero, F.J., 2013. Chem. Soc. Rev. 42 (9), 3913.

Yang, P.D., Deng, T., Zhao, D.Y., Feng, P.Y., Pine, D., Chmelka, B.F., Whitesides, G.M., Stucky, G.D., 1998. Science 282 (5397), 2244.

Zeleňàk, V., Hornebecq, V., Mornet, S., Schäf, O., Llewellyn, P.L., 2006. Chem. Mater. 18, 3184.

Zeleňàk, V., Halamova, D., Gaberova, L., Bloch, E., Llewellyn, P., 2008. Micro. Meso. Mat. 116 (1–3), 358.

Zhao, D.Y., Feng, J.L., Huo, Q.S., Melosh, N., Fredrickson, G.H., Chmelka, B.F., Stucky, G.D., 1998a. Science 279 (5350), 548.

Zhao, D., Huo, Q., Feng, J., Chmelka, B.F., Stucky, G.D., 1998b. J. Am. Chem. Soc. 120 (24), 6024.

Zukal, A., Siklova, H., Čejka, J., 2008. Langmuir. 24 (17), 9837.

Zukal, A., Mayerová, J., Čejka, J., 2010. Phys. Chem. Chem. Phys. 12, 5240.

14 Adsorption by Metal-Organic Frameworks

Philip Llewellyn*, Guillaume Maurin†, Jean Rouquerol*

*Aix Marseille University-CNRS, MADIREL Laboratory, Marseille, France,
†University of Montpellier 2, Institute Charles Gerhardt, Montpellier, France

---

Chapter Contents

14.1 Introduction

The family of metal-organic frameworks (MOFs) or porous coordination polymers (PCPs) (Kitagawa et al., 2004; Rowsell and Yaghi, 2004; Tranchemontagne et al., 2009; Farha and Hupp, 2010; Janiak and Vieth, 2010; Meek et al., 2011; Stock and Biswas, 2012) encompasses a very wide range of materials of which subsets can be known as ZIFs (zeolite imadzole frameworks) (Phan et al., 2010), COFs (covalent

Adsorption by Powders and Porous Solids. http://dx.doi.org/10.1016/B978-0-08-097035-6.00014-0

organic frameworks) (Ding and Wang, 2013), MOPs (microporous organic polymers) (Jiang and Cooper, 2010) and many others. Thousands of different structures have been prepared of which many do not show permanent porosity observed with nitrogen at 77 K and these will not be treated in this chapter.

The realm of PCPs itself is not particularly new. A report of a potentially PCP material dates back to the 1950s (Kinoshita et al., 1959), although no sorption results are given. Recent interest in MOFs dates from the mid-1990s (Hoskins and Robson, 1989; Yaghi and Li, 1995; Kondo et al., 1997; Férey, 2001) where porosity and significant gas uptakes were reported.

MOFs are constructed of metals or metal cluster nodes which are linked together through organic moieties (see Figure 14.1). The metals that can be used are di-valent (Cu, Zn, Mg, etc.), tri-valent (Al, Cr, Ga, Fe, In, etc.) or tetra-valent (V, Zr, Ti, Hf, etc.) with almost every metal in the periodic table having been incorporated in one MOF structure or another. A very large number of organic linker moieties have equally been introduced with the carboxylate, imidazolate, phosphonate, pyrazolate families of note. The length and degree of functionalisation of these linkers have been varied enormously.

The building block approach schematised in Figure 14.1 suggests that the synthesis of this family of materials is relatively simple. This can be far from the case, as the chemistry needs to be adapted for each node and linker pair. In some cases, the activation of the solid can be a problem in itself. One can appreciate therefore, the talent of those specialised in the synthesis of this family of materials.

Many thousands of different MOF structures that have been reported so far which can be understood by the wealth of possibilities to vary the metal node and organic linker. The coordination of both node and linker can be varied to give different structure geometries (Eddaoudi et al., 2002; Yaghi et al., 2003; Kitagawa et al., 2004; Férey, 2008). The linker can be modified in size to tailor the available pore volume (Rowsell and Yaghi, 2004). The metals can be partially unsaturated to give specific adsorption sites. Furthermore, the nodes and linkers can be functionalised with groups of varying degrees of polarity including methyl, halide, acid and amines. This can lead to a very large spectrum of materials which able to be prepared with an almost infinite variation of physical and chemical properties.

Some MOFs are of quite low density and thus have been reported to have remarkable specific pore volumes and BET specific areas. This has largely contributed to the interest in these materials for adsorption based applications.

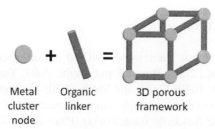

Metal   Organic   3D porous
cluster   linker   framework
node

Figure 14.1 Schematic representation of the construction of a metal-organic framework.

A more unique property of some MOFs is their large flexibility in response to external stimuli such as temperature change (Liu et al., 2008a), mechanical pressure (Beurroies et al., 2010; Ghoufi et al., 2012) and probe molecule uptake (Seki, 2002; Bourrelly et al., 2005). Indeed, as an example, the unit cell variation of the carboxylate-based MIL-88d (MIL stands from Materials of the Institut Lavosisier) can be over 200% upon the uptake of some fluids (Serre et al., 2007a). Further to this, the organic linkers themselves can show some significant rotational movement which can lead to some exceptional findings with respect to the uptake of specific molecules (Fairen-Jimenez et al., 2011).

Functionalisation of the metal nodes and organic linkers is possible with moieties that can specifically attract target species (Yang et al., 2011). Further to this, clusters or nanoparticles have been occluded in the pores, again, in order to improve the uptake of given species, notably hydrogen (Dybtsev et al., 2010; Zlotea et al., 2011).

Adsorption studies on these materials have largely been focused on applications with comparisons of the adsorption properties of hydrogen, methane and carbon dioxide for example. These studies have been carried out in a range of temperatures and often up to very high pressures. In the case of flexible MOFs, their adsorption performance may indeed be lowered at a low temperature where they have lost their flexibility: some MOFs seem to be non-porous at 77 K but show significant uptake at room temperature.

Because of their regular and well-defined porous structures, MOFs have lent themselves well to the development of adsorption simulation studies (see Chapter 6). Indeed, this field offers an exceptional opportunity for profitable synergy between simulation and experiment, so that sometimes, the experiments have come first and other times the modelling predictions have preceded those made experimentally. Furthermore, due to the complexity of these solids, a wide variety of *in situ* and *ex situ* approaches has been adopted to understand the adsorption phenomena. Thus, microcalorimetry has been used to gain energetic information, IR spectroscopy has been employed to characterise the adsorption sites and strength of adsorbate/adsorbent interactions, and X-ray diffraction experiments have been performed to follow the structural changes of the solids as well as the preferential location of the adsorbates.

It can thus be appreciated that with the numerous structures and wide spectrum of chemistry that can be employed in this fascinating family of materials, the number of adsorption studies are equally extremely diverse. It is an impossible task to relate all of these studies and we were therefore obliged to select a number – unfortunately limited – of examples that are able to highlight the adsorbent properties of these materials.

For the moment, although there is not yet any universally accepted nomenclature for MOFs or their structure types as in the case for zeolites, we should stress that the IUPAC division of inorganic chemistry is proposing, at the time when this book goes for printing, a provisional recommendation on 'Terminology of Metal-Organic Frameworks and Coordination Polymers' (Batten et al., 2012). They propose the following definition for a MOF: 'A Metal-Organic Framework, abbreviated to MOF, is a Coordination Polymer (or alternatively Coordination Network) with an open framework containing potential voids'. They explicitly discourage the use of the term 'hybrid organic–inorganic material'. Also, in their comments, they 'agree with the common practice of giving important new compounds trivial or nick names based on their place of origin followed by a number such as HKUST-1, MIL-101 and NOTT-112'. This is what we shall do in this chapter. Thus we shall use HKUST-1 to name the structure discovered at the Hong Kong University of Science and Technology (Chui et al., 1999) rather than Cu-btc (btc, benzene tricarboxylate) by which this material is also known. Indeed, the combination of a metal with the same organic linker can give rise to several different structures and so the latter nomenclature can lead to ambiguities. Furthermore, several metals have been used to make the same structure, and in this case, the metal is represented in brackets (e.g. MIL-100(Cr) and MIL-100(Fe) (Férey et al., 2005) or CPO-27(Ni) and CPO-27(Mn) (Dietzel et al., 2008a,b)).

Prior to discussing the particularities of MOFs, some remarks need to be made concerning the assessment of surface area. The BET method being the most used, since 70 years, for a first characterisation of adsorbents, it was natural that it is also broadly used in the case of MOFs (Düren et al., 2007; Farha et al., 2012; Walton and Snurr, 2007). Nevertheless, as it will be developed in the next section, the frequent complexity of their adsorption isotherms and porous structures, together with the usual presence of micropores, raise special problems which must be considered in the application and interpretation of the BET method.

14.2   Assessment and Meaning of the BET Area of MOFs

14.2.1   Assessment of the BET Area of MOFs

Like for many microporous adsorbents, the BET plot for N_2 adsorption isotherms of MOFs at 77 K often offers several pressure ranges where linearity is observed. This issue is normally addressed by applying, *in addition to linearity, the four criteria* (a) to (d) listed in Section 7.2.2 for the selection of a single, appropriate, pressure range.

The example shown in Figures 14.2a and 14.2b is that of the N_2 adsorption isotherm at 77 K of MOF HKUST-1. Application of criterion (b) given in Section 7.2.2 is made through the 'consistency plot' of Figure 14.2c, where $n(p^\circ - p)$ is plotted versus p/p°. This criterion requires limiting the BET calculation to the pressure range under the maximum, which is here at $p/p^\circ = 0.011$. This means that the relative pressure range for the calculation will be totally out of and beneath the common range of application of the BET method, namely 0.05–0.35. With this upper pressure limitation, we obtain the BET plot represented in Figure 14.2d. The corresponding BET n_m is 18.09 mmol g^{-1}, which occurs for a $(p/p^\circ)_m$ of 0.005. Criterion (c) requires this relative pressure to fall within the range used for the calculation (here, 0.0001–0.0108), which is the case. Finally, criterion (d) requires this same relative pressure to be equal, within 20%, to $1/(\sqrt{C} + 1)$. This latter term is here equal to 0.0052 and $(p/p^\circ)_m$ differs from it by 4%. All criteria are therefore fulfilled and the BET n_m to retain is 18.09 mmol g^{-1}, hence a '(N_2)BET area' (derived with the conventional σ of 0.162 nm^2 g^{-1}) of 1764 m^2 g^{-1}.

Figure 14.3a shows the N_2 experimental adsorption isotherm for MOF NU 109, taken from Farha et al. (2012), with an important step in the range of relative pressure from 0.22 to 0.3. This corresponds to, at least a bimodal micropore-size distribution,

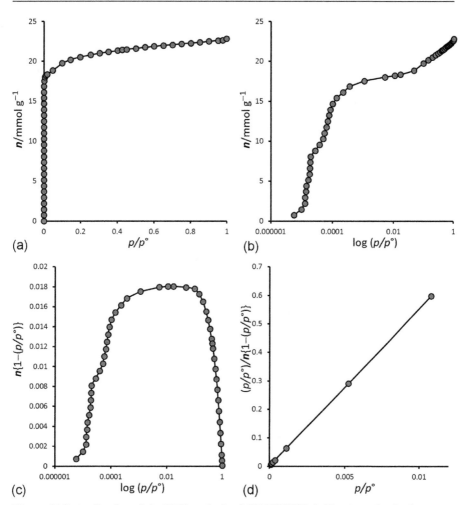

Figure 14.2 Application of the BET method to MOF HKUST-1: N_2 adsorption isotherm at 77 K (a) normal plot, (b) semi-log plot, (c) consistency plot and (d) BET plot.

with most probably, the presence of a significant quantity of wide micropores. The 'consistency plot' in Figure 14.3b shows a maximum at approximately $p/p° = 0.28$ but the BET plot in Figure 14.3c indicates that this criterion is not sufficient, as three straight lines can be drawn, for different pressure ranges, as shown with the BET plots of Figures 14.3d–f. The BET areas derived from these three BET plots are 2442, 4050 and 7864 m^2 g^{-1}, respectively, whereas the corresponding values of C are 143, 38 and 4: the span is considerable! Actually, for the BET plots shown in Figures 14.3d and f, the derived BET n_m corresponds to a $(p/p°)_m$ which is above the actual pressure range of the BET plot, so that criterion (c) is not fulfilled. It is only for the BET plot of Figure 14.3e that both criteria (c) and (d) are fulfilled. The corresponding

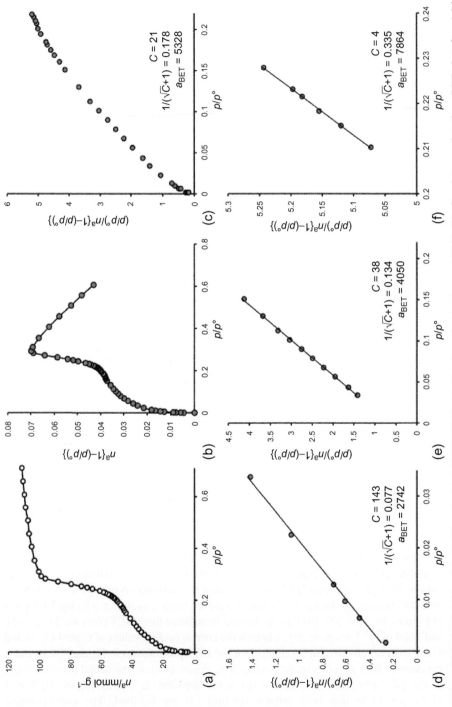

Figure 14.3 Application of the BET method to MOF NU 109. (a): N_2 adsorption isotherm at 77 K; (b): Consistency plot; then, BET plots for $p/p°$ in the following ranges: 0.01–0.3 (c), 0.001–0.03 (d), 0.03–0.15 (e) and 0.21–0.23 (f).

(N_2) BET n_m of 42 mmol g^{-1}, (N_2) BET area of 4050 m^2 g^{-1} and C value of 38 are therefore those it makes sense to consider characteristic of this system.

In the still more delicate case of flexible MOFs, also giving rise to one or several steps, it seems that the above criteria tend to select a pressure region where the porous structure is completely open and becomes stable. Nevertheless, in case, the criteria would be fulfilled for several regions of the adsorption isotherms it would be essential to always indicate the $p/p°$ range used for the calculation of the BET n_m and derived (N_2) BET area.

In any case, when providing a BET area, it is highly recommended to systematically indicate (i) whether all criteria listed in Section 7.2.2 are fulfilled, (ii) what is the selected pressure range? and (iii) what is the corresponding value of C?

14.2.2 Meaning of the BET Area of MOFs

In the presence of micropores, which most often happens with MOF's, the BET n_m includes the capacity of the micropores which are already filled at that point of the adsorption isotherm. The conversion of this capacity into a surface area, after assuming a single cross-sectional area of 0.162 nm^2 for all N_2 molecules is then simply an exercise to guide (or to mislead, with our agreement, etc.) our imagination, as it will only be by chance that the resulting BET area will be equal to the 'probe-accessible area' (see Section 6.4.2). This is because, by assuming that the N_2 molecules cover the same area as if they were on a flat surface, we are forgetting that they are covering an almost 4 × larger area in a narrow micropore able to accommodate only one molecule in width and that they are covering a much smaller area in a wide micropore in which part of the adsorbed molecules do not even touch the micropore walls.

This lack of direct relationship, in the presence of micropores, between the BET area and the probe-accessible surface area, is stressed by the voluntary use of the shortened term '*BET area*', instead of 'BET surface area' which should better be reserved to the case when, notably because of the known absence of micropores, it is considered that the BET method gives a reasonable access to the probe-accessible surface area.

Nevertheless, experience shows that, even when different from the probe-accessible surface area, the BET area still has some meaning and usefulness.

The *meaning* of the BET n_m is indeed, as often shown by gas adsorption calorimetry, that it embraces *the major part of the amount of adsorptive in energetic interaction with the surface*, that is whose energy of adsorption exceeds the energy of condensation. It can also be considered as an adsorption capacity in the pressure range of the BET calculation, as it includes the capacity of the micropores completely filled at that point of the isotherm, together with the capacity of the monolayer on the rest of the surface. It would probably be wiser not to convert the BET n_m into a BET area, but we must accept that the latter is much more suggestive and easy to remember, so that we shall simply have to be careful never to mix the 'BET area' with a 'probe-accessible area'.

The *usefulness* of the BET area follows from the strict procedure used for its assessment, by means of the five criteria considered (i.e. the four criteria listed in Section 7.2.2 in addition to the linearity of the BET plot). It can indeed be considered

as a characteristic data of the adsorbent, well suited for comparison purposes: study of
activation or ageing, study of the effect of changing a parameter in the synthesis procedure or, still, comparison between a predicted structure model with that of the actual
adsorbent. In the latter case, indeed, it makes sense to compare the BET areas calculated, with an identical procedure, from the simulated and experimental adsorption
isotherms, respectively (Düren et al., 2007; Walton and Snurr, 2007). Although after
the latter studies, the r-distant surface area (i.e. the 'accessible surface area' of sim-
ulators) and the BET area may seem to compare well, their definition and nature is
different, so that their comparison is less meaningful (see also Chapter 6 for more
details).

Although '*Langmuir areas*' are also often calculated in the case of MOF's, one
should be aware of their meaning. In the case of a real Type-I isotherm (i.e. with a
plateau which has been experimentally reached), the Langmuir area corresponds to
an experimental saturation capacity, artificially converted into an area. In the case
of a supposedly Type-I isotherm (i.e. for which the plateau has not been experimentally reached), the Langmuir equation (in the form of Equations 5.13 or 5.14) allows
assessing a Langmuir so-called monolayer capacity n_m corresponding to a hypothetical plateau, necessarily located above the last experimental point. The resulting
Langmuir area is then always higher than the BET area and its physical meaning is
rather weak.

14.3   Effect of Changing the Nature of the Organic Ligands

14.3.1   Changing the Ligand Length

One of the possibilities to vary the adsorption properties is to change the length of the
organic linker unit. This possibility was initially shown by Yaghi and co-workers
(Rowsell and Yaghi, 2004) for the case of the IRMOF series.

The IRMOF series contains ZrO_4 centres which are linked together via different
carboxylate linkers of which some are given in Figure 14.4.

Increasing the linker length leads to an increase in pore size and pore volume which
is reflected in the change in the pressure of gas uptake as it has been evidenced in the
case of N_2 in IRMOF-1, -10 and -16, as shown in Figure 14.5 (Walton and Snurr, 2007;
Walton et al., 2008). This phenomenon should also be observed in other MOF series
with different ligands such as the NOTT materials, although this has not been experimentally confirmed at present (Lin et al., 2009).

As expected, a change in the pore size also affects the isotherm shape (Figure 14.6).
Indeed, depending on the probe and the temperature, one observes a transformation in
shape from Type-I to Type-IV as one evolves from more microporous systems to
mesoporous samples. Some modelling studies also suggest the possibility to obtain
Type-V isotherms (Fairen-Jimenez et al., 2010), which is considered as a consequence
of relatively weak fluid–solid interactions. While this is certainly a consequence of the
chemical nature of many MOFs, it is also suggested that this is due to the relatively
open nature of these MOF frameworks, compared with other microporous materials,

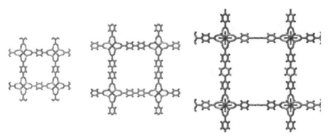

Figure 14.4 Some samples from the IRMOF series: IRMOF-1, IRMOF-10 and IRMOF-16. The nodes consist of ZrO4 units and the linker units are benzene-1,4-dicarboxylate, biphenyl-4,4′-dicarboxylate and terphenyl-4,4″-dicarboxylate respectively.
Figure courtesy of B. Borah, Northwestern University.

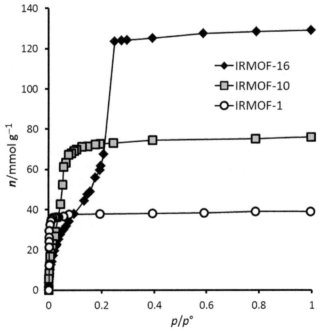

Figure 14.5 Nitrogen isotherms, obtained by GCMC on various IRMOFs.
Adapted from Walton and Snurr (2007).

such as zeolites or active carbons. Thus, as the temperature of the adsorption decreases, it is possible to adsorb more than one layer of fluid, increasing the fluid–fluid interactions, and thus inducing this transition, in the case shown in Figure 14.6 for IRMOF-16, from Type-I (200 K) to Type-V (150 K) and Type-IV (125 K) methane adsorption behaviour.

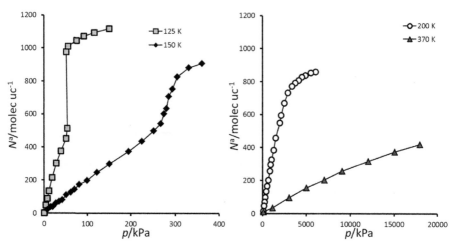

Figure 14.6 Methane isotherms, obtained by Grand Canonical Monte Carlo simulations (see Chapter 6 for details) on IRMOF-16 at various temperatures.
Adapted from Fairen-Jimenez et al. (2010).

An open question concerns the feasibility to apply the BET method for the characterization of such materials. This is especially the case for the IRMOF series (IRMOF-1, IRMOF-6, IRMOF-10, IRMOF-16) where it is difficult to imagine a continuous pore wall or surface. Here, molecular simulations suggest that adsorption either occurs first in the corners of the structure and then around the linkers or follows a pore-filling mechanism in case of narrow micropores (Walton and Snurr, 2007). This study further compares experimental BET calculations with those obtained from isotherms simulated using a Grand Canonical Monte Carlo (GCMC) strategy. This modelling study further allows assessing the r-distant surface areas and these can also be compared to the two BET areas (see Chapter 6). These results show rather good correlations between the simulated and experimental BET areas (except for an IRMOF-14 sample, whose low experimental BET area is explained by crystal defects or imperfect outgassing) and also between the BET areas and the r-distant surface areas. Here, the BET areas were calculated with the help of criteria (a) and (b) given in Section 7.2.2. In any case, for such adsorbents, the r-distant surface area and the BET area have in common that they do not measure a probe-accessible area but, rather, an adsorption capacity whose translation into an area gives a rather high value, only weakly connected to the probe-accessible surface area. Indeed, such an approach to compare the BET areas obtained from simulations on model structures with those obtained with experimentally synthesised samples is of good use to the experimentalist in order to ascertain whether the sample made is optimal in terms of structure and, more importantly, in terms of activation and potential subsequent degradation.

A second example of a series of MOFs, where the ligand can be systematically modulated, is that of the porous zirconium dicarboxylate solids denoted MIL-140A, B, C and D. These materials have a general $ZrO[O_2\text{---}R\text{---}CO_2]$ formula ($R = C_6H_4$ (A), $C_{10}H_6$ (B), $C_{12}H_8$ (C) and $C_{12}N_2H_8Cl_2$ (D)) (Guillerm et al., 2012).

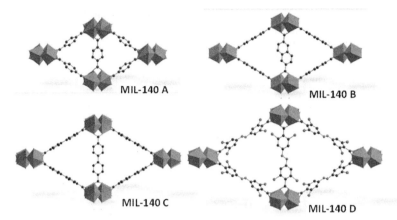

Figure 14.7 View of the crystal structure of the series of MIL-140(Zr) solids along the *c*-axis with polyhedral of zirconium atoms.

These materials present an inorganic sub-unit based on complex zirconium oxide chains, oriented along the *c*-axis, related to six other chains through dicarboxylate linkers, that delimits triangular channels along the crystallographic *c*-axis (Figure 14.7).

The peculiarity of these solids is to contain narrow micropores (<0.7 nm), whose sizes estimated following the methodology reported by Gelb and Gubbins (see Chapter 6) are 0.32, 0.40, 0.57 and 0.63 nm for the MIL-140A, B, C and D respectively. Indeed, the strategy to theoretically assess the surface areas of such microporous solids is questionable. Figure 14.8 reports the surface areas calculated from different methods. The BET areas estimated from the Grand Canonical Monte Carlo (GCMC) simulated nitrogen isotherms at 77 K using the first two consistency criteria recommended above, agree very well with the calculated *r*-distant surface areas (see Chapter 6) in the case of the MIL-140C and D solids. The BET areas extracted from the experimental nitrogen isotherms using the same criteria are however significantly lower than the *r*-distant surface areas, this discrepancy being mainly explained by the presence of residual Zr oxide and/or incomplete activation of the sample.

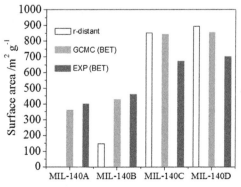

Figure 14.8 Surface areas of the MIL-140 series: Comparison of two theoretical approaches with the experimental BET method.

The scenario drastically differs for the narrowest pore MIL-140A and B structures. Here, estimating the r-distant surface areas via the geometric method which considers all atoms as hard spheres, leads to unphysical values with for instance the absence of any surface area for MIL-140A. This may be due either to the different meaning of the BET and r-distant surface areas or to the fact that the narrow micropores of such solids cannot be or only partially accessible by the nitrogen-sized (0.368 nm) probe molecule, thus rendering this method inappropriate for characterising the surface areas of solids with pore dimensions very close to those of the nitrogen molecule. Here, the only possible way to carry out a meaningful comparison consists of calculating the adsorption isotherms via GCMC simulations which allow a full exploration of the available surface as the atoms are treated as soft Lennard Jones spheres and further applying the BET method. The resulting 'theoretical' BET areas thus reproduce reasonably well the experimental BET values (Figure 14.8).

Based on this typical MIL-140 series, we can consider the BET method as a useful means for characterising materials, even with narrow micropores, as long as one carefully considers the appropriate pressure range based on the consistency criteria.

The pore volume seems to be an alternative way to circumvent the limitations discussed above about the 'surface areas' assessed by various ways and further characterise the narrow pore sizes structures such as MIL-140A and B. The pore volumes calculated by means of the method developed by Myers and Monson (2002), lead to values of 0.10, 0.21, 0.35 and 0.36 $cm^3 g^{-1}$ for MIL-140A, B, C and D, respectively, which fit well with experimental adsorption data.

14.3.2   Changing the Ligand Functionalisation

Functionalising the organic linkers with diverse chemical groups is an efficient route to tune the physico-chemical properties of the MOF frameworks (polarity, acidity, basicity, etc.) with the aim to modulate their adsorption capacities and selectivities. This is achieved either starting simply from linkers already containing the desired functional group or by post-synthetic modifications (Cohen, 2012). One of the active endeavours on MOFs is to graft —NH_2 polar functions on the organic ligands while keeping their original topologies with the aim to design more efficient CO_2 capture materials. The molar enthalpy of adsorption for CO_2 is in general increased by up to 10 kJ mol^{-1}, resulting from a strong electrostatic CO_2/—NH_2 interactions, while the values for CH_4 and N_2 remain almost unchanged compared to the parent material (Yang et al., 2011).

The ligand functionalisation equally affects the selectivity which can be significantly enhanced in the low pressure domain as was experimental observed or predicted for a wide range of MOFs including some MILs and the ZIF, IRMOF and DMOF series. The CO_2/N_2 selectivity of the 3D carboxylate-based MIL-68(Al) is for instance predicted to increase from 38.0 to 85.0 once —NH_2 groups are incorporated to the phenyl ring (Yang et al., 2012). With increasing pressure, the impact of the functionalisation becomes much less pronounced as illustrated in Figure 14.9, as the major part of the amino functions is not anymore available.

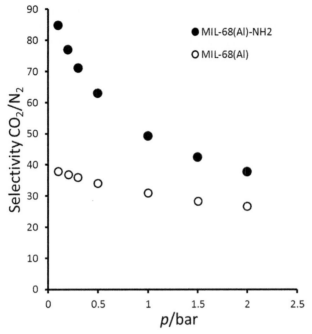

Figure 14.9 GCMC predicted selectivity for CO_2 over N_2 for a binary CO_2/N_2 gas mixture (bulk composition 15–85) in the MIL-68(Al) and MIL-68(Al)-NH_2 solids as a function of the bulk pressure at 303 K.
Adapted from Yang et al. (2012).

Some MOFs have proven to show even higher selectivity with more polar functions such as —SO_3H and —CO_2H groups paving the way to tune materials as highly selective as the most commonly used zeolite 13X in industrial applications. The functionalised zirconium oxide UiO-66(Zr) system is a typical example as described in Chapter 6 and by Yang et al. (2011).

The scenario significantly differs for the flexible MOFs, the functionalisation induces not only modification of their physico-chemical properties but also more importantly major changes in the pore opening that can also impact their adsorption/separation properties. Indeed, for the flexible carboxylate MIL-53(Fe), the presence of functional groups affects strongly its initial pore size and shape leading to different adsorption behaviours with respect to either liquid or gas molecules (Devic et al., 2012). Typically, while the functional group has generally only a minor impact on the storage capacity of rigid MOFs, a significant enhancement of both properties is obtained for the apolar n-alkane in the halogenated (-Br, -Cl) functionalised MIL-53(Fe) solids due to their higher degree of initial pore opening compared to the parent structure which is controlled by the steric hindrance of the functional group. In the case of the MIL-53(Al) solid, the —NH_2 functional group has still a positive effect on its selective adsorption of CO_2 but this is not related to the expected additional CO_2/—NH_2 interactions but rather to its indirect impact on the flexibility of the framework (Stavitski et al., 2011), leading to a narrow pore form inaccessible to CH_4 and thus to an increase of the CO_2/CH_4 by orders of magnitude whilst maintaining a relatively high CO_2 adsorption capacity. In contrast, the same polar function can negatively affect the adsorption properties of flexible MOFs by forming strong intra-framework interactions in such a way to render the pores inaccessible for the molecules of interest. This is for instance the case of the MIL-53(Fe), where the CO_2/framework interactions are not strong enough to overcome the intra-framework hydrogen bonds involving $-NH_2$ leading to the absence of any CO_2 adsorption uptake in this material. This later observation also holds true for the adsorption of even more polar molecule such as water and ethanol in the liquid phase (Devic et al., 2012).

14.4   Effect of Changing the Metal Centre

As we have seen above, the possibility to change the chemistry of MOFs and related materials is vast. One of the parameters that can be varied is the metal centre. For a given structure, this metal centre can affect the rigidity/flexibility of the system as for the MIL-47(V)/MIL-53(Cr, Al, Sc, Fe) solid. Here, if vanadium (IV) is considered then the material is rather rigid in presence of guest molecules but if aluminium, chromium, scandium or iron is used then a large variety of flexibility is introduced into the structure. The question of MOF structural flexibility is discussed further below. The metal used also seems to confer an inherent stability of the material. Indeed, in general terms, the use of tetravalent (Zr, Ti, etc.) or to a lesser extent tri-valent metals (Cr, Al, Fe, Sc, etc.) leads to structures which are usually more thermal stable and humidity resistant than the structures prepared with divalent metals (Mg, Zr, Cu, etc.) (Low et al., 2009).

In most cases, the metal centres are fully coordinated and are often shielded by their surrounding environment and thus have a little impact on the adsorption *per se*. However, some MOFs are constructed of metal centres which are not fully coordinated to the structural organic ligands. This therefore leads to coordinative unsaturated sites (CUS) which are also commonly termed open metal sites. These CUS can act as specific adsorption sites as well as being of interest in other domains such as catalysis. As typical MOFs with CUS, the HKUST-1 (Chui et al., 1999), CPO-27 (Dietzel et al., 2008a,b) and MIL-100 (Férey et al., 2005) structures have been most documented to date.

HKUST-1 or Cu-btc was first synthesised by Chui et al. at the Hong Kong University of Science and Technology in 1999 (Chui et al., 1999). The structure is built from two copper atoms bound to the oxygen of four btc linkers (benzene-1,3,5-tricarboxylate) to create the assembly of $[Cu_3(btc)_2(H_2O)_3]$. This material presents a bimodal distribution of pores: a large central pore with a diameter of 0.9 nm (Figure 14.10) surrounded by pockets with a diameter of 0.5 nm. Pores and pockets are interconnected by triangular windows with a diameter of 0.35 nm. The fine pore structure is visible in the paper by Rubes et al. (2012).

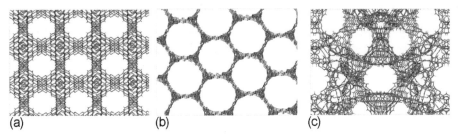

(a)   (b)   (c)

Figure 14.10 Structures of HKUST-1 (a), CPO-27 (b) and MIL-100 (c).

On dehydration, a reduction in cell volume is observed due to a shrinking of the $[Cu_2C_4O_8]$ cage. Infra-red studies suggest that the water removal results in the Cu (II) sites becoming labile although a shielding of the copper by the four oxygen atoms is apparent which would infer a rather low coordinative unsaturation of these metal sites (Prestipino et al., 2006).

These Cu(II) coordinative unsaturated sites can thus act as specific centres for the adsorption of different probe molecules although the shielding of the copper by the oxygen molecules may limit the interactions observed. Carbon monoxide can be considered as a molecule of choice for the characterisation of such sites. Infra-red studies suggest the formation of Cu(II)—CO adducts at 77 K. This is confirmed by quantum chemical calculations (Rubes et al., 2012; see Chapter 6 for more details on the computational method) which predict a preferential adsorption of CO on Cu(II) via its carbon atom with a corresponding molar enthalpy of adsorption around 30 kJ mol^{-1} (for the sake of simplification, in the text of this chapter we shall only give the absolute values of the enthalpies of adsorption, knowing that they are of course always negative). If we envisage the scenario at higher loading, the simulations suggest that the coordination of a second CO molecule would lead to a slightly lower molar enthalpy of adsorption of about 28 kJ mol^{-1}.

The adsorption data of carbon monoxide at 303 K on HKUST-1 is shown in Figure 14.11. The adsorption isotherm is obtained using a point by point method. Two dose sizes are used and both lead to overlapping isotherms. The experimentally measured differential enthalpies of adsorption are equally shown in Figure 14.11. The first point shows a quite high value of around 63 kJ mol^{-1}, after which the differential enthalpy of adsorption drops immediately to 29 kJ mol^{-1} at a coverage of 0.1 mmol g^{-1}. This higher initial value can be explained by the presence of Cu(I) defect sites which can be present in relatively small amounts. The rest of the enthalpy curve shows a slight decrease from 29 to around 25 kJ mol^{-1} at a coverage of around 3 mmol g^{-1}. This decreasing profile would be assigned to the gradual increase of the coordination sphere around Cu(II) as suggested by the quantum chemical calculations.

A second probe molecule of interest here can be carbon dioxide. The CO_2 adsorption isotherm obtained at 303 K is given in Figure 14.12 which again shows gradual uptake with pressure. The differential enthalpies of adsorption are relatively constant at around 29 kJ mol^{-1} up to a coverage of about 10 mmol g^{-1}. Above this value, a slight increase in enthalpy of 2–3 kJ mol^{-1} is observed. The horizontal region of

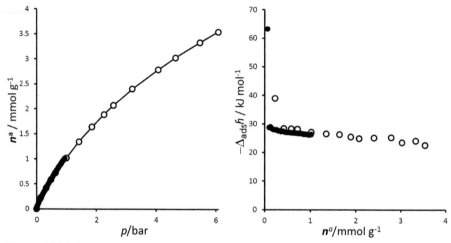

Figure 14.11 CO adsorption at 303 K on CuBTC: Large doses (o) and small doses (•).
Left, adsorption isotherm. Right, corresponding differential enthalpies of adsorption
(Rubes et al., 2012).

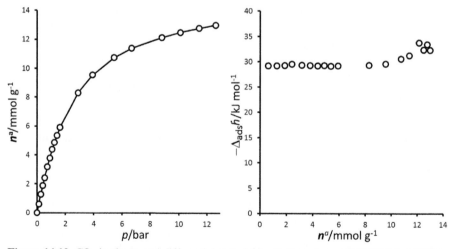

Figure 14.12 CO_2 isotherm and differential enthalpies of adsorption on HKUST-1 at 303 K.

the enthalpy curve can suggest interactions on an energetically homogeneous surface
as in the case of argon adsorption on graphite for example (see Chapter 10). Again, the
use of quantum chemical calculations can shed some light on the adsorption mechanisms. The strongest interaction calculated is that of the CO_2 with the Cu(II) site
(Grajciar et al., 2011). However, it is suggested that a tilt of the adsorbed molecule
occurs with a Cu—O—C angle of 123°. This allows for further interaction of
the CO_2 with a second Cu(II) site as well as dispersion interactions with the organic
linkers. The calculated enthalpy value for this site is $\Delta_{ads}u = 28.2$ kJ mol^{-1}. Other

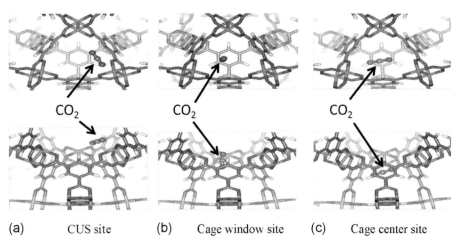

(a)   CUS site   (b)   Cage window site   (c)   Cage center site

Figure 14.13 Geometries of the CO_2 adsorption complexes on the (a) CUS, (b) small cage window and (c) small cage centre sites issued from quantum chemical energy optimization. The upper and lower parts of the figure represent the views perpendicular and parallel, respectively, to one of the small cage openings.
Adapted from Grajciar et al., 2011.

adsorption sites could be at the cage centre or cage window (see Figure 14.13) which are calculated to be around 5 kJ mol^{-1} less stable than the CO_2—Cu(II) interaction. It is suggested that all the Cu(II) sites are initially coordinated before the filling of the rest of the cages. This further filling leads to interactions of the CO_2 with the MOF surface but also to lateral interactions between the adsorbed CO_2 molecules. These additional lateral interactions initially result in a compensation of the decreasing CO_2-surface interactions. The fact that this leads to a flat enthalpy profile is coincidental. At higher loading however, the increased lateral interactions do lead to an increasing differential enthalpy until the end of pore filling which is equally observed in the experiment.

It would thus seem that the CO and CO_2 interactions with the Cu(II) CUS sites are similar. The fact that the oxygen molecules surrounding the Cu(II) partially shield the metal probably results in a non-optimal interaction of the linear CO_2 with the Cu (II). Indeed, the optimal interaction would be of the Cu(II) with the C of the CO_2 as is the case with CO. This would explain the relatively high CO interactions when compared to CO_2.

The accessibility of CUS sites is imperative if they are to be of interest in adsorption or catalysis for example. One structure that seems to have quite accessible CUS sites is known as CPO-27(M=Co, Ni, Mg) (Dietzel et al., 2005) or MOF-74 (M=Zn, Mn) (Rosi et al., 2005). The pore structure of this material resembles a honeycomb with the metal forming the corners and the organic linker, 2,5-dihydroxyterephthalic acid, forming the walls. The six-coordinated metal forms a threefold helix parallel to the pore wall with five bonds to organic ligand oxygens and a sixth bond to water which can be removed on heating resulting in a very high density of coordinative unsaturated sites available for specific adsorption interactions.

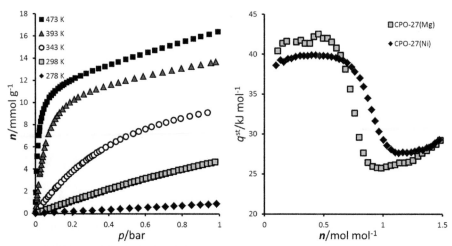

Figure 14.14 CO_2 adsorption isotherms at different temperatures on CPO-27(Ni) (left) and isosteric adsorption enthalpies relative to the amount of metal atoms (right) for the adsorption of CO_2 on CPO-27(Ni) and CPO-27(Mg) (Dietzel et al., 2009).

Once again, carbon oxides are interesting probe molecules for this structure. As an illustration, the CO_2 isotherms and differential enthalpies are given in Figure 14.14.

Once can note that the adsorption energies calculated for the adsorption of carbon dioxide in the CPO-27 structures are much higher than for the HKUST-1 previously described. Indeed, absolute values of the isosteric enthalpies of around 39 and 42 kJ mol^{-1} are obtained for CO_2 adsorption on the Ni and Mg forms of CPO-27 (Figure 14.14) which is around 10 kJ mol^{-1} higher than for the HKUST-1. One can explain this difference by the more 'active' metal centres in the CPO-27 structure (Valenzano et al., 2011). This 'activity' is probably due to the fact that the metals are more exposed in the CPO-27 structure than in the HKUST-1 where the oxygen molecules partially shield the Cu. The carbon monoxide isosteric enthalpy at low coverage is around 59 kJ mol^{-1} for the CPO-27(Ni) (Figure 14.15) which compares to 29 kJ mol^{-1} for HKUST-1. Quantum chemical calculations (Valenzano et al., 2011) made with CPO-27 predict that the carbon of the CO molecule points to the metal centre with a perpendicular alignment of the probe (see Chapter 6). In the case of the nickel sample, π-back donation charge transfer may occur. These simulations further show that the carbon dioxide molecule adsorbs at a tilted position due to interactions of the CO_2 oxygen atoms to both the metal centre and oxygen of the organic linker. The authors further suggest that the enhanced interaction of these probe molecules with the CPO-27 may be also due to the curvature of the pore surface compared to a planar one.

The quantity and strength of these metal sites make them of interest for other probe molecules. One probe of current interest is hydrogen with respect to potential hydrogen storage applications. In terms of adsorption energy, it has been suggested that a hydrogen molar enthalpy of adsorption of between 15 and 20 kJ mol^{-1} would be ideal (Bhatia and Myers, 2006). Most MOFs provide with hydrogen molar adsorption

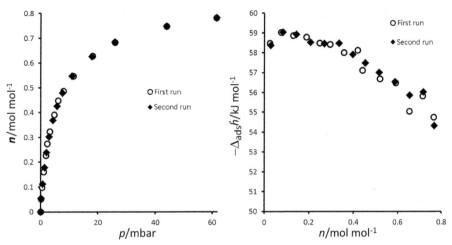

Figure 14.15 CO isotherms (left) and corresponding isosteric adsorption enthalpies (right) obtained at 303 K on CPO-27(Ni). The uptakes are given in CO molecules per metal site (Chavan et al., 2009).

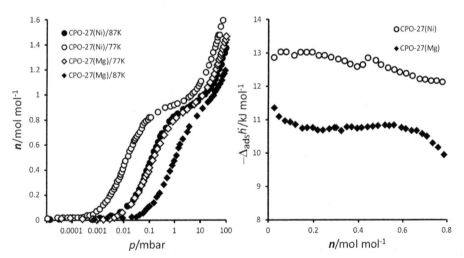

Figure 14.16 Isotherms (in semi-log plot) and isosteric enthalpies for H_2 adsorption CPO-27(Ni) and CPO-27(Mg), as a function of H_2 adsorbed per formula unit per metal atom (Dietzel et al., 2010).

enthalpies in the range 6–8 kJ mol^{-1}. In the case of CPO-27 (Figure 14.16), the isosteric enthalpies at low loading are 13.0, 11.5 and 10.9 kJ mol^{-1} for the Ni, Co and Mg forms of CPO-27 (Dietzel et al., 2010). Interestingly, these values remain relatively constant until the adsorption of about one molecule per metal atom suggesting the strong influence of the metal centres although neutron scattering suggests that two adsorption sites are present: close to the metal centre and further along the chain close to three oxygen atoms (Liu et al., 2008b). Nevertheless, the influence of the available metal centres as specific adsorption sites is of considerable interest here for further development in this domain.

The metal sites of some MOFs can thus form quite strong specific adsorption sites whose strength can depend on their position in the structure, with various effects of shielding by the organic linkers, and on the metal chosen, depending on their electronic properties for instance. However, it may also be of interest to modulate the interactions with a given site. This is possible with MOFs whose metal centres can be reduced, via a thermal treatment for example. An example of this effect can be observed with the MIL-100 structure with iron as the metal centre.

The MIL-100 structure (Figure 14.10c) is a 3D framework made up of hybrid supertetrahedra units consisting of inorganic subunits of trimers of iron (III) octahedra connected together by oxygen atoms from the organic linker benzene-1,3,5-tricarboxylate. Two types of mesoporous cages are created by the assembly of supertetrahedra with diameters of 2.5 and 2.9 nm, respectively, accessible through microporous windows of 0.47×0.55 and 0.86 nm. This sample can be synthesised with a number of metal centres: Cr, Al, Sc, Mn, V and Fe. (Férey et al., 2004, 2005; Horcajada et al., 2007; Volkringer et al., 2009b; Mowat et al., 2011). Interestingly, a significant amount of iron, Fe(III) sites can be reduced to give Fe(II) Lewis acid sites. This can be achieved by thermal treatment; outgassing at 150 °C under a secondary vacuum renders the MIL-100(Fe(III)) sample whereas treatment to 250 °C can lead to a maximum of 1/3 of the Fe(III) sites being reduced to Fe(II) (Yoon et al., 2010). Thermal treatment between these two extremes leads to various amounts of Lewis acid sites.

The variation in amount of Lewis sites has little effect on the adsorption of CO_2 (Yoon et al., 2010). Indeed, the differential enthalpies of adsorption for the sample outgassed to 150 or to 250 °C increase only by about 3 kJ mol^{-1}. In contrast, the differential enthalpies of adsorption for CO increase by more than 10 kJ mol^{-1} (Figure 14.17).

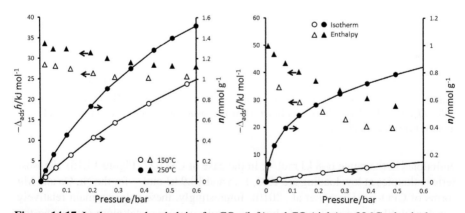

Figure 14.17 Isotherms and enthalpies for CO_2 (left) and CO (right) at 30 °C adsorbed on MIL-100(Fe) after outgassing to 150 and 250 °C.
From Yoon et al. (2010).

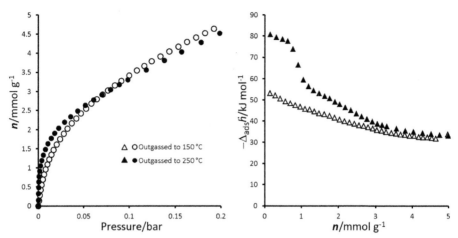

Figure 14.18 Isotherms (left) and differential enthalpies (right) for propyne adsorbed on MIL-100(Fe) after outgassing to 150 and 250 °C.
From Leclerc et al. (2011).

This increased interaction of a given probe molecule with a variation in the strength of adsorption site can be extended to other types of interaction. Indeed, it is possible to follow the metal-π interactions with probes such as propane, propene and propyne (Figure 14.18). Indeed, while the differential enthalpy of adsorption of propane does not change between the samples outgassed at 150 or 250 °C, there is a significant increase in the differential enthalpy of adsorption of propene (Yoon et al., 2010) and propyne (Leclerc et al., 2011) with the increased outgassing temperature. This highlights the π-back donation effects of the double or triple bonds with the metal centres.

14.5 Effect of Changing the Nature of Other Surface Sites

The influence of the bridging μ_2-OH or μ_2-O on the adsorption properties of gases can be highlighted when comparing the isostructural MIL-53(Cr)/MIL-47(V) series.

The MIL-53(Cr) structure is constructed of corner-sharing $CrO_4(OH)_2$ octahedra chains, interconnected by *trans*-benzene-dicarboxylate ligands creating a 3D framework defining 1D diamond-shaped pores of free diameter close to 0.85 nm (Serre et al., 2002). The MIL-53(Cr^{III}) contains hydroxyl groups (μ_2-OH groups) located at the metal–oxygen–metal links, whereas the isostructural MIL-47(V^{IV}) material presents oxo bridges (μ_2-O groups) (Barthelet et al., 2002) (Figure 14.19). As these sites could act as potential attractive sites for the guest molecules and/or block out their mobility, one can expect different adsorption mechanisms in both solids.

The adsorption isotherms recorded at room temperature for the MIL-53(Cr) compound exhibit unexpected shapes for a series of probe molecules including CO_2, hydrocarbons, H_2S, H_2O and different alcohols with the presence of sub-steps (Serre et al., 2002; Bourrelly et al., 2005, 2010; Trung et al., 2008; Hamon et al., 2009).

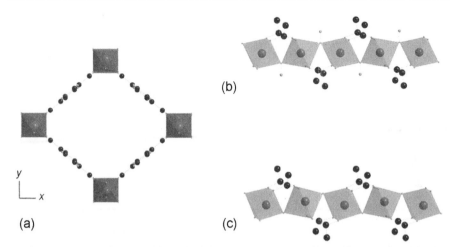

Figure 14.19 (a) View of the MIL-53(Cr)/MIL-47(V) structures along the chain (z-axis), highlighting the 1D. View perpendicular to the pores of the MIL-53(Cr) (b) and MIL-47(V) (c) with μ_2-OH and μ_2-O atoms linked to the metal (M) atom, respectively.

Such an unusual behaviour for a microporous solid is related to a breathing of the structure in presence of such adsorbate molecules that potentially give rise to specific interactions with the pore wall. This spectacular phenomenon will be discussed in more detail in Section 14.4. It has been shown by molecular simulations coupled with IR measurements that the interactions between the mentioned adsorbed molecules and the dipole of the μ_2-OH group are at the origin of the onset of flexibility during the adsorption in the MIL-53 structure (Serre et al., 2007b). This observation is supported by the fact that the MIL-47(V) solid which only contains oxo bridges (μ_2-O groups) does not breathe in presence of guest molecules and its resulting adsorption isotherm does not present any distinct step with a characteristic I-type shape usually observed for microporous solids. Such a difference is illustrated in Figure 14.20 which reports the H_2S adsorption isotherms for both solids.

There is also a clear signature in the microcalorimetry profile to distinguish both types of materials when adsorbed polar molecules such as CO_2 (Bourrelly et al., 2005; Serre et al., 2007b). While the curve for MIL-53(Cr) is non-monotonous with rather high enthalpy value at the initial stage of adsorption, the one for MIL-47(V) is almost flat when the loading increases. It means that the adsorption surface of MIL-47(V) experienced by the probe molecules can be described as being energetically homogeneous with no preferential adsorption sites while the presence of the μ_2-OH group in the MIL-53(Cr) leads to a heterogeneous energetic surface. This is confirmed by quantum chemical calculations which evidence the absence of any specific adsorption sites within the pore of the MIL-47(V) (Figure 14.21a) solid although some associations with the framework may be seen with minimum distances close to 0.35 nm. In contrast in the MIL-53(Cr) solid, the main interaction involves a 'head-on' interaction between CO_2 and the hydrogen atom with much shorter distances centred around 0.23 nm (Figure 14.21b).

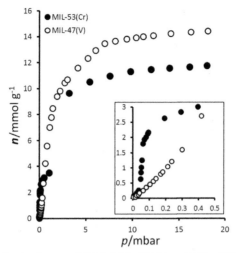

Figure 14.20 Adsorption isotherms of H_2S in MIL-53(Cr) and MIL-47(V) structure obtained at 303 K.
From Hamon et al. (2009).

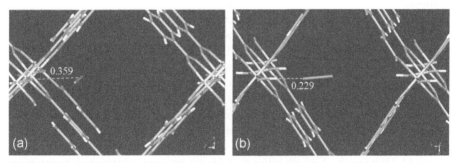

Figure 14.21 Arrangement of the CO_2 molecule within the porosity of (a) the MIL-47(V) and (b) the MIL-53 (Cr) solids issued from quantum chemical energy optimization.

The presence of the μ_2-OH group also affects the dynamics of the polar molecules within the pores as can be evidenced by an appropriate coupling of Molecular Dynamics (see Chapter 6 for more details) and Quasi-elastic neutrons scattering measurements. While for instance, the displacements of the CO_2 molecules remain mainly restricted along the direction of the tunnel in the MIL-53(Cr) solid, the most favourable single interactions between CO_2 and the μ_2-OH groups leading to a global 1D normal diffusion mechanism (Salles et al., 2009), a 3D diffusion mechanism characterised by random motions occurs in MIL-47(V) due to the absence of any CO_2—OH preferential interaction (Salles et al., 2010b).

Substituting the μ_2-OH sites of the MIL-53(Cr) solid by the μ_2-O groups with a lower affinity for water also induces a more pronounced hydrophobic character of the MIL-47(V) surface. This effect was confirmed by Grand Canonical Monte

Carlo simulations (Devautour-Vinot et al., 2010). Indeed, the distribution of water molecules within the pores of MIL-47(V) was predicted to be strongly heterogeneous, with either completely empty or filled pores (Salles et al., 2011), consistent with the behaviour previously pointed out in several zeolites including silicalite-1 known as hydrophobic solids (Desbiens et al., 2005; Caillez et al., 2008). The water molecules are then arranged in such a way to form clusters within the centre of the pore instead of interacting each with the pore wall, which results from stronger H_2O/H_2O interactions compared to the $H_2O/MIL-47(V)$ surface. These clusters are constituted by up to 12 H_2O at high loading as illustrated in Figure 14.22a, behaviour very similar to those observed in the liquid phase where up to 14 water molecules are implicated in the formation of the tetrahedral clusters (Ludwig, 2001). The situation slightly differs in the case of the MIL-53(Cr) solid wherein all the pores contain water molecules at low pressure even if the distribution is heterogeneous. Indeed, it is predicted that some pores are partially filled but never completely empty with water molecules localised in the region around the μ_2-OH groups, consistent with a soft hydrophobic character of this structure compared to MIL-47(V) (Salles et al., 2011). Further, typical configurations at high loadings reported in Figure 14.22b shows that the μ_2-OH group plays a similar role as the chemical defects present at the surface of silicalite-1 (Desbiens et al., 2005) allowing the water molecules to initiate the formation of clusters and condense.

Such a different predicted water adsorption in these materials have been further confirmed by (i) the adsorption isotherms which showed that, at 303 K, the hydration process starts only at around 30 mbar in MIL-47(V) (vs. 15 mbar in the MIL-53(Cr)), that is closer to the saturation pressure of water and (ii) the differential enthalpies of adsorption comprised in the range (35; 39 kJ mol^{-1}) that lie below the molar enthalpy of vaporisation ($\sim$44 kJ mol^{-1}) (Bourrelly et al., 2010).

Regarding the interactions with apolar molecules such as CH_4 and H_2, the adsorption behaviour at ambient temperature is not affected by the presence of the μ_2-OH group (Bourrelly et al., 2005; Rosenbach et al., 2008; Salles et al., 2008). Indeed, Figure 14.23 shows that for both MILs, the CH_4 adsorption isotherms and the resulting absolute differential enthalpies are very similar ($\sim$17 kJ mol^{-1}). This holds also true

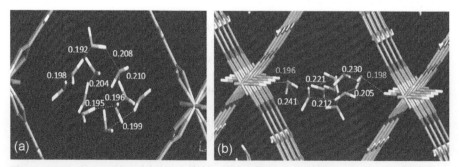

Figure 14.22 Typical arrangements of the water molecules at high loading in (a) the MIL-47(V) and (b) MIL-53(Cr) obtained at 303 K from Grand Canonical Monte Carlo simulations. The characteristic distances are reported in nm.

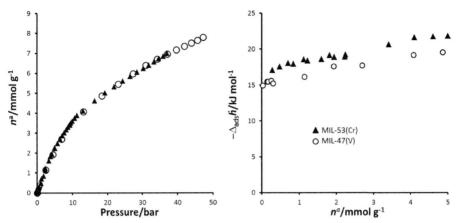

Figure 14.23 Adsorption isotherms (left) and differential enthalpies (right) of adsorption for CH_4 at 303 K on MIL-53(Cr) and MIL-47(V).
From Bourrelly et al. (2005).

for the dynamic behaviour with the motions of CH_4 predominantly orientated along the direction of the tunnel following a 1D diffusion mechanism in both cases. The only variant comes from the microscopic displacements of the molecules via a sequence of jumps between two consecutive μ_2-OH groups and displacements mainly centred in the middle of the pore for MIL-53(Cr) and MIL-47(V), respectively (Rosenbach et al., 2008; Salles et al., 2008).

14.6 Influence of Extra-Framework Species

In an analogous manner to zeolites, the pore chemistry of some MOFs has been modified by the inclusion of extra-framework species. This has been with the aim to enhance the adsorption capacity for a given molecule such as hydrogen as well as for other areas including catalysis.

Metal nanoparticles such as Pt and Pd have been occluded in the pores of MOFs most often with the aim to increase the hydrogen adsorption (Zlotea et al., 2010). Other groups have prepared complexes inside large pore MOFs which have included $[Mo_6Br_8F_6]^{2-}$, $[Re_4S_4F_{12}]^{4-}$ and $[SiW_{11}O_{39}]^{7-}$ (Dybtsev et al., 2010; Klyamkin et al., 2011). These studies all seem to show a reduced capacity for hydrogen at 77 K due to loss of pore volume, but in some cases, an increase in adsorption at room temperature has been observed, possibly due to spill-over type effects.

However, in terms of systematically varying the adsorption properties of MOFs, several studies have focussed on the addition and variation of extra-framework cations in MOFs. Indeed, some porous MOFs with anionic frameworks charge compensated by extra-framework cations inside the cavity have been synthesised during the past 5 years mainly by conducting a direct synthesis route considering the presence of a given linker, a metallic salt (most of the time a transition metal cation) and an inorganic base (alkali hydroxide). The incorporation of extra-framework cations in MOFs

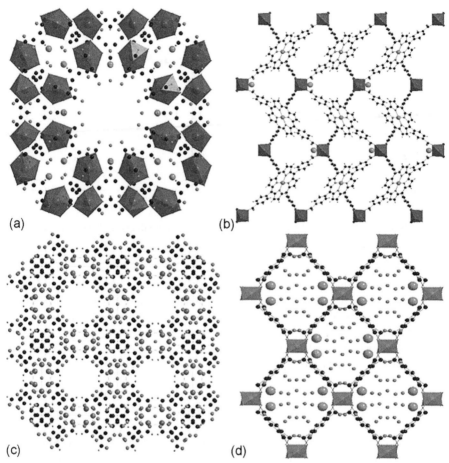

Figure 14.24 Crystal structures of the porous MOFs with alkali cations within their pores:
(a) sod-ZMOF (Chen et al., 2011), (b) Fe^{III}(NiPp-TC)Ksolv (NiPp-TC, nickel(II) tetra
(carboxyphenyl)porphyrine) (Fateeva et al., 2011), (c) [Cu$_3$-(OH)(CPZ)$_3$]NH$_4$ solv (CPZ,
4-carboxypyrazolate) (Quartapelle Procopio et al., 2010) and (d) M^{II}$_2$(BTeC)C(H$_2$O)$_n$
(M = Zn, Co; C = Na, K, NH$_4$; BTeC = 1,2,4,5-benzenetetracarboxylate) (Wu et al., 2008).
(For colour version of this figure, the reader is referred to the online version of this chapter.)

was also envisaged via a post-synthetic strategy. They include the series of MOFs with
zeolite-analogous framework topologies first reported by Eddaoudi and co-workers
(Liu et al., 2006; Sava et al., 2008; Alkordi et al., 2009; Nouar et al., 2009; Chen
et al., 2012), the tetrazolate-based MOFs from Long (Dinca and Long, 2007) and other
3D type MOFs (Horike et al., 2006; Wu et al., 2008; Yang et al., 2008; Quartapelle
Procopio et al., 2010; Fateeva et al., 2011; as illustrated in Figure 14.24 that provides
some structures with a 'true' accessible porous structure, either *permanent* or
dynamic. This relatively recent class of MOFs is expected to show higher performance
in adsorption/separation processes that are governed by the direct interactions

between the adsorbates and the extra-framework cations. In this sense, compared to the most commonly used zeolite systems, such MOFs can potentially provide significant improvements on several points including a higher accessibility of the cations, that are usually less bounded to the framework, thus leading to an enhancement of the strength of interactions with the guest molecules and a higher chemical versatility that allows to tune the electronic features/polarities of the cations by controlling the nature of the organic linkers or/and the metal centres in the framework.

A few studies focused on the H_2 sorption properties of such materials depending on the nature of the extra-framework cations (Dinca and Long, 2007; Nouar et al., 2009). It has been thus evidenced that introducing Li^+ cations in the pores of such MOFs either through acid–basic or redox reactions, enhance both H_2 enthalpy of adsorption and capacity at low-pressure/ambient temperature. As a typical illustration, it was shown that the adsorption uptake and differential enthalpy at low coverage for H_2 in a series of cation containing tetrazolate-based MOFs (Dinca and Long, 2007) strongly depend on the nature of the exchanged cations (Figure 14.25), the highest differential enthalpies of adsorption being obtained for the Co^{2+} exchanged form $(10.5 \text{ kJ mol}^{-1})$.

It has been also predicted by molecular simulations that rho-ZMOF in its sodium version can be a very promising candidate for the separations of various mixtures including CO_2/N_2, CO_2/CH_4 and CO_2/H_2 by a predominant electrostatic CO_2/Na^+ interactions (Yang et al., 2007; Jiang et al., 2011; Chen et al., 2012). These predictions were further complemented by experimental CO_2 adsorption/desorption measurements performed on various alkali-exchanged rho-ZMOF and sod-ZMOF materials which evidenced CO_2/N_2 selectivity higher than 20, the K^+ exchanged form showing the highest increase in adsorption capacity associated with a reversible and sustainable performance in five recycling runs and a complete regeneration under milder

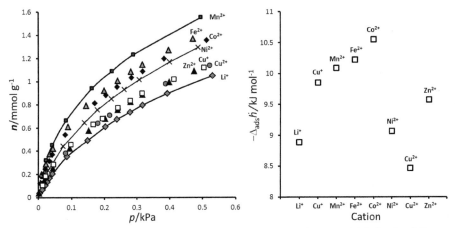

Figure 14.25 (Left) Low pressure region of the adsorption isotherms at 77 K for the different cations exchanged $M_3[(M_4Cl)_3(BTT)_8(CH_3OH)_{10}]_2$ forms and (right) the corresponding differential enthalpy of adsorption at low coverage.
Taken from Dinca and Long (2007).

conditions (40 °C) (Chen et al., 2011). A more systematic exploration of the impact of mono-, di- and tri-valent cations on the interactions between CO_2 and the rho-ZMOF computationally evidenced that the isosteric heat and Henry's constant increase following the sequence $Cs^+ < Rb^+ < K^+ < Na^+ < Ca^{2+} < Mg^{2+} < Al^{3+}$ in accordance with the increasing order of the charge-to-diameter ratio of the cation used in a first approximation to quantify the electric field generated by the cation (Chen et al., 2012). Such a correlation is consistent with what has been previously evidenced for the adsorption of N_2 in various cation containing X-Faujasites which was also governed by strong electrostatic N_2/cation interactions (Maurin et al., 2005).

14.7 Special Case of the Flexibility of MOFs

One of the most fascinating properties of some MOFs is their significant structural flexibility under the influence of external stimuli (Férey and Serre, 2009; Férey et al., 2011). As we will see below, the flexibility of these materials is different from that observed in other porous solids. The amplitude of the flexibility can be over 200% in the case of MIL88 (Serre et al., 2007a) whereas in zeolites phase changes tend to be of fractions of a nm as for example the orthorhombic to monoclinic transition in silicalite-1 (Wu et al., 1979). The flexibility in many MOFs involves structure transitions between ordered crystalline phases (Figure 14.26) which differ from the flexibility observed in porous polymers which remain amorphous.

In terms of flexibility, it has been shown that several MOFs show local rearrangement of their ligands in presence of guest molecules corresponding to π flips and rotational librations which have been characterised by complementary experimental tools including neutron scattering, NMR and dielectric relaxation (Devautour-Vinot et al., 2012; Kolokolov et al., 2012). This behaviour is usually referred to a gate opening process. However, we will restrict our discussion to structural flexibility. A summary of different types of gas induced flexibility in MOFs was given by Kitagawa (Kitagawa and Uemura, 2005) as shown in (Figure 14.26) along with some (nonexhaustive) representative publications for each type. There are numerous examples of the gas induced flexibility in MOFs, with the 'jungle jim' (Kitaura et al., 2003) and breathing (Bourrelly et al., 2005) types having probably been most studied to date. The MIL-53 solid which belongs to this last category is of interest due to the crystalline nature of each structural form and the possibility to observe distinct behaviours for different molecules which has made it fascinating to characterise and so is detailed below. While we are concerned with gas adsorption as a stimulus for the MOF flexibility, it can be noted that in the case of MIL-53, both temperature (Liu et al., 2008a) and pressure (Beurroies et al., 2010; Ghoufi et al., 2012) can equally act as stimuli for some transitions.

To follow the structure changes in these flexible solids, *in situ* X-ray powder diffraction (XRPD) investigations are required to establish unambiguously a correspondence between variations in the adsorption isotherms and different degrees of pore opening (Bourrelly et al., 2010). For this, a highly leak tight dosing apparatus

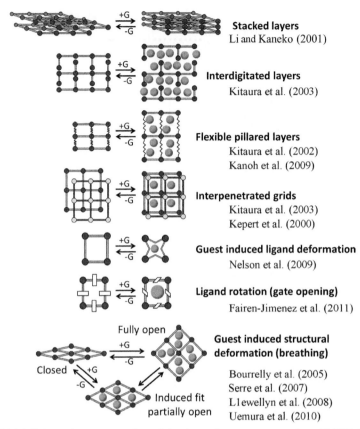

Stacked layers
Li and Kaneko (2001)

Interdigitated layers
Kitaura et al. (2003)

Flexible pillared layers
Kitaura et al. (2002)
Kanoh et al. (2009)

Interpenetrated grids
Kitaura et al. (2003)
Kepert et al. (2000)

Guest induced ligand deformation
Nelson et al. (2009)

Ligand rotation (gate opening)
Fairen-Jimenez et al. (2011)

Guest induced structural
deformation (breathing)

Bourrelly et al. (2005)
Serre et al. (2007)
Llewellyn et al. (2008)
Uemura et al. (2010)

Figure 14.26 Schematic representation of the dynamic porous properties of MOFs in the presence of gases. The references contain some examples of these different types of flexibility.

has been developed that enables well-controlled doses to be introduced to the sample at both low pressures down to 10^{-2} bar and up to 60 bar. The gas dosing system is schematised in Figure 14.27. The MIL-53(Fe) sample is thus introduced inside a quartz capillary of 0.7 mm external diameter which is itself mounted on a T-piece which can be attached to a goniometer head (G). The T-piece was then attached to the system via PEEK capillary (0.8 mm ext. diameter) to a stainless steel manifold via two manual valves.

The valve setup itself is mounted on a single rail which is attached to one of the bar mounts around the detector. This allows a minimal manifold encumbrance in the experimental hutch. The pneumatic valves used (V1–V8, ref. Swagelok) along with the rest of the VCR fittings are leak tight and can withstand an over-pressure of 250 bar. The pneumatic valves are operated via an electrovalve block (Joucoumatic) which is connected to an electronic control box (built in house). This controller gives the possibility to action the valves either manually or via a computer.

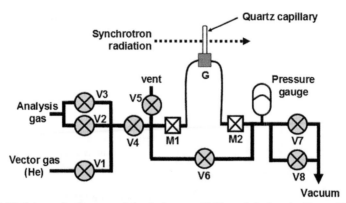

Figure 14.27 Schematic diagram of the dosing manifold used during the X-ray diffraction investigations. (For colour version of this figure, the reader is referred to the online version of this chapter.)

In a typical experiment, the sample is outgassed separately under a vacuum pressure of 10^{-3} mbar at 250 °C for a few hours. The sample is then cooled down to room temperature and transferred under vacuum to the dosing manifold and placed onto the goniometer head. A cryostream is thus used to adjust the temperature of the sample. The gas is dosed to the manifold and then to the sample. Typical X-Ray diffractogram is collected 1 min after the gas introduction, with an acquisition time of 30 s (rotation rate $1° s^{-1}$). New X-ray diffraction patterns are recorded at the same pressure every 5 min, and equilibrium (at a given pressure) is assumed when no change is observed between two successive patterns. When such equilibrium condition is reached, the pressure is increased and the same procedure applied once again.

14.7.1   MIL-53(Al, Cr)

This metal terephtalate MIL-53(Cr, Al) solid is among the most spectacular breathing MOFs upon adsorption at room temperature of some guest molecules including CO_2, H_2O, H_2S and alkanes which correspond to a highly flexibility of its framework (Serre et al., 2002; Bourrelly et al., 2005, 2010; Trung et al., 2008; Hamon et al., 2009; Férey et al., 2011). Such an unusual phenomenon has been assigned to a reversible structural transformation between two distinct forms, namely the large pore (LP) (orthorhombic symmetry, unit cell volume $V \approx 1.5$ nm^3) and the narrow pore (NP) (monoclinic symmetry, $V \approx 1$–1.2 nm^3 depending on the guest molecules) structures (Figure 14.28) that implies for instance in the case of the quadrupolar CO_2 molecules, a large unit cell volume contraction/expansion of about 40%. Such behaviour leads to an unusual adsorption isotherm as illustrated for CO_2 in MIL-53(Cr) with the presence of a step at around 6 bar accompanied by distinct enthalpy changes. Indeed, *in situ* XRPD measurements (Serre et al., 2007a) and molecular simulations based on a sophisticated Hybrid Osmotic Monte Carlo (HOMC) (Ghoufi and Maurin, 2010; see Chapter 6) approach showed that when the CO_2 is introduced, the structure initially in its LP form

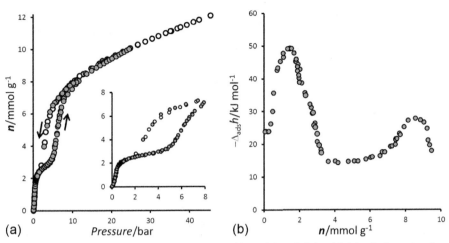

Figure 14.28 Adsorption isotherms (left) and differential enthalpies (right) of adsorption for CO_2 at 303 K on MIL-53(Cr).

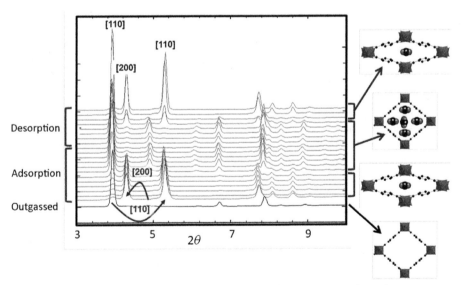

Figure 14.29 *In* situ X-ray powder diffraction patterns of MIL-53(Cr) under various pressures of CO_2 at 293 K (Serre et al., 2007a).

rapidly shrinks due to the CO_2/μ_2-OH interactions, leading to a NP version. This situation persists up to the end of the plateau and a second structural transition occurs from the NP to the LP forms at a pressure above. Figure 14.29 reports the corresponding evolution of the *in situ* XRPD patterns as a function of the pressure that concurs well with the profile of the adsorption isotherm.

The structural transition can be also detected from the microcalorimetry profile recorded along the adsorption process. At very low pressure, one observes in Figure 14.28 an exothermic contribution which has been assigned to the structural switching from the LP to the NP form, the energy barrier being roughly estimated to be around 20 kJ mol^{-1} that agrees well with (i) the value obtained by DSC measurements on the water/MIL-53(Cr) system (Devautour-Vinot et al., 2009) and (ii) some of the modelling predictions which claimed that at room temperature, a minimum enthalpy of adsorption of 20 kJ mol^{-1} needs to be provided by any type of guest molecules in the initial LP form, to induce the structural transition towards the NP form (Llewellyn et al., 2009). Further, above 6 bar, there is a sharp decrease in the enthalpies indicating a change in the mode of interaction between the hybrid porous framework and the CO_2 molecule. This drop in enthalpies occurs at the same pressure as the increase in adsorbed amount in the experimental isotherm. The simulated evolutions of the differential enthalpies of adsorption as a function of the pressure help the interpretation of such a microcalorimetry profile. An almost flat enthalpy profile was simulated for the NP form, in the wide range of pressure with value centred around 37.0 kJ mol^{-1} that concur well with the experimental energetic behaviour in the region below 6 bar where the NP form is predominant (Bourrelly et al., 2005; Serre et al., 2007a), whereas a significant increase of the enthalpy of adsorption with the pressure is predicted for the LP version with absolute values ranging from 20 to 24 kJ mol^{-1} (Ramsahye et al., 2007, 2008; Serre et al., 2007a,b). This latter simulated value and trend fit well with the experimental data recorded in the same range of pressure where the MIL-53(Cr) is in its LP form (Bourrelly et al., 2005).

Monte Carlo and quantum chemical calculations (see Chapter 6) are valuable tools for providing insights into the microscopic adsorption mechanism in such breathing materials. Indeed, the CO_2 molecules are lined up along the direction of the tunnel almost parallel to each other, with a strong double interaction with the opposing walls of the same pore via both their carbon and oxygen atoms with the μ_2-OH groups present at the MIL-53(Cr) surface (Figure 14.30) (Ramsahye et al., 2007, 2008; Serre et al., 2007a; Ghoufi and Maurin, 2010; Ghoufi et al., 2012). At the opposite, the arrangements of the CO_2 molecules are less ordered in the LP form with preferential interactions between their oxygens and the hydrogen atoms of the μ_2-OH group, suggesting a drastic change of the order parameter of the adsorbate during the structural

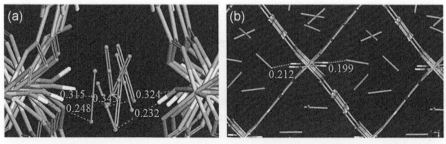

Figure 14.30 Preferential arrangements of the CO_2 molecules within the pores of the (a) NP and (b) LP forms of MIL-53(Cr) obtained by Grand Canonical Monte Carlo simulations.

switching, the highest value in the NP form being attributed to the high degree of confinement in the NP structure compared to that in the LP form, which tends to force the CO_2 molecules to be strictly aligned along the tunnel (Figure 14.30). One should mention that such predictive geometries can be further validated by a careful analysis of the Infra-red spectra obtained in the same condition (Serre et al., 2007a).

Beyond the description of the microscopic arrangements of the adsorbate molecules within the pores, molecular simulations allow a subtle analysis of the factors and mechanisms that govern the structural transition of the framework in play. The HOMC simulations (Hybrid Osmotic Monte Carlo, see Section 6.5.1) clearly emphasised that probing the interplay between the host and the guest molecules is crucial to capture the physics behind such a transition, suggesting that the host/guest interactions preliminary induce a soft mode in the host framework which is a pre-requirement for initiating the structural transition of the MIL-53(Cr) solid (Ghoufi and Maurin, 2010; Ghoufi et al., 2010, 2012). Modelling effort based on thermodynamic and analytical models has been also deployed to both predict and rationalise the spectacular behaviour of such dynamic frameworks (Coudert et al., 2008, 2011; Ghysels et al., 2013; Neimark et al., 2011).

14.7.2 MIL-53(Fe)

The iron form of the MIL-53 system shows a more complex breathing behaviour with a three-stepped pore-opening process as observed in the case of water (Bourrelly et al., 2010). In contrast with the Cr and Al analogues, its anhydrous state labelled as MIL-53 (Fe)VNP (VNP for *very narrow pore* form; monoclinic symmetry, $V \approx 0.900$ nm^3) exhibits closed pores with no accessible pores to most gases (Millange et al., 2008; Volkringer et al., 2009a). Upon an increase of the gas pressure, the structure switches first to an *intermediate* (MIL-53(Fe)INT; triclinic symmetry, $V \approx 0.892$ nm^3) phase where half of the pores are filled with guest molecules while the others are closed and empty before attaining a narrow pore form (MIL-53(Fe)NP; monoclinic symmetry, $V \approx 0.990$ nm^3). The pores nevertheless completely reopen at higher pressure (MIL-53(Fe)LP; orthorhombic symmetry, $V \approx 1.560$ nm^3) with pore dimensions identical to those observed for the *large pore* form of Al- and Cr-based MIL-53LP solids. This structural behaviour is summarised in Figure 14.31.

As a typical illustration, the adsorption isotherms measured at 303 K for a series of short linear alkanes in the MIL-53(Fe) are reported in Figure 14.32 both in standard and in semi-log forms in order to highlight the low pressure steps. Indeed, for all the hydrocarbon molecules, steps in the isotherms have been clearly distinguished. The smaller molecule, methane, shows a negligible uptake up to 10 bar, followed by a step. The adsorption of the other investigated hydrocarbons (ethane, propane, n-butane) leads to a first step in the isotherm occurring for a similar amount of adsorbed species. A second step is also detected for propane and n-butane corresponding to an uptake of about 2.8 mmol g^{-1}, while a final step is also observable for the C2–C4 hydrocarbons although not always complete in these isotherms.

In situ XRPD patterns collected on MIL-53(Fe) during the adsorption of the linear alkanes (C1–C4) showed unambiguous changes in the crystalline structure as a

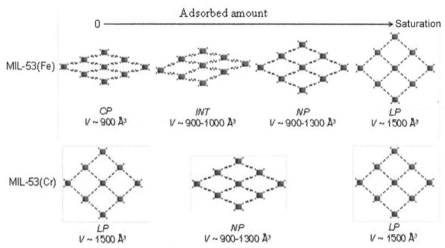

Figure 14.31 Structural evolution of MIL-53(Fe) (top) and MIL-53(Cr, Al) (bottom) with an increase in the amount of guest molecules inside the system (from left to right). (For colour version of this figure, the reader is referred to the online version of this chapter.)

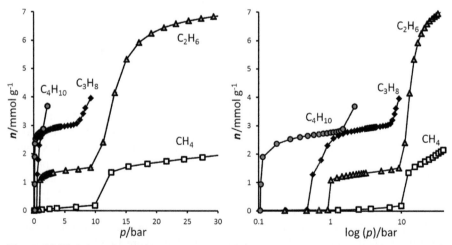

Figure 14.32 Adsorption isotherms of the C1–C4 linear hydrocarbons at 303 K in the MIL-53 (Fe) solid: standard coordinates (left), semi-log plot (right).

function of the gas adsorption into the channels. Except for methane, where only one structural change was observed in a wide range of pressure up to 40 bars, three structural transformations were present for the longer alkanes (ethane, propane, butane), with different XRPD signatures according to the gas pressure. Moreover, the pressures associated with these structural changes were on the whole in very good agreement with the ones corresponding to the steps observed in the adsorption isotherms. These small differences may be related to the difference in equilibrium criteria used for each

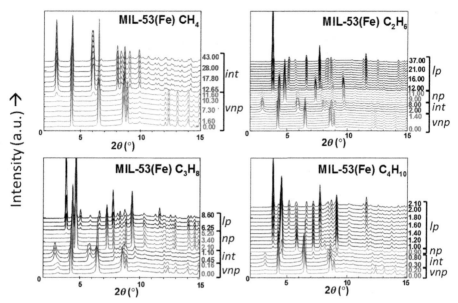

Figure 14.33 Adsorption of linear alkanes on MIL-53(Fe) at 303 K followed by XRPD. The numbers correspond to the adsorption pressures (bar) and the different phases are indicated on the right of each figure (*vnp*, very narrow pore; *int*, intermediate; *np*, narrow pore; *lp*, arge pore. See main text for details).

study, the one chosen for the isotherms measurements being the most severe. The comparison between both results nevertheless clearly proved that the steps are associated with structural modifications.

It was thus observed that starting with the VNP form (see Figure 14.33) upon increasing the gas pressure, a first characteristic diffraction pattern appears which can be assigned to the *intermediate* phase whose unit cell volume varies depending on the size of the hydrocarbon. When the pressure increases, except for methane, the *narrow pore* form was then observed. Finally, at much higher pressures, a new pattern was observed, which is assigned to the *large pore* form.

14.7.3   Co(BDP)

Compared to the previous two- or three-stepped behaviour of the typical flexible MIL-53(Cr, Al, Fe) MOFs presented above, the five steps observed in the N_2 adsorption isotherms collected at 77 K on the Co(BDP) ($BDP^{2-} = 1,4$-benzenedipyrazolate) solid is without precedent and has piqued a large curiosity to understand the origins of such phenomena (Figure 14.34) (Choi et al., 2008; Salles et al., 2010a).

Such a 3D solid which is composed of tetrahedrally coordinated Co^{II} centres along the plane [001] linked in two orthogonal directions by bridging BDP^{2-} ligands to form

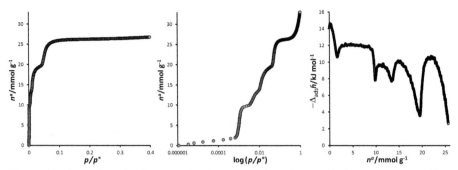

Figure 14.34 N_2 adsorption isotherm for Co(BDP) measured at 77 K (left, normal plot; middle, semi-log plot), indicating the five pressure-dependent phases. Corresponding differential enthalpy of adsorption measured at 77 K (right) (Salles et al., 2010a).

square channels along the plane [001], represents a rare example of a MOF solid exhibiting flexibility upon adsorption of non-polar molecules. To probe this intriguing five-step isotherm, an *in situ* X-ray powder diffraction study was performed at various increasing N_2 pressures using a custom-built apparatus to control the N_2 pressure above the sample. This investigation provided a remarkable confirmation that the steps detected on the adsorption isotherms are assigned to consecutive structural transitions occurring reversibly and reflecting the feasibility of repeated pore-opening and pore-closing cycles. A computational assisted structure determination strategy was able to determine the structures of the successive phases, denoted dry, Int.1, Int.2, Int.3 and filled, revealing a remarkable shrinkage of the cell volumes upon loss of guest molecules as summarised in Figure 14.35. A remarkable phase evolution was discovered, which is associated with the change in coordination geometry at the Co^{II} centres. Notably, the topology and symmetry of the framework in the monoclinic cells remains unaltered from dry to Int.3, demonstrating that Co^{II} adopts a square planar coordination geometry in all phases except filled, when it suddenly transforms to a tetrahedral geometry.

As an additional probe of the phase changes, the differential enthalpy of N_2 adsorption was measured at 77 K employing an *in situ* pressure-controlled Tian–Calvet type microcalorimeter. As shown in Figure 14.34, the five phases are distinctly discriminated with the differential enthalpies of each step estimated as 14.1, 12.0, 9.5, 10.0 and 10.5 kJ mol^{-1} for dry, Int.1, Int.2, Int.3 and filled, respectively. The result is obviously distinguishable from a typical plot of differential enthalpy of adsorption for a rigid microporous material. The low-coverage differential enthalpy of 14.1 kJ mol^{-1} with its subsequent drop is presumably attributed to a weak N_2 binding at open Co^{II} coordination sites and is somewhat smaller in magnitude than those reported for zeolites (19–27 kJ mol^{-1}) (Maurin et al., 2005). This was confirmed by Monte Carlo simulations which evidenced that N_2 binds Co^{II} in a side-on fashion with a relatively long Co$\cdots$N distance of 0.26 nm. Significant valleys appear between each step, which is indicative of additional heat required for the phase transition concomitant with adsorption. While that energy is estimated as approximately 2 kJ mol^{-1} for the dry through Int.3 transitions, it significantly increases to 7 kJ mol^{-1} for the last transition, reflecting the higher activation barrier associated with the transformation of the Co^{II}

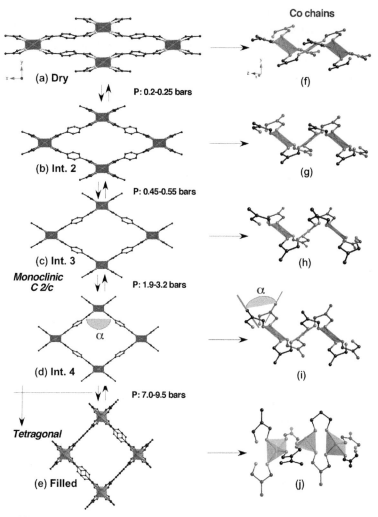

Co chains

(a) Dry

P: 0.2-0.25 bars

(b) Int. 2

P: 0.45-0.55 bars

(c) Int. 3

Monoclinic C 2/c

P: 1.9-3.2 bars

(d) Int. 4

P: 7.0-9.5 bars

Tetragonal

(e) Filled

(f)

(g)

(h)

(i)

(j)

Figure 14.35 (a-e) Structural evolution of Co(BDP) under increasing N_2 pressures via a five-step phase (f-j) perspective view of the cobalt chains showing the connectivity of the square coordination planes of CO atoms with the pyrazolate ligands (Salles et al., 2010a).

coordination geometry from square planar to tetrahedral. It is a rare example where such a complex transformation involving significant enthalpy changes has been measured directly during a gas adsorption at low temperature.

14.8 Towards Application Performances

The domain of MOFs has rapidly expanded due to the wide spectrum of structures prepared and to the expectation that some of them should necessarily have a specific application. It is indeed already possible to list a large number of potential applications for MOFs, though very few of them have reached an industrial scale (Kuppler et al., 2009).

Before examining a few of these applications, it may be worthwhile to mention a few difficulties encountered in the use of MOFs which can suffer from one of the following shortcomings: limited resistance to thermal degradation, instability in the presence of moisture (which may drastically limit the lifetime of the material), high cost of synthesis (which makes that they hardly compete with zeolites or activated carbons.) and low apparent density, which is a consequence of their most open structure. This low density is to be taken into account in most industrial applications where the volume to be occupied by the adsorbent is limited or imposed, for example, in gas storage containers or even in gas separation units. This makes that a MOF can provide high values of its specific properties (i.e. per unit mass), though quite ordinary values of its volumic properties (i.e. per unit volume). For this reason, a fair comparison of MOFs with other adsorbents should often be made on a volumic basis.

On the other hand, there are many MOFs which are thermally robust (e.g. the aluminium based MOFs, which can usually withstand temperatures up to 450 °C), water-resistant, chemically resistant (even to contaminants like H_2S) and not too expensive to prepare. Their main applications can be listed as follows:

14.8.1   Gas Storage

Natural gas storage (i.e. essentially *methane storage*) is probably the most advanced industrial application of MOFs (Yilmaz et al., 2012). This application can draw benefit from the robustness of Al-based MOFs and their stability towards water vapour. Such a MOF (aluminium-fumarate, trade name: Basolite A520, from BASF) is prepared through a cheap water route, with an amazing 'Space-Time-Yield'(STY) of 3600 kg/(m^3 day), to be compared with the STY of 50–150 kg/(m^3 day) of synthetic zeolites, which makes its cost competitive (Gaab et al., 2012). For a given storage volume and pressure, it allows increasing by 40% the autonomy of a vehicle operating with natural gas.

Hydrogen storage is more difficult to achieve at room temperature, because the enthalpy of adsorption is too weak (usually around 4–10 kJ mol^{-1}) to allow a satisfactory hydrogen adsorption capacity at room temperature: as shown by Bhatia and Myers (2006), an enthalpy of adsorption of ca. 15 kJ mol^{-1} would be required to ensure the desired adsorption and delivery of hydrogen at room temperature. An effort is currently being done to increase this enthalpy of adsorption by reducing the pore width, for instance by catenation, the optimum width being considered to be ca. 0.6 nm (Wang and Johnson, 1999). In the meanwhile, it is at 77 K that hydrogen storage can be obtained on MOFs (Suh et al., 2012).

14.8.2   Gas Separation or Purification

An efficient gas separation requires both a good adsorption capacity and a good selectivity of the adsorbent, two qualities that MOFs are expected to be able to fulfil, due to the possibility of tuning their pore size and windows size, and to the variety of possible active adsorption sites (Li et al., 2012).

The tuning is indeed so fine that it was possible, with MOF PCN-17, to separate, by simple size-exclusion, N_2 (kinetic diameter: 0.364 nm) from O_2 (kinetic diameter: 0.346 nm), by adsorbing $10\times$ as much O_2 as N_2 (Ma et al., 2008).

Promising materials are the 'Mesh-Adjustable Molecular Sieves' (MAMS) in which the tuning is obtained by simply varying the temperature of operation (Ma et al., 2007).

A selectivity based on different adsorbate–adsorbent interaction is also possible, as with the MOF Cu (bdt) which contains unsaturated metal coordination sites and is again efficient for the O_2/N_2 separation (Dinca et al., 2006). Much effort has been devoted to the use of MOFs for CO_2 capture (Sumida et al., 2012). The use of flexible MOFs or those with specific grafting could be of interest. One example is the efficient CO_2/CH_4 separation obtained with an amino-functionalised MIL-53 (Al) (Couck et al., 2009). This however leads to a reduced capacity and in such cases, large pore MOFs such as MIL100 may be more of interest (Llewellyn et al., 2008).

The large amount of work done on the performance of MOFs for the removal of hazardous chemicals (NO_x, SO_x, CO_x, H_2S, volatile organic compounds (VOC), dyes, drugs) was reviewed by Khan et al. (2013).

While many MOFs seem to be more selective than active carbons, only a few are more selective than zeolites. At the present time, MOFs seem to be a compromise between active carbons and zeolites (Wiersum et al., 2013a). In many cases, high working capacities can be attained with moderate selectivities, which would be acceptable for bulk separations. Highly selective MOFs, such as those used in sensor applications, are more suitable in the case of purifications where small quantities of gas need to be removed.

14.8.3 Catalysis

With their high metal content, which can be of any catalytically active transition metal, and with the broad adaptability of their pore structure to the needs of a shape-selective catalytic nanoreactor, MOFs are expected to have a great future in catalysis.

The synthesis, properties (including a thermal stability up to 450 °C) and direct catalytic activity of IRMOF-8 (Zn_4O (naphtalenedicarboxylate)$_3$) for Friedel–Crafts acylation reactions was thoroughly studied by Nguyen et al. (2011).

The coordinatively unsaturated metal sites of MOFs offer the possibility to graft active groups; for instance, amine-grafted MIL-101(Cr) exhibits a high activity for the Knoevenagel condensation, where an active hydrogen compound is brought to condense on the carbonyl group of a ketone or aldehyde (Hwang et al., 2008).

Another possibility also examined is to use a MOF as a host for large catalytically active molecules like metalloporphyrins (Alkordi et al., 2008).

14.8.4 Drug Delivery

What is expected from MOFs in controlled drug delivery is a large capacity, due to their high porosity, together with a slow, controlled, release due to an appropriate windows size. Nevertheless, they should also be biocompatible (making use of metals like Fe, Zn,

Mg, Ca, Ti, etc.) and, because the drug molecules are often large molecules, the pore windows should be wider than in the majority of MOFs (Horcajada et al., 2012).

The above conditions were fulfilled by MIL-101, whose pores are 0.29–0.34 nm wide, with specially wide windows of 1.2 (pentagonal) and 1.6 nm (hexagonal): each gram of dry MOF is able to slowly release, over 3 days, ca. 1.25 g of ibuprofen, whose molecule size is 0.6×1.03 nm. The delivery mechanism seems to be two-stepped, (i) by simple diffusion of weakly bound molecules and (ii) by slow desorption of the molecules bound by π–π interactions to the aromatic rings (Horcajada et al., 2006).

MIL-53 only uptakes ca. 20% ibuprofen in weight and, due to the flexibility of its framework, which continuously maintains a close contact between the molecule and the adsorbent, it is able to extend to 3 weeks the time for a total delivery (Horcajada et al., 2008; Loiseau et al., 2004)

Nanoparticles of MIL-100(Fe) were also used to entrap nucleoside analogues such as the azidothymidine tri-phosphate (AZT-Tp), used in the treatment of AIDS and HIV infection, or the cidofovir, an anti-smallpox agent, with unprecedented encapsulation loadings, up to fifty times higher than those reported so far with polymeric and mesoporous silica nanocarriers (Horcajada et al., 2012).

All of these materials offer the possibility of controlling *in vitro* the release of their cargo over the long period desired for improving the administration conditions for the patient.

14.8.5   Sensors

Luminescent MOFs were prepared (e.g. with lanthanide metal ion or stilbene ligand), whose luminescence can be influenced by the nature and amount of the adsorbed species, allowing to envisage an application as gas sensors (Chen et al., 2009).

A sensor with a *mechanical response* resulting from the stress induced by adsorption on a thin film of MOF HKUST-1 stuck to a microcantilever was also devised and found to respond to the presence of methanol and ethanol vapour (Allendorf et al., 2009).

14.8.6   Comparing MOFs with Other Adsorbents

We already mentioned the need, when comparing MOFs to other materials for industrial applications, to consider their bulk density and to compare amounts adsorbed per unit volume of material.

One can also make use of comparison factors such as those proposed by Rege and Yang (2001) or Wiersum et al. (2013a), which can take into account the working capacity, the selectivity and even the adsorption enthalpy.

The cost of the MOF is of course an important parameter which, at present, is not often in favour of MOFs. Nevertheless, some of them can be prepared with cheap linkers (eg. terephthalic or trimesic acid) and, as seen above for Basolite A520, with extremely effective and cost-saving processes, which shows that this issue of cost can be circumvented once the needs of the market are evident, that is once the interest of the MOF for a given application is clearly seen.

Before any comparison is carried out with another adsorbent, a selection of the best MOF available for a given application is necessary. So many structures are continuously appearing (either in theory or through an experimental synthesis) that such a screening has become an important and compulsory work. The theoretical approaches that are used to screen materials are discussed in Chapter 6. From an experimental standpoint, several setups have been developed either for specific applications, such as carbon dioxide recovery (Pirngruber et al., 2012) stability to water and acid gases at low pressures (Han et al., 2012) or more general parallelisation of isotherm measurement to high pressure (Wiersum et al., 2013b).

References

Alkordi, M.H., Liu, Y.L., Larsen, R.W., Eubank, J.F., Eddaoudi, M., 2008. J. Am. Chem. Soc. 131, 17753.

Alkordi, M.H., Brant, J.H., Wojtas, L., Kratsov, V.C., Cairns, A.J., Eddaoudi, M., 2009. J. Am. Chem. Soc. 130, 12639.

Allendorf, M.D., Bauer, C.A., Bhakta, R.K., Houk, R.J.T., 2009. Chem. Soc. Rev. 38, 1330.

Barthelet, K., Marrot, J., Riou, D., Férey, G., 2002. Angew. Chem. Int. Ed. 41, 281.

Batten, S.R., Champness, N.R., Chen, X.-M., Garcia-Martinez, J., Kitagawa, S., Öhrström, L., O'Keeffe, M., Suh, M.P., Reedijk, J., 2012. CrystEngComm 14, 3001.

Beurroies, I., Boulhout, M., Llewellyn, P.L., Kuchta, B., Férey, G., Serre, C., Denoyel, R., 2010. Angew. Chem. Int. Ed. 49 (41), 7526.

Bhatia, S.K., Myers, A.L., 2006. Langmuir 22, 1688.

Bourrelly, S., Llewellyn, P.L., Serre, C., Millange, F., Loiseau, T., Férey, G., 2005. J. Am. Chem. Soc. 127 (39), 13519.

Bourrelly, S., Moulin, B., Rivera, A., Maurin, G., Devautour-Vinot, S., Serre, C., Devic, T., Horcajada, P., Vimont, A., Clet, G., Daturi, M., Lavalley, J.-C., Loera-Serna, S., Denoyel, R., Llewellyn, P.L., Férey, G., 2010. J. Am. Chem. Soc. 132 (27), 9488.

Caillez, F., Stirnemann, G., Boutin, A., Demachy, I., Fuchs, A.H., 2008. J. Phys. Chem. C. 112, 10435.

Chavan, S., Vitillo, J.G., Groppo, E., Bonino, F., Lamberti, C., Dietzel, P.D.C., Bordiga, S., 2009. J. Phys. Chem. C. 113 (8), 3292.

Chen, B.L., Wang, L.B., Xiao, Y.Q., Fronczek, F.R., Xue, M., Cui, Y.J., Qian, G.D., 2009. Angew. Chem. Int. Ed. Eng. 48, 500.

Chen, C., Kim, J., Yang, D., Ahn, W., 2011. Chem. Eng. J. 168 (3), 1134.

Chen, Y.F., Nalaparaju, A., Eddaoudi, M., Jiang, J.W., 2012. Langmuir 28 (8), 3903.

Choi, H.J., Dincă, M., Long, J.R., 2008. J. Am. Chem. Soc. 130 (25), 7848.

Chui, S.S.Y., Lo, S.M.F., Charmant, J.P.H., Orpen, A.G., Williams, I.D., 1999. Science 283 (5405), 1148.

Cohen, S.M., 2012. Chem. Rev. 112 (2), 970.

Couck, S., Denayer, J.F.M., Baron, G.V., Rémy, T., Gascon, J., Kapteijn, F., 2009. J. Am. Chem. Soc. 131, 6326.

Coudert, F.X., Jeffroy, M., Fuchs, A.H., Boutin, A., Mellot-Draznieks, C., 2008. J. Am. Chem. Soc. 130 (43), 14294.

Coudert, F.X., Boutin, A., Jeffroy, M., Mellot-Draznieks, C., Fuchs, A.H., 2011. Chem. Phys. Chem. 12, 247.

Desbiens, N., Demachy, I., Fuchs, A.H., Kirsch-Rodeschini, H., Soulard, M., Patarin, J., 2005. Angew. Chem. Int. Ed. 44, 5310.

Devautour-Vinot, S., Maurin, G., Henn, F., Serre, C., Devic, T., Férey, G., 2009. Chem. Commun. 19, 2733.

Devautour-Vinot, S., Maurin, G., Henn, F., Serre, C., Férey, G., 2010. Phys. Chem. Chem. Phys. 12 (39), 12478.

Devautour-Vinot, S., Maurin, G., Serre, C., Horcajada, P., Paula da Cunha, D., Guillerm, V., Taulelle, F., Martineau, C., 2012. Chem. Mater. 24 (11), 2168.

Devic, T., Salles, F., Bourrelly, S., Moulin, B., Maurin, G., Horcajada, P., Serre, C., Vimont, A., Lavalley, J.C., Leclerc, H., Clet, G., Daturi, M., Llewellyn, P.L., Filinchuk, Y., Férey, G., 2012. J. Mater. Chem. 22 (20), 10266.

Dietzel, P.D.C., Morita, Y., Blom, R., Fjellvag, H., 2005. Angew. Chem. Int. Ed. 44, 6354.

Dietzel, P.D.C., Johnsen, R.E., Blom, R., Fjellvag, H., 2008a. Chem. Eur. J. 14 (8), 2389.

Dietzel, P.D.C., Johnsen, R.E., Fjellvag, H., Bordiga, S., Groppo, E., Chavan, S., Blom, R., 2008b. Chem. Commun. 41, 5125.

Dietzel, P.D.C., Besikiotis, V., Blom, R., 2009. J. Mater. Chem. 19, 7362.

Dietzel, P.D.C., Georgiev, P.A., Eckert, J., Blom, R., Strässle, T., Unruh, T., 2010. Chem. Commun. 46, 4962.

Dinca, M., Long, J.R., 2007. J. Am. Chem. Soc. 129 (36), 11172.

Dinca, M., Yu, A.F., Long, J.R., 2006. J. Am. Chem. Soc. 128, 8904.

Ding, S.-Y., Wang, W., 2013. Chem. Soc. Rev. 42 (2), 548.

Düren, T., Millange, F., Férey, G., Walton, K.S., Snurr, R.Q., 2007. J. Phys. Chem. C. 111 (42), 15350.

Dybtsev, D., Serre, C., Schmitz, B., Panella, B., Hirscher, M., Latroche, M., Llewellyn, P.L., Cordier, S., Molard, Y., Haouas, M., Taulelle, F., Férey, G., 2010. Langmuir 26 (13), 11283.

Eddaoudi, M., Kim, J., Rosi, N., Vodak, D., Wachter, J., O'Keeffe, M., Yaghi, O.M., 2002. Science. 295, 469.

Fairen-Jimenez, D., Seaton, N.A., Duren, T., 2010. Langmuir 26 (18), 14694.

Fairen-Jimenez, D., Moggach, S.A., Wharmby, M.T., Wright, P.A., Parsons, S., Düren, T., 2011. J. Am. Chem. Soc. 133 (23), 8900.

Farha, O.K., Hupp, J.T., 2010. Acc. Chem. Res. 43 (8), 1166.

Farha, O.K., Eryazici, I., Jeong, N.C., Haauser, B.G., Wilmer, C.E., Sarjeant, A.A., Snurr, R.Q., Nguyen, S.B.T., Yazaydin, A.O., Hupp, J.T., 2012. J. Am. Chem. Soc. 134, 15016.

Fateeva, A., Devautour-Vinot, S., Heymans, N., Devic, T., Grenèche, J.M., Wuttke, S., Miller, S., Lago, A., Serre, C., De Weireld, G., Maurin, G., Vimont, A., Férey, G., 2011. Chem. Mater. 23 (20), 4641.

Férey, G., 2001. Chem. Mater. 13, 3084.

Férey, G., 2008. Chem. Soc. Rev. 37, 191.

Férey, G., Serre, C., 2009. Chem. Soc. Rev. 38, 1380.

Férey, G., Serre, C., Mellot-Draznieks, C., Millange, F., Surblé, S., Dutour, J., Margiolaki, I., 2004. Angew. Chem. Int. Ed. 43 (46), 6296.

Férey, G., Mellot-Draznieks, C., Serre, C., Millange, F., Dutour, J., Surble, S., Margiolaki, I., 2005. Science. 309 (5743), 2040.

Férey, G., Serre, C., Devic, T., Maurin, G., Jobic, H., Llewellyn, P.L., De Weireld, G., Vimont, A., 2011. Chem. Soc. Rev. 40 (2), 550.

Gaab, M., Trukhan, N., Maurer, S., Gummaraju, R., Müller, U., 2012. Micropor. Mesopor. Mater. 157, 131.

Getman, R.B., Bae, Y.-S., Wilmer, C.E., Snurr, R.Q., 2012. Chem. Rev. 112 (2), 703.

Ghoufi, A., Maurin, G., 2010. J. Phys. Chem. C. 114 (14), 6496.

Ghoufi, A., Férey, G., Maurin, G., 2010. J. Phys. Chem. Lett. 1, 2810.

Ghoufi, A., Subercaze, A., Ma, Q., Yot, P., Yang, K., Puente, O.I., Devic, T., Guillerm, V., Zhong, C., Serre, C., Férey, G., Maurin, G., 2012. J. Phys. Chem. C. 116 (24), 13289.

Ghysels, A., Vanduyfhuys, L., Vandichel, M., Waroquier, M., Van Speybroeck, V., Smit, B., 2013. J. Phys. Chem. C. 117, 11540.

Grajciar, L., Wiersum, A.D., Llewellyn, P.L., Chang, J.-S., Nachtigall, P., 2011. J. Phys. Chem. C 115, 17925.

Guillerm, V., Ragon, F., Dan-Hardi, M., Devic, T., Vishnuvarthan, M., Campo, B., Vimont, A., Clet, G., Yang, Q., Maurin, G., Férey, G., Vittadini, A., Gross, S., Serre, C., 2012. Angew. Chem. Int. Ed. 51 (37), 9267.

Hamon, L., Serre, C., Devic, T., Loiseau, T., Millange, F., Férey, G., De Weireld, G., 2009. J. Am. Chem. Soc. 131 (25), 8775.

Han, S., Huang, Y., Watanabe, T., Dai, Y., Walton, K.S., Nair, S., Sholl, D.S., Meredith, J.C., 2012. ACS Comb. Sci. 14, 263.

Horcajada, P., Serre, C., Vallet-Regi, M., Sebban, M., Taulelle, F., Férey, G., 2006. Angew. Chem. Int. Ed. Engl. 45, 5974.

Horcajada, P., Surble, S., Serre, C., Hong, D.-Y., Seo, Y.-K., Chang, J.-S., Greneche, J.-M., Margiolaki, I., Férey, G., 2007. Chem. Commun. 27, 2820.

Horcajada, P., Serre, C., Maurin, G., Ramsahye, N.A., Balas, F., Vallet-Regi, M., Sebban, M., Taulelle, F., Férey, G., 2008. J. Am. Chem. Soc. 130, 6774.

Horcajada, P., Gref, R., Baati, T., Allan, P.K., Maurin, G., Couvreur, P., Férey, G., Morris, R.E., Serre, C., 2012. Chem. Rev. 112 (2), 1232.

Horike, S., Matsuda, R., Tanaka, D., Mizuno, M., Endo, K., Kitagawa, S., 2006. J. Am. Chem. Soc. 128, 4222.

Hoskins, B.F., Robson, R., 1989. J. Am. Chem. Soc. 111, 5962.

Hupp, J.T., 2012. J. Amer. Chem. Soc. 134 (36), 15016.

Hwang, Y.K., Hong, D.Y., Chang, J.S., Jhung, S.H., Seo, Y.K., Kim, J., Vimont, A., Daturi, M., Serre, C., Férey, G., 2008. Angew. Chem. Int. Ed. Engl. 47, 4144.

Janiak, C., Vieth, J.K., 2010. New J. Chem. 34 (11), 2366.

Jiang, J.-X., Cooper, A.I., 2010. In: Schröder, M. (Ed.), Functional Metal-Organic Frameworks: Gas Storage, Separation and Catalysis. Topics in Current Chemistry, Vol. 293. Springer Verlag, Berlin, p. 1.

Jiang, J., Babarao, R., Hu, Z., 2011. Chem. Soc. Rev. 40, 3599.

Khan, N.A., Hasan, Z., Jhung, S.H., 2013. J. Hazard. Mater. 244–245, 444.

Kinoshita, Y., Matsubara, I., Higuchi, T., Saito, Y., 1959. Bull. Chem. Soc. Jpn. 32, 1221.

Kitagawa, S., Uemura, K., 2005. Chem. Soc. Rev. 34, 109.

Kitagawa, S., Kitaura, R., Noro, S.-I., 2004. Angew. Chem. Int. Ed. 43 (18), 2334.

Kitaura, R., Seki, K., Akiyama, G., Kitagawa, S., 2003. Angew. Chem. Int. Ed. 42 (4), 428.

Klyamkin, S.N., Berdonosova, E.A., Kogan, E.V., Kovalenko, K.A., Dybtsev, D.N., Fedin, V.P., 2011. Chem. Asian J. 6 (7), 1854.

Kolokolov, D.I., Stepanov, A.G., Guillerm, V., Serre, C., Frick, B., Jobic, H., 2012. J. Phys. Chem. C 116 (22), 12131.

Kondo, M., Yoshitomi, T., Seki, K., Matsuzaka, H., Kitagawa, S., 1997. Angew. Chem. Int. Ed. 36, 1725.

Kuppler, R.J., Timmons, D.J., Fang, Q.R., Li, J.R., Makal, T.A., Young, M.D., Yuan, D., Zhao, D., Zhuang, W., Zhou, H.C., 2009. Coordinat. Chem. Rev. 253, 3042.

Leclerc, H., Vimont, A., Lavalley, J.-C., Daturi, M., Wiersum, A.D., Llewellyn, P.L., Horcajada, P., Férey, G., Serre, C., 2011. Phys. Chem. Chem. Phys. 13 (24), 11748.

Li, J.R., Sculley, J., Zhou, H.-C., 2012. Chem. Rev. 112 (2), 869.

Lin, X., Telepeni, I., Blake, A.J., Dailly, A., Brown, C.M., Simmons, J.M., Zoppi, M., Walker, G.S., Thomas, K.M., Mays, T.J., Hubberstey, P., Champness, N.R., Schröder, M., 2009. J. Am. Chem. Soc. 131, 2159.

Liu, Y., Kravtsov, V.C., Larsen, R., Eddaoudi, M., 2006. Chem. Commun., 1488.

Liu, Y., Her, J.-H., Dailly, A., Ramirez-Cuesta, A.J., Neumann, D.A., Brown, C.M., 2008a. J. Am. Chem. Soc. 130 (35), 11813.

Liu, Y., Kabbour, H., Brown, C.M., Neumann, D.A., Ahn, C.C., 2008b. Langmuir 24, 4772.

Llewellyn, P.L., Bourrelly, S., Serre, C., Vimont, A., Daturi, M., Hamon, L., De Weireld, G., Chang, J.-S., Hong, D.-Y., Hwang, Y.K., Jhung, S.H., Férey, G., 2008. Langmuir. 24 (14), 7245.

Llewellyn, P.L., Horcajada, P., Maurin, G., Devic, T., Rosenbach, N., Bourrelly, S., Serre, C., Vincent, D., Loera-Serna, S., Filinchuk, Y., Ferey, G., 2009. J. Am. Chem. Soc. 131 (36), 13002.

Loiseau, T., Serre, C., Huguenard, C., Fink, G., Taulelle, F., Henry, M., Bataille, T., Férey, G., 2004. Chem. Eur. J. 10, 1373.

Low, J.J., Benin, A.I., Jakubczak, P., Abrahamian, J.F., Faheem, S.A., Willis, R.R., 2009. J. Am. Chem. Soc. 131 (43), 15834.

Ludwig, R., 2001. Angew. Chem. Int. Ed. Engl. 40, 1808.

Ma, S.Q., Wang, X.S., Collier, C.D., Manis, E.S., Zhou, H.C., 2007. Inorg. Chem. 46, 8499.

Ma, S.Q., Wang, X.S., Yuan, D.Q., Zhou, H.C., 2008. Angew. Chem. Int. Ed. Engl. 47, 4130.

Maurin, G., Llewellyn, P.L., Poyet, T., Kuchta, B., 2005. Micropor. Mesopor. Mater. 79 (1–3), 53.

Meek, S.T., Greathouse, J.A., Allendorf, M.D., 2011. Adv. Mater. 23, 249.

Millange, F., Guillou, N., Walton, R.I., Grenèche, J.M., Margiolaki, I., Férey, G., 2008. Chem. Commun., 4732.

Mowat, J.P.S., Miller, S.R., Slawin, A.M.Z., Seymour, V.R., Ashbrook, S.E., Wright, P.A., 2011. Micropor. Mesopor. Mater. 142 (1), 322.

Myers, A.L., Monson, P.A., 2002. Langmuir. 18 (26), 10261.

Neimark, A.V., Coudert, F.X., Triguero, C., Boutin, A., Fuchs, A.H., Beurroies, I., Denoyel, R., 2011. Langmuir 27, 4734.

Nelson, A.P., Parrish, D.A., Cambrea, L.R., Baldwin, L.C., Trivedi, N.J., Mulfort, K.L., Farha, O.K., Hupp, J.T., 2009. Cryst. Growth Des. 9 (11), 4588.

Nguyen, L.T.L., Nguyen, C.V., Dang, G.H., Le, K.K.A., Phan, N.T.S., 2011. J. Mol. Cat. A Chem. 349, 28.

Nouar, F., Eckert, J., Eubank, J.F., Forster, P., Eddaoudi, M., 2009. J. Am. Chem. Soc. 131 (8), 2864.

Phan, A., Doonan, C.J., Uribe-Romo, F.J., Knobler, C.B., O'Keeffe, M., Yaghi, O.M., 2010. Acc. Chem. Res. 43 (1), 58.

Pirngruber, G.D., Hamon, L., Bourrelly, S., Llewellyn, P.L., Lenoir, E., Guillerm, V., Serre, C., Devic, T., 2012. ChemSusChem 5 (4), 762–776.

Prestipino, C., Regli, L., Vitillo, J.G., Bonino, F., Damin, A., Lamberti, C., Zecchina, A., Solari, P.L., Kongshaug, K.O., Bordiga, S., 2006. Chem. Mater. 18, 1337.

Quartapelle Procopio, E., Linares, F., Montoro, C., Colombo, V., Maspero, C., Barea, E., Navarro, J.A.R., 2010. Angew. Chem. Int. Ed. 49, 7308.

Ramsahye, N., Maurin, G., Bourrelly, S., Llewellyn, P.L., Loiseau, T., Serre, C., Férey, G., 2007. Chem. Commun. 31, 3261.

Ramsahye, N., Maurin, G., Bourrelly, S., Llewellyn, P.L., Loiseau, T., Serre, C., Férey, G., 2008. J. Phys. Chem. C 112, 514.

Rege, S.U., Yang, R.T., 2001. Sep. Sci. Technol. 36, 3355.

Rosenbach, N., Jobic, H., Ghoufi, A., Salles, F., Maurin, G., Bourrelly, S., Llewellyn, P.L., Devic, T., Serre, C., Férey, G., 2008. Angew. Chem. Int. Ed. 47, 6611.

Rosi, N.L., Kim, J., Eddaoudi, M., Chen, B.L., O'Keeffe, M., Yaghi, O.M., 2005. J. Am. Chem. Soc. 127, 1504.

Rowsell, J.L.C., Yaghi, O.M., 2004. Micropor. Mesopor. Mater. 73, 3.

Rubes, M., Grajciar, L., Bludsky, O., Wiersum, A.D., Llewellyn, P.L., Nachtigall, P., 2012. Chem. Phys. Chem. 13 (2), 488–495.

Salles, F., Jobic, H., Maurin, G., Llewellyn, P.L., Devic, T., Serre, C., Férey, G., 2008. Phys. Rev. Lett. 100, 245901.

Salles, F., Jobic, H., Ghoufi, A., Llewellyn, P.L., Serre, C., Bourrelly, S., Férey, G., Maurin, G., 2009. Angew. Chem. Int. Ed. 48 (44), 8335.

Salles, F., Maurin, G., Serre, C., Llewellyn, P.L., Knöfel, C., Choi, H.J., Filinchuk, Y., Oliviero, L., Vimont, A., Long, J.R., Férey, G., 2010a. J. Am. Chem. Soc. 132 (39), 13782.

Salles, F., Jobic, H., Devic, T., Llewellyn, P.L., Serre, C., Férey, G., Maurin, G., 2010b. ACS Nano 4 (1), 143.

Salles, F., Bourrelly, S., Jobic, H., Devic, T., Guillerm, V., Llewellyn, P.L., Serre, C., Férey, G., Maurin, G., 2011. J. Phys. Chem. C 115 (21), 10764.

Sava, D.F., Kratsov, V.C., Nouar, F., Wotjas, L., Eubank, J.F., Eddaoudi, M., 2008. J. Am. Chem. Soc. 130, 3768.

Seki, K., 2002. Phys. Chem. Chem. Phys. 4, 1968.

Serre, C., Millange, F., Thouvenot, C., Noguès, M., Marsolier, G., Louër, D., Férey, G., 2002. J. Am. Chem. Soc. 124 (45), 13519.

Serre, C., Mellot-Draznieks, C., Surble, S., Audebrand, N., Filinchuk, Y., Férey, G., 2007a. Science. 315 (5820), 1828.

Serre, C., Bourrelly, S., Vimont, A., Ramsahye, N.A., Maurin, G., Llewellyn, P.L., Daturi, M., Filinchuk, Y., Leynaud, O., Barnes, P., Férey, G., 2007b. Adv. Mater. 19 (17), 2246.

Stavitski, E., Pidko, E.A., Couck, S., Rémy, T., Hensen, E.J.M., Weckhuysen, B.M., Denayer, J., Gascon, J., Kaptein, F., 2011. Langmuir 27 (7), 3970.

Stock, N., Biswas, S., 2012. Chem. Soc. Rev. 112 (2), 933.

Suh, P.M., Park, H.J., Prasad, T.K., Lim, D.W., 2012. Chem. Rev. 112 (2), 782.

Sumida, K., Rogow, D.L., Mason, J.A., McDonld, T.M., Bloch, E.D., Herm, Z.R., Bae, T.H., Long, J.R., 2012. Chem. Rev. 112 (2), 724.

Tranchemontagne, D.J., Mendoza-Cortés, J.L., O'Keefe, M., Yaghi, O.M., 2009. Chem. Soc. Rev. 38 (5), 1257.

Trung, T.K., Trens, P., Tanchoux, N., Bourrelly, S., Llewellyn, P.L., Loera-Serna, S., Serre, C., Loiseau, T., Fajula, F., Férey, G., 2008. J. Am. Chem. Soc. 130 (50), 16926.

Uemura, K., Yamasaki, Y., Onishi, F., Kita, H., Ebihara, M., 2010. Inorg. Chem. 49 (21), 10133.

Valenzano, L., Civalleri, B., Sillar, K., Sauer, J., 2011. J. Phys. Chem. C. 115, 21777.

Volkringer, C., Loiseau, T., Guillou, N., Férey, G., Elkaim, E., Vimont, A., 2009a. Dalton Trans. 12, 2241.

Volkringer, C., Popov, D., Loiseau, T., Férey, G., Burghammer, M., Riekel, C., Haouas, M., Taulelle, F., 2009b. Chem. Mater. 21 (24), 5695.

Walton, K.S., Snurr, R.Q., 2007. J. Am. Chem. Soc. 129 (27), 8552.

Walton, K.S., Millward, A.R., Dubbeldam, D., Frost, H., Low, J.J., Yaghi, O.M., Snurr, R.Q., 2008. J. Am. Chem. Soc. 130, 406.

Wang, Q., Johnson, J.K., 1999. J. Chem. Phys. 110, 577.

Wiersum, A.D., Chang, J.-S., Serre, C., Llewellyn, P.L., 2013a. Langmuir 29 (10), 3301.

Wiersum, A.D., Giovannangeli, C., Bloch, E., Reinsch, H., Stock, N., Lee, J.S., Chang, J.-S., Llewellyn, P.L., 2013b. ACS Comb. Sci. 15 (2), 111.

Wu, E.L., Lawton, S.L., Olson, D.H., Rohrman, A.C., Kokotailo, G.T., 1979. J. Phys. Chem. 83 (21), 2777.

Wu, J.-Y., Yang, S.L., Luo, T.-T., Liu, Y.-H., Cheng, Y.-W., Chen, Y.-F., Wen, Y.-S., Lin, L.-G., Lu, K.-L., 2008. Chem. Eur. J. 14, 7136.

Yaghi, O.M., Li, H.L., 1995. J. Am. Chem. Soc. 117, 10401.

Yaghi, O.M., O'Keeffe, M., Ockwig, N.W., Chae, H.K., Eddaoudi, M., Kim, J., 2003. Nature 423 (6941), 705.

Yang, Q.Y., Xue, C.Y., Zhong, C.L., Chen, J.F., 2007. AIChE J. 53, 2832.

Yang, S., Lin, X., Blake, A.J., Thoms, K.M., Hubberstey, P., Champness, N.R., Schröder, M., 2008. Chem. Commun., 6108.

Yang, Q.Y., Wiersum, A.D., Llewellyn, P.L., Guillerm, V., Serre, C., Maurin, G., 2011. Chem. Commun. 47 (34), 9603.

Yang, Q.Y., Vaesen, S., Vishnuvarthan, M., Ragon, F., Serre, C., Vimont, A., Daturi, M., De Weireld, G., Maurin, G., 2012. J. Mater. Chem. 22 (20), 10210.

Yilmaz, B., Trukhan, N., Müller, U., 2012. Chim. J. Catal. 33 (1), 3.

Yoon, J.W., Seo, Y.-K., Hwang, Y.K., Chang, J.-S., Leclerc, H., Wuttke, S., Bazin, P., Vimont, A., Daturi, M., Bloch, E., Llewellyn, P.L., Serre, C., Horcajada, P., Greneche, J.-M., Rodrigues, A.E., Férey, G., 2010. Angew. Chem. Int. Ed. 49, 5949.

Zlotea, C., Cuevas, F., Paul-Boncour, V., Leroy, E., Dibandjo, P., Gadiou, R., Vix-Guterl, C., Latroche, M., 2010. J. Am. Chem. Soc. 132 (22), 7720.

Zlotea, C., Phanon, D., Mazaj, M., Heurtaux, D., Guillerm, V., Serre, C., Horcajada, P., Devic, T., Magnier, E., Cuevas, F., Férey, G., Llewellyn, P.L., Latroche, M., 2011. Dalton Trans. 40 (18), 4879.

Index

Printed and bound by CPI Group (UK) Ltd, Croydon, CR0 4YY

08/05/2025

01864786-0002